The Fractional Fourier Transform

WILEY SERIES IN PURE AND APPLIED OPTICS

Founded by Stanley S. Ballard, University of Florida

EDITOR: Joseph W. Goodman, Stanford University

BEISER • *Holographic Scanning*
BERGER-SCHUNN • *Practical Color Measurement*
BOYD • *Radiometry and The Detection of Optical Radiation*
BUCK • *Fundamentals of Optical Fibres*
CATHEY • *Optical Information Processing and Holography*
CHUANG • *Physics of Optoelectronic Devices*
DELONE AND KRAINOV • *Fundamentals of Nonlinear Optics of Atomic Gases*
DERENIAK AND BOREMAN • *Infrared Detectors and Systems*
DERENIAK AND CROWE • *Optical Radiation Detectors*
DE VANY • *Master Optical Techniques*
GASKILL • *Linear Systems, Fourier Transform, and Optics*
GOODMAN • *Statistical Optics*
HOBBS • *Building Electro-Optical Systems: Optics That Work*
HUDSON • *Infrared Systems Engineering*
JUDD AND WYSZECKI • *Color in Business, Science and Industry*, Third Edition
KAFRI AND GLATT • *The Physics of Moire Metrology*
KAROW • *Fabrication Methods for Precision Optics*
KLEIN AND FURTAK • *Optics*, Second Edition
MALACARA • *Optical Shop Testing*, Second Edition
MILONNI AND EBERLY • *Lasers*
NASSAU • *The Physics and Chemistry of Color*
NIETO-VESPERINAS • *Scattering and Diffraction in Physical Optics*
O'SHEA • *Elements of Modern Optical Design*
SALEH AND TEICH • *Fundamentals of Photonics*
SCHUBERT AND WILHELMI • *Nonlinear Optics and Quantum Electronics*
SHEN • *The Principles of Nonlinear Optics*
UDD • *Fiber Optic Sensors: An Introduction for Engineers and Scientists*
UDD • *Fiber Optic Smart Structures*
VANDERLUGT • *Optical Signal Processing*
VEST • *Holographic Interferometry*
VINCENT • *Fundamentals of Infrared Detector Operation and Testing*
WILLIAMS AND BECKLUND • *Introduction to the Optical Transfer Function*
WYSZECKI AND STILES • *Color Science: Concepts and Methods, Quantitative Data and Formulae*, Second Edition
XU AND STROUD • *Acousto-Optic Devices*
YAMAMOTO • *Coherence, Amplification and Quantum Effects in Semiconductor Lasers*
YARIV AND YEH • *Optical Waves in Crystals*
YEH • *Optical Waves in Layered Media*
YEH • *Introduction to Photorefractive Nonlinear Optics*
YEH AND GU • *Optics of Liquid Crystal Displays*

The Fractional Fourier Transform
with Applications in Optics and Signal Processing

Haldun M. Ozaktas
Bilkent University, Ankara, Turkey

Zeev Zalevsky
Tel Aviv University, Tel Aviv, Israel

M. Alper Kutay
TÜBİTAK-UEKAE, Ankara, Turkey

JOHN WILEY & SONS, LTD
Chichester New York Weinheim Brisbane Singapore Toronto

National   01243 779777
International (+44) 1243 779777
e-mail (for orders and customer service enquiries): cs-books@wiley.co.uk
Visit our Home Page on http://www.wiley.co.uk

Other Wiley Editorial Offices

New York, Weinheim, Brisbane, Singapore, Toronto

Library of Congress Cataloging-in-Publication Data

Ozaktas, Haldun M.
  The Fractional Fourier transform with applications in optics and signal processing /
Haldun M. Ozaktas, M. Alper Kutay, Zeev Zalevsky.
     p.cm.
  Includes bibliographical references and index.
  ISBN 0-471-96346-1 (acid-free paper)
  1.Fourier transformations. 2.Fourier transform optics. 3.Control theory. I. Kutay, M.
Alper. II. Zalevsky, Zeev. III. Title.

  QC20.7.F67 O93 2000
  530.15'5723--dc21
                                                          00-043350

British Library Cataloguing in Publication Data

A catalogue record for this book is available from the British Library

ISBN   0 471 96346 1

Typeset by the author
Printed and bound in Great Britain by Antony Rowe Ltd, Chippenham, Wiltshire
This book is printed on acid-free paper responsibly manufactured from sustainable forestry,
in which at least two trees are planted for each one used for paper production.

Cover design based on an idea by Haldun M. Ozaktas and Tayfun Akgül.
The image is of Jean-Baptiste-Joseph Fourier (1768–1830).

To our families
Past, present, and future

Contents

Preface

The fractional Fourier transform has received considerable interest since the early nineties, finding itself a place in standard texts and handbooks such as Bracewell 1999 and *The Transforms and Applications Handbook* 2000. Our primary purpose in writing this book has been to provide a widely accessible account of the transform covering both theory and applications.

Little need be said of the importance and ubiquity of the ordinary Fourier transform and frequency-domain concepts and techniques in many diverse areas of science and engineering. As a generalization of the ordinary Fourier transform, the fractional Fourier transform is only richer in theory and more flexible in applications—but not more costly in implementation. Therefore the transform is likely to have something to offer in every area in which Fourier transforms and related concepts are used. So far applications of the transform have been studied mostly in the areas of optics and wave propagation, and signal analysis and processing. These applications are discussed extensively in this book. However, we expect the transform to find applications in many other areas, and hope that this book will contribute to this end.

This text should primarily be of interest to graduate students, academics, and researchers in branches of mathematics, science, and engineering where Fourier transforms and related concepts are used. A partial list of these areas is operator theory, harmonic analysis and integral transforms, linear algebra, group representation theory, phase-space methods, time- and space-frequency representations, transform theory and techniques, signal analysis and processing, wave propagation, and many areas of optics. We have made an effort to make the book accessible to such a cross-disciplinary audience, entailing a number of compromises. The emphasis is mostly on elucidating the basic concepts from different perspectives and showing as many of the relationships between them as possible. Although most arguments and results are analytical in nature, we did not hesitate to employ suggestive physical arguments where we felt this was appropriate. Mathematical rigor is delegated to the references, as are most experimental and practical considerations. Discussion of optics has been strictly segregated so that readers with no interest in optics can simply ignore chapters 7, 8, and 9.

The fractional Fourier transform is intimately related to several indispensable concepts appearing in diverse areas. We have tried to present the transform in a broad context, showing its relationship to as many of these different concepts as possible. This has required the inclusion of a considerable amount of background and review material to ensure that the book is reasonably self-contained. Nevertheless, we have assumed the reader has at least elementary undergraduate-level exposure to signals and systems and linear algebra. A similar exposure to optics is assumed for those wishing to study the chapters on optics. For instance, we define the ordinary Fourier transform and list its properties, but we do not attempt to develop the insight and intuition that would constitute the focus of an elementary text. Specific suggestions for background reading are provided in the introductory chapter, as well as at the end of certain chapters.

The background material contained in chapters 2 and 7, and especially chapters 3 and 8, is an important feature of the book which we hope readers not very familiar with these topics will find useful in their own right. In these chapters, we occasionally go beyond providing background and preliminary material, to present a self-complete exposition of certain topics which have been neglected in other texts. As such, these chapters may be useful as primary or supplementary material for a variety of different courses. Substantial parts of this book have been used as the primary material for courses on time-frequency analysis, advanced signal processing, and as the theoretical core material of a course on optical information processing at Bilkent University. The material may be especially useful for courses in advanced Fourier optics or information optics emphasizing phase-space concepts and the Wigner distribution, or as supplementary material for introductory courses in these areas. The book can also form the basis of a specialized course on the fractional Fourier transform or the fractional Fourier transform and time-frequency representations, and their applications in optics and/or signal processing. However, depending on the emphasis of the course, and especially if optics is excluded, it may be useful to use the book together with one of the excellent tutorials or books on time-frequency representations (see the end of chapter 3).

A detailed overview of the book, including the relationships of the chapters to each other and suggestions for using the book for self-study, is presented in section 1.3.

Acknowledgments

One summer day in 1992, Adolf W. Lohmann, David Mendlovic, and Haldun M. Ozaktas were having a meeting at the Physics Institute of the University of Erlangen-Nürnberg. Haldun M. Ozaktas had been pondering the fact that in an optical system with several lenses employing a point source for illumination, one observes the Fourier transform of the object at the images of the point source (section 8.4.3). In such a system one typically observes first a Fourier transform, then an inverted image, then an inverted Fourier transform, then an erect image, then another Fourier transform, and so on. Inspired by other fractional operations he was familiar with, this led him to propose that the distributions of light at intermediate planes may be interpreted as "fractional Fourier transforms" of continually increasing orders, only if one could figure out how to define such a transform. David Mendlovic noted the irregular and nonuniform nature of such optical systems, and suggested that it might be more illuminating to consider propagation in quadratic graded-index media. Since a certain length d_0 of such a medium was known to produce an ordinary Fourier transform, and since the medium is uniform in the direction of propagation, he suggested that a length ad_0 should produce a distribution of light which can be interpreted as the ath order "fractional Fourier transform." This led us to define the transform as an operator power, essentially according to definition B in chapter 4. It was known that rays propagating in quadratic graded-index media exhibit circular trajectories when viewed in geometrical-optical phase space. Grasping the implications of this from a wave-optical perspective, Adolf W. Lohmann suggested that since ordinary Fourier transformation corresponds to $\pi/2$ rotation of the Wigner distribution, then rotation of the Wigner distribution by $a\pi/2$ should correspond to ath order fractional Fourier transformation. Although the initial motivation came from optics, mathematical and signal processing aspects of the transform were soon to intrigue us. Our journey exploring the transform and its applications, soon joined by M. Alper Kutay and Zeev Zalevsky, was a memorable one. Here we would like to express our most heartfelt and warm appreciation of having had the privilege of interacting and collaborating with David Mendlovic and Adolf W. Lohmann. The idea of writing a book on the fractional Fourier transform was originally conceived together with David Mendlovic; unfortunately, his obligations prevented him from actively participating in the preparation of this book.

We have had the benefit of collaborating and interacting with many students and colleagues throughout the years. We would like to thank the following both for their contributions to the area and for many fruitful discussions: Özer K. Akdemir, Orhan Arıkan, Orhan Aytür, Laurence Barker, Billur Barshan, A. Ümit Batur, Yigal Bitran,

Gözde Bozdağı, Çağatay Candan, John H. Caulfield, Rainer G. Dorsch, Nilgün Erkaya, Carlos Ferreira, Javier García, Ruşen Güldoğan, Özgür Güleryüz, Tuğrul Hakioğlu, Levent Onural, Hakan Özaktaş, Ayşegül Şahin, Hakan Ürey, K. Bernardo Wolf, and İ. Şamil Yetik. M. Fatih Erden deserves to be singled out for his many contributions to the area and the many inspiring discussions we have had.

We are most grateful to several individuals for reading and commenting on substantial parts of the manuscript, for answering our many questions, or providing information: İsrafil Bahçeci, Laurence Barker, Çağatay Candan, Dmitry B. Karp, David Mendlovic, David Mustard, Sergei M. Sitnik, K. Bernardo Wolf, and İ. Şamil Yetik. We would also like to extend our gratitude to many others who commented on various parts of the manuscript, answered our questions, made suggestions or pointed out errors in early drafts, or otherwise lent us their expertise: Zafer Aydın, Billur Barshan, Volkan Cevher, Yamaç Dikmelik, Lütfiye Durak, Joseph W. Goodman, Bahadır K. Güntürk, Tuğrul Hakioğlu, Ataç İmamoğlu, David A. B. Miller, Hakan Özaktaş, A. Kemal Özdemir, Rafael Piestun, Michael Raymer, and A. Pınar Tan. Finally, we would like to thank those we have failed to mention, hoping they will forgive us.

1

Introduction

The fractional Fourier transform is a generalization of the ordinary Fourier transform with an order parameter a. Mathematically, the ath order fractional Fourier transform $\mathcal{F}^a$ is the ath power of the ordinary Fourier transform operation $\mathcal{F}$. The first-order ($a = 1$) fractional transform is the ordinary Fourier transform operation, and the zeroth-order fractional transform is the identity operation. Thus the first-order fractional transform of a function is its ordinary Fourier transform, and the zeroth-order transform is the function itself. With the development of the fractional Fourier transform and related concepts, we see that the ordinary frequency domain is merely a special case of a continuum of fractional Fourier domains, which are intimately related to time-frequency (or space-frequency) representations such as the Wigner distribution. Every property and application of the ordinary Fourier transform becomes a special case of the fractional Fourier transform. In every area in which Fourier transforms and frequency-domain concepts are used, there exists the potential for generalization and improvement by using the fractional transform. For instance, the theory of optimal Wiener filtering in the ordinary Fourier domain can be generalized to optimal filtering in fractional domains, resulting in smaller signal recovery errors at practically no additional cost. The well-known result stating that the far-field optical diffraction pattern of an aperture is in the form of the Fourier transform of the aperture can be generalized to state that at closer distances, one observes the fractional Fourier transform of the aperture. (Ozaktas and Mendlovic 1995b, 1996b)

The fractional Fourier transform has been found to play an important role in the study of optical systems known as *Fourier optics*, with applications in *optical information processing*, allowing a reformulation of this area in a much more general way. It has also allowed a generalization of the notion of the frequency domain and increased our understanding of the time-frequency plane, two central concepts in *signal analysis* and *signal processing*, and is expected to have an impact in the form of deeper understanding or new applications in every area in which the Fourier transform plays a significant role, and to take its place among the standard mathematical tools of physics and engineering. We hope that this book will make possible the discovery of new applications by introducing the subject to new audiences.

1.1 Fractional operations and the fractional Fourier transform

Going from the whole of an entity to fractions of it represents a relatively major conceptual leap. The third power of 5 may be defined as $5^3 = 5 \times 5 \times 5$, but it is not obvious from this definition how one might define $5^{2.5}$. It must have taken some time before the common definition $5^{2.5} = 5^{5/2} \equiv \sqrt{5^5}$ emerged. The derivative of the function $f(u)$ is commonly denoted by $df(u)/du$. The second derivative of $f(u)$ is simply defined as the derivative of the derivative of $f(u)$, and denoted by $d^2f(u)/du^2 = d[df(u)/du]/du = (d/du)[df(u)/du] = (d/du)^2 f(u)$. Higher-order derivatives are defined similarly. Once again, it is not obvious from this definition what the 1.5th derivative of a function might mean. Let $F(\mu)$ denote the Fourier transform of $f(u)$. The Fourier transform of $d^n f(u)/du^n$, the nth derivative of $f(u)$, is known to be given by $(i2\pi\mu)^n F(\mu)$, for integer n. Now, let us generalize this property by replacing n with the real order a and take it as the definition of the ath derivative of $f(u)$. Thus, to find $d^a f(u)/du^a$, the ath derivative of $f(u)$, we simply find the inverse Fourier transform of $(i2\pi\mu)^a F(\mu)$. In both of these examples, we are dealing with fractions of an operation performed on an entity, rather than fractions of the entity itself. $2^{0.5}$ is the square root of the integer 2. The function $[f(u)]^{0.5}$ is the square root of the function $f(u)$. But $d^{0.5} f(u)/du^{0.5}$ is the 0.5th derivative of $f(u)$, with $(d/du)^{0.5}$ being the square root of the derivative operator d/du. Bracewell (1986) shows how fractional derivatives can be used to characterize the discontinuities of certain functions. Applications of so-called fractional calculus (fractional differentiation and integration) to electromagnetic theory are discussed by Engheta (1995, 1996a, b, c, 1997, 1998). Many historical and general references on this subject may be found in Engheta 1997. Fractional differences are discussed in Miller and Ross 1989.

The process of going from the whole of an entity to fractions of it underlies several of the more important conceptual developments of the last decades. One example is the study of *fractal objects*, which are essentially sets characterized by fractional, rather than integral, dimensions. Another example is *fuzzy logic*, where the binary 1 and 0 (logical true and false) are replaced by continuous values representing our certainty or uncertainty of a proposition.

Integer powers of the Fourier transform can be defined through repeated application, much like higher-order derivatives. Thus the Fourier transform of the Fourier transform of the function $f(u)$ can be defined to be the second-order Fourier transform. In other words, the second Fourier transform is obtained by applying the second power of the Fourier transform operation (or operator) to the function $f(u)$. Once again, it is not obvious how one might define the 0.5th or 1.3rd fractional Fourier transform of a function. The mathematical definition of the fractional Fourier transform and its properties will be undertaken in chapter 4. Our purpose here will be to provide some general motivation.

The ath order fractional Fourier transform operation is the ath power of the Fourier transform operation. This is most easily understood in the discrete case. Let the $N \times 1$ column vector $\mathbf{f}$ denote a discrete signal, the $N \times 1$ column vector $\mathbf{f}_1$ its discrete Fourier transform (DFT), and $\mathbf{F}$ the $N \times N$ discrete Fourier transform matrix with elements $F_{jl} = N^{-1/2} \exp[-i2\pi(j-1)(l-1)/N]$ so that

$$\mathbf{f}_1 = \mathbf{F}\mathbf{f}. \qquad (1.1)$$

Note that $\mathbf{F}$ denotes the DFT matrix and not the DFT of $\mathbf{f}$. Now, the ath discrete fractional Fourier transform of $\mathbf{f}$, which we will denote by $\mathbf{f}_a$, can be defined by multiplying $\mathbf{f}$ with the ath power of the discrete Fourier transform matrix:

$$\mathbf{f}_a = \mathbf{F}^a \mathbf{f}. \tag{1.2}$$

Clearly, when $a = 1$, $\mathbf{F}^a = \mathbf{F}^1 = \mathbf{F}$ so that $\mathbf{f}_a = \mathbf{f}_1$; the first-order Fourier transform is the ordinary Fourier transform. When $a = 0$, $\mathbf{F}^a = \mathbf{F}^0$ is the identity matrix so that $\mathbf{f}_0 = \mathbf{f}$. It is particularly easy to understand the meaning of $\mathbf{F}^a$ when a is a rational number expressible as the ratio of two integers: $a = p/q$. Then, $\mathbf{F}^{(p/q)}$ is simply the pth power of $\mathbf{F}^{(1/q)}$, where the latter is the qth root of $\mathbf{F}$. The qth root of $\mathbf{F}$ is simply the matrix whose qth power is $\mathbf{F}$. Thus the $(1/q)$th fractional Fourier transform operation is that operation which, when applied q times in succession, is equivalent to the ordinary Fourier transform operation. The (p/q)th fractional Fourier transform is simply the $(1/q)$th transform applied p times in succession.

Another viewpoint is available for those already familiar with time-frequency or space-frequency representations (or phase-space representations), such as the Wigner distribution. It is well known that the Wigner distribution of the Fourier transform of $f(u)$ is simply obtained by rotating the Wigner distribution of $f(u)$ by $\pi/2$ in the clockwise direction. The Wigner distribution of the ath order fractional Fourier transform of $f(u)$ is likewise obtained by rotating the Wigner distribution of $f(u)$ by $a\pi/2$.

The fractional Fourier transform is intimately related to several ubiquitous and indispensable concepts appearing in diverse contexts. One of these is, of course, the ordinary Fourier transform of which the fractional Fourier transform is a generalization. Another is wave and beam propagation and diffraction, which are the most basic physical embodiments of fractional Fourier transformation. In many different contexts, it is possible to interpret the solution of the wave equation and the propagation of waves as an act of continuously unfolding fractional Fourier transformation. Somewhat less stressed in this book but certainly not less important is simple harmonic motion. The harmonic oscillator is of central importance in classical and quantum mechanics not only because it represents the simplest and most basic vibrational system, but also because systems disturbed from equilibrium, no matter what the underlying forces are, can often be satisfactorily modeled in terms of harmonic oscillations. The kernel of the fractional Fourier transform corresponds to the Green's function associated with the quantum-mechanical harmonic oscillator differential equation (Schrödinger's equation). It is known that harmonic oscillations correspond to circular (or elliptic) motions in the space-frequency plane (or phase space), and so does fractional Fourier transformation, with the ordinary Fourier transformation corresponding to a $\pi/2$ rotation. Many physical phenomena are described by a family of transforms known as linear canonical transforms or one of its subclasses. While fractional Fourier transforms constitute one of the subclasses of the class of linear canonical transforms, it is possible to fruitfully interpret any linear canonical transform as a scaled fractional Fourier transform with residual quadratic phase factor, an interpretation which is sometimes more meaningful and insightful from a physical perspective. Of course, all of the concepts discussed in this paragraph are closely related to each other to begin with, and together with other time-frequency (or space-frequency, or phase-space) concepts, the fractional Fourier transform plays an important role in providing a basis for their unified study.

1.2   Applications of the fractional Fourier transform

It will be shown in chapter 9 that the fractional Fourier transform can be used
to provide an alternative statement of the law of wave propagation and to analyze
and describe a rather general class of optical systems. One of the central results of
diffraction theory is that the far-field diffraction pattern is the Fourier transform
of the diffracting object. It is possible to generalize this result by showing that
the field patterns at closer distances are the fractional Fourier transforms of the
diffracting object. More generally, in an optical system involving many lenses separated
by arbitrary distances, it is possible to show that the amplitude distribution is
continuously fractional Fourier transformed as it propagates through the system.
The order $a(z)$ of the fractional transform observed at the distance z along the
optical axis is a continuous monotonic increasing function. As light propagates, its
distribution evolves through fractional transforms of increasing orders. Wherever the
order of the transform $a(z)$ is equal to $4j + 1$ for any integer j, we observe the Fourier
transform of the input. Wherever the order is equal to $4j + 2$, we observe an inverted
image, and so on. Propagation in graded-index media and Gaussian beam propagation
are also intimately related to the fractional Fourier transform. The transform has
also been employed in the study of spherical mirror resonators (used in lasers) and
optical systems, has found use in quantum optics and statistical optics, and has been
used for lens design, beam synthesis and shaping, optical and quantum wave field
reconstruction and phase retrieval, and perspective correction. The transform has
even found its way to the journal *Zoological Studies* (Abe and Sheridan 1995c), in the
context of microscopy applications. (Ozaktas and Mendlovic 1995b, 1996b)

So-called fractional Fourier domains correspond to oblique axes in the time-
frequency plane, and the transform is directly related to the Radon transforms of
the Wigner distribution and the ambiguity function. Of particular interest from a
signal processing perspective is the concept of filtering in fractional Fourier domains,
which consists of multiplicatively modifying the signal in the fractional Fourier domain,
rather than the ordinary Fourier domain. Since the fractional Fourier transform
can be digitally implemented in $O(N \log N)$ time, just like the ordinary Fourier
transform, the additional degree of freedom associated with fractional Fourier domain
filtering comes without any additional cost. ($O(N \log N)$ means "order of $N \log N$.")
Furthermore, combining several fractional Fourier domain filtering blocks in the form
of series, parallel, or more general *filtering circuits* has led to flexible and cost-effective
time- or space-variant filtering configurations with applications in signal and image
restoration, reconstruction, denoising, and recovery, signal and system synthesis, and
signal extraction. The applications of the transform to correlation, detection, and
pattern recognition have also received significant attention. Other signal processing
applications which have been investigated or suggested include phase retrieval, radar,
tomography, multiplexing, data compression, and linear FM detection.

The fact that the fractional Fourier transform can be optically realized in a similar
manner as the ordinary Fourier transform means that the many signal and image
processing applications of the transform can also be implemented optically. Indeed
several of the mentioned applications have been first studied in an optical context.

The transform has also been employed to define new time-frequency representations
and solve certain kinds of differential equations. It has been related to a certain class of

wavelet transforms, to neural networks, and has also inspired the study of the fractional versions of many other transforms employed in signal analysis and processing.

We believe that these are only a fraction of the possible applications. Despite the fact that most of the publications in this area have so far appeared in mathematics, optics, and signal processing journals, we believe that the fractional Fourier transform will have a significant impact also in other areas of science and engineering where Fourier concepts are used.

1.3   Overview of the book

Chapter 2 is a review and collection of several basic concepts and results that will be employed throughout the book. It establishes notation and consolidates several concepts, some of which all readers may not be familiar with. All readers are assumed to have prior understanding of at least the very elementary notions of signals and systems and the Fourier transform (Papoulis 1977, Bracewell 1986, 1999). Readers with a more solid background in these areas should be able to quickly go through this chapter, perhaps spending more time on any concepts which might be new to them. The same comments apply to chapter 3, which deals with time-frequency (or space-frequency) analysis and linear canonical transforms. For readers never formally exposed to these topics, however, this chapter can serve as a tutorial which can be read from beginning to end.

Chapter 4 is the "cover story" and constitutes a self-complete, relatively comprehensive exposition of the fractional Fourier transform from a mathematical and signal analysis perspective. It brings together, for the first time, a broad array of viewpoints and results. Readers with a reasonable amount of exposure to the concepts discussed in chapters 2 and 3 (or anxious to learn about the fractional Fourier transform) should be able to directly begin with this chapter, and fall back to the earlier chapters as the need arises. Although chapters 2 and 3 aim to provide a relatively broad conceptual framework within which the content of later chapters can be better appreciated, most of the more advanced material is not essential and should not deter the reader from moving on to chapter 4 and beyond. Of the six equivalent definitions of the fractional Fourier transform discussed in chapter 4, the last three may be omitted at first reading. Chapter 5 may be considered to be an extension of chapter 4. It deals with so-called *time-order representations*, which are new time-frequency representations based on interpreting the order of the fractional Fourier transform as another dimension. Chapter 6 deals with the discrete fractional Fourier transform. Research on the discrete transform is currently in a state of flux, so that this chapter may soon become incomplete. Chapters 5 and 6 are not necessary for understanding later chapters and can be omitted without loss of continuity.

Chapter 7 is an introduction to the basic concepts of wave and geometrical optics. As in chapter 2, the purpose is to establish notation and consolidate results that will be used in later chapters; a number of less encountered perspectives and results are also presented. Readers with a solid background in elementary wave, geometrical, and Fourier optics should be able to quickly go through this chapter. Free-space propagation is treated at some length from three alternative perspectives. Readers comfortable with (or willing to take for granted) the Fresnel integral may skip this section. Chapter 8 is to chapter 7, what chapter 3 is to chapter 2. It deals with optical signals and systems in phase space (in the space-frequency plane), and provides a self-complete treatment of first-order optical systems as linear canonical transforms. *ABCD* matrix algebra is employed in a unified manner for both wave and geometrical optical perspectives. Quite a number of results not found in standard treatments have been included. Considerable space is devoted to the relationship between first-order wave and geometrical optics, and an extended treatment of optical invariants is included, since this important topic is neglected in most works. However, neither of these two topics is a prerequisite for understanding later chapters and may be omitted. Taken together, chapters 7 and 8 (and also chapter 9), supplemented with the relevant mathematical tools from earlier chapters, may be viewed as a short, self-contained course on advanced Fourier optics emphasizing space-frequency concepts. For those with no previous exposure to elementary wave, geometrical, and Fourier optics this material should be complemented with standard texts such as Papoulis 1968, Saleh and Teich 1991, or Goodman 1996, which devote a greater amount of space to explaining the fundamentals. Chapters 2, 3, 7, and 8 previously appeared as Ozaktas and Kutay 2000b.

Chapter 9 presents a broad treatment of the fractional Fourier transform in optics. This not only includes the role of the transform in understanding and analyzing optical systems, but how to implement the transform optically. Chapters 7, 8, and 9 can be omitted by those with no interest in optics, wave and beam propagation, and their applications.

Chapters 10 and 11 deal with signal and image processing applications of the transform. Either of these chapters can be read before the other. No doubt these chapters will require updating as investigations in this area progress. However, by providing several examples of how applications of the ordinary Fourier transform are generalized to the fractional Fourier transform, we hope to stimulate the development of new applications.

There are two bibliographies. One is devoted to the fractional Fourier transform and its applications. The other includes other works cited in this book which we judged are not directly related to the fractional Fourier transform. While we decided that providing separate bibliographies would be more useful, the reader will be burdened by having to look in two places to locate a given reference.

As a final word on terminology, we believe that ultimately the term "Fourier transform" should in general mean "fractional Fourier transform," and that the ordinary Fourier transform should be referred to as the "first-order Fourier transform." Likewise, DFT should stand for the discrete (fractional) Fourier transform, and instead of speaking of "fractional Fourier optics," we should be able to speak once again simply of "Fourier optics." Although we discourage the invention of new acronyms and abbreviations, in contexts where it is necessary to distinguish the fractional Fourier transform from the ordinary Fourier transform, FRT or fFt may be used to denote the fractional Fourier transform and DFRT or dfFt may be used to denote the discrete fractional Fourier transform. Less elegant alternatives include FRFT or fFT and DFRFT or DfFT. (FFT is, of course, already widely employed for the fast Fourier transform.)

2

Signals, Systems, and Transformations

2.1 Signals

2.1.1 Signals

Signals are information-bearing entities which are usually represented by one or more functions of one or more independent variables. We will mostly deal with signals represented by a scalar function of one or two real variables. For instance, a voltage signal $\hat{v}$ may be represented as a function of time t by the function $\hat{v}(t) = 2\cos(2\pi 10t)\,\mathrm{V}$, with t being measured in seconds. It is common to think of and to refer to this function as the signal itself, as a consequence of the primacy we attach to time as an independent variable. However, the same information can be equally well represented by other functions of other variables. For instance, the same information can be represented as a function of temporal frequency f (measured in hertz) in the form $\hat{V}(f) = [\delta(f-10)+\delta(f+10)]\,\mathrm{V}$, where $\hat{V}(f)$ is the Fourier transform of $\hat{v}(t)$. Thus, it will be more useful to use the term *signal* to refer to the signal $\hat{v}$ as an information-bearing entity in the abstract, and to refer to the *function* $\hat{v}(t)$ as the *time-domain representation of the signal*. When there is possibility of confusion, we will write $\hat{v}_t(t)$ instead of $\hat{v}(t)$, $\hat{v}_f(f)$ instead of $\hat{V}(f)$, and so on to identify the different functional representations of the signal $\hat{v}$. This notion of a signal is similar to the notion of a vector in classical mechanics and the notion of a ket vector in quantum mechanics (Cohen-Tannoudji, Diu, and Laloë 1977). A vector $\mathbf{r}$ is a geometrical entity independent of any coordinate system. One way of representing it is with respect to a particular rectangular coordinate system in the form $\mathbf{r} = x\hat{\mathbf{u}}_x + y\hat{\mathbf{u}}_y + z\hat{\mathbf{u}}_z$ or $\mathbf{r} = (x,y,z)$, where $\hat{\mathbf{u}}_x, \hat{\mathbf{u}}_y, \hat{\mathbf{u}}_z$ are unit vectors along the coordinate axes. Many other representations with respect to many other coordinate systems are possible. When the distinction is not crucial, we will simply say "the signal $\hat{f}(t)$" rather than "the time-domain representation $\hat{f}(t)$ of the signal $\hat{f}$."

2.1.2 Notation

Throughout this book, u, v, w will be used as generic dimensionless coordinate variables, which may be referred to as time or space depending on the context. The associated frequency-domain variables will be denoted by u_1, v_1, w_1, for reasons that will become apparent in chapter 4. The same variables will also be denoted by

$\mu \equiv u_1, \nu \equiv v_1, \eta \equiv w_1$ in simpler contexts. Dimensionless coordinate and frequency vectors will be denoted by $\mathbf{q}$ and $\mathbf{q}_1$ respectively, with $\varsigma \equiv \mathbf{q}_1$ in simpler contexts. Functions which take dimensionless arguments will simply be denoted by lowercase letters, such as $f(u), g(u, v)$. Their Fourier transforms will be denoted by $F(\mu), G(\mu, \nu)$.

The symbol t will be used for the time coordinate, $f \equiv t_1$ for temporal frequency, $r_x \equiv x, r_y \equiv y, r_z \equiv z$ for the space coordinates, and $\sigma_x \equiv x_1, \sigma_y \equiv y_1, \sigma_z \equiv z_1$ for spatial frequencies. (f will also commonly be used to represent a generic signal $f(\cdot)$ and the focal length of a lens, but this will cause no confusion.) Spatial coordinate and frequency vectors will be denoted by $\mathbf{r}$ and $\boldsymbol{\sigma}$ respectively. Functions whose arguments have the dimensions of time or distance will be denoted as $\hat{f}(t), \hat{g}(x, y), \hat{F}(f), \hat{G}(\sigma_x, \sigma_y)$.

Integrals whose limits are not indicated will denote integrals from minus to plus infinity. Likewise, summations whose limits are not indicated will denote summations over the complete range of the indices, which is often either from minus to plus infinity or from zero to plus infinity.

When we use the square root function and unless we indicate otherwise, $\sqrt{z}$ will mean the square root of z whose argument lies in the interval $(-\pi/2, \pi/2]$. We denote the imaginary unit by $i = \sqrt{-1}$.

We will use $\equiv$ instead of simply $=$ when it is important to emphasize that the left-hand side is defined as the expression on the right-hand side.

2.1.3   Some commonly used functions

The rectangle function $\mathrm{rect}(u)$ is defined to be equal to 1 in the interval $(-0.5, 0.5)$, 0.5 at $u = \pm 0.5$, and 0 elsewhere. The unit step function $\mathrm{step}(u)$ is defined to be equal to 1 when $u > 0$, 0.5 at $u = 0$, and 0 when $u < 0$ and the sign function $\mathrm{sgn}(u)$ is defined as $\mathrm{sgn}(u) = 2\,\mathrm{step}(u) - 1$. The sinc (or interpolation) function $\mathrm{sinc}(u)$ is defined as $\mathrm{sinc}(u) = \sin(\pi u)/(\pi u)$. The Gaussian function $\mathrm{gauss}(u)$ is given by $\mathrm{gauss}(u) = \exp(-\pi u^2)$, the harmonic function $\mathrm{har}(u)$ by $\mathrm{har}(u) = \exp(i 2\pi u)$, and the chirp function $\mathrm{chirp}(u)$ by $\mathrm{chirp}(u) = e^{-i\pi/4} \exp(i\pi u^2)$.

The Dirac delta function $\delta(u)$ is a generalized function which is zero everywhere except at $u = 0$, such that its integral over any interval including $u = 0$ is equal to unity. It may be defined as the limit of parametric continuous functions:

$$\delta(u) = \lim_{c \to 0} c^{-1} \mathrm{rect}(u/c), \tag{2.1}$$

$$\delta(u) = \lim_{c \to 0} c^{-1} \mathrm{sinc}(u/c), \tag{2.2}$$

$$\delta(u) = \lim_{c \to 0} c^{-1} \mathrm{gauss}(u/c), \tag{2.3}$$

$$\delta(u) = \lim_{c \to 0} c^{-1} \mathrm{chirp}(u/c), \tag{2.4}$$

where $c > 0$. The last two equations can be rewritten for all real values of c as

$$\delta(u) = \lim_{c \to 0} \frac{1}{\sqrt{|c|}} e^{-\pi u^2/|c|}, \tag{2.5}$$

$$\delta(u) = \lim_{c \to 0} e^{-i\pi \mathrm{sgn}(c)/4} \frac{1}{\sqrt{|c|}} e^{i\pi u^2/c} = \lim_{c \to 0} e^{-i\pi/4} \sqrt{\frac{1}{c}} e^{i\pi u^2/c}. \tag{2.6}$$

Table 2.1.   Properties of the Dirac delta function.

1.	$\delta(Mu) = \delta(u)/|M|$
2.	$f(u)\delta(u - \xi) = f(\xi)\delta(u - \xi)$
3.	$\int \delta(u - \xi)f(u)\,du = f(\xi)$
4.	$\int \delta(u - \xi)\delta(u - \xi')\,du = \delta(\xi - \xi')$
5.	$\int e^{\pm i2\pi(u-\xi)u'}\,du' = \delta(u - \xi)$
6.	$\int \delta'(u - u')f(u')\,du' = df(u)/du$
7.	$\int_{0-}^{\infty} \delta(u - u')\,du' = \text{step}(u)$
8.	$\delta(u) = d[\text{step}(u)]/du$

M, ξ, ξ' are real numbers.

Alternatively, the delta function may be defined through its effect under the integral sign. For every continuous function $f(u)$,

$$f(u) = \int_{-\infty}^{\infty} \delta(u - u')f(u')\,du', \tag{2.7}$$

which is known as the *sifting property*. Table 2.1 is a list of some of the common properties of the delta function. Of particular importance is the following identity:

$$\delta(u) = \int_{-\infty}^{\infty} e^{\pm i2\pi u\mu}\,d\mu. \tag{2.8}$$

The first and higher-order derivatives of the delta function are denoted as $\delta'(u)$, $\delta''(u)$, and so on. The comb function is defined as $\text{comb}(u) = \sum_{n=-\infty}^{\infty} \delta(u - n)$. The Kronecker delta $\delta_{ll'}$ is defined to be 0 when $l \neq l'$ and 1 when $l = l'$. A common identity valid for arbitrary real δu is

$$\sum_{n=-\infty}^{\infty} \delta(u + n\delta u) = \frac{1}{\delta u} \sum_{n=-\infty}^{\infty} e^{i2\pi nu/\delta u}, \tag{2.9}$$

whose right-hand side can be interpreted as the Fourier series of the comb function.

2.1.4   Analytic signals and the Hilbert transform

We will mostly deal with complex signals bearing the same information as the real physical signal. A real physical signal $f(u)$ and the associated complex signal $f_{\text{as}}(u)$, also known as the *analytic signal*, are related by

$$f_{\text{as}}(u) = f(u) + if_{\text{H}}(u), \tag{2.10}$$
$$f(u) = \Re[f_{\text{as}}(u)], \tag{2.11}$$

where $\Re[\cdot]$ denotes the real part of a complex entity and

$$f_{\text{H}}(u) = \int_{-\infty}^{\infty} \frac{1}{\pi(u - u')} f(u')\,du' \tag{2.12}$$

is the *Hilbert transform* of $f(u)$. The Fourier transform of $f_H(u)$ is given by $-i\,\text{sgn}(\mu)F(\mu)$ and $f(u)$ is orthogonal to its Hilbert transform: $\int f^*(u)f_H(u)\,du = 0$. The Fourier transform $\mathcal{F}[f_{as}(u)](\mu)$ of $f_{as}(u)$ is obtained from $F(\mu)$ according to

$$\mathcal{F}[f_{as}(u)](\mu) = 2\,\text{step}(\mu)F(\mu). \tag{2.13}$$

In the event that $f_{as}(u)$ is a narrowband signal whose spectrum is centered around some center frequency μ_0, it is convenient to express it in the form

$$f_{as}(u) = A_c(u)e^{i2\pi\mu_0 u}, \tag{2.14}$$

where $A_c(u)$ is known as the *complex envelope*, and is a lowpass (slowly varying) function. In general, the analytic signal of $A(u)\cos[2\pi\mu_0 u + \phi(u)]$ is not $A(u)\exp[i2\pi\mu_0 u + i\phi(u)]$, where $A(u)$ and $\phi(u)$ are real functions. However, this is approximately true when $A(u)$ and $\phi(u)$ are slowly varying functions (Cohen 1989).

A *monochromatic* signal is one which consists of only a single frequency: $f(u) = A_0\cos(2\pi\mu_0 u + \phi_0)$ for some particular μ_0, A_0, and ϕ_0. The associated complex representation (analytic signal) is given by $A_0\exp(i2\pi\mu_0 u + i\phi_0)$ and the complex envelope is simply $A_c(u) = A_0\exp(i\phi_0)$. In this case the complex envelope is also known as the *phasor* of $f(u)$. The real signal is recovered by multiplying the phasor by $\exp(i2\pi\mu_0 u)$ and taking the real part. Since we will mostly deal with complex representations, we will often omit the subscript "as" and simply write $f(u)$ instead of $f_{as}(u)$.

2.1.5   Signal spaces

A vector space is a set of entities for which addition and scalar multiplication have been defined such that certain axioms are satisfied (see the appendix to this chapter). A set of signals which constitute a vector space is referred to as a *signal space*. Although we will not rigorously specify what this means mathematically, we will restrict ourselves to the space of signals whose members are "physically realizable," which in particular implies that they have finite energy, and that their representations are smooth, and negligible outside some finite interval (Cohen-Tannoudji, Diu, and Laloë 1977:94). We will also use certain physically unrealizable signals which have infinite energy, but which nevertheless serve as useful intermediaries (such as the delta and harmonic functions), and occasionally deal with discontinuous functions. The fact that these signals are not "physically realizable" will not overly concern us in this book.

For concreteness, we will concentrate on the space of functions consisting of the representations of the members of a signal space in discrete and continuous domains (such as the time domain). We will mostly use $l, m, \ldots$ to denote the independent variable(s) in a discrete domain. Likewise, we will mostly use $u, v, \ldots$ to denote the independent variable(s) in a continuous domain. The addition of two functions is defined as ordinary arithmetic addition and scalar multiplication is defined as ordinary arithmetic multiplication with a complex number. The *inner product* $\langle f, g \rangle$ of two signals f and g may be defined in terms of their discrete or continuous functional representations in the l or u domain as

$$\langle f, g \rangle = \sum_l f^*(l)g(l), \tag{2.15}$$

$$\langle f, g \rangle = \int f^*(u)g(u)\,du, \tag{2.16}$$

respectively. The *energy* $\|f\|^2$ and *norm* $\|f\|$ of a signal f are defined by $\|f\|^2 = \langle f, f \rangle$. Two signals whose inner product is zero are called *orthogonal* to each other. The distance between two signals f and g is defined to be the norm of their difference: $\|f - g\|$. We will later show that the definitions of the inner product, norm, and energy of a signal are independent of the particular functional representation or domain in which we calculate it (by using equation 2.15 or 2.16).

2.2 Systems

2.2.1 Systems

A *system* is a process, event, mechanism or the like that maps a given signal into another signal. Mathematically, a system is a rule for assigning to any element f of some set of signals, an element g of (another or the same) set of signals. The signal f is referred to as the input, and the signal g is referred to as the output of the system. In other words, a system is a mapping from the input set of signals to the output set of signals (Papoulis 1977).

The rule relating the output signal g to the input signal f is denoted as

$$g = S[f]. \tag{2.17}$$

In this notation, f becomes the argument of the (possibly many-to-one) relation $S[\cdot]$ which characterizes the system. Alternatively, by interpreting S as an operator that operates on objects to its right, we may write the above in the form

$$g = Sf. \tag{2.18}$$

The same relationships can be written more explicitly in terms of the time-domain (or space-domain) representations of the signals in the form

$$g(u) = \{S[f(u)]\}(u) \tag{2.19}$$

or more simply as $g(u) = \{S[f]\}(u)$, or even $g(u) = S[f(u)]$ when there is no room for confusion. In operator notation one may write

$$g(u) = \{Sf\}(u) \tag{2.20}$$

or more simply $g(u) = Sf(u)$.

2.2.2 Linearity and superposition integrals

For a linear system $\mathcal{L}$, the output corresponding to a linear superposition of a sequence of inputs f_j, is the same linear superposition of the corresponding sequence of outputs g_j:

$$g_j = \mathcal{L}[f_j] \quad \text{for all } j \quad \Rightarrow \quad \sum_j \alpha_j g_j = \mathcal{L}\left[\sum_j \alpha_j f_j\right], \qquad (2.21)$$

where α_j are arbitrary complex coefficients. If we have continuously many inputs f_v, the summations above should be replaced with integrals over v.

$$g_v = \mathcal{L}[f_v] \quad \text{for all } v \quad \Rightarrow \quad \int \alpha_v g_v \, dv = \mathcal{L}\left[\int \alpha_v f_v \, dv\right], \qquad (2.22)$$

where α_v are arbitrary complex coefficients.

Let us express the input function $f(u)$ as a linear superposition of shifted delta functions as

$$f(u) = \int_{-\infty}^{\infty} \delta(u - u') f(u') \, du', \qquad (2.23)$$

and let $h(u, u')$ denote the output of a linear system when the input is $\delta(u - u')$:

$$h(u, u') = \mathcal{L}[\delta(u - u')]. \qquad (2.24)$$

It follows that the output $g(u)$ is related to the input $f(u)$ by the relation

$$g(u) = \int_{-\infty}^{\infty} h(u, u') f(u') \, du'. \qquad (2.25)$$

Now, let the signal f be input to the linear system $\mathcal{L}_1$ and the output be input to a second linear system $\mathcal{L}_2$:

$$\mathcal{L}_2\left[\mathcal{L}_1[f]\right], \qquad (2.26)$$

or simply $\mathcal{L}_2\mathcal{L}_1[f]$ or $\mathcal{L}_2\mathcal{L}_1 f$. The kernel $h(u, u')$ corresponding to the composite system $\mathcal{L} = \mathcal{L}_2\mathcal{L}_1$ can be given by

$$h(u, u') = \int h_2(u, u'') h_1(u'', u') \, du''. \qquad (2.27)$$

2.2.3 Some special linear systems

A number of systems that will be of special interest are tabulated in table 2.2 together with their linear transform kernels $h(u, u')$ and inverses. The inverse $\mathcal{L}^{-1}$ of a system $\mathcal{L}$, if it exists, satisfies $\mathcal{L}\mathcal{L}^{-1} = \mathcal{L}^{-1}\mathcal{L} = \mathcal{I}$. The kernel of a system $h(u, u')$ and the kernel of its inverse $h^{-1}(u, u')$ satisfy the relation

$$\int h(u, u'') h^{-1}(u'', u') \, du'' = \delta(u - u'). \qquad (2.28)$$

Table 2.2.   Special linear systems and their kernels.

Symbol	Kernel	Inverse kernel
$\mathcal{I}$	$\delta(u - u')$	$\delta(u - u')$
$\mathcal{P}$	$\delta(u + u')$	$\delta(u + u')$
$\mathcal{M}_M$	$\sqrt{|M|}\,\delta(u - Mu')$	$(1/\sqrt{|M|})\delta(u - u'/M)$
$\mathcal{SH}_\xi$	$\delta(u - u' + \xi)$	$\delta(u - u' - \xi)$
$\mathcal{PH}_\xi$	$\exp(i2\pi\xi u)\delta(u - u')$	$\exp(-i2\pi\xi u)\delta(u - u')$
Λ_h	$h(u)\delta(u - u')$	$[1/h(u)]\delta(u - u')$
$\mathcal{Q}_q$	$\exp(-i\pi q u^2)\delta(u - u')$	$\exp(i\pi q u^2)\delta(u - u')$
Λ_H	$h(u - u')$	$h^{-1}(u - u')$
$\mathcal{R}_r$	$e^{-i\pi/4}\sqrt{1/r}\,\exp[i\pi(u - u')^2/r]$	$e^{i\pi/4}(1/\sqrt{r})\exp[-i\pi(u - u')^2/r]$
$\mathcal{U}$	$u\delta(u - u')$	$u^{-1}\delta(u - u')$
$\mathcal{D}$	$(i2\pi)^{-1}\delta'(u - u')$	$(i2\pi)\text{step}(u - u')$
$\mathcal{F}$	$\exp(-i2\pi uu')$	$\exp(i2\pi uu')$

$\mathcal{I}$: identity, $\mathcal{P}$: parity, $\mathcal{M}_M$: scaling, $\mathcal{SH}_\xi$: shift or translation, $\mathcal{PH}_\xi$: phase shift, Λ_h: multiplicative filter, $\mathcal{Q}_q$: chirp multiplication, Λ_H: convolutive filter, $\mathcal{R}_r$: chirp convolution, $\mathcal{U}$: coordinate multiplication, $\mathcal{D}$: differentiation, $\mathcal{F}$: Fourier transform. M, ξ, q, r are real parameters and $\delta'(u - u') = d[\delta(u - u')]/du$. $h^{-1}(u)$ is related to $h(u)$ through $\int h(u - u')h^{-1}(u')\,du' = \delta(u)$.

Among the systems listed in the table we comment only on chirp convolution, also known as the Fresnel transform or the Fresnel integral:

$$g(u) = \sqrt{\frac{1}{r}}\,\text{chirp}(u/\sqrt{r}) * f(u) = e^{-i\pi/4}\sqrt{\frac{1}{r}}\int_{-\infty}^{\infty} f(u')e^{i\pi(u-u')^2/r}\,du'. \qquad (2.29)$$

The Fresnel transform satisfies many properties (Gori 1994) of which we will need to know

$$h^{-1}(u, u'; r) = h(u, u'; -r) = h^*(u, u', r), \qquad (2.30)$$

$$\mathcal{R}_{r_1}\mathcal{R}_{r_2} = \mathcal{R}_{r_1 + r_2}, \qquad (2.31)$$

$$\lim_{r \to 0} \mathcal{R}_r = \mathcal{I}. \qquad (2.32)$$

Here $h(u, u'; r)$ explicitly shows the dependence of the Fresnel transform kernel on the parameter r. We also note that as $r \to 0$, the transform approaches the identity transform characterized by the kernel $\delta(u - u')$.

Another important class of systems, which includes most of the above as special cases, is the class of linear canonical transforms, which we will discuss at length in chapter 3.

2.2.4   Shift invariance and convolution

Let the output of a system corresponding to the input $\delta(u)$ be denoted by $h(u)$. This system is called shift-invariant (or time-invariant or space-invariant) if the output $h(u, u')$ corresponding to the input $\delta(u - u')$ is equal to $h(u - u')$ for all u'. In this

Table 2.3. Properties of the convolution and correlation operations.

1.	$f(u) * h(u) = h(u) * f(u)$
2.	$f(-u) * h(-u) = g(-u)$
3.	$f(u) * [h_1(u) * h_2(u)] = [f(u) * h_1(u)] * h_2(u)$
4.	$f(u) * [h_1(u) + h_2(u)] = f(u) * h_1(u) + f(u) * h_2(u)$
5.	$f(u - \xi) * h(u) = g(u - \xi)$
6.	$R_{fh}(u) = f(u) \star h(u) = h^*(-u) \star f^*(-u) = R_{h^* f^*}(-u)$
7.	$f(-u) \star h(-u) = R_{fh}(-u)$
8.	$f(u) * [h_1(u) \star h_2(u)] = [f * h_1(u)] \star h_2(u)$
9.	$f(u) \star [h_1(u) + h_2(u)] = f(u) \star h_1(u) + f(u) \star h_2(u)$
10.	$f(u - \xi) \star h(u) = R_{fh}(u - \xi)$
11.	$R_{ff}(u) = R_{ff}^*(-u)$
12.	$\max[R_{ff}] = R_{ff}(0) = \int |f(u)|^2 \, du$

$g(u) = f(u) * h(u)$, $R_{fh} = f(u) \star h(u)$, and ξ is real.

case $h(u)$ is called the impulse response and the relation between the output and the input becomes

$$g(u) = f(u) * h(u) \equiv \int_{-\infty}^{\infty} h(u - u') f(u') \, du'. \tag{2.33}$$

We say that $g(u)$ is the *convolution* of the two functions $f(u)$ and $h(u)$. The *correlation* of $f(u)$ and $h(u)$ is denoted by $R_{fh}(u)$ or $f(u) \star h(u)$ and is defined as

$$R_{fh}(u) \equiv f(u) \star h(u) \equiv f(u) * h^*(-u) = \int_{-\infty}^{\infty} f(u + u') h^*(u') \, du'$$

$$= \int_{-\infty}^{\infty} f(u' + u/2) h^*(u' - u/2) \, du'. \tag{2.34}$$

Some properties of the convolution and correlation operations are summarized in table 2.3. We might also recall that the Fourier transform of $f(u) * h(u)$ is $F(\mu)H(\mu)$ and the Fourier transform of $f(u) \star h(u)$ is $F(\mu)H^*(\mu)$.

Let us now consider the eigenvalue equation for a linear shift-invariant system with impulse response $h(u)$:

$$\{\mathcal{L}[f(u)]\}(u) = \lambda f(u), \tag{2.35}$$

which we may simply write as $\mathcal{L}f(u) = h(u) * f(u) = \lambda f(u)$ in operator notation. Rewriting the right-hand side of equation 2.33 as $\int f(u - u')h(u') \, du'$, it is easy to show that $f(u) = \exp(i2\pi\mu u)$ is a solution for all real μ with eigenvalue λ_μ given by

$$\lambda_\mu = \int_{-\infty}^{\infty} h(u) e^{-i2\pi\mu u} \, du. \tag{2.36}$$

Interpreted as a function of μ, we see that λ_μ is nothing but $H(\mu)$, the Fourier transform of $h(u)$. Thus we see that harmonic functions are eigenfunctions of linear shift-invariant systems, with the eigenvalues being given by the Fourier transform of the impulse response.

2.3   Representations and transformations

2.3.1   Systems versus transformations

Signals can be represented in many different ways which are distinct in appearance but nevertheless contain the same information. A common example is given by the time- and frequency-domain representations of a signal. Another common example is given by the two functional forms of an image with respect to two coordinate systems which are rotated with respect to each other. In both cases the two representations both contain the same information and either representation can be obtained from the other. The act of obtaining one representation from the other is called a *transformation*.

Different representations of a signal correspond to different coordinate systems or basis sets. Once an appropriate basis set is chosen, the signals may be expressed as a linear superposition of the elements of the basis set. In other words, the signal may be expanded in terms of the elements of the basis set. The coefficients appearing in this superposition or expansion, which uniquely specify the signal, constitute the representation of the signal with respect to this basis set.

It is important to distinguish clearly between systems and transformations, although mathematically they can take similar forms. A system is a rule that maps an input signal into an output signal. A system is usually a mathematical abstraction of a physical system which alters a physical input in a certain way to produce a physical output. For instance, a live television broadcasting system tries to reproduce the event in front of the camera as faithfully as possible on the retinas of human observers watching their televisions at home. A careful study of its physical components will enable characterization of this system and how it departs from this ideal. Other systems will intentionally alter the input, such as a pattern recognition system whose inputs are images and outputs are labels of recognized images. A system can alter the information content of the input signal in producing the output signal, so that systems need not always be invertible. A system, much like a signal, is an abstract entity whose existence is independent of which coordinate system or basis set we choose to work with. The output may be represented in either the same representation as the input, or a different one, without affecting the nature of the system.

A transform(ation), on the other hand, is merely a change of the coordinate system or basis set used, with which we move from one representation of a signal to another. The signal is not altered, but expressed in another form bearing the same information. As such, transformations are usually invertible. Despite this clear distinction between systems and transformations, they are often mathematically expressed in the same way, and both are often represented by abstract operators (which we denote by calligraphic letters). As an example, consider an image signal f. Let this image be input to a system $\mathcal{L}$ which rotates the input image by $\pi/4$ in the clockwise direction to obtain the output image. Notice that the definition of the system is not tied to any particular coordinate system or representation. To relate the output image to the input image mathematically, we may choose a particular rectangular coordinate system in which the image f is represented by the function $f(u,v)$. Then, the output image will be represented by the function $g(u,v)$ which is related to the input image by

$$g(u,v) = f\left(\cos(\pi/4)u - \sin(\pi/4)v,\ \sin(\pi/4)u + \cos(\pi/4)v\right). \qquad (2.37)$$

Now, let us set the system $\mathcal{L}$ aside and consider a transformation $\mathcal{T}$ which rotates the coordinate axes by $\pi/4$ in the counterclockwise direction. The new coordinates u', v' are related to u, v as follows:

$$u' = +\cos(\pi/4)u + \sin(\pi/4)v,$$
$$v' = -\sin(\pi/4)u + \cos(\pi/4)v, \tag{2.38}$$

and the representation of the signal with respect to the new coordinate axes, which we denote by $f'(u', v')$, is given by

$$f'(u', v') = f\left(\cos(\pi/4)u' - \sin(\pi/4)v', \sin(\pi/4)u' + \cos(\pi/4)v'\right), \tag{2.39}$$

which we see is identical in form to the output $g(u, v)$ of system $\mathcal{L}$.

One can always define a system based on a transformation (but not necessarily the other way around). For instance, consider the Fourier transformation which relates the frequency-domain representation $F(\mu)$ of a signal to its space-domain representation $f(u)$ as follows:

$$F(\mu) = \int f(u)e^{-i2\pi\mu u}\, du. \tag{2.40}$$

Let us rewrite the same with a change in dummy variables as

$$g(u) = \int f(u')e^{-i2\pi u u'}\, du'. \tag{2.41}$$

This latter equation can be interpreted as the rule relating the output of a system g to its input f, expressed in a particular coordinate system. Thus a transformation is employed as the rule that relates the output of the system to the input of the system in a particular representation.

In other words, although we more often think of the Fourier transform as a transformation, it can also be interpreted as a system. This is particularly useful in physical contexts. For instance, the simple $2f$ setup used to implement Fourier transforms optically (chapter 7) is a physical system which alters an input distribution of light in a particular way to produce an output distribution of light. This physical *system* can be characterized by (a scaled version of) equation 2.41. However, the purpose for which this system is most often employed is to compute the Fourier *transformation* of the input and present it as the output; that is, to obtain the frequency-domain representation of a signal from its space-domain representation.

In physics the distinction between systems and transformations is often referred to as the distinction between *active* and *passive* transformations (Wolf 1979). Active transformations are produced by operators which bodily move the vectors or signals, and correspond to what we have called "systems." Passive transformations arise from a change in the basis used for the description of the space, and correspond to what we have simply called "transformations."

2.3.2   Basis sets and representations

A set of discretely (countably) many signals, denoted by $\{\psi_l\}$, is said to be *orthonormal* if all of its members have unit norm and are orthogonal to each other:

$$\langle \psi_l, \psi_{l'} \rangle = \delta_{ll'}. \tag{2.42}$$

A set of continuously (uncountably) many signals, denoted by $\{\Psi_v\}$, is likewise orthonormal if

$$\langle \Psi_v, \Psi_{v'} \rangle = \delta(v - v'). \tag{2.43}$$

Using the inner product definition given in equation 2.16, these conditions may be written in the time or space domain as

$$\int \psi_l^*(u)\psi_{l'}(u)\,du = \delta_{ll'}, \tag{2.44}$$

$$\int \Psi_v^*(u)\Psi_{v'}(u)\,du = \delta(v - v'). \tag{2.45}$$

The set $\{\psi_l\}$ (or $\{\Psi_v\}$) is said to constitute a *basis* for a signal space if every member of the signal space can be expanded in one and only one way in terms of the elements of this set. In the discrete case, this expansion may be expressed as

$$f = \sum_l f_\psi(l)\,\psi_l, \tag{2.46}$$

where $f_\psi(l)$ are the expansion coefficients of the signal f with respect to the basis $\{\psi_l\}$. These expansion coefficients, interpreted as a function of the discrete variable l, constitute the representation of the signal f in the basis $\{\psi_l\}$. Alternatively, we may say that they represent the signal f in this basis, or with respect to this basis. Likewise, in the continuous case,

$$f = \int f_\Psi(v)\,\Psi_v\,dv, \tag{2.47}$$

where $f_\Psi(v)$ are the expansion coefficients of the signal f with respect to the basis $\{\Psi_v\}$. These coefficients constitute the representation of signal f in the basis $\{\Psi_v\}$. The above equations which are written in terms of abstract signals may be specialized to a particular domain, such as the time domain:

$$f(u) = \sum_l f_\psi(l)\psi_l(u), \tag{2.48}$$

$$f(u) = \int f_\Psi(v)\Psi_v(u)\,dv. \tag{2.49}$$

If every function $f(u)$ belonging to the function space of interest can be expanded in terms of the elements of the set of functions $\{\psi_l(u)\}$, this set of functions is said to constitute a *complete* set of functions. Alternatively, it is sometimes said that the set of functions *spans* the space of interest. A linearly independent set of functions which spans the space of interest, such that the expansion not only exists but is unique, constitutes a basis for that space. Similar statements can be made for the set of functions $\{\Psi_v(u)\}$. The act of going from the $f(u)$ representation to the $f_\psi(l)$ or $f_\Psi(v)$ representation is referred to as a *transformation*.

In order to obtain the expansion coefficients for an orthonormal basis set, we take the inner product of both sides of equations 2.46 or 2.47 with a particular member of the basis set. For the discrete and continuous cases respectively,

$$\langle \psi_{l'}, f \rangle = \sum_l f_\psi(l) \langle \psi_{l'}, \psi_l \rangle = \sum_l f_\psi(l) \delta_{l'l} = f_\psi(l'), \tag{2.50}$$

$$\langle \Psi_{v'}, f \rangle = \int f_\Psi(v) \langle \Psi_{v'}, \Psi_v \rangle \, dv = \int f_\Psi(v) \delta(v' - v) \, dv = f_\Psi(v'), \tag{2.51}$$

so that

$$f_\psi(l) = \langle \psi_l, f \rangle = \int \psi_l^*(u) f(u) \, du, \tag{2.52}$$

$$f_\Psi(v) = \langle \Psi_v, f \rangle = \int \Psi_v^*(u) f(u) \, du, \tag{2.53}$$

where the rightmost forms are expressed in the time domain using the inner product definition given in equation 2.16. The orthonormality conditions given by equations 2.42 and 2.43 have been used in deriving these results.

Now, let us substitute equations 2.52 and 2.53 in equations 2.48 and 2.49, respectively, and use the orthonormality relations to obtain

$$\sum_l \psi_l^*(u') \psi_l(u) = \delta(u - u'), \tag{2.54}$$

$$\int \Psi_v^*(u') \Psi_v(u) \, dv = \delta(u - u'), \tag{2.55}$$

for all u, u'. Treating u and u' as parameters, the summation and integral can be interpreted as inner products so that we can formally write the above equations in the form of inner products:

$$\langle \psi_l(u'), \psi_l(u) \rangle = \delta(u - u'), \tag{2.56}$$
$$\langle \Psi_v(u'), \Psi_v(u) \rangle = \delta(u - u'), \tag{2.57}$$

for all u, u'. These relations are known as *closure* or *completeness* conditions. It is worth comparing and contrasting the closure conditions with the orthonormality conditions given by equations 2.42 and 2.43. These conditions are indeed a statement of completeness of the set of functions and their constituting an orthonormal basis set. The *braket notation* employed in quantum mechanics provides a very elegant means of expressing such relations; see Cohen-Tannoudji, Diu, and Laloë 1977.

Let us summarize by repeating the following two key relations for discrete bases:

$$f_\psi(l) = \int \psi_l^*(u) f(u) \, du, \tag{2.58}$$

$$f(u) = \sum_l f_\psi(l) \psi_l(u). \tag{2.59}$$

The first of these equations gives the coefficient $f_\psi(l)$ appearing in the expansion of $f(u)$ given in the second equation. If this second equation is a given; that is, if we know that $f(u)$ *can* be expanded in terms of the orthonormal set $\{\psi_l(u)\}$, then the first equation for $f_\psi(l)$ is derived simply by orthonormality. The second equation expresses the somewhat more subtle fact that the projections $f_\psi(l)\psi_l(u)$ for all l add up to $f(u)$ itself. This is what we mean when we say that the set $\{\psi_l(u)\}$ is a basis. (For example, in $\mathbf{R}^3$, any two of the common unit vectors $\hat{\mathbf{u}}_x$, $\hat{\mathbf{u}}_y$, $\hat{\mathbf{u}}_z$ do not constitute a basis but all three of them do.)

Let us now assume that we are given the representations $f_\psi(l)$ and $g_\psi(l)$ of two signals f and g in the basis $\{\psi_l\}$, or the representations $f_\Psi(v)$ and $g_\Psi(v)$ in the basis $\{\Psi_v\}$, and that we wish to calculate the inner product $\langle f, g \rangle$ directly in terms of these representations. By substituting equations 2.46 and 2.47 in equations 2.15 and 2.16 and using the orthonormality conditions we can easily show that

$$\langle f, g \rangle = \sum_l f_\psi^*(l) g_\psi(l), \tag{2.60}$$

$$\langle f, g \rangle = \int f_\Psi^*(v) g_\Psi(v) \, dv, \tag{2.61}$$

which we see are identical in form to equations 2.15 and 2.16. Thus, the expression for the inner product and hence the norm $\|f\| = \sqrt{\langle f, f \rangle}$ is independent of the particular basis set in which we represent the signal. In other words, no matter which representation of the signal we use in equations 2.15 and 2.16, we will always obtain the same result. This justifies our previous assertion that inner products and norms are properties of the signals in the abstract, and not tied to any particular representation. When it comes to actually calculating them, we can calculate inner products and norms in any representation we find convenient.

We now turn our attention to the representation of systems with respect to particular bases. Let a linear system $\mathcal{L}$ mapping an input signal f to an output $g = \mathcal{L}f$ be defined with respect to the basis set $\{\psi_l\}$ as

$$g_\psi(l) = \sum_{l'} L_\psi(l, l') f_\psi(l'), \tag{2.62}$$

where $L_\psi(l, l')$ is the representation of this linear system with respect to the basis set $\{\psi_l\}$. This equation is the most general linear relation between the representation of g and the representation of f. To see how $L_\psi(l, l')$ can be expressed in terms of the members of the basis set $\{\psi_l\}$, let us start from the system equation in abstract form $g = \mathcal{L}f$ and substitute the expansions of g and f to obtain

$$\sum_l g_\psi(l)\, \psi_l = \mathcal{L} \left[\sum_{l'} f_\psi(l')\, \psi_{l'} \right] = \sum_{l'} f_\psi(l')\, \mathcal{L}\psi_{l'}. \tag{2.63}$$

Now, taking the inner product of both sides from the left with ψ_l we obtain

$$g_\psi(l) = \sum_{l'} \langle \psi_l, \mathcal{L}\psi_{l'} \rangle f_\psi(l'), \tag{2.64}$$

from which we recognize

$$L_\psi(l, l') = \langle \psi_l, \mathcal{L}\psi_{l'} \rangle. \tag{2.65}$$

This expression shows how the representation of a system with respect to a particular basis is related to the abstract system operator $\mathcal{L}$ and the members of the basis set.

The trace of a system is a representation-invariant quantity defined by

$$\text{Tr}[\mathcal{L}] = \sum_l L_\psi(l,l), \tag{2.66}$$

$$\text{Tr}[\mathcal{L}] = \int L_\Psi(u,u)\, du, \tag{2.67}$$

for discrete and continuous bases respectively. It is easy to show that the trace is the same no matter which representation it is calculated in. That is, $\sum_l L_\psi(l,l) = \sum_l L_\phi(l,l)$ for any two discrete basis sets $\{\psi_l\}$ and $\{\phi_l\}$. It is also known that the trace is equal to the summation of the eigenvalues. Corresponding results hold for continuous bases.

We finally note that if we have a set of signals spanning a certain space, it is possible to obtain an orthonormal basis set by using a process known as *Gram-Schmidt orthogonalization* (Naylor and Sell 1982).

2.3.3 Impulse and harmonic bases

We will now illustrate some of the above concepts with two familiar examples. First, we consider the set of signals $\Psi_v = \delta_v$ which are defined through their representations in the time domain as $\delta_v(u) = \delta(u-v)$. This set of signals constitute an orthonormal basis set as they obviously satisfy the orthonormality and closure conditions:

$$\langle \delta_v, \delta_{v'} \rangle = \int \delta(u-v)\delta(u-v')\, du = \delta(v-v'), \tag{2.68}$$

$$\langle \delta_v(u), \delta_v(u') \rangle = \int \delta(u-v)\delta(u'-v)\, dv = \delta(u-u'). \tag{2.69}$$

This basis set will be referred to as the impulse basis. The expansion of a signal f in this basis takes the form

$$f = \int f_\delta(v)\, \delta_v\, dv, \tag{2.70}$$

$$f_\delta(v) = \langle \delta_v, f \rangle. \tag{2.71}$$

Expressed in the time domain, these expression take the form

$$f(u) = \int f_\delta(v)\delta(u-v)\, dv, \tag{2.72}$$

$$f_\delta(v) = \int \delta(u-v)f(u)\, du, \tag{2.73}$$

from which we see that $f_\delta(v) = f(v)$. The expansion coefficients of f corresponding to the impulse basis set is simply $f(v)$, the representation of the signal in the time domain. Alternatively, we may say that what we conventionally call the time-domain representation of the signal f and denote by $f(u)$, is nothing but the representation of the signal in the impulse basis. Equivalently, the impulse basis $\{\delta_v\}$ is the basis set associated with what is conventionally called the time domain. Thus the time-domain representation is no different from any other representation in terms of the status accorded to it in our framework. It does not have a special place and is on an equal footing with other representations.

As a second example, we consider the set of signals $\Psi_v = \text{har}_v$ which are defined to correspond in the time domain to the set of eigenfunctions $\text{har}_v(u) = e^{i2\pi vu}$ of linear shift-invariant systems. This set of signals constitutes an orthonormal basis set as they satisfy the orthonormality and closure conditions:

$$\langle \text{har}_v, \text{har}_{v'} \rangle = \int e^{-i2\pi vu} e^{i2\pi v'u} \, du = \delta(v - v'), \tag{2.74}$$

$$\langle \text{har}_v(u), \text{har}_v(u') \rangle = \int e^{-i2\pi vu} e^{i2\pi vu'} \, dv = \delta(u - u'). \tag{2.75}$$

These equations are simply two different instances of equation 2.8. This basis set will be referred to as the harmonic basis. The expansion of a signal f in this basis takes the form

$$f = \int f_{\text{har}}(v) \, \text{har}_v \, dv, \tag{2.76}$$

$$f_{\text{har}}(v) = \langle \text{har}_v, f \rangle. \tag{2.77}$$

Expressed in the time domain, these expression take the form

$$f(u) = \int f_{\text{har}}(v) e^{i2\pi vu} \, dv, \tag{2.78}$$

$$f_{\text{har}}(v) = \int e^{-i2\pi vu} f(u) \, du, \tag{2.79}$$

from which we see that $f_{\text{har}}(v)$ is equal to $F(v)$, the Fourier transform of $f(u)$. The expansion coefficients of f corresponding to the harmonic basis set are simply $F(v)$, the representation of the signal in the frequency domain. Alternatively, we may say that what we conventionally call the frequency-domain representation of the signal f and denote by $F(\mu)$, is nothing but the representation of the signal in the harmonic basis. Equivalently, the harmonic basis $\{\text{har}_v\}$ is the basis set associated with what is conventionally called the frequency domain.

Notice that members of both the impulse set and the harmonic set have infinite norms and energies; they are not square integrable. They are not physically realizable signals but are mathematical idealizations which are found to be quite indispensable as intermediaries. We cannot physically realize impulse or harmonics functions but we can expand physically realizable functions in terms of them. Both of them are examples of continuous bases; an example of a discrete basis set will be given in section 2.5.2.

2.3.4 Transformations between representations

We have seen that the coefficients appearing in the expansion of a signal in terms of an orthonormal basis set constitute the representation of the signal with respect to that basis set. We will now examine more closely the relations between different representations and transformations between them.

Let us assume that a new orthonormal basis $\{\phi_l\}$ is defined in terms of the orthonormal basis $\{\psi_l\}$ through the relation

$$\psi_l = \mathcal{T}\phi_l \qquad \text{for all } l, \tag{2.80}$$

where $\mathcal{T}$ is a linear system. To ensure that the set $\{\phi_l\}$ as defined is indeed an orthonormal basis, $\mathcal{T}$ must satisfy certain properties. A linear system $\mathcal{T}$ which maps any orthonormal basis into another orthonormal basis, is called a *unitary* system. (Conversely, a unitary system will always map an orthonormal basis into another orthonormal basis.) A unitary system always has an inverse $\mathcal{T}^{-1}$ so that we can write

$$\phi_l = \mathcal{T}^{-1}\psi_l \qquad \text{for all } l. \tag{2.81}$$

In the continuous case, a new orthonormal basis $\{\Phi_v\}$ may be defined in terms of the orthonormal basis $\{\Psi_v\}$ through

$$\Psi_v = \mathcal{T}\Phi_v \qquad \text{for all } v. \tag{2.82}$$

Let us now consider two discrete orthonormal basis sets $\{\psi_l\}$ and $\{\phi_l\}$ and consider the expansion of f in terms of both of these basis sets:

$$f = \sum_{l'} f_\psi(l')\psi_{l'}, \tag{2.83}$$

$$f = \sum_{l} f_\phi(l)\phi_l. \tag{2.84}$$

We wish to find the relation between $f_\psi(l)$ and $f_\phi(l)$. One way is to expand each member of one of the sets in terms of members of the other:

$$\psi_{l'} = \sum_{l} \langle \phi_l, \psi_{l'} \rangle \phi_l, \tag{2.85}$$

and substitute this in equation 2.83 to recognize

$$f_\phi(l) = \sum_{l'} \langle \phi_l, \psi_{l'} \rangle f_\psi(l') \tag{2.86}$$

from equation 2.84. This equation allowing us to compute $f_\phi(l)$ in terms of $f_\psi(l)$ is an explicit expression of the transformation from the $\{\psi_l\}$ basis to the $\{\phi_l\}$ basis. The inner products $\langle \phi_l, \psi_{l'} \rangle = \langle \psi_{l'}, \phi_l \rangle^*$ constitute a two-dimensional array of coefficients, which we will define as the *transformation coefficients* $T(l, l')$ of the transformation from the $\{\psi_l\}$ representation to the $\{\phi_l\}$ representation: $T(l, l') \equiv \langle \phi_l, \psi_{l'} \rangle$. Using equation 2.81, these inner products can also be written as $\langle \mathcal{T}^{-1}\psi_l, \psi_{l'} \rangle = \langle \phi_l, \mathcal{T}\phi_{l'} \rangle$. Equation 2.86 can now be rewritten as

$$f_\phi(l) = \sum_{l'} T(l, l') f_\psi(l'). \tag{2.87}$$

The coefficients $T^{-1}(l, l')$ of the inverse transformation (equation 2.81) from the $\{\phi_l\}$ representation to the $\{\psi_l\}$ representation are likewise given by $T^{-1}(l, l') \equiv \langle \psi_l, \phi_{l'} \rangle = \langle \phi_{l'}, \psi_l \rangle^*$, from which we conclude that

$$T^{-1}(l, l') = T^*(l', l). \tag{2.88}$$

To obtain the array of transformation coefficients for the inverse transformation, we simply take the conjugate transpose of the array of coefficients for the forward transformation. Equation 2.88 also implies

$$\sum_{l'} T^*(l, l') T(j, l') = \delta_{lj}, \tag{2.89}$$

$$\sum_{l} T^*(l, l') T(l, j') = \delta_{l'j'}, \tag{2.90}$$

from which we see that the rows and columns of the array of coefficients are orthogonal to each other.

We have showed that if $\mathcal{T}$ is unitary; that is, if $\mathcal{T}$ maps any orthonormal basis set into another orthonormal basis set, it satisfies equation 2.88. Conversely, it is also possible to show that an array of coefficients satisfying this equation does indeed map any orthonormal basis into another orthonormal basis. (This amounts to showing that if the basis $\{\psi_l\}$ satisfies orthonormality and closure relations, then so does the basis $\{\phi_l\}$, a task which we leave to the reader.) For this reason, equation 2.88 is often taken to be the defining property of a unitary system. Likewise, the transformation expressed by equation 2.87, with $T(l, l')$ satisfying equation 2.88, is called a unitary transformation.

For continuous orthonormal basis sets we can analogously write

$$f_\Phi(u) = \int \langle \Phi_u, \Psi_{u'} \rangle f_\Psi(u') \, du'. \tag{2.91}$$

Defining the transformation coefficients $T(u, u') \equiv \langle \Phi_u, \Psi_{u'} \rangle = \langle \Psi_{u'}, \Phi_u \rangle^*$, we can write

$$f_\Phi(u) = \int T(u, u') f_\psi(u') \, du', \tag{2.92}$$

and so on. We note that again $T^{-1}(u, u') = T^*(u', u)$. Transformations from a discrete set to a continuous set and the other way around are similarly handled.

A transformation is linear if the relation between the two representations is linear, as in equations 2.92 and 2.87. This implies that if $f_{j_\phi}(l) = \sum_{l'} T(l, l') f_{j_\psi}(l')$ for some sequence of signals f_j, then

$$\sum_{l'} T(l, l') \left[\sum_j \alpha_j f_{j_\psi}(l') \right] = \sum_j \alpha_j f_{j_\phi}(l), \tag{2.93}$$

where α_j are arbitrary complex coefficients.

We now turn our attention to the transformation of the representations of systems, rather than signals, from one basis set to another. Let the output g of a linear system $\mathcal{L}$ be related to the input f through the relation $g = \mathcal{L}f$. This can be expressed as

$$g_\psi(l) = \sum_{l'} L_\psi(l, l') f_\psi(l'), \tag{2.94}$$

$$g_\phi(l) = \sum_{l'} L_\phi(l, l') f_\phi(l'), \tag{2.95}$$

in the $\{\psi_l\}$ and $\{\phi_l\}$ representations respectively. Our aim is to find the relation between $L_\phi(l,l')$ and $L_\psi(l,l')$. We can write two instances of equation 2.87 as

$$f_\psi(l) = \sum_{l'} T^{-1}(l,l') f_\phi(l'), \qquad (2.96)$$

$$g_\phi(l) = \sum_{l'} T(l,l') g_\psi(l'), \qquad (2.97)$$

and use these in equation 2.94 to obtain

$$g_\phi(l) = \sum_{l'} \sum_{l''} \sum_{l'''} T(l,l') L_\psi(l',l'') T^{-1}(l'',l''') f_\phi(l'''), \qquad (2.98)$$

from which we recognize the desired result as

$$L_\phi(l,l') = \sum_{l''} \sum_{l'''} T(l,l'') L_\psi(l'',l''') T^{-1}(l''',l'), \qquad (2.99)$$

or, since $T^{-1}(l,l') = T^*(l',l)$,

$$L_\phi(l,l') = \sum_{l''} \sum_{l'''} T(l,l'') L_\psi(l'',l''') T^*(l',l'''). \qquad (2.100)$$

For the continuous case we can likewise derive

$$L_\Phi(u,u') = \iint T(u,u'') L_\Psi(u'',u''') T^*(u',u''') \, du'' \, du'''. \qquad (2.101)$$

An alternative derivation of the above result, which we leave to the reader, takes as a starting point the closure relation and equation 2.7.

As a simple example, consider the kernel $h(u,u') = h(u-u')$ in the time domain $(g(u) = \int h(u,u') f(u') \, du')$, which becomes the kernel $H(\mu,\mu') = H(\mu)\delta(\mu-\mu')$ in the frequency domain $(G(\mu) = \int H(\mu,\mu') F(\mu') \, d\mu')$, where $F(\mu)$, $G(\mu)$, $H(\mu)$ are the Fourier transforms of $f(u)$, $g(u)$, $h(u)$. The reader may illustrate the above general results for this special case where the Fourier transform plays the role of the unitary transformation.

2.4 Operators

2.4.1 Operators

Operators are mathematical objects that can be used to denote either systems or transformations. They are denoted by calligraphic letters such as $\mathcal{S}$ or $\mathcal{T}$. For instance, the clockwise rotation system or the counterclockwise coordinate transformation discussed on page 15 may both be denoted by the symbol $\mathcal{ROT}_{\pi/4}$. In the case of systems, they denote a system in the abstract, without reference to any particular representation or basis set. In the case of transformations, they denote the underlying system through which the new basis set is related to the old (equation 2.80).

A linear operator is an operator denoting a linear system or transformation. In the case of systems, an operator relates an output signal g to an input signal f in the form $g = \mathcal{L}f$. If we choose to represent signals with respect to a particular continuous basis set $\{\Psi_u\}$ in the form $f(u)$, $g(u)$, and so forth, we may write this relation more explicitly in one of several forms:

$$g(u) = \mathcal{L}f(u) \equiv (\mathcal{L}f)(u) \equiv \mathcal{L}[f](u) \equiv \{\mathcal{L}[f]\}(u). \qquad (2.102)$$

We are writing simply $f(u)$, $g(u)$ instead of $f_\Psi(u)$, $g_\Psi(u)$ since only one representation is involved in this context. Also note that u is a dummy variable; we could have written v or some other symbol instead. The rightmost form in the above equation explicitly denotes the Ψ-representation of the abstract signal $g = \mathcal{L}[f]$. The form preceding it is essentially the same but the brackets have been omitted. In the two forms preceding these, we have employed the convention that $\mathcal{L}f$ stands for $\mathcal{L}[f]$. The form $\mathcal{L}f(u)$ is interpreted as the Ψ-representation of the abstract signal $\mathcal{L}f$. This form also allows another interpretation. Let us define $\mathcal{L}$ such that it acts on the function $f(u)$, rather than the abstract signal f, to result in another function $g(u) = \mathcal{L}[f(u)]$ in the obvious manner:

$$\mathcal{L}[f(u)] \equiv (\mathcal{L}f)(u). \qquad (2.103)$$

Thus the action of a system, on a function which is the representation of a signal with respect to some basis, will be defined to be the representation (with respect to the same basis) of the output of that system when the input is the underlying signal f. In other words, since a signal is fully characterized by any of its representations, we may write one of these representations in place of the signal and agree that this means that the result is also expressed in the same representation. Sometimes it will be useful to denote the output explicitly:

$$g(u) = \mathcal{L}f(u) \equiv \mathcal{L}[f(u)] \equiv \mathcal{L}[f(u)](u) \equiv \{\mathcal{L}[f(u)]\}(u). \qquad (2.104)$$

The form $\mathcal{L}f(u)$ is ambiguous in that it allows $\mathcal{L}$ to be interpreted both as an operator that acts on abstract signals and as an operator that acts on functions. Since both interpretations are consistent and useful, this expression will be used to denote the Ψ-representation of $g = \mathcal{L}f$, or equivalently, $\mathcal{L}[f(u)]$.

Likewise, an expression such as $\langle f(u), g(u) \rangle$ will simply be interpreted as the inner product of f and g evaluated in the Ψ-representation. (Of course, the inner product is always the same no matter what representation it is evaluated in.)

The representation of the system $\mathcal{L}$ with respect to this basis set will be denoted by $L(u, u')$ and the output $g(u)$ will be related to the input $f(u)$ as in equation 2.25 or the continuous version of equation 2.62:

$$g(u) = \int L(u, u') f(u') \, du'. \qquad (2.105)$$

When there is possibility of confusion, we will employ explicit labels such as $f_\Psi(u)$, $L_\Psi(u, u')$, and so on, as introduced earlier.

The interpretation of $\mathcal{L}$ as a mapping from functions to functions (representations to representations) is convenient also in the case of transformations. Let us repeat the above forms for the transformation $\mathcal{T}$:

$$f_\Phi(v) = \mathcal{T}f_\Psi(u) \equiv \mathcal{T}[f_\Psi(u)] \equiv \mathcal{T}[f_\Psi(u)](v) \equiv \{\mathcal{T}[f_\Psi(u)]\}(v). \qquad (2.106)$$

We have used distinct variables u and v for the two representations since this reminds us that the functions inhabit distinct spaces, but this is not of any deeper significance since both u and v are dummy variables. For a transformation between a basis set $\{\Psi_u\}$ to another basis set $\{\Phi_v\}$, the representation of the transformation $T(v, u)$ with respect to these basis sets will be denoted by $T(v, u)$ and $f_\Phi(v)$ will be related to $f_\Psi(u)$ as in equation 2.92:

$$f_\Phi(v) = \mathcal{T} f_\Psi(u) = \int T(v, u) f_\Psi(u) \, du. \tag{2.107}$$

When there is possibility of confusion, we will employ explicit labels such as $T_{\Psi \to \Phi}(v, u)$, and so on. Analogous expressions may be written for discrete representations.

To summarize, we have now defined the effect of operators (whether they represent systems or transformations) on signals and on functions in a consistent manner. The expression $\mathcal{L} f(u)$ can be interpreted as the Ψ-representation of the abstract signal $\mathcal{L} f$, or the action of the operator $\mathcal{L}$ on the function $f(u)$. The expression $\mathcal{T} f_\Psi(u)$ denotes the Φ-representation of f.

The formal similarity between systems and transformations is sometimes useful, although their physical interpretations are distinct. It is sometimes useful to think of transformations as if they were systems in mathematical manipulations. $f_\Psi(u)$ is interpreted as the input, and $f_\Phi(v)$ as the output. For instance, the system which rotates the input by $\pi/4$ in the clockwise direction is associated with the transformation which corresponds to rotation of the rectangular coordinate system by $\pi/4$ in the counterclockwise direction. The operator notation embodies the formal similarity between systems and transformations and allows them to be treated in a unified manner, so that in the course of symbolic manipulations we do not need to distinguish between systems and transformations.

The *Hermitian conjugate* (or *Hermitian transpose* or *conjugate transpose* or *adjoint*) $\mathcal{L}^H$ of a linear operator $\mathcal{L}$ with representation $L(u, u')$ (or $L(l, l')$) is defined as the operator whose representation is $L^H(u, u') = L^*(u', u)$ (or $L^H(l, l') = L^*(l', l)$). Hermitian conjugation satisfies the following properties:

$$(\mathcal{L}^H)^H = \mathcal{L}, \tag{2.108}$$

$$(\mathcal{L}_1 \mathcal{L}_2 \cdots \mathcal{L}_n)^H = \mathcal{L}_n^H \cdots \mathcal{L}_2^H \mathcal{L}_1^H, \tag{2.109}$$

$$\langle f, \mathcal{L} g \rangle = \langle \mathcal{L}^H f, g \rangle, \tag{2.110}$$

$$\langle \mathcal{L} f, g \rangle = \langle f, \mathcal{L}^H g \rangle, \tag{2.111}$$

where f and g are any two signals. It can be shown, for instance by using equations 2.100 or 2.101, that this definition is independent of the representation in which the conjugate transpose is taken. In fact, it is also possible to take the representation-independent equation 2.110 or equation 2.111 as the definition of Hermitian conjugation, which can be readily shown to be equivalent to the definition we have given above. (Choosing a particular basis $\{\psi_l\}$ and using equation 2.65, $L^H(l, l') = \langle \psi_l, \mathcal{L}^H \psi_{l'} \rangle = \langle \mathcal{L} \psi_l, \psi_{l'} \rangle = \langle \psi_{l'}, \mathcal{L} \psi_l \rangle^* = L^*(l', l)$, and similarly for the continuous case.)

An operator $\mathcal{H}$ is called *Hermitian* if it is equal to its Hermitian conjugate: $\mathcal{H}^H = \mathcal{H}$. Thus the representation of such an operator satisfies the relation $H(u, u') = H^*(u', u)$.

For an operator denoting a system, being Hermitian is a property of the system in the abstract, and is not tied to a particular representation $H(u, u')$. It is easy to see from equation 2.110 or equation 2.111 that Hermitian operators satisfy $\langle f, \mathcal{H}g \rangle = \langle \mathcal{H}f, g \rangle$. In fact, this equality can be taken as a representation-independent definition of Hermitian operators. $(H(l, l') \equiv \langle \psi_l, \mathcal{H}\psi_{l'} \rangle = \langle \mathcal{H}\psi_l, \psi_{l'} \rangle = \langle \psi_{l'}, \mathcal{H}\psi_l \rangle^* = H^*(l', l)$, and similarly for the continuous case.) If two operators are Hermitian so is their sum and difference. The quantity $\langle f, \mathcal{H}f \rangle$ for arbitrary f is always real as can be seen easily by writing $\langle \mathcal{H}f, f \rangle^* = \langle f, \mathcal{H}f \rangle = \langle \mathcal{H}f, f \rangle$.

An operator $\mathcal{T}$ is called *unitary* if its inverse equals its Hermitian conjugate: $\mathcal{T}^H = \mathcal{T}^{-1}$, or equivalently $\mathcal{T}\mathcal{T}^H = \mathcal{T}^H\mathcal{T} = \mathcal{I}$ where $\mathcal{I}$ is the identity operator. Thus the representation of such an operator satisfies the relation $T^*(u, u') = T^{-1}(u', u)$. Operators denoting linear transformations (in the sense of expressing a signal with respect to a new orthonormal basis set) are always unitary, as we have seen in association with equations 2.87 and 2.92. If two operators are unitary then so is their product. It is easy to verify that if $\mathcal{H}$ is Hermitian and $\mathcal{T}$ is unitary, then $\mathcal{T}^{-1}\mathcal{H}\mathcal{T}$ is also Hermitian, $\langle \mathcal{T}f, g \rangle = \langle f, \mathcal{T}^{-1}g \rangle$, $\langle \mathcal{T}f, \mathcal{T}g \rangle = \langle f, \mathcal{T}^{-1}\mathcal{T}g \rangle = \langle f, g \rangle$, and $\|\mathcal{T}f\|^2 = \|f\|^2$, for arbitrary f and g. The latter properties mean that inner products and norms are conserved when the signals in question are acted upon by unitary operators. This property is what underlies their being interpretable as transformations from one basis to another, as we have already seen. A kernel $T(l, l')$ whose columns (or rows) constitute an orthonormal set is unitary since it can be directly shown that $\mathcal{T}^H\mathcal{T} = \mathcal{I} = \mathcal{T}\mathcal{T}^H = \mathcal{I}$. Conversely, the columns (or rows) of a unitary kernel constitute an orthonormal set.

Particularly important properties of Hermitian and unitary operators are those regarding their eigenfunctions, which will be discussed later on.

Another correspondence between systems and transformations is that between equation 2.65, which we repeat for convenience,

$$L_\psi(l, l') = \langle \psi_l, \mathcal{L}\psi_{l'} \rangle, \tag{2.112}$$

and the corresponding

$$T_{\psi \to \phi}(l, l') \equiv T(l, l') = \langle \phi_l, \mathcal{T}\phi_{l'} \rangle = \langle \psi_l, \mathcal{T}\psi_{l'} \rangle, \tag{2.113}$$

$$T_{\phi \to \psi}(l, l') \equiv T^{-1}(l, l') = \langle \psi_l, \mathcal{T}^{-1}\psi_{l'} \rangle = \langle \phi_l, \mathcal{T}^{-1}\phi_{l'} \rangle, \tag{2.114}$$

which can be derived from the definition of the transformation matrix $T(l, l') = \langle \phi_l, \psi_{l'} \rangle$ and $\psi_l = \mathcal{T}\phi_l$. Let us now write $f = \sum_l f_\phi(l)\, \phi_l$ and apply the unitary operator $\mathcal{T}$ on both sides to obtain

$$\mathcal{T}f = \sum_l f_\phi(l)\, \psi_l, \tag{2.115}$$

implying $\langle \psi_l, \mathcal{T}f \rangle = f_\phi(l)$ or

$$[\mathcal{T}f]_\psi(l) = f_\phi(l). \tag{2.116}$$

How is this last equation to be interpreted? It says that the ψ-representation of the signal $\mathcal{T}f$ is functionally identical in appearance to the ϕ-representation of the signal f. Looking at the same equation from the other way around, the ϕ-representation of the signal f may be found by finding the ψ-representation of the signal $\mathcal{T}f$. The operator $\mathcal{T}$ interpreted as a system, is related to the operator $\mathcal{T}$ interpreted as a transformation, in the same way that the rotational system and rotational transformation discussed on page 15 are related. $\mathcal{T}f$ is that signal whose ψ-representation looks exactly like the ϕ-representation of the signal f. Thus, if we wish to find the ϕ-representation of the signal f, we might obtain the ψ representation of the signal $\mathcal{T}f$ instead.

To show the utility of this formalism, let us rederive two previous results. We have already shown that the inner products and norms of signals are independent of which representation they are calculated in. This is particularly easy to see by using the properties of unitary operators. Let $\mathcal{T}$ denote the unitary transformation between two representations of the signals f and g as $f_\Phi(u) = \mathcal{T}f_\Psi(u)$ and $g_\Phi(u) = \mathcal{T}g_\Psi(u)$. Then

$$\langle f_\Psi(u), g_\Psi(u) \rangle = \langle \mathcal{T}^{-1}f_\Phi(u), \mathcal{T}^{-1}g_\Phi(u) \rangle = \langle f_\Phi(u), \mathcal{T}\mathcal{T}^{-1}g_\Phi(u) \rangle = \langle f_\Phi(u), g_\Phi(u) \rangle, \tag{2.117}$$

proving the desired result. Now, let us consider the transformation of the representation of a linear system from one basis to another, which is also particularly transparent in operator notation. With $f_\Phi(u) = \mathcal{T}f_\Psi(u)$, $g_\Phi(u) = \mathcal{T}g_\Psi(u)$ and $g_\Psi(u) = \int L_\Psi(u, u')f_\Psi(u')\,du'$ we obtain

$$L_\Phi(u, u') = \mathcal{T}L_\Psi(u, u')\mathcal{T}^{-1}, \tag{2.118}$$

whose explicit form was derived earlier (equation 2.101).

2.4.2 Eigenvalue equations

Let us consider the eigenvalue equation for the linear operator $\mathcal{L}$:

$$\mathcal{L}f = \lambda f. \tag{2.119}$$

f is called the eigenvector or eigensignal, and λ the eigenvalue. The representation of an eigensignal with respect to a particular basis is referred to as an eigenfunction. To solve this abstract equation, we must first write it in a particular representation. For instance, in the discrete ψ-representation we have

$$\mathcal{L}f_\psi(l) = \lambda f_\psi(l), \tag{2.120}$$

or more explicitly

$$\sum_{l'} L_\psi(l, l')f_\psi(l') = \lambda f_\psi(l), \tag{2.121}$$

and in the continuous Ψ-representation we have

$$\mathcal{L}f_\Psi(u) = \lambda f_\Psi(u), \tag{2.122}$$

or more explicitly

$$\int L_\Psi(u, u')f_\Psi(u')\,du' = \lambda f_\Psi(u). \tag{2.123}$$

Let us consider the discrete case and let $\mathcal{T}$ represent the (unitary) transformation from the ψ-representation to the ϕ-representation so that $f_\phi(u) = \mathcal{T} f_\psi(u)$ and $L_\phi(u, u') = \mathcal{T} L_\psi(u, u') \mathcal{T}^{-1}$. With these equation 2.121 becomes

$$\sum_{l'} \mathcal{T}^{-1} L_\phi(l, l') \, \mathcal{T} \mathcal{T}^{-1} f_\phi(l) = \lambda \mathcal{T}^{-1} f_\phi(l), \tag{2.124}$$

$$\sum_{l'} \mathcal{T}^{-1} L_\phi(l, l') f_\phi(l') = \lambda \mathcal{T}^{-1} f_\phi(l), \tag{2.125}$$

$$\sum_{l'} L_\phi(l, l') f_\phi(l') = \lambda f_\phi(l), \tag{2.126}$$

which is of the same form as equation 2.121. We have simply rewritten the eigenvalue equation in another representation. Clearly, if we have a solution $f_\psi(l)$ of equation 2.121 with eigenvalue λ, then $f_\phi(l)$ will be a solution of equation 2.126 *with the same eigenvalue*. Thus the eigenvalues and eigensignals of a system are properties of the system in the abstract, and are not tied to the particular representation in which we solve the eigenvalue equation.

From now on we restrict our attention to Hermitian or unitary operators. The eigenvalues of Hermitian operators are always real and the eigenvalues of unitary operators are always of unit magnitude, as can be easily verified. In general there will be several values of λ for which a solution to the eigenvalue equation can be found. For such operators, the eigensignals corresponding to distinct eigenvalues are always orthogonal to each other, but this cannot be said for two eigensignals which share the same eigenvalue. However, it is possible to show that within the subspace spanned by all eigensignals which share the same m-degenerate eigenvalue, it is always possible to find m linearly independent eigensignals. These m linearly independent eigensignals can be orthogonalized among themselves so that for such operators it is always possible to find an orthogonal set of eigensignals. When speaking of the eigensignals of Hermitian or unitary operators, we will always assume that the eigensignals have been chosen so that they constitute an orthonormal set. In general, it may not always be the case that this orthonormal set constitutes a basis for the space of signals we are interested in (Cohen-Tannoudji, Diu, and Laloë 1977:137). We will, however, assume that this is the case for the operators we are dealing with.

2.4.3 Diagonalization and spectral expansion

We will now assume that $\{\psi_l\}$ is an orthonormal set of eigensignals of the operator $\mathcal{L}$, constituting a basis for the signal space we are interested in. Eigensignal decompositions are often a convenient way of finding the output of a linear system in response to an arbitrary input. If we expand the input signal in terms of the eigensignals of the system in the form

$$f = \sum_l f_\psi(l) \, \psi_l, \tag{2.127}$$

we can easily obtain the output g corresponding to this input by applying the linear system operator $\mathcal{L}$ to both sides of the above equation to obtain

$$g = \mathcal{L}f = \sum_l f_\psi(l)\,\mathcal{L}\psi_l, \qquad (2.128)$$

and since ψ_l is an eigensignal of $\mathcal{L}$ with eigenvalue λ_l, we have

$$g = \mathcal{L}f = \sum_l f_\psi(l)\,\lambda_l\psi_l, \qquad (2.129)$$

which we can compare with $g = \sum_l g_\psi(l)\,\psi_l$ to recognize

$$g_\psi(l) = f_\psi(l)\,\lambda_l. \qquad (2.130)$$

We see that the output g of the system is simply a signal whose expansion coefficient is $f_\psi(l)\,\lambda_l$. The effect of a linear system on an input signal turns out to be particularly simple if we know the representation of the signal in the eigensignal basis. Let us also find the representation of $\mathcal{L}$ with respect to the eigensignal basis. We earlier showed that $L_\psi(l,l') = \langle \psi_l, \mathcal{L}\psi_{l'} \rangle$. But since $\psi_{l'}$ is an eigensignal of $\mathcal{L}$ we have

$$L_\psi(l,l') = \lambda_l \delta_{ll'}. \qquad (2.131)$$

Likewise, for a continuous set of eigensignals,

$$L_\Psi(u,u') = \lambda_u \delta(u - u'). \qquad (2.132)$$

We see that the representation of $\mathcal{L}$ is diagonal in the eigensignal basis. The representation of $\mathcal{L}$ in another basis will not be diagonal; the act of transforming to the eigensignal basis is thus referred to as *diagonalization* and takes the form given in equation 2.100. The transformation kernel $T_{\phi \to \psi}(l,l')$ from an arbitrary basis $\{\phi_l\}$ to the eigensignal basis $\{\psi_l\}$ is given by $\langle \psi_l, \phi_{l'} \rangle = \langle \psi_l, T_{\phi \to \psi}\psi_{l'} \rangle$. Interpreted as a function of l with l' a parameter, we recognize this kernel as the representation of $\phi_{l'}$ in the $\{\psi_l\}$ basis. Interpreted as a transformation matrix (with infinite dimensions), we see that $T_{\phi \to \psi}(l,l')$ consists of the orthonormal eigensignals of $\mathcal{L}$ as its columns. The orthonormality of the ψ_l is consistent with the unitarity of $T_{\phi \to \psi}(l,l')$. If $\mathcal{L}$ is Hermitian, a set of orthonormal eigensignals always exists, so that $\mathcal{L}$ can always be diagonalized by a unitary transformation whose columns consist of the orthonormal set of eigensignals of the Hermitian matrix.

If the eigensignals of $\mathcal{L}$ constitute an orthonormal basis for the signal space of interest, then knowing the eigensignals and eigenvalues of $\mathcal{L}$ is sufficient to completely characterize the system. Let us start by expanding the input f in terms of the eigensignal basis $\{\psi_l\}$ in the form $f = \sum_l f_\psi(l)\psi_l$ where $f_\psi(l) = \langle \psi_l, f \rangle$ and write

$$g = \mathcal{L}f = \sum_l f_\psi(l)\mathcal{L}\psi_l = \sum_l \langle \psi_l, f \rangle \mathcal{L}\psi_l = \sum_l \langle \psi_l, f \rangle \lambda_l \psi_l. \qquad (2.133)$$

Now, let us represent this abstract equation in any representation we find convenient to work with. For instance, in the time domain,

$$g(u) = \sum_l \langle \psi_l, f \rangle \lambda_l \psi_l(u) = \int \left[\sum_l \lambda_l \psi_l(u)\psi_l^*(u') \right] f(u')\,du', \qquad (2.134)$$

from which we can recognize the time-domain kernel $L(u, u')$ as

$$L(u, u') = \sum_l \lambda_l \psi_l(u) \psi_l^*(u'). \tag{2.135}$$

Likewise, with respect to a continuous set of eigensignals we can show that

$$L(u, u') = \int \lambda_v \psi_v(u) \psi_v^*(u') \, dv. \tag{2.136}$$

Such expansions of a kernel are known as *spectral expansions* (or *spectral decompositions* or *singular value decompositions*).

Another way of interpreting the spectral expansion is as follows. Note that the effect of $\psi_v(u)\psi_v^*(u')$ under the integral $\int du'$ is precisely to find the projection of $f(u)$ along $\psi_v(u)$. Letting $\mathcal{PR}_v$ denote the projection operators whose kernels are $\mathcal{PR}_v = \psi_v(u)\psi_v^*(u')$, we can write the spectral expansion as

$$\mathcal{L} = \int \lambda_v \mathcal{PR}_v \, dv, \tag{2.137}$$

where we have assumed that all eigenvalues are distinct. (If there are multiple eigenvalues, then we employ projection operators onto the distinct eigenspaces.) Thus each term in equation 2.134 is the projection of the signal onto one of the eigensignals, multiplied with the corresponding eigenvalue. These are added back together to obtain $g(u)$. (Also note that the projection operators satisfy $\mathcal{PR}_v \mathcal{PR}_{v'} = \mathcal{PR}_v$ if $v = v'$ and $\mathcal{PR}_v \mathcal{PR}_{v'} = 0$ if $v \neq v'$.)

As an example, let us apply the above procedure to a linear shift-invariant system $\mathcal{L}$ whose eigenfunctions we saw were $\exp(i2\pi\mu u)$. An arbitrary input $f(u)$ may be expanded in terms of these eigenfunctions as

$$f(u) = \int F(\mu) e^{i2\pi\mu u} \, d\mu, \tag{2.138}$$

$$F(\mu) = \int e^{-i2\pi\mu u} f(u) \, du. \tag{2.139}$$

The eigenfunction representation of $\mathcal{L}$ is diagonal; its kernel has the particularly simple form $L_{\text{har}}(\mu, \mu') = H(\mu)\delta(\mu - \mu')$ where $H(\mu)$ is the eigenvalue given by equation 2.36. The effect of $\mathcal{L}$ in this representation is simply expressed as $G(\mu) = \int L_{\text{har}}(\mu, \mu')F(\mu') \, d\mu' = H(\mu)F(\mu)$. This final relation is nothing but the so-called convolution property of the Fourier transform. An interesting exercise is to start from equation 2.136 and specialize it to obtain the spectral expansion of a linear shift-invariant system with impulse response $h(u)$. With $\psi_{\text{har}}(u) = \exp(i2\pi\mu u)$ and $\lambda_{\text{har}} = H(\mu)$ we obtain

$$h(u) = \int H(\mu) e^{i2\pi\mu u} d\mu, \tag{2.140}$$

which is nothing but the inverse of equation 2.36. Note that the expansion coefficient $H(\mu)$, which is nothing but the frequency-domain representation of $h(u)$, is also equal to the eigenvalue associated with the eigenfunction $\exp(i2\pi\mu u)$ (page 14). If we expand the impulse response of a particular system in terms of the eigenfunctions of that system, the expansion coefficients will correspond to the eigenvalues.

As another example, let us consider a system with time-domain representation $L(u, u')$ and whose eigenfunctions constitute a discrete set $\{\psi_l\}$. (We will later see systems whose eigensignals constitute such a discrete set, most notably a system defined by the fractional Fourier transform.) The eigenfunction representation of the system will be of the form $L_\psi(l, l') = \lambda_l \delta_{ll'}$, as we now show. The unitary transformation from the time-domain representation to the eigenfunction representation is given by $T(l, u) = \langle \psi_l(u'), \delta(u' - u) \rangle = \int \psi_l^*(u') \delta(u' - u) \, du' = \psi_l^*(u)$ and exhibits one continuous and one discrete variable. The eigenfunction representation of the system is found as

$$L_\psi(l, l') = \int \int \psi_l^*(u) L(u, u') \psi_{l'}(u') \, du \, du', \tag{2.141}$$

which can be shown without difficulty to be simply equal to

$$L_\psi(l, l') = \lambda_l \delta_{ll'}. \tag{2.142}$$

2.4.4 Functions of operators

Integer powers of operators are simply defined as their repeated application. Thus $\mathcal{L}^2 = \mathcal{L}\mathcal{L}$, $\mathcal{L}^3 = \mathcal{L}\mathcal{L}^2$, and so on. $\mathcal{L}^{-1}$ is defined as the inverse of $\mathcal{L}$, so that this definition can be easily generalized to negative integers. If $\mathcal{L}$ is Hermitian or unitary, $\mathcal{L}^n$ is also so.

Let $\Upsilon(z)$ denote a function of a complex variable whose polynomial series is defined everywhere:

$$\Upsilon(z) = \sum_{n=0}^{\infty} \Upsilon_n z^n. \tag{2.143}$$

Then we can take

$$\Upsilon(\mathcal{L}) \equiv \sum_{n=0}^{\infty} \Upsilon_n \mathcal{L}^n \tag{2.144}$$

as the definition of $\Upsilon(\mathcal{L})$. It can be shown that if λ_l is an eigenvalue of $\mathcal{L}$ and ψ_l the corresponding eigensignal, then $\mathcal{L}^n$ has the same eigensignal with the eigenvalue λ_l^n. This property can also be made the basis of an alternative definition. We have earlier seen that, provided they constitute a basis set for the set of signals under consideration, the eigenvalues and eigensignals of an operator are sufficient to fully characterize it. Thus we may define $\Upsilon(\mathcal{L})$ by specifying its eigensignals and eigenvalues as follows:

$$\Upsilon(\mathcal{L}) \psi_l = \Upsilon(\lambda_l) \psi_l \qquad \text{for all } l. \tag{2.145}$$

In order to calculate $\Upsilon(\mathcal{L}) f$, we first decompose f in terms of the eigensignals of $\mathcal{L}$

and use the preceding equation:

$$f = \sum_l f_\psi(l)\,\psi_l, \tag{2.146}$$

$$\Upsilon(\mathcal{L})\,f = \sum_l f_\psi(l)\Upsilon(\lambda_l)\,\psi_l. \tag{2.147}$$

Equation 2.52 allows us to write this in the time domain as

$$\Upsilon(\mathcal{L})\,f(u) = \int \Upsilon(u,u')f(u')\,du', \tag{2.148}$$

$$\text{with} \quad \Upsilon(u,u') = \sum_l \Upsilon(\lambda_l)\psi_l(u)\psi_l^*(u'). \tag{2.149}$$

Alternatively, we may start from the diagonalized $L_\psi(l,l')$ given in equation 2.142, replace the eigenvalues λ_l with $\Upsilon(\lambda_l)$, and then transform back to the time domain:

$$\Upsilon(u,u') = \sum_l \sum_{l'} \psi_l(u)\Upsilon(l,l')\psi_{l'}^*(u'), \tag{2.150}$$

$$\text{with} \quad \Upsilon(l,l') = \Upsilon(\lambda_l)\delta_{ll'}, \tag{2.151}$$

which can be shown to yield equation 2.149.

If the operator $\mathcal{H}$ is Hermitian, the operator $\exp(ia\mathcal{H})$ with real a, is unitary. (In different applications a may represent time, the axis of propagation, or the order of a parametric transform such as the fractional Fourier transform.)

The derivative of an operator $\mathcal{L}$ is the operator whose representation is the common derivative of the representation of $\mathcal{L}$. Thus if the time-domain representation of $\mathcal{L}$ is $L(u,u')$, the time-domain representation of $d\mathcal{L}/da$, where a is some real parameter implicit in $\mathcal{L}$, is $dL(u,u')/da$. Manipulations involving derivatives of operators and functions of operators are easily carried out by considering series expansion of the functions. One can show, for instance, that some common rules of differentiation apply to operators and their functions:

$$\frac{d}{da}e^{\mathcal{A}a} = \mathcal{A}e^{\mathcal{A}a}, \tag{2.152}$$

where $\mathcal{A}$ is assumed not to depend on a.

The *commutator* of two operators $\mathcal{A}$ and $\mathcal{B}$ is another operator denoted by $[\mathcal{A},\mathcal{B}]$ and defined as

$$[\mathcal{A},\mathcal{B}] \equiv \mathcal{A}\mathcal{B} - \mathcal{B}\mathcal{A}. \tag{2.153}$$

Two operators whose commutator is the zero operator are said to commute: $\mathcal{A}\mathcal{B} = \mathcal{B}\mathcal{A}$. It is always the case that $[\mathcal{A},\Upsilon(\mathcal{A})] = 0$; operators commute with functions of themselves. If two operators $\mathcal{A}$ and $\mathcal{B}$ commute, then $[\mathcal{B},\Upsilon(\mathcal{A})] = 0$ and

$$e^{\mathcal{A}}e^{\mathcal{B}} = e^{\mathcal{A}+\mathcal{B}} = e^{\mathcal{B}+\mathcal{A}} = e^{\mathcal{B}}e^{\mathcal{A}}. \tag{2.154}$$

If $[\mathcal{A},\mathcal{B}] = \pm i\mathcal{I}$, then we have

$$[\mathcal{A},\mathcal{B}^n] = \pm in\mathcal{B}^{n-1}, \qquad [\mathcal{A},\Upsilon(\mathcal{B})] = \pm i\Upsilon'(\mathcal{B}), \tag{2.155}$$

where $\Upsilon'(\cdot)$ is the derivative of the function $\Upsilon(\cdot)$. The latter equality can be shown by expanding $\Upsilon(\cdot)$ into a power series. Two operators satisfying equation 2.155 are the coordinate multiplication and differentiation operators $\mathcal{U}$ and $\mathcal{D}$.

We conclude with some additional results applying to two operators $\mathcal{A}$ and $\mathcal{B}$ which commute ($[\mathcal{A}, \mathcal{B}] = 0$). It is possible to show that if f is an eigensignal of $\mathcal{A}$ with eigenvalue λ, then $\mathcal{B}f$ is also an eigensignal of $\mathcal{A}$ with the same eigenvalue, since $\mathcal{A}(\mathcal{B}f) = \lambda(\mathcal{B}f)$. Furthermore, if λ is a nondegenerate eigenvalue, then $\mathcal{B}f \propto f$ so that f is also an eigensignal of $\mathcal{B}$. For any two commuting operators which are Hermitian or unitary, it is always possible to find a common set of orthonormal eigensignals. Thus, in this case one can find a representation in which both of these operators are diagonal. (Cohen-Tannoudji, Diu, and Laloë 1977)

2.5   The Fourier transform

2.5.1   Definition and properties

The Fourier transform(ation) $F(\mu)$ of the function $f(u)$ is defined as (Bracewell 1986)

$$F(\mu) = \int_{-\infty}^{\infty} f(u)e^{-i2\pi\mu u}\, du. \tag{2.156}$$

The function $f(u)$ can be recovered from its Fourier transform by

$$f(u) = \int_{-\infty}^{\infty} F(\mu)e^{i2\pi\mu u}\, d\mu. \tag{2.157}$$

To see this, substitute either of these equations in the other and use equation 2.8. These relations are valid for the set of finite-energy functions mentioned on page 10, as well as a more general set of functions which include pure harmonic functions, delta functions, chirp functions, and so on, whose exact nature we will not precisely define (Dym and McKean 1972).

The Fourier transform can be interpreted either as a system or a transformation. As a system, the Fourier transform of a signal f will be denoted by $F \equiv \mathcal{F}f \equiv \mathcal{F}[f]$. This relation may be expressed in the time domain as

$$F(u) = \mathcal{F}f(u) \equiv (\mathcal{F}f)(u) \equiv \mathcal{F}[f](u) \equiv \{\mathcal{F}[f]\}(u). \tag{2.158}$$

As a transformation from the time domain to the frequency domain we write

$$F(\mu) = \mathcal{F}f(u) \equiv \mathcal{F}[f(u)] \equiv \mathcal{F}[f(u)](\mu) \equiv \{\mathcal{F}[f(u)]\}(\mu) \equiv \{\mathcal{F}[f]\}(\mu) \equiv \{\mathcal{F}f\}(\mu), \tag{2.159}$$

where $\mathcal{F}[f] \equiv \mathcal{F}f$ simply stand for F.

The inverse Fourier transform operator will be denoted by $\mathcal{F}^{-1}$ and satisfies $\mathcal{F}^{-1}\mathcal{F} = \mathcal{F}\mathcal{F}^{-1} = \mathcal{I}$. Integer powers of the Fourier transform are defined through repeated application: $\mathcal{F}^0 = \mathcal{I}$ and $\mathcal{F}^j = \mathcal{F}\mathcal{F}^{j-1}$ for integer j. Properties of the Fourier transform are summarized in tables 2.4 and 2.5.

Table 2.4.   Properties of the Fourier transform, part I.

	$f(u)$	$F(\mu)$
1.	$f(-u)$	$F(-\mu)$
2.	$|M|^{-1}f(u/M)$	$F(M\mu)$
3.	$f(u-\xi)$	$\exp(-i2\pi\mu\xi)F(\mu)$
4.	$\exp(i2\pi\xi u)f(u)$	$F(\mu-\xi)$
5.	$u^n f(u)$	$(-i2\pi)^{-n}d^n F(\mu)/d\mu^n$
6.	$(i2\pi)^{-n}d^n f(u)/du^n$	$\mu^n F(\mu)$
7.	$f^*(u)$	$F^*(-\mu)$
8.	$f^*(-u)$	$F^*(\mu)$
9.	$[f(u)+f(-u)]/2$	$[F(\mu)+F(-\mu)]/2$
10.	$[f(u)-f(-u)]/2$	$[F(\mu)-F(-\mu)]/2$
11.	$f(u)*h(u)$	$F(\mu)H(\mu)$
12.	$f(u)h(u)$	$F(\mu)*H(\mu)$
13.	$f(u)\star h(u)$	$F(\mu)H^*(\mu)$
14.	$R_{ff}(u)=f(u)\star f(u)$	$|F(\mu)|^2$

The expressions on the right are Fourier transforms of the
expressions on the left. M, ξ are real but $M \neq 0, \pm\infty$, and
n is a positive integer.

Table 2.5.   Properties of the Fourier transform, part II.

1.	$\mathcal{F}[\sum_j \alpha_j f_j] = \sum_j \alpha_j \mathcal{F}f_j$	$\mathcal{F}[\sum_j \alpha_j f_j(u)] = \sum_j \alpha_j F_j(\mu)$
2.	$\mathcal{F}^2 = \mathcal{P}$	$\mathcal{F}^2 f(u) = \mathcal{F}F(u) = f(-u)$
3.	$\mathcal{F}^3 = \mathcal{P}\mathcal{F} = \mathcal{F}\mathcal{P}$	$\mathcal{F}^3 f(u) = \mathcal{F}f(-u) = \mathcal{F}^2 F(u) = F(-u)$
4.	$\mathcal{F}^4 = \mathcal{F}^0 = \mathcal{I}$	$\mathcal{F}^4 f(u) = f(u)$
5.	$\langle f, g \rangle = \langle F, G \rangle$	$\int f^*(u)g(u)\,du = \int F^*(\mu)G(\mu)\,d\mu$
6.	$\text{En}[f] = \text{En}[F]$	$\int |f(u)|^2\,du = \int |F(u)|^2\,d\mu$

The same properties are expressed in abstract signal and operator form on the left
and explicitly on the right, where the Fourier transform has been interpreted as a
system. α_j are arbitrary complex constants.

Properties 1 to 6 in table 2.4 can be expressed in operator notation as well (see
table 2.2):

$$\mathcal{F}\mathcal{P} = \mathcal{P}\mathcal{F}, \tag{2.160}$$

$$\mathcal{F}\mathcal{M}_M = \mathcal{M}_{1/M}\mathcal{F}, \tag{2.161}$$

$$\mathcal{F}\,\mathcal{S}\mathcal{H}_\xi = \mathcal{P}\mathcal{H}_\xi\mathcal{F}, \tag{2.162}$$

$$\mathcal{F}\,\mathcal{P}\mathcal{H}_\xi = \mathcal{S}\mathcal{H}_{-\xi}\mathcal{F}, \tag{2.163}$$

$$\mathcal{F}\mathcal{U}^n = (-\mathcal{D})^n\mathcal{F}, \tag{2.164}$$

$$\mathcal{F}\mathcal{D}^n = \mathcal{U}^n\mathcal{F}. \tag{2.165}$$

Examination of the kernel $\exp(-i2\pi\mu u)$ of the Fourier transform and the kernel
$\exp(i2\pi\mu u)$ of the inverse Fourier transform reveals that the Fourier transform

Table 2.6. Some common Fourier pairs.

	$f(u)$	$F(\mu)$
1.	$\delta(u - \xi)$	$\mathrm{har}(-\xi\mu) = \exp(-i2\pi\xi\mu)$
2.	$\mathrm{har}(\xi u) = \exp(i2\pi\xi u)$	$\delta(\mu - \xi)$
3.	$\mathrm{rect}(u)$	$\mathrm{sinc}(\mu)$
4.	$\mathrm{sinc}(u)$	$\mathrm{rect}(\mu)$
5.	$\mathrm{gauss}(u) = \exp(-\pi u^2)$	$\mathrm{gauss}(\mu) = \exp(-\pi\mu^2)$
6.	$e^{\mp i\pi/4}\exp(\pm i\pi u^2)$	$\exp(\mp i\pi\mu^2)$
7.	$\exp[i\pi(\chi u^2 + 2\xi u)]$	$(e^{i\pi/4}/\sqrt{\chi})\exp[-i\pi(u - \xi)^2/\chi]$
8.	$\mathrm{comb}(u) = \sum_{n=-\infty}^{\infty}\delta(u - n)$	$\mathrm{comb}(\mu) = \sum_{n=-\infty}^{\infty}\delta(\mu - n)$
9.	$i/\pi u$	$\mathrm{sgn}(\mu)$
10.	$\sum_{n=-N}^{N}\delta(u - n\delta u)$	$\sin[\pi(2N + 1)\delta u\,\mu]/\sin[\pi\delta u\,\mu]$

$\xi, \chi, \delta u$ are real, N is a positive integer.

operator is unitary. Parseval's relation (table 2.5, property 5) is a direct consequence of this fact.

Further properties may be derived from those given. For instance, if $f(u)$ is an analytic signal expressed in the form $A(u)\exp[i\phi(u)]$ where $A(u)$ and $\phi(u)$ are real functions, then we have (Cohen 1989)

$$\int \mu|F(\mu)|^2\,d\mu = \frac{1}{2\pi}\int \frac{d\phi(u)}{du}\,|f(u)|^2\,du; \qquad (2.166)$$

we leave the proof to the reader. $d\phi(u)/du$ can be interpreted as the instantaneous frequency of $f(u)$, if $f(u)$ is a narrowband signal. Two other properties concern the magnitudes of the function and its derivatives (Bracewell 1986):

$$|f(u)| \leq \int |F(\mu)|\,d\mu, \qquad (2.167)$$

$$\frac{df(u)}{du} \leq 2\pi \int |\mu F(\mu)|\,d\mu. \qquad (2.168)$$

The derivatives of a function provide information on its rate of change. So does the frequency spectrum. For instance, if the Fourier transform of $f(u)$ is zero for $|\mu| \geq \mu_{\max}$, and the function is bounded by $f_{\max}$ such that $|f(u)| \leq f_{\max}$ for all u, then $|df(u)/du| \leq \mu_{\max}f_{\max}$ for all u (Papoulis 1968:131). This result shows the relationship between the frequency spectrum and derivatives of the function.

Common Fourier transform pairs are given in table 2.6. Pair number 6 in table 2.6 is intimately related to equation 2.6, as can be seen by writing it as

$$\mathcal{F}\left[e^{-i\pi/4}\frac{1}{|s|}e^{i\pi(u/s)^2}\right] = e^{-i\pi(s\mu)^2}, \qquad (2.169)$$

where s is real and considering $s \to 0$. The finite delta train appearing in pair 10 is equal to $\mathrm{rect}[u/(2N + 1)\delta u]\,\delta u\,\mathrm{comb}(u/\delta u)$, whose Fourier transform is given by

$(2N+1)\delta u\operatorname{sinc}[(2N+1)\delta u\,\mu]*\operatorname{comb}(\delta u\,\mu)$. Thus the right-hand side of pair 10 can also be written as

$$\frac{\sin[\pi(2N+1)\delta u\,\mu]}{\sin[\pi\delta u\,\mu]}=\sum_{n=-\infty}^{\infty}(2N+1)\delta u\operatorname{sinc}\left[(2N+1)\delta u(\mu-n/\delta u)\right],\qquad(2.170)$$

showing that it consists of periodically replicated sinc functions. This function is illustrated in figure 2.1.

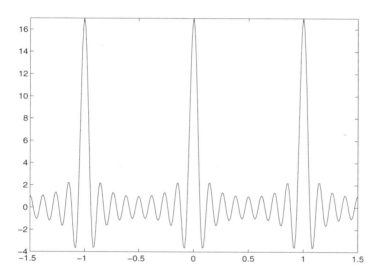

Figure 2.1.   The function given in equation 2.170 with $N=8$ and $\delta u=1$.

We might also mention that the Fourier transform of $P_n(u)\exp(-\pi u^2)$ where $P_n(u)$ is a polynomial of degree n is always of the same form $R_n(\mu)\exp(-\pi\mu^2)$ where $R_n(\mu)$ is another polynomial of nth degree. To prove this, one first writes the polynomial as a series and then takes the Fourier transform term by term, noting that the Fourier transform of $u^n f(u)$ corresponds to the nth derivative of $F(\mu)$. Of course, the nth order derivative of $\exp(-\pi u^2)$ is simply a polynomial of that order times $\exp(-\pi u^2)$, which upon collecting terms results in the form $R_n(u)\exp(-\pi u^2)$. The Hermite polynomials to be introduced in section 2.5.2 are special in that they reproduce themselves: $R_n(u)\propto P_n(u)$.

Equation 2.157 is a linear superposition of functions of the form $\exp(i2\pi\mu u)$. Thus the Fourier transform $F(\mu)$ is essentially the expansion coefficient when we expand the function $f(u)$ in terms of these functions. Equation 2.156, which shows us how to calculate the expansion coefficients, is essentially an inner product between the function $f(u)$ and the basis functions $\exp(i2\pi\mu u)$. Thus equations 2.156 and 2.157 are a special case of equations 2.49 and 2.53.

This expansion is of special interest when we are dealing with linear shift-invariant systems. Since complex exponential functions are the eigenfunctions of such systems, application of such a system with impulse response $h(u)$ on both sides of equation 2.157 gives

$$\mathcal{L}f(u) = \int H(\mu)F(\mu)e^{i2\pi\mu u}\,d\mu. \qquad (2.171)$$

The right-hand side is simply the function whose Fourier transform is $H(\mu)F(\mu)$, which we know is $f(u) * h(u)$. This is nothing but a derivation of the convolution property given in table 2.4, property 11. The process of calculating the convolution $f(u) * h(u)$ by multiplying its Fourier transform with a Fourier domain filter function $H(\mu)$ is referred to as *multiplicative filtering in the Fourier domain*.

An interesting consequence of equation 2.9 is

$$\sum_{n=-\infty}^{\infty} f(u + n\delta u) = \frac{1}{\delta u} \sum_{n=-\infty}^{\infty} F(n/\delta u)e^{i2\pi nu/\delta u}. \qquad (2.172)$$

Equation 2.172 and equation 2.9 are referred to as Poisson's sum(mation) formulas.

The discrete Fourier transform (DFT) is a mapping from $\mathbf{R}^N$ to $\mathbf{R}^N$. If we let $f(l)$ denote the lth component of a vector in $\mathbf{R}^N$, then its discrete Fourier transform is defined by

$$F(j) \equiv \frac{1}{\sqrt{N}} \sum_{l=0}^{N-1} W(j,l)f(l), \qquad j = 0, 1, \ldots, N-1, \qquad (2.173)$$

$$W(j,l) \equiv W^{jl}, \qquad W \equiv e^{-i2\pi/N},$$

where $F(j)$ denotes the jth component of the discrete Fourier transform of $f(l)$. The DFT can be computed in $O(N \log N)$ time on a serial computer, and $O(\log N)$ time on a parallel computer by using the fast Fourier transform (FFT) algorithm (Bracewell 1986, Iizuka 1987, Oppenheim and Shafer 1989). We will discuss the relationship between the continuous and discrete Fourier transforms in chapter 3.

2.5.2   Eigenfunctions of the Fourier transform

In this section we will examine the eigenfunctions of the Fourier transform, interpreted either as a system or a transformation. That is, we are looking for solutions of equation 2.119 for the Fourier transform operator. The solutions are denoted by $\psi_n(u)$ for $n = 0, 1, 2, \ldots$ and are known as the Hermite-Gaussian functions:

$$\mathcal{F}\psi_n(u) = e^{-in\pi/2}\psi_n(u). \qquad (2.174)$$

These functions may be written explicitly as

$$\psi_n(u) = A_n H_n(\sqrt{2\pi}\,u)\,e^{-\pi u^2}, \qquad A_n = \frac{2^{1/4}}{\sqrt{2^n n!}}, \qquad (2.175)$$

where $H_n(u)$ denotes the nth order Hermite polynomial. (How it can be proved that these are indeed eigenfunctions of the Fourier transform will be mentioned in a broader context later on page 124.) $H_n(u)$ is an nth degree polynomial; it is an odd function when n is odd and an even function when n is even. $H_n(u)$ has n real zeros between which are interposed the $n-1$ zeros of $H_{n-1}(u)$. Properties of the Hermite polynomials are given in table 2.7. Using property 8, we can see that the Hermite-Gaussian functions $\psi_n(u)$ constitute an orthonormal set. An excellent introduction to Hermite polynomials and Hermite-Gaussian functions may be found in Cohen-Tannoudji, Diu, and Laloë 1977 and Wolf 1979, the former being of a more elementary nature.

Table 2.7. Properties of Hermite polynomials.

1.	$H_0(u) = 1,\ H_1(u) = 2u,\ H_2(u) = 4u^2 - 2,\ H_3(u) = 8u^3 - 12u,\ \ldots$
2.	$H_n(u) = (-1)^n e^{u^2} \frac{d^n}{du^n} e^{-u^2}$
3.	$e^{-\xi^2 + 2\xi u} = \sum_{n=0}^{\infty} \frac{\xi^n}{n!} H_n(u) \qquad H_n(u) = \left[\frac{\partial^n}{\partial \xi^n} e^{-\xi^2 + 2\xi u} \right]_{\xi=0}$
4.	$\frac{dH_n(u)}{du} = 2n H_{n-1}(u)$
5.	$H_{n+1}(u) = 2u H_n(u) - 2n H_{n-1}(u)$
6.	$H_n(u) = (2u - \frac{d}{du}) H_{n-1}(u)$
7.	$(\frac{d^2}{du^2} - 2u\frac{d}{du} + 2n) H_n(u) = 0$
8.	$\int e^{-u^2} H_n(u) H_{n'}(u)\, du = 2^n n! \sqrt{\pi}\, \delta_{nn'}$
9.	$\int e^{-(u-u')^2} H_n(u')\, du' = 2^n \pi^{1/2} u^n$
10.	$\lim_{n \to \infty} \frac{(-1)^n \sqrt{n}}{4^n n!} H_{2n}(u/2\sqrt{n}) = \frac{1}{\sqrt{\pi}} \cos(u)$
11.	$\lim_{n \to \infty} \frac{(-1)^n}{4^n n!} H_{2n+1}(u/2\sqrt{n}) = \frac{2}{\sqrt{\pi}} \sin(u)$
12.	$\sum_{j=0}^{\infty} \frac{1}{2^{n/2}} \binom{n}{j} H_j(\sqrt{2}\,u) H_{n-j}(\sqrt{2}\,u') = H_n(u + u')$
13.	$\sum_{j=0}^{n} \frac{H_j(u) H_j(u')}{2^j j!} = \frac{H_{n+1}(u) H_n(u') - H_n(u) H_{n+1}(u')}{2^{n+1} n! (u-u')}$
14.	$\pi^{-1/2} e^{-(u^2+u'^2)/2} \sum_{n=0}^{\infty} \frac{1}{2^n n!} H_n(u) H_n(u') = \delta(u - u')$
15.	$\sum_{n=0}^{\infty} \frac{e^{in\alpha}}{2^n n!} H_n(u) H_n(u') = (1 - e^{2i\alpha})^{-1/2} \exp\left[\frac{2uu' e^{i\alpha} - e^{2i\alpha}(u^2+u'^2)}{1 - e^{2i\alpha}} \right]$
16.	$\sum_{n=0}^{\infty} \frac{z^n}{2^n n!} H_n(u) H_n(u') = (1 - z^2)^{-1/2} \exp\left[u'^2 - \frac{(u'-zu)^2}{1-z^2} \right]$

ξ, α are real, z is complex.

The Hermite-Gaussian functions are often recognized as the solutions of the differential equation

$$\frac{d^2 f(u)}{du^2} + 4\pi^2 \left(\frac{2n+1}{2\pi} - u^2 \right) f(u) = 0, \tag{2.176}$$

associated with the quantum-mechanical harmonic oscillator or propagation in quadratic graded-index media. For the moment, we treat $(2n+1)/2\pi$ as a single constant. By taking the Fourier transform of this equation, and using elementary identities regarding the transforms of $d^2 f(u)/du^2$ and $(-i2\pi u)^2 f(u)$, we can show

$$\frac{d^2 F(\mu)}{d\mu^2} + 4\pi^2 \left(\frac{2n+1}{2\pi} - \mu^2 \right) F(\mu) = 0. \tag{2.177}$$

Since equations 2.176 and 2.177 are identical in form, it is easy to accept the well-known fact that solutions of this equation, the Hermite-Gaussian functions, are eigenfunctions of the Fourier transform operation (Wiener 1933, Dym and McKean 1972). It is also possible to show directly that the Hermite-Gaussian functions are indeed eigenfunctions of the Fourier transform by directly substituting them in the definition of the Fourier transform. We will do this later in chapter 4 for a more general case.

It is also not difficult to see that the Hermite-Gaussian functions $\psi_n(u) \propto H_n(\sqrt{2\pi}\,u)\exp(-\pi u^2)$ are indeed solutions of the above equation. When we substitute them, we obtain

$$\frac{d^2 H_n(u)}{du^2} + (-2u)\frac{dH_n(u)}{du} + (2n)H_n(u) = 0, \tag{2.178}$$

which is nothing but item 7 in table 2.7. (This equation is sometimes taken as the defining equation of the Hermite polynomials.)

The Hermite-Gaussian functions constitute an orthonormal basis for the set of finite-energy functions:

$$\langle \psi_n, \psi_{n'} \rangle = \int \psi_n(u)\psi_{n'}(u)\,du = \delta_{nn'}, \tag{2.179}$$

$$\sum_{n=0}^{\infty} \psi_n(u)\psi_n(u') = \delta(u - u'). \tag{2.180}$$

Thus any finite-energy signal can be expanded in the form

$$f = \sum_{n=0}^{\infty} f_\psi(n)\,\psi_n, \tag{2.181}$$

$$f_\psi(n) = \langle \psi_n, f \rangle = \int \psi_n(u)f(u)\,du. \tag{2.182}$$

The transformation from $f(u)$ to $f_\psi(n)$ is unitary. In the ψ-representation, the Fourier transform has a particularly simple form. The representation of the Fourier transform of $f_\psi(n)$ is simply $\exp(-in\pi/2)f_\psi(n)$; that is, the kernel for Fourier transformation is diagonal in this representation.

The coefficients $f_\psi(n)$ constitute the representation of the signal f in the Hermite-Gaussian basis set. The act of obtaining these coefficients from the time-domain representation of the signal $f(u)$ is an example of a transformation from a continuous to a discrete representation. Clearly, as with any orthonormal basis expansion,

$$\sum_{n=0}^{\infty} |f_\psi(n)|^2 = \int |f(u)|^2\,du. \tag{2.183}$$

Other properties of the Hermite-Gaussian functions are given in table 2.8.

Another interesting property is that "most" of the energy of the nth order Hermite-Gaussian function is concentrated between the bounds $-\sqrt{(n + 1/2)/\pi}$ and $\sqrt{(n + 1/2)/\pi}$. It is within these bounds that the function shows oscillatory behavior. Outside of these bounds, the exponential factor dominates and its value decays quickly. See Ozaktas and Mendlovic 1993b for a physical justification. The first four Hermite-Gaussian functions are shown in figure 2.2.

Table 2.8.   Properties of Hermite-Gaussian functions.

1. $\psi_n(u) = \frac{(-1)^n A_n}{(\sqrt{2\pi})^n} e^{\pi u^2} \frac{d^n}{du^n} e^{-2\pi u^2}$

2. $\psi_n(u) = A_n \left[\frac{\partial^n}{\partial \xi^n} e^{-\xi^2} e^{u(2\xi - \pi u)} \right]_{\xi=0}$

3. $(\sqrt{2\pi}\, u + \frac{1}{\sqrt{2\pi}} \frac{d}{du}) \psi_n(u) = 2n \frac{A_n}{A_{n-1}} \psi_{n-1}(u)$

4. $\psi_n(u) = 2\sqrt{2\pi}\, u \frac{A_n}{A_{n-1}} \psi_{n-1}(u) - 2(n-1) \frac{A_n}{A_{n-2}} \psi_{n-2}(u)$

5. $\psi_n(u) = \frac{A_n}{A_{n-1}} (\sqrt{2\pi}\, u - \frac{1}{\sqrt{2\pi}} \frac{d}{du}) \psi_{n-1}(u)$

6. $[\frac{d^2}{du^2} + 4\pi^2 (\frac{2n+1}{2\pi} - u^2)] \psi_n(u) = 0$

7. $\sum_{n=0}^{\infty} \psi_n(u) \psi_n(u') = \delta(u - u')$

8. $\sum_{n=0}^{\infty} e^{-in\pi/2} \psi_n(u) \psi_n(u') = e^{-i2\pi uu'}$

9. $\sum_{n=0}^{\infty} e^{-in\alpha} \psi_n(u) \psi_n(u') = \sqrt{1 - i \cot \alpha}\, e^{i\pi(\cot \alpha\, u^2 - 2\csc \alpha\, uu' + \cot \alpha\, u'^2)}$

ξ, α are real.

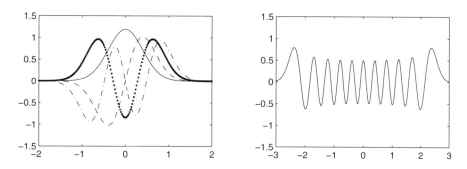

Figure 2.2.   Hermite-Gaussian functions of order (left panel) $n = 0$ (solid), $n = 1$ (dashed), $n = 2$ (dotted), $n = 3$ (chain-dotted), and (right panel) $n = 20$.

As a final comment on the Hermite-Gaussian functions, we note that if one starts with the set of functions $u^n \exp(-\pi u^2)$ for $n = 0, 1, 2, \ldots$ and uses the Gram-Schmidt orthonormalization process to construct an orthonormal set of functions $Q_n(u)$, what one obtains is precisely the Hermite-Gaussian functions; that is, $Q_n(u) = \psi_n(u)$ (Wiener 1933:54).

The Fourier transform of any function can be obtained by first expanding it in terms of the Hermite-Gaussian functions. Upon application of the eigenvalue equation and substituting for $f_\psi(n)$, we can show that

$$e^{-i2\pi\mu u} = \sum_{n=0}^{\infty} e^{-in\pi/2} \psi_n(\mu) \psi_n(u). \tag{2.184}$$

This is called the spectral expansion of the kernel of the Fourier transformation.

The eigenfunctions of the Fourier transform have received special interest under the name "self-Fourier functions" or "self-reciprocal Fourier functions" (Caola 1991, Cincotti, Gori, and Santarsiero 1992, Lohmann and Mendlovic 1992a, b,

1994a, Lakhtakia 1993, Lipson 1993, Coffey 1994, Choudhury, Puntambekar, and Chakraborty 1995). The eigenvalues and eigenvectors of the discrete Fourier transform will be discussed in chapter 6.

2.6   Some important operators

2.6.1   Coordinate multiplication and differentiation operators

We will now investigate the operators $\mathcal{U}$ and $\mathcal{D}$ which may be defined by specifying their effect in the time domain (Cohen-Tannoudji, Diu, and Laloë 1977:149):

$$g(u) = (\mathcal{U}f)(u) = uf(u), \tag{2.185}$$

$$g(u) = (\mathcal{D}f)(u) = (i2\pi)^{-1}\frac{df(u)}{du}, \tag{2.186}$$

corresponding to the kernels

$$h_{\mathcal{U}}(u, u') = u\delta(u - u'), \tag{2.187}$$
$$h_{\mathcal{D}}(u, u') = (i2\pi)^{-1}\delta'(u - u'). \tag{2.188}$$

These operators are duals in the sense that their effects in the frequency domain are

$$G(\mu) = (\mathcal{U}f)(\mu) = (-i2\pi)^{-1}\frac{dF(\mu)}{d\mu}, \tag{2.189}$$

$$G(\mu) = (\mathcal{D}f)(\mu) = \mu F(\mu), \tag{2.190}$$

corresponding to the kernels

$$H_{\mathcal{U}}(\mu, \mu') = (-i2\pi)^{-1}\delta'(\mu - \mu'), \tag{2.191}$$
$$H_{\mathcal{D}}(\mu, \mu') = \mu\delta(\mu - \mu'). \tag{2.192}$$

If we had defined a frequency multiplication operator in the frequency domain, this would be identical to the derivative operator defined in the time domain. Likewise if we had defined a derivative operator in the frequency domain, this would be identical to the coordinate multiplication operator defined in the time domain.

   The kernels in the frequency domain are related to the kernels in the time domain through

$$H(\mu, \mu') = \int\int e^{-i2\pi\mu u} h(u, u') e^{i2\pi\mu' u'}\, du\, du', \tag{2.193}$$

which is a special case of equation 2.101.

   The impulse signals δ_v defined earlier are the eigensignals of $\mathcal{U}$. This is most easily seen in the time domain:

$$(\mathcal{U}\delta_v)(u) = u\delta(u - v) = \lambda\delta(u - v). \tag{2.194}$$

with the eigenvalue $\lambda = v$. The set of eigensignals of $\mathcal{U}$, namely the impulse set, was earlier shown to constitute an orthonormal basis. It is also instructive to write the eigensignal equation in the frequency domain:

$$(\mathcal{U}\delta_v)(\mu) = \frac{1}{-i2\pi}\frac{d}{d\mu}e^{-i2\pi v\mu} = \lambda e^{-i2\pi v\mu}. \tag{2.195}$$

with $\lambda = v$.

Likewise, the harmonic signals har_v are the eigensignals of $\mathcal{D}$. This is most easily seen in the frequency domain:

$$(\mathcal{D}\,\mathrm{har}_v)(\mu) = \mu\delta(\mu - v) = \lambda\delta(\mu - v) \tag{2.196}$$

with the eigenvalue $\lambda = v$. The set of eigensignals of $\mathcal{D}$, namely the harmonic set, was earlier shown to constitute an orthonormal basis. In the time domain

$$(\mathcal{D}\,\mathrm{har}_v)(u) = \frac{1}{i2\pi}\frac{d}{du}e^{i2\pi vu} = \lambda e^{i2\pi vu} \tag{2.197}$$

with $\lambda = v$.

Since we know that the time-domain representation and the frequency-domain representation are associated with the impulse set and the harmonic set respectively, we are not surprised by the fact that the $\mathcal{U}$ and $\mathcal{D}$ operators have such simple expressions in these domains.

It is easy to show that the $\mathcal{U}$ and $\mathcal{D}$ operators are Hermitian by examining their kernels. Their commutator is given by

$$[\mathcal{U}, \mathcal{D}] = \frac{i}{2\pi}\mathcal{I}. \tag{2.198}$$

Some properties of Hermitian operators satisfying this relation can be found in Cohen-Tannoudji, Diu, and Laloë 1977; also see equation 2.155 and the following paragraphs.

We will now consider functions of $\mathcal{U}$ and $\mathcal{D}$. The effect of a function $\Upsilon(\mathcal{U})$ on a signal f can be found easily by considering the power series of $\Upsilon(\cdot)$. Thus

$$g = \Upsilon(\mathcal{U})f = \sum_{n=0}^{\infty}\Upsilon_n\mathcal{U}^n f. \tag{2.199}$$

For instance, in the time domain,

$$g(u) = \left[\sum_{n=0}^{\infty}\Upsilon_n u^n\right]f(u) = \Upsilon(u)f(u), \tag{2.200}$$

where we have used $(\mathcal{U}^n f)(u) = u^n f(u)$, which can be easily derived by considering repeated application of equation 2.185. The kernel corresponding to $\Upsilon(\mathcal{U})$ in the time domain is likewise easily shown to be given by

$$h_{\Upsilon(\mathcal{U})}(u, u') = \Upsilon(u)\delta(u - u'). \tag{2.201}$$

Let us also consider the effect of $\Upsilon(\mathcal{U})$ in the frequency domain:

$$G(\mu) = \left[\sum_{n=0}^{\infty}\Upsilon_n(-i2\pi)^{-n}\frac{d^n}{d\mu^n}\right]F(\mu) = \Upsilon\left[(-i2\pi)^{-1}\frac{d}{d\mu}\right]F(\mu). \tag{2.202}$$

The corresponding kernel is difficult to derive from this result, but it can be obtained from a common property of the Fourier transform. Since we know that $G(\mu) = (\mathcal{F}\Upsilon)(\mu) * F(\mu)$, we have

$$H_{\Upsilon(\mathcal{U})}(\mu, \mu') = (\mathcal{F}\Upsilon)(\mu - \mu'). \tag{2.203}$$

The effect of $\Upsilon(\mathcal{D})$ can be found similarly. In the frequency domain

$$G(\mu) = \left[\sum_{n=0}^{\infty} \Upsilon_n \mu^n\right] F(\mu) = \Upsilon(\mu)F(\mu), \tag{2.204}$$

with the corresponding kernel

$$H_{\Upsilon(\mathcal{D})}(\mu, \mu') = \Upsilon(\mu)\delta(\mu - \mu'). \tag{2.205}$$

In the time domain

$$g(u) = \left[\sum_{n=0}^{\infty} \Upsilon_n (i2\pi)^{-n} \frac{d^n}{du^n}\right] f(u) = \Upsilon\left[(i2\pi)^{-1}\frac{d}{du}\right] f(u), \tag{2.206}$$

where we have used $(\mathcal{D}^n f)(u) = (i2\pi)^{-n} d^n f(u)/du^n$, which can be easily derived by considering repeated application of equation 2.186. The corresponding kernel is difficult to derive from this result, but can be obtained from a common property of the Fourier transform. Since we know that $g(u) = (\mathcal{F}^{-1}\Upsilon)(u) * f(u)$, we have

$$h_{\Upsilon(\mathcal{D})}(u, u') = (\mathcal{F}^{-1}\Upsilon)(u - u'). \tag{2.207}$$

It is not difficult to show that $\Upsilon(\mathcal{U})$ and $\Upsilon(\mathcal{D})$ have the same eigensignals as $\mathcal{U}$ and $\mathcal{D}$ respectively:

$$\Upsilon(\mathcal{U})\delta_v = \Upsilon(v)\,\delta_v, \tag{2.208}$$
$$\Upsilon(\mathcal{D})\mathrm{har}_v = \Upsilon(v)\,\mathrm{har}_v. \tag{2.209}$$

The various kernels are summarized in table 2.9.

Table 2.9.   Kernels associated with the $\mathcal{U}$ and $\mathcal{D}$ operators and their functions.

	$\mathcal{U}$	$h(\mathcal{U})$	$\mathcal{D}$	$H(\mathcal{D})$
u domain	$u\delta(u - u')$	$h(u)\delta(u - u')$	$(i2\pi)^{-1}\delta'(u - u')$	$h(u - u')$
μ domain	$(-i2\pi)^{-1}\delta'(\mu - \mu')$	$H(\mu - \mu')$	$\mu\delta(\mu - \mu')$	$H(\mu)\delta(\mu - \mu')$

In this context it is also worth discussing the so-called moment theorem, which is closely related to the above considerations (Papoulis 1977). It can be stated in various forms and derived directly from the properties of the Fourier transform without resorting to operator concepts. Consider the expansion of $F(\mu)$:

$$F(\mu) = \sum_{n=0}^{\infty} \frac{\mu^n}{n!} \left. \frac{d^n F(\mu)}{d\mu^n} \right|_{\mu=0}.$$ (2.210)

Using

$$m_f^n \equiv \int u^n f(u) \, du = \frac{1}{(-i2\pi)^n} \left. \frac{d^n F(\mu)}{d\mu^n} \right|_{\mu=0},$$ (2.211)

where m_f^n denotes the nth moment of $f(u)$, we obtain the moment theorem:

$$F(\mu) = \sum_{n=0}^{\infty} \frac{(-i2\pi\mu)^n}{n!} m_f^n.$$ (2.212)

This result can also be derived by starting from

$$F(\mu) = \int f(u) e^{-i2\pi\mu u} \, du$$ (2.213)

and replacing $\exp(-i2\pi\mu u)$ with its series expansion $\sum_{n=0}^{\infty}(-i2\pi\mu u)^n/n!$. The dual of the moment theorem is

$$f(u) = \sum_{n=0}^{\infty} \frac{(i2\pi u)^n}{n!} m_F^n,$$ (2.214)

where m_F^n denotes the nth moment of $F(\mu)$.

As an additional exercise, consider the shift-invariant system characterized in the time and frequency domains respectively as

$$g(u) = h(u) * f(u) = \int h(u - u') f(u') \, du',$$ (2.215)

$$G(\mu) = H(\mu) F(\mu).$$ (2.216)

Let us consider the series expansion of $H(\mu)$,

$$H(\mu) = \sum_{n=0}^{\infty} m_h^n \frac{(-i2\pi\mu)^n}{n!},$$ (2.217)

so that

$$G(\mu) = \sum_{n=0}^{\infty} m_h^n \frac{(-i2\pi\mu)^n}{n!} F(\mu),$$ (2.218)

$$g(u) = \sum_{n=0}^{\infty} \frac{(-1)^n}{n!} m_h^n \frac{d^n f(u)}{du^n},$$ (2.219)

an expression which, though not obvious from its appearance, is equivalent to the convolution $g(u) = h(u) * f(u)$.

An alternative approach will be instructive. In table 2.9, the kernel associated with the operator $H(\mathcal{D})$ is denoted by $h(u - u')$ in the u domain. But we can also expand the series of $H(\mathcal{D})$ to obtain

$$(H(\mathcal{D}))\,(u, u') = \sum_{n=0}^{\infty} K_n (i2\pi)^{-n} \delta^{(n)}(u - u'), \qquad (2.220)$$

where K_n is the series coefficient given by

$$K_n = \frac{1}{n!} \left.\frac{d^n H(\mu)}{d\mu^n}\right|_{\mu=0} = \frac{1}{n!}(-i2\pi)^n m_h^n, \qquad (2.221)$$

so that

$$(H(\mathcal{D}))\,(u, u') = h(u - u') = \sum_{n=0}^{\infty} \frac{(-1)^n}{n!} m_h^n \delta^{(n)}(u - u'), \qquad (2.222)$$

$$h(u) = \sum_{n=0}^{\infty} \frac{(-1)^n}{n!} m_h^n \delta^{(n)}(u). \qquad (2.223)$$

When equation 2.223 is convolved with $f(u)$, we obtain equation 2.219.

2.6.2 Phase shift, translation, chirp multiplication, and chirp convolution operators

We now consider the phase shift operator $\mathcal{PH}_\xi = \exp(i2\pi\xi\mathcal{U})$ and the translation operator $\mathcal{SH}_\xi = \exp(i2\pi\xi\mathcal{D})$ (Cohen-Tannoudji, Diu, and Laloë 1977:187). Their effect in the time domain can be shown to be

$$(\mathcal{PH}_\xi f)(u) = e^{i2\pi\xi u} f(u), \qquad (2.224)$$

$$(\mathcal{SH}_\xi f)(u) = \exp\left[i2\pi\xi(i2\pi)^{-1}d/du\right] f(u) = f(u + \xi). \qquad (2.225)$$

The last equality can be shown by expanding $\exp(\xi d/du)$ in a series, applying it to $f(u)$, and comparing it to the series of $f(u + \xi)$ with respect to the variable ξ. Since this is important, it is worth going through carefully:

$$e^{\xi d/du} f(u) = \sum_{n=0}^{\infty} \frac{\xi^n}{n!} \frac{d^n f(u)}{du^n} = f(u + \xi), \qquad (2.226)$$

where the last equality follows from the fact that the summation is simply the series expansion of $f(u + \xi)$ around u. Another method is to consider $\exp(\xi d/du)f(u)$ as a function $k(u, \xi)$ of two variables. Now, differentiate $k(u, \xi)$ with respect to ξ. Since $\partial \exp(\xi d/du)/\partial\xi = (d/du)\exp(\xi d/du)$, we have

$$\frac{\partial k(u, \xi)}{\partial \xi} = \frac{\partial k(u, \xi)}{\partial u}. \qquad (2.227)$$

Noting that $k(u, 0) = f(u)$, we find the solution of this equation as $k(u, \xi) = f(u + \xi)$, which is again the desired result. However, perhaps the most direct way of seeing this result is to note that the effect of the $\exp(i2\pi\xi\mathcal{D})$ operator in the frequency domain is to multiply with $\exp(i2\pi\xi\mu)$, which we know corresponds to a shift by ξ in the time domain: $f(u + \xi)$.

The associated kernels are

$$h_{\mathcal{PH}_\xi}(u, u') = e^{i2\pi\xi u}\delta(u - u'), \tag{2.228}$$

$$h_{\mathcal{SH}_\xi}(u, u') = \delta(u + \xi - u'). \tag{2.229}$$

That these two operators are also duals is easily seen by examining their effect in the frequency domain:

$$(\mathcal{PH}_\xi F)(\mu) = \exp\left[i2\pi\xi(-i2\pi)^{-1}d/d\mu\right] F(\mu) = F(\mu - \xi), \tag{2.230}$$

$$(\mathcal{SH}_\xi F)(\mu) = e^{i2\pi\xi\mu} F(\mu), \tag{2.231}$$

with associated kernels

$$H_{\mathcal{PH}_\xi}(\mu, \mu') = \delta(\mu - \xi - \mu'), \tag{2.232}$$

$$H_{\mathcal{SH}_\xi}(\mu, \mu') = e^{i2\pi\xi\mu}\delta(\mu - \mu'). \tag{2.233}$$

The kernels in the time and frequency domains are related by equation 2.193. Since $\mathcal{U}$ and $\mathcal{D}$ are Hermitian, $\mathcal{PH}_\xi$ and $\mathcal{SH}_\xi$ are unitary. The commutator of $\mathcal{PH}_\xi$ and $\mathcal{SH}_{\xi'}$ is

$$[\mathcal{PH}_\xi, \mathcal{SH}_{\xi'}] = [1 - e^{i2\pi\xi\xi'}]\mathcal{PH}_\xi\mathcal{SH}_{\xi'}. \tag{2.234}$$

Obviously these operations do not commute.

Now we consider the chirp multiplication operator $\mathcal{Q}_q = \exp(-i\pi q\mathcal{U}^2)$ and the chirp convolution (Fresnel) operator $\mathcal{R}_r = \exp(-i\pi r\mathcal{D}^2)$. Their effect in the time domain is

$$(\mathcal{Q}_q f)(u) = e^{-i\pi qu^2} f(u), \tag{2.235}$$

$$(\mathcal{R}_r f)(u) = \exp\left[-i\pi r[(i2\pi)^{-2}d^2/du^2]\right] f(u)$$
$$= e^{-i\pi/4}\sqrt{1/r}\, e^{i\pi u^2/r} * f(u). \tag{2.236}$$

The last equality is a special case of equation 2.207. The associated kernels are

$$h_{\mathcal{Q}_q}(u, u') = e^{-i\pi qu^2}\delta(u - u'), \tag{2.237}$$

$$h_{\mathcal{R}_r}(u, u') = e^{-i\pi/4}\sqrt{1/r}\, e^{i\pi(u-u')^2/r}. \tag{2.238}$$

That these two operators are also duals is easily seen by examining their effect in the frequency domain:

$$(\mathcal{R}_r F)(\mu) = e^{-i\pi r\mu^2} F(\mu), \tag{2.239}$$

$$(\mathcal{Q}_q F)(\mu) = \exp\left[-i\pi q[(-i2\pi)^{-2}d^2/d\mu^2]\right] F(\mu)$$
$$= e^{-i\pi/4}\sqrt{1/q}\, e^{i\pi\mu^2/q} * F(\mu), \tag{2.240}$$

with associated kernels

$$H_{\mathcal{R}_r}(\mu, \mu') = e^{-i\pi r\mu^2}\delta(\mu - \mu'), \tag{2.241}$$

$$H_{\mathcal{Q}_q}(\mu, \mu') = e^{-i\pi/4}\sqrt{1/q}\, e^{i\pi(\mu-\mu')^2/q}. \tag{2.242}$$

The kernels in the time and frequency domains are once again related by equation 2.193. Since $\mathcal{U}$ and $\mathcal{D}$ are Hermitian, $\mathcal{R}_r$ and $\mathcal{Q}_q$ are unitary. The time-domain kernel of their commutator is given by

$$h_{[\mathcal{R}_r, \mathcal{Q}_q]}(u, u') = e^{-i\pi/4} \sqrt{\frac{1}{r}} \, e^{i\pi(u-u')^2/r} \left(e^{-i\pi qu'^2} - e^{-i\pi qu^2} \right). \tag{2.243}$$

It is important to note that since these two operators do not commute, an expression such as

$$e^{i\pi(r\mathcal{D}^2 + q\mathcal{U}^2)} = e^{i\pi(q\mathcal{U}^2 + r\mathcal{D}^2)} \tag{2.244}$$

cannot be written as $\exp(i\pi r\mathcal{D}^2)\exp(i\pi q\mathcal{U}^2)$ or $\exp(i\pi q\mathcal{U}^2)\exp(i\pi r\mathcal{D}^2)$. However, several relations can be used to manipulate such expressions, of which one important example is

$$e^{-i\theta\pi(\mathcal{U}^2 + \mathcal{D}^2)} = e^{-i\pi \tan\theta \, \mathcal{U}^2} e^{-i\pi \ln(\cos\theta)\,(\mathcal{U}\mathcal{D} + \mathcal{D}\mathcal{U})} e^{-i\pi \tan\theta \, \mathcal{D}^2}. \tag{2.245}$$

Such relationships are in general known as Baker-Campbell-Hausdorff formulas (Wilcox 1967, Gilmore 1974, Wolf 1979, Stoler 1981). Similar considerations apply to and similar relations exist for the phase shift and translation operators discussed previously. In general, such a formula exists to correspond with each one of the decompositions we will come across in section 3.4.4, from which they can be derived. Another celebrated example is Glauber's formula (Cohen-Tannoudji, Diu, and Laloë 1977:174). If $\mathcal{A}$ and $\mathcal{B}$ both commute with their commutator ($[\mathcal{A}, [\mathcal{A}, \mathcal{B}]] = 0$ and $[\mathcal{B}, [\mathcal{A}, \mathcal{B}]] = 0$), then $[\mathcal{A}, \Upsilon(\mathcal{B})] = [\mathcal{A}, \mathcal{B}]\Upsilon'(\mathcal{B})$ and

$$e^{\mathcal{A}} e^{\mathcal{B}} = e^{\mathcal{A} + \mathcal{B}} e^{[\mathcal{A}, \mathcal{B}]/2}. \tag{2.246}$$

At this point the reader may wish to look back at tables 2.2 and 2.4. Most properties of the Fourier transform are examples of finding the effect of a linear system on the Fourier transform. In the left-hand column of table 2.4, there is a linear alteration of the function $f(u)$. The right-hand column shows how the same alteration looks in the Fourier domain. It is in particular interesting to examine the effects of the six operations of coordinate multiplication, differentiation, phase shift, translation, chirp multiplication, and chirp convolution. The latter four of these operators are sometimes referred to as *hyperdifferential operators*. A useful source is Wolf 1979. For a self-consistent operational calculus based on such operators, see Yosida 1984. Such operators have been made the basis of a study of optical systems in a series of papers by Nazarathy and Shamir (1980, 1982a, b).

We end by mentioning the eigenvalues and eigenfunctions of these operators. The eigenfunctions of the $\mathcal{Q}_q$ operator in the space domain are of course $\delta(u - v)$ (as for all functions of the $\mathcal{U}$ operator) with eigenvalue $\exp(-i\pi qv^2)$. Likewise, the eigenfunctions of the $\mathcal{R}_r$ operator in the frequency domain are $\delta(\mu - v)$ (as for all functions of the $\mathcal{D}$ operator) with eigenvalue $\exp(-i\pi rv^2)$. Notice that these latter eigenfunctions are harmonic functions in the space domain.

2.6.3 Annihilation and creation operators

Properties 3 and 5 of table 2.8 can be rewritten as

$$\mathcal{A}\psi_n(u) = \sqrt{n}\,\psi_{n-1}(u), \tag{2.247}$$

$$\mathcal{A}^H\psi_{n-1}(u) = \sqrt{n}\,\psi_n(u), \tag{2.248}$$

$$\text{or}\quad \mathcal{A}^H\psi_n(u) = \sqrt{n+1}\,\psi_{n+1}(u),$$

where

$$\mathcal{A} \equiv \sqrt{2\pi}\,\frac{\mathcal{U} + i\mathcal{D}}{\sqrt{2}}, \tag{2.249}$$

$$\mathcal{A}^H \equiv \sqrt{2\pi}\,\frac{\mathcal{U} - i\mathcal{D}}{\sqrt{2}}. \tag{2.250}$$

These operators are respectively referred to as the *annihilation* and *creation* operators, because of their effect on $\psi_n(u)$. Their commutator can be easily shown to be $[\mathcal{A}, \mathcal{A}^H] = \mathcal{I}$ by using $[\mathcal{U}, \mathcal{D}] = (i/2\pi)\mathcal{I}$.

Now, let us form the products

$$\mathcal{A}^H\mathcal{A} = \pi(\mathcal{U}^2 + \mathcal{D}^2) - \frac{1}{2}, \tag{2.251}$$

$$\mathcal{A}\mathcal{A}^H = \pi(\mathcal{U}^2 + \mathcal{D}^2) + \frac{1}{2}, \tag{2.252}$$

by using the definitions given in equations 2.249 and 2.250. Alternately, using equations 2.247 and 2.248, we can obtain

$$(\mathcal{A}^H\mathcal{A})\psi_n(u) = n\psi_n(u), \tag{2.253}$$

$$(\mathcal{A}\mathcal{A}^H)\psi_n(u) = (n+1)\psi_n(u), \tag{2.254}$$

from which we see that $\psi_n(u)$ is an eigenfunction of $(\mathcal{A}^H\mathcal{A})$ with eigenvalue n. Comparing the first of these with the first of the preceding pair of equations, we obtain

$$\left[\pi(\mathcal{U}^2 + \mathcal{D}^2) - \frac{1}{2}\right]\psi_n(u) = n\psi_n(u), \tag{2.255}$$

$$\left[\pi(\mathcal{U}^2 + \mathcal{D}^2)\right]\psi_n(u) = (n+1/2)\psi_n(u), \tag{2.256}$$

from which we note that $\psi_n(u)$ is also an eigenfunction of $\pi(\mathcal{U}^2 + \mathcal{D}^2)$ with eigenvalue $(n + 1/2)$. The reader may also wish to note that the final eigenvalue equation is the same as equation 2.176.

It is also possible to show $[(\mathcal{A}^H\mathcal{A}), \mathcal{A}] = -\mathcal{A}$ and $[(\mathcal{A}^H\mathcal{A}), \mathcal{A}^H] = \mathcal{A}^H$. Using $(\mathcal{A}^H\mathcal{A})\psi_n(u) = n\psi_n(u)$, these commutation relations can be shown to imply

$$(\mathcal{A}^H\mathcal{A})\mathcal{A}\psi_n(u) = (n-1)\mathcal{A}\psi_n(u), \tag{2.257}$$

$$(\mathcal{A}^H\mathcal{A})\mathcal{A}^H\psi_n(u) = (n+1)\mathcal{A}^H\psi_n(u), \tag{2.258}$$

which state that $\mathcal{A}\psi_n(u)$ and $\mathcal{A}^H\psi_n(u)$ are also eigenfunctions of $(\mathcal{A}^H\mathcal{A})$ with the eigenvalues $(n-1)$ and $(n+1)$ respectively, consistent with equations 2.247 and 2.248.

Finally, we note that use of equation 2.248 in the form $\psi_n(u) = n^{-1/2}\mathcal{A}^H\psi_{n-1}(u)$ leads to the expression $\psi_n(u) = (n!)^{-1/2}(\mathcal{A}^H)^n\psi_0(u)$, which allows the calculation of Hermite-Gaussian functions of arbitrary order.

These operators are commonly used in quantum mechanics texts to solve the equation of the quantum-mechanical harmonic oscillator (Cohen-Tannoudji, Diu, and Laloë 1977).

2.7   Uncertainty relations

In the context of deterministic signals, uncertainty relations are bounds on the concentration or spread of the energy of a signal in two domains, commonly the time and frequency domains. In this section we will present a rather general result and then discuss some of its special cases. First, however, we introduce some definitions which may already be familiar to readers who have studied quantum mechanics (Cohen-Tannoudji, Diu, and Laloë 1977). Means η_A, mean squares m_A^2, and squared standard deviations (variances) σ_A^2 of an operator $\mathcal{A}$ are defined as the weighted averages of $\mathcal{A}$, $\mathcal{A}^2$ and $(\mathcal{A} - \eta_A)^2$, normalized by the energy $\text{En}[f] = \|f\|^2 = \int |f(u)|^2\, du$ of the signal f under consideration:

$$\eta_A \equiv \left[\int f^*(u)\mathcal{A}f(u)\, du\right]/\|f\|^2, \tag{2.259}$$

$$m_A^2 \equiv \left[\int f^*(u)\mathcal{A}^2 f(u)\, du\right]/\|f\|^2, \tag{2.260}$$

$$\sigma_A^2 \equiv \left[\int f^*(u)(\mathcal{A} - \eta_A)^2 f(u)\, du\right]/\|f\|^2 = m_A^2 - \eta_A^2. \tag{2.261}$$

For example, if $\mathcal{A} = \mathcal{U}$, the coordinate multiplication operator, then $\eta_\mathcal{U}$ is simply $\int u|f(u)|^2\, du/\|f\|^2$, the center of gravity of $|f(u)|^2$, and $\sigma_\mathcal{U}$ is a measure of the spread of $|f(u)|^2$. Likewise, if $\mathcal{A} = \mathcal{D}$, the derivative operator, Parseval's relation allows us to write $\int f^*(u)[\mathcal{D}f(u)]\, du = \int F^*(\mu)[\mu F(\mu)]\, d\mu$, so that $\eta_\mathcal{D}$ will give us the center of gravity of $|F(\mu)|^2$ and similarly $\sigma_\mathcal{D}$ is a measure of the spread of $|F(\mu)|^2$.

Let us denote the commutator of two operators $\mathcal{A}$ and $\mathcal{B}$ as $[\mathcal{A}, \mathcal{B}] = i\mathcal{C}$. Then the standard deviations σ_A and σ_B of $\mathcal{A}$ and $\mathcal{B}$ satisfy

$$\sigma_A^2 \sigma_B^2 \geq \frac{|\eta_C|^2}{4}, \tag{2.262}$$

where η_C is the mean of $\mathcal{C}$ (Dym and McKean 1972:119). As an example we again consider the coordinate multiplication and differentiation operators $\mathcal{U}$ and $\mathcal{D}$ whose commutator was given before as $[\mathcal{U}, \mathcal{D}] = (i/2\pi)\mathcal{I}$, so that $\mathcal{C} = (1/2\pi)\mathcal{I}$. These lead to an uncertainty relation of the form $\sigma_\mathcal{U}\sigma_\mathcal{D} \geq 1/4\pi$ or more explicitly

$$\sigma_\mathcal{U}\sigma_\mathcal{D} = \frac{\left[\int (u - \eta_\mathcal{U})^2|f(u)|^2\, du\right]^{1/2} \left[\int (\mu - \eta_\mathcal{D})^2|F(\mu)|^2\, d\mu\right]^{1/2}}{\left[\int |f(u)|^2\, du\right]^{1/2} \left[\int |F(\mu)|^2\, d\mu\right]^{1/2}} \geq \frac{1}{4\pi}, \tag{2.263}$$

which means that the spread of $|f(u)|^2$ and the spread of $|F(\mu)|^2$ cannot simultaneously be very small.

If we define measures of spread Δu and $\Delta \mu$ as $\sqrt{4\pi}$ times the standard deviations of $|f(u)|^2$ and $|F(\mu)|^2$ respectively, then the uncertainty relation expressed in terms of these measures of spread takes the form

$$\Delta u \, \Delta \mu \geq 1. \tag{2.264}$$

The absolute square of the unit-energy Gaussian function $2^{1/4}\Delta u^{-0.5}\exp(-\pi u^2/\Delta u^2)$ has standard deviation $\Delta u/\sqrt{4\pi}$. In the case of Gaussian functions, the above inequality is satisfied with equality: $\Delta u \, \Delta \mu = 1$. (At $u = \Delta u/2$ the Gaussian function drops to 0.46 of its value at $u = 0$ and 92.5% of the energy of the Gaussian is contained in the interval $[-\Delta u/2, \Delta u/2]$.)

We can further manipulate the numerator of the central term in equation 2.263 by first letting $f_1(u) \equiv f(u + \eta_{\mathcal{U}})$, and noting that $F_1(\mu)$ has the same magnitude as $F(\mu)$:

$$\left[\int u^2 |f_1(u)|^2 \, du \right]^{1/2} \left[\int (\mu - \eta_{\mathcal{D}})^2 |F_1(\mu)|^2 \, d\mu \right]^{1/2}. \tag{2.265}$$

Now, with $F_2(\mu) \equiv F_1(\mu + \eta_{\mathcal{D}})$, it becomes

$$\left[\int u^2 |f_2(u)|^2 \, du \right]^{1/2} \left[\int \mu^2 |F_2(\mu)|^2 \, d\mu \right]^{1/2}, \tag{2.266}$$

which, by eliminating the subscripts, leads to

$$\left[\int u^2 |f(u)|^2 \, du \right] \left[\int \mu^2 |F(\mu)|^2 \, d\mu \right] \geq (16\pi^2)^{-1} \|f\|^4, \tag{2.267}$$

with equality for $f(u)$ a constant multiple of $\exp(-\chi u^2)$ where $\chi > 0$ (Dym and McKean 1972:117–118). This is an alternate form of the uncertainty relation given in equation 2.263.

We will now take a slightly different approach to arrive at yet a third form. Let us first note that

$$m_{\mathcal{D}}^2 \|f\|^2 = \int f^*(u) \mathcal{D}^2 f(u) \, du = \langle f, \mathcal{D}^2 f \rangle = \langle \mathcal{D}^{\mathrm{H}} f, \mathcal{D} f \rangle. \tag{2.268}$$

In the time domain we know that $\mathcal{D}f(u) = (i2\pi)^{-1} df(u)/du$ and $\mathcal{D}^{\mathrm{H}} f(u) = (-i2\pi)^{-1} df(u)/du$, so that

$$m_{\mathcal{D}}^2 \|f\|^2 = \int \mathcal{D}^{\mathrm{H}} f^*(u) \mathcal{D} f(u) \, du = \int |(i2\pi)^{-1} df(u)/du|^2 \, du. \tag{2.269}$$

Likewise, it is not difficult to show that $m_{\mathcal{U}}^2 \|f\|^2 = \int |uf(u)|^2 \, du$. Noting that $|\eta_{\mathcal{I}}|^2/4 = 1/4$, we can write

$$(m_{\mathcal{U}}^2 - \eta_{\mathcal{U}}^2)(m_{\mathcal{D}}^2 - \eta_{\mathcal{D}}^2) \geq \left(\frac{1}{2\pi} \right)^2 \frac{1}{4}. \tag{2.270}$$

Now, noting that $m_U^2 - \eta_U^2 \leq m_U^2$ and $m_D^2 - \eta_D^2 \leq m_D^2$, and using the above we ultimately obtain

$$\|uf(u)\| \, \|df(u)/du\| \geq \|f(u)\|^2/2, \qquad (2.271)$$

again with equality when $f(u)$ is proportional to $\exp(-\chi u^2)$ (Born and Wolf 1980:773).

Intuitively, the uncertainty relation states that both a function and its Fourier transform cannot be simultaneously concentrated. Generally speaking, if one of them is narrow then the other must be broad. A function $f(u)$ of approximate duration Δu will necessarily exhibit nonnegligible frequency components around $1/\Delta u$ so that its Fourier transform will exhibit a spread of $\Delta \mu \simeq 1/\Delta u$, implying $\Delta u \, \Delta \mu \geq 1$.

There are other kinds of uncertainty relations which embody the same basic concepts. For instance, assume $\|f\| = 1$ and define

$$\alpha^2 = \int_{-\Delta u/2}^{\Delta u/2} |f(u)|^2 \, du, \qquad (2.272)$$

$$\beta^2 = \int_{-\Delta \mu/2}^{\Delta \mu/2} |F(\mu)|^2 \, d\mu. \qquad (2.273)$$

Let us fix $\Delta u, \Delta \mu > 0$. Now, $\alpha = \beta = 1$ is clearly not possible. But how close to unity can they simultaneously be? Such bounds on α and β for given Δu and $\Delta \mu$ are discussed by Slepian and Pollak (1961), Landau and Pollak (1961, 1962), Slepian (1964, 1978), and more briefly by Dym and McKean (1972:122).

A simpler but related result states that a function and its Fourier transform cannot both be compact (unless they are identically zero). A compact function is one which is zero outside a finite interval around the origin. Such a function can be expressed as itself multiplied by a sufficiently wide rectangle function. Upon Fourier transforming the product, we see that the Fourier transform of the function is equal to itself convolved with a sinc function, which cannot be compact.

An instructive treatment of uncertainty relations may be found in Vakman 1968.

2.8   Random processes

2.8.1   Fundamental definitions

A random process can be considered as a parametric random variable (Papoulis 1991). That is, if $f(u)$ is a random process then $f(u_0)$, for a particular value of u_0, is a random variable with a probability density function $P_{f(u_0)}[f(u_0)]$. Thus, the probability that $f(u_0)$ will lie in $[f(u_0), f(u_0) + \Delta f(u_0)]$ is given by $P_{f(u_0)}[f(u_0)]\Delta f(u_0)$ and $\int P_{f(u_0)}[f(u_0)] \, df(u_0) = 1$. (In this section u_0, u_1, u_2 will denote particular instances of u.)

The mean of a random process $f(u)$ is defined through the ensemble average

$$\eta_f(u) \equiv \langle f(u) \rangle \equiv \int f(u) \, P_{f(u)}[f(u)] \, df(u), \qquad (2.274)$$

where u is interpreted as a parameter. The autocorrelation is defined as

$$R_{ff}(u_1, u_2) \equiv \langle f(u_1) f^*(u_2) \rangle, \qquad (2.275)$$

where the calculation of the expectation value will this time involve the joint probability density of $f(u_1)$ and $f(u_2)$. Since the instantaneous power of a deterministic signal is given by $|f(u)|^2$, we interpret $R_{ff}(u, u) = \langle |f(u)|^2 \rangle \equiv m_f^2(u)$ as the expected power of $f(u)$. The autocorrelation is a nonnegative definite function. It is Hermitian: $R_{ff}(u_1, u_2) = R_{ff}^*(u_2, u_1)$. It also satisfies $R_{ff}(u_1, u_2) \leq R_{ff}(u_1, u_1)$. The cross-correlation of f and h is defined as

$$R_{fh}(u_1, u_2) = \langle f(u_1)h^*(u_2) \rangle. \tag{2.276}$$

$f(u)$ and $h(u)$ are orthogonal if $R_{fh}(u_1, u_2) = 0$ for all u_1, u_2.

A process $f(u)$ is said to be *wide-sense stationary* if and only if (i) $m_f(u)$ is finite, (ii) $\eta_f(u)$ is a constant independent of u, (iii) the autocorrelation is a function of $u_1 - u_2$ only and not a function of u_1 and u_2 separately: $R_{ff}(u_1, u_2) = R_{ff}(u_1 - u_2)$. It follows from the properties of the autocorrelation that $R_{ff}(u) = R_{ff}^*(-u)$, $R_{ff}(u) \leq R_{ff}(0)$, and $R_{ff}(0) = m_f^2(0)$.

The time-averaged mean and autocorrelation of a wide-sense stationary process are denoted by $\overline{f(u)}$ and $\overline{f(u + v/2)f^*(u - v/2)}$ and defined as

$$\overline{f(u)} = \lim_{T \to \infty} \frac{1}{2T} \int_{-T}^{T} f(u)\, du, \tag{2.277}$$

$$\overline{f(u + v/2)f^*(u - v/2)} = \lim_{T \to \infty} \frac{1}{2T} \int_{-T}^{T} f(u + v/2)f^*(u - v/2)\, du. \tag{2.278}$$

The wide-sense stationary process $f(u)$ will be called ergodic in the mean if $\langle f(u) \rangle = \overline{f(u)}$ and equals a constant. Likewise, the wide-sense stationary process $f(u)$ will be called ergodic in the autocorrelation if $\langle f(u + v/2)f^*(u - v/2) \rangle = \overline{f(u + v/2)f^*(u - v/2)}$ and is a function of v only. Thus, ergodicity is a concept involving interchangeability of time and ensemble averages. The present discussion of ergodicity is not rigorous or completely accurate; the reader should consult standard texts on random processes for further details.

2.8.2 Power spectral density

Let $f(u)$ be a wide-sense stationary random process with mean η_f and autocorrelation $R_{ff}(u)$. The power spectral density $S_{ff}(\mu)$ is defined as the Fourier transform of the autocorrelation:

$$S_{ff}(\mu) \equiv \int R_{ff}(u)e^{-i2\pi\mu u}\, du, \tag{2.279}$$

$$R_{ff}(u) = \int S_{ff}(\mu)e^{i2\pi\mu u}\, d\mu. \tag{2.280}$$

It can be shown from the properties of the autocorrelation that $S_{ff}(\mu)$ is real, always ≥ 0, and that

$$\langle |f(u)|^2 \rangle = R_{ff}(0) = \int S_{ff}(\mu)\, d\mu, \tag{2.281}$$

which suggests that $S_{ff}(u)$ deserves its name in that it can indeed be interpreted as the spectral density of power. Further confirmation to this end comes from the relation between $S_{ff}(\mu)$ and the expectation value of $|F(\mu)|^2$, discussed in many standard texts.

2.8.3 Linear systems with random inputs

Let the random process $f(u)$ be input to the linear system $\mathcal{L}$ whose output $g(u)$ is given by

$$g(u) = \int h(u, u') f(u') \, du'. \tag{2.282}$$

It is possible to show that the mean of the output η_g is given by $\eta_g = \eta_f \int h(u, u') \, du'$. As for the autocorrelation of the output process $R_{gg}(u_1, u_2)$, it is given by

$$R_{gg}(u_1, u_2) = \iint R_{ff}(u'_1, u'_2) h(u_1, u'_1) h^*(u_2, u'_2) \, du'_1 \, du'_2. \tag{2.283}$$

In the event that the processes are wide-sense stationary, then this simplifies to

$$R_{gg}(v) = R_{ff}(v) * h(v) * h^*(-v), \tag{2.284}$$

which implies the following relation between input and output power spectral densities:

$$S_{gg}(\mu) = |H(\mu)|^2 S_{ff}(\mu). \tag{2.285}$$

2.9 Generalization to two dimensions

Most of the definitions and results given in this and other chapters can be generalized easily to two and higher dimensions in a trivial manner. Expressions involving abstract signals are of course not affected in any way. An expression of the form $g = \mathcal{L}f$ does not imply any dimensionality. As for expressions involving functional representations of signals, it is often possible to write two-dimensional versions of them by simply replacing the variables u and μ with the pairs of variables u, v and μ, ν, and by replacing integrals and summations with double integrals and summations. For instance, equations 2.15 and 2.16 become

$$\langle f, g \rangle = \sum_l \sum_m f^*(l, m) g(l, m), \tag{2.286}$$

$$\langle f, g \rangle = \iint f^*(u, v) g(u, v) \, du \, dv. \tag{2.287}$$

The output $g(u, v)$ of a linear system is related to its input by a relation of the form

$$g(u, v) = \iint h(u, v; u', v') f(u', v') \, du' \, dv'. \tag{2.288}$$

Most of the commonly used functions introduced at the beginning of this chapter are generalized such that they are separable. For instance, $\text{rect}(u, v) \equiv \text{rect}(u)\text{rect}(v)$ and so on for $\text{sinc}(u, v)$, $\text{gauss}(u, v)$, $\text{har}(u, v)$, $\text{chirp}(u, v)$, and $\delta(u, v) \equiv \delta(u)\delta(v)$. However, some new functions which are not separable can also be defined. Introducing polar coordinates $q^2 = u^2 + v^2$ and $\tan \phi = u/v$, it is possible to define the function $\text{rect}(q)$ which is unity inside the circle of radius $1/2$ centered at the origin and zero outside. It is also convenient to define the jinc function as $\text{jinc}(q) = J_1(\pi q)/2q$, where $J_1(\cdot)$ is the first-order Bessel function of the first kind. Furthermore, nonseparable versions of the chirp and Gaussian functions also exist which have terms in the exponent not only in u^2 and v^2 but also in uv.

Likewise, some of the systems defined in table 2.2 may be generalized in a separable manner. For instance, the kernels of $\mathcal{I}$ and $\mathcal{P}$ become $\delta(u \pm u', v \pm v') \equiv \delta(u \pm u')\delta(v \pm v')$. The multiplicative filter now has the more general kernel $h(u, v)\delta(u - u', v - v')$. The kernels of the shift, phase shift, chirp multiplication, and chirp convolution operators may be generalized as

$$\delta(u - u' + \xi_u, v - v' + \xi_v), \tag{2.289}$$

$$\exp[i2\pi(\xi_u u + \xi_v v)]\delta(u - u', v - v'), \tag{2.290}$$

$$\exp[-i\pi(q_u u^2 + q_v v^2)]\delta(u - u', v - v'), \tag{2.291}$$

$$e^{-i\pi/2}\sqrt{\frac{1}{r_u}}\sqrt{\frac{1}{r_v}}\exp[i\pi((u - u')^2/r_u + (v - v')^2/r_v)]. \tag{2.292}$$

The coordinate multiplication and differentiation operators, as already defined, will act along one dimension only. It is possible to define a complementary $\mathcal{V}$ operator with kernel $v\delta(v - v')$ which acts along the v axis and it is possible to distinguish differentiation operators in the two dimensions as $\mathcal{D}_u$ and $\mathcal{D}_v$.

The two-dimensional Fourier transform is defined as

$$F(\mu, \nu) = \int\int f(u, v)e^{-i2\pi(\mu u + \nu v)}\, du\, dv. \tag{2.293}$$

The two-dimensional Fourier transforms of common separable functions of the form $f(u, v) = f_u(u)f_v(v)$ are obtained easily from their one-dimensional counterparts by using the result

$$F(\mu, \nu) = F_\mu(\mu)F_\nu(\nu), \tag{2.294}$$

where $F_\mu(\mu)$ and $F_\nu(\nu)$ are the one-dimensional Fourier transforms of $f_u(u)$ and $f_v(v)$ respectively. (Some common Fourier pairs are not separable and cannot be obtained this way; for instance, the two-dimensional Fourier transform of $\text{rect}(q)$ is $\text{jinc}(\varsigma)$, where ς is the polar coordinate variable in the frequency plane.) Since the Fourier transform kernel is separable, its eigenfunctions are also separable. Denoting the two-dimensional Hermite-Gaussian functions by $\psi_{lm}(u, v)$, we have

$$\psi_{lm}(u, v) \equiv \psi_l(u)\psi_m(v). \tag{2.295}$$

Most properties of the two-dimensional Fourier transform are straightforward generalizations of the one-dimensional property. A notable exception is the following. If

$f(u, v)$ has a two-dimensional Fourier transform $F(\mu, \nu)$, then $f(au+bv+e, cu+dv+f)$ has the two-dimensional Fourier transform

$$\frac{1}{|\Delta|} \exp\left[\frac{i2\pi[(de - bf)\mu + (af - ec)\nu]}{\Delta}\right] F\left(\frac{d\mu - c\nu}{\Delta}, \frac{-b\mu + a\nu}{\Delta}\right), \qquad (2.296)$$

where $\Delta = ac - bd$.

As far as this book is concerned, the reader who has grasped the one-dimensional version of a result or concept should have no difficulty generalizing it to two dimensions. Special discussion of two-dimensional signals and systems may be found in Bracewell 1995 and Dudgeon and Mersereau 1984.

One often encounters rotationally symmetric two-dimensional functions and systems, especially in optics. These depend only on $q = (u^2 + v^2)^{1/2}$ but not ϕ when expressed in polar coordinates. Referring the reader once again to Bracewell 1995 for a more extensive treatment, we will satisfy ourselves by noting that the two-dimensional Fourier transform $F(\varsigma)$ of a rotationally symmetric function $f(q)$ is also rotationally symmetric and is given by

$$F(\varsigma) = 2\pi \int_0^\infty f(q) J_0(2\pi\varsigma q) q \, dq, \qquad (2.297)$$

$$f(q) = 2\pi \int_0^\infty F(\varsigma) J_0(2\pi\varsigma q) \varsigma \, d\varsigma, \qquad (2.298)$$

where $J_0(q)$ is the zeroth-order Bessel function of the first kind. The relationship between $f(q)$ and $F(\varsigma)$ is known as a Hankel transform.

2.10   Some additional definitions and results

2.10.1   The Radon transform and projection-slice theorem

The Radon transform $\mathcal{RDN}_\phi[f(u, v)](u')$ of a two-dimensional function $f(u, v)$ is defined as the integral projection of the function onto an axis making angle ϕ with the u axis:

$$\mathcal{RDN}_\phi[f(u, v)](u') \equiv \int f(u' \cos\phi - v' \sin\phi, u' \sin\phi + v' \cos\phi) \, dv'. \qquad (2.299)$$

(If $h(u, v)$ is the two-dimensional impulse response of a two-dimensional system, then the Radon transform $\mathcal{RDN}_\phi[h(u, v)](u')$ can be interpreted as the line response to the input $\delta(u \cos\phi + v \sin\phi)$. For example, the integral projection at angle $\phi = 0$ is the response to the input $\delta(u)$, and is given by $\int h(u, v) \, dv$.) Sometimes, $\mathcal{RDN}_\phi[f](u')$ is interpreted as a function of plane polar coordinates, with ϕ being the common polar angle and u' corresponding to the radial variable.

Let us also define the slice $\mathcal{SLC}_\phi[F(\mu, \nu)](\mu')$ of a two-dimensional function $F(\mu, \nu)$ through the relation

$$\mathcal{SLC}_\phi[F(\mu, \nu)](\mu') \equiv F(\mu' \cos\phi, \mu' \sin\phi). \qquad (2.300)$$

The slice of a two-dimensional function $F(\mu, \nu)$ at angle ϕ is a one-dimensional function which takes the values of $F(\mu, \nu)$ along the radial line making angle ϕ with the μ axis. This radial line has the parametric form $\mu = \mu' \cos\phi$, $\nu = \mu' \sin\phi$.

Of particular interest is the projection-slice theorem, which states how a function $f(u,v)$ can be recovered from its Radon transform. According to this theorem, the one-dimensional Fourier transform of the integral projection at angle ϕ is equal to the slice of the two-dimensional Fourier transform at angle ϕ:

$$\mathcal{F}\left\{\mathcal{RDN}_\phi[f]\right\}(\mu') = \mathcal{SLC}_\phi[F(\mu,\nu)](\mu'), \tag{2.301}$$

where $F(\mu,\nu)$ is the two-dimensional Fourier transform of $f(u,v)$. If we think of ϕ as a parameter, both sides of the last equation are functions of the single variable μ'. The above relation can be written more explicitly as

$$\int [\mathcal{RDN}_\phi[f(u,v)](u')]\, e^{-i2\pi\mu' u'}\, du' = \iint f(u,v) e^{-i2\pi(\mu'\cos\phi\, u + \mu'\sin\phi\, v)}\, du\, dv. \tag{2.302}$$

The left-hand side is the one-dimensional Fourier transform of the integral projection at angle ϕ. The right-hand side is the two-dimensional Fourier transform of $f(u,v)$ expressed in polar coordinates (μ', ϕ). The proof of the theorem follows immediately upon substitution of equation 2.299 in the above relation.

More on the Radon transform and projection-slice theorem can be found, for instance, in Bracewell 1995 or Barrett 1984, among many other references.

2.10.2 Complex exponential integrals

Here we list a number of complex Gaussian integrals that will be needed in later chapters:

$$\int e^{-p^2 u^2 \pm qu}\, du = \frac{\sqrt{\pi}}{p}\, e^{q^2/4p^2}, \tag{2.303}$$

$$\int u e^{-p^2 u^2 + 2qu}\, du = \sqrt{\frac{\pi}{p}}\frac{q}{p}\, e^{q^2/p}. \tag{2.304}$$

The square roots are taken so that the arguments lie in the interval $(-\pi/2, \pi/2]$. These results are, of course, not valid when the integrals do not converge.

With the above, care must be exercised if p^2 or p is complex or pure imaginary. Thus it will be more convenient to write a number of special forms. Let us start with the Fourier pair $\exp(i\pi u^2)$ and $\exp(i\pi/4)\exp(-i\pi\mu^2)$, write the Fourier transform relations between them, and employ variable substitutions to obtain (McBride and Kerr 1987)

$$\int e^{i\pi(\chi u^2 \pm 2\xi u)}\, du = \frac{1}{\sqrt{\chi}} e^{i\pi/4} e^{-i\pi\xi^2/\chi} \qquad \xi \text{ real}, \ \chi > 0, \tag{2.305}$$

$$\int e^{-i\pi(\chi u^2 \pm 2\xi u)}\, du = \frac{1}{\sqrt{\chi}} e^{-i\pi/4} e^{i\pi\xi^2/\chi} \qquad \xi \text{ real}, \ \chi > 0. \tag{2.306}$$

Both formulas are consistent with equation 2.303 if we use $\sqrt{i} = \exp(i\pi/4)$ and $\sqrt{-i} = \exp(-i\pi/4)$ when extracting p from p^2. Equation 2.305 is also valid for $\chi < 0$, provided we employ the same square root convention. This can be shown by writing $\chi = -|\chi|$ and using $\sqrt{-1} = i = \exp(i\pi/2)$.

2.10.3 Stationary-phase integral

If $f(u)$ is continuous and the derivative of $\kappa(u)$ vanishes at only a single point $u = \xi$ in $(-\infty, \infty)$ such that $\kappa'(\xi) = 0$ and $\kappa''(\xi) \neq 0$, then for sufficiently large μ,

$$\int f(u)e^{i2\pi\mu\kappa(u)}\,du \simeq e^{i2\pi\mu\kappa(\xi)}f(\xi)\sqrt{\frac{i}{\mu\kappa''(\xi)}}. \qquad (2.307)$$

Further discussion may be found in Papoulis 1968. Applications of the stationary phase integral in optics are particularly well discussed in Lohmann 1986.

2.10.4 Schwarz's inequality

The general form of this inequality is given in the appendix to this chapter. A commonly used form for two functions $f(u)$ and $h(u)$ is

$$\left|\int f^*(u)h(u)\,du\right|^2 \leq \left[\int |f(u)|^2\,du\right]\left[\int |h(u)|^2\,du\right]. \qquad (2.308)$$

2.11 Further reading

Relatively elementary texts which may be useful for background reading include Bracewell 1995, 1999, Cohen-Tannoudji, Diu, and Laloë 1977, Papoulis 1968, 1977, and Strang 1988. Texts which may be useful for further study include Dym and McKean 1972, Naylor and Sell 1982, and Wolf 1979.

2.12 Appendix: Vector spaces and function spaces

Here we provide a basic review of vector and function spaces, assuming familiarity only with elementary linear algebra and vectors in $\mathbf{R}^N$ (Strang 1988). Our main purpose is to enable the reader to grasp the various parts of our presentation in a deeper and more unified way. More extensive discussions and greater rigor is to be found in, for instance, Wolf 1979, Naylor and Sell 1982, Debnath and Mikusiński 1990, and Roman 1992. Wolf 1979 will particularly suit those with a mathematical physics bent and covers in greater detail many of the other topics discussed in this chapter as well. An excellent exposition to most of the basic concepts used here for the simpler case of finite-dimensional vectors and matrices may be found in Strang 1988.

2.12.1 Vector spaces

The most familiar example of a vector space is the common three-dimensional space consisting of the set of position vectors denoted as $\mathbf{r} = x\hat{\mathbf{u}}_x + y\hat{\mathbf{u}}_y + z\hat{\mathbf{u}}_z$ or (x, y, z), where $\hat{\mathbf{u}}_x, \hat{\mathbf{u}}_y, \hat{\mathbf{u}}_z$ are unit vectors along the coordinate axes, and addition and multiplication with a scalar number are defined in the obvious way.

A vector space is a set of objects for which addition and scalar multiplication operations have been defined such that, for any vectors f, g, and h which are members of this vector space, and for any scalars a and b:

1. the sum $f + g$ is also a member of the vector space,

2. $f + g = g + f$,
3. $f + (g + h) = (f + g) + h$,
4. there exists a zero vector denoted by 0 such that $f + 0 = f$,
5. there exists the negative of f denoted by $-f$ such that $f + (-f) = 0$,
6. the product af is also a member of the vector space,
7. $(ab)f = a(bf)$,
8. $a(f + g) = af + ag$,
9. $(a + b)f = af + bf$,
10. there exists a scalar unity denoted by 1 such that $1f = f$.

The following properties of a vector space can be derived directly from the above defining axioms:

1. for any set of vectors f_j and any set of scalars a_j, labeled by the integer index j, the vector $\sum_j a_j f_j$ is also a member of the vector space,
2. for any set of vectors f_v and any set of scalars a_v, labeled by the real index v, the vector $\int_v a_v f_v$ is also a member of the vector space,
3. $0 + f = f + 0 = f$, $(-f) + f = f + (-f) = 0$, $-0 = 0$, $-(-f) = f$,
4. $f1 = 1f = f$, $-1f = -f$.

It is easy to see that the three-dimensional space of position vectors satisfies the axioms and hence the properties listed above. There are many other vector spaces whose elements or structure may be less familiar, but which still satisfy all of the above axioms and properties. These include vector spaces with infinite but still countably (discretely) many dimensions. The simplest example is the extension of the three-dimensional vector space considered above to infinite dimensions. Members of such spaces will be represented by functions $f(l)$, $g(l)$, and so on, where each value of the integer variable l corresponds to one of the discretely many dimensions. Of greater interest to us will be spaces which have not only infinite, but also uncountably (continuously) many dimensions. Members of such spaces will be represented by functions $f(u)$, $g(u)$, and so on, where each value of the real variable u corresponds to one of the continuously many dimensions. Such vector spaces are also referred to as *function spaces*.

In this book we will usually deal with complex vector spaces whose members are complex-valued functions and in which the scalars are complex numbers.

2.12.2 Inner products and norms

An *inner product* associates a scalar $\langle f, g \rangle$ with any two elements f and g of a vector space, such that, for any vectors f, g, and h, and scalar a:

1. $\langle f, f \rangle$ is real and ≥ 0, with equality if and only if $f = 0$,
2. $\langle f, g \rangle = \langle g, f \rangle^*$,
3. $\langle f, g + h \rangle = \langle f, g \rangle + \langle f, h \rangle$,
4. $\langle f, ag \rangle = a \langle f, g \rangle$.

A vector space for which an inner product is defined is called an inner product space. Two vectors whose inner product is zero are called *orthogonal* to each other. We will mostly employ the inner product definitions

$$\langle f, g \rangle \equiv \sum_l f^*(l)g(l), \tag{2.309}$$

$$\langle f, g \rangle \equiv \int f^*(u)g(u)\, du, \tag{2.310}$$

for the discrete and continuous cases respectively. These can be shown to satisfy the listed axioms.

A *norm* associates a scalar real number $\|f\|$ with every element f of a vector space such that, for any vectors f and g, and any scalar a:

1. $\|f\| \geq 0$, with equality if and only if $f = 0$,
2. $\|af\| = |a|\,\|f\|$,
3. $\|f + g\| \leq \|f\| + \|g\|$ (triangle inequality).

A vector space for which a norm is defined is known as a normed vector space. If an inner product has already been defined, a norm can be defined as

$$\|f\| \equiv \sqrt{\langle f, f \rangle}, \tag{2.311}$$

which, in the event that the inner product is defined through equations 2.310 or 2.309, is known as the L_2 norm. It is possible to show that this definition satisfies the axioms listed above. The energy of f is defined as $\langle f, f \rangle = \|f\|^2$ so that with the definitions in equations 2.309 and 2.310 we have

$$\langle f, f \rangle = \|f\|^2 = \sum_l |f(l)|^2, \tag{2.312}$$

$$\langle f, f \rangle = \|f\|^2 = \int |f(u)|^2\, du, \tag{2.313}$$

for the discrete and continuous cases respectively.

An inner product satisfying the axioms given above satisfies the following properties:

1. $\langle f, 0 \rangle = \langle 0, f \rangle = 0$,
2. $\langle af, g \rangle = a^*\langle f, g \rangle$,
3. $\langle f + g, h \rangle = \langle f, h \rangle + \langle g, h \rangle$,
4. $|\langle f, g \rangle|^2 \leq \langle f, f \rangle\langle g, g \rangle$ (Cauchy-Schwarz inequality),

and with the norm as defined by equation 2.311 the further properties:

1. $|\langle f, g \rangle| \leq \|f\|\,\|g\|$ (Cauchy-Schwarz inequality),
2. $\|f + g\|^2 = \|f\|^2 + \|g\|^2 + 2\Re[\langle f, g \rangle]$,
3. $\|f + g\|^2 \leq \|f\|^2 + \|g\|^2 + 2|\langle f, g \rangle| \leq \|f\|^2 + \|g\|^2 + 2\|f\|\,\|g\|$,

where $\Re[\cdot]$ denotes the real part of a complex entity.

The distance $d(f, g)$ between two vectors f and g can be defined as

$$d(f, g) = \|f - g\|, \tag{2.314}$$

which is always nonzero if $f \neq g$. This relation associating a real number with every pair of vectors f and g in the vector space defines a *metric* for the inner product space. A space with a defined metric is known as a *metric space*. A definition of distance is often expected to satisfy the following axioms:

1. $d(f, g) \geq 0$, with equality if and only if $f = g$,
2. $d(f, g) = d(g, f)$,
3. $d(f, g) + d(g, h) \geq d(f, h)$,

as the definition given by equation 2.314 indeed does.

We will mostly, but not exclusively, deal with vectors whose energies and norms are finite. This is because, in most physical applications, the energy as defined here corresponds to actual physical energy. However, this will not exclude us from employing certain idealized unphysical functions with infinite energy as intermediaries in our calculations.

An inner product space (with norm given by equation 2.311 and all of whose members have finite energy), which satisfies an additional condition known as *completeness* which we do not discuss here, is known as a *Hilbert space* (Naylor and Sell 1982). The Hilbert space of complex-valued functions $f(u)$, u real, for which the inner product is defined by equation 2.310, is known as L_2, whereas the Hilbert space of complex-valued functions $f(l)$, l integer, for which the inner product is defined by equation 2.309, is known as ℓ_2. Both of these spaces have discretely (countably) many dimensions (or degrees of freedom). Self-evident in the case of ℓ_2, this is also true for L_2 whose members can always be represented by discretely (countably) many coefficients, for instance as when expanded in terms of a discrete basis such as the Hermite-Gaussian functions. The space of "physically realizable" signals introduced on page 10 is somewhat more restricted than these spaces.

3

Wigner Distributions and Linear Canonical Transforms

3.1 Time-frequency and space-frequency representations

The Fourier transform of a signal gives the relative weights of the various frequency components that make up the signal. It tells us which frequencies exist in the signal and their strengths. However, since the Fourier transform $F(\mu)$ involves integration of the time-domain representation of the signal $f(u)$ from minus to plus infinity, it is difficult to tell by just looking at $F(\mu)$ where these frequencies are located in $f(u)$. The value of $F(\mu)$ at each frequency μ depends on the value of $f(u)$ at all values of u. This character of the Fourier transform is sometimes found to be at odds with common physical intuition and experience. For instance, music scores tell the musician which frequencies to generate at particular time intervals, embodying the notion of particular frequencies being localized around particular instances. When we change the frequency setting of a sinusoidal signal generator from 1 MHz to 2 MHz, we would be reluctant to say that the output waveform continues to contain a frequency component at 1 MHz; we would rather say that it used to contain this frequency but it no longer does. As another example, let us consider a linear FM (frequency modulation) signal of the form $\exp(i\pi u^2)$, whose instantaneous frequency $(2\pi)^{-1}d(\pi u^2)/du = u$ is linearly increasing with time u. The most pronounced frequency in this signal increases with passing time. Looking at a picture of a dressed person, we may say that their checkered jacket exhibits high spatial frequencies, whereas their white shirt exhibits low spatial frequencies, and their striped tie exhibits equally spaced discrete frequency components.

Clearly, we are well accustomed to the concept of time- or space-dependent frequency content; we often speak about the frequency content of signals at different times or locations. Time- or space-frequency distributions are functions of time (or space) and temporal (or spatial) frequency which display the frequency content of signals for different times (or locations).

3.1.1 Short-time or windowed Fourier transform

One way of obtaining the time-dependent frequency content of a signal is to take the Fourier transform of $f(u')$ over an interval around a point u, where u is a variable parameter. This is called the short-time or windowed Fourier transform $WF_f^{(w)}(u, \mu)$

and may be defined as follows (Hlawatsch and Boudreaux-Bartels 1992):

$$WF_f^{(w)}(u,\mu) \equiv \int [f(u')w^*(u'-u)]e^{-i2\pi\mu u'} \, du', \tag{3.1}$$

where $w(u')$ is a suitably chosen lowpass unit-energy window function centered around the origin, which suppresses $f(u')$ outside an interval centered around u. A common choice is the unit-energy Gaussian function $2^{1/4}\Delta_u^{-0.5}\exp(-\pi u^2/\Delta_u^2)$, for which $WF_f^{(w)}(u,\mu)$ is essentially the Fourier transform of the function over the interval $[u-\Delta_u/2, u+\Delta_u/2]$, and thus gives us the distribution of frequencies in $f(u)$ in this interval. (At $u = \Delta_u/2$ the Gaussian function drops to 0.46 of its value at $u = 0$, and 92.5% of the energy of the Gaussian is contained in this interval.) Thus, the short-time Fourier transform allows us to be specific about the location of certain frequencies with a time resolution of $\sim \Delta_u$. If we want to be able to specify the distribution of frequencies as a function of time with greater temporal accuracy, we must choose shorter (narrower) windows (smaller Δ_u).

It is possible to show that one can express $WF_f^{(w)}(u,\mu)$ in terms of $F(\mu)$ as well:

$$WF_f^{(w)}(u,\mu) = e^{-i2\pi\mu u} \int [F(\mu')W^*(\mu'-\mu)]e^{i2\pi\mu'u} \, d\mu'$$

$$= \int [F(\mu''+\mu)W^*(\mu'')]e^{i2\pi\mu''u} \, d\mu'', \tag{3.2}$$

where $W(\mu)$ is the Fourier transform of $w(u)$. For the Gaussian function above, $W(\mu) = 2^{1/4}\Delta_\mu^{-0.5}\exp(-\pi\mu^2/\Delta_\mu^2)$, where $\Delta_\mu = 1/\Delta_u$. Multiplying $F(\mu')$ with the bandpass filter $W^*(\mu'-\mu)$ essentially suppresses all frequencies other than those in the interval $[\mu-\Delta_\mu/2, \mu+\Delta_\mu/2]$, resulting in a bandpass signal with center frequency μ. Multiplying with $\exp(-i2\pi\mu u)$ in the time domain amounts to a frequency shift which converts this bandpass signal into a lowpass signal. Thus $WF_f^{(w)}(u,\mu)$, interpreted as a function of time u with frequency μ as a parameter, is a lowpass signal whose frequency distribution is given by $F(\mu''+\mu)W^*(\mu'')$, which is simply the frequency distribution of $f(u)$ in an interval of width $\sim \Delta_\mu$ around μ, shifted down to zero frequency. If we want to be able to specify μ with greater accuracy, the width $\sim \Delta_\mu$ of $W(\mu)$ must be made smaller. Since $\Delta_u\Delta_\mu = 1$ for a Gaussian window, we conclude that choosing a shorter window $w(u)$ increases temporal resolution while decreasing frequency resolution, whereas choosing a longer window decreases temporal resolution while increasing frequency resolution. More generally, the product of the temporal extent Δ_u and spectral extent Δ_μ of an arbitrary window must always be greater or equal than (approximately) unity. Thus, this trade-off between temporal and frequency resolution always exists. The extreme cases of $w(u) = \delta(u)$ and $w(u) = 1$ correspond to perfect time resolution and perfect frequency resolution respectively. (Hlawatsch and Boudreaux-Bartels 1992)

It is possible to recover $f(u)$ from $WF_f^{(w)}(u,\mu)$ by using the following easily derived result (Hlawatsch and Boudreaux-Bartels 1992):

$$f(u) = \int\int WF_f^{(w)}(u',\mu')w(u-u')e^{i2\pi\mu'u} \, du' \, d\mu', \tag{3.3}$$

which may be expressed as

$$f(u) = \iint W F_f^{(w)}(u', \mu') w_{u', \mu'}(u)\, du'\, d\mu', \tag{3.4}$$

$$w_{u', \mu'}(u) \equiv w(u - u') e^{i2\pi\mu' u},$$

where $w_{u', \mu'}(u)$ is interpreted as a basis signal centered at the time-frequency point (u', μ'), since $w(u)$ is a lowpass signal centered around the origin. If $w(u)$ is taken as the Gaussian function used above, the time-frequency extent of this basis signal is $\sim \Delta_u \times \Delta_\mu$. Thus, we see that we can interpret $W F_f^{(w)}(u', \mu')$ as the weighting coefficient of the basis signal concentrated around the time-frequency point (u', μ'), indicating the relative strength in $f(u)$ of certain frequencies μ' at certain times u'.

The absolute square of the windowed or short-time Fourier transform is known as the spectrogram $SP_f^{(w)}(u, \mu) \equiv |W F_f^{(w)}(u, \mu)|^2$. It can be interpreted as an indicator of the energy of the signal at the time and frequency point (u, μ) in the sense that $SP_f^{(w)}(u, \mu)\, du\, d\mu$ gives the energy of the signal in the time-frequency region $[u - du/2, u + du/2] \times [\mu - d\mu/2, \mu + d\mu/2]$. We also note that equations 3.1 and 3.2 can be made symmetrical by redefining the windowed Fourier transform as follows (Almeida 1994):

$$W F_f^{(w)}(u, \mu) = e^{i\pi\mu u} \int f(u') w^*(u' - u) e^{-i2\pi\mu u'}\, du', \tag{3.5}$$

$$W F_f^{(w)}(u, \mu) = e^{-i\pi\mu u} \int F(\mu') W^*(\mu' - \mu) e^{i2\pi\mu' u}\, d\mu'. \tag{3.6}$$

We will mostly remain with the standard definitions, however.

3.1.2 Gabor expansion

The short-time Fourier transform is closely related to the Gabor expansion, which is an expansion of $f(u)$ in term of discretely many basis functions which are localized in time and frequency (Hlawatsch and Boudreaux-Bartels 1992):

$$f(u) = \sum_l \sum_m G_f^{(w)}(l, m) w_{lm}(u), \tag{3.7}$$

$$w_{lm}(u) \equiv w(u - l\,\delta u) e^{i2\pi(m\,\delta\mu)u},$$

where $w(u)$ is a suitably chosen lowpass unit-energy window function centered around the origin, so that $w_{lm}(u)$ is a function localized in time and frequency around the time-frequency point $(l\,\delta u, m\,\delta\mu)$. The time and frequency spacings δu and $\delta\mu$ define a lattice in the time-frequency plane. The coefficients $G_f^{(w)}(l, m)$ are referred to as Gabor coefficients, and the $w_{lm}(u)$ are known as Gabor logons. Note that this expansion is in terms of basis signals which are discretely spaced in the time-frequency plane, as opposed to the expansion in equation 3.4, which is in terms of basis signals which are continuously spaced in the time-frequency plane. A necessary condition for the discretely many logons to be sufficient to expand an arbitrary finite-energy signal $f(u)$ is $\delta u\, \delta\mu \leq 1$ (Daubechies 1990, 1992). This condition indicates the minimum density of the logons that is needed for them to constitute a basis set for finite-energy signals. If the temporal extent of the signal is $\sim \Delta u$ and its bandwidth is $\sim \Delta \mu$, then we would expect the number of Gabor coefficients which are nonnegligible to be given by $\sim (\Delta u/\delta u)(\Delta \mu/\delta \mu)$. If $\delta u\, \delta \mu > 1$ the number of these coefficients would be less than $\Delta u\, \Delta \mu$, the time-bandwidth product of the signal. Clearly, we cannot expect the signal to be characterized by fewer coefficients than its time-bandwidth product. On the other hand, if $\delta u\, \delta \mu < 1$ the number of Gabor coefficients will be larger than the time-bandwidth product of the signal, indicating that in this case the logons are not linearly independent, and that the representation is redundant: it has more coefficients than needed. When $\delta u\, \delta \mu = 1$ the logons are linearly independent and the coefficients contain no redundancy; the number of coefficients equals the time-bandwidth product of the signal (Hlawatsch and Boudreaux-Bartels 1992). We will later return to these concepts and see that each coefficient corresponds to one degree of freedom of the signal, hence the number of nonnegligible Gabor coefficients corresponds to the number of degrees of freedom of the signal.

We now assume $\delta u\, \delta \mu = 1$ and also that $w(u)$ is chosen so that the $w_{lm}(u)$ constitute a basis set for all finite-energy signals (completeness). In general, the $w_{lm}(u)$ will not be orthogonal to each other, so that we cannot find the coefficient $G_f^{(w)}(l, m)$ by taking the inner product of $f(u)$ with $w_{lm}(u)$. However, the coefficients can be found by taking the inner product of $f(u)$ with a new set of signals $v_{lm}(u)$ as follows:

$$G_f^{(w)}(l, m) = \langle v_{lm}(u), f(u) \rangle = \int f(u) v_{lm}^*(u)\, du, \tag{3.8}$$

$$v_{lm}(u) \equiv v(u - l\, \delta u) e^{i2\pi(m\, \delta \mu)u},$$

where the $v_{lm}(u)$ and $w_{lm}(u)$ satisfy the biorthonormality condition

$$\langle v_{l'm'}(u), w_{lm}(u) \rangle = \int v_{l'm'}^*(u) w_{lm}(u)\, du = \delta_{ll'} \delta_{mm'}. \tag{3.9}$$

Determination of an appropriate function $v(u)$ such that the biorthonormality condition is satisfied is discussed in Bastiaans 1994, where it is shown how $v(u)$ may be easily determined by employing the Zak transform.

The biorthonormality condition above ensures that if we start with the coefficients $G_f^{(w)}(l, m)$ and construct the signal $f(u)$ using equation 3.7, then we can obtain the original coefficients from this $f(u)$ by using equation 3.8. It is also possible to show that the above biorthonormality condition implies the dual biorthonormality condition (Bastiaans 1994)

$$\sum_l \sum_m v_{lm}^*(u) w_{lm}(u') = \delta(u - u'). \tag{3.10}$$

This condition ensures that if we start from a signal $f(u)$ and find its Gabor coefficients by using equation 3.8, then we can reconstruct the signal by using equation 3.7.

From equation 3.8 we conclude that the Gabor coefficients are in fact samples of the windowed Fourier transform with window $v(u)$:

$$G_f^{(w)}(l, m) = WF_f^{(v)}(l\, \delta u, m\, \delta \mu). \tag{3.11}$$

Thus Gabor's expansion can also be viewed as a way of recovering a signal from the samples of its windowed Fourier transform, rather than the continuous windowed Fourier transform as in equation 3.3.

Choosing $w(u) = 2^{1/4}\Delta_u^{-0.5}\exp(-\pi u^2/\Delta_u^2)$ ensures the greatest possible simultaneous concentration in time and frequency of the logons. (Actually the choice of a Gaussian function is not compatible with $\delta u\,\delta\mu = 1$ but requires that $\delta u\,\delta\mu$ be smaller than unity, even if only slightly so; see Daubechies 1992:107.) A natural choice for the parameters Δ_u and $\Delta_\mu = 1/\Delta_u$ characterizing the time and frequency extent of this Gaussian is $\Delta_u = 1\,\delta u$ and $\Delta_\mu = 1\,\delta\mu$. (The factor 1 is somewhat arbitrarily chosen; slightly different values may also be used.) In this case each $w_{lm}(u)$ snugly occupies the time-frequency cell on which it is centered; there is little overlap with adjacent cells. In this case the interpretation of the Gabor coefficients as an indicator of the time-frequency content of a signal around given time-frequency points becomes especially transparent. The function $v(u)$ corresponding to a Gaussian $w(u)$ is determined in Bastiaans 1994.

Gabor's expansion is a special case of what are more generally referred to as phase-space expansions (Landau 1993). What makes such expansions of interest is the fact that when appropriately defined, the coefficients $G_f^{(w)}(l, m)$ indicate how the energy of $f(u)$ is distributed over time and frequency; the larger this coefficient, the larger the contribution of that frequency at that time. If $f(u)$ is limited approximately to some time-frequency region such that its energy outside this region is small, then $f(u)$ can be reconstructed to the same degree of approximation from only those components $w_{lm}(u)$ which lie inside that region (Landau 1993). With Gabor's expansion, we have seen that we must have $\delta u\,\delta\mu \le 1$; that is, at least one sampling point per unit time-frequency area is required. Similar conditions exist for more general classes of expansions (Daubechies 1990, 1992, Landau 1993). In general, at least one coefficient is required per unit time-frequency area. This is consistent with an argument based on the Nyquist sampling theorem, which requires sampling at a rate of $\delta u = 1/\Delta\mu$ over the extent Δu of $f(u)$, implying a total of $\Delta u\,\Delta\mu$ samples, one sample per unit time-frequency area. Taken together, these facts support interpreting the minimum number of (nonredundant) expansion coefficients (or the time-bandwidth product) as the number of degrees of freedom of a signal. These concepts will be further discussed in section 3.3.

Original work underlying the Gabor expansion is scattered through many references. The reader may refer to the references found in the above cited works or to the useful papers by Bastiaans (1980, 1981a, 1982a, b, 1985, 1991a, 1994). The 1994 paper containing many references may constitute a useful starting point. A useful exposition of the Zak transform is Janssen 1988.

3.1.3 Wavelet transforms

Until now we exclusively took the width of the window function Δ_u and thus its dimensions in time-frequency $\Delta_u \times \Delta_\mu$ to be the same for all values of u and μ. Thus the absolute resolution obtained for all frequency components, high or low, was the same. However, since relative resolution is sometimes considered to be more relevant than absolute resolution (10 Hz resolution for 10 MHz is as good as 1 Hz resolution for 1 MHz), time-frequency representations involving windows of variable width have been invented. These so-called wavelet transforms are based on a complete set of orthogonal child wavelets $w_{u,\mu}(u')$ which are generated from a parent wavelet $w(u')$ through scaling and shift operations, where $w(u')$ is a unit-energy bandpass function with center frequency μ_0 centered around the origin (Hlawatsch and Boudreaux-Bartels 1992):

$$w_{u,\mu}(u') = |\mu/\mu_0|^{1/2} w\left(\frac{\mu}{\mu_0}(u' - u)\right). \tag{3.12}$$

We note that $w_{u,\mu}(u')$ is centered around u and has center frequency μ. The bandwidth of $w_{u,\mu}(u')$ is μ/μ_0 times the bandwidth of $w(u')$. Thus we see that the bandwidth of $w_{u,\mu}(u')$ is proportional to its center frequency. This directly translates into frequency-proportionate frequency resolution in the wavelet transform defined as

$$WT_f^{(w)}(u, \mu) = \int f(u')w_{u,\mu}^*(u')\, du', \tag{3.13}$$

which may be compared to equation 3.1. A more detailed comparison of the wavelet transform to the short-time Fourier transform may be found in Hlawatsch and Boudreaux-Bartels 1992. Wavelet transforms are often expressed as time-scale representations, rather than time-frequency representations. Defining $\xi \equiv \mu_0/\mu$, we can rewrite the above definition as

$$TS_f^{(w)}(u, \xi) = \int f(u')w_{u,\xi}^*(u')\, du', \tag{3.14}$$

$$w_{u,\xi}(u') = \frac{1}{|\xi|^{1/2}}\, w\left(\frac{u' - u}{\xi}\right).$$

Introductory sources on wavelet transforms and time-scale representations include Mallat 1989, Daubechies 1990, 1992, Rioul and Vetterli 1991, Akansu and Haddad 1992, Chui 1992, Walter 1994, Vetterli and Kovacevic 1995, Strang and Nguyen 1996, Suter 1997, and Mallat 1998. There is also the edited volume *Wavelets: Mathematics and Applications* 1993.

We will be particularly interested in wavelet transforms generated from the parent wavelet $w(u) = \exp(i\pi u^2)$, resulting in the wavelet transform

$$TS_f^{(\text{chirp})}(u, \xi) = \frac{1}{|\xi|^{1/2}} \int f(u')e^{-i\pi(u'-u)^2/\xi^2}\, du', \tag{3.15}$$

which we recognize to be essentially the Fresnel transform. This class of wavelet transforms has been discussed in Onural 1993 and Onural and Kocatepe 1995.

3.1.4   Remarks

A very large number of different time-frequency representations have been suggested for their particular properties and suitability for different applications (Cohen 1989, 1995, Hlawatsch and Boudreaux-Bartels 1992). We will mostly concentrate on the Wigner distribution and a number of other closely related representations which are discussed in detail in the following sections.

It is tempting to view time-frequency representations as alternative representations of a signal, just as the time-domain representation, frequency-domain representation, and so on, as we discussed in chapter 2. This is justified by the fact that they often contain the same (or almost the same) information as these other representations. However, there are a number of differences. (i) Time-frequency representations are not always linearly related to other representations, such as the time-domain representation. (ii) It is not always possible to interpret a time-frequency representation as the coefficient of expansion in terms of a basis set, and even when this is the case, the basis may not be orthonormal. (iii) Time-frequency representations of functions of one variable are functions of two variables (time and frequency).

We end by noting that the discussion of this section is not totally precise and far from rigorous. Readers desiring greater rigor and more precise versions of the various statements we have made should consult the references cited.

3.2   The Wigner distribution and the ambiguity function

3.2.1   The Wigner distribution

The Wigner distribution $W_f(u, \mu)$ of a signal f can be defined in terms of the time-domain representation $f(u)$ of the signal as follows (Claasen and Mecklenbräuker 1980a, b, c):

$$W_f(u, \mu) \equiv \int f(u + u'/2) f^*(u - u'/2) e^{-i2\pi \mu u'} \, du'. \tag{3.16}$$

Roughly speaking, $W_f(u, \mu)$ is a function which gives the distribution of signal energy over time and frequency, a fact which is not immediately evident from the above definition. However, it is possible to show directly from the above definition that

$$\int W_f(u, \mu) \, d\mu = |f(u)|^2, \tag{3.17}$$

$$\int W_f(u, \mu) \, du = |F(\mu)|^2, \tag{3.18}$$

$$\iint W_f(u, \mu) \, du \, d\mu = \|f\|^2 = \text{En}[f] = \text{signal energy}. \tag{3.19}$$

Note that since the signal energy is given by the integral of either $|f(u)|^2$ over time or the integral of $|F(\mu)|^2$ over frequency, the first two equations are consistent and imply the third. These properties are consistent with, but do not imply, the interpretation of $W_f(u, \mu)$ as the energy density at time-frequency point (u, μ). Indeed, there are intrinsic difficulties associated with the notion of the energy density of a signal at a specific time and frequency point, stemming from the uncertainty relation (section 2.7). However, we will later justify the interpretation of local averages of $W_f(u, \mu)$ approximately as the time-frequency energy density of a signal. As a consequence, the energy of the signal in any extended time-frequency region can be found by integrating $W_f(u, \mu)$ over that region.

A number of characteristics make the Wigner distribution a very attractive time-frequency representation. The Wigner distribution is completely symmetric with

Table 3.1. Wigner distributions of some common signals.

$f(u)$	$W_f(u, \mu)$
1.	$\exp(i2\pi\xi u)$	$\delta(\mu - \xi)$
2.	$\delta(u - \xi)$	$\delta(u - \xi)$
3.	$\exp[i\pi(\chi u^2 + 2\xi u + \zeta)]$	$\delta(\mu - \chi u - \xi)$
4.	$(2\chi)^{1/4}\exp(-\pi\chi u^2)$	$2\exp[-2\pi(\chi u^2 + \mu^2/\chi)]$
5.	$\mathrm{rect}(u)$	$2(1 - |2u|)\,\mathrm{rect}(u)\,\mathrm{sinc}\,[2(1 - |2u|)\mu]$

ξ, χ, ζ are real.

respect to the time and frequency domains, as evidenced by its expression in terms of the frequency-domain representation $F(\mu)$ of the signal:

$$W_f(u, \mu) = \int F(\mu + \mu'/2)F^*(\mu - \mu'/2)e^{i2\pi\mu' u}\,d\mu', \qquad (3.20)$$

which can also be derived from equation 3.16. We will see in chapter 4 that this symmetry is preserved also with respect to a continuum of domains we will refer to as fractional Fourier domains. Then it will become particularly apparent that the geometric shape of the Wigner distribution has a reality and significance independent of the particular coordinate system in the time-frequency plane in which it is expressed. The Wigner distribution should be considered as an abstract geometric entity associated with the signal f in the abstract, not being tied to a particular representation of f in a particular domain.

The Wigner distributions of some common signals are given in table 3.1. The Wigner distributions of the impulse and harmonic functions are easily interpreted in terms of the expected distribution of signal energy of these functions. The Wigner distribution of the chirp function, which includes these two as special cases, is found to be concentrated along the line giving the instantaneous frequency of the chirp: $(2\pi)^{-1}d[\pi(\chi u^2 + 2\xi u + \zeta)]/du = \chi u + \xi$. The Wigner distribution of a Gaussian signal is a Gaussian in u and μ whose time and frequency profiles match the time and frequency representations of the Gaussian signal. We observe that the Wigner distribution of the rectangle function is negative for certain values of u and μ. In fact, it is more the norm than the exception for the Wigner distribution of a signal to exhibit negative values for some values of u and μ, a fact which complicates its interpretation as an energy density. We also give the Wigner distribution of the scaled Hermite-Gaussian function $\psi_n(u/M)$:

$$W_{\psi_n(u/M)}(u, \mu) = 2(-1)^n \exp\left[-2\pi(u^2/M^2 + M^2\mu^2)\right] L_n\left[4\pi(u^2/M^2 + M^2\mu^2)\right],$$
$$(3.21)$$

where $L_n(\cdot)$ denotes the Laguerre polynomials (Bastiaans 1997). We finally note the Wigner distribution of the delta train or comb function defined as $f(u) = \sum_{n=-\infty}^{\infty} \delta(u - n)$, which is given by

$$W_f(u, \mu) = \frac{1}{2}\sum_{n=-\infty}^{\infty}\sum_{n'=-\infty}^{\infty}(-1)^{nn'}\delta(u - n/2)\delta(\mu - n'/2). \qquad (3.22)$$

Table 3.2. Properties of the Wigner distribution (Hlawatsch and Boudreaux-Bartels 1992).

1.	$W_f^*(u,\mu) = W_f(u,\mu)$
2.	$f(u-\xi)$ has Wigner distribution $W_f(u-\xi,\mu)$
3.	$\exp(i2\pi\xi u)f(u)$ has Wigner distribution $W_f(u,\mu-\xi)$
4.	$\int W_f(u,\mu)\,d\mu = |f(u)|^2$
5.	$\int W_f(u,\mu)\,du = |F(\mu)|^2$
6.	$\iint u^n W_f(u,\mu)\,du\,d\mu = \int u^n |f(u)|^2\,du$
7.	$\iint \mu^n W_f(u,\mu)\,du\,d\mu = \int \mu^n |F(\mu)|^2\,d\mu$
8.	$\sqrt{|1/M|}\,f(u/M)$ has Wigner distribution $W_f(u/M, M\mu)$
9.	$\dfrac{\int \mu W_f(u,\mu)\,d\mu}{\int W_f(u,\mu)\,d\mu} = \dfrac{1}{2\pi}\dfrac{d\angle[f(u)]}{du}$
10.	$\dfrac{\int u W_f(u,\mu)\,du}{\int W_f(u,\mu)\,du} = -\dfrac{1}{2\pi}\dfrac{d\angle[F(\mu)]}{d\mu}$
11.	If $f(u) = 0$ outside $[\xi_1,\xi_2]$, then $W_f(u,\mu) = 0$ outside $[u=\xi_1, u=\xi_2]$
12.	If $F(\mu) = 0$ outside $[\xi_1,\xi_2]$, then $W_f(u,\mu) = 0$ outside $[\mu=\xi_1,\mu=\xi_2]$
13.	$|\int f^*(u)h(u)\,du|^2 = \iint W_f(u,\mu)W_h(u,\mu)\,du\,d\mu$
14.	$h(u) * f(u)$ has Wigner distribution $\int W_h(u-u',\mu)W_f(u',\mu)\,du'$
15.	$h(u)f(u)$ has Wigner distribution $\int W_h(u,\mu-\mu')W_f(u,\mu')\,d\mu'$
16.	$F(u) = \{\mathcal{F}[f]\}(u)$ has Wigner distribution $W_f(-\mu,u)$

ξ,ξ_1,ξ_2,M are real and n is a positive integer. Adapted by permission of IEEE.

A pictorial discussion of this Wigner distribution may be found in Testorf and Ojeda-Castañeda 1996.

To recover $f(u)$ from its Wigner distribution $W_f(u,\mu)$, we note that equation 3.16 is a Fourier transform relation which can be inverted as

$$f(u+u'/2)f^*(u-u'/2) = \int W_f(u,\mu)e^{i2\pi\mu u'}\,d\mu. \qquad (3.23)$$

If and only if upon evaluation of the right-hand side of this equation we arrive at a function expressible in the form indicated by the left-hand side, is the given two-dimensional function $W_f(u,\mu)$ a legitimate Wigner distribution of some function $f(u)$. Otherwise, the given two-dimensional function does not correspond to the Wigner distribution of any function, as is the case for most two-dimensional functions. Assuming that $W_f(u,\mu)$ is indeed a legitimate Wigner distribution, we can show

$$f(u) = \frac{1}{f^*(0)}\int W_f(u/2,\mu)e^{i2\pi\mu u}\,d\mu. \qquad (3.24)$$

We see that the original function can be recovered only up to a complex constant of unit magnitude. In other words, any function of the form $f(u)\exp(i\pi\zeta)$ where ζ is a real constant has the same Wigner distribution as $f(u)$.

Important properties of the Wigner distribution are listed in table 3.2. We briefly comment on these properties. (1) The Wigner distribution is everywhere real but not always positive, an issue we will further discuss below. (2,3) It is time and frequency shift-invariant in the sense that shifting the time- and/or frequency-domain representations of a signal results in corresponding shifts in the Wigner distribution.

(4,5) Integrating out time or frequency returns the energy distribution with respect to frequency or time. (6,7) The time and frequency moments can be calculated as weighted averages in the time-frequency plane directly. Here these follow directly from properties 4 and 5, but there are other time-frequency distributions which satisfy 6 and 7 without satisfying 4 and 5. (8) Time and frequency scale inversely. (9,10) The weighted average of frequency for a given time is equal to the instantaneous frequency at that time, and the weighted average of time for a given frequency is equal to the group delay for that frequency. (However, instantaneous frequency and group delay can be meaningfully interpreted for only certain classes of signals.) (11,12) If the time- or frequency-domain representation of a signal is identically zero outside a certain interval, so is its Wigner distribution. However, if the time- or frequency-domain representation is zero inside a finite interval with nonzero values outside this interval, the Wigner distribution will not in general be zero inside that interval (Cohen 1989). (13) This property, known as Moyal's formula, is some kind of Parseval's relation between the time- (or frequency-) domain representation and the Wigner distribution. It basically states that the overlap integral or inner product of two Wigner distributions is equal to the absolute square of the inner product of the two original signals. (14,15) Convolving $f(u)$ with another function $h(u)$ corresponds to convolving $W_f(u,\mu)$ with $W_h(u,\mu)$ in the time coordinate. Multiplying $f(u)$ with another function $h(u)$ corresponds to convolving $F(\mu)$ with $H(\mu)$, which further corresponds to convolving $W_f(u,\mu)$ with $W_h(u,\mu)$ in the frequency coordinate. (16) The Wigner distribution of the Fourier transform of a function is the Wigner distribution of the original function rotated clockwise by a right angle.

Certain properties of the Wigner distribution are sometimes considered undesirable. Since it is not linear but quadratic in the signal, the Wigner distribution of the sum of two signals will not be equal to the sum of their Wigner distributions, resulting in often undesired cross terms. This has motivated dealing with projections of the Wigner distribution, which we will see correspond to linear representations of the signal in different fractional Fourier domains (chapter 4). From a fundamental viewpoint, since the Wigner distribution is meant to have an energetic interpretation, and since energy is quadratic in the signal, the existence of cross terms should not be considered a defect. However, it does lead to difficulty when visually interpreting signals with multiple components, which we might want to be able to separately identify in a time-frequency plot. Experience with and recognition of the nature of the interference terms (Hlawatsch and Boudreaux-Bartels 1992) is of great benefit in interpreting such signals. Certain smoothed Wigner distributions, such as the Choi-Williams distribution (Choi and Williams 1989, Cohen 1989), allow the suppression of the interference terms at the expense of time-frequency concentration in a controlled manner by including an adjustable parameter.

The fact that the Wigner distribution can be negative for certain time-frequency values is often considered undesirable because it conflicts with the interpretation of the Wigner distribution as the distribution of signal energy. Such negative values will also tend to disappear with smoothing. We will return to this issue further below.

It is instructive to write the Wigner distribution in terms of the Hermite-Gaussian expansion of a function. The Wigner distribution of a function $f(u) = \sum_{n=0}^{\infty} C_n \psi_n(u)$ becomes

$$W_f(u,\mu) = \int \sum_{n=0}^{\infty} C_n \psi_n(u + u'/2) \sum_{n'=0}^{\infty} C_{n'}^* \psi_{n'}(u - u'/2) e^{-i2\pi\mu u'} \, du'$$

$$= \sum_{n=0}^{\infty} \sum_{n'=0}^{\infty} C_n C_{n'}^* \int \psi_n(u + u'/2) \psi_{n'}(u - u'/2) e^{-i2\pi\mu u'} \, du'$$

$$= \sum_{n=0}^{\infty} \sum_{n'=0}^{\infty} C_n C_{n'}^* W_{\psi_n,\psi_{n'}}(u,\mu), \tag{3.25}$$

where $W_{\psi_n,\psi_{n'}}(u,\mu)$ is the *cross Wigner distribution* of $\psi_n(u)$ and $\psi_{n'}(u)$ (Claasen and Mecklenbräuker 1980a). If the final equation is considered as the expansion of a two-dimensional function in terms of the cross Wigner distributions $W_{\psi_n,\psi_{n'}}(u,\mu)$, we see that arbitrary expansion coefficients of the form $C_{nn'}$ are not possible, but that only outer products of the form $C_n C_{n'}^*$ can appear. This is consistent with the fact that the Wigner distribution contains the information of a one-dimensional signal; very few two-dimensional functions are Wigner distributions of some signal.

A useful edited collection on the Wigner distribution and its applications in signal processing is *The Wigner Distribution: Theory and Applications in Signal Processing* 1997. An extension of the Wigner distribution to signal spaces, rather than a single signal is given in Hlawatsch and Kozek 1993. The original work of Wigner is Wigner 1932.

3.2.2 The ambiguity function

Another time-frequency distribution which is closely related to the Wigner distribution is known as the ambiguity function $A(\bar{u}, \bar{\mu})$ defined as

$$A_f(\bar{u}, \bar{\mu}) \equiv \int f(u' + \bar{u}/2) f^*(u' - \bar{u}/2) e^{-i2\pi\bar{\mu}u'} \, du'$$

$$= \int F(\mu' + \bar{\mu}/2) F^*(\mu' - \bar{\mu}/2) e^{i2\pi\mu'\bar{u}} \, d\mu', \tag{3.26}$$

where the second equality is a consequence of the definition given in the first line. This definition should be carefully compared to that of the Wigner distribution given in equation 3.16. Whereas the Wigner distribution is the prime example of an *energetic* time-frequency representation, the ambiguity function is the prime example of a *correlative* time-frequency representation (Hlawatsch and Boudreaux-Bartels 1992). (The term "time-frequency distribution" is sometimes used interchangeably with the term "time-frequency representation." However, it is more appropriate to reserve the term "distribution" for those representations which have an energetic interpretation.) The ambiguity function deserves this by virtue of the following properties:

$$A_f(\bar{u}, 0) = R_{ff}(\bar{u}) \equiv \int f(u' + \bar{u}) f^*(u') \, du' = \int |F(\mu')|^2 e^{i2\pi\mu'\bar{u}} \, d\mu', \tag{3.27}$$

$$A_f(0, \bar{\mu}) = R_{FF}(\bar{\mu}) \equiv \int F(\mu' + \bar{\mu}) F^*(\mu') \, d\mu' = \int |f(u')|^2 e^{-i2\pi\bar{\mu}u'} \, du', \tag{3.28}$$

$$A_f(\bar{u}, \bar{\mu}) \le A_f(0, 0) = \|f\|^2 = \mathrm{En}[f] = \text{signal energy}, \tag{3.29}$$

Table 3.3.   Ambiguity functions of some common signals.

	$f(u)$	$A_f(\bar{u}, \bar{\mu})$
1.	$\exp(i2\pi\xi u)$	$\exp(i2\pi\xi\bar{u})\delta(\bar{\mu})$
2.	$\delta(u - \xi)$	$\exp(i2\pi\xi\bar{\mu})\delta(\bar{u})$
3.	$\exp[i\pi(\chi u^2 + 2\xi u + \zeta)]$	$\exp(i2\pi\xi\bar{u})\delta(\bar{\mu} - \chi\bar{u})$
4.	$(2\chi)^{1/4}\exp(-\pi\chi u^2)$	$\exp[-\pi(\chi\bar{u}^2 + \bar{\mu}^2/\chi)/2]$
5.	$\mathrm{rect}(u)$	$(1 - |\bar{u}|)\,\mathrm{rect}(2\bar{u})\,\mathrm{sinc}\,[\bar{\mu}(1 - |\bar{u}|)]$

ξ, χ, ζ are real.

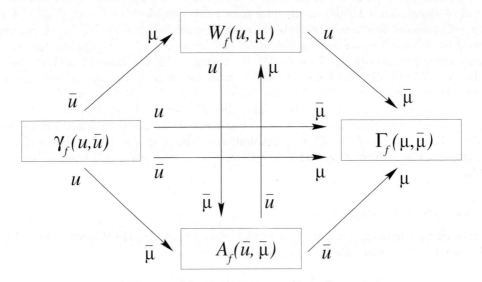

Figure 3.1.   Relationships between $\gamma(u, \bar{u})$, $\Gamma(\mu, \bar{\mu})$, $A_f(\bar{u}, \bar{\mu})$, and $W_f(u, \mu)$; the arrows indicate a Fourier transform with respect to the variables shown (Bamler and Glünder 1983, Hlawatsch and Boudreaux-Bartels 1992). Adapted by permission of Taylor & Francis (www.tandf.co.uk/journals) and IEEE.

which say that the on-axis profiles of the ambiguity function are equal to the autocorrelation of the signal in the time and frequency domains respectively.

The ambiguity functions of some common signals are given in table 3.3.

The Wigner distribution and ambiguity function are Fourier transforms of two auxiliary functions defined as (Claasen and Mecklenbräuker 1980c)

$$\gamma(u, \bar{u}) \equiv f(u + \bar{u}/2)f^*(u - \bar{u}/2), \tag{3.30}$$

$$\Gamma(\mu, \bar{\mu}) \equiv F(\mu + \bar{\mu}/2)F^*(\mu - \bar{\mu}/2), \tag{3.31}$$

such that

$$W_f(u, \mu) = \int \gamma(u, \bar{u})e^{-i2\pi\mu\bar{u}}\, d\bar{u} = \int \Gamma(\mu, \bar{\mu})e^{i2\pi\bar{\mu}u}\, d\bar{\mu}, \tag{3.32}$$

$$A_f(\bar{u}, \bar{\mu}) = \int \gamma(u, \bar{u}) e^{-i2\pi \bar{\mu} u} \, du = \int \Gamma(\mu, \bar{\mu}) e^{i2\pi \mu \bar{u}} \, d\mu. \tag{3.33}$$

Combining these relationships, we find that the ambiguity function is related to the Wigner distribution by what is essentially a two-dimensional Fourier transform:

$$A_f(\bar{u}, \bar{\mu}) = \iint W_f(u, \mu) e^{-i2\pi(\bar{\mu} u - \bar{u} \mu)} \, du \, d\mu, \tag{3.34}$$

consistent with the energetic nature of the Wigner distribution and the correlative nature of the ambiguity function. It is also possible to show that $\Gamma(\mu, \bar{\mu})$ is the two-dimensional Fourier transform of $\gamma(u, \bar{u})$. These relationships are summarized in figure 3.1. It is also instructive to see the relationship between the (approximate) supports of these four functions. By the support of a function we mean the region where the value of the function is nonnegligible. Let us assume that the signal f has negligible energy outside the time interval $[-\Delta u/2, \Delta u/2]$ and the frequency interval $[\Delta \mu/2, \Delta \mu/2]$. Clearly, the Wigner distribution will (approximately) have a rectangular support defined by these intervals. The supports of the remaining three functions are shown in figure 3.2.

The properties of the ambiguity function are summarized in table 3.4 which has been prepared parallel to table 3.2. The properties of the ambiguity function are easily interpreted by virtue of the fact that the ambiguity function is the Fourier transform of the Wigner distribution.

Table 3.4.   Properties of the ambiguity function (Hlawatsch and Boudreaux-Bartels 1992).

1.	$A_f^*(-\bar{u}, -\bar{\mu}) = A_f(\bar{u}, \bar{\mu})$
2.	$f(u - \xi)$ has ambiguity function $A_f(\bar{u}, \bar{\mu}) \exp(-i2\pi \xi \bar{\mu})$
3.	$\exp(i2\pi \xi u) f(u)$ has ambiguity function $A_f(\bar{u}, \bar{\mu}) \exp(i2\pi \xi \bar{u})$
4.	$A_f(0, \bar{\mu}) = \int F(\mu + \bar{\mu}) F^*(\mu) \, d\mu$
5.	$A_f(\bar{u}, 0) = \int f(u + \bar{u}) f^*(u) \, du$
6.	$\dfrac{1}{(-i2\pi)^n} \dfrac{d^n A_f(0, \bar{\mu})}{d\bar{\mu}^n} \bigg|_{\bar{\mu}=0} = \int u^n |f(u)|^2 \, du$
7.	$\dfrac{1}{(i2\pi)^n} \dfrac{d^n A_f(\bar{u}, 0)}{d\bar{u}^n} \bigg|_{\bar{u}=0} = \int \mu^n |F(\mu)|^2 \, d\mu$
8.	$\sqrt{|1/M|} \, f(u/M)$ has ambiguity function $A_f(\bar{u}/M, M\bar{\mu})$
9.	$\dfrac{1}{i2\pi} \dfrac{\int [dA_f(\bar{u}, \bar{\mu})/d\bar{u}]_{\bar{u}=0} e^{i2\pi u \bar{\mu}} \, d\bar{\mu}}{\int A_f(0, \bar{\mu}) e^{i2\pi u \bar{\mu}} \, d\bar{\mu}} = \dfrac{1}{2\pi} \dfrac{d\angle[f(u)]}{du}$
10.	$-\dfrac{1}{i2\pi} \dfrac{\int [dA_f(\bar{u}, \bar{\mu})/d\bar{\mu}]_{\bar{\mu}=0} e^{-i2\pi \mu \bar{u}} \, d\bar{u}}{\int A_f(\bar{u}, 0) e^{-i2\pi \mu \bar{u}} \, d\bar{u}} = -\dfrac{1}{2\pi} \dfrac{d\angle[F(\mu)]}{d\mu}$
11.	If $f(u) = 0$ outside $[\xi_1, \xi_2]$, then $A_f(\bar{u}, \bar{\mu}) = 0$ for $|\bar{u}| > \xi_2 - \xi_1$
12.	If $F(\mu) = 0$ outside $[\xi_1, \xi_2]$, then $A_f(\bar{u}.\bar{\mu}) = 0$ for $|\bar{\mu}| > \xi_2 - \xi_1$
13.	$|\int f^*(u) h(u) \, du|^2 = \iint A_f^*(\bar{u}, \bar{\mu}) A_h(\bar{u}, \bar{\mu}) \, d\bar{u} \, d\bar{\mu}$
14.	$h(u) * f(u)$ has ambiguity function $\int A_h(\bar{u} - \bar{u}', \bar{\mu}) A_f(\bar{u}', \bar{\mu}) \, d\bar{u}'$
15.	$h(u) f(u)$ has ambiguity function $\int A_h(\bar{u}, \bar{\mu} - \bar{\mu}') A_f(\bar{u}, \bar{\mu}') \, d\bar{\mu}'$
16.	$F(u) = \{\mathcal{F}[f]\}(u)$ has ambiguity function $A_f(-\bar{\mu}, \bar{u})$

ξ, ξ_1, ξ_2, M are real and n is a positive integer. Adapted by permission of IEEE.

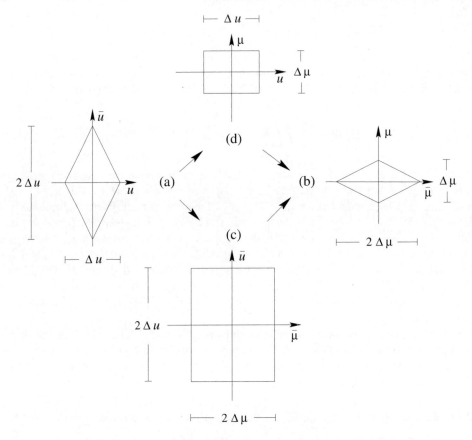

Figure 3.2. Supports of (a) $\gamma(u,\bar{u})$, (b) $\Gamma(\mu,\bar{\mu})$, (c) $A_f(\bar{u},\bar{\mu})$, and (d) $W_f(u,\mu)$ for an approximately time- and band-limited signal f (Bamler and Glünder 1983). Adapted by permission of Taylor & Francis (www.tandf.co.uk/journals).

The special cases of properties 13 in tables 3.2 and 3.4 when $g = f$ are worth noting:

$$\|f\|^4 = (\text{En}[f])^2 = \int\int [W_f(u,\mu)]^2 \, du \, d\mu = \int\int |A_f(\bar{u},\bar{\mu})|^2 \, d\bar{u} \, d\bar{\mu}. \qquad (3.35)$$

The consistency of these two properties follows from the fact that Parseval's relation implies that the square integrals of the Wigner distribution and ambiguity function, essentially a two-dimensional Fourier transform pair, must be equal.

A rather different and useful exposition of the ambiguity function and further discussion of its relation to the second-order moments of a signal may be found in Papoulis 1977:284–295.

3.2.3   Cohen's class of shift-invariant distributions

A relatively broad class of energetic time-frequency distributions which includes the Wigner distribution and spectrogram as special cases is known as Cohen's class of shift-invariant time-frequency distributions (Cohen 1966, 1976, 1989, 1995, Hlawatsch and Boudreaux-Bartels 1992). These distributions may be defined in terms of the Wigner distribution through the two-dimensional convolution relation

$$TFE_f(u,\mu) = \psi_{TFE}(u,\mu) * * W_f(u,\mu) = \iint \psi_{TFE}(u - u', \mu - \mu') W_f(u', \mu') \, du' \, d\mu'.$$
(3.36)

$\psi_{TFE}(u,\mu)$ is a kernel uniquely corresponding to the distribution $TFE_f(u,\mu)$. A distribution is a member of this class if and only if the distribution corresponding to $f(u - \xi_u) \exp(i2\pi\xi_\mu u)$ is equal to $TFE_f(u - \xi_u, \mu - \xi_\mu)$ (as in properties 2 and 3 in table 3.2), as can be shown from equation 3.36. This is what is meant by shift invariance. The fact that we have defined this class in terms of the Wigner distribution does not by itself confer a privileged status to the Wigner distribution among other members of the class. It is also possible to define the Cohen class in terms of members other than the Wigner distribution (Cohen 1989).

In analogy with equation 3.34, we can define the correlative dual time-frequency representation $TFC_f(\bar{u}, \bar{\mu})$ of the energetic time-frequency distribution $TFE_f(u, \mu)$ as follows (Hlawatsch and Boudreaux-Bartels 1992):

$$TFC_f(\bar{u}, \bar{\mu}) = \iint TFE_f(u, \mu) e^{-i2\pi(\bar{\mu}u - \bar{u}\mu)} \, du \, d\mu.$$
(3.37)

Then equation 3.36 can also be written as

$$TFC_f(\bar{u}, \bar{\mu}) = \Psi_{TFE}(\bar{u}, \bar{\mu}) A_f(\bar{u}, \bar{\mu}),$$
(3.38)

where $\Psi_{TFE}(\bar{u}, \bar{\mu})$ is the two-dimensional Fourier transform of $\psi_{TFE}(u, \mu)$. In this case the representation corresponding to $f(u - \xi_u) \exp(i2\pi\xi_\mu u)$ is equal to $TFC_f(u, \mu) \exp[i2\pi(\xi_\mu\bar{u} - \xi_u\bar{\mu})]$ (as in properties 2 and 3 in table 3.4). Table 3.5 lists the names of a number of time-frequency distributions together with their defining kernels $\Psi_{TFE}(\bar{u}, \bar{\mu})$. The correlative duals of these distributions can also be deduced from these kernels.

Table 3.5. Selected shift-invariant time-frequency distributions and their defining kernels (Hlawatsch and Boudreaux-Bartels 1992).

Distribution	Kernel $\Psi_{TFE}(\bar{u}, \bar{\mu})$
Born-Jordan	$\mathrm{sinc}(\bar{u}\bar{\mu})$
Choi-Williams (exponential)	$\exp[-(2\pi\bar{u}\bar{\mu})^2/\zeta]$
Generalized Wigner	$\exp(i2\pi\chi\bar{u}\bar{\mu})$
Page	$\exp(-i\pi|\bar{u}|\bar{\mu})$
Pseudo Wigner	$\eta(\bar{u}/2)\eta^*(-\bar{u}/2)$
Rihaczek	$\exp(i\pi\bar{u}\bar{\mu})$
Spectrogram	$\int w(u' - \bar{u}/2)w^*(u' + \bar{u}/2)e^{i2\pi\bar{\mu}u'} \, du'$
Wigner	1

ζ, χ are real parameters and $\eta(\cdot), w(\cdot)$ are suitably selected functions. Adapted by permission of IEEE.

Since the kernels $\psi_{TFE}(u, \mu)$ or $\Psi_{TFE}(\bar{u}, \bar{\mu})$ fully characterize a distribution which is a member of the Cohen class, the properties of the distribution can often be determined by examining these kernels. Table 3.6 has been prepared in parallel with table 3.2 and states the constraints that these kernels must satisfy in order for the distribution to exhibit a given property. The kernel $\psi_{TFE}(u, \mu) = \delta(u, \mu)$ or $\Psi_{TFE}(\bar{u}, \bar{\mu}) = 1$ of the Wigner distribution satisfies all of these constraints so that the Wigner distribution exhibits all of these properties.

Table 3.6. Kernel constraints corresponding to various properties of shift-invariant time-frequency distributions (Hlawatsch and Boudreaux-Bartels 1992).

	Property	Kernel constraint
1.	Real-valued	$\Psi^*_{TFE}(-\bar{u}, -\bar{\mu}) = \Psi_{TFE}(\bar{u}, \bar{\mu})$
2.	Time shift	Always satisfied
3.	Frequency shift	Always satisfied
4.	Time marginal	$\Psi_{TFE}(0, \bar{\mu}) = 1$
5.	Frequency marginal	$\Psi_{TFE}(\bar{u}, 0) = 1$
6.	Time moments	$\Psi_{TFE}(0, \bar{\mu}) = 1$
7.	Frequency moments	$\Psi_{TFE}(\bar{u}, 0) = 1$
8.	Time-frequency scaling	$\Psi_{TFE}(\bar{u}/M, M\bar{\mu}) = \Psi_{TFE}(\bar{u}, \bar{\mu})$
9.	Instantaneous frequency	$\Psi_{TFE}(0, \bar{\mu}) = 1$ and $\left. \frac{d\Psi_{TFE}(\bar{u}, \bar{\mu})}{d\bar{u}} \right\|_{\bar{\mu}=0} = 0$
10.	Group delay	$\Psi_{TFE}(\bar{u}, 0) = 1$ and $\left. \frac{d\Psi_{TFE}(\bar{u}, \bar{\mu})}{d\bar{\mu}} \right\|_{\bar{\mu}=0} = 0$
11.	Finite time support	$\int \Psi_{TFE}(\bar{u}, \bar{\mu}) e^{i2\pi\bar{\mu}u} \, d\bar{\mu} = 0$ for $|u/\bar{u}| > 1/2$
12.	Finite frequency support	$\int \Psi_{TFE}(\bar{u}, \bar{\mu}) e^{-i2\pi\mu\bar{u}} \, d\bar{u} = 0$ for $|\mu/\bar{\mu}| > 1/2$
13.	Moyal's formula (unitarity)	$|\Psi_{TFE}(\bar{u}, \bar{\mu})| = 1$
14.	Convolution	$\Psi_{TFE}(\bar{u} + \bar{u}', \bar{\mu}) = \Psi_{TFE}(\bar{u}, \bar{\mu}) \, \Psi_{TFE}(\bar{u}', \bar{\mu})$
15.	Multiplication	$\Psi_{TFE}(\bar{u}, \bar{\mu} + \bar{\mu}') = \Psi_{TFE}(\bar{u}, \bar{\mu}) \, \Psi_{TFE}(\bar{u}, \bar{\mu}')$
16.	Fourier transform	$\Psi_{TFE}(-\bar{\mu}, \bar{u}) = \Psi_{TFE}(\bar{u}, \bar{\mu})$

The listed properties correspond to those in table 3.2. The constraints are given in terms of the kernel Ψ_{TFE}. $M, \bar{u}, \bar{u}', \bar{\mu}, \bar{\mu}'$ are arbitrary real numbers. Adapted by permission of IEEE.

3.2.4   Smoothing of the Wigner distribution

The existence of interference terms is usually considered an undesirable property of the Wigner distribution which often makes its visual interpretation difficult. Since such terms are often of oscillatory nature (Hlawatsch and Boudreaux-Bartels 1992), it is possible to attenuate them considerably by smoothing the Wigner distribution. This can be achieved by convolving the Wigner distribution with a smooth function. Looking back to equation 3.36, we conclude that such smoothed Wigner distributions are time-frequency distributions belonging to the Cohen class. (However, not all distributions belonging to the Cohen class are smoothed Wigner distributions; this requires that the kernel $\psi_{TFE}(u, \mu)$ be a smooth function.) Naturally, such smoothing will result in some loss of time-frequency resolution, and also in a loss of some of the desirable properties of the Wigner distribution (as can be determined from table 3.6).

As we have mentioned before, another undesirable property of the Wigner distribution is that it can be negative for certain time-frequency values, which is troubling because it conflicts with the interpretation of the Wigner distribution as the distribution of signal energy. We will now argue that this is not a fundamental flaw, but rather only an inconvenience. We must first recognize that the value of the Wigner distribution of a signal at a certain time-frequency point, mathematically defined through equation 3.16, does not correspond to a physically measurable quantity. (This is in contrast to the spectrogram which corresponds to a physically measurable quantity, as one can physically window a function and then observe its spectrum.) This has to do with the fact that it is not possible to resolve or isolate a part of a signal which is concentrated at a single time-frequency point. By applying a narrowband filter to the signal in the frequency domain, we can isolate as narrow a band of frequencies as we wish, but application of this filter will also inevitably result in a broadening of the signal in the time domain (since the signal will be convolved with a broad function in the time domain). Alternatively, by multiplying the signal with a narrow window in the time domain, we can isolate as short an interval of the signal as we wish, but this process will also inevitably result in a broadening of the signal in the frequency domain (since the signal will be convolved with a broad function in the frequency domain). According to the uncertainty relation, the product of the duration of the impulse response of a filter and its bandwidth must be greater than (approximately) unity. Thus, the smallest part of the signal we can isolate and subject to an energy measurement has time-frequency area which must also be greater than (approximately) unity. The energy of such an isolated part of the signal is given by an (appropriately weighted) integral of the Wigner distribution over the relevant time-frequency region and it is this quantity that we would expect to be positive.

Measuring the energy of a signal over a unit time-frequency area is closely related to the concept of smoothing the Wigner distribution with a kernel of unit time-frequency area. If it does not make sense to speak about the energy of the signal in time-frequency areas smaller than unity, then it also does not make sense to specify time-frequency points with joint accuracy exceeding that suggested by the uncertainty relation.

From table 3.5 we see that by smoothing the Wigner distribution of a signal with the coordinate-inverted Wigner distribution of a window function $w(u)$, we obtain the spectrogram of the signal based on the same window function (Hlawatsch and Boudreaux-Bartels 1992):

$$SP_f^{(w)}(u,\mu) = |WF_f^{(w)}(u,\mu)|^2 = \int\int \psi_{SP}(u - u', \mu - \mu') W_f(u', \mu') \, du' \, d\mu', \quad (3.39)$$

$$\psi_{SP}(u,\mu) = W_w(-u, -\mu),$$

where $W_w(u,\mu)$ is the Wigner distribution of the window function $w(u)$. From this result we can conclude that if the smoothing kernel $\psi_{TFE}(u,\mu)$ is the Wigner distribution of some unit-energy function, the resulting distribution will be a spectrogram and hence nonnegative (and also a measurable quantity). In particular, let us consider a Gaussian window $2^{1/4}\Delta_u^{-0.5} \exp(-\pi u^2/\Delta_u^2)$ whose Wigner distribution is

$$W_{\text{Ga}}(u,\mu) = 2\exp\left[-2\pi\left(\frac{u^2}{\Delta_u^2} + \mu^2\Delta_u^2\right)\right]. \quad (3.40)$$

The Wigner distribution of any function extends over a time-frequency region whose area is at least unity. The Wigner distribution of the Gaussian function satisfies this condition with approximate equality, extending roughly over a region of unity area in the time-frequency plane. More generally, it is known that choosing a Gaussian smoothing function $\psi_{\mathrm{Ga}}(u, \mu)$ of the form

$$\psi_{\mathrm{Ga}}(u, \mu) = \frac{2}{\sqrt{\Delta_u \Delta_\mu}} \exp\left[-2\pi \left(\frac{u^2}{\Delta_u^2} + \frac{\mu^2}{\Delta_\mu^2}\right)\right] \tag{3.41}$$

will result in a positive distribution $TFE_f(u, \mu)$ if $\Delta_u \Delta_\mu \geq 1$ (Cohen 1989). Notice that the Gaussian here is not necessarily the Wigner distribution of anything. The same result generalizes to oblique kernels of the form (Cohen 1989)

$$\psi_{\mathrm{Ga}}(u, \mu) \propto \exp\left[-2\pi \left(\frac{u^2}{\Delta_u^2} + \frac{2u\mu}{\Delta_{u\mu}^2} + \frac{\mu^2}{\Delta_\mu^2}\right)\right], \tag{3.42}$$

provided $\Delta_u^2 \Delta_\mu^2 \geq \Delta_{u\mu}^2/(1 + \Delta_{u\mu}^2)$. The quantities Δ_u and Δ_μ appearing in equation 3.41 are approximate measures of the spread of $\psi_{\mathrm{Ga}}(u, \mu)$ in time and frequency respectively. The factor $\Delta_{u\mu}$ appearing in the cross term is related to the obliqueness of the ellipsoidal contours of $\psi_{\mathrm{Ga}}(u, \mu)$. These examples suggest that smoothing the Wigner distribution with a kernel $\psi_{TFE}(u, \mu)$ whose time-frequency area is equal to or greater than unity will result in a positive distribution. Since convolution with this kernel effectively corresponds to a weighted average of the Wigner distribution, one is tempted to state that averages or integrals of the Wigner distribution over regions of at least unit time-frequency area are always positive. However, despite being common wisdom, these statements are not true in general (Cohen 1989).

While not being true in general, this common wisdom indeed holds for a broad range of interesting cases, particularly when the kernel itself is a smooth localized function. Thus, allowing ourselves to be imprecise, we will take it to be an approximate truth that localized averages of Wigner distributions over time-frequency regions of area greater than or equal to unity are always positive.

We emphasize that the above considerations are of theoretical interest, having to do with the interpretation of the negative values of the Wigner distribution. As far as visual interpretation of plots of Wigner distributions are of concern, smoothing with functions which are not the Wigner distribution of anything and/or which have time-frequency area less than unity, may be just as much or more effective if chosen properly. These may result in a display which still exhibits some negative values, but may nevertheless offer an attractive and meaningful visual result. For practical purposes, the choice of a smoothing kernel is often governed by the need to find a compromise between the two goals of maximum interference suppression and maximum time-frequency resolution. Kernels with an adjustable parameter, such as the Choi-Williams distribution already mentioned, are particularly suited to this purpose because of the tuning they allow through their free parameter (Cohen 1989).

3.2.5   Effect of linear systems on the Wigner distribution

If $g(u)$ is related to $f(u)$ through the linear relation

$$g(u) = \int h(u, u') f(u') \, du', \tag{3.43}$$

then $W_g(u, \mu)$ is related to $W_f(u, \mu)$ through the relation (Bastiaans 1978, 1979a)

$$W_g(u, \mu) = \iint K_h(u, \mu; u', \mu') W_f(u', \mu') \, du' \, d\mu', \tag{3.44}$$

$$K_h(u, \mu; u', \mu') = \iint h(u + u''/2, u' + u'''/2) h^*(u - u''/2, u' - u'''/2)$$

$$\times e^{-i2\pi(u''\mu - u'''\mu')} \, du'' \, du'''.$$

It is possible to show that the kernel $K_h(u, \mu; u', \mu')$ is always real. If two systems with kernels $K_{h_1}(u, \mu; u', \mu')$ and $K_{h_2}(u, \mu; u', \mu')$ are cascaded, the kernel $K_{h_3}(u, \mu; u', \mu')$ of the resulting system is given by

$$K_{h_3}(u, \mu; u', \mu') = \iint K_{h_2}(u, \mu; u'', \mu'') K_{h_1}(u'', \mu''; u', \mu') \, du'' \, d\mu''. \tag{3.45}$$

We recall that unitary systems conserve norm and energy; that is, the energy of the output g is equal to the energy of the input f. It is also possible to show that a linear system which conserves energy is necessarily unitary. (Systems which conserve energy are also referred to as lossless and gainless systems.) We know that the kernel of a unitary system satisfies $h^{-1}(u, u') = h^*(u', u)$. The same condition can be expressed in terms of the kernel $K_h(u, \mu; u', \mu')$ as follows (Bastiaans 1978):

$$\iint K_h(u, \mu; u', \mu') \, du \, d\mu = 1. \tag{3.46}$$

This can be derived either from the condition $h^{-1}(u, u') = h^*(u', u)$ or more instructively as

$$\mathrm{En}[g] = \iint W_g(u, \mu) \, du \, d\mu = \iiiint K_h(u, \mu; u'\mu') W_f(u', \mu') \, du' \, d\mu' \, du \, d\mu$$

$$= \iint W_f(u', \mu') \left[\iint K_h(u, \mu; u', \mu') \, du \, d\mu \right] du' \, d\mu'. \tag{3.47}$$

Clearly, if equation 3.46 is satisfied, then $\mathrm{En}[g] = \mathrm{En}[f]$. The converse is also true; the only way for $\mathrm{En}[g] = \mathrm{En}[f]$ for all f is for equation 3.46 to be satisfied for all u', μ'.

We will now focus our attention to the special linear systems given earlier in table 2.2. In table 3.7 we have summarized the associated kernels $K_h(u, \mu; u'\mu')$ and the Wigner distribution of the output $W_g(u, \mu)$ for these systems.

Most of the items in the table correspond to items appearing in table 3.2; the reader will have no difficulty matching these up. Convolution of $f(u)$ with $h(u)$ corresponds to convolution of their Wigner distributions in u. This is denoted as

$$W_g(u, \mu) = W_h(u, \mu) \overset{u}{*} W_f(u, \mu) \equiv \int W_h(u - u', \mu) W_f(u', \mu) \, du'. \tag{3.48}$$

Likewise, multiplication of $f(u)$ with $h(u)$, which corresponds to convolution of $F(\mu)$ with $H(\mu)$, corresponds to convolution of their Wigner distributions in μ. This is denoted as

$$W_g(u,\mu) = W_h(u,\mu) \overset{\mu}{*} W_f(u,\mu) \equiv \int W_h(u,\mu-\mu')W_f(u,\mu')\,d\mu'. \qquad (3.49)$$

It is worth examining the geometric distortions in the u-μ plane to which some of the systems appearing in table 3.7 correspond. The region bounded by the rectangle shown in figure 3.3a represents the Wigner distribution of $f(u)$, within which a large fraction of the energy of $f(u)$ is assumed to be contained. The effect of the scaling operation, of which the identity and parity operations are a special case, is shown in figure 3.3b, which should be interpreted in the light of the fact that the Fourier transform of $f(u/M)$ is $|M|F(M\mu)$ and the fact that the projections of the Wigner distribution of $f(u)$ on the u and μ axes correspond to $|f(u)|^2$ and $|F(\mu)|^2$ respectively. The effects of the coordinate shift and phase shift operations, shown in figure 3.3c and d, are simply to shift the Wigner distribution in the u and μ directions respectively, in line with the corresponding properties of the Fourier transform (table 2.4, properties 3 and 4). The effect of the Fourier transform, not shown in the figure, is to rotate the Wigner distribution by a right angle in the clockwise direction, essentially resulting in an interchange of the u and μ axes. Chirp convolution results in shearing of the Wigner distribution in the u direction and chirp multiplication results in shearing of the Wigner distribution in the μ direction. Notice that all of the above geometric transformations, including the shearing operations, are area preserving.

When we represent the Wigner distribution of a signal by such a closed curve or rectangle, we assume that a certain large fraction, say 95%, of the signal energy is contained inside the region bounded by that curve. Thus the geometric transformations illustrated in the figure show how the signal energy is redistributed in the u-μ plane. The fact that the area of the region containing 95% of the signal energy does not change, means that while energy is redistributed and mapped to different time- or space-frequency points, the concentration of energy in the u-μ plane does not change under the action of these systems. We may also finally note that any system composed by concatenating any number of the above systems will also result in an area-preserving geometric transformation in the u-μ plane.

3.2.6 Time-frequency filtering

Here we briefly mention the concept of filtering in the time-frequency plane. In analogy with conventional time-invariant filtering where we modify the Fourier transform of a function with a multiplicative filter to alter its frequency content in the desired manner, time-frequency filtering is based on the idea of modifying the Wigner distribution (or other time-frequency representation) to alter the time-frequency content of the signal in the desired manner, or to construct signals with desired time-frequency content. The procedure is complicated by the fact that even the most reasonable modifications on the Wigner distribution of a signal, such as requiring it to be zero over a certain interval, may result in two-dimensional functions which are not the Wigner distributions of anything. This problem can be remedied in a number of ways.

Table 3.7. The effect of some special linear systems on the Wigner distribution.

Symbol	$h(u,u')$ $g(u)$	$K_h(u,\mu;u',\mu')$ $W_g(u,\mu)$
$\mathcal{I}$	$\delta(u-u')$ $f(u)$	$\delta(u-u')\delta(\mu-\mu')$ $W_f(u,\mu)$
$\mathcal{P}$	$\delta(u+u')$ $f(-u)$	$\delta(u+u')\delta(\mu+\mu')$ $W_f(-u,-\mu)$
$\mathcal{M}_M$	$\sqrt{\lvert M\rvert}\,\delta(u-Mu')$ $(1/\sqrt{\lvert M\rvert})f(u/M)$	$\delta(u-Mu')\delta(\mu-\mu'/M)$ $W_f(u/M,M\mu)$
$\mathcal{SH}_\xi$	$\delta(u+\xi-u')$ $f(u+\xi)$	$\delta(u+\xi-u')\delta(\mu-\mu')$ $W_f(u+\xi,\mu)$
$\mathcal{PH}_\xi$	$\exp(i2\pi\xi u)\delta(u-u')$ $\exp(i2\pi\xi u)f(u)$	$\delta(u-u')\delta(\mu-\xi-\mu')$ $W_f(u,\mu-\xi)$
Λ_h	$h(u)\delta(u-u')$ $h(u)f(u)$	$W_h(u,\mu-\mu')\delta(u-u')$ $\int W_h(u,\mu-\mu')W_f(u,\mu')\,d\mu'$
$\mathcal{Q}_q$	$\exp(-i\pi qu^2)\delta(u-u')$ $\exp(-i\pi qu^2)f(u)$	$\delta(\mu+qu-\mu')\delta(u-u')$ $W_f(u,\mu+qu)$
Λ_H	$h(u-u')$ $\int h(u-u')f(u')\,du'$	$W_h(u-u',\mu)\delta(\mu-\mu')$ $\int W_h(u-u',\mu)W_f(u',\mu)\,du'$
$\mathcal{R}_r$	$e^{-i\pi/4}\sqrt{1/r}\,\exp[i\pi(u-u')^2/r]$ $e^{-i\pi/4}\sqrt{1/r}\,\exp(i\pi u^2/r)*f(u)$	$\delta(u-r\mu-u')\delta(\mu-\mu')$ $W_f(u-r\mu,\mu)$
$\mathcal{U}$	$u\delta(u-u')$ $uf(u)$	$u^2\delta(\mu-\mu')+(16\pi^2)^{-1}\delta''(\mu-\mu')$ $u^2W_f(u,\mu)+(16\pi^2)^{-1}d^2W_f(u,\mu)/d\mu^2$
$\mathcal{D}$	$(i2\pi)^{-1}\delta'(u-u')$ $(i2\pi)^{-1}df(u)/du$	$\mu^2\delta(u-u')+(16\pi^2)^{-1}\delta''(u-u')$ $\mu^2W_f(u,\mu)+(16\pi^2)^{-1}d^2W_f(u,\mu)/du^2$
$\mathcal{F}$	$\exp(-i2\pi uu')$ $F(u)$	$\delta(\mu+u')\delta(u-\mu')$ $W_f(-\mu,u)$

$\mathcal{I}$: identity, $\mathcal{P}$: parity, $\mathcal{M}_M$: scaling, $\mathcal{SH}_\xi$: translation, $\mathcal{PH}_\xi$: phase shift, Λ_h: multiplicative filter, $\mathcal{Q}_q$: chirp multiplication, Λ_H: convolutive filter, $\mathcal{R}_r$: chirp convolution, $\mathcal{U}$: coordinate multiplication, $\mathcal{D}$: differentiation, $\mathcal{F}$: Fourier transform. M,ξ,q,r are real parameters, $\delta'(u-u')\equiv d[\delta(u-u')]/du$ and $\delta''(u-u')\equiv d^2[\delta(u-u')]/du^2$.

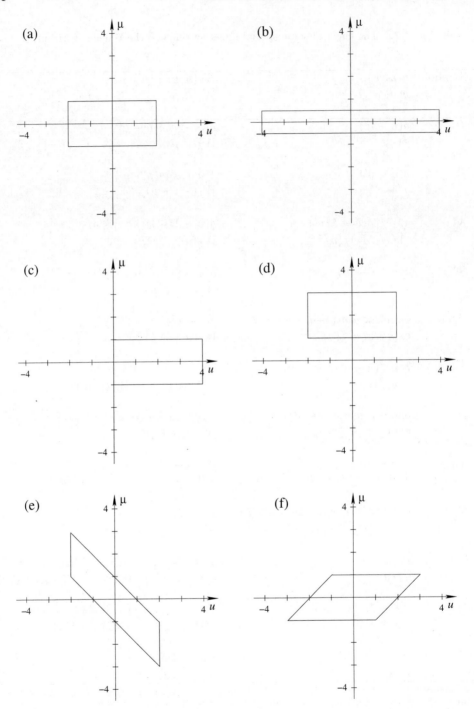

Figure 3.3. (a) Original signal; (b) scaling with $M = 2$; (c) translation with $\xi = -2$; (d) phase shift with $\xi = 2$; (e) chirp multiplication with $q = 1$; (f) chirp convolution with $r = 1$.

For instance, we may seek the signal which has the distribution closest to the one at hand, where closeness may be defined in the mean square sense. See Saleh and Subotic 1985 and Cohen 1989 for further discussion.

3.2.7 Wigner distribution of random signals

The Wigner distribution of a random signal $f(u)$ is defined as the expectation value of the Wigner distribution defined in equation 3.16:

$$
\begin{aligned}
W_f(u, \mu) &= \left\langle \int f(u + u'/2) f^*(u - u'/2) e^{-i2\pi\mu u'} \, du' \right\rangle \\
&= \int \langle f(u + u'/2) f^*(u - u'/2) \rangle \, e^{-i2\pi\mu u'} \, du' \\
&= \int R_{ff}(u + u'/2, u - u'/2) e^{-i2\pi\mu u'} \, du',
\end{aligned}
\tag{3.50}
$$

where $R_{ff}(u, u') \equiv \langle f(u) f^*(u') \rangle$ is the ensemble-averaged autocorrelation of $f(u)$. We see that the Wigner distribution of a random signal is essentially the Fourier transform of its autocorrelation with respect to the delay variable. Indeed, a similar interpretation is possible for the deterministic case. If we interpret $\gamma(u, u') = f(u + u'/2) f^*(u - u'/2)$ as some kind of time-dependent autocorrelation function (Cohen 1989), then the Wigner distribution is simply the Fourier transform of this function:

$$
W_f(u, \mu) = \int \gamma(u, u') e^{-i2\pi\mu u'} \, du',
\tag{3.51}
$$

which can be considered to be a time-dependent generalization of the common result

$$
|F(\mu)|^2 = \int R_{ff}(u') e^{-i2\pi\mu u'} \, du',
\tag{3.52}
$$

where $R_{ff}(u) \equiv \int f(u + u') f^*(u') \, du'$ here is the deterministic autocorrelation function.

If the signal is wide-sense stationary, then the ensemble-averaged autocorrelation $R_{ff}(u, u') = R_{ff}(u - u')$ so that $R_{ff}(u + u'/2, u - u'/2) = R_{ff}(u')$ and the Wigner distribution becomes independent of u and reduces to the conventional power spectral density $S_{ff}(\mu)$. Thus, in the general case where the signal is not necessarily stationary, the Wigner distribution is readily interpreted as a time-varying power spectral density. For finite-energy signals, most of the original properties hold:

$$
\int W(u, \mu) \, du = \langle |F(\mu)|^2 \rangle,
\tag{3.53}
$$

$$
\int W(u, \mu) \, d\mu = \langle |f(u)|^2 \rangle.
\tag{3.54}
$$

Readers wishing to learn more may consult the references in, for instance, Hlawatsch and Boudreaux-Bartels 1992 and Cohen 1995.

3.2.8   Wigner distribution of analytic signals

All of the results presented so far are valid for complex as well as real signals $f(u)$. In most physical applications it is more common practice to work with the Wigner distribution of the real signal. However, some authors have argued that it is more meaningful to work with the Wigner distribution of the analytic signal (Boashash 1988, Zhu, Peyrin, and Goutte 1989). The Wigner distribution of the analytic signal of a real signal $f(u)$ is not simply the upper ($\mu > 0$) part of the Wigner distribution $W_f(u, \mu)$ of the real signal (Cohen 1989:969). For instance, the Wigner distribution of the signal $\exp(i\pi u^2)$, given by $\delta(\mu - u)$, is concentrated along the line $\mu = u$, with a simple interpretation in terms of the instantaneous frequency. However, the real signal $\cos(\pi u^2)$ will exhibit interference terms in addition to the line deltas $\delta(\mu - u)$ and $\delta(\mu + u)$.

The relationship between the Wigner distribution of a signal $f(u)$ and the Wigner distribution of its analytic signal $f_{\text{as}}(u)$ is given by

$$W_{f_{\text{as}}}(u, \mu) = 16 \int W_f(u - u', \mu) \, \mu \, \text{sinc}(4\mu u') \, du', \tag{3.55}$$

if $\mu \geq 0$, and 0 if $\mu < 0$ (Claasen and Mecklenbräuker 1980a).

3.2.9   Other properties

There are many other interesting properties of the Wigner distribution and ambiguity function that we do not discuss here. Of particular interest are properties relating the moments of a signal to the moments of its Wigner distribution and properties involving instantaneous frequencies and group delays (Claasen and Mecklenbräuker 1980a, b, c, Cohen 1989, Bastiaans 1989, 1991b). An inspiring treatment of the ambiguity function is Vakman 1968.

3.3   Sampling and the number of degrees of freedom

The support of a function is the subset of the real axis in which the function is not equal to zero. This subset is said to be compact if and only if its members are confined to a finite interval around the origin. A function will be referred to as compact if its support is so. In other words, a function is compact if and only if its nonzero values are confined to a finite interval around the origin. A signal is said to be compact in the u domain if it is zero outside a finite interval around the origin in that domain. For instance, the function $\text{rect}(u)$ is compact, but the function $\exp(-\pi u^2)$ is not (although the latter may be considered to be approximately compact because its values are very small for larger u). If the Fourier transform of a function is compact, being zero outside the interval $(-\Delta\mu/2, \Delta\mu/2)$, it is said to be band-limited with bandwidth $\Delta\mu$. Such a signal can be recovered from its samples taken at intervals $\delta u \leq 1/\Delta\mu$, a result known as Nyquist's sampling theorem. Taking the fewest possible samples ($\delta u = 1/\Delta\mu$), the sampled function $f_{\text{samp}}(u)$ becomes

$$f_{\text{samp}}(u) = f(u) \, \text{comb}(u/\delta u) = \sum_{l=-\infty}^{\infty} f(l \, \delta u) \, \delta(u/\delta u - l)$$

$$= \delta u \sum_{l=-\infty}^{\infty} f(l\,\delta u)\,\delta(u - l\,\delta u) = \delta u \sum_{l=-\infty}^{\infty} f(l/\Delta\mu)\,\delta(u - l/\Delta\mu). \qquad (3.56)$$

Taking the Fourier transform of both sides

$$\mathcal{F}[f_{\text{samp}}(u)](\mu) = F(\mu) * \delta u\,\text{comb}(\delta u\,\mu) = F(\mu) * \delta u \sum_{l=-\infty}^{\infty} \delta(\delta u\,\mu - l)$$

$$= F(\mu) * \sum_{l=-\infty}^{\infty} \delta(\mu - l\,\Delta\mu) = \sum_{l=-\infty}^{\infty} F(\mu - l\,\Delta\mu). \qquad (3.57)$$

The last expression tells us that sampling in the time domain results in periodic replication in the frequency domain. The original signal can be recovered by multiplying $\mathcal{F}[f_{\text{samp}}(u)](\mu)$ with a rectangular window $\text{rect}(\mu/\Delta\mu)$ which will single out the original spectrum $F(\mu)$. The same operation can be written in the time domain as a convolution of the form

$$f_{\text{samp}}(u) * \Delta\mu\,\text{sinc}(u\,\Delta\mu) = \sum_{l=-\infty}^{\infty} f(l/\Delta\mu)\,\text{sinc}(\Delta\mu\,u - l)$$

$$= \sum_{l=-\infty}^{\infty} f(l\,\delta u)\,\text{sinc}(u/\delta u - l) = \sum_{l=-\infty}^{\infty} f(l\,\delta u)\,\text{sinc}\left(\frac{u - l\,\delta u}{\delta u}\right), \qquad (3.58)$$

which is the formula allowing us to reconstruct $f(u)$ from its samples and is known as the interpolation formula. Analogous results hold for sampling of the frequency-domain representation $F(\mu)$. Further discussion on the fundamentals of sampling may be found in Bracewell 1986, Marks 1991, and the edited book *Advanced Topics in Shannon Sampling and Interpolation Theory* 1993.

We already know that a function and its Fourier transform cannot both be compact (unless they are identically zero). That is, Δu and $\Delta\mu$ as defined above cannot both be finite. In practice, however, it seems that we are always working with both a finite time (or space) interval and a finite bandwidth. Thus we will find it useful to abandon the above definitions of Δu and $\Delta\mu$ in favor of less well defined yet more meaningful ones. A large percentage of the energy of most finite-energy signals arising in physical applications will be concentrated in a finite interval both in the time domain and in the frequency domain, although neither $f(u)$ nor $F(\mu)$ may be identically zero outside of these intervals. For instance, the Gaussian function $\text{gauss}(u)$ is clearly well concentrated around the origin in both the time and frequency domains, although it is not identically zero anywhere and its tails extend to infinity. Strictly speaking, both the temporal extent and the bandwidth of this signal are infinite. Since this is clearly counterintuitive, other measures of spread are often employed. One of these is to take the temporal extent or bandwidth of the signal as the standard deviation (or a certain number of standard deviations) of the time- and frequency-domain representations of the signal respectively. Another measure of spread, appropriate for certain functions which tend to diminish as we move away from their center of gravity, is the distance between the points at which the function has dropped to a certain fraction of its peak value, say $1/e$ or 0.05. A somewhat more generally applicable measure of spread is the length of the interval which contains a certain fraction, say 0.95, of the total energy of the signal. (All three of these measures are appropriate for the particularly well-behaved Gaussian function.). When we speak of the temporal extent Δu and bandwidth (spectral extent) $\Delta \mu$ of a signal, we will usually be speaking of the length of an interval containing a sufficiently large fraction of the total energy of the signal. Having ensured that the signal energy outside this interval is negligible, we may thus assume that the signal is (approximately) confined to that interval. The Nyquist sampling theorem and interpolation formula will hold approximately in this case. This definition of spread works well especially when we are dealing with signals of large time-bandwidth product, a notion which is defined further below.

We recall that we take u and μ to be dimensionless variables. Let us assume that the time-domain representations of the signals we are dealing with are approximately confined to the interval $[-\Delta t/2, \Delta t/2]$ and that their frequency-domain representations are approximately confined to the interval $[-\Delta f/2, \Delta f/2]$ in real physical units. With this statement we mean that a sufficiently large percentage of the energies of the signals are confined to these intervals in the respective domains. This can be ensured by choosing Δt and Δf sufficiently large. (If the time or frequency representations of the signals are confined to intervals which are not centered around the origin, we may simply shift the origin of time and frequency so that this becomes the case.)

Let us now introduce the scaling parameter s with the dimension of time and introduce dimensionless coordinates $u = t/s$ and $\mu = fs$. With these new coordinates, the time and frequency domain representations will be confined to intervals of length $\Delta t/s$ and $s\Delta f$. If we choose $s = \sqrt{\Delta t/\Delta f}$, the lengths of both intervals will now be equal to the dimensionless quantity $\sqrt{\Delta f \Delta t}$ which we may denote by Δu. It is often convenient to assume that such a dimensional normalization has been performed on the signals we work with so that the spreads of the signal in the time and frequency domains are comparable in dimensionless coordinates.

We now define the time-bandwidth product N for a set of signals, whose members we assume are approximately confined to an interval of length Δu in the time domain and to an interval of length $\Delta \mu$ in the frequency domain. The time-bandwidth product (or space-bandwidth product) is defined as

$$N \equiv \Delta u \Delta \mu. \tag{3.59}$$

N is always greater than or equal to unity by virtue of the uncertainty relation. The time-bandwidth product is the minimum number of samples needed to characterize or identify a signal out of all possible signals whose energies are confined to time and frequency intervals of length Δu and $\Delta \mu$. If we sample the time-domain representation of a signal at the Nyquist rate of $\delta u = 1/\Delta \mu$, the total number of samples lying in the interval Δu is given by $\Delta u/(1/\Delta \mu) = \Delta u \Delta \mu$, which is simply the time-bandwidth product N. Alternatively, if we sample the frequency-domain representation of a signal at the Nyquist rate of $\delta \mu = 1/\Delta u$, the total number of samples lying in the interval $\Delta \mu$ is given by $\Delta \mu/(1/\Delta u) = \Delta u \Delta \mu$, which is again the time-bandwidth product N. (With the dimensional normalization above which results in $\Delta u = \Delta \mu$, the number of samples is $N = \Delta u^2$ with the samples being spaced $\Delta u^{-1} = N^{-1/2}$ apart in both domains.)

The time-bandwidth product of the set of signals we are dealing with can often be interpreted as the *number of degrees of freedom* or *dimensionality* of the set of signals.

Since signals whose energies are (approximately) confined to intervals of length Δu and $\Delta \mu$ in the time and frequency domains can be fully characterized by N numbers, there is a one-to-one correspondence between these signals and N-dimensional vectors $\mathbf{r} = [r_1 \ r_2 \ \ldots \ r_N]^{\mathrm{T}}$. We saw above that the number of samples needed to fully characterize a signal is the same in both the time domain and the frequency domain, and is given by the time-bandwidth product N. The time-bandwidth product will remain invariant under transformations to other representations as well, as the information content of the signal is not altered under invertible unitary transformations of the type we discussed in chapter 2. Thus if a signal can be uniquely characterized in a particular representation by N complex numbers, this will also be the case in any other representation. Much like the norm and energy, the number of degrees of freedom is a property of the signals in the abstract, and not tied to their representations in a particular domain.

We can summarize by saying that the set of time- and band-limited signals in question has approximately $\Delta u \Delta \mu$ degrees of freedom. (Of course, strictly speaking, $f(u) = 0$ is the only such signal, so that we are continuing to talk about approximate time- and band-limitedness.) Our argument has been based on sampling theory. We sample every $1/\Delta \mu$ over Δu so that the total number of samples is $\Delta u/(1/\Delta \mu) = \Delta u \Delta \mu$, which we interpret as the number of degrees of freedom. For a more rigorous account, see Dym and McKean 1972:129–131.

Great insight into the concept of the number of degrees of freedom can be gained through time-frequency representations. For instance, let us consider Gabor's expansion whose definition we repeat (equation 3.7):

$$f(u) = \sum_l \sum_m G_f^{(w)}(l, m) w_{lm}(u), \tag{3.60}$$

where $w_{lm}(u)$ are basis signals centered at the time-frequency point $(l \, \delta u, m \, \delta \mu)$, with $\delta u \, \delta \mu = 1$. Here we will assume that the $w_{lm}(u)$ are well concentrated in their respective time-frequency cells (of dimensions $\delta u \times \delta \mu$) centered around $(l \, \delta u, m \, \delta \mu)$. (For instance, this will be the case if we choose $w(u) = 2^{1/4} \Delta_u^{-0.5} \exp(-\pi u^2/\Delta_u^2)$ with $\Delta_u = \delta u$, as on page 67.) The number of degrees of freedom of a set of signals can be defined as the number of Gabor coefficients $G_f^{(w)}(l, m)$ which are not negligibly small for all of the signals in this set, since these nonnegligible coefficients are sufficient—to a good degree of approximation—to completely characterize and distinguish a particular signal in this set from the others. The region in the time-frequency plane in which the Gabor coefficients are not negligible may be referred to as the *time-frequency support* of the set of signals. The coefficients corresponding to points lying in this region are, roughly speaking, the time-frequency samples of the signal. The number of these samples corresponds to the number of degrees of freedom of the set of signals.

Notice that since a time-frequency area of unity is associated with each coefficient, the number of degrees of freedom thus defined is also equal to the area of the time-frequency support. If we assume that these nonnegligible coefficients lie neatly in a rectangular region of dimensions $\Delta u \times \Delta \mu$, we see that the number of these coefficients is simply $(\Delta u/\delta u)(\Delta \mu/\delta \mu) = \Delta u \Delta \mu$, which is equal to the time-bandwidth product. Thus, when the time-frequency content of a signal is confined to a rectangular region, the number of degrees of freedom is equal to the time-bandwidth product. However, when the time-frequency content of the signal is not confined to a rectangular region,

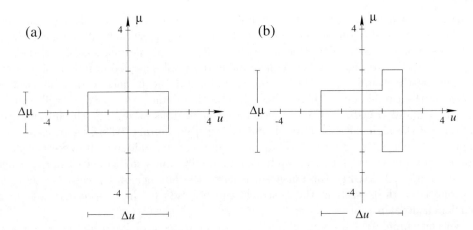

Figure 3.4. (a) Rectangular time-frequency support; the time-frequency area is equal to the time-bandwidth product $\Delta u \Delta \mu = 8$. (b) Irregular time-frequency support; the time-frequency area is 10, which is smaller than the time-bandwidth product $\Delta u \Delta \mu = 16$.

the actual number of degrees of freedom is less than the time-bandwidth product. In such cases the two concepts must be clearly distinguished, as will be further discussed below.

Although the above argument has been based on the Gabor expansion, similar arguments are possible with any reasonably well-localized energetic time-frequency distribution. The picture is somewhat more complicated for time-frequency distributions exhibiting interference terms (such as the Wigner distribution); nevertheless, similar concepts and arguments are found useful in these cases as well. As above, we may define the time-frequency support of the set of signals we are dealing with as the time-frequency region outside of which the values of the Wigner distributions are negligible. Alternatively, we may define this region by requiring that the integral of the Wigner distributions over this region should be equal to a certain significant fraction (say 0.95) of the energies of the signals. (Remember that the integral of the Wigner distribution over the whole plane is equal to the energy of the signal.) Throughout this book, when we say that we assume the Wigner distribution of a signal or a set of signals to be confined to a certain region, we will be referring to the time-frequency support and implying that such a significant fraction of the energy or energies are confined to that region (as we already did in figure 3.3). A recent paper dealing in a more rigorous manner with the number of degrees of freedom concept in the context of time-frequency representations is Landau 1993. An older reference dealing with related concepts is Vakman 1968.

Thus, in the general case, we define the number of degrees of freedom as the time-frequency support. In general, this will be smaller than the the time-bandwidth product, unless the time-frequency support is a rectangle perpendicular to the time-frequency axes (figure 3.4) (Lohmann and others 1996a). Quite commonly the time-bandwidth product is simply taken to be equal to the number of degrees of freedom of a set of signals, without regard to the shape of the time-frequency support. As we have seen, this may overstate the number of degrees of freedom of the set of signals in question. A simple analogy may be useful. Consider the set of points in three-dimensional space which are confined to some particular plane. The elements of the coordinate vector $\mathbf{r} = (x, y, z)$ will each assume values over the complete interval $[-\infty, \infty]$, but not independently so; the number of degrees of freedom is 2 and not 3. A particularly striking example is the set of chirp signals $A\,\mathrm{chirp}(u)$ whose single degree of freedom is represented by the number A. However, both the temporal extent and the spectral extent of these signals are infinite.

It is important to note that the number of degrees of freedom is an approximate concept in the sense that the number of degrees of freedom will depend on the amplitude accuracy we are working with, since this determines what is negligible and what is nonnegligible. If we are working with greater accuracy, then a larger region in the u-μ plane will have Gabor coefficients which are nonnegligible and the signal and its Fourier transform will have nonnegligible values over larger intervals. In the case of signals with approximately rectangular time-frequency content, the values of Δu and $\Delta \mu$ outside of which these signals are negligible in the time and frequency domains, respectively, again depend on the accuracy we are working with. The important point is to choose the smallest intervals outside of which the signals are truly negligible with that accuracy. Of course, if the values of the signals are negligible outside an interval of length Δu, they will also be negligible outside a larger interval. However, unless we choose the smallest possible values of Δu and $\Delta \mu$, we will be overstating the number of degrees of freedom of the set of signals.

Until now, we spoke of the temporal and spectral extent and time-bandwidth product of signals without discussing the origin of these finite extents. In the real world a signal is always represented in some physical form in some physical system. These systems which carry or process the signals always limit their temporal duration (or spatial extent) and bandwidths to certain finite values. A physical system cannot allow the existence of frequencies outside a certain band because there is always some limit to the resolution that can be supported. Likewise, since all physical events of interest have a beginning and an end, or because all physical events or systems have a finite extent, the temporal duration or spatial extent of the signal will also be finite. For instance, a computer display with a certain number of pixels cannot represent an image of greater space-bandwidth product. In an optical system the size of the lenses will limit both the spatial extent of the images that can be dealt with and their spatial bandwidths. It is these physical limitations that determine the temporal (or spatial) and spectral extent of the signals and thus their time-bandwidth product. Just as these may be undesirable physical limitations which limit the performance of the system, they may also be deliberate limitations with the purpose of limiting the set of signals we are dealing with. When a signal previously represented by a system with greater time-bandwidth product is input into a system with smaller time-bandwidth product, an uninvertible process in which information is lost takes place. (It is important to understand that signals have no physical existence outside of a system. Even the "wire" connecting two "systems," is a system itself. Thus signals always move from one system to another.)

The fact that all physical systems support only a finite time-bandwidth product, means that their effect on signals can be simulated with discrete-time systems with the same degree of accuracy that is inherent in the continuous systems or measurement devices from which the signals originate. Further insight on these matters may be

obtained by discussing how the discrete Fourier transform provides an approximation
to the continuous Fourier transform. The discrete Fourier transform $F(j)$ of $f(l)$ has
been defined in equation 2.173 and is repeated here:

$$F(j) \equiv \frac{1}{\sqrt{N}} \sum_{l=0}^{N-1} f(l) e^{-i2\pi jl/N} \qquad j = 0, 1, \dots, N-1. \qquad (3.61)$$

We shall now see that provided N is chosen to be *at least* equal to the time-bandwidth
product of the set of signals we are dealing with, the discrete Fourier transform, which
can be efficiently computed on a digital computer using the fast Fourier transform
(FFT) algorithm, can be used to obtain a good approximation to the continuous
Fourier transform. The approximation improves with increasing N.

Let us consider a function $f(u)$ and its Fourier transform $F(\mu)$ and define the
periodically replicated functions

$$f_{\mathrm{pr}}(u) \equiv \sum_{n=-\infty}^{\infty} f(u - n\,\Delta u), \qquad (3.62)$$

$$F_{\mathrm{pr}}(\mu) \equiv \sum_{n=-\infty}^{\infty} F(\mu - n\,\Delta\mu), \qquad (3.63)$$

where Δu and $\Delta\mu$ are arbitrary. It is possible to show that samples of these functions
constitute a discrete Fourier transform pair as follows (Papoulis 1977:74):

$$F_{\mathrm{pr}}(j/\Delta u) = \frac{1}{\Delta\mu} \sum_{l=0}^{\Delta u \Delta\mu - 1} f_{\mathrm{pr}}(l/\Delta\mu) \exp(-i2\pi jl/\Delta u\Delta\mu). \qquad (3.64)$$

Now, let us assume that a significant fraction of the energy of the signal is confined to
the intervals $[-\Delta u/2, \Delta u/2]$ and $[-\Delta\mu/2, \Delta\mu/2]$ in the time and frequency domains
respectively. In this case, $f(u) \approx f_{\mathrm{pr}}(u)$ and $F(\mu) \approx F_{\mathrm{pr}}(\mu)$ in the respective intervals.
Thus, $f_{\mathrm{pr}}(l/\Delta\mu) \approx f(l/\Delta\mu)$ and $F_{\mathrm{pr}}(j/\Delta u) \approx F(j/\Delta u)$. As before, $\Delta u\Delta\mu \equiv N$ will
denote the time-bandwidth product. To further simplify we may assume scaling such
that $\Delta u = \Delta\mu$ so that $\Delta u = \Delta\mu = 1/\sqrt{N}$. Under these circumstances, we see that
the DFT of the samples of a function are the samples of the Fourier transform of the
function; or, in other words, the DFT maps the samples of $f(u)$ to the samples of
$F(\mu)$. The sampling interval in the time domain is $1/\Delta\mu$ and the sampling interval in
the frequency domain is $1/\Delta u$ and the total number of samples in both domains is N.

In the interest of easier interpretation with respect to the continuous case, the
discrete Fourier transform is sometimes expressed as

$$F(j) \equiv \frac{1}{\sqrt{N}} \sum_{l=N/2-1}^{N/2} f(l) e^{-i2\pi jl/N} \qquad j = N/2 - 1, N/2, \dots, N/2, \qquad (3.65)$$

where N is assumed to be even. This is easily seen to be identical to equation 3.61 if
we think of $f(l)$ and $F(j)$ as periodic functions.

3.4   Linear canonical transforms

The class of linear canonical transforms is a three-parameter class of linear integral transforms which includes Fresnel transforms, fractional Fourier transforms, and simple scaling and chirp multiplication operations, as well as certain other transforms among its members.

Linear canonical transforms have been reinvented or reconsidered by many authors under many different names at different times in different contexts, a fact which we consider a tribute to their ubiquity. They have been referred to as quadratic-phase systems (Bastiaans 1979a), generalized Huygens integrals (Siegman 1986), generalized Fresnel transforms (James and Agarwal 1996, Palma and Bagini 1997), special affine Fourier transforms (Abe and Sheridan 1994a, b), extended fractional Fourier transforms (Hua, Liu, and Li 1997c), and Moshinsky-Quesne transforms (Wolf 1979), among other things.

An excellent and alternative exposition to linear canonical transforms may be found in Wolf 1979: chapter 9; which also contains an account of the history of these transforms. Among the important works in this area we may mention Moshinsky and Quesne 1971; Quesne and Moshinsky 1971; Wolf 1974a, b, 1976; García-Calderón and Moshinsky 1980; and Basu and Wolf 1982. Further references may be found in Wolf 1979: chapter 9.

3.4.1   Definition and properties

The linear canonical transform $f_{\mathbf{M}}(u) = (\mathcal{C}_{\mathbf{M}}f)(u)$ of $f(u)$ with parameter $\mathbf{M}$ is most conveniently defined as

$$(\mathcal{C}_{\mathbf{M}}f)(u) = \int C_{\mathbf{M}}(u, u')f(u')\, du', \tag{3.66}$$

$$C_{\mathbf{M}}(u, u') = A_{\mathbf{M}} \exp\left[i\pi(\alpha u^2 - 2\beta u u' + \gamma u'^2)\right],$$

$$A_{\mathbf{M}} = \sqrt{\beta}\, e^{-i\pi/4},$$

where α, β, and γ are real parameters independent of u and u'. $\mathcal{C}_{\mathbf{M}}$ is the linear canonical transform operator. The label $\mathbf{M}$ represents the three parameters α, β, and γ which completely specify the transform. As with the Fourier transform, a linear canonical transform can be interpreted both as a system and as a transformation to another representation. In the former case, $f(u)$ is the input and $(\mathcal{C}_{\mathbf{M}}f)(u)$, or simply $\mathcal{C}_{\mathbf{M}}f(u)$, is the output. In the latter case, the transform gives us the representation of the signal f in another "domain." Linear canonical transforms are unitary; that is, the inverse transform kernel is the Hermitian conjugate of the original transform kernel: $C_{\mathbf{M}}^{-1}(u, u') = C_{\mathbf{M}}^{*}(u', u)$, or more explicitly

$$(\mathcal{C}_{\mathbf{M}}^{-1}f)(u) = \int C_{\mathbf{M}}^{-1}(u, u')f(u')\, du, \tag{3.67}$$

$$C_{\mathbf{M}}^{-1}(u, u') = A_{\mathbf{M}}^{*} \exp\left[-i\pi(\gamma u^2 - 2\beta u u' + \alpha u'^2)\right],$$

$$A_{\mathbf{M}}^{*} = (1/\sqrt{1/\beta})e^{i\pi/4} = \sqrt{-\beta}\, e^{-i\pi/4}.$$

$1/\sqrt{1/\beta}$ is equal to the complex conjugate of $\sqrt{\beta}$ (see the square root convention on page 8). That the transform given in equation 3.67 is indeed the inverse transform can be verified by confirming the identity

$$\int C_{\mathbf{M}}^{-1}(u, u'') C_{\mathbf{M}}(u'', u')\, du'' = \delta(u - u'),\qquad(3.68)$$

with the help of equation 2.8. The inverse of a linear canonical transform is also a linear canonical transform, so we can write $(C_{\mathbf{M}}^{-1} f)(u) = (C_{\mathbf{M}^{-1}} f)(u)$ where $\mathbf{M}^{-1}$ denotes the set of parameters of the inverse transform $\alpha_{\mathrm{inv}} = -\gamma$, $\beta_{\mathrm{inv}} = -\beta$, $\gamma_{\mathrm{inv}} = -\alpha$.

We will now examine the consecutive application (also referred to as composition or concatenation) of two linear canonical transforms with arbitrary parameters. We will start with a signal $f(u)$, apply a transform with the parameters α_1, β_1, γ_1 and then apply to the result a second transform with the parameters α_2, β_2, γ_2 to obtain finally a signal $g(u)$:

$$g(u) = \int C_{\mathbf{M}_2}(u, u'') \left[\int C_{\mathbf{M}_1}(u'', u') f(u')\, du' \right] du''$$

$$= \int \left[\int C_{\mathbf{M}_2}(u, u'') C_{\mathbf{M}_1}(u'', u')\, du'' \right] f(u')\, du',\qquad(3.69)$$

so that the kernel $h(u, u')$ of the composite operation relating $g(u)$ to $f(u)$ is given by

$$g(u) = \int h(u, u') f(u')\, du',\qquad(3.70)$$

$$h(u, u') = \int C_{\mathbf{M}_2}(u, u'') C_{\mathbf{M}_1}(u'', u')\, du''.$$

Upon evaluating the final integral we find that $h(u, u')$ can be expressed in the following form (Wolf 1979:387):

$$h(u, u') = \mathrm{sgn}(\beta_1 \beta_2 / \beta_3) \sqrt{\beta_3}\, e^{-i\pi/4} \exp\left[i\pi(\alpha_3 u^2 - 2\beta_3 uu' + \gamma_3 u'^2) \right],\qquad(3.71)$$

where

$$\alpha_3 = \alpha_2 - \frac{\beta_2^2}{\alpha_1 + \gamma_2},$$

$$\beta_3 = \frac{\beta_1 \beta_2}{\alpha_1 + \gamma_2},$$

$$\gamma_3 = \gamma_1 - \frac{\beta_1^2}{\alpha_1 + \gamma_2}.\qquad(3.72)$$

If the factor $\mathrm{sgn}(\beta_1 \beta_2 / \beta_3)$ did not appear in the result, we could conclude simply that the composition of any two linear canonical transforms is another linear canonical transform. This is not strictly true because of this sign factor. However, we will refer to transforms which differ from equation 3.66 by a minus sign also as linear canonical transforms, so we can speak of the composition of two linear canonical transforms as being another linear canonical transform. While a more precise formulation is possible, diverting into this technicality would not serve our purpose; we keep our present definition to maintain simplicity. (The ± 1 is related to the fact that the class of linear canonical transforms involves a so-called double or twofold cover of the circle, as will be briefly discussed later.)

The composition we have just examined is not in general commutative; that is, in general $\mathcal{C}_{\mathbf{M}_1}\mathcal{C}_{\mathbf{M}_2} \neq \mathcal{C}_{\mathbf{M}_2}\mathcal{C}_{\mathbf{M}_1}$. However, such compositions are associative:

$$(\mathcal{C}_{\mathbf{M}_1}\mathcal{C}_{\mathbf{M}_2})\mathcal{C}_{\mathbf{M}_3} = \mathcal{C}_{\mathbf{M}_1}(\mathcal{C}_{\mathbf{M}_2}\mathcal{C}_{\mathbf{M}_3}). \tag{3.73}$$

Until now, we let the symbol $\mathbf{M}$ denote the three parameters α, β, γ characterizing a linear canonical transform. Now, we will more specifically define $\mathbf{M}$ as a matrix of the form

$$\mathbf{M} \equiv \begin{bmatrix} A & B \\ C & D \end{bmatrix} \equiv \begin{bmatrix} \gamma/\beta & 1/\beta \\ -\beta + \alpha\gamma/\beta & \alpha/\beta \end{bmatrix} = \begin{bmatrix} \alpha/\beta & -1/\beta \\ \beta - \alpha\gamma/\beta & \gamma/\beta \end{bmatrix}^{-1}, \tag{3.74}$$

with determinant $AD - BC = 1$. (Such matrices are called unit-determinant or unimodular matrices.) The matrix elements are fully equivalent to the three independent parameters α, β, and γ, which can be recovered in terms of the matrix elements as follows:

$$\alpha = \frac{D}{B} = \frac{1}{A}\left(\frac{1}{B} + C\right),$$

$$\beta = \frac{1}{B},$$

$$\gamma = \frac{A}{B} = \frac{1}{D}\left(\frac{1}{B} + C\right). \tag{3.75}$$

The reason why we define $\mathbf{M}$ in this manner, rather than simply as a parameter vector $[\alpha \ \beta \ \gamma]$, is because it leads to some attractive properties. The primary rationale behind the definition of this matrix is that the matrix corresponding to the composition of two systems is the matrix product of the matrices corresponding to the individual systems. That is,

$$\mathbf{M}_3 = \mathbf{M}_2\mathbf{M}_1, \tag{3.76}$$

a result which can be proved by using equations 3.72. (This result is oblivious to the ± 1 that might appear in front of the kernel.) Furthermore, it is easy to show that the matrix corresponding to the inverse of a transform is the inverse of the matrix corresponding to the original transform, as we have already built into our notation:

$$\mathcal{C}_{\mathbf{M}}^{-1} = \mathcal{C}_{\mathbf{M}^{-1}}, \tag{3.77}$$

$$C_{\mathbf{M}}^{-1}(u, u') = C_{\mathbf{M}^{-1}}(u, u') = C_{\mathbf{M}}^{*}(u', u). \tag{3.78}$$

If desired, the defining equation 3.66 can be rewritten in terms of the matrix parameters as follows:

$$f_{\mathbf{M}}(u) = (\mathcal{C}_{\mathbf{M}}f)(u) = \int C_{\mathbf{M}}(u, u')f(u')\,du', \tag{3.79}$$

$$C_{\mathbf{M}}(u, u') = A_{\mathbf{M}}\exp\left[i\pi\left(\frac{D}{B}u^2 - 2\frac{1}{B}uu' + \frac{A}{B}u'^2\right)\right],$$

$$A_{\mathbf{M}} = \sqrt{1/B}\,e^{-i\pi/4}.$$

The set of linear canonical transforms satisfy all the axioms of a noncommutative group (closure, associativity, existence of identity, inverse of each element), just like the set of all unit-determinant 2×2 matrices. (Again, this is true to the extent that we are willing to be flexible with signs in front of the transform integrals.) Certain subsets (or subclasses) of the set (class) of linear canonical transforms are groups in themselves and thus are subgroups. Several of them will be discussed further below. For example, we will see that the fractional Fourier transform is a subgroup with one real parameter. Integer powers of the Fourier transform are a subgroup with one integer parameter.

A rather trivial extension of linear canonical transforms are transforms which include not only quadratic terms such as u^2, u'^2, uu', but also linear terms such as u, u' in the exponent. Any second-order expression including such terms can be expressed as a quadratic form of $(u - \xi)$ and $(u' - \xi')$, where ξ and ξ' are constants. Thus such transforms can be obtained by shifting the input and output of linear canonical transforms as defined above. We will not develop this extension, with the understanding that it can be readily introduced when necessary. A more involved extension is to allow elements of the matrix $\mathbf{M}$ to be complex.

Some of the operational properties of linear canonical transforms are listed in table 3.8. Of course, one must add to these all properties associated with unitarity, such as Parseval's relation: $\|f_{\mathbf{M}}(u)\| = \|f(u)\|$.

We also note here the linear canonical transforms of the eigenfunctions of the Fourier transform (Wolf 1979):

$$(\mathcal{C}_{\mathbf{M}}\psi_n)(u) = \left[\left(2\frac{A+iB}{A-iB} \right)^n n! \pi^{1/2} (A+iB) \right]^{-1/2}$$
$$\times \exp\left(-\frac{D-iC}{A+iB} \pi u^2 \right) H_n \left[(A^2+B^2)^{-1/2} \sqrt{2\pi} \, u \right]. \quad (3.80)$$

3.4.2 Effect on Wigner distributions

We now discuss the effect of a linear canonical transform on the Wigner distribution or ambiguity function of a signal. We let f denote a signal and $f_{\mathbf{M}}$ its linear canonical transform, where $\mathbf{M}$ is the unit-determinant matrix of coefficients characterizing the transform. Then

$$W_{f_{\mathbf{M}}}(Au + B\mu, Cu + D\mu) = W_f(u, \mu), \quad (3.81)$$
$$W_{f_{\mathbf{M}}}(u, \mu) = W_f(Du - B\mu, -Cu + A\mu). \quad (3.82)$$

That is, the Wigner distribution of the transformed signal is simply a linearly distorted form of the Wigner distribution of the original signal. This result can be demonstrated directly from the definition of linear canonical transforms and the definition of the Wigner distribution, although the algebra is somewhat involved. This distortion can also be interpreted in terms of a coordinate transformation with new coordinates u_{new}, μ_{new} defined in terms of the old u, μ as

$$u_{\text{new}} = +Du - B\mu,$$
$$\mu_{\text{new}} = -Cu + A\mu. \quad (3.83)$$

Table 3.8. Properties of linear canonical transforms.

	$f(u)$	$f_M(u)$
1.	$\sum_j \alpha_j f_j(u)$	$\sum_j \alpha_j f_{jM}(u)$
2.	$f(-u)$	$f_M(-u)$
3.	$|M|^{-1} f(u/M)$	$f_{M'}(u)$
4.	$f(u - \xi)$	$\exp[i\pi(2u\xi C - \xi^2 AC)] f_M(u - A\xi)$
5.	$\exp(i2\pi\xi u) f(u)$	$\exp[i\pi\xi D(2u - \xi B)] f_M(u - B\xi)$
6.	$\exp[i\pi(\chi u^2 + 2\xi u)] f(u)$	$\exp[i\pi\xi D(2u - \xi B)] f_{M''}(u - B\xi)$
7.	$u^n f(u)$	$[Du - B(i2\pi)^{-1} d/du]^n f_M(u)$
8.	$(i2\pi)^{-n} d^n f(u)/du^n$	$[-Cu + A(i2\pi)^{-1} d/du]^n f_M(u)$
9.	$f^*(u)$	$f^*_{M^{-1}}(u)$
10.	$[f(u) + f(-u)]/2$	$[f_M(u) + f_M(-u)]/2$
11.	$[f(u) - f(-u)]/2$	$[f_M(u) - f_M(-u)]/2$

The expressions on the right are linear canonical transforms of the expressions on the left. α_j are arbitrary complex constants, M, ξ, χ are real, and n is a positive integer. $\mathbf{M'}$ is the matrix that corresponds to the parameters $\alpha' = \alpha$, $\beta' = M\beta$, $\gamma' = M^2\gamma$ and $\mathbf{M''}$ is the matrix that corresponds to the parameters $\alpha'' = \alpha$, $\beta'' = \beta$, $\gamma'' = \gamma + \chi$.

In this interpretation we see that the Wigner distribution $W_{f_M}(u, \mu)$ of the transformed signal f_M is of the same functional form $W_f(\cdot, \cdot)$ with respect to the newly defined axes: $W_f(u_{new}, \mu_{new})$. The Jacobian of the above two-dimensional coordinate transformation is simply the determinant of $\mathbf{M}$, which is by definition equal to unity. Thus this transformation is area preserving. It distorts but does not concentrate or deconcentrate the Wigner distribution. For instance, consider the time-frequency support of a signal, defined as the time-frequency region which contains a certain significant percentage of the signal energy. Then the similarly defined region corresponding to the transformed signal, containing the same percentage of the signal energy, will have the same support area. That is, the time-frequency area in which this percentage of the signal energy is contained remains invariant under a linear canonical transform. Figure 3.5 illustrates these concepts for a Wigner distribution with approximately rectangular time-frequency support.

Remembering that the Wigner distribution gives us the distribution of signal energy over time and frequency, the time-frequency area-preserving nature of linear canonical transforms means that such transforms do not concentrate or deconcentrate energy in the time-frequency plane. (Since linear canonical transforms are unitary, the total signal energy is conserved to begin with.) This property has many interpretations in physics, some of which we will discuss in chapter 8. For the time being, we note that since the area of the time-frequency support for a set of signals can be interpreted as the number of degrees of freedom, conservation of this area also implies conservation of the number of degrees of freedom and thus information, which is consistent with the fact that linear canonical transforms are invertible.

The results just mentioned are of sufficient importance to warrant an alternative treatment (Bastiaans 1979a). Earlier we presented the general form of the kernel transforming the Wigner distribution for any given linear transform (equation 3.44).

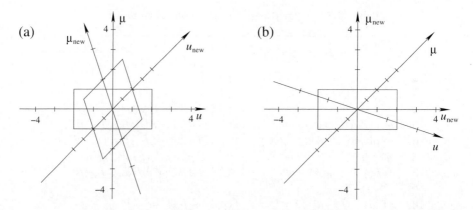

Figure 3.5. (a) Rectangular time-frequency support of the Wigner distribution of $f(u)$ and the time-frequency support of the Wigner distribution of its linear canonical transform $f_{\mathbf{M}}(u)$. (b) Time-frequency support of the Wigner distribution of $f_{\mathbf{M}}(u)$ with respect to the axes u_{new} and μ_{new}, which is seen to be of the same rectangular form as the support of the original Wigner distribution with respect to the axes u and μ.

Specializing this to linear canonical transforms we find that

$$K_{\mathbf{M}}(u, \mu; u', \mu') = \beta\delta[\mu - (\alpha u - \beta u')]\delta[\mu' - (\beta u - \gamma u')]. \qquad (3.84)$$

The value of the Wigner distribution $W_f(u, \mu)$ at a certain time-frequency point is mapped to another point which is determined by setting the arguments of the delta functions equal to zero:

$$\begin{bmatrix} \mu \\ \mu' \end{bmatrix} = \begin{bmatrix} \alpha & -\beta \\ \beta & -\gamma \end{bmatrix} \begin{bmatrix} u \\ u' \end{bmatrix}. \qquad (3.85)$$

This equation can be algebraically rewritten as

$$\begin{bmatrix} u' \\ \mu' \end{bmatrix} = \begin{bmatrix} A & B \\ C & D \end{bmatrix}^{-1} \begin{bmatrix} u \\ \mu \end{bmatrix}, \qquad (3.86)$$

where

$$A = \gamma/\beta, \qquad B = 1/\beta, \qquad C = -\beta + \alpha\gamma/\beta, \qquad D = \alpha/\beta, \qquad (3.87)$$

just as defined previously in equation 3.74. In the light of these equations, the kernel can be rewritten as

$$K_{\mathbf{M}}(u, \mu; u'\mu') = \delta(u' - Du + B\mu)\,\delta(\mu' + Cu - A\mu). \qquad (3.88)$$

This can also be derived directly by substituting equation 3.66 in equation 3.44. This form of the kernel clearly indicates the pointwise mapping involved. The value of the Wigner distribution at time-frequency point (u, μ) is mapped into the time-frequency point $(Au + B\mu, Cu + D\mu)$ so that

$$W_{f_{\mathbf{M}}}(u, \mu) = W_f(Du - B\mu, -Cu + A\mu), \qquad (3.89)$$

$$W_{f_{\mathbf{M}}}(Au + B\mu, Cu + D\mu) = W_f(u, \mu). \qquad (3.90)$$

These equations are the same as equations 3.81 and 3.82.

To summarize, when a function undergoes a linear canonical transform, its Wigner distribution undergoes a pointwise geometrical distortion or deformation: the value of the Wigner distribution at each time-frequency point is mapped to another time-frequency point. (We will see in chapter 8 that these points can be interpreted as optical rays.) It is often easier to visualize this by concentrating on the boundary of the region to which the Wigner distribution is approximately confined—the region in which most of the energy of the signal lies. However, the area-preserving nature of the distortion is more general. For instance, let us concentrate on any particular closed contour of the Wigner distribution, which does not necessarily contain a large percentage of the signal energy. The region defined by such a contour will be distorted and deformed but its area will remain the same regardless of the parameters of the linear canonical transform. Even more generally, if we take any region R in (u, μ) space, and find the image of this region under a linear canonical transform, their areas will always be the same. (If R is a given region in (u, μ) space, then its image is simply the set of points $(Au + B\mu, Cu + D\mu)$ such that $(u, \mu) \in R$.)

Until now we limited our attention to the Wigner distribution. The ambiguity function is affected in a similar way. We remember that the ambiguity function is essentially the two-dimensional Fourier transform of the Wigner distribution (equation 3.34). Thus, the two-dimensional Fourier transform property given in equation 2.296 allows us to show that the ambiguity function of the linear canonical transform of a function is related to that of the original function through the relation

$$A_{f_\mathbf{M}}(\bar{u}, \bar{\mu}) = A_f(D\bar{u} - B\bar{\mu}, -C\bar{u} + A\bar{\mu}). \tag{3.91}$$

More generally, time-frequency distributions of the Cohen class will exhibit a similar distortion property if the kernel satisfies $\psi_{TFE}(u, \mu) = \psi_{TFE}(Du - B\mu, -Cu + A\mu)$ for all u, μ and all A, B, C, D satisfying $AD - BC = 1$ (Ozaktas, Erkaya, and Kutay 1996).

3.4.3 Special linear canonical transforms

We now consider a number of important special cases of linear canonical transforms. We will see that most of these are operations that we have already defined in table 2.2. We will present the forward kernel $C_\mathbf{M}(u, u')$, the inverse kernel $C_\mathbf{M}^{-1}(u, u') = C_{\mathbf{M}^{-1}}(u, u')$, the result of the transform $f_\mathbf{M}(u)$, the matrix $\mathbf{M}$, and the Wigner distribution of the transformed signal $W_{f_\mathbf{M}}(u, \mu)$.

First, we consider simple scaling which is characterized by

$$M_M(u, u') = \sqrt{M}\, \delta(u - Mu'), \tag{3.92}$$

$$M_M^{-1}(u, u') = M_{(1/M)}(u, u') = \sqrt{1/M}\, \delta(u - u'/M), \tag{3.93}$$

$$\mathcal{C}_{\mathbf{M}_M} f(u) = \mathcal{M}_M f(u) = \sqrt{1/M}\, f(u/M), \tag{3.94}$$

$$\mathbf{M}_M = \begin{bmatrix} M & 0 \\ 0 & 1/M \end{bmatrix} = \begin{bmatrix} 1/M & 0 \\ 0 & M \end{bmatrix}^{-1}, \tag{3.95}$$

$$W_{\mathcal{M}_M f}(u, \mu) = W_f(u/M, M\mu), \tag{3.96}$$

where $M > 0$. Here $\mathbf{M}_M$ is the transform matrix corresponding to the scaling operation

$\mathcal{M}_M$ with kernel $M_M(u, u')$. (It is unfortunate that we use $\mathbf{M}$ also to denote a generic transform matrix, and that we use M to denote both the functional form of the kernel and its parameter.) The effect on the Wigner distribution is illustrated in figure 3.6b. To see that this is indeed a special case of equation 3.66 requires some care. First notice that $A = M = 1/D$ and $B = 0 = C$ implies $M\alpha = \beta = \gamma/M \to \infty$ from equations 3.75. Now the desired results can be obtained by virtue of equation 2.4. The identity operation is a further special case of the scaling operation with $M = 1$.

We now turn to chirp multiplication for which

$$Q_q(u, u') = e^{-i\pi q u^2} \delta(u - u'), \tag{3.97}$$

$$Q_q^{-1}(u, u') = Q_{-q}(u, u') = e^{i\pi q u^2} \delta(u - u'), \tag{3.98}$$

$$C_{\mathbf{Q}_q} f(u) = \mathcal{Q}_q f(u) = e^{-i\pi q u^2} f(u), \tag{3.99}$$

$$\mathbf{Q}_q = \begin{bmatrix} 1 & 0 \\ -q & 1 \end{bmatrix} = \begin{bmatrix} 1 & 0 \\ q & 1 \end{bmatrix}^{-1}, \tag{3.100}$$

$$W_{\mathcal{Q}_q f}(u, \mu) = W_f(u, \mu + qu). \tag{3.101}$$

Here $\mathbf{Q}_q$ is the transform matrix corresponding to the chirp multiplication operation $\mathcal{Q}_q$ with kernel $Q_q(u, u')$. Chirp multiplication is characterized by a lower triangular matrix. The effect on the Wigner distribution is a special case of property 15 in table 3.2 (figure 3.6c). It is also of some interest to note that the matrix $\mathbf{Q}_q$ can be written as $\mathbf{Q}_q = \exp(-q\,\mathbf{lower})$ where $\mathbf{lower}$ is a 2×2 matrix $[0\ 0; 1\ 0]$.

The dual of chirp multiplication is chirp convolution for which

$$R_r(u, u') = e^{-i\pi/4} \sqrt{1/r}\, \exp[i\pi(u - u')^2/r], \tag{3.102}$$

$$R_r^{-1}(u, u') = R_{-r}(u, u') = e^{i\pi/4}(1/\sqrt{r}) \exp[-i\pi(u - u')^2/r], \tag{3.103}$$

$$C_{\mathbf{R}_r} f(u) = \mathcal{R}_r f(u) = f(u) * e^{-i\pi/4} \sqrt{1/r}\, \exp(i\pi u^2/r), \tag{3.104}$$

$$\mathbf{R}_r = \begin{bmatrix} 1 & r \\ 0 & 1 \end{bmatrix} = \begin{bmatrix} 1 & -r \\ 0 & 1 \end{bmatrix}^{-1}, \tag{3.105}$$

$$W_{\mathcal{R}_r f}(u, \mu) = W_f(u - r\mu, \mu). \tag{3.106}$$

Here $\mathbf{R}_r$ is the transform matrix corresponding to the chirp convolution operation $\mathcal{R}_r$ with kernel $R_r(u, u')$. The effect on the Wigner distribution is a special case of property 14 in table 3.2 (figure 3.6d). Chirp convolution is characterized by an upper triangular matrix. It is also of some interest to note that the matrix $\mathbf{R}_q$ can be written as $\mathbf{R}_q = \exp(r\,\mathbf{upper})$ where $\mathbf{upper}$ is a 2×2 matrix $[0\ 1; 0\ 0]$.

The geometrical effect of chirp multiplication and chirp convolution on the Wigner distribution, or more precisely the support of the Wigner distribution, is referred to as "shearing." This geometrical distortion is a special case of that described by equation 3.81, and like all distortions described by this equation, is area preserving. This is easily seen to be the case by considering the effect of shearing on a rectangle. Shearing may be viewed as cutting this rectangle into narrow strips and sliding them with respect to each other (figure 3.7).

We will see later that a general area-preserving distortion, characterized by a unit-determinant matrix, can be decomposed in terms of these two kinds of area-preserving vertical and horizontal shearing operations. Also note that shearing the

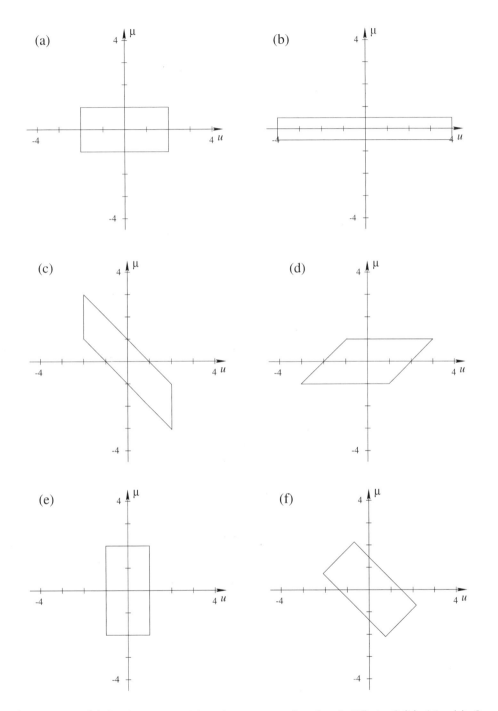

Figure 3.6.   (a) Region representing the support of a signal. Effect of (b) $\mathcal{M}_2$, (c) $\mathcal{Q}_1$, (d) $\mathcal{R}_1$, (e) $\mathcal{F}$, (f) $\mathcal{F}^{0.5}$.

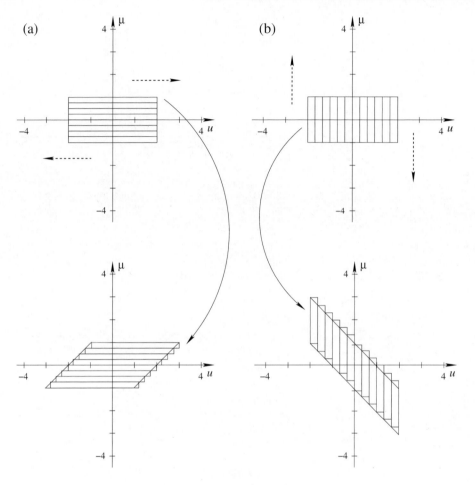

Figure 3.7. Horizontal (a) and vertical (b) shearing as sliding of narrow strips.

Wigner distribution in a certain direction will not change the projection of the Wigner distribution (its Radon transform) along that direction; this is also evident from figure 3.7. (As an additional exercise, the reader may want to show that such general area-preserving distortions always map a bundle of parallel lines to another bundle of parallel lines, and ellipses to ellipses.)

The Fourier transform is also a special case with

$$F_{\text{lc}}(u, u') = e^{-i\pi/4} e^{-i2\pi uu'}, \tag{3.107}$$

$$F_{\text{lc}}^{-1}(u, u') = F_{\text{lc}}^{*}(u, u') = e^{i\pi/4} e^{i2\pi uu'}, \tag{3.108}$$

$$\mathcal{C}_{\mathbf{F}_{\text{lc}}} f(u) = \mathcal{F}_{\text{lc}} f(u) = e^{-i\pi/4} \int f(u) e^{-i2\pi uu'} \, du', \tag{3.109}$$

$$\mathbf{F}_{\text{lc}} = \begin{bmatrix} 0 & 1 \\ -1 & 0 \end{bmatrix} = \begin{bmatrix} 0 & -1 \\ 1 & 0 \end{bmatrix}^{-1}, \tag{3.110}$$

$$W_{\mathcal{F}_{1c}f}(u,\mu) = W_f(-\mu, u). \tag{3.111}$$

The effect on the Wigner distribution is seen to be property 16 in table 3.2 (figure 3.6e). The factor $\exp(-i\pi/4)$ is not normally a part of the definition of the Fourier transform. However, it is this form which is the special case of linear canonical transforms (Wolf 1979). This usually does not cause any trouble since it differs from the standard definition only by a unit-magnitude constant. The reader may also have noticed that $\mathcal{F}_{1c}$ as defined above satisfies $\mathcal{F}_{1c}^4 = -\mathcal{I}$ rather than $\mathcal{F}^4 = \mathcal{I}$ satisfied by the standard definition. This fact is related to the sign factor discussed on page 94. (Similar comments also apply to the fractional Fourier transform introduced below.)

Note that all of these special cases, with the exception of the Fourier transform, are one-parameter subgroups of the group of linear canonical transforms. A further such group is the fractional Fourier transform, characterized by the matrix

$$\mathbf{F}_{1c}^a = \begin{bmatrix} \cos(a\pi/2) & \sin(a\pi/2) \\ -\sin(a\pi/2) & \cos(a\pi/2) \end{bmatrix} = \begin{bmatrix} \cos(a\pi/2) & -\sin(a\pi/2) \\ \sin(a\pi/2) & \cos(a\pi/2) \end{bmatrix}^{-1}, \tag{3.112}$$

and the rotational time-frequency distortion (figure 3.6f)

$$W_{\mathcal{F}_{1c}^a f}(u,\mu) = W_f\left[\cos(a\pi/2)\, u - \sin(a\pi/2)\,\mu, \sin(a\pi/2)\, u + \cos(a\pi/2)\,\mu\right]. \tag{3.113}$$

The fractional Fourier transform will be discussed in detail in chapter 4. This one-parameter subgroup is also referred to as the *elliptic subgroup* (Wolf 1979). It is also of some interest to note that the matrix $\mathbf{F}_{1c}^a$ can be written as $\mathbf{F}_{1c}^a = \exp[(a\pi/2)\mathbf{F}_{1c}]$ where $\mathbf{F}_{1c}$ is the 2×2 matrix $[0\ 1; -1\ 0]$. We emphasize that here $\mathbf{F}$ and $\mathbf{F}^a$ denote the 2×2 linear canonical transform matrices associated with the continuous Fourier and fractional Fourier transforms; they do *not* denote the discrete Fourier and fractional Fourier transform matrices for which these symbols are used elsewhere in this book.

Another one-parameter subgroup which we will not further discuss has the matrix

$$\mathbf{HYP}_a = \begin{bmatrix} \cosh(a\pi/2) & \sinh(a\pi/2) \\ \sinh(a\pi/2) & \cosh(a\pi/2) \end{bmatrix} = \begin{bmatrix} \cosh(a\pi/2) & -\sinh(a\pi/2) \\ -\sinh(a\pi/2) & \cosh(a\pi/2) \end{bmatrix}^{-1} \tag{3.114}$$

and is referred to as the *hyperbolic subgroup*. We might also add for completeness that the one-parameter subgroup corresponding to the scaling operation is sometimes written with $M = \exp(a\pi/2)$ and referred to as the *parabolic subgroup* (Wolf 1979).

We will briefly mention a number of additional special cases. First we consider the transform characterized by an arbitrary unit-determinant lower triangular matrix

$$\begin{bmatrix} M & 0 \\ -qM & 1/M \end{bmatrix} = \begin{bmatrix} 1 & 0 \\ -q & 1 \end{bmatrix}\begin{bmatrix} M & 0 \\ 0 & 1/M \end{bmatrix}, \tag{3.115}$$

which we have written as the composition of a scaling operation followed by a chirp multiplication, so that the effect of the linear canonical transform corresponding to this matrix is to take $f(u)$ to

$$e^{-i\pi q u^2}\sqrt{1/M}\, f(u/M). \tag{3.116}$$

Also, we may consider the transform characterized by an arbitrary unit-determinant upper triangular matrix

$$\begin{bmatrix} M & r/M \\ 0 & 1/M \end{bmatrix} = \begin{bmatrix} 1 & r \\ 0 & 1 \end{bmatrix}\begin{bmatrix} M & 0 \\ 0 & 1/M \end{bmatrix}, \tag{3.117}$$

which we have written as the composition of a scaling operation followed by a chirp convolution, so that the effect of the linear canonical transform corresponding to this matrix is to take $f(u)$ to

$$e^{-i\pi/4}\sqrt{\frac{1}{r}}\, e^{i\pi u^2/r} * \sqrt{1/M}\, f(u/M). \tag{3.118}$$

Matrices of the remaining two triangular forms can be decomposed as follows:

$$\begin{bmatrix} -r/M & M \\ -1/M & 0 \end{bmatrix} = \begin{bmatrix} 1 & r \\ 0 & 1 \end{bmatrix}\begin{bmatrix} M & 0 \\ 0 & 1/M \end{bmatrix}\begin{bmatrix} 0 & 1 \\ -1 & 0 \end{bmatrix}, \tag{3.119}$$

$$\begin{bmatrix} 0 & M \\ -1/M & -qM \end{bmatrix} = \begin{bmatrix} 1 & 0 \\ -q & 1 \end{bmatrix}\begin{bmatrix} M & 0 \\ 0 & 1/M \end{bmatrix}\begin{bmatrix} 0 & 1 \\ -1 & 0 \end{bmatrix}. \tag{3.120}$$

The second of these corresponds to a scaled Fourier transform with a residual phase factor.

Lastly we mention that other simple operations, such as coordinate shifting or multiplication by a phase factor, also become special cases of linear canonical transforms if linear terms are allowed in the exponent of the kernel (as discussed on page 96).

3.4.4 Decompositions

We have already given above a number of decompositions for triangular matrices. Here we list a number of further decompositions for the general case:

$$\begin{bmatrix} A & B \\ C & D \end{bmatrix} = \begin{bmatrix} 1 & (A-1)/C \\ 0 & 1 \end{bmatrix}\begin{bmatrix} 1 & 0 \\ C & 1 \end{bmatrix}\begin{bmatrix} 1 & (D-1)/C \\ 0 & 1 \end{bmatrix} \tag{3.121}$$

$$= \begin{bmatrix} 1 & 0 \\ (D-1)/B & 1 \end{bmatrix}\begin{bmatrix} 1 & B \\ 0 & 1 \end{bmatrix}\begin{bmatrix} 1 & 0 \\ (A-1)/B & 1 \end{bmatrix} \tag{3.122}$$

$$= \begin{bmatrix} 1 & 0 \\ C/A & 1 \end{bmatrix}\begin{bmatrix} A & 0 \\ 0 & 1/A \end{bmatrix}\begin{bmatrix} 1 & B/A \\ 0 & 1 \end{bmatrix} \tag{3.123}$$

$$= \begin{bmatrix} 1 & B/D \\ 0 & 1 \end{bmatrix}\begin{bmatrix} 1/D & 0 \\ 0 & D \end{bmatrix}\begin{bmatrix} 1 & 0 \\ C/D & 1 \end{bmatrix} \tag{3.124}$$

$$= \begin{bmatrix} B & 0 \\ D & 1/B \end{bmatrix}\begin{bmatrix} 0 & 1 \\ -1 & 0 \end{bmatrix}\begin{bmatrix} 1 & 0 \\ A/B & 1 \end{bmatrix} \tag{3.125}$$

$$= \begin{bmatrix} 1 & 0 \\ D/B & 1 \end{bmatrix}\begin{bmatrix} B & 0 \\ 0 & 1/B \end{bmatrix}\begin{bmatrix} 0 & 1 \\ -1 & 0 \end{bmatrix}\begin{bmatrix} 1 & 0 \\ A/B & 1 \end{bmatrix} \tag{3.126}$$

$$= \begin{bmatrix} -1/C & -A \\ 0 & -C \end{bmatrix}\begin{bmatrix} 0 & 1 \\ -1 & 0 \end{bmatrix}\begin{bmatrix} 1 & D/C \\ 0 & 1 \end{bmatrix} \tag{3.127}$$

$$= \begin{bmatrix} 1 & A/C \\ 0 & 1 \end{bmatrix}\begin{bmatrix} -1/C & 0 \\ 0 & -C \end{bmatrix}\begin{bmatrix} 0 & 1 \\ -1 & 0 \end{bmatrix}\begin{bmatrix} 1 & D/C \\ 0 & 1 \end{bmatrix}. \tag{3.128}$$

Care must be exercised when a term appearing in the denominator of any of the matrix elements is zero. The decompositions will remain valid if the limits are evaluated carefully. Some of these decompositions are from Nazarathy and Shamir 1982a.

These decompositions are very easy to demonstrate when stated as matrix products in the above manner. The reader should nevertheless see how these decompositions work out in terms of integral transforms to gain familiarity with them. Remember that lower triangular matrices correspond to chirp multiplication, upper triangular matrices to chirp convolution, diagonal matrices to scaling, and skew diagonal matrices to Fourier transforming. Thus, these decompositions show the many ways in which a general linear canonical transform can be decomposed in terms of these more elementary operations. The decompositions given in equations 3.121 and 3.122 are referred to as canonical decompositions type I and type II, respectively (Papoulis 1974, 1977). (They can be obtained from each other by taking the transpose of both sides of these decompositions.) It is also useful to interpret these decompositions in terms of "shear diagrams," which show how the overall distortion of the Wigner distribution can be broken down into simpler shearing operations. We illustrate this concept with equation 3.121 in figure 3.8.

Equations 3.121 and 3.122 show how any unit-determinant matrix can be written as the product of lower and upper triangular matrices. We have already seen that these lower and upper triangular matrices (which correspond to chirp multiplication and convolution) geometrically correspond to vertical and horizontal shears. Since it is easy to see that these shears preserve area, and since the distortion associated with any unit determinant matrix can be decomposed into such shears, it follows that such distortions are also area preserving. As previously stated, this result also follows more directly from the facts that the determinant of the matrix is the Jacobian of the geometric distortion and that distortions with unity Jacobian are area preserving.

It is also possible to write equations 3.121 through 3.128 in operator form as follows:

$$\mathcal{R}_{(A-1)/C}\, \mathcal{Q}_{-C}\, \mathcal{R}_{(D-1)/C}, \tag{3.129}$$

$$\mathcal{Q}_{(1-D)/B}\, \mathcal{R}_B\, \mathcal{Q}_{(1-A)/B}, \tag{3.130}$$

$$\mathcal{Q}_{-C/A}\, \mathcal{M}_A\, \mathcal{R}_{B/A}, \tag{3.131}$$

$$\mathcal{R}_{B/D}\, \mathcal{M}_{1/D}\, \mathcal{Q}_{-C/D}, \tag{3.132}$$

$$(\mathcal{Q}_{-D/B}\, \mathcal{M}_B)\, \mathcal{F}\, \mathcal{Q}_{-A/B}, \tag{3.133}$$

$$\mathcal{Q}_{-D/B}\, \mathcal{M}_B\, \mathcal{F}\, \mathcal{Q}_{-A/B}, \tag{3.134}$$

$$(\mathcal{R}_{A/C}\, \mathcal{M}_{-1/C})\, \mathcal{F}\, \mathcal{R}_{D/C}, \tag{3.135}$$

$$\mathcal{R}_{A/C}\, \mathcal{M}_{-1/C}\, \mathcal{F}\, \mathcal{R}_{D/C}. \tag{3.136}$$

The last two pairs have been repeated so as to maintain parallelism with the list of matrix decompositions. These formulas are analogous to what are known as Baker-Campbell-Hausdorff formulas; see the discussion surrounding equation 2.245.

At this point, the reader should be able to interpret any decomposition in all of the following ways and understand the relationships between these interpretations:

- Decomposition into matrices.
- Decomposition into consecutive integral transforms.
- Decomposition into abstract operators.

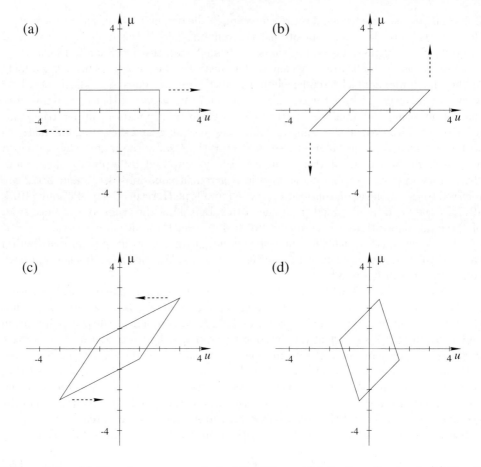

Figure 3.8. (a) Time-frequency support of the Wigner distribution of a signal. (b) Result of first horizontal shear. (c) Result of vertical shear. (d) Result of second horizontal shear, which is the time-frequency support of the linear canonical transform of the signal. $A = 0.5$, $B = -0.5$, $C = 0.5$, $D = 1.5$.

- Decomposition into chirp multiplications and chirp convolutions, and scaling operations and Fourier transforms.
- Decomposition of the overall geometric distortion of the time-frequency support of the Wigner distribution into horizontal and vertical shears, and scalings and rotations.

Decomposition into hyperdifferential operators will be added to the list later in this chapter. Further interpretations will be possible in an optical context: decomposition into lenses, sections of free space, imaging systems, and Fourier transforming systems. Any of these interpretations can be used to verify a particular decomposition, but matrix multiplication is often the simplest. Of course, we remember that these correspondences are valid within a factor of ±1.

Any linear canonical transform can be obtained from another by properly scaling it and appending chirp multiplications to both ends:

$$\begin{bmatrix} A & B \\ C & D \end{bmatrix} = \begin{bmatrix} 1 & 0 \\ (DB - D'B')/B^2 & 1 \end{bmatrix} \begin{bmatrix} B/B' & 0 \\ 0 & B'/B \end{bmatrix}$$
$$\times \begin{bmatrix} A' & B' \\ C' & D' \end{bmatrix} \begin{bmatrix} 1 & 0 \\ (AB' - A'B)/BB' & 1 \end{bmatrix} \quad (3.137)$$

This result can be specialized by replacing either or both linear canonical transforms by special transforms such as Fourier, scaling, fractional Fourier, chirp multiplication, chirp convolution, and so forth. Thus, for example, a Fourier transform can be expressed as a chirp multiplication followed by a scaling operation followed by another chirp multiplication. A dual of this result based on appending chirp convolutions rather than chirp multiplications to both ends also exists. Further variations may be obtained by changing the position of the scaling matrix in the above equation.

It is worth explicitly writing some of the special cases of the above equation for future reference:

$$\mathcal{C_M} = \mathcal{Q}_{q_1} \mathcal{M}_{M_1} \mathcal{F}^a \mathcal{Q}_{q_2}, \quad (3.138)$$
$$\mathcal{C_M} = \mathcal{Q}_{q_3} \mathcal{M}_{M_2} \mathcal{F} \mathcal{Q}_{q_4}, \quad (3.139)$$

which hold for suitable choices of the parameters.

We will most commonly use the canonical decompositions given by equations 3.121 and 3.122. However, if both B and C are zero, it may be more convenient to employ the alternative decomposition (Papoulis 1977):

$$\begin{bmatrix} M & 0 \\ 0 & 1/M \end{bmatrix} = \begin{bmatrix} 1 & 0 \\ (M^{-1} - M^{-2})/X & 1 \end{bmatrix} \begin{bmatrix} 1 & -XM \\ 0 & 1 \end{bmatrix}$$
$$\times \begin{bmatrix} 1 & 0 \\ (M^{-1} - 1)/X & 1 \end{bmatrix} \begin{bmatrix} 1 & X \\ 0 & 1 \end{bmatrix}, \quad (3.140)$$

where X is a parameter of our choosing.

We finally note that the operators $\mathcal{Q}_q$ and $\mathcal{R}_r$ can be referred to as being *duals* of each other when $q = r$, since they correspond to each other under Fourier transformation (chirp multiplication corresponds to chirp convolution in the Fourier domain and vice versa):

$$\mathcal{Q}_q = \mathcal{F}^{-1} \mathcal{R}_r \mathcal{F}, \quad (3.141)$$
$$\mathcal{R}_r = \mathcal{F} \mathcal{Q}_q \mathcal{F}^{-1}. \quad (3.142)$$

These equations are nothing but the convolution and product theorems for the Fourier transform.

3.4.5   Transformation of moments

The matrix **M** also allows one to write simple expressions for the transformation of the first- and second-order moments of a signal, relating the moments of the linear canonical transform to the moments of the original signal. Let $\bar{u}$ and $\bar{\mu}$ denote the first-order moments of the Wigner distribution of $f(u)$:

$$\bar{u} = \frac{\iint u W_f(u,\mu)\,du\,d\mu}{\iint W_f(u,\mu)\,du\,d\mu}, \tag{3.143}$$

$$\bar{\mu} = \frac{\iint \mu W_f(u,\mu)\,du\,d\mu}{\iint W_f(u,\mu)\,du\,d\mu}, \tag{3.144}$$

and define the first-order time-frequency moment vector $\bar{\mathbf{u}} \equiv [\bar{u}\ \bar{\mu}]^{\mathrm{T}}$. Then, it is possible to show that this vector transforms as follows (Bastiaans 1989, 1991b):

$$\bar{\mathbf{u}}_{f_{\mathbf{M}}} = \mathbf{M}\bar{\mathbf{u}}_f. \tag{3.145}$$

That is, the first-order moment vector of the transformed signal is obtained simply by multiplying the moment vector of the original signal by $\mathbf{M}$. Now, let us define the nonnegative-definite centralized second-order moment matrix (moment of inertia tensor) as

$$\mathbf{P} \equiv \begin{bmatrix} \overline{(u-\bar{u})^2} & \overline{(u-\bar{u})(\mu-\bar{\mu})} \\ \overline{(u-\bar{u})(\mu-\bar{\mu})} & \overline{(\mu-\bar{\mu})^2} \end{bmatrix}, \tag{3.146}$$

where the various second-order moments are defined in a manner similar to the first-order moments. It is possible to show that these are transformed as (Bastiaans 1989, 1991b)

$$\mathbf{P}_{f_{\mathbf{M}}} = \mathbf{M}\mathbf{P}_f\mathbf{M}^{\mathrm{T}}, \tag{3.147}$$

a result which preserves the nonnegative-definiteness of the moment matrix.

Since Gaussian signals are completely characterized by their first- and second-order moments, these results are particularly useful when dealing with such signals. We note that the determinant of $\mathbf{P}_{f_{\mathbf{M}}}$ is the same as that of $\mathbf{P}_f$. On the other hand, the trace of $\mathbf{P}_{f_{\mathbf{M}}}$ will be the same as that of $\mathbf{P}_f$ if and only if $\mathbf{M}$ is a rotation matrix. (The square root of the determinant is a measure of the support area whereas the trace corresponds to the moment of inertia of the Wigner distribution.) Further details and discussion with applications in optics may be found in Bastiaans 1989, 1991b.

3.4.6   Linear fractional transformations

Let us consider the linear canonical transform $f_{\mathbf{M}}(u)$ of a unit-energy *complex Gaussian function* $f(u)$ with *complex radius* r_c:

$$f(u) = (2\Im[1/r_c])^{1/4}\, e^{i\pi u^2/r_c}, \qquad \Im[1/r_c] > 0, \tag{3.148}$$

$$W_f(u,\mu) = 2\exp\left\{-2\pi\left[\Im[1/r_c]u^2 + \frac{(\mu - \Re[1/r_c]u)^2}{\Im[1/r_c]}\right]\right\}, \tag{3.149}$$

$$f_{\mathbf{M}}(u) = (2\Im[1/r_c'])^{1/4}\, e^{i\pi u^2/r_c'}, \tag{3.150}$$

where

$$r_c' = \left[\alpha - \frac{\beta^2 r_c}{\gamma r_c + 1}\right]^{-1}, \tag{3.151}$$

is the complex radius of $f_M(u)$ and $\Re[1/r_c]$ and $\Im[1/r_c]$ are the real and imaginary parts of $1/r_c$. (The term complex radius comes from optics, where r_c is interpreted as the radius of a wavefront.) We see that the linear canonical transform of a complex Gaussian function with given complex radius r_c is always another complex Gaussian function with complex radius r_c'. The relationship between the complex radius of the transformed function and that of the original function can be cast in a simpler form if we use the A, B, C, D parameters instead of α, β, γ. Using equation 3.75 we obtain the following result (Bastiaans 1989, 1991b):

$$r_c' = \frac{Ar_c + B}{Cr_c + D}. \tag{3.152}$$

Such functional relationships are referred to as linear fractional transformations. (When $\Im[1/r_c] = 0$, we expect to recover the simple chirp function $\propto \exp(i\pi u^2/r)$ with $\Re[1/r_c] = 1/r_c = 1/r$. However, since the chirp function does not have unit energy, this does not follow as a special case of equation 3.148 by letting $\Im[1/r_c] \to 0$. Nevertheless, the result $r' = (Ar + B)/(Cr + D)$ still holds for such chirp functions.)

Linear fractional transformations constitute an alternative to the matrix formulation of linear canonical transforms. Equation 3.152 can be generalized to arbitrary finite-energy functions (Bastiaans 1989, 1991b). Let us define the complex quantities Z and $Y = 1/Z$ as

$$Z = \frac{\overline{u\mu}}{\overline{\mu^2}} + i\, \frac{-\sqrt{\overline{u^2}\,\overline{\mu^2} - \overline{u\mu}^2}}{\overline{\mu^2}}, \tag{3.153}$$

$$Y = \frac{\overline{u\mu}}{\overline{u^2}} + i\, \frac{\sqrt{\overline{u^2}\,\overline{\mu^2} - \overline{u\mu}^2}}{\overline{u^2}}, \tag{3.154}$$

where the moments are those of the Wigner distribution of the arbitrary function (see page 107). (If desired, Z and Y can also be expressed in terms of moments defined in the time and frequency domains.) Now, let Z denote the complex parameter associated with a function $f(u)$. The complex parameter Z' associated with the linear canonical transform of $f(u)$ is given by

$$Z' = \frac{AZ + B}{CZ + D}. \tag{3.155}$$

Likewise, the complex parameters Y and Y' are related through

$$Y' = \frac{DY + C}{BY + A}. \tag{3.156}$$

We see that the Z parameter transforms in the same way as the complex radius and is thus a generalization of this concept for arbitrary signals. Let us consider some special cases. First, consider chirp convolution. In this case $A = D = 1$, $B = r$, and $C = 0$. Then

$$Z' = Z + r, \tag{3.157}$$

$$\frac{1}{Y'} = \frac{1}{Y} + r. \tag{3.158}$$

Second, consider chirp multiplication for which $A = D = 1$, $C = -q$, and $B = 0$. Then

$$Y' = Y - q, \tag{3.159}$$

$$\frac{1}{Z'} = \frac{1}{Z} - q. \tag{3.160}$$

We will see in chapter 7 that chirp convolution corresponds to free-space propagation and that r represents the distance of propagation. Thus, the effect of propagation on the Z parameter is simply to increase it by the distance of propagation. We will also see that chirp multiplication corresponds to passage through a thin lens and that $1/q$ corresponds to the focal length of the lens. This leads to the well-recognized lens formula $1/Z' = 1/Z - 1/(\text{focal length})$. If we make an analogy with electrical circuits, the Z parameter is like an impedance and propagation through free space is like the addition of a series element; the Y parameter is like an admittance and passage through a lens is like the addition of a parallel element.

We finally add that it is also possible to generalize these results to noncentered Gaussian or quadratic-phase signals passing through systems which distort the Wigner distribution in an affine manner (that is, including time/space and frequency shifts).

3.4.7 Coordinate multiplication and differentiation operators

The coordinate multiplication operator $\mathcal{U}$ and the differentiation operator $\mathcal{D}$ switch roles when we transform to the Fourier domain. Functions related to each other through linear canonical transforms may be considered representations of the same abstract signal in different linear canonical transform domains. We now seek the relationship between the coordinate multiplication and differentiation operators associated with these different domains. $\mathcal{U}_{\mathbf{M}}$ will denote the operator which multiplies by the coordinate variable and $\mathcal{D}_{\mathbf{M}}$ will denote the operator which differentiates with respect to the coordinate variable, both in the domain represented by the matrix $\mathbf{M}$. Since the effect of $\mathcal{U}_{\mathbf{M}}$ (or $\mathcal{D}_{\mathbf{M}}$) amounts to first taking the linear canonical transform, then coordinate multiplying (or differentiating), and then going back to the original domain, we can write the relationship between these operators as

$$\mathcal{U}_{\mathbf{M}} = \mathcal{C}_{\mathbf{M}}^{-1} \mathcal{U} \mathcal{C}_{\mathbf{M}},$$
$$\mathcal{D}_{\mathbf{M}} = \mathcal{C}_{\mathbf{M}}^{-1} \mathcal{D} \mathcal{C}_{\mathbf{M}}. \tag{3.161}$$

By using properties 7 and 8 in table 3.8, it is possible to show that

$$\begin{bmatrix} \mathcal{U}_{\mathbf{M}} \\ \mathcal{D}_{\mathbf{M}} \end{bmatrix} = \begin{bmatrix} A & B \\ C & D \end{bmatrix} \begin{bmatrix} \mathcal{U} \\ \mathcal{D} \end{bmatrix}. \tag{3.162}$$

An alternative development of linear canonical transforms may be found in Wolf 1979, where these equations are taken as the defining characteristics of linear canonical transforms and the integral form is subsequently derived. (It should be noted that Wolf poses the same idea in a different form by concentrating on the transformation between the different representations of the same operator $\mathcal{U}$, rather than the relationship between the different operators $\mathcal{U}_{\mathbf{M}}$ and $\mathcal{U}$ which both coordinate multiply in their respective domains.)

3.4.8   Uncertainty relation

The standard deviations of $f(u)$ and $f_{\mathbf{M}}(u)$, denoted by Δu and $\Delta u_{\mathbf{M}}$ respectively, satisfy

$$\Delta u \Delta u_{\mathbf{M}} \geq |B|/4\pi, \tag{3.163}$$

which reduces to the ordinary Fourier uncertainty relation when the linear canonical transform in question is the ordinary Fourier transform (Wolf 1979).

3.4.9   Invariants and hyperdifferential forms

A number of Hermitian operators are invariant under some of the special canonical transforms we have discussed. An operator $\mathcal{H}$ is invariant under $\mathcal{C}_{\mathbf{M}}$ if and only if $\mathcal{C}_{\mathbf{M}}^{-1}\mathcal{H}\mathcal{C}_{\mathbf{M}} = \mathcal{H}$, so that $g = \mathcal{H}f$ implies $(\mathcal{C}_{\mathbf{M}}g) = \mathcal{H}(\mathcal{C}_{\mathbf{M}}f)$. According to Wolf 1979:391, these invariant operators are

$$
\begin{array}{llll}
\mathcal{H}_M & \equiv\ 2\pi\tfrac{1}{2}(\mathcal{U}\mathcal{D} + \mathcal{D}\mathcal{U}) & \equiv\ 2\mathcal{J}_2, & \text{under scaling or parabolic,} \\
\mathcal{H}_R & \equiv\ 2\pi\tfrac{1}{2}\mathcal{D}^2, & & \text{under chirp convolution,} \\
\mathcal{H}_Q & \equiv\ 2\pi\tfrac{1}{2}\mathcal{U}^2, & & \text{under chirp multiplication,} \\
\mathcal{H}_{HYP} & \equiv\ 2\pi\tfrac{1}{2}(\mathcal{D}^2 - \mathcal{U}^2) & \equiv\ 2\mathcal{J}_1, & \text{under hyperbolic,} \\
\mathcal{H}_{fF} & \equiv\ 2\pi\tfrac{1}{2}(\mathcal{D}^2 + \mathcal{U}^2) & \equiv\ 2\mathcal{J}_0, & \text{under fractional Fourier transform.}
\end{array}
$$

The invariance of these operators can be demonstrated directly by using equations 3.161 and equation 3.162. Invariance can equivalently be stated as a commutation relation since $\mathcal{C}_{\mathbf{M}}^{-1}\mathcal{H}\mathcal{C}_{\mathbf{M}} = \mathcal{H}$ implies $\mathcal{H}\mathcal{C}_{\mathbf{M}} = \mathcal{C}_{\mathbf{M}}\mathcal{H}$. For instance, we can write $\mathcal{Q}_q^{-1}\mathcal{U}^2\mathcal{Q}_q = \mathcal{U}^2$ which means that $\mathcal{U}^2$ commutes with $\mathcal{Q}_q$. Thus, eigenfunctions of these operators are also eigenfunctions of the indicated special linear canonical transforms. We also note that $\mathcal{H}_{fF} = \mathcal{A}^{\mathrm{H}}\mathcal{A} + 1/2 = \mathcal{A}\mathcal{A}^{\mathrm{H}} - 1/2$, where $\mathcal{A}$ was defined in equation 2.249.

We have already discussed several one-parameter subgroups of the group of linear canonical transforms. Noting that most of these have the property that when their parameter is zero, they reduce to the identity transform, we will now seek hyperdifferential forms for these operators of the form $\exp(-ip\mathcal{H})$, where p is the relevant parameter. We know that $\exp(-ip\mathcal{H})$ is unitary if and only if $\mathcal{H}$ is Hermitian. The hyperdifferential forms of the various special linear canonical transform operators we have discussed can be given in terms of the above Hermitian operators as follows (Wolf 1979:408):

$$
\begin{aligned}
\mathcal{M}_M &= \exp(-i \ln M\, \mathcal{H}_M) \\
&= \exp[-i 2\pi \ln M (\mathcal{U}\mathcal{D} + \mathcal{D}\mathcal{U})/2] = M^{[-i\pi(\mathcal{U}\mathcal{D}+\mathcal{D}\mathcal{U})]}, \tag{3.164} \\
\mathcal{R}_r &= \exp(-ir\mathcal{H}_R) = \exp(-i\pi r\mathcal{D}^2), \tag{3.165} \\
\mathcal{Q}_q &= \exp(-iq\mathcal{H}_Q) = \exp(-i\pi q\mathcal{U}^2), \tag{3.166} \\
\mathcal{HYP}_a &= \exp\left[-i(a\pi/2)\mathcal{H}_{HYP}\right], \tag{3.167} \\
\mathcal{F}_{\mathrm{lc}}^a &= \exp\left[-i(a\pi/2)\mathcal{H}_{fF}\right] = \exp[-i(a\pi^2/2)(\mathcal{U}^2 + \mathcal{D}^2)]. \tag{3.168}
\end{aligned}
$$

The unitary operators given above are sufficient to construct any linear canonical transform operator by concatenation. This follows from the matrix decompositions given in section 3.4.4. By using the hyperdifferential forms given above, we can also obtain a Baker-Campbell-Hausdorff formula corresponding to each matrix decomposition appearing in section 3.4.4. These abstract operator formulas can then be specialized to any domain or representation. For example, using equation 3.123 for the matrix given in equation 3.112, we obtain

$$
\begin{bmatrix} \cos\alpha & \sin\alpha \\ -\sin\alpha & \cos\alpha \end{bmatrix} = \begin{bmatrix} 1 & 0 \\ -\tan\alpha & 1 \end{bmatrix} \begin{bmatrix} \cos\alpha & 0 \\ 0 & 1/\cos\alpha \end{bmatrix} \begin{bmatrix} 1 & \tan\alpha \\ 0 & 1 \end{bmatrix}. \tag{3.169}
$$

Now, we recognize this decomposition as a chirp convolution followed by scaling followed by chirp multiplication (see page 99 onwards). Identifying $r = q = \tan\alpha$ and $M = \cos\alpha$ and using the hyperdifferential forms given above, one immediately obtains equation 2.245.

We should perhaps also note that if $\mathcal{C}_{\mathbf{M}} = \exp(-ip\mathcal{H})$ is the hyperdifferential form of a particular linear canonical transform $\mathcal{C}_{\mathbf{M}}$, it is *necessarily* the case that $\mathcal{H}$ is invariant under that transform. That is,

$$
e^{ip\mathcal{H}}\mathcal{H}e^{-ip\mathcal{H}} = \mathcal{H}, \tag{3.170}
$$

which is true since $e^{ip\mathcal{H}}$ and $\mathcal{H}$ commute.

Thus, to the list of alternative interpretations of decompositions given on page 105, we may add decomposition into hyperdifferential operators. Each interpretation is associated with a different set of objects. These sets of objects are matrices; integral transforms; abstract operators; chirp multiplication, convolution and other operations; geometric distortion operations; and hyperdifferential operators. In later chapters we will add optical components to this list.

3.4.10 Differential equations

We now consider differential equations of the form

$$
\mathcal{H}f_p(u) = i\,\frac{\partial f_p(u)}{\partial p}, \tag{3.171}
$$

where $\mathcal{H}$ is a quadratic Hermitian operator in $\mathcal{U}$ and $\mathcal{D}$. Since the Hermitian conjugate of $\mathcal{D}\mathcal{U}$ is $\mathcal{U}\mathcal{D}$ and vice versa, such an operator may only contain terms proportional to $\mathcal{U}^2$, $\mathcal{D}^2$, and $\mathcal{U}\mathcal{D} + \mathcal{D}\mathcal{U}$. The solution of this equation is

$$
f_p(u) = e^{-ip\mathcal{H}}f_0(u), \tag{3.172}
$$

where $f_0(u)$ serves as the initial or boundary condition. (To prove that this satisfies equation 3.171, it is sufficient to expand the exponential into a series.) Alternatively, the solution can be expressed as a one-parameter canonical transform (Wolf 1979:410):

$$
\mathcal{C}_p = e^{-ip\mathcal{H}}, \tag{3.173}
$$

$$
f_p(u) = (\mathcal{C}_p[f_0(u)])(u) = \int \mathcal{C}_p(u, u')f_0(u')\,du'. \tag{3.174}
$$

The operator $\mathcal{C}_p$ may be referred to as the time-evolution or Green's operator for the system governed by the differential equation in question. The associated kernel is known as the Green's function $C_p(u, u')$, which is simply the response of the system to $f_0(u) = \delta(u - u')$. (Wolf 1979)

Certain functions will preserve their forms under the one-parameter linear canonical transform described above. Let $\psi_\lambda(u)$ be an eigenfunction of $\mathcal{H}$ with eigenvalue λ. Then

$$\mathcal{C}_p[\psi_\lambda(u)](u) = \int C_p(u, u')\psi_\lambda(u')\, du' = e^{-ip\mathcal{H}}\psi_\lambda(u) = e^{-ip\lambda}\psi_\lambda(u). \qquad (3.175)$$

We see that the dependence of $\mathcal{C}_p[\psi_\lambda(u)](u)$ on u and p is separable. The eigenfunctions corresponding to some of the operators we have discussed are

$$\psi_{R_\lambda}(u) = e^{\pm i2\pi\sqrt{\lambda/\pi}\,u} \qquad\qquad \lambda \geq 0 \qquad\qquad \text{chirp conv.,}$$

$$\psi_{Q_\lambda}(u) = \delta(u \pm \sqrt{\lambda/\pi}\,) \qquad\qquad \lambda \geq 0 \qquad\qquad \text{chirp mult.,}$$

$$\psi_{fF_\lambda}(u) = \psi_{\lambda-1/2}(u) \qquad \lambda = n + 1/2, \quad n = 0, 1, \ldots \qquad \text{frac. Four. trans.,}$$

where $\psi_{\lambda-1/2}(u)$ is the $(\lambda - 1/2)$th order Hermite-Gaussian function (also see page 49). Further discussion of the eigenfunctions of linear canonical transforms may be found in Wolf 1977.

3.4.11   Symplectic systems

We will begin by mentioning the concept of a *form*. The most common example of a form is the scalar inner product of two signals or vectors. It is often of interest to inquire into the nature of the set of systems under which such a form is preserved (is invariant). We know that the inner product of two signals remains the same under passage through a unitary system. We also know that physically this corresponds to a system conserving power or energy (since the norm is a special case of the inner product and signal energy is the square of the norm, and since signal energy usually corresponds to physical energy or power). The inner product is an example of a *symmetric form*.

An example of an *antisymmetric form* is the *symplectic form*. Here we will consider the symplectic form $\prec[u_1, \mu_1]^T, [u_2, \mu_2]^T \succ$ of two space-frequency (or time-frequency) vectors $[u_1, \mu_1]^T$ and $[u_2, \mu_2]^T$, which we define as follows (Folland 1989):

$$\prec[u_1, \mu_1]^T, [u_2, \mu_2]^T \succ \equiv [u_1, \mu_1] \begin{bmatrix} 0 & 1 \\ -1 & 0 \end{bmatrix} \begin{bmatrix} u_2 \\ \mu_2 \end{bmatrix} = u_1\mu_2 - u_2\mu_1. \qquad (3.176)$$

The antisymmetric 2×2 matrix appearing in the definition will be denoted by the symbol $\mathbf{J}$ and satisfies $\mathbf{J}^T = -\mathbf{J} = \mathbf{J}^{-1}$. We see that the symplectic form of two space-frequency vectors gives the (signed) area of the parallelogram defined by those two vectors. (To see this, recall that the signed area of a parallelogram is given by the vector cross product of the vectors defining it, and that $u_1\mu_2 - u_2\mu_1$ is nothing but the value of this cross product. The "sign" here refers to the direction of the resultant cross product vector.) We also note that the symplectic form can be interpreted as the space-frequency hypervolume for higher-dimensional spaces as well. Thus, physically, preservation of the symplectic form corresponds to invariance of space-frequency area or volume, a fact that will become important in chapter 8.

Now, let us consider a matrix $\mathbf{M} = [A\ B;\ C\ D]$ mapping the space-frequency vectors $[u_1, \mu_1]^T$ and $[u_2, \mu_2]^T$ into $\mathbf{M}[u_1, \mu_1]^T$ and $\mathbf{M}[u_2, \mu_2]^T$. Equating the symplectic form of these new vectors to the original symplectic form as

$$\prec \mathbf{M}[u_1, \mu_1]^\mathrm{T}, \mathbf{M}[u_2, \mu_2]^\mathrm{T} \succ \; = \; \prec [u_1, \mu_1]^\mathrm{T}, [u_2, \mu_2]^\mathrm{T} \succ, \tag{3.177}$$

$$\left(\mathbf{M} \begin{bmatrix} u_1 \\ \mu_1 \end{bmatrix} \right)^\mathrm{T} \mathbf{J} \, \mathbf{M} \begin{bmatrix} u_2 \\ \mu_2 \end{bmatrix} = [u_1, \mu_1] \, \mathbf{J} \begin{bmatrix} u_2 \\ \mu_2 \end{bmatrix}, \tag{3.178}$$

leads us to the condition

$$\mathbf{M}^\mathrm{T} \mathbf{J} \mathbf{M} = \mathbf{J}, \tag{3.179}$$

for the preservation of the symplectic form under the mapping represented by $\mathbf{M}$. Matrices $\mathbf{M}$ satisfying this relation are called symplectic matrices; they preserve the symplectic form. In the one-dimensional case we are considering, it is easy to show that this condition is fully equivalent to the unit-determinant condition $AD - BC = 1$. Thus, the matrices characterizing linear canonical transforms introduced in equation 3.74 are symplectic. Although trivial in the one-dimensional case, similar results hold for higher dimensions as well (Folland 1989). In general, matrices are called symplectic if they preserve the symplectic form (which corresponds to preservation of space-frequency area or volume), and the condition for this turns out to be of the same form as equation 3.179, which is not simply equivalent to a unit-determinant condition in higher dimensions. The matrices characterizing multi-dimensional linear canonical transforms are always symplectic.

Moving further in this direction is beyond what we can achieve in this book. The interested reader is referred to Folland 1989, particularly chapters 1 and 4, and the introduction to Guillemin and Sternberg 1984. Another interesting work is Turski 1998. We will revisit these issues in an optical context in section 8.8.

We finally note that the symplecticity of matrices characterizing linear canonical transforms can be alternatively shown as follows. First, we can easily show that the matrices for chirp multiplication and chirp convolution satisfy the symplecticity definition given above. Then, we can show that if two matrices satisfy this definition, all their products and inverses also do. Since any linear canonical transform can be expressed as the product of chirp multiplications and chirp convolutions, we can then conclude that the matrix associated with any linear canonical transform is symplectic.

3.4.12   Connections to group theory

By now we have talked about quite a number of different sets of objects in association with linear canonical transforms. These sets of objects are matrices; integral transforms; abstract operators; chirp multiplication, convolution and other operations; geometric distortion operations; hyperdifferential operators; and optical operators (to be introduced later).

Each of these sets of objects together with a rule of composition satisfy the axioms of a group. (i) The composition of any two elements of the set is also an element of the set (closure). (ii) There exists an identity element such that the composition of any element with that identity is again the same element. (iii) Each element has an inverse such that the composition of any element with its inverse is the identity element. (iv) Composition is associative.

It is not difficult to show that any of the sets of objects we considered and their composition rules constitute a group. Furthermore, for every element in one of these groups, there are corresponding elements in each of the other groups, and these correspondences are preserved under composition (save for a possible sign) (Wolf 1979). If we ignore differences caused by $\pm$ signs, these correspondences become one-to-one. Such groups are called isomorphic to each other.

Some groups are given special names. In particular, the group of real 2×2 matrices of determinant $+1$ is referred to as the $SL(2, \mathbf{R})$ group (Dym and McKean 1972:273). All of our groups are isomorphic to this group. The notation and definition offered in Folland 1989:171 is slightly different but equivalent: the symplectic group $Sp(2, \mathbf{R})$ is the group of 2×2 real matrices which preserve the symplectic form. (A matrix "preserves the symplectic form" if and only if it is symplectic.) Dym and McKean (1972:275) also define the one-parameter subgroup $SO(2)$ of matrices of the form given by equation 3.112. Thus fractional Fourier transforms as a group are isomorphic to the group denoted as $SO(2)$.

As we have noted on several occasions, the above correspondences are true only to the extent that we are willing to be flexible with $\pm$ signs. While it is beyond the scope of this book to present a more precise formulation (Guillemin and Sternberg 1984, Folland 1989), we mention that linear canonical transforms more precisely constitute a metaplectic group $Mp(2, \mathbf{R})$, rather than a symplectic group. The group $Mp(2, \mathbf{R})$ is what is known as a *double cover* (or *twofold cover*) of the group $Sp(2, \mathbf{R})$. Linear canonical transforms are not strictly isomorphic to the symplectic group, but are rather *locally isomorphic*.

An example of the application of linear canonical transforms and group theoretical methods to optical imaging and image processing may be found in Seger and Lenz 1992 and Seger 1993.

3.5   Generalization to two and higher dimensions

Most of the concepts presented in this chapter can be generalized to two and higher dimensions in a straightforward manner. Here we will present generalizations of only some of the more important concepts. The Wigner distribution of $f(u, v)$ is defined as

$$W_f(u, v; \mu, \nu) = \iint f(u + u'/2, v + v'/2) f^*(u - u'/2, v - v'/2) e^{-i2\pi(\mu u' + \nu v')} \, du' \, dv',$$

$$(3.180)$$

in two dimensions and similarly in higher dimensions. The separable two-dimensional linear canonical transform of $f(u, v)$ is defined as

$$f_\mathbf{M}(u, v) = \iint C_\mathbf{M}(u, v; u', v') f(u', v') \, du' \, dv', \qquad (3.181)$$

$$C_\mathbf{M}(u, v; u', v') = C_{\mathbf{M}_u}(u, u') C_{\mathbf{M}_v}(v, v').$$

Here $\mathbf{M}$ is a four-dimensional parameter matrix given by

$$\mathbf{M} = \begin{bmatrix} A_u & 0 & B_u & 0 \\ 0 & A_v & 0 & B_v \\ C_u & 0 & D_u & 0 \\ 0 & C_v & 0 & D_v \end{bmatrix}. \qquad (3.182)$$

Thus defined, we may multiply matrices associated with two transforms and find the matrix associated with their concatenation. The Wigner distribution of $f_M(u,v)$ is related to that of $f(u,v)$ according to

$$W_{f_M}(u,v;\mu,\nu) = W_f(D_u u - B_u \mu, D_v v - B_v \nu; -C_u u + A_u \mu, -C_v v + A_v \nu). \quad (3.183)$$

It is also possible to define nonseparable linear canonical transforms in which the cross terms are not zero (Folland 1989).

Just as Hankel transforms correspond to two-dimensional Fourier transforms under circular symmetry, it is possible to derive transforms which correspond to two-dimensional linear canonical transforms under circular symmetry. Such transforms are discussed in Moshinsky, Seligman, and Wolf 1972 and Zalevsky, Mendlovic, and Lohmann 1998.

3.6 Further reading

General tutorial references on time-frequency representations with a large number of additional references include Hlawatsch and Boudreaux-Bartels 1992, Cohen 1989, 1995, Flandrin 1993, and Qian and Chen 1996. A classic exposition of the Wigner distribution is Claasen and Mecklenbräuker 1980a, b, c. A general reference which also includes time-frequency transformations is *The Transforms and Applications Handbook* 2000. Time- and space-frequency representations are widely used in physics, and in fact have their origin in physics. For such a perspective, see Hillery and others 1984.

Chapter 9 of Wolf 1979 is an excellent discussion of linear canonical transforms.

4

The Fractional Fourier Transform

4.1 Definitions of the fractional Fourier transform

It is possible to define the fractional Fourier transform in several different ways. Any of these definitions can be taken as a starting point, and the others then derived as properties. Rather than biasing our presentation in favor of one definition over the others, we will present all of them, and show that they are equivalent to each other. This will enable us to obtain a more complete understanding of the transform. Furthermore, the different definitions lead to different physical interpretations which become useful in a variety of applications, and readers with different backgrounds may feel more comfortable with different definitions. While definitions A, B, and C are essential, definitions D, E, and F may be omitted at first reading.

We have already seen in chapter 3 that fractional Fourier transforms are a one-parameter subclass of the class of linear canonical transforms. As such, most of the properties of fractional Fourier transforms are special cases of those of linear canonical transforms. In this chapter we will treat the important special case of fractional Fourier transforms in greater depth than we treated the broader class of linear canonical transforms in chapter 3.

We precede the definitions by introducing some notation and general assumptions. The ath order fractional Fourier transform of the function $f(u)$ will be denoted in any of the following ways, depending on the context and requirements of clarity. Most commonly, we will simply denote the fractional transform by $f_a(u)$ or equivalently $\mathcal{F}^a f(u)$. The latter expression may be interpreted in two ways, either of which amounts to the same. First, we may interpret it as the operator $\mathcal{F}^a$ acting on the abstract signal f, the result of which is expressed in the u domain:

$$f_a(u) \equiv \mathcal{F}^a f(u) \equiv (\mathcal{F}^a f)(u) \equiv \mathcal{F}^a[f](u) \equiv (\mathcal{F}^a[f])(u). \qquad (4.1)$$

Second, we may interpret $\mathcal{F}^a f(u)$ as the operator $\mathcal{F}^a$ acting on the function $f(u)$, with the result again being expressed in the u domain:

$$f_a(u) \equiv \mathcal{F}^a f(u) \equiv \mathcal{F}^a[f(u)](u) \equiv (\mathcal{F}^a[f(u)])(u). \qquad (4.2)$$

This second interpretation is appropriate regardless of whether the operator $\mathcal{F}^a$ denotes a system or a transformation, whereas the first is appropriate only when $\mathcal{F}^a$ denotes a system. We have used the same dummy variable u both for the original function in the time (or space) domain, and its fractional Fourier transform. Both explicitly appear in the last two forms in equation 4.2, whereas either is absent in the first two. In equation 4.1 only the argument of the fractional Fourier transform is shown. In equation 4.2 the argument of the fractional Fourier transform is implicit in the expression $\mathcal{F}^a f(u)$; it is understood that it is the same as the argument of the original function $f(u)$.

The last form in equation 4.2 is the most explicit, but it is often possible to employ the simpler ones without confusion. When the fractional Fourier transform operator is interpreted as a system acting on an input signal f, it is possible to suppress both dummy variables and write

$$f_a \equiv \mathcal{F}^a f \equiv \mathcal{F}^a[f]. \tag{4.3}$$

The time-domain representation of the abstract signal f_a is given by any of the forms in equation 4.1 or 4.2. We will refer to $\mathcal{F}^a[\cdot]$, or simply $\mathcal{F}^a$, as the ath order fractional Fourier transform operator. This operator transforms a signal f or a function $f(\cdot)$ into its fractional Fourier transform f_a or $f_a(\cdot)$, respectively. (The notation $f(\cdot)$ will be used when we want to explicitly denote the representation of the signal f in a particular domain expressed as a function of a single variable, but do not want to specify a particular dummy variable.)

Sometimes, especially when the fractional Fourier transform operation is interpreted as a transformation from one representation of a signal to another, it will be useful to distinguish the argument of the transformed function from that of the original function. In this case we will let u_a denote the argument of the ath order fractional Fourier transform: $f_a(u_a) = (\mathcal{F}^a[f(u)])(u_a)$. We will see that with this convention, u_0 corresponds to u, the time (or space) coordinate, and u_1 corresponds to the temporal (or spatial) frequency coordinate. We will also have $u_2 = -u_0$ and $u_3 = -u_1$.

We will restrict ourselves to the case where the order parameter a is a real number. Complex-ordered fractional Fourier transforms may be treated as a special case of complex-parametered linear canonical transforms (Wolf 1979). We generally assume that f is a finite-energy signal which is well behaved in the sense described in chapter 2. We also deal with sets of signals which do not satisfy this assumption but serve as useful intermediaries in our formulations. As before, we interpret u as a dimensionless variable.

4.1.1 Definition A: Linear integral transform

The first definition we present is the most direct and concrete one, although it will not be immediately evident why this transform deserves to be called the fractional Fourier transform. We define the transform by explicitly specifying its linear transform kernel.

Definition A: The ath order fractional Fourier transform is a linear operation defined by the integral

$$f_a(u) \equiv \int_{-\infty}^{\infty} K_a(u, u') f(u') \, du', \tag{4.4}$$

$$K_a(u, u') \equiv A_\alpha \exp\left[i\pi(\cot \alpha \, u^2 - 2\csc \alpha \, uu' + \cot \alpha \, u'^2)\right],$$

$$A_\alpha \equiv \sqrt{1 - i \cot \alpha}, \qquad \alpha \equiv \frac{a\pi}{2}$$

when $a \neq 2j$ and $K_a(u, u') \equiv \delta(u - u')$ when $a = 4j$ and $K_a(u, u') \equiv \delta(u + u')$ when $a = 4j \pm 2$, where j is an integer. The ath order transform is sometimes referred to as the αth order transform, a practice which will occasionally be found convenient when no confusion can arise. The square root is defined such that the argument of the result lies in the interval $(-\pi/2, \pi/2]$. For $0 < |a| < 2$ $(0 < |\alpha| < \pi)$, A_α can be rewritten without ambiguity as

$$A_\alpha = \frac{e^{-i[\pi \operatorname{sgn}(\alpha)/4 - \alpha/2]}}{\sqrt{|\sin \alpha|}}, \tag{4.5}$$

where $\operatorname{sgn}(\cdot)$ is the sign function. When a is outside the interval $0 \leq |a| \leq 2$, we need simply replace a by its modulo 4 equivalent lying in this interval and use this value in equation 4.5.

At first sight, this definition does not offer much insight into the nature of the fractional Fourier transform, unless one is very well versed with the class of linear canonical transforms, of which fractional Fourier transforms are easily seen to constitute a one-parameter subclass upon comparison with equation 3.66 (a slight discrepancy will be discussed on page 125). Nevertheless, equation 4.4 is the most direct way of defining the transform. To obtain the ath order fractional Fourier transform of a function $f(u)$, we simply substitute it in equation 4.4. (Note that the parameter α here does not correspond to that appearing in equation 3.66. The α appearing in equation 3.66 corresponds to the $\cot \alpha$ appearing here. We will remain with this unfortunate use of symbols to maintain consistency with the literature.)

The transform is by definition linear, but it is not shift-invariant (unless $a = 4j$), since the kernel is not a function of $(u - u')$ only. Let us first examine the case where a is an integer. Letting j denote an arbitrary integer, we immediately note that by definition $\mathcal{F}^{4j}$ and $\mathcal{F}^{4j \pm 2}$ correspond to the identity operator $\mathcal{I}$ and the parity operator $\mathcal{P}$ respectively. For $a = 1$ we find $\alpha = \pi/2$, $A_\alpha = 1$, and

$$f_1(u) = \int_{-\infty}^{\infty} e^{-i2\pi u u'} f(u') \, du'. \tag{4.6}$$

We see that $f_1(u)$ is equal to the ordinary Fourier transform of $f(u)$, which until now we denoted as $F(u)$. Likewise, it is possible to see that $f_{-1}(u)$ is the ordinary inverse Fourier transform of $f(u)$. It is further possible to conclude that the above definition of the fractional Fourier transform is consistent with our definition of integer powers of the Fourier transform in section 2.5.1. Since $\alpha = a\pi/2$ appears in equation 4.4 only in the argument of trigonometric functions, the definition is periodic in a (or α) with period 4 (or 2π). Thus we will often limit our attention to the interval $a \in (-2, 2]$ (or $\alpha \in (-\pi, \pi]$), and sometimes $a \in [0, 4)$ (or $\alpha \in [0, 2\pi)$). These facts can be restated in operator notation:

$$\mathcal{F}^0 = \mathcal{I}, \tag{4.7}$$

$$\mathcal{F}^1 = \mathcal{F}, \tag{4.8}$$

$$\mathcal{F}^2 = \mathcal{P}, \tag{4.9}$$

$$\mathcal{F}^3 = \mathcal{F}\mathcal{P} = \mathcal{P}\mathcal{F}, \tag{4.10}$$

$$\mathcal{F}^4 = \mathcal{F}^0 = \mathcal{I}, \tag{4.11}$$

$$\mathcal{F}^{4j+a} = \mathcal{F}^{4j'+a}, \tag{4.12}$$

where j, j' are arbitrary integers.

According to equation 4.4, the zeroth-order transform of a function is equal to the function itself by definition. Likewise, the $\pm$2nd order transform is equal to $f(-u)$ by definition. This piecewise definition would be rather artificial if it did not exhibit some kind of continuity with respect to a for all values of a. It is not difficult to see by examining the kernel that a slight change in a results in only a slight change in $f_a(u)$ when a is not close to an integer multiple of 2. To see that this is true when a approaches an integer multiple of 2 as well, we first consider the behavior of the kernel as $a \to 0$. For infinitesimal $|a| > 0$ the kernel can be rewritten as

$$K_a(u, u') = \frac{e^{-i\pi\operatorname{sgn}(\alpha)/4}}{\sqrt{|\alpha|}} \exp[i\pi(u - u')^2/\alpha]. \tag{4.13}$$

Now, using equation 2.6 this is seen to indeed reduce to $\delta(u - u')$ in the limit $a \to 0$. Alternatively, noting that $K_a(u, u')$ is a function only of $(u - u')$, we can define the function

$$K_a(u) \equiv \frac{e^{-i\pi\operatorname{sgn}(\alpha)/4}}{\sqrt{|\alpha|}} \exp(i\pi u^2/\alpha), \tag{4.14}$$

which is convolved with $f(u)$ to obtain $K_a(u) * f(u) = f_a(u)$. The Fourier transform of $K_a(u)$, which is given in table 2.6, property 6 as $\exp(-i\pi\alpha\mu^2)$, approaches unity as α approaches zero, which in turn implies that $K_a(u)$ approaches a delta function. Thus we see that the definition of the transform is indeed continuous with respect to a around $a = 0$. A similar discussion is possible when a approaches other integer multiples of 2. A rigorous discussion of the continuity of the transform with respect to the order parameter may be found in McBride and Kerr 1987.

We now discuss a very important property of the fractional Fourier transform operator, the index additivity property, which can be stated in the alternative forms

$$\mathcal{F}^{a_1}\left[\mathcal{F}^{a_2}[f]\right] = \mathcal{F}^{a_1+a_2}[f] = \mathcal{F}^{a_2}\left[\mathcal{F}^{a_1}[f]\right],$$

$$\mathcal{F}^{a_1}\mathcal{F}^{a_2}f = \mathcal{F}^{a_1+a_2}f = \mathcal{F}^{a_2}\mathcal{F}^{a_1}f,$$

$$\mathcal{F}^{a_1}\mathcal{F}^{a_2} = \mathcal{F}^{a_1+a_2} = \mathcal{F}^{a_2}\mathcal{F}^{a_1}. \tag{4.15}$$

This can be proved by repeated application of equation 4.4, a process which is complicated by the square root appearing in the coefficient A_α. This process amounts to showing

$$\int K_{a_2}(u, u'')K_{a_1}(u'', u')\, du'' = K_{a_1+a_2}(u, u') \tag{4.16}$$

by direct integration, which can be accomplished by using Gaussian integrals. We do not present this proof, which may be found in McBride and Kerr 1987, since definitions B and C both allow much simpler proofs.

For instance, the 0.3rd fractional Fourier transform of the 0.6th transform is the 0.9th transform. Repeated application allows generalization to any number of consecutive transformations. For instance, the 1.7th transform of the 2.5th transform of the 2nd transform is the 6.2th transform (which is the same as the 2.2th (or -1.8th) transform. Transforms of different orders commute with each other so that their order can be freely interchanged. From the index additivity property, we deduce that the inverse $(\mathcal{F}^a)^{-1}$ of the ath order fractional Fourier transform operator $\mathcal{F}^a$ is simply equal to the operator $\mathcal{F}^{-a}$ (because $\mathcal{F}^{-a}\mathcal{F}^a = \mathcal{I}$). (This can also be shown by directly demonstrating that $\int K_a(u, u'')K_{-a}(u'', u')\, du'' = \delta(u - u')$, so that $K_a^{-1}(u, u') = K_{-a}(u, u')$.) Thus we see that we can freely manipulate the order parameter a as if it denoted a power of the Fourier transform operator $\mathcal{F}$.

Since fractional Fourier transforms are linear canonical transforms, they also satisfy the associative property, as well as other properties of linear canonical transforms. In particular, fractional Fourier transforms are unitary, as we can see by examining the kernel of the inverse transform obtained by replacing a with $-a$:

$$K_a^{-1}(u, u') = K_{-a}(u, u') = K_a^*(u, u') = K_a^*(u', u) = K_a^{\mathrm{H}}(u, u'). \qquad (4.17)$$

Note that the kernel $K_a(u, u')$ is symmetric, but not Hermitian. Unitarity implies that the fractional Fourier transform can be interpreted as a transformation from one representation to another, and that inner products and norms are not changed under the transform(ation).

We are now in a position to offer our first interpretation of the fractional Fourier transform. Let us concentrate on the interval $0 \leq a \leq 1$. We know that when $a = 0$ the fractional Fourier transform is the original function and when $a = 1$ it is the ordinary Fourier transform. As a varies from 0 to 1, the transform evolves smoothly from the original function to the ordinary Fourier transform. The reader may look forward to figures 4.3, 4.4, and 4.5 for a graphic illustration of the evolution of the $\mathrm{rect}(u)$ function into the $\mathrm{sinc}(u)$ function. The fact that the fractional Fourier transform interpolates between the original function and its ordinary Fourier transform with the continuous parameter a, offers some justification for its name.

Almeida (1994) has noted that since chirp functions such as $\exp(i\pi u^2)$ have constant magnitude, it is possible to make a rather general statement about the existence of the fractional Fourier transform. If $f(u)$ is a member of the function space L_2 or is a generalized function, its product with $\exp(i\pi u^2)$ is also in L_2 or is also a generalized function. Therefore, the fractional Fourier transform of $f(u)$ exists under the same conditions under which its Fourier transform exists (Almeida 1994). For further details on such matters, the reader is referred to McBride and Kerr 1987, Kerr 1988a, b, and Zayed 1998c.

Some final comments are in order before we move on to the other definitions. Remember that the transform can be interpreted as a system mapping "input" functions $f(u)$ to "output" functions $f_a(u)$. In this interpretation we use the same dummy variable u for both the inputs and the outputs. We will usually find this notation to be the most convenient. On the other hand, the functions $f_a(\cdot)$ for different values of a may be interpreted as different representations of the same abstract signal f, and we may speak of $f_a(\cdot)$ as being the representation of the signal f in the ath order *fractional Fourier domain*. In this case it is often useful to distinguish the variables associated with each domain by labelling them as u_a. Thus $f_a(u_a)$ is the representation in the ath domain, $f_0(u_0)$ is the representation in the time domain, and $f_1(u_1)$ is the frequency-domain representation. The axis u_a may be referred to as the ath fractional Fourier domain, so that u_0 and u_1 are the conventional time and frequency domains u and μ. The representation of the signal in the a'th domain can be obtained from its representation in the ath domain through an $(a' - a)$th order fractional Fourier transformation:

$$f_{a'}(u_{a'}) = \int K_{a'-a}(u_{a'}, u_a) f_a(u_a) \, du_a. \tag{4.18}$$

The concept of fractional Fourier domains will be further discussed later in this chapter.

4.1.2   Definition B: Fractional powers of the Fourier transform

Now, we will actually define the ath order fractional Fourier transform operation as the ath fractional power of the Fourier transform operation. We have already discussed functions of operators in section 2.4.4. We will use equation 2.145 as the basis of our definition. This was the motivation of Namias (1980a). Kober (1939) had obtained the fractional Fourier transform by an essentially equivalent approach.

> **Definition B:** Let ψ_l (or $\psi_l(u)$) denote the Hermite-Gaussian signals (or functions) which are eigensignals (or eigenfunctions) of the ordinary Fourier transform operation with respective eigenvalues λ_l, and which are known to constitute an orthonormal basis for the space of well-behaved finite-energy signals (functions). The fractional Fourier transform operation is defined to be linear and to satisfy
>
> $$\mathcal{F}^a \psi_l = \lambda_l^a \psi_l = \left(e^{-il\pi/2} \right)^a \psi_l = e^{-ial\pi/2} \psi_l, \tag{4.19}$$
>
> or
>
> $$\mathcal{F}^a \psi_l(u) = \lambda_l^a \psi_l(u) = \left(e^{-il\pi/2} \right)^a \psi_l(u) = e^{-ial\pi/2} \psi_l(u). \tag{4.20}$$

This statement completely defines the fractional Fourier transform by specifying its eigenfunctions and eigenvalues. The definition depends on the particular set of eigenfunctions that have been chosen as well as on the particular way in which the ath powers of the eigenvalues λ_l are chosen. Different choices would lead to different definitions. We should also note that, strictly speaking, the operator $\mathcal{F}^a$ is not the function $(\cdot)^a$ of the ordinary Fourier operator in the sense of section 2.4.4. These matters will be carefully discussed in section 4.3.

Several properties of the fractional Fourier transform discussed in terms of definition A are much more easily deduced from the present definition, such as the special cases $a = 0$, $a = 1$, and the index additivity property. The latter can be shown by applying $\mathcal{F}^{a'}$ to both sides of equation 4.20.

To find the fractional transform of a given function $f(u)$, we first expand it as a linear superposition of the eigenfunctions of the fractional Fourier transform:

$$f(u) = \sum_{l=0}^{\infty} C_l \psi_l(u) \tag{4.21}$$

$$C_l = \int \psi_l(u') f(u') \, du'.$$

Applying $\mathcal{F}^a$ to both sides of equation 4.21 and using equation 4.20 we obtain

$$\mathcal{F}^a f(u) = \sum_{l=0}^{\infty} e^{-ial\pi/2} C_l \psi_l(u) = \int \sum_{l=0}^{\infty} e^{-ial\pi/2} \psi_l(u) \psi_l(u') f(u') \, du'. \tag{4.22}$$

Upon comparison with equation 4.4, the kernel $K_a(u, u')$ can be identified as

$$K_a(u, u') = \sum_{l=0}^{\infty} e^{-ial\pi/2} \psi_l(u) \psi_l(u'). \tag{4.23}$$

This is called the spectral expansion (or singular-value decomposition) of the kernel of the fractional Fourier transform, and should be compared with equation 2.184, which is a special case of the above with $a = 1$. The kernel given in equation 4.23 can be directly shown to be identical to that given in equation 4.4 by using property 9 in table 2.8. This demonstrates the equivalence of definitions A and B.

It is also possible to directly show that the Hermite-Gaussian functions are indeed eigenfunctions of the transform defined in equation 4.4 with the eigenvalues given in equation 4.20 (McBride and Kerr 1987). This can be established by induction. First, we can show that $\psi_0(u)$ and $\psi_1(u)$ are eigenfunctions with eigenvalues 1 and $\exp(-ia\pi/2)$, by evaluating the complex exponential integrals. (Such integrals have been reviewed in section 2.10.2.) Then, by using the recurrence relations given in table 2.8, it is possible to *assume* that the result to be shown holds for $l-1$ and l, and to show that as a consequence it holds for $l+1$. This completes the induction. The recurrence relations used are

$$\psi_{l+1}(u) = 2\sqrt{2\pi} \frac{A_{l+1}}{A_l} u \psi_l(u) - 2l \frac{A_{l+1}}{A_{l-1}} \psi_{l-1}(u),$$

$$\frac{d\psi_l(u)}{du} = \frac{A_l}{A_{l-1}} 2l \sqrt{2\pi} \, \psi_{l-1}(u) - 2\pi u \psi_l(u), \tag{4.24}$$

where $A_l = 2^{1/4}/\sqrt{2^l l!}$. These are derived from the corresponding Hermite polynomial recurrence relations

$$H_{l+1}(u) = 2u H_l(u) - 2l H_{l-1}(u),$$

$$\frac{dH_l(u)}{du} = 2l H_{l-1}(u). \tag{4.25}$$

Also needed in the course of these calculations is the property

$$\mathcal{F}^a[u f(u)](u) = \left[\cos\alpha \, u + i \sin\alpha \, (2\pi)^{-1} \frac{d}{du} \right] f_a(u), \tag{4.26}$$

which readily follows from definition A (equation 4.4) if one differentiates $f_a(u)$ with respect to u.

The above demonstration of the fact that Hermite-Gaussian functions are eigenfunctions of the fractional Fourier transform as defined by equation 4.4, reduces to the well-known fact that Hermite-Gaussian functions are eigenfunctions of the ordinary Fourier transform when $a = 1$, since equation 4.4 reduces to the definition of the ordinary Fourier transform when $a = 1$.

Of course, once we have shown that the Hermite-Gaussian functions are eigenfunctions of the transform with kernel $K_a(u, u')$ given by equation 4.4, we may immediately write the spectral expansion of this kernel as equation 4.23, as this expansion follows directly from a knowledge of the complete set of orthonormal eigenfunctions and eigenvalues. This procedure amounts to a derivation of property 9 in table 2.8.

It is also instructive to demonstrate index additivity (equation 4.16) directly from the spectral expansion (equation 4.23) by using the orthonormality of the Hermite-Gaussian functions, a task we leave to the reader.

4.1.3   Definition C: Rotation in the time-frequency plane

We have already mentioned that the fractional Fourier transform is a linear canonical transform, a fact which was evident upon examination of its kernel as defined in equation 4.4. We now define the fractional Fourier transform directly as a special one-parameter subclass of the class of linear canonical transforms.

> **Definition C.1:** The ath order fractional Fourier transform is a linear canonical transform defined by the transform matrix
>
> $$\mathbf{M} = \begin{bmatrix} A & B \\ C & D \end{bmatrix} \equiv \begin{bmatrix} \cos\alpha & \sin\alpha \\ -\sin\alpha & \cos\alpha \end{bmatrix}, \qquad (4.27)$$
>
> where $\alpha = a\pi/2$.

We immediately recognize the defining matrix as the two-dimensional rotation matrix in the time-frequency plane. The effect of a linear canonical transform on the Wigner distribution of a function was stated in equation 3.81. Thus this definition can also be stated as follows.

> **Definition C.2:** The ath order fractional Fourier transform corresponds to rotation of the Wigner distribution of the signal (or function) in the clockwise direction by angle $\alpha = a\pi/2$ in the time-frequency plane:
>
> $$W_{f_a}(u, \mu) \equiv W_f(u\cos\alpha - \mu\sin\alpha, u\sin\alpha + \mu\cos\alpha). \qquad (4.28)$$

This latter version of definition C is slightly weaker than the other definitions in that it defines the fractional Fourier transform up to a unit magnitude complex constant. Apart from this, definition C.2 is equivalent to definition C.1 by virtue of equation 3.81.

When $a = 0$ ($\alpha = 0$) we simply obtain the identity matrix and the identity operation. When $a = 1$ ($\alpha = \pi/2$) we obtain the matrix given in equation 3.110 and the ordinary Fourier transform. In this case the Wigner distribution is rotated by $\alpha = \pi/2$, consistent with property 16 in table 3.2. Indeed, interpolating this $\pi/2$ rotation was the motivation of Lohmann (1993b), who gave definition C.2 as the primary definition of the transform. The index additivity property follows immediately from the angle additivity of the rotation matrix.

Definition C.1 can be alternately phrased in terms of the parameters α, β, γ appearing in the definition of linear canonical transforms (equation 3.66), rather than specifying A, B, C, D as we have done in equation 4.27. This can be slightly confusing since we use the parameter α appearing in equation 3.66 for a different purpose in this chapter: $\alpha = a\pi/2$. Bearing this in mind, and using the relations between these two sets of parameters (equation 3.74), equation 4.27 can be equivalently written as

$$\alpha = \gamma = \cot \alpha,$$
$$\beta = \csc \alpha. \tag{4.29}$$

The α appearing on the left-hand side is the same parameter appearing in equation 3.66, whereas that appearing on the right-hand side is the one defined in this chapter as $a\pi/2$. In this chapter we will use α solely to denote $a\pi/2$.

Definition C differs from the previous ones by a unit magnitude complex constant. This situation is somewhat inelegant, though usually not harmful. The discrepancy arises because we wish to comply with the commonly accepted definition of both linear canonical transforms (for instance, as given by Wolf 1979) and fractional Fourier transforms (for instance, as given by McBride and Kerr 1987). (If we were to allow ourselves to deviate from either of these commonly accepted definitions, we could redefine the complex constant appearing in front of either linear canonical transforms or fractional Fourier transforms to make them consistent with each other.) This is hardly ever an issue in applications, since complex constants of unit and nonunit magnitude are commonly of no major consequence, and it is common to refer to an integral differing from the mathematical definitions by a complex constant as still being a linear canonical or fractional Fourier transform anyway. Nevertheless, the precise relationship between the fractional Fourier transform $\mathcal{F}^a$ as we use it throughout this book, and the linear canonical transform $\mathcal{C}_{\mathbf{M}}$ with matrix $\mathbf{M}$ as given in definition C.1 is

$$\mathcal{C}_{\mathbf{M}} = e^{-ia\pi/4}\mathcal{F}^a, \tag{4.30}$$

for $-2 \leq a \leq 2$.

The equivalence of definition C to definition A within the factor $e^{-ia\pi/4}$ is established easily. Definition C.1 defines the fractional Fourier transform as a linear canonical transform by specifying its transform matrix. The equivalent set of parameters α, β, γ have already been obtained in equation 4.29. These allow us to immediately obtain equation 4.4 by using equation 3.66. Alternatively, we can start from definition A, identify the parameters α, β, γ, and then find the corresponding matrix coefficients by using equation 3.74, and finally see that these are the same as given in equation 4.27.

These are sufficient to establish the equivalence of all three definitions. However, we will point out one of the more instructive direct relationships between definitions B and C. Naturally, we expect substitution of the kernel of the fractional Fourier transform into equation 3.44 to yield

$$K_M(u, \mu; u', \mu') = \delta(u' - \cos\alpha\, u + \sin\alpha\, \mu)\, \delta(\mu' - \sin\alpha\, u - \cos\alpha\, \mu), \quad (4.31)$$

which is essentially a restatement of definition C.2 (also see equation 3.88). Although this can be demonstrated by using the kernel as given by equation 4.4, it is much more instructive to use the kernel as given by equation 4.23 and use the orthonormality of the Hermite-Gaussian functions, a task we leave to the reader.

Within sign changes, the fractional Fourier transform is the only linear canonical transform which commutes with the Fourier transform. This can be established by showing that the only transform matrix that commutes with the transform matrix of the Fourier transform is a rotation matrix. Geometrically, this means that the only linear distortions in the time-frequency plane which commute with rotation by $\pi/2$ are rotations by arbitrary angles.

4.1.4 Definition D: Transformation of coordinate multiplication and differentiation operators

This approach is essentially due to Condon (1937). Wolf (1979) uses a similar approach to define linear canonical transforms. Yurke, Schleich, and Walls (1990) derive the same transform kernel in a very similar manner, though without being aware that what they obtain is the fractional Fourier transform.

It will be useful to recall the following properties of the $\mathcal{U}$ and $\mathcal{D}$ operators in the time and frequency domains:

$$\mathcal{U}f(u) = uf(u), \quad (4.32)$$

$$\mathcal{D}f(u) = \frac{1}{i2\pi}\frac{d}{du}f(u), \quad (4.33)$$

$$-\mathcal{U}F(\mu) = \frac{1}{i2\pi}\frac{d}{d\mu}F(\mu), \quad (4.34)$$

$$\mathcal{D}F(\mu) = \mu F(\mu). \quad (4.35)$$

Throughout this section we will use u_a as the variable associated with the ath fractional Fourier domain representation $f_a(u_a)$ of a signal f. As usual, u will denote the variable associated with the ordinary time domain.

Definition D.1: The two operators $\mathcal{U}_a$ and $\mathcal{D}_a$ are defined as

$$\begin{bmatrix} \mathcal{U}_a \\ \mathcal{D}_a \end{bmatrix} \equiv \begin{bmatrix} \cos\alpha & \sin\alpha \\ -\sin\alpha & \cos\alpha \end{bmatrix} \begin{bmatrix} \mathcal{U} \\ \mathcal{D} \end{bmatrix}, \qquad \alpha \equiv \frac{a\pi}{2}, \quad (4.36)$$

such that $\mathcal{U}_0 = \mathcal{U}$, $\mathcal{U}_1 = \mathcal{D}$, $\mathcal{D}_0 = \mathcal{D}$, $\mathcal{D}_1 = -\mathcal{U}$. Now, we define the fractional Fourier domain representation $f_a(u_a)$ of a signal f such that it satisfies the properties

$$\mathcal{U}_a f_a(u_a) = u_a f_a(u_a),$$

$$\mathcal{D}_a f_a(u_a) = \frac{1}{i2\pi}\frac{d}{du_a}f_a(u_a), \quad (4.37)$$

which are generalizations of equations 4.32, 4.35 and 4.33, 4.34. The operator $\mathcal{U}_a$ corresponds to multiplication by the coordinate variable and the operator $\mathcal{D}_a$ corresponds to differentiation with respect to the coordinate variable in the ath order fractional Fourier domain representation. The unitary transformation from the time-domain representation $f(u)$ to the fractional Fourier domain representation $f_a(u_a)$ is the fractional Fourier transform operation.

Before showing that this definition is equivalent to earlier ones, we consider an alternative approach which will provide a different perspective. The inverse of the kernel $K_0(u_0, u) = \delta(u_0 - u)$ of the identity operation $\mathcal{I}$, interpreted as a function of u, is the eigenfunction of the coordinate multiplication operator $\mathcal{U}$ with eigenvalue u_0, and the inverse of the kernel $K_1(u_1, u) = \exp(-i2\pi u_1 u)$ of the Fourier transform operation $\mathcal{F}$, interpreted as a function of u, is the eigenfunction of the differentiation operator $\mathcal{D}$ with eigenvalue u_1. That is,

$$\mathcal{U}K_0^{-1}(u, u_0) = u_0 K_0^{-1}(u, u_0), \tag{4.38}$$

$$\mathcal{D}K_1^{-1}(u, u_1) = u_1 K_1^{-1}(u, u_1). \tag{4.39}$$

It is also possible to verify the duals of these equations:

$$-\mathcal{D}K_0^{-1}(u, u_0) = \frac{1}{i2\pi}\frac{d}{du_0}K_0^{-1}(u, u_0), \tag{4.40}$$

$$\mathcal{U}K_1^{-1}(u, u_1) = \frac{1}{i2\pi}\frac{d}{du_1}K_1^{-1}(u, u_1). \tag{4.41}$$

Equations 4.38 and 4.39 suggest an association between the identity operator and the coordinate multiplication operator, and the Fourier transform operator and the differentiation operator. (Likewise, $-\mathcal{U}$ may be associated with the parity operator $\mathcal{P}$, and $-\mathcal{D}$ may be associated with the inverse Fourier transform operator, by writing equations similar to 4.38 and 4.39 for $K_2(u_2, u) = \delta(u_2 + u)$ and $K_{-1}(u_{-1}, u) = \exp(i2\pi u_{-1}u)$.) Referring back to section 2.6.1, the reader will recall that the kernel $K_0(u_0, u)$ or $K_1(u_1, u)$ of a transformation from one representation to another is simply the representation of the members of the old basis set in the new basis set. (Equivalently, the inverse kernel $K_0^{-1}(u, u_0)$ or $K_1^{-1}(u, u_1)$ is the representation of the members of the new basis set in the old basis set.) We have seen that the eigensignals of the operators $\mathcal{U}$ and $\mathcal{D}$, the impulse and harmonic sets respectively, constitute orthonormal basis sets. The representation of a signal with respect to the impulse set associated with the operator $\mathcal{U}$ is the time-domain representation and the representation of the signal with respect to the harmonic set associated with the operator $\mathcal{D}$ is the frequency-domain representation. The bottom line is that there is an association between the identity operator, the impulse basis, the time-domain representation of a signal, and the $\mathcal{U}$ operator; and likewise, an association between the ordinary Fourier transform operator, the harmonic basis, the frequency-domain representation of a signal, and the $\mathcal{D}$ operator.

Definition D.2: Again let $\mathcal{U}_a$ and $\mathcal{D}_a$ be defined by equation 4.36. Now, we associate the ath order fractional Fourier transform operator $\mathcal{F}^a$ with $\mathcal{U}_a$, in the same way as we associate the ordinary Fourier transform operator with $\mathcal{D}$ and the identity operator with $\mathcal{U}$. That is,

$$\mathcal{U}_a K_a^{-1}(u, u_a) = u_a K_a^{-1}(u, u_a),$$

$$-\mathcal{D}_a K_a^{-1}(u, u_a) = \frac{1}{i2\pi} \frac{d}{du_a} K_a^{-1}(u, u_a), \qquad (4.42)$$

which are generalizations of equations 4.38, 4.39 and 4.40, 4.41, where $K_a^{-1}(u, u_a)$ is interpreted as a function of u. In other words, we form an association between the ath order fractional Fourier domain representation $f_a(u_a)$ of a signal f, and the $\mathcal{U}_a$ operator, and the unitary transformation from the time-domain representation $f(u)$ to the fractional Fourier domain representation $f_a(u_a)$ is the fractional Fourier transform operation.

We now show the equivalence of definitions D.1 and D.2. Since $f_a(u_a)$ is related to $f(u)$ through the relations $f_a(u_a) = \int K_a(u_a, u) f(u) \, du$ or $f(u) = \int K_a^{-1}(u, u_a) f_a(u_a) \, du_a$, we can rewrite $\mathcal{U}_a f_a(u_a) = u_a f_a(u_a)$ (equation 4.37) in the time domain as

$$\mathcal{U}_a \int K_a^{-1}(u, u_a) f_a(u_a) \, du_a = \int K_a^{-1}(u, u_a) u_a f_a(u_a) \, du_a. \qquad (4.43)$$

The left-hand side is simply the operator $\mathcal{U}_a$ applied to $f(u)$. The right-hand side is simply the time-domain representation of $u_a f_a(u_a)$. We can now immediately see that $\mathcal{U}_a K_a^{-1}(u, u_a) = u_a K_a^{-1}(u, u_a)$ (equation 4.42) implies equation 4.43. Conversely, if equation 4.43 is true for all $f_a(u)$, then it follows that $\mathcal{U}_a K_a^{-1}(u, u_a) = u_a K_a^{-1}(u, u_a)$. We can also rewrite $\mathcal{D}_a f_a(u_a) = (i2\pi)^{-1} df_a(u_a)/du_a$ (equation 4.37) in the time domain as

$$\mathcal{D}_a \int K_a^{-1}(u, u_a) f_a(u_a) \, du_a = \int K_a^{-1}(u, u_a) \frac{1}{i2\pi} \frac{df_a(u_a)}{du_a} \, du_a. \qquad (4.44)$$

Integrating the right-hand side by parts and using the fact that the well-behaved finite-energy functions we deal with go to zero as their arguments go to infinity, we obtain

$$\mathcal{D}_a \int K_a^{-1}(u, u_a) f_a(u_a) \, du_a = \int \frac{-1}{i2\pi} \frac{dK_a^{-1}(u, u_a)}{du_a} f_a(u_a) \, du_a. \qquad (4.45)$$

Now, we can immediately see that $-\mathcal{D}_a K_a^{-1}(u, u_a) = (i2\pi)^{-1}(d/du_a) K_a^{-1}(u, u_a)$ (equation 4.42) implies equation 4.45. Conversely, if equation 4.45 is to hold for all $f_a(u)$, it follows that $-\mathcal{D}_a K_a^{-1}(u, u_a) = (i2\pi)^{-1}(d/du_a) K_a^{-1}(u, u_a)$. Thus we have established the equivalence of definitions D.1 and D.2.

There are a number of ways of showing that this constructive definition is equivalent to definition A. Using equation 4.36 and equations 4.42 we can write

$$\left(\cos\alpha \, u + \sin\alpha \, \frac{1}{i2\pi} \frac{d}{du} \right) K_a^{-1}(u, u_a) = u_a K_a^{-1}(u, u_a), \qquad (4.46)$$

$$-\left(-\sin\alpha \, u + \cos\alpha \, \frac{1}{i2\pi} \frac{d}{du} \right) K_a^{-1}(u, u_a) = \frac{1}{i2\pi} \frac{d}{du_a} K_a^{-1}(u, u_a), \qquad (4.47)$$

where $K_a^{-1}(u, u_a)$ is interpreted as a function of u. The solution of this pair of differential equations can be shown to be

$$K_a^{-1}(u, u_a) = (\text{constant}) \exp\left[-i\pi(\cot\alpha\, u^2 - 2\csc\alpha\, u u_a + \cot\alpha\, u_a^2)\right]. \qquad (4.48)$$

The angle additivity inherent in equation 4.36 translates into index additivity of the kernel (equation 4.16) within a constant factor. Requiring that this factor disappear is sufficient to uniquely determine the constant appearing in equation 4.48. The resulting $K_a^{-1}(u, u_a)$ turns out to be identical to the inverse of the kernel $K_a(u_a, u)$ appearing in equation 4.4, establishing the equivalence of the present definition to definition A. Alternatively, we could have imposed the condition that the transform be a linear canonical transform to obtain the slightly different kernel $C_{\mathbf{M}}(u_a, u)$ (see equation 4.30).

The relationship of the present definition to definition C is immediate from equation 4.36 in which the rotation matrix explicitly appears. Just as the time-domain representation is associated with the $\mathcal{U}$ operator and the horizontal u axis, and the frequency-domain representation is associated with the $\mathcal{D}$ operator and the vertical μ axis, the ath order fractional Fourier domain representation is associated with the $\mathcal{U}_a$ operator and the u_a axis making angle $\alpha = a\pi/2$ with the u axis.

Equation 4.36 is a special case of equation 3.162 so that we can also write

$$\mathcal{U}_a = \mathcal{F}^{-a}\mathcal{U}\mathcal{F}^a, \qquad (4.49)$$

$$\mathcal{D}_a = \mathcal{F}^{-a}\mathcal{D}\mathcal{F}^a, \qquad (4.50)$$

as in equation 3.161.

4.1.5 Definition E: Differential equation

Here the fractional Fourier transform $f_a(u)$ of a function $f_0(u)$ is defined as the solution of a differential equation, with $f_0(u)$ acting as the initial condition. This is the quantum-mechanical harmonic oscillator differential equation and also the equation governing optical propagation in quadratic graded-index media. (In the former case the order parameter a corresponds to time and in the latter case it corresponds to the coordinate along the direction of propagation.) The solution of this equation is well known and may be found in books on quantum mechanics (Cohen-Tannoudji, Diu, and Laloë 1977) and optical electronics (Yariv 1989, 1997). In fact, in some sources the solution is written in the form of an integral transform whose kernel is sometimes referred to as the harmonic oscillator Green's function, without the authors knowing that this is the fractional Fourier transform. (To be precise, we must note that the differential equations governing these physical phenomena differ slightly from the equation we discuss below; see page 133 for a discussion.)

Definition E: Consider the differential equation

$$\left[-\frac{1}{4\pi}\frac{\partial^2}{\partial u^2} + \pi u^2 - \frac{1}{2}\right]f_a(u) = i\frac{2}{\pi}\frac{\partial f_a(u)}{\partial a}, \qquad (4.51)$$

with the initial condition $f_0(u) = f(u)$. The solution $f_a(u)$ of the equation is the ath order fractional Fourier transform of $f(u)$.

It is possible to show by direct substitution that the solution of this equation is

$$f_a(u) = \int K_a(u, u') f_0(u') \, du', \tag{4.52}$$

with $K_a(u, u')$ being explicitly given by equation 4.4. Namias (1980a) solves equation 4.51 in two steps by substituting $f_a(u) = \int K_a(u, u') f_0(u') \, du'$ in equation 4.51 and obtaining a differential equation for $K_a(u, u')$:

$$\left[-\frac{1}{4\pi} \frac{\partial^2}{\partial u^2} + \pi u^2 - \frac{1}{2} \right] K_a(u, u') = i \frac{2}{\pi} \frac{\partial K_a(u, u')}{\partial a}, \tag{4.53}$$

with the initial condition $K_0(u, u') = \delta(u - u')$. The kernel Namias obtains again coincides with that in equation 4.4. This establishes the equivalence of this definition with definition A.

Alternatively, one may first seek the eigensolutions of equation 4.51 and then construct arbitrary solutions as linear superpositions of these eigensolutions. Substituting the eigensolution form $f_a(u) = \lambda_a f_0(u)$ in equation 4.51, we obtain

$$\frac{d^2 f_0(u)}{du^2} + 4\pi^2 \left(\frac{1}{2\pi} + \frac{i2}{\pi^2 \lambda_a} \frac{d\lambda_a}{da} - u^2 \right) f_0(u) = 0. \tag{4.54}$$

Comparing this equation with equation 2.176, we see that its solutions are the Hermite-Gaussian functions provided

$$\frac{2l + 1}{2\pi} = \frac{1}{2\pi} + \frac{i2}{\pi^2 \lambda_a} \frac{d\lambda_a}{da}, \tag{4.55}$$

which implies

$$\frac{d\lambda_a}{da} = -il \left(\frac{\pi}{2} \right) \lambda_a, \tag{4.56}$$

yielding

$$\lambda_a = e^{-ial\pi/2} \tag{4.57}$$

as the eigenvalue associated with the lth Hermite-Gaussian function. (We would have obtained equation 2.176 directly from equation 4.51 if we had assumed this form for λ_a to begin with.) This demonstrates the equivalence of the present definition with definition B.

We have shown that the eigensolutions of equation 4.51 are the Hermite-Gaussian functions; that is, if the initial condition $f_0(u)$ is $\psi_l(u)$, the solution $f_a(u)$ is $\exp(-ial\pi/2) \, \psi_l(u)$. Given an arbitrary initial condition $f_0(u)$, we can expand it in terms of the Hermite-Gaussian functions as

$$f_0(u) = \sum_{l=0}^{\infty} C_l \psi_l(u), \tag{4.58}$$

$$C_l = \int \psi_l(u) f_0(u) \, du.$$

Since equation 4.51 is linear, the solution corresponding to this initial condition is readily obtained as

$$f_a(u) = \sum_{l=0}^{\infty} C_l e^{-ial\pi/2} \psi_l(u), \tag{4.59}$$

from which one can obtain

$$f_a(u) = \int K_a(u, u') f_0(u') \, du', \tag{4.60}$$

$$K_a(u, u') = \sum_{l=0}^{\infty} e^{-ial\pi/2} \psi_l(u) \psi_l(u'),$$

in the same manner as with equation 4.23.

The reader may have noted that equation 4.51 is of the same form as equation 3.171. Specializing equation 3.171 to $\mathcal{H} = \pi(\mathcal{U}^2 + \mathcal{D}^2)$ and with $p = \alpha = a\pi/2$ we obtain

$$\pi(\mathcal{U}^2 + \mathcal{D}^2) f_a(u) = i \frac{\partial f_a(u)}{\partial(\pi a/2)}. \tag{4.61}$$

To see how this is related to equation 4.51, let us first determine the effect of $\mathcal{U}^2 + \mathcal{D}^2$ on $f_a(u)$. If we interpret $f_a(u)$ to represent different functions of u for different values of a, then we simply use $\mathcal{U} f_a(u) = u f_a(u)$ and $\mathcal{D} f_a(u) = (i2\pi)^{-1} \partial f_a(u)/\partial u$ to obtain

$$\left[-\frac{1}{4\pi} \frac{\partial^2}{\partial u^2} + \pi u^2 \right] f_a(u) = i \frac{2}{\pi} \frac{\partial f_a(u)}{\partial a}. \tag{4.62}$$

Alternatively, if we interpret $f_a(u_a)$ to stand for the representation of $f_0(u)$ in different fractional domains, then we must use $\mathcal{U}^2 f_a(u_a) = (\cos\alpha \ u_a - \sin\alpha \ (i2\pi)^{-1}\partial/\partial u_a)^2 f_a(u_a)$ and $\mathcal{D}^2 f_a(u_a) = (\sin\alpha \ u_a + \cos\alpha \ (i2\pi)^{-1}\partial/\partial u_a)^2 f_a(u_a)$, which again leads to equation 4.62. The fact that both interpretations are consistent is a necessary consequence of the invariance of the operator $\mathcal{U}_a^2 + \mathcal{D}_a^2$, which means that this operator has the same effect and the same kernel for all values of a. This rotational invariance is easily shown from equation 4.36: $\mathcal{U}_a^2 + \mathcal{D}_a^2 = \mathcal{U}^2 + \mathcal{D}^2$ for all a. This result relates the present definition to definition D and also more indirectly to definition C.

Equation 4.62 differs from equation 4.51 only by the term $(-1/2)f_a(u)$. We will see below in discussing definition F that this is the same discrepancy discussed on page 125.

It immediately follows from the specified initial condition of equation 4.51 that the zeroth-order transform is equal to the original function. The special case $a = 1$ follows from the fact that the eigenfunctions and eigenvalues of equation 4.51 coincide with those of the ordinary Fourier transform when $a = 1$. The periodicity with respect to a and the index additivity property follow from the corresponding properties of the eigenvalues as given by equation 4.57. Index additivity also follows from the form of equation 4.51 by noting that the term in square brackets on the left-hand side is independent of a. This will become clearer when we discuss definition F.

More on the relationship of the fractional Fourier transform to differential equations and their solutions is found in Namias 1980a, McBride and Kerr 1987, and Kerr

1988a, b. The fractional Fourier transformation is a useful technique for solving various kinds of differential equations. Just as ordinary Fourier (or sometimes Laplace) transformation enables one to convert differential equations into algebraic equations, taking the fractional Fourier transformation of certain differential equations enables one to convert them into a form which is easier to solve. Once the solution of this new equation is obtained, an inverse fractional Fourier transformation gives the solution of the original equation. The order parameter serves as a useful degree of freedom; we need not commit ourselves to a specific order before taking the fractional Fourier transformation of the original equation, we can select it afterwards to result in as much simplification as possible. For instance, in dealing with second-order differential equations, appropriate choice of the order allows the second-order term to disappear in the transformed equation, leaving us with a first-order equation which is easy to solve. While the ordinary Fourier transformation is useful in dealing with constant-coefficient equations (corresponding to time- or space-invariant systems), the fractional Fourier transform is able to handle a variety of equations whose coefficients are functions of the coordinate variable.

4.1.6 Definition F: Hyperdifferential operator

Lastly, we show how the fractional Fourier transform can be defined in hyperdifferential form.

> **Definition F:** The fractional Fourier transform is defined by the hyperdifferential operator
>
> $$\mathcal{F}^a = e^{-i(a\pi/2)\mathcal{H}}, \tag{4.63}$$
>
> $$\mathcal{H} = \pi\left(\mathcal{D}^2 + \mathcal{U}^2\right) - \frac{1}{2}.$$

In the time domain this becomes

$$\mathcal{F}^a f(u) = \exp\left[-i\left(\frac{a\pi}{2}\right)\left(-\frac{1}{4\pi}\frac{d^2}{du^2} + \pi u^2 - \frac{1}{2}\right)\right] f(u). \tag{4.64}$$

This definition is closely related to definition E. To see this, let us recall that the solution of the differential equation $\mathcal{H}f_a(u) = i(2/\pi)\partial f_a(u)/\partial a$ (equation 3.171) is given by $f_a(u) = \exp(-i(a\pi/2)\mathcal{H})f_0(u)$, where $f_0(u)$ serves as the initial or boundary condition. Thus we see that definition F is simply the solution of the differential equation given in definition E, expressed in hyperdifferential form. The operator $\mathcal{F}^a$ given in equation 4.63 generates $f_a(u)$ for all values of a from $f_0(u) = f(u)$.

A related derivation (Namias 1980a) demonstrates the equivalence of this definition to definition B, which was based on the eigenvalue equation

$$\mathcal{F}^a \psi_l(u) = e^{-ial\pi/2}\psi_l(u) = e^{-i\alpha l}\psi_l(u), \tag{4.65}$$

where $\alpha = a\pi/2$ and $\psi_l(u)$ are the Hermite-Gaussian functions satisfying the differential equation

$$\left[\frac{d^2}{du^2} + 4\pi^2\left(\frac{2l+1}{2\pi} - u^2\right)\right]\psi_l(u) = 0. \tag{4.66}$$

Now, starting from the last two equations, let us seek a hyperdifferential representation for $\mathcal{F}^a$ of the form $\exp(-i\alpha\mathcal{H})$. Differentiating

$$\exp(-i\alpha\mathcal{H})\psi_l(u) = e^{-i\alpha l}\psi_l(u) \tag{4.67}$$

with respect to α and setting $\alpha = 0$ we obtain

$$\mathcal{H}\psi_l(u) = l\psi_l(u), \tag{4.68}$$

which upon comparison with equation 4.66 leads to

$$\mathcal{H}\psi_l(u) = \left(-\frac{1}{4\pi}\frac{d^2}{du^2} + \pi u^2 - \frac{1}{2}\right)\psi_l(u). \tag{4.69}$$

By expanding arbitrary $f(u)$ in terms of the $\psi_l(u)$, we obtain

$$\mathcal{H}f(u) = \left(-\frac{1}{4\pi}\frac{d^2}{du^2} + \pi u^2 - \frac{1}{2}\right)f(u), \tag{4.70}$$

by virtue of the linearity of $\mathcal{H}$. Now, in abstract operator form, we may write

$$\mathcal{H} = \pi\left(\mathcal{D}^2 + \mathcal{U}^2\right) - \frac{1}{2}, \tag{4.71}$$

precisely corresponding to the present definition (equation 4.63).

The definition of $\mathcal{H}$ given in equation 4.63, which we saw is fully consistent with definition E, differs from $\mathcal{H}_{fF}$ given on page 111 by $-1/2$. This is the same discrepancy already discussed on page 125. It is possible to show that, if the $-1/2$ is eliminated from equation 4.51, definition E of the fractional Fourier transform is modified by a factor $\exp(-i\alpha\pi/4)$, consistent with equation 4.30. Likewise, precisely the same happens to definition F if the $-1/2$ is eliminated from equation 4.63. The physical differential equations governing the quantum-mechanical harmonic oscillator and propagation in quadratic graded-index media do not include this $-1/2$ and are thus best described by $\mathcal{C}_M$ rather than $\mathcal{F}^a$ (equation 4.30). $\mathcal{H}$ as defined in equation 4.63 is simply the Hamiltonian of the quantum-mechanical harmonic oscillator $\mathcal{H}_{fF}$ minus the ground-state (or zero-point) energy: $\mathcal{H} = \mathcal{H}_{fF} - 1/2$. Recalling equations 2.251 and 2.252, we further note that $\mathcal{H}$ is precisely given by $\mathcal{H} = \mathcal{A}^H\mathcal{A}$, so that equations 2.253 and 4.68 are the same.

To establish the equivalence of the present definition to definition D, we will use the hyperdifferential form of $\mathcal{F}^a$ in equations 4.49 and 4.50 to obtain explicit expressions for $\mathcal{U}_a$ and $\mathcal{D}_a$, which we will see are precisely equivalent to those given in equation 4.36. We will employ the identity

$$\mathcal{F}^a = e^{-i\alpha\mathcal{H}} = e^{-i\alpha[\pi(\mathcal{U}^2+\mathcal{D}^2)-1/2]}$$

$$= e^{-i\pi(\csc\alpha-\cot\alpha)\mathcal{U}^2}e^{-i\pi(\sin\alpha)\mathcal{D}^2}e^{-i\pi(\csc\alpha-\cot\alpha)\mathcal{U}^2}e^{i\alpha/2}. \tag{4.72}$$

This identity corresponds to the decomposition of the fractional Fourier transform into a chirp multiplication followed by a chirp convolution followed by another chirp multiplication (equation 3.122 specialized to the rotation matrix). Substituting this decomposition of $\mathcal{F}^a$ in equation 4.49 and using the commutation relations $[\mathcal{U}, \Upsilon(\mathcal{U})] = 0$ and $[\mathcal{U}, \Upsilon(\mathcal{D})] = (i/2\pi)\Upsilon'(\mathcal{D})$ (equation 2.155) to move $\mathcal{U}$ towards the left, we obtain

$$\mathcal{U}_a = \mathcal{U} + \sin\alpha \; e^{i\pi(\csc\alpha - \cot\alpha)\mathcal{U}^2} \; \mathcal{D} \; e^{-i\pi(\csc\alpha - \cot\alpha)\mathcal{U}^2}. \tag{4.73}$$

Now, using $[\mathcal{D}, \Upsilon(\mathcal{U})] = (-i/2\pi)\Upsilon'(\mathcal{U})$ to move $\mathcal{D}$ towards the left, we obtain

$$\mathcal{U}_a = \mathcal{U} + \sin\alpha[\mathcal{D} - (\csc\alpha - \cot\alpha)\mathcal{U}] = \cos\alpha \, \mathcal{U} + \sin\alpha \, \mathcal{D}, \tag{4.74}$$

precisely as in equation 4.36. Substituting the dual of equation 4.72 (in which the roles of $\mathcal{U}$ and $\mathcal{D}$ are exchanged) in equation 4.50, it is possible to similarly show that

$$\mathcal{D}_a = -\sin\alpha \, \mathcal{U} + \cos\alpha \, \mathcal{D}, \tag{4.75}$$

again precisely as in equation 4.36. Thus we have shown that the hyperdifferential form of $\mathcal{F}^a$ implies equation 4.36, which constitutes the heart of definition D. Since equation 4.36 shows that the coordinate multiplication and differentiation operators corresponding to different values of a are related by the rotation matrix, this derivation also relates the present definition to definition C. Readers with some exposure to quantum mechanics will be familiar with the idea of operators replacing classical variables such as position and momentum (which correspond to our time or space variable u and frequency variable μ, and the operators $\mathcal{U}$ and $\mathcal{D}$). In such operator contexts, equation 4.36 serves as the concrete expression of rotation in the time- or space-frequency plane, corresponding to equation 4.28 written in terms of the Wigner distribution $W_f(u, \mu)$. Readers familiar with the use of the Wigner distribution in quantum mechanics may already know that the effect of the Hamiltonian $\mathcal{H} = \pi(\mathcal{U}^2 + \mathcal{D})^2 - 1/2$ is to rotate the Wigner distribution. That the operation represented by the hyperdifferential form (or differential equation) associated with this Hamiltonian should correspond to rotation in the time- or space-frequency plane, is closely related to the fact that this Hamiltonian is rotationally invariant. Rotational invariance simply means that $\mathcal{H}_a \equiv \pi(\mathcal{U}_a^2 + \mathcal{D}_a^2) - 1/2 = \pi(\mathcal{U}^2 + \mathcal{D}^2) - 1/2 \equiv \mathcal{H}$, as a result of equation 4.36. Stated differently, the fact that the hyperdifferential operator associated with this Hamiltonian is the fractional Fourier transform operator in function space, is related to the fact that this Hamiltonian is invariant under the fractional Fourier transform; see the discussion surrounding equation 3.170.

Readers uncomfortable with operator algebra may prefer the following alternative derivation of equation 4.36 from the hyperdifferential form of $\mathcal{F}^a$. We again substitute $\mathcal{F}^a = \exp(-i\alpha\mathcal{H})$ in equations 4.49 and 4.50 to obtain the equations

$$\mathcal{U}_a = e^{i\alpha\mathcal{H}}\mathcal{U}e^{-i\alpha\mathcal{H}}, \tag{4.76}$$

$$\mathcal{D}_a = e^{i\alpha\mathcal{H}}\mathcal{D}e^{-i\alpha\mathcal{H}}, \tag{4.77}$$

which also constituted our starting point in the operator-algebraic derivation above. Let us express $\alpha = a\pi/2$ in the form $\alpha = (\alpha/\epsilon)\epsilon$ where ϵ is infinitesimally small and rewrite the above equations as

$$\mathcal{U}_{2\alpha/\pi} = e^{i\epsilon\mathcal{H}} \dots e^{i\epsilon\mathcal{H}}\mathcal{U}e^{-i\epsilon\mathcal{H}} \dots e^{-i\epsilon\mathcal{H}}, \tag{4.78}$$

$$\mathcal{D}_{2\alpha/\pi} = e^{i\epsilon\mathcal{H}} \dots e^{i\epsilon\mathcal{H}}\mathcal{D}e^{-i\epsilon\mathcal{H}} \dots e^{-i\epsilon\mathcal{H}}, \tag{4.79}$$

where the $\exp(\pm i\epsilon\mathcal{H})$ factors are repeated α/ϵ times. First, let us evaluate $\mathcal{U}_{2\epsilon/\pi}$ and $\mathcal{D}_{2\epsilon/\pi}$ by expanding the exponentials to first order:

$$\mathcal{U}_{2\epsilon/\pi} = e^{i\epsilon\mathcal{H}}\mathcal{U}e^{-i\epsilon\mathcal{H}} = (1 + i\epsilon\mathcal{H})\mathcal{U}(1 - i\epsilon\mathcal{H}), \tag{4.80}$$

$$\mathcal{D}_{2\epsilon/\pi} = e^{i\epsilon\mathcal{H}}\mathcal{D}e^{-i\epsilon\mathcal{H}} = (1 + i\epsilon\mathcal{H})\mathcal{D}(1 - i\epsilon\mathcal{H}). \tag{4.81}$$

The right-hand sides can be evaluated by substituting $\mathcal{H} = \pi(\mathcal{U}^2 + \mathcal{D}^2) - 1/2$ and using $[\mathcal{U}, \mathcal{D}] = (i/2\pi)\mathcal{I}$, $[\mathcal{U}, \mathcal{D}^2] = (i/\pi)\mathcal{D}$, and $[\mathcal{D}, \mathcal{U}^2] = (-i/\pi)\mathcal{U}$. Ignoring higher-order terms in ϵ we obtain

$$\mathcal{U}_{2\epsilon/\pi} = \mathcal{U} + \epsilon\mathcal{D}, \tag{4.82}$$

$$\mathcal{D}_{2\epsilon/\pi} = -\epsilon\mathcal{U} + \mathcal{D}. \tag{4.83}$$

Now let us rewrite equations 4.78 and 4.79 in the form

$$\mathcal{U}_{2\alpha/\pi} = e^{i\epsilon\mathcal{H}} \ldots e^{i\epsilon\mathcal{H}}\mathcal{U}_{2\epsilon/\pi}e^{-i\epsilon\mathcal{H}} \ldots e^{-i\epsilon\mathcal{H}}, \tag{4.84}$$

$$\mathcal{D}_{2\alpha/\pi} = e^{i\epsilon\mathcal{H}} \ldots e^{i\epsilon\mathcal{H}}\mathcal{D}_{2\epsilon/\pi}e^{-i\epsilon\mathcal{H}} \ldots e^{-i\epsilon\mathcal{H}}, \tag{4.85}$$

where the $\exp(\pm i\epsilon\mathcal{H})$ factors are now repeated $(\alpha/\epsilon - 1)$ times. Now let us evaluate $\mathcal{U}_{4\epsilon/\pi}$ and $\mathcal{D}_{4\epsilon/\pi}$. Using equations 4.82 and 4.83, it can be easily shown that $\mathcal{U}_{2\epsilon/\pi}$ and $\mathcal{D}_{2\epsilon/\pi}$ satisfy the same commutation relations as $\mathcal{U}$ and $\mathcal{D}$. Thus we obtain

$$\mathcal{U}_{4\epsilon/\pi} = e^{i\epsilon\mathcal{H}}\mathcal{U}_{2\epsilon/\pi}e^{-i\epsilon\mathcal{H}} = (1 + i\epsilon\mathcal{H})\mathcal{U}_{2\epsilon/\pi}(1 - i\epsilon\mathcal{H}) = \mathcal{U}_{2\epsilon/\pi} + \epsilon\mathcal{D}_{2\epsilon/\pi}, \tag{4.86}$$

$$\mathcal{D}_{4\epsilon/\pi} = e^{i\epsilon\mathcal{H}}\mathcal{D}_{2\epsilon/\pi}e^{-i\epsilon\mathcal{H}} = (1 + i\epsilon\mathcal{H})\mathcal{D}_{2\epsilon/\pi}(1 - i\epsilon\mathcal{H}) = -\epsilon\mathcal{U}_{2\epsilon/\pi} + \mathcal{D}_{2\epsilon/\pi}. \tag{4.87}$$

Repeating this nested procedure, we finally obtain

$$\begin{bmatrix} \mathcal{U}_{2\alpha/\pi} \\ \mathcal{D}_{2\alpha/\pi} \end{bmatrix} = \begin{bmatrix} 1 & \epsilon \\ -\epsilon & 1 \end{bmatrix}^{\alpha/\epsilon} \begin{bmatrix} \mathcal{U} \\ \mathcal{D} \end{bmatrix}. \tag{4.88}$$

The matrix power appearing in this equation may be evaluated either by using standard eigenvalue techniques or by induction, which leads to a power series form that we immediately recognize as the rotation matrix:

$$\lim_{\epsilon\to 0} \begin{bmatrix} 1 & \epsilon \\ -\epsilon & 1 \end{bmatrix}^{\alpha/\epsilon} = \begin{bmatrix} 1 - \alpha^2/2 + \cdots & \alpha - \alpha^3/3 + \cdots \\ -(\alpha - \alpha^3/3 + \cdots) & 1 - \alpha^2/2 + \cdots \end{bmatrix} = \begin{bmatrix} \cos\alpha & \sin\alpha \\ -\sin\alpha & \cos\alpha \end{bmatrix}. \tag{4.89}$$

Given the direct relation between the hyperdifferential form of $\mathcal{F}^a = \exp[-i(a\pi/2)\mathcal{H}]$ and the differential equation $\mathcal{H}f_a(u) = i(2/\pi)\partial f_a(u)/\partial a$, the above discussions also enable one to quite directly relate definition E to definitions D and C.

The two derivations above both yielded the result

$$e^{i\alpha\mathcal{H}}\mathcal{U}e^{-i\alpha\mathcal{H}} = \cos\alpha\,\mathcal{U} + \sin\alpha\,\mathcal{D}, \tag{4.90}$$

$$e^{i\alpha\mathcal{H}}\mathcal{D}e^{-i\alpha\mathcal{H}} = -\sin\alpha\,\mathcal{U} + \cos\alpha\,\mathcal{D}. \tag{4.91}$$

It will be instructive to engage in a final check of consistency. Equality of both sides is equivalent to equality of their derivatives with respect to α and equality at a particular value of α, say $\alpha = 0$. Equality at $\alpha = 0$ is immediate. Differentiating the left-hand side of equation 4.90 with respect to α, we obtain

$$\left(i\mathcal{H}e^{i\alpha\mathcal{H}}\right)\mathcal{U}e^{-i\alpha\mathcal{H}} + e^{i\alpha\mathcal{H}}\mathcal{U}\left(-i\mathcal{H}e^{-i\alpha\mathcal{H}}\right) = e^{i\alpha\mathcal{H}}\left(i[\mathcal{H},\mathcal{U}]\right)e^{-i\alpha\mathcal{H}}. \qquad (4.92)$$

Using $[\mathcal{H},\mathcal{U}] = -i\mathcal{D}$, we obtain

$$e^{i\alpha\mathcal{H}}\mathcal{D}e^{-i\alpha\mathcal{H}}, \qquad (4.93)$$

which we recognize as the left-hand side of equation 4.91. But the derivative of the right-hand side of equation 4.90 is also the same as the right-hand side of equation 4.91. One can similarly differentiate equation 4.91 to obtain equation 4.90, which shows that both equations are indeed consistent.

Finally, we discuss the relation of the present definition to definition A. Readers familiar with the solution of the quantum-mechanical harmonic oscillator (or the solution of propagation in quadratic graded-index media) may already be familiar with the fact that the Green's function appearing in these problems has the form of equation 4.4 (Agarwal and Simon 1994). Here we derive this by employing equation 2.245:

$$e^{-i\alpha\mathcal{H}} = e^{-i\alpha[\pi(\mathcal{U}^2+\mathcal{D}^2)-1/2]} = e^{-i\pi\tan\alpha\,\mathcal{U}^2}e^{-i\pi\ln(\cos\alpha)\,(\mathcal{U}\mathcal{D}+\mathcal{D}\mathcal{U})}e^{-i\pi\tan\alpha\,\mathcal{D}^2}e^{i\alpha/2}. \qquad (4.94)$$

The operators on the right-hand side are recognized as chirp multiplication, scaling, and chirp convolution respectively (see page 111). Thus the kernel $K_a(u,u')$ of the fractional Fourier transform can be obtained by concatenating the kernels associated with these operations. Although we do not provide the details, this indeed yields the kernel given in equation 4.4. (It is also possible to multiply the matrices associated with these operations, but this will not exactly reveal the phase of A_α.)

The index additivity property and the special case $a = 0$ immediately follow from the exponential form $\exp(-i\alpha\mathcal{H})$. The special case $a = 1$ which is $\mathcal{F} = \exp[-i(\pi/2)\mathcal{H}]$ is nothing but the hyperdifferential form of the ordinary Fourier transform (although we did not derive this form of the ordinary Fourier transform independently in this book). Alternatively, the special case $a = 1$ can be seen to reduce to the ordinary Fourier transform by noting that it corresponds to a $\pi/2$ rotation in the time- or space-frequency plane. The periodicity with respect to a can be deduced from equation 4.94 or indirectly from the rotation matrix.

The hyperdifferential form is especially useful when one considers fractional Fourier transforms of very small order a, as we did in one of the derivations above. Since it will be seen in chapter 9 that the fractional Fourier transform models wave propagation, this corresponds to finding how the amplitude distribution of a wave is altered upon propagation over a very small distance. When a is infinitesimal, the operator $\mathcal{F}^a = \exp[-i(a\pi/2)\mathcal{H}]$ is called an *infinitesimal operator*. Such operators can be expressed as a first-order expansion of the form $\mathcal{F}^a = \mathcal{I} + i(a\pi/2)\mathcal{H}$. This implies

$$f_a(u) - f(u) = i(a\pi/2)\mathcal{H}f(u) = i\left(\frac{a\pi}{2}\right)\left(-\frac{1}{4\pi}\frac{d^2}{du^2} + \pi u^2 - \frac{1}{2}\right)f(u), \qquad (4.95)$$

so that we can determine the change in the wave amplitude without integration (also see Wawrzyńczyk 1990). Likewise, when a is infinitesimal, the mapping of an operator

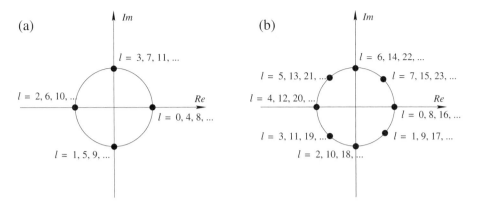

Figure 4.1. (a) Eigenvalues of $\mathcal{F}$ on the unit circle in the complex plane. (b) Eigenvalues of $\mathcal{F}^{1/2}$ on the unit circle in the complex plane.

$\mathcal{A}$ into the new operator $\mathcal{A}' = \mathcal{F}^{-a}\mathcal{A}\mathcal{F}^a$ takes the following simple form (Cohen-Tannoudji, Diu, and Laloë 1977:181):

$$\mathcal{A}' - \mathcal{A} = i\alpha[\mathcal{H}, \mathcal{A}]. \tag{4.96}$$

The hyperdifferential form of the fractional Fourier transform is also discussed by Mustard (1987a, c). We have already mentioned the relationship between the fractional Fourier transform and the quantum-mechanical harmonic oscillator; the transform is the time-evolution operator of the harmonic oscillator. This viewpoint is stressed in Dattoli, Torre, and Mazzacurati 1998, where multidimensional extensions are also considered.

4.2 Eigenvalues and eigenfunctions

The eigenvalue equation for the fractional Fourier transform is (definition B)

$$\mathcal{F}^a\psi_l(u) = e^{-ial\pi/2}\psi_l(u), \qquad l = 0, 1, 2, \ldots, \tag{4.97}$$

where $\psi_l(u)$ are the Hermite-Gaussian functions. The eigenvalue $e^{-ial\pi/2}$ was motivated in definition B as the ath power of the eigenvalue $e^{-il\pi/2}$ of the ordinary Fourier transform. The reader might have noted that the ath power λ_l^a of $\lambda_l = e^{-il\pi/2}$, is not unique. It is the particular power $e^{-ial\pi/2}$ which is the eigenvalue of the fractional Fourier transform, by definition. We will denote this particular power by $\lambda_{a,l}$ to avoid confusion. When $a = 1$ we simply have $\lambda_{1,l} = \lambda_l$.

Let us now examine the number of distinct eigenvalues of $\mathcal{F}^a$. When $a = 1$, corresponding to the ordinary Fourier transform, the eigenvalues are $1, -i, -1, i, 1, \ldots$ respectively for $l = 0, 1, 2, 3, 4, \ldots$ as shown in figure 4.1a. There are four distinct eigenvalues which are the fourth roots of unity: $\lambda_{1,l}^4 = \lambda_l^4 = 1$. If a is a rational number of the form $a = p/q$ where p and q are integers which have no common divisor, then there are $4q$ distinct eigenvalues, since $(\mathcal{F}^{p/q})^{4q} = \mathcal{I}$ implies $\lambda_{p/q,l}^{4q} = 1$. For example, if $a = 1/2$, the eight eigenvalues satisfying $\lambda_{1/2,l}^8 = 1$ are as shown in figure 4.1b. The fact that there are only $4q$ distinct eigenvalues is also easily seen by noting that $e^{-i(p/q)l\pi/2}$ repeats itself after $l = 4q$, having made p trips around the unit circle.

It is important to note that the functions $\psi_l(u)$ which had the same eigenvalues as eigenfunctions of $\mathcal{F}$, no longer do so for $\mathcal{F}^a$. For example, $\psi_1(u), \psi_5(u), \psi_9(u), \ldots$ have the same degenerate eigenvalue $-i$ as eigenfunctions of $\mathcal{F}$. However, only $\psi_1(u), \psi_9(u), \psi_{17}(u), \ldots$ have the same degenerate eigenvalue $e^{-i\pi/4}$ as eigenfunctions of $\mathcal{F}^{1/2}$. The complementary set of functions $\psi_5(u), \psi_{13}(u), \psi_{21}(u), \ldots$, on the other hand, share the other square root of $-i$ as a degenerate eigenvalue. With $\mathcal{F}$ the eigenvalue 1 is shared by eigenfunctions $l = 0, 4, 8, 12, 16, 20, \ldots$, whereas with $\mathcal{F}^{1/2}$ the eigenvalue 1 is shared by $l = 0, 8, 16, \ldots$ and the eigenvalue -1 is shared by $l = 4, 12, 20, \ldots$. We again see that the eigenfunctions are split into two groups such that one group shares one of the square roots of the original eigenvalue 1, and the other group shares the other. The same happens with the other eigenvalues: the eigenvalue $-i$ for $l = 1, 5, 9, 13, \ldots$ is split into $e^{-i\pi/4}$ for $l = 1, 9, \ldots$ and $e^{i3\pi/4}$ for $l = 5, 13, \ldots$; the eigenvalue -1 for $l = 2, 6, 10, 14, \ldots$ is split into $-i$ for $l = 2, 10, \ldots$ and i for $l = 6, 14, \ldots$; the eigenvalue i for $l = 3, 7, 11, 15, \ldots$ is split into $e^{-i3\pi/4}$ for $l = 3, 11, \ldots$ and $e^{i\pi/4}$ for $l = 7, 15, \ldots$. Thus, going from $a = 1$ to $a = 1/2$, we find that the two square roots of each eigenvalue are distributed alternately among the eigenfunctions which previously shared that eigenvalue. More generally, the set of eigenfunctions of $\mathcal{F}$ which share the same degenerate eigenvalue are split into q disjoint sets such that each set now shares one of the q distinct qth roots of 1, $-i$, -1, or i. When a is irrational, the degeneracy is totally eliminated: each eigenfunction has a distinct eigenvalue.

Let us recall from section 2.4.4 the result stating that if an operator $\mathcal{A}$ has eigenvalues λ_l with degeneracy g_l, then a function $\Upsilon(\mathcal{A})$ of that operator will have eigenvalues $\Upsilon(\lambda_l)$ with the same degeneracy. The careful reader will note that this flatly contradicts the above discussion regarding the eigenvalues of the fractional Fourier transform. In section 2.4.4 we implicitly assumed that $\Upsilon(z)$ is a single-valued function. On the other hand, the function $\Upsilon(z) = z^a$ is not single-valued unless we specify which fractional power of z is to be chosen. Once the function $\Upsilon(z)$ has been defined unambiguously in such a manner, then the above statement regarding degeneracy is valid. Of course, for different ways of specifying which fractional power of z is to be chosen, we end up with different operators $\Upsilon(\mathcal{A})$. One choice might be to choose the principal powers of z and define the fractional Fourier transform accordingly. This approach has been considered by a number of authors and is discussed further below. However, this definition does not correspond to the definition of the fractional Fourier transform employed in this book. Furthermore, the definition in this book *does not correspond to any unambiguous way of specifying the power function z^a.* For instance, consider the function $z^{1/2}$ which is ambiguous. The ambiguity may be removed by specifying which square root of z is to be chosen. But our definition does not correspond to choosing a specific square root of z. In our definition, we take all possible powers of z and distribute them among the eigenfunctions in a specific manner. For example, one square root of λ_l is assigned to half of the eigenfunctions which formerly shared the eigenvalue λ_l, and the other square root is assigned to the others. In other words, our definition does not favor one way of resolving the ambiguity of the power function over the others. Consequently, *the fractional Fourier transform operator $\mathcal{F}^a$ we have defined cannot be expressed in the form $\Upsilon(\mathcal{F})$ where $\Upsilon(z) = z^a$*

is a single-valued power function. Finally, we note that although our definition does not favor one way of resolving the ambiguity of the power function over the others, it does depend on the choice of the Hermite-Gaussian functions as the complete set of eigenfunctions, and also on the way in which we assign the newly created eigenvalues to these eigenfunctions. These issues will be further discussed below.

The relationship between self-Fourier functions (page 41) and fractional Fourier transforms has been discussed in Mendlovic, Ozaktas, and Lohmann 1994b, Alieva 1996a, Zhang, Gu, and Yang 1998, and Alieva and Barbé 1997, 1999.

4.3   Distinct definitions of the fractional Fourier transform

Our particular definition of the fractional Fourier transform is justified by its many desirable properties and its usefulness in many applications. Nevertheless, it will be instructive to consider some of the distinct definitions of the fractional Fourier transform which have been proposed, definitions which are not equivalent to the transform discussed in this book.

First, let us consider a definition appearing in Dickinson and Steiglitz 1982, and also later in Shih 1995b and Santhanam and McClellan 1995, 1996. In this approach, one defines the power function $\Upsilon(z) = z^a$ uniquely by choosing the principal power of z. The principal ath power of a unit-magnitude complex number $z = e^{-i\phi}$ where $\phi \in [0, 2\pi)$ is defined as $z = e^{-ia\phi}$. (In some texts the principal power of a unit-magnitude complex number $z = e^{i\phi}$ where $\phi \in [0, 2\pi)$ is defined differently as $z = e^{ia\phi}$. Another variation is to work with the interval $(0, 2\pi]$ instead of $[0, 2\pi)$. The present definition is most suitable for our purposes.)

One way of explicitly deriving the kernel of the fractional Fourier transform defined through the principal power function, is to use annihilating polynomials (Roman 1992). We will prefer a related but more simple and transparent approach (Shih 1995b). Consider the series expansion of the principal ath fractional power $\mathcal{F}^a_{\mathrm{pp}}$ of the Fourier operator $\mathcal{F}$:

$$\mathcal{F}^a_{\mathrm{pp}} = \sum_{k=0}^{\infty} \beta'_k(a) \, \mathcal{F}^k. \tag{4.98}$$

Since we know that $\mathcal{F}^4 = \mathcal{F}^0$, $\mathcal{F}^5 = \mathcal{F}^1$, $\mathcal{F}^6 = \mathcal{F}^2$, and in general $\mathcal{F}^k = \mathcal{F}^{k \bmod 4}$, this summation can be reduced to one with only four terms:

$$\mathcal{F}^a_{\mathrm{pp}} = \sum_{k=0}^{3} \beta_k(a) \, \mathcal{F}^k. \tag{4.99}$$

Now we apply both sides of this equation to the eigensignals ψ_l and—in the spirit of definition B—require $\mathcal{F}^a_{\mathrm{pp}} \psi_l = \lambda^a_l \psi_l$, where $\lambda_l = e^{-il\pi/2}$ are the eigenvalues of the ordinary Fourier transform (equation 2.174). We then obtain

$$\sum_{k=0}^{3} \beta_k(a) \lambda^k_l \psi_l = \lambda^a_l \psi_l, \qquad \text{for all } l. \tag{4.100}$$

Since $\lambda_l = \lambda_{l\mathrm{mod}4}$, only four of these equations are independent:

$$\lambda_0^a = e^{-ia0} = \beta_0(a) + \beta_1(a) + \beta_2(a) + \beta_3(a) = \sum_{k=0}^{3} \beta_k(a)e^{-ik2\pi}, \tag{4.101}$$

$$\lambda_1^a = e^{-ia\pi/2} = \beta_0(a) - i\beta_1(a) - \beta_2(a) + i\beta_3(a) = \sum_{k=0}^{3} \beta_k(a)e^{-ik\pi/2}, \tag{4.102}$$

$$\lambda_2^a = e^{-ia\pi} = \beta_0(a) - \beta_1(a) + \beta_2(a) - \beta_3(a) = \sum_{k=0}^{3} \beta_k(a)e^{-ik\pi}, \tag{4.103}$$

$$\lambda_3^a = e^{-ia3\pi/2} = \beta_0(a) + i\beta_1(a) - \beta_2(a) - i\beta_3(a), = \sum_{k=0}^{3} \beta_k(a)e^{-ik3\pi/2}, \tag{4.104}$$

where we used the principal powers of the eigenvalues of the ordinary Fourier transform, as defined above. The above equations are easily solved for the coefficients $\beta_k(a)$:

$$\beta_k(a) = \frac{1}{4} \sum_{n=0}^{3} e^{-i(a-k)n\pi/2}. \tag{4.105}$$

Some of the references mentioned above give slightly different expressions for $\beta_k(a)$ due to slightly different ways of defining the principal power. Also, whereas our derivation is posed in terms of abstract operators and is thus applicable for both the discrete and continuous cases, some of the mentioned authors have chosen to work directly in a discrete framework, which makes it possible to employ standard techniques for obtaining functions of matrices. Using these techniques, it is possible to arrive at the same result given above by evaluating the ath power of the discrete Fourier transform matrix.

To understand how this definition differs from the definition employed in this book, we carefully examine the eigenvalues for both cases. In the definition just discussed, the eigenvalues assigned to Hermite-Gaussian functions with orders $l = 0, 1, 2, 3, 4, 5, 6, 7, 8, \ldots$ are

$$e^{-ia0\pi/2}, e^{-ia1\pi/2}, e^{-ia2\pi/2}, e^{-ia3\pi/2}, e^{-ia0\pi/2}, e^{-ia1\pi/2}, e^{-ia2\pi/2}, e^{-ia3\pi/2}, e^{-ia0\pi/2}, \ldots \tag{4.106}$$

respectively. Note that there are only four distinct eigenvalues of the fractional Fourier transform thus defined, just as for the ordinary Fourier transform. These four eigenvalues are simply the principal ath powers of the four eigenvalues of the ordinary Fourier transform $(1, -i, -1, i)$. Hermite-Gaussian functions with orders $0, 4, 8, \ldots$ share the same eigenvalue $e^{-ia0\pi/2} = 1$. Likewise, those with orders $1, 5, 9, \ldots$, those with orders $2, 6, 10, \ldots$, and $3, 7, 11, \ldots$ share the remaining eigenvalues.

This is in contrast with the eigenvalue splitting associated with the definition of the fractional Fourier transform employed in this book, as we have discussed in detail in section 4.2. In that case the eigenvalues assigned to Hermite-Gaussian functions with orders $l = 0, 1, 2, 3, 4, 5, 6, 7, 8, \ldots$ are given by the formula $e^{-ial\pi/2}$:

$$e^{-ia0\pi/2}, e^{-ia1\pi/2}, e^{-ia2\pi/2}, e^{-ia3\pi/2}, e^{-ia4\pi/2}, e^{-ia5\pi/2}, e^{-ia6\pi/2}, e^{-ia7\pi/2}, e^{-ia8\pi/2}, \ldots \tag{4.107}$$

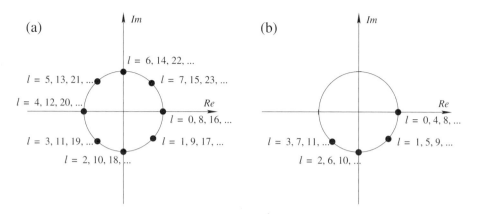

Figure 4.2. Eigenvalues for the definition employed in this book (a) and for the alternative definition discussed in this section (b).

As an example, let us take $a = 1/2$ for which the principal square roots of 1, $-i$, -1, and i are 1, $e^{-i\pi/4}$, $-i$, and $e^{-i3\pi/4}$ respectively. The eigenvalues for both cases are shown in figure 4.2.

Returning to equation 4.99, we emphasize that the present approach defines the fractional Fourier transform operator as the linear combination of four integer powers of the ordinary Fourier transform. Since $\mathcal{F}^0 f(u) = f(u)$, $\mathcal{F}^1 f(u) = F(u)$, $\mathcal{F}^2 f(u) = f(-u)$, and $\mathcal{F}^3 f(u) = F(-u)$, the ath fractional Fourier transform of a function is simply a linear combination of the function itself, its Fourier transform, its mirror reflection, and the mirror reflection of its Fourier transform:

$$\mathcal{F}^a_{pp} f(u) = \beta_0(a) f(u) + \beta_1(a) F(u) + \beta_2(a) f(-u) + \beta_3(a) F(-u). \tag{4.108}$$

Given the form which this definition boils down to, we do not believe it is of special interest. Furthermore, when the function $f(u)$ is even, we simply obtain

$$\mathcal{F}^a_{pp} f(u) = e^{-ia\pi/2} \left[\cos(a\pi/2) f(u) + i \sin(a\pi/2) F(u)\right], \tag{4.109}$$

which implies an operator relationship of the form

$$\mathcal{F}^a_{pp} = e^{-ia} \left[\cos\alpha\, \mathcal{I} + i \sin\alpha\, \mathcal{F}\right], \tag{4.110}$$

from which we see that when $f(u)$ is even, $\mathcal{F}^a_{pp}$ is simply a linear combination of the identity operator and the ordinary Fourier transform operator. The reader may also wish to contrast this result with equation 4.36.

The careful reader may be troubled by the fact that the particular way in which the ath power was chosen played no part in going from equation 4.98 to equation 4.99, which would seem to imply that the definition employed in this book is also reducible to a summation of the form given in equation 4.99. However, for the definition employed in this book, it is not possible to write an expansion of the form given by equation 4.98 to begin with. This is not because there is a certain power function and that function does not have a series expansion, but because there can be no such function associated with our definition in the first place. In other words, the fractional Fourier transform operator $\mathcal{F}^a$ as we define and use it, cannot be expressed as $\Upsilon(\mathcal{F})$ for any function $\Upsilon(\cdot)$. There is no particular way of resolving the ambiguity of the power function such that it results in our definition. As discussed in section 4.2, rather than choosing a particular branch of the power function, we use all possible powers, distributing them among the eigenfunctions in a certain manner. We must also note that we were being imprecise in this regard when we motivated definition B, where we spoke of the fractional Fourier transform as if it were the function $(\cdot)^a$ of the ordinary Fourier operator.

To see that there is indeed no function $\Upsilon(\cdot)$ such that our fractional Fourier transform operator satisfies $\mathcal{F}^a = \Upsilon(\mathcal{F})$, we recall from section 2.4.4 that the function of an operator was defined in terms of the power series of the function in question, which was assumed to be defined everywhere:

$$\Upsilon(\mathcal{F}) = \sum_{k=0}^{\infty} \Upsilon'_k \mathcal{F}^k. \tag{4.111}$$

Since $\mathcal{F}^4 = \mathcal{I}$, this summation collapses down to

$$\Upsilon(\mathcal{F}) = \sum_{k=0}^{3} \Upsilon_k \mathcal{F}^k. \tag{4.112}$$

Now, we assume that $\mathcal{F}^a = \Upsilon(\mathcal{F})$ for some a and show that this results in a contradiction. Recalling the eigenvalue equation $\mathcal{F}^a \psi_l(u) = e^{-ial\pi/2}\psi_l(u)$, we apply both sides of the above equation to $\psi_l(u)$:

$$\mathcal{F}^a \psi_l(u) = e^{-ial\pi/2}\psi_l(u) = \sum_{k=0}^{3} \Upsilon_k \mathcal{F}^k \psi_l(u) = \left[\sum_{k=0}^{3} \Upsilon_k e^{-ikl\pi/2}\right]\psi_l(u). \tag{4.113}$$

The factor preceding $\psi_l(u)$ in the rightmost expression is periodic in l with period 4. However, the factor $e^{-ial\pi/2}$ is not periodic in l with period 4 unless a is an integer. This contradiction shows that $\mathcal{F}^a$ is not the function $(\cdot)^a$ of $\mathcal{F}$ in the strict sense. (Ozaktas and others 1996)

We may further note that any unambiguously defined function $\Upsilon(\cdot)$ which can be expanded into a power series, when applied to the Fourier operator $\mathcal{F}$, will result in an operator which is a linear superposition of the form given by equation 4.112. This implies that the effect of $\Upsilon(\mathcal{F})$ on a given function $f(u)$ is simply a linear superposition of $f(u)$, $f(-u)$, $F(u)$, and $F(-u)$. Thus, functions of the Fourier operator such as $\exp(\mathcal{F})$ or $\sin(\mathcal{F})$ are not very interesting and do not seem to deserve special study.

A number of authors have explored in detail the multiplicity of possible fractional Fourier transform definitions based on different ways of dealing with the ambiguity associated with the power function (Shamir and Cohen 1995, Liu, Zhang, and Zhang 1997, Liu and others 1997a, Cariolaro and others 1998, Kraniauskas, Cariolaro, and Erseghe 1998). As discussed above, resolving the ambiguity of the power function in one way or another so as to obtain an unambiguous function $\Upsilon(\cdot) = (\cdot)^a$, leads to a relatively uninteresting definition of the fractional Fourier transform. Definitions based on different assignments of the various powers of the eigenvalues to the Hermite-Gaussian functions (or to members of a different set of eigenfunctions) may lead to more interesting definitions. Much remains to be explored in this direction. What distinguishes our definition is its many interesting properties, especially in the time-frequency plane, and also the fact that it is particularly suited for modeling wave propagation problems.

An alternative approach has been taken by Marhic (1995), who explores the roots of the identity operator. This work considers fractional identity operators, rather than fractional Fourier operators. Not surprisingly, one arrives at the same class of fractional operators; the index used in this work is simply related to the index a we use. Marhic argues that this approach is purer: "I believe, however, that by giving a central role to $\mathcal{I}$ rather than to one of its fourth roots, however important it might be, we obtain a broader perspective that naturally leads to properties and possible implementations that are not readily apparent otherwise."

Here we also mention Várilly and Gracia-Bondía 1987 and Rashid 1989, where an entity whose sixth power is equal to the inverse Fourier transform appears.

A totally distinct definition, which we mention only because it shares the same name, has been given by Bailey and Swarztrauber (1991) and further discussed in Bailey and Swarztrauber 1994. Their definition is given in a discrete context. Instead of the discrete Fourier transform defined in terms of the integer roots of unity $e^{-i2\pi/N}$, they allow fractional roots of unity $e^{-i2\pi b}$ in the definition:

$$f_{\mathrm{BS}b}(j) = \frac{1}{\sqrt{N}} \sum_{l=0}^{N-1} f(l)e^{-i2\pi jlb}, \qquad j = 0, 1, \ldots, N-1. \qquad (4.114)$$

Although this interesting definition may have many useful applications, in a continuous context it would correspond merely to a scaled Fourier transform, so that it is not directly comparable to our definition.

4.4   Transforms of some common functions

The fractional Fourier transforms of some common functions are given in table 4.1. Most of them can be derived by straightforward use of the defining integral given in equation 4.4, with the help of the definite integrals given in section 2.10.2. Pair 8 is a special case of pair 9, as well as a special case of pair 7. Pair 7 gives the eigenfunctions and eigenvalues of the fractional Fourier transform, as discussed in definition B. Pairs 4 and 5 are special cases of pair 6, and pair 9 is a special case of pair 10. Similar tables have appeared in several sources, including Namias 1980a, Almeida 1994, and Ozaktas, Kutay, and Mendlovic 1998, 1999.

The fractional Fourier transforms of most other common functions do not have simple closed-form expressions. But they may be obtained numerically, as discussed in chapter 6. Figure 4.3 shows the real and imaginary parts of the fractional Fourier transforms of $\mathrm{rect}(u)$ for various values of a. Notice that the $a = 1$ transform corresponds to the $\mathrm{sinc}(u)$ function. Figure 4.4 shows the magnitudes and phases of the fractional Fourier transforms of $\mathrm{rect}(u)$ for the same values of a. The phase plots wrap around at $\pm\pi$. In these figures we observe how the transform evolves from a rectangle function to the sinc function continuously as a ranges from 0 to 1. This evolution can be seen more graphically in figure 4.5, where $f_a(u)$ has been plotted as a function of both a and u. The profile of this graph for any value of a is simply

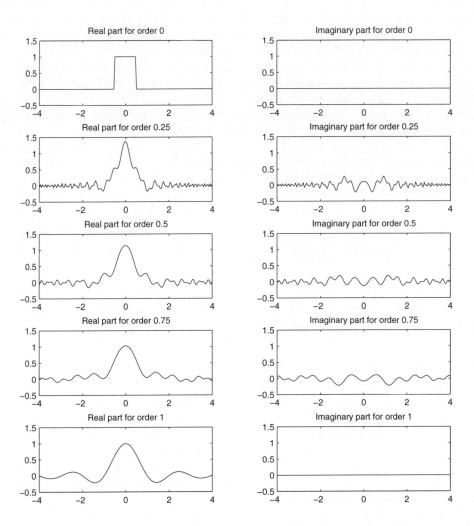

Figure 4.3. Real and imaginary parts of the fractional Fourier transforms of the rectangle function.

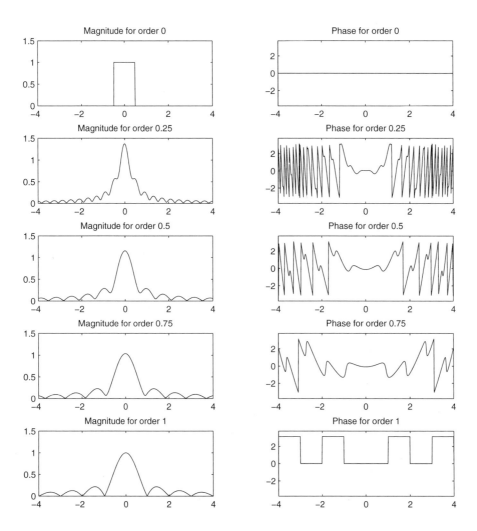

Figure 4.4. Magnitudes and phases of the fractional Fourier transforms of the rectangle function (Ozaktas and others 1996).

Table 4.1. Fractional Fourier transforms of some common functions.

	$f(u)$	$f_a(u)$
1.	$\delta(u)$	$\sqrt{1 - i\cot\alpha}\, \exp(i\pi u^2 \cot\alpha)$
2.	$\delta(u - \xi)$	$\sqrt{1 - i\cot\alpha}\, \exp[i\pi(u^2 \cot\alpha - 2u\xi \csc\alpha + \xi^2 \cot\alpha)]$
3.	1	$\sqrt{1 + i\tan\alpha}\, \exp(-i\pi u^2 \tan\alpha)$
4.	$\exp(i2\pi\xi u)$	$\sqrt{1 + i\tan\alpha}\, \exp[-i\pi(u^2 \tan\alpha - 2u\xi \sec\alpha + \xi^2 \tan\alpha)]$
5.	$\exp(i\pi\chi u^2)$	$\sqrt{\dfrac{1 + i\tan\alpha}{1 + \chi\tan\alpha}}\, \exp\left[i\pi u^2 \dfrac{\chi - \tan\alpha}{1 + \chi\tan\alpha}\right]$
6.	$\exp[i\pi(\chi u^2 + 2\xi u)]$	$\sqrt{\dfrac{1 + i\tan\alpha}{1 + \chi\tan\alpha}}\, \exp\left[i\pi \dfrac{u^2(\chi - \tan\alpha) + 2u\xi \sec\alpha - \xi^2 \tan\alpha}{1 + \chi\tan\alpha}\right]$
7.	$\psi_l(u)$	$\exp(-il\alpha)\, \psi_l(u)$
8.	$\exp(-\pi u^2)$	$\exp(-\pi u^2)$
9.	$\exp(-\pi\chi u^2)$	$\sqrt{\dfrac{1 - i\cot\alpha}{\chi - i\cot\alpha}}\, \exp\left[i\pi u^2 \dfrac{\cot\alpha(\chi^2 - 1)}{\chi^2 + \cot^2\alpha}\right] \exp\left[-\pi u^2 \dfrac{\chi \csc^2\alpha}{\chi^2 + \cot^2\alpha}\right]$
10.	$\exp[-\pi(\chi u^2 + 2\xi u)]$	$\sqrt{\dfrac{1 - i\cot\alpha}{\chi - i\cot\alpha}}\, \exp\left[i\pi \cot\alpha \dfrac{u^2(\chi^2 - 1) + 2u\xi \sec\alpha + \xi^2}{\chi^2 + \cot\alpha}\right]$
		$\times \exp\left[-\pi \csc^2\alpha \dfrac{u^2\chi + 2u\xi \cos\alpha - \chi\xi^2 \sin^2\alpha}{\chi^2 + \cot\alpha}\right]$

ξ and χ are real. $\psi_n(u)$ are the Hermite-Gaussian functions. (1) and (2) are valid when $a \neq 2j$, (3) and (4) are valid when $a \neq 2j+1$, and (5) and (6) are valid when $a - (2/\pi)\arctan\chi \neq 2j+1$, where j is an arbitrary integer. The transform of $\delta(u - \xi)$ is $\delta(u - \xi)$ when $a = 4j$ and $\delta(u + \xi)$ when $a = 4j + 2$. The transform of $\exp(i2\pi\xi u)$ is $\delta(u - \xi)$ when $a = 4j + 1$ and $\delta(u + \xi)$ when $a = 4j + 3$. The tranform of $\exp(i\pi\chi u^2)$ is $\sqrt{1/(1 - i\chi)}\,\delta(u)$ when $[a - (2/\pi)\arctan\chi] = 2j + 1$ and $\sqrt{1/(1 - i\chi)}$ when $[a - (2/\pi)\arctan\chi] = 2j$. In (9) and (10) $\chi > 0$ is required for convergence.

the ath fractional Fourier transform. We note from these figures that the transform undergoes especially fast changes when a is close to zero. Figure 4.6 illustrates the nature of these changes.

Figure 4.7 (left column) and figure 4.8 show the fractional Fourier transforms of the triangle function $\text{rect}(u) * \text{rect}(u)$. When $a = 1$ we simply obtain the Fourier transform of the triangle function $\text{sinc}^2(u)$. The right column of figure 4.7 shows the fractional Fourier transforms of the exponential function $\exp(-2\pi|u|)$.

Figure 4.9 shows the real and imaginary parts of the fractional Fourier transforms of the Dirac delta function $\delta(u)$. We note that for orders close to zero, the transform of the delta function is highly oscillatory, and thus will approximately behave like the delta function under the integral sign. It is in this sense that the fractional Fourier transform of the delta function approaches the delta function itself as $a \to 0$. Figure 4.10 shows the real and imaginary parts of the fractional Fourier transforms of the harmonic function $\exp(i2\pi u)$. This time the transform is highly oscillatory for orders close to unity, and the transform approaches $\delta(u - 1)$ as $a \to 0$.

The fractional Fourier transform of a delta function is in general a chirp function unless the order is an even number, when it is again a delta function, or an odd number, when it is a harmonic function (table 4.1, pair 2). Likewise, the fractional Fourier transform of a harmonic function is also in general a chirp function unless the order is an even number, when it is again a harmonic function, or an odd number, when it is a delta function (table 4.1, pair 4). More generally, the transform of a chirp function $\exp[i\pi(\chi u^2 + 2\xi u)]$ is in general another chirp function unless $[a - (2/\pi)\arctan\chi]$ is

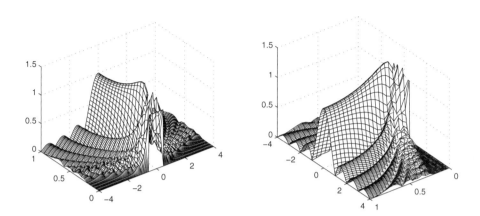

Figure 4.5. Magnitude of the fractional Fourier transform of the rectangle function as a function of the transform order from two different perspectives.

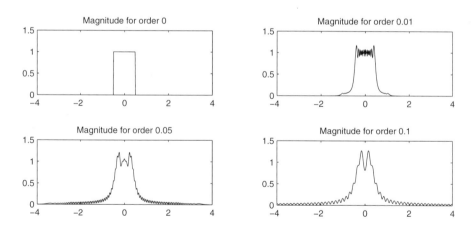

Figure 4.6. Magnitudes of the fractional Fourier transforms of the rectangle function for small orders.

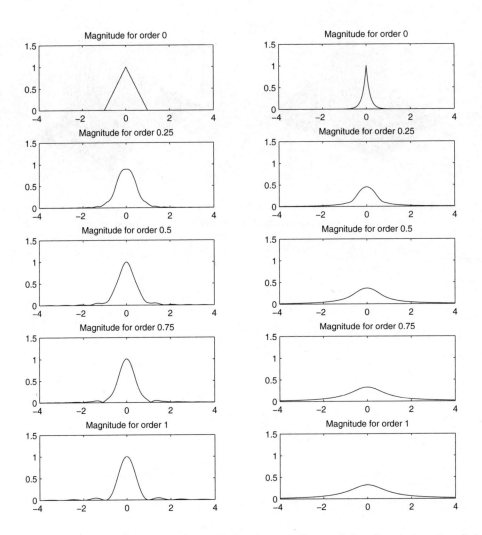

Figure 4.7. Magnitudes of the fractional Fourier transforms of the triangle function (left column) and the exponential function (right column).

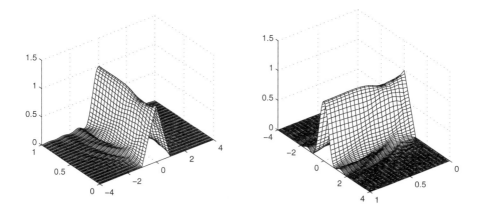

Figure 4.8. Magnitude of the fractional Fourier transform of the triangle function as a function of the transform order from two different perspectives.

an odd number, when the transform is a delta function, or an even number, when it is a harmonic function (table 4.1, pair 6). Since harmonic functions and delta functions can be considered degenerate or limiting cases of chirp functions, it is possible to make the general statement that the fractional Fourier transform of a chirp function is always another chirp function. These facts are particularly easy to observe in the time-frequency plane. The Wigner distribution of the chirp function $\exp[i\pi(\chi u^2 + 2\xi u)]$ is $\delta(\mu - \chi u - \xi)$, which is simply a line delta in the time-frequency plane along the line $\mu = \chi u + \xi$ making angle $\arctan \chi$ with the time axis (figure 4.11). The Wigner distribution of $\exp(i2\pi \xi u)$ is $\delta(\mu - \xi)$ and is seen to be a special case in which the line delta is concentrated along the horizontal line $\mu = \xi$. Similarly, the Wigner distribution of $\delta(u - \xi)$ is $\delta(u - \xi)$ and can also be shown to be a special case through a limiting argument. In this case the line delta is concentrated along the vertical line $u = \xi$. Recalling from definition C that fractional Fourier transformation of order a corresponds to rotation of the Wigner distribution by angle $\alpha = a\pi/2$, one can easily verify the facts discussed at the beginning of this paragraph.

It is also of interest to write pair 6 of table 4.1 for the more general case of complex constants ξ and χ. In this case the fractional Fourier transform of $\exp[i\pi(\chi u^2 + 2\xi u)]$ may be expressed as

$$e^{i\pi/4} \frac{\sqrt{1 - i \cot \alpha}}{\sqrt{\chi + \cot \alpha}} \exp\left[i\pi \frac{u^2(\chi - \tan \alpha) + 2u\xi \sec \alpha - \xi^2 \tan \alpha}{1 + \chi \tan \alpha}\right]. \tag{4.115}$$

Specializing the discussion associated with equation 3.148, we can also write the fractional Fourier transform of the complex Gaussian function $(2\Im[1/r_c])^{1/4} \exp(i\pi u^2/r_c)$ with complex radius r_c as

$$(2\Im[1/r'_c])^{1/4} e^{i\pi u^2/r'_c}, \tag{4.116}$$

$$r'_c = \frac{r_c + \tan \alpha}{1 - r_c \tan \alpha},$$

which agrees with equation 3.151 if we substitute $\alpha, \gamma \to \cot \alpha$ and $\beta \to \csc \alpha$.

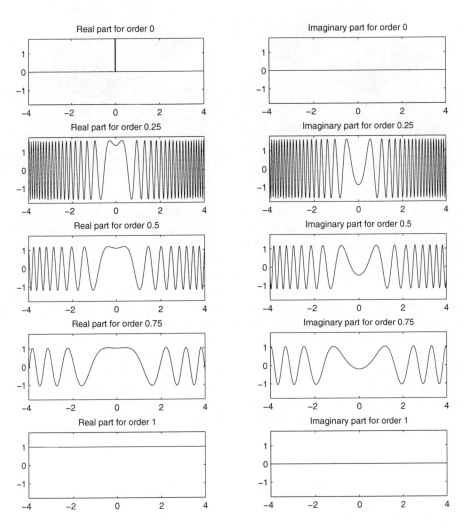

Figure 4.9. Real and imaginary parts of the fractional Fourier transforms of the delta function.

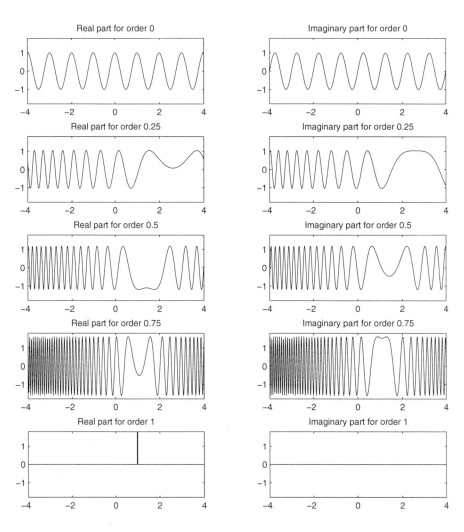

Figure 4.10. Real and imaginary parts of the fractional Fourier transforms of a harmonic function.

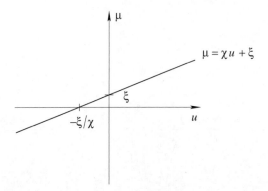

Figure 4.11.   Wigner distribution of a chirp function.

The fractional Fourier transforms of the Hermite-Gaussian functions, and in particular the Gaussian function $\exp(-\pi u^2)$, are equal to themselves multiplied by a complex constant, by virtue of being eigenfunctions of the transform. Scaled versions of these functions are not eigenfunctions in the precise sense, although they will still be transformed into functions with similar magnitude profiles. The ath order transform of $\psi_l(u/M)$ is

$$\left[e^{-il\alpha'} \sqrt{\frac{1 - i\cot\alpha}{M^{-2} - i\cot\alpha}} \right] e^{i\pi q' u^2} \psi_l(u/M'), \qquad (4.117)$$

where

$$\alpha' = \arctan(M^{-2}\tan\alpha),$$

$$q' = \cot\alpha \left(1 - \frac{\cos^2\alpha'}{\cos^2\alpha}\right),$$

$$M' = \frac{\sin\alpha}{M\sin\alpha'}. \qquad (4.118)$$

This result is readily derived by using the scaling property to be given in the next section (table 4.3). We see that the transform of $\psi_l(u/M)$ is a scaled and chirp modulated version of itself. To better understand its behavior, it is useful to examine the dependence of the chirp parameter q' and scale parameter M' on a (figure 4.12). These curves behave as expected when a approaches an integer value. In particular, as a approaches 0, we see that $\alpha' \to 0$, $q' \to 0$, and $M' \to M$, so that we are left with the original function $\psi_l(u/M)$. On the other hand, as a approaches 1, we see that $\alpha' \to \pi/2$, $q' \to 0$, and $M' \to 1/M$, so that we are left with $(-i)^l|M|\psi_l(Mu)$.

4.5   Properties

Several properties of the fractional Fourier transform have been mentioned or derived previously; they are restated in table 4.2 for convenience. The fractional Fourier transform is evidently linear but not shift-invariant. Property 2 states that when a is equal to an integer j, the ath order fractional Fourier transform is equivalent

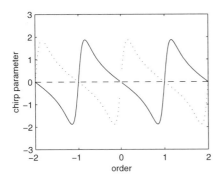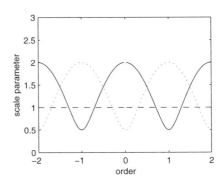

Figure 4.12. Chirp parameter q' and scale parameter M' for $M = 2$ (solid), $M = 1$ (dashed), $M = 1/2$ (dotted).

to the jth integer power of the ordinary Fourier transform, as defined by repeated application. (The way this property is written in the table looks trivial, but this is because this property is already built into our notation.) It also follows that $\mathcal{F}^2 = \mathcal{P}$ (the parity operator), $\mathcal{F}^3 = \mathcal{F}^{-1} = (\mathcal{F})^{-1}$ (the inverse transform operator), $\mathcal{F}^4 = \mathcal{F}^0 = \mathcal{I}$ (the identity operator), and $\mathcal{F}^j = \mathcal{F}^{j \bmod 4}$ (see table 2.5). When a is equal to $4j + 1$, we simply have $K_a(u, u') = K_1(u, u') = \exp(-i2\pi u u')$, and when a is equal to $4j + 3$, we simply have $K_a(u, u') = K_{-1}(u, u') = \exp(i2\pi u u')$. When $a = 4j$, we have $K_a(u, u') = K_0(u, u') = \delta(u - u')$, and when $a = 4j \pm 2$, we have $K_a(u, u') = K_{\pm 2}(u, u') = \delta(u + u')$. Property 3 allows us to associate positive orders with forward transforms and negative orders with inverse transforms. In terms of the kernel, this property is stated as $K_a^{-1}(u, u') = K_{-a}(u, u')$. Property 4 may likewise be stated as $K_a^{-1}(u, u') = K_a^*(u', u)$, which when combined with the previous property implies $(\mathcal{F}^a)^{\mathrm{H}} = \mathcal{F}^{-a}$ or $K_{-a}(u, u') = K_a^*(u', u) = K_a^{\mathrm{H}}(u, u')$. Property 5 is expressed in terms of the kernels as follows:

$$K_{a_2 + a_1}(u, u') = \int K_{a_2}(u, u'') K_{a_1}(u'', u') \, du''. \tag{4.119}$$

Table 4.2. Properties of the fractional Fourier transform, part I.

1.	Linearity	$\mathcal{F}^a[\sum_j \alpha_j f_j(u)] = \sum_j \alpha_j[\mathcal{F}^a f_j(u)]$
2.	Integer orders	$\mathcal{F}^j = (\mathcal{F})^j$
3.	Inverse	$(\mathcal{F}^a)^{-1} = \mathcal{F}^{-a}$
4.	Unitarity	$(\mathcal{F}^a)^{-1} = (\mathcal{F}^a)^{\mathrm{H}}$
5.	Index additivity	$\mathcal{F}^{a_2} \mathcal{F}^{a_1} = \mathcal{F}^{a_2 + a_1}$
6.	Commutativity	$\mathcal{F}^{a_2} \mathcal{F}^{a_1} = \mathcal{F}^{a_1} \mathcal{F}^{a_2}$
7.	Associativity	$\mathcal{F}^{a_3}(\mathcal{F}^{a_2} \mathcal{F}^{a_1}) = (\mathcal{F}^{a_3} \mathcal{F}^{a_2})\mathcal{F}^{a_1}$
8.	Eigenfunctions	$\mathcal{F}^a \psi_l = \exp(-ial\pi/2)\psi_l$
9.	Wigner distribution	$W_{f_a}(u, \mu) = W_f(u \cos\alpha - \mu \sin\alpha, u \sin\alpha + \mu \cos\alpha)$
10.	Parseval	$\langle f(u), g(u) \rangle = \langle f_a(u), g_a(u) \rangle$

α_j are arbitrary complex constants and j is an arbitrary integer.

Table 4.3. Properties of the fractional Fourier transform, part II.

	$f(u)$	$f_a(u)$
1.	$f(-u)$	$f_a(-u)$
2.	$\|M\|^{-1}f(u/M)$	$\sqrt{\frac{1-i\cot\alpha}{1-iM^2\cot\alpha}}\,\exp\left[i\pi u^2\cot\alpha\left(1-\frac{\cos^2\alpha'}{\cos^2\alpha}\right)\right]f_{a'}\left(\frac{Mu\sin\alpha'}{\sin\alpha}\right)$
3.	$f(u-\xi)$	$\exp(i\pi\xi^2\sin\alpha\cos\alpha)\exp(-i2\pi u\xi\sin\alpha)f_a(u-\xi\cos\alpha)$
4.	$\exp(i2\pi\xi u)f(u)$	$\exp(-i\pi\xi^2\sin\alpha\cos\alpha)\exp(i2\pi u\xi\cos\alpha)f_a(u-\xi\sin\alpha)$
5.	$u^n f(u)$	$[\cos\alpha\,u-\sin\alpha\,(i2\pi)^{-1}d/du]^n f_a(u)$
6.	$[(i2\pi)^{-1}d/du]^n f(u)$	$[\sin\alpha\,u+\cos\alpha\,(i2\pi)^{-1}d/du]^n f_a(u)$
7.	$f(u)/u$	$-i\csc\alpha\,\exp(i\pi u^2\cot\alpha)\int_{-\infty}^{2\pi u}f_a(u')\exp(-i\pi u'^2\cot\alpha)\,du'$
8.	$\int_\xi^u f(u')\,du'$	$\sec\alpha\,\exp(-i\pi u^2\tan\alpha)\int_\xi^u f_a(u')\exp(i\pi u'^2\tan\alpha)\,du'$
9.	$f^*(u)$	$f_{-a}^*(u)$
10.	$f^*(-u)$	$f_{-a}^*(-u)$
11.	$[f(u)+f(-u)]/2$	$[f_a(u)+f_a(-u)]/2$
12.	$[f(u)-f(-u)]/2$	$[f_a(u)-f_a(-u)]/2$

ξ and M are real but $M\neq0,\pm\infty$. $\alpha'=\arctan(M^{-2}\tan\alpha)$ where α' is taken to be in the same quadrant as α. Property 7 is not valid when a is an even integer and property 8 is not valid when a is an odd integer.

Property 6 follows immediately from property 5. Property 7 is not special to the fractional Fourier transform and is valid for all linear canonical transforms. Property 10 is equivalent to unitarity. Energy or norm conservation ($\mathrm{En}[f]=\mathrm{En}[f_a]$ or $\|f\|=\|f_a\|$) is a special case. We also note a slightly more general way of writing this property:

$$\langle f_{a_1}(u), g_{a_2}(u)\rangle = \langle f_{a_3}(u), g_{a_4}(u)\rangle \quad\text{when}\quad a_1-a_2=a_3-a_4, \tag{4.120}$$

which is seen to be true by noting that f_{a_3} is the (a_3-a_1)th transform of f_{a_1} and g_{a_4} is the (a_4-a_2)th transform of g_{a_2}. Property 8 was discussed in detail as definition B. It simply states that the eigenfunctions of the fractional Fourier transform are the Hermite-Gaussian functions $\psi_l(u)$. Property 9 was discussed as part of definition C and will also be further discussed below (section 4.6). This property states that the Wigner distribution of the fractional Fourier transform of a function is a rotated version of the Wigner distribution of the original function. Another important property is the uncertainty relation; this is discussed on page 169. We may finally note that the transform is continuous in the order a. That is, small changes in the order a correspond to small changes in the transform $f_a(u)$. In other words, as a is changed gradually, we also observe the transform $f_a(u)$ to change gradually. Nevertheless, care is always required in dealing with cases where a approaches an even integer, since in this case the kernel approaches a delta function. A more rigorous discussion of continuity with respect to a may be found in McBride and Kerr 1987.

Various operational properties of the transform are listed in table 4.3 (Namias 1980a, McBride and Kerr 1987, Mendlovic and Ozaktas 1993a, Almeida 1994, Ozaktas, Kutay, and Mendlovic 1998, 1999). Most of them are most readily derived or verified by using definition A or the symmetry properties of the kernel. Properties 5 and 6 can be first shown for $n=1$ and then generalized by repeated application. Properties 3 to 6 will be discussed in greater detail in sections 4.7 and 4.8. When a is an even integer, property 7 is trivial: if $a=4j$ for any integer j then $f_a(u)=f(u)/u$, and if $a=4j+2$ then $f_a(u)=f(-u)/(-u)$. When $a=4j+1$ or $a=4j+3$, the right-hand side of property 8 reduces to the forward or inverse ordinary Fourier transforms of $\int_{\xi}^{u} f(u')\, du'$ respectively. Property 1 is a special case of property 2, which will be further discussed below. Operations satisfying property 1 are referred to as even operations, so that the fractional Fourier transform is an even operation. Properties 9 and 10 imply that if $f(u)$ is real then $f_a(u) = f_{-a}^*(u)$ or $f_{-a}(u) = f_a^*(u)$, and if $f(u)$ is purely imaginary then $f_a(u) = -f_{-a}^*(u)$ or $f_{-a}(u) = -f_a^*(u)$. Properties 11 and 12 imply that the transform of every even function is even and the transform of every odd function is odd. Similar facts can be stated in operator form: all even operators, and in particular the fractional Fourier transform operator, commute with the parity operator $\mathcal{P}$: $\mathcal{F}^a \mathcal{P} = \mathcal{P} \mathcal{F}^a$ or $\mathcal{F}^a = \mathcal{P} \mathcal{F}^a \mathcal{P}$. It is also known that the eigenfunctions of even operations can always be chosen to be of definite (even or odd) parity, as the Hermite-Gaussian functions indeed are. Let us finally note that since $\mathcal{F}^{a\pm 2} = \mathcal{F}^a \mathcal{F}^{\pm 2} = \mathcal{F}^a \mathcal{P} = \mathcal{P} \mathcal{F}^a$, it follows that $f_{a\pm 2}(u) = f_a(-u)$.

Property 2 is the generalization of the ordinary Fourier transform property stating that the Fourier transform of $f(u/M)$ is $|M| F(M\mu)$. Notice that the fractional Fourier transform of $f(u/M)$ cannot be expressed as a scaled version of $f_a(u)$ for the same order a. Rather, the fractional Fourier transform of $f(u/M)$ turns out to be a scaled and chirp modulated version of $f_{a'}(u)$ where $a' \neq a$ is a different order given by

$$\frac{a'\pi}{2} \equiv \alpha' = \arctan\left(\frac{\tan \alpha}{M^2}\right), \tag{4.121}$$

where α' is chosen to be in the same quadrant as α. (Actually, the final expression for the transform of $f(u/M)$ does not depend on which quadrant we chose α' in. Nevertheless, we specify the quadrant chosen for concreteness.) This result is most easily grasped by considering a numerical example in the time-frequency plane. Let us consider the function $f(u) = \exp(-\pi \chi u^2)$ with $\chi = 3$. The ath order fractional Fourier transform of $f(u/M)$ with $a = 0.7$ and $M = 3$ can be calculated by using pair 9 of table 4.1 in property 2 of table 4.3. Figure 4.13a shows a contour of the Wigner distribution $W_f(u, \mu)$ of the original function $f(u)$. Figure 4.13b shows the Wigner distribution of $f(u/3)$ which we know is given by $3W_f(u/3, 3\mu)$ (table 3.2, property 8). (Since $M = \chi$ in our example, this distribution also corresponds to the distribution of the ordinary Fourier transform of $f(u)$; however, this would not be true for other functions or other values of M and has no bearing on our discussion.) Figure 4.13d shows the distribution of the 0.7th order fractional Fourier transform of $f(u/3)$, which is obtained by rotating part b clockwise by an angle $0.7\pi/2$ ($= 63°$). On the other hand, figure 4.13c shows the distribution of the 0.7th transform of the unscaled function $f(u)$, obtained by rotating part a by the same angle. Now, it is not possible to arrive at part d by scaling part c, corresponding to the fact that $\mathcal{F}^a[f(u/M)]$ is not a scaled version of $f_a(u)$. Figure 4.13e shows the distribution of the $a' = 0.14$th transform of $f(u)$, which is obtained by rotating part a by an angle $\alpha' = 12.6°$. Figure 4.13f shows the distribution of the scaled version $\propto f_{a'}(Mu \sin \alpha'/\sin \alpha) = f_{0.14}(u/1.39)$ of this a'th transform. Now, it is possible to obtain part d by shearing this figure in the vertical direction, corresponding to the fact that the ath order fractional Fourier transform of a scaled function is a scaled and chirp multiplied version of the a'th transform of the original function (anchor the center of the contour in part f at the origin, pull the lower left corner up and pull the upper right corner down until you coincide with part d). For completeness, we note

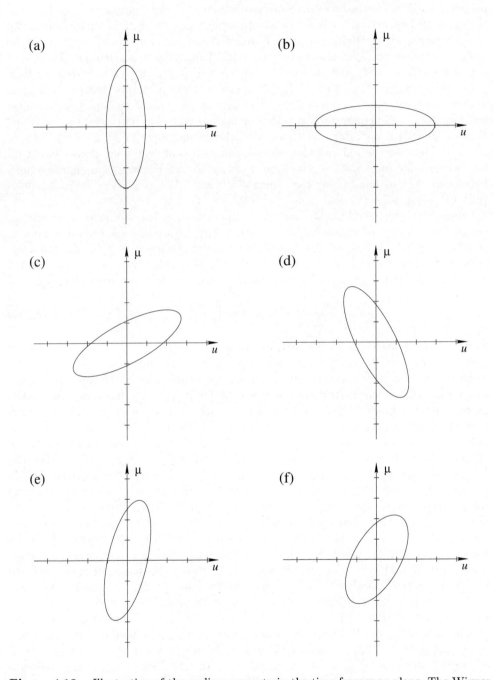

Figure 4.13. Illustration of the scaling property in the time-frequency plane. The Wigner distribution drops to $\exp(-2\pi)$ of its peak value at the contours shown. Each tick mark represents a distance of $1/\sqrt{3}$.

that $M' = \sin\alpha/M\sin\alpha' = 1.39$ and $q' = \cot\alpha(1 - \cos^2\alpha'/\cos^2\alpha) = -1.85$ in our example (see equations 4.118). It is also possible to interpret the scaling theorem in terms of the matrix decompositions discussed on page 104. Since fractional Fourier transforms, scaling, and chirp multiplication are all linear canonical transforms, the above theorem corresponds to the following matrix equality:

$$\begin{bmatrix} \cos\alpha & \sin\alpha \\ -\sin\alpha & \cos\alpha \end{bmatrix} \begin{bmatrix} M & 0 \\ 0 & 1/M \end{bmatrix} = \begin{bmatrix} 1 & 0 \\ q' & 1 \end{bmatrix} \begin{bmatrix} M' & 0 \\ 0 & 1/M' \end{bmatrix} \begin{bmatrix} \cos\alpha' & \sin\alpha' \\ -\sin\alpha' & \cos\alpha' \end{bmatrix}. \tag{4.122}$$

The left-hand side of this equation corresponds to the ath order fractional Fourier transform of the scaled function, whereas the right-hand side corresponds to the a'th order fractional Fourier transform of the original function followed by scaling followed by chirp multiplication.

The fractional Fourier transform does not have a convolution or multiplication property of comparable simplicity to that of the ordinary Fourier transform. The basic result leading to these properties relates the fractional Fourier transform of the product $h(u) = f(u)g(u)$ of two functions $f(u)$ and $g(u)$ to their individual fractional Fourier transforms (Wolf 1979):

$$h_a(u) = \iint f_a(u')g_a(u'')K^{(a)}(u, u', u'')\, du'\, du'', \tag{4.123}$$

$$K^{(a)}(u, u', u'') = \int K_a(u, u''')K_{-a}(u''', u')K_{-a}(u''', u'')\, du'''$$

$$= \frac{\mathrm{sgn}(\cot\alpha)\sqrt{1 - i\tan\alpha}}{|\sin\alpha|} e^{i\pi\cot\alpha(u^2 - u'^2 - u''^2)} e^{i\pi\sec\alpha\csc\alpha(u - u' - u'')^2},$$

where $K^{(a)}(u, u', u'')$ serves as a coupling coefficient (whose explicit form has been derived using the integrals in section 2.10.2). It is possible to show that this reduces to the familiar convolution and multiplication properties for $a = -1$ and $a = 1$, respectively, using equation 2.6. These properties are also discussed in Almeida 1997, Mustard 1998, and Zayed 1998a.

It is instructive to discuss the convolution and multiplication properties also from a more general perspective. Let us consider the multiplication of $f_a(u_a)$ with $h(u_a)$ to produce $g_a(u_a)$, where all functions reside in the ath fractional Fourier domain:

$$g_a(u_a) = h(u_a)f_a(u_a). \tag{4.124}$$

To find what this corresponds to in the u domain, let us write it as $g_a(u_a) = h(\mathcal{U}_a)f_a(u_a)$ and then in the representation-independent abstract form $g = h(\mathcal{U}_a)f$. Now, this equation can be written in the u domain as

$$g(u) = h(\cos\alpha\,\mathcal{U} + \sin\alpha\,\mathcal{D})f(u). \tag{4.125}$$

Likewise, the convolution

$$g_a(u_a) = h(u_a) * f_a(u_a) \tag{4.126}$$

can be written as $g_{a+1}(u_{a+1}) = h_1(\mathcal{U}_{a+1})f_{a+1}(u_{a+1})$, $g = h_1(\mathcal{D}_a)f$, and

$$g(u) = h_1(-\sin \alpha\, \mathcal{U} + \cos \alpha\, \mathcal{D})f(u). \qquad (4.127)$$

While equations 4.125 and 4.127 may not be useful when it comes to actually evaluating them for given functions, they exhibit the convolution and multiplication properties in their simplest and most transparent forms.

Although we did not include them in the table, here we note a final pair of properties which are especially useful in wave propagation applications. The fractional Fourier transform of the multiplication of $f(u)$ with a chirp of the form $\exp(-i\pi q u^2)$ is given by

$$\sqrt{\frac{1 - i\cot \alpha}{1 - i\cot \alpha'}}\, \exp\left[i\pi u^2 \cot \alpha \left(1 - \frac{\sin \alpha' \cos \alpha'}{\sin \alpha \cos \alpha}\right)\right] f_{a'}\left(\frac{u \sin \alpha'}{\sin \alpha}\right), \qquad (4.128)$$

where $a'\pi/2 = \alpha' = \operatorname{arccot}(\cot \alpha - q)$. Here we choose to invert the cotangent such that if $\alpha \in [0, \pi)$ then α' is also in $[0, \pi)$, and if $\alpha \in [-\pi, 0)$ then α' is also in $[-\pi, 0)$. Similarly, the fractional Fourier transform of the convolution of $f(u)$ with a chirp of the form $e^{-i\pi/4}\sqrt{1/r}\, \exp(i\pi u^2/r)$ is given by

$$\sqrt{\frac{1 + i\tan \alpha}{1 - i\cot \alpha'}}\, \exp\left[-i\pi u^2 \tan \alpha \left(1 + \frac{\sin \alpha' \cos \alpha'}{\sin \alpha \cos \alpha}\right)\right] f_{a'}\left(\frac{-u \sin \alpha'}{\cos \alpha}\right), \qquad (4.129)$$

where $a'\pi/2 = \alpha' = \operatorname{arccot}(-\tan \alpha - r)$. Here we choose to invert the cotangent such that if $\alpha \in [\pi/2, 3\pi/2)$ then α' is also in $[\pi/2, 3\pi/2)$, and if $\alpha \in [-\pi/2, \pi/2)$ then α' is also in $[-\pi/2, \pi/2)$.

As a final property, we note that the trace of the fractional Fourier transform kernel is

$$\int K_a(u, u)\, du = \frac{e^{-i\alpha/2}}{2\sin(\alpha/2)}, \qquad (4.130)$$

for $0 < |\alpha| < \pi$.

Zayed (1996) and Xia (1996) have discussed sampling theorems for the fractional Fourier transform. In particular, they show that if $f_a(u)$ is compact for any a, then it is possible to recover $f_{a'}(u)$ from appropriately spaced samples of $f_{a'}(u)$ for any $a' \neq a \pm 2j$ where j is an integer. The required sampling rate is proportional to $\csc(\alpha' - \alpha)$. When $(\alpha' - \alpha)$ is equal to an odd integer multiple of $\pi/2$ the sampling rate is as small as possible, corresponding to the conventional Fourier sampling theorem. As $(\alpha' - \alpha)$ approaches even integer multiples of $\pi/2$, larger and larger sampling rates are required. Compactness of the ordinary Fourier transform of $f(u)$—commonly referred to as band-limitedness—puts the strongest constraint on how fast the variations of $f(u)$ can be and thus allows the samples to be spaced as widely as possible. On the other hand, as a becomes smaller and smaller, the constraint becomes weaker and weaker and $f(u)$ is able to make faster variations, so that a narrower sample spacing is required. When $a = 0$, compactness of the function $f(u)$ itself puts no constraint on how quickly it varies.

4.6   Rotations and projections in the time-frequency plane

4.6.1   Rotation of the Wigner distribution

We now reconsider one of the most important properties of the fractional Fourier transform (Mustard 1989, 1996, Lohmann 1993a, b, Özaktaş and Mendlovic 1993, Ozaktas, Mendlovic, and Lohmann 1993, Almeida 1993, 1994, Mendlovic, Ozaktas, and Lohmann 1994a, Ozaktas and others 1994a, Stankovic and Djurovic 1998). Some authors have chosen to make this property the definition of the fractional Fourier transform (Lohmann 1993b), defining the transform as that operation which corresponds to rotation of the Wigner distribution (definition C.2).

If $W_f(u, \mu)$ denotes the Wigner distribution of $f(u)$, then the Wigner distribution of the ath fractional Fourier transform of $f(u)$, denoted by $W_{f_a}(u, \mu)$, is given by equation 4.28:

$$W_{f_a}(u, \mu) = W_f(u \cos \alpha - \mu \sin \alpha, u \sin \alpha + \mu \cos \alpha), \qquad (4.131)$$

so that the Wigner distribution of $W_{f_a}(u, \mu)$ is obtained from $W_f(u, \mu)$ by rotating it clockwise by an angle α. Let us define $\mathcal{ROT}_\alpha$ to be the operator which rotates a function of (u, μ) by angle α in the conventional counterclockwise direction:

$$\mathcal{ROT}_\alpha[W_f(u, \mu)] \equiv W_f(u \cos \alpha + \mu \sin \alpha, -u \sin \alpha + \mu \cos \alpha). \qquad (4.132)$$

With this definition equation 4.131 becomes

$$W_{f_a}(u, \mu) = \mathcal{ROT}_{-\alpha}[W_f(u, \mu)], \qquad \alpha = a\pi/2. \qquad (4.133)$$

It also follows that

$$W_{f_{a_2}}(u, \mu) = \mathcal{ROT}_{-(\alpha_2 - \alpha_1)}[W_{f_{a_1}}(u, \mu)]. \qquad (4.134)$$

We now show how equation 4.131 can be derived directly from definition A (Ozaktas and others 1994a). A similar derivation is given by Mustard (1996) and by Almeida (1994). Lohmann's derivation goes in the reverse direction, starting from the rotation property and arriving at what is essentially equation 4.4 (Lohmann 1993b). The derivation is straightforward yet somewhat lengthy, so that we only sketch the steps. The fractional Fourier transform was defined in equation 4.4 as a linear integral transform with kernel $K_a(u, u')$:

$$f_a(u) = \int K_a(u, u') f(u') \, du'. \qquad (4.135)$$

The Wigner distribution of $f_a(u)$ can be evaluated by substituting this integral expression in the definition of the Wigner distribution (equation 3.16):

$$W_{f_a}(u, \mu) = \int f_a(u + u'/2) f_a^*(u - u'/2) e^{-i2\pi\mu u'} \, du'. \qquad (4.136)$$

Writing

$$f_a(u + u'/2) = \int K_a(u + u'/2, u'') f(u'') \, du'', \qquad (4.137)$$

$$f_a^*(u - u'/2) = \int K_a^*(u - u'/2, u''') f^*(u''') \, du''', \qquad (4.138)$$

we obtain an expression involving three integrals over the dummy variables u'', u''', u'. After some manipulation, we are faced with an integral of the form

$$\int e^{-i2\pi u'[\mu - u\cot\alpha + \csc\alpha(u'' + u''')/2]} \, du' \qquad (4.139)$$

inside the expression we are working with. This integral is equal to $\delta(\mu - u\cot\alpha + \csc\alpha(u'' + u''')/2)$. This delta function will eliminate the integral over u''' by virtue of the sifting property, leaving us with

$$W_{f_a}(u, \mu) = 2\,e^{-i4\pi u\csc\alpha(\mu\sin\alpha - u\cos\alpha)} e^{-i4\pi\cot\alpha(\mu\sin\alpha - u\cos\alpha)^2}$$

$$\times \int e^{-i4\pi u''(u\sin\alpha + \mu\cos\alpha)} f(u'') f^*(-u'' - 2(\mu\sin\alpha - u\cos\alpha)) \, du''. \quad (4.140)$$

We wish to show that this is the Wigner distribution of $f(u)$ rotated by α (equation 4.131). The Wigner distribution of $f(u)$ is given by equation 3.16. Rotating this clockwise by α we obtain

$$W_f(u\cos\alpha - \mu\sin\alpha, u\sin\alpha + \mu\cos\alpha)$$

$$= \int f(u\cos\alpha - \mu\sin\alpha + u'/2) f^*(u\cos\alpha - \mu\sin\alpha - u'/2)$$

$$\times\, e^{-i2\pi(u\sin\alpha + \mu\cos\alpha)u'} \, du'. \quad (4.141)$$

Now, with the change of integration variable $u\cos\alpha - \mu\sin\alpha + u'/2 = u''$, it is easily shown that the right-hand sides of the last two equations are identical, demonstrating the equality of the left-hand sides, which is the result we sought to show.

4.6.2 Projections of the Wigner distribution

An at least equally important form of equation 4.133 follows easily (Mustard 1989, 1996, Lohmann and Soffer 1993, 1994, Ozaktas and others 1994a). Let us recall equations 3.17 and 3.18:

$$\int W_f(u, \mu) \, d\mu = |f(u)|^2, \qquad (4.142)$$

$$\int W_f(u, \mu) \, du = |F(\mu)|^2, \qquad (4.143)$$

which state that the integral projection of $W_f(u, \mu)$ onto the u axis is the squared magnitude of the u-domain representation of the signal and that the integral projection of $W_f(u, \mu)$ onto the μ axis is the squared magnitude of the μ-domain representation of the signal. Now, let us rewrite the first of these equations for $f_a(u)$, the ath order fractional Fourier transform of $f(u)$:

$$\int W_{f_a}(u, \mu) \, d\mu = |f_a(u)|^2. \qquad (4.144)$$

Since $W_{f_a}(u, \mu)$ is simply $W_f(u, \mu)$ rotated clockwise by angle α, the integral projection of $W_{f_a}(u, \mu)$ onto the u axis is geometrically identical to the integral projection of

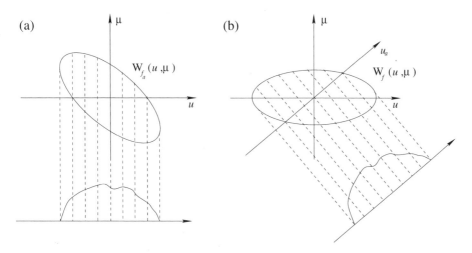

Figure 4.14. The integral projection of $W_{f_a}(u,\mu)$ onto the u axis (a) is equal to the integral projection of $W_f(u,\mu)$ onto the u_a axis (b). The u_a axis makes an angle $\alpha = a\pi/2$ with the u axis.

$W_f(u,\mu)$ onto an axis making angle α with the u axis (figure 4.14). We will refer to this new axis making angle $\alpha = a\pi/2$ with the u axis as the u_a axis. Let $\mathcal{RDN}_\alpha$ denote the *Radon transform* operation, which maps a two-dimensional function of (u,μ), to its integral projection onto an axis making angle α with the u axis (page 56). Thus

$$\int W_{f_a}(u,\mu)\,d\mu = \mathcal{RDN}_0[W_{f_a}(u,\mu)] = \mathcal{RDN}_\alpha[W_f(u,\mu)], \qquad (4.145)$$

which in turn implies

$$\mathcal{RDN}_\alpha[W_f(u,\mu)] = |f_a(u)|^2. \qquad (4.146)$$

Thus *the integral projection of the Wigner distribution $W_f(u,\mu)$ of a function $f(u)$ onto an axis making angle $\alpha = a\pi/2$ with the u axis, is equal to the squared magnitude of the ath order fractional Fourier transform $f_a(u)$ of the function.* When $a = 0$ this property reduces to equation 4.142 and when $a = 1$ this property reduces to equation 4.143. Thus, equation 4.146 is a generalization of these two equations.

In a series of papers, Wood and Barry discussed what they referred to as the "Radon-Wigner transform" without realizing its relation to (and most likely without being aware of) the fractional Fourier transform (Wood and Barry 1994a, b, c). The above discussion demonstrates that the Radon-Wigner transform is simply the squared magnitude of the fractional Fourier transform. Wood and Barry discussed the properties and applications of the Radon transform of the Wigner distribution and showed that it could be evaluated directly as the magnitude squared of a dechirping integral, without first evaluating the Wigner distribution. These dechirping integrals are of the form

$$\left|\int e^{i\pi(\cot\alpha\,u'^2 - 2\csc\alpha\,uu')} f(u')\,du'\right|^2, \qquad (4.147)$$

which we recognize as being equal to $|\sin\alpha|\,|f_a(u)|^2$ from equation 4.4. They also recognized that Radon-Wigner transforms can be expressed as the integral projections of sheared (and scaled) Wigner distributions, results which are readily interpretable in terms of the various decomposition we studied in section 3.4.4. They also noted that for certain values of the parameters, dechirping the Fourier transform of the function rather than the function itself, resulted in better numerical implementation. A similar issue will come up in section 6.7 when we discuss the numerical computation of the fractional Fourier transform.

Before we continue, we note that equation 4.146 is easily generalized to cross Wigner distributions. The cross Wigner distribution W_{fg} of two functions f and g is defined by an equation similar to equation 3.16, but with the second appearance of f replaced by g. Then it can be shown that (Raveh and Mendlovic 1999, Özdemir and Arıkan 2000)

$$\mathcal{RDN}_\alpha[W_{fg}(u,\mu)] = f_a(u)g_a^*(u).\tag{4.148}$$

A number of additional results follow immediately from the projection-slice theorem (page 56). We saw in chapter 3 that the ambiguity function is essentially the two-dimensional Fourier transform of the Wigner distribution:

$$\iint W_f(u,\mu)e^{-i2\pi(\bar{\mu}u+\bar{u}\mu)}\,du\,d\mu = A_f(-\bar{u},\bar{\mu}).\tag{4.149}$$

The projection-slice theorem implies that the slice of $A_f(-\bar{u},\bar{\mu})$ at angle α is equal to the Fourier transform of the integral projection of $W_f(u,\mu)$ at angle α, which we know to be equal to $|f_a(u)|^2$ (equation 4.146). Thus the slices of the ambiguity function are equal to the Fourier transforms of $|f_a(u)|^2$:

$$\mathcal{F}[|f_a(u)|^2] = \mathcal{SLC}_{\alpha+\pi/2}[A_f(\bar{u},\bar{\mu})],\tag{4.150}$$

The slice of $A_f(-\bar{u},\bar{\mu})$ appearing in the projection-slice theorem must be taken at angle α with respect to a horizontal axis $\bar{\mu}$ and vertical axis $\bar{u}$, since $\bar{\mu}$ is the Fourier variable corresponding to u and $\bar{u}$ is the Fourier variable corresponding to μ. This is equivalent to the standard slice $\mathcal{SLC}$ of $A_f(\bar{u},\bar{\mu})$ at angle $\alpha+\pi/2$ with respect to the horizontal axis $\bar{u}$ and vertical axis $\bar{\mu}$. Equation 4.150 is a generalization of its special cases for $a=0$ and $a=1$ given by equations 3.27 and 3.28.

The results of this section, and in particular equation 4.146, continue to hold when $W_f(u,\mu)$ is the Wigner distribution of a random process since the expectation value operation can move inside the Radon transform and rotation operators:

$$\langle W_{f_a}(u,\mu)\rangle = \mathcal{ROT}_{-\alpha}[\langle W_f(u,\mu)\rangle],\tag{4.151}$$

$$\mathcal{RDN}_\alpha[\langle W_f(u,\mu)\rangle] = \langle|f_a(u)|^2\rangle.\tag{4.152}$$

4.6.3   Other time-frequency representations

The Wigner distribution is not the only time-frequency representation satisfying a rotation property having the form of equation 4.133. A similar relation holds for the ambiguity function (Ozaktas and others 1994a, Almeida 1994), by virtue of the fact that the ambiguity function is essentially the two-dimensional Fourier transform of the Wigner distribution, and the fact that the two-dimensional Fourier transform of the rotated version of a function is the rotated version of the two-dimensional Fourier transform of the original function. Almeida (1994) further showed that a rotation property also holds for (a modified form of) the short-time Fourier transform and spectrogram. Consider the modified short-time Fourier transform (MSTFT) as defined in equations 3.5 and 3.6. It is shown in Almeida 1994 that

$$WF_f^{(w)}(u, \mu) = e^{i\pi\mu_a u_a} \int f_a(u'_a) w_a^*(u'_a - u_a) e^{-i2\pi\mu_a u'_a} \, du'_a, \tag{4.153}$$

where $u_a = \cos\alpha \, u + \sin\alpha \, \mu$ and $\mu_a = -\sin\alpha \, u + \cos\alpha \, \mu$ are the rotated axes. (The u_a and μ_a axes both make positive angle α with the u and μ axes respectively.) We quote Almeida: "The right-hand side of this equation is the modified STFT of f_a computed with window w_a and with arguments (u_a, μ_a). The left-hand side is the modified STFT of f computed with window w and with arguments (u, μ). As in the case with the Wigner distribution, this equation shows that the MSTFT of f_a is the same as the MSTFT of f, again taking into account the rotation, i.e., that it is simply a rotated version of the MSTFT of f or that it is the MSTFT of f expressed in the rotated axes (u_a, μ_a). ... The spectrogram is simply the squared magnitude of the STFT and, therefore, of the MSTFT as well. The results we obtained on the MSTFT immediately lead us to conclude that the effect of the fractional Fourier transform on the spectrogram is identical to the one it has on the MSTFT: the spectrogram of f_a computed with window w_a is a rotated version of the spectrogram of f computed with window w."

It is also possible to show that the Radon transforms of the ambiguity function and the cross ambiguity function are given by (Özdemir and Arıkan 2000)

$$\mathcal{RDN}_\alpha[A_f(\bar{u}, \bar{\mu})] = f_a(u/2) f_a^*(-u/2), \tag{4.154}$$

$$\mathcal{RDN}_\alpha[A_{fg}(\bar{u}, \bar{\mu})] = f_a(u/2) g_a^*(-u/2). \tag{4.155}$$

The cross ambiguity function is defined by an equation similar to equation 3.26 by replacing the second occurrence of f with g. The Radon transforms of the Wigner distribution and ambiguity function will be further explored in chapter 5.

These results lead one to inquire whether a rotational relation similar to equation 4.133 is valid for a more general class of time-frequency representations. We will now see that the rotation property indeed generalizes to certain other time-frequency representations belonging to the Cohen class. The Cohen class of time-frequency representations are obtained by convolving the Wigner distribution with a kernel characterizing each representation. We show below that the time-frequency representation of the fractional Fourier transform of a function is a rotated version of the representation of the original function, if the kernel is rotationally symmetric. Thus the fractional Fourier transform corresponds to rotation of a relatively large class of time-frequency representations (phase-space representations), confirming the important role this transform plays in the study of such representations.

We denote by $TFE_f(u, \mu)$ time-frequency representations which are members of the Cohen class. These representations can be derived from the Wigner distribution through equation 3.36 (Cohen 1989, 1995, Hlawatsch and Boudreaux-Bartels 1992):

$$TFE_f(u,\mu) = \iint \psi_{TFE}(u-u',\mu-\mu')W_f(u',\mu')\,du'\,d\mu'. \tag{4.156}$$

$\psi_{TFE}(u,\mu)$ is a kernel uniquely corresponding to the representation $TFE_f(u,\mu)$. We show below that $TFE_f(u,\mu)$ will satisfy

$$TFE_{f_a}(u,\mu) = \mathcal{ROT}_{-\alpha}[TFE_f(u,\mu)] \tag{4.157}$$

for all f and $\alpha = a\pi/2$, if $\psi_{TFE}(u,\mu)$ is rotationally symmetric around the origin; that is, if $\psi_{TFE}(u,\mu)$ is a function of $(u^2+\mu^2)^{1/2}$ only. (The same condition can also be stated in terms of the alternative kernel function $\Psi_{TFE}(\bar{u},\bar{\mu})$, which is also employed in the study of the Cohen class (equation 3.38). Since $\Psi_{TFE}(\bar{u},\bar{\mu})$ and $\psi_{TFE}(u,\mu)$ constitute a two-dimensional Fourier transform pair, rotational symmetry of either implies rotational symmetry of the other.)

We only sketch the main features of the proof since the operations involved are elementary (Ozaktas, Erkaya, and Kutay 1996). First, apply $\mathcal{ROT}_{-\alpha}$, as defined in equation 4.132, to both sides of equation 4.156 to obtain

$$\mathcal{ROT}_{-\alpha}[TFE_f(u,\mu)]$$
$$= \iint \psi_{TFE}(u\cos\alpha - \mu\sin\alpha - u'', u\sin\alpha + \mu\cos\alpha - \mu'')W_f(u'',\mu'')\,du''\,d\mu'', \tag{4.158}$$

where we have also switched from the dummy variables (u',μ') to (u'',μ''). Now, consider an instance of equation 4.156 for the function $f_a(u)$, rather than $f(u)$. Use equations 4.133 and 4.132 to replace $W_{f_a}(u',\mu')$ by $W_f(u'\cos\alpha - \mu'\sin\alpha, u'\sin\alpha + \mu'\cos\alpha)$. Finally, put $u'' = u'\cos\alpha - \mu'\sin\alpha$, $\mu'' = u'\sin\alpha + \mu'\cos\alpha$, to obtain

$$TFE_{f_a}(u,\mu)$$
$$= \iint \psi_{TFE}(u - u''\cos\alpha - \mu''\sin\alpha, \mu + u''\sin\alpha - \mu''\cos\alpha)W_f(u'',\mu'')\,du''\,d\mu''. \tag{4.159}$$

Equation 4.157 will be true if the right-hand sides of the last two equations are equal for all f and for all α (or a). This will be the case if

$$\psi_{TFE}(u\cos\alpha - \mu\sin\alpha - u'', u\sin\alpha + \mu\cos\alpha - \mu'')$$
$$= \psi_{TFE}(u - u''\cos\alpha - \mu''\sin\alpha, \mu + u''\sin\alpha - \mu''\cos\alpha) \tag{4.160}$$

for all of the appearing variables. By treating u,μ as coordinate variables and u'',μ'',α as parameters, it is easy to see that this equality can hold only if $\psi_{TFE}(u,\mu)$ is rotationally symmetric. (The left-hand side is shifted and rotated, whereas the right-hand side is merely shifted. Thus rotational symmetry is necessary for both sides to be equal.) It is also not difficult to show, by the transformation $u' = u - u''\cos\alpha - \mu''\sin\alpha$, $\mu' = \mu + u''\sin\alpha - \mu''\cos\alpha$, that this is also a sufficient condition. This completes the proof.

Thus, fractional Fourier transformation corresponds to rotation of not only the Wigner distribution, ambiguity function, and spectrogram, but of a much larger class of time-frequency representations (phase-space representations). This not only confirms the important role this transform plays in the study of such representations, but also further supports the notion of referring to the axis making angle $\alpha = a\pi/2$ with the u axis as the *ath fractional Fourier domain*, as will be further discussed in section 4.9. Originally, applications of the fractional Fourier transform have been predominantly posed in terms of the Wigner distribution. This result indicates that it is also possible to deal with other time-frequency representations which might be more appropriate for particular applications.

Despite this generalization, the only representation which satisfies a relation of the form of equation 4.146 is the Wigner distribution: there is only one representation whose integral projections are equal to the squared magnitudes of the fractional Fourier transforms at all angles, and that is the Wigner distribution (Mustard 1989, 1996). This follows from the fact that a complete set of integral projections uniquely determines a representation by virtue of the projection-slice theorem (page 56). That is, knowledge of $\mathcal{RDN}_\alpha[TFE_f(u,\mu)](u_a)$ for all α and u_a uniquely determines $TFE_f(u,\mu)$. In particular, if $\mathcal{RDN}_\alpha[TFE_f(u,\mu)](u_a) = |f_a(u_a)|^2$, then $TFE_f(u,\mu)$ is equal to the Wigner distribution $W_f(u,\mu)$. There is no other representation all of whose integral projections are equal to $|f_a(u_a)|^2$. (This is the basis of so-called phase-space tomography, which will be discussed in chapter 9.)

The integral projections of a time-frequency representation on the time and frequency axes are often referred to as the time and frequency *marginals*, borrowing the terminology used for probability distributions. Thus, integral projections onto oblique axes may be referred to as *generalized marginals*. The integral projection of many time-frequency representations belonging to the Cohen class, onto the time and frequency axes, are equal to the squared magnitudes of the time and frequency representations of the signal (they are said to "satisfy the time and frequency marginals," see table 3.6). Likewise, many members of the Cohen class have rotationally symmetric $\psi_{TFE}(u,\mu)$ and $\Psi_{TFE}(\bar{u},\bar{\mu})$. However, the Wigner distribution is the only representation combining both properties. Representations which satisfy the time and frequency marginals are characterized by $\Psi_{TFE}(0,\bar{\mu}) = 1$ and $\Psi_{TFE}(\bar{u},0) = 1$ (table 3.6, properties 4 and 5). If $\Psi_{TFE}(\bar{u},\bar{\mu})$ is also rotationally symmetric, this implies $\Psi_{TFE}(\bar{u},\bar{\mu}) = 1$, which corresponds to the Wigner distribution (table 3.5). Rotational symmetry combined with satisfaction of the time and frequency marginals leads to satisfaction of all generalized marginals. Thus the Wigner distribution is the only representation all of whose marginals are equal to the squared magnitudes of the fractional Fourier transforms, although many other representations satisfy the time and frequency marginals.

Mustard also discussed these in relation to the shift invariance property of the Cohen class of representations (1989, 1996). This class of representations is characterized by time-frequency invariance; time and frequency shifts of the original function $f(u)$ are simply and directly reflected as corresponding shifts in the time-frequency plane (table 3.6, properties 2 and 3). The concept of shift invariance can be generalized by considering shifts along arbitrary oblique axes: we can require that a shift of the *ath* order fractional Fourier transform correspond to a shift of the representation by the same amount along the u_a axis in the time-frequency plane (for all a).

Mustard (1989, 1996) concluded that "there is exactly one member of the Cohen class of phase-space distributions that satisfies the generalized translation property and the generalized marginal distribution (or Radon transform) property and that is the Wigner distribution." Requiring either the generalized translation property or the generalized marginal distribution property is sufficient to single out the Wigner distribution.

Cohen (1989) argued that there is nothing special about the Wigner distribution among other members of the Cohen class, since all members of this class (including the Wigner distribution) are derivable from each other through relations such as equation 4.156 (the role of the Wigner distribution in this equation can be taken by any other member of this class). However, we have seen above that only the Wigner distribution satisfies equation 4.146 and generalized shift invariance. This has led Mustard to suggest that the Wigner distribution *is* a specially distinguished member of the Cohen class (Mustard 1989, 1996). Although their discussion is not posed in terms of the fractional Fourier transform, Bertrand and Bertrand (1987) have also argued that the Wigner distribution has a prominent position among the others, since "it is the only one to have the correct marginalizations whatever the direction of integration in the u-μ plane."

The concept of generalized marginals has found use in the study and evaluation of time-frequency distributions. For instance, see Bertrand and Bertrand 1987, 1992, Fonollosa and Nikias 1994, Fonollosa 1996, Sang, Williams, and O'Neill 1996, and Xia and others 1996. The fractional Fourier transform has also been found to play an important role in transforming from a rectangular to a quincunx Gabor lattice (Bastiaans and van Leest 1998). Other works dealing with the application of the fractional Fourier transform to the study of time-frequency representations include Akay and Boudreaux-Bartels 1998c and Owechko 1998.

4.7   Coordinate multiplication and differentiation operators

In this and the following section we discuss in greater detail properties 3 to 6 of table 4.3. First we focus on properties 5 and 6 in table 4.3 with $n = 1$:

$$\mathcal{F}^a \left[u f(u) \right] = \cos\alpha \; u_a f_a(u_a) - \sin\alpha \; \frac{1}{i2\pi} \frac{df_a(u_a)}{du_a}, \qquad (4.161)$$

$$\mathcal{F}^a \left[\frac{1}{i2\pi} \frac{df(u)}{du} \right] = \sin\alpha \; u_a f_a(u_a) + \cos\alpha \; \frac{1}{i2\pi} \frac{df_a(u_a)}{du_a}, \qquad (4.162)$$

where we have used u_a as the dummy variable on the right-hand side to emphasize that $\mathcal{F}^a$ represents a transformation from the zeroth domain to the ath domain. We first note that these reduce to the corresponding pair of properties for the ordinary Fourier transform when $a = 1$ (table 2.4, properties 5 and 6 with $n = 1$), and to simple identities when $a = 0$. When $a = 1$ the transform of a coordinate multiplied function $uf(u)$ is the derivative of the transform of the original function $f(u)$, and the transform of the derivative of a function $df(u)/du$ is the coordinate-multiplied transform of the original function. Thus coordinate multiplication in the time or frequency domain corresponds to differentiation in the other domain. For arbitrary values of a, we see that the transform of a coordinate-multiplied function $uf(u)$ is a linear combination of the coordinate-multiplied transform of the original function and the derivative of the transform of the original function. The coefficients in the linear combination are $\cos \alpha$ and $-\sin \alpha$. As a approaches 0, there is more $uf(u)$ and less $df(u)/du$ in the linear combination. As a approaches 1, there is more $df(u)/du$ and less $uf(u)$. The same is true for the transform of the derivative of a function $df(u)/du$, but this time the coefficients in the linear combination are $\sin \alpha$ and $\cos \alpha$ and the roles of $uf(u)$ and $df(u)/du$ are reversed as a approaches 0 or 1.

The above system of two equations, with $u_a f_a(u_a)$ and $(i2\pi)^{-1} df_a(u_a)/du_a$ interpreted as unknowns, can be solved by direct matrix inversion to give the following equations (McBride and Kerr 1987):

$$u_a f_a(u_a) = +\cos \alpha \; \mathcal{F}^a \left[uf(u) \right] + \sin \alpha \; \mathcal{F}^a \left[\frac{1}{i2\pi} \frac{df(u)}{du} \right], \qquad (4.163)$$

$$\frac{1}{i2\pi} \frac{df_a(u_a)}{du_a} = -\sin \alpha \; \mathcal{F}^a \left[uf(u) \right] + \cos \alpha \; \mathcal{F}^a \left[\frac{1}{i2\pi} \frac{df(u)}{du} \right]. \qquad (4.164)$$

We see from this equation that knowing the derivative and coordinate-multiplied version of a function in any fractional Fourier domain allows us to obtain the same in any other domain.

Equations 4.161 and 4.162 can be written in operator form by recalling that $\mathcal{U}f(u) = uf(u)$ and $\mathcal{D}f(u) = (i2\pi)^{-1} df(u)/du$ and by similarly defining the operators $\mathcal{U}_a$ and $\mathcal{D}_a$ through the relations $\mathcal{U}_a f_a(u_a) = u_a f_a(u_a)$ and $\mathcal{D}_a f_a(u_a) = (i2\pi)^{-1} df_a(u_a)/du_a$ (see equation 4.37). The result is

$$\mathcal{F}^a \left[\mathcal{U}f(u) \right] = \cos \alpha \; \mathcal{U}_a f_a(u_a) - \sin \alpha \; \mathcal{D}_a f_a(u_a), \qquad (4.165)$$
$$\mathcal{F}^a \left[\mathcal{D}f(u) \right] = \sin \alpha \; \mathcal{U}_a f_a(u_a) + \cos \alpha \; \mathcal{D}_a f_a(u_a), \qquad (4.166)$$

or

$$(\mathcal{U}f)_a(u_a) = \cos \alpha \; (\mathcal{U}_a f)_a(u_a) - \sin \alpha \; (\mathcal{D}_a f)_a(u_a), \qquad (4.167)$$
$$(\mathcal{D}f)_a(u_a) = \sin \alpha \; (\mathcal{U}_a f)_a(u_a) + \cos \alpha \; (\mathcal{D}_a f)_a(u_a). \qquad (4.168)$$

These equations can be written in representation-independent form in terms of the abstract signal f as follows:

$$\mathcal{U}f = \cos \alpha \; \mathcal{U}_a f - \sin \alpha \; \mathcal{D}_a f, \qquad (4.169)$$
$$\mathcal{D}f = \sin \alpha \; \mathcal{U}_a f + \cos \alpha \; \mathcal{D}_a f. \qquad (4.170)$$

Furthermore, since these equations are true for an arbitrary signal f, we may write

$$\left[\begin{array}{c} \mathcal{U} \\ \mathcal{D} \end{array} \right] = \left[\begin{array}{cc} \cos \alpha & -\sin \alpha \\ \sin \alpha & \cos \alpha \end{array} \right] \left[\begin{array}{c} \mathcal{U}_a \\ \mathcal{D}_a \end{array} \right], \qquad (4.171)$$

which we see to be equivalent to equation 4.36. There this equation was used to define $\mathcal{U}_a$ and $\mathcal{D}_a$ and thus the fractional Fourier transform. Here we assumed the fractional Fourier transform was already defined and we showed that $\mathcal{U}_a$ and $\mathcal{D}_a$ satisfy this equation. The reader may also wish to start from equations 4.36 and 4.37 and derive equations 4.161 and 4.162 and also take another look at equations 4.46 and 4.47. Thus, properties 5 and 6 of table 4.3 are equivalent to the operator property given by equation 4.171 or equation 4.36, which formed the basis of definition D. It also follows that properties 5 and 6 are sufficient to define the fractional Fourier transform (within a constant coefficient). We also note that equation 4.171 can also be written between two domains a and a' as follows:

$$\begin{bmatrix} \mathcal{U}_{a'} \\ \mathcal{D}_{a'} \end{bmatrix} = \begin{bmatrix} \cos(\alpha - \alpha') & -\sin(\alpha - \alpha') \\ \sin(\alpha - \alpha') & \cos(\alpha - \alpha') \end{bmatrix} \begin{bmatrix} \mathcal{U}_a \\ \mathcal{D}_a \end{bmatrix}. \tag{4.172}$$

Equations 4.161 and 4.162 may be generalized to functions $f(u)$ multiplied by arbitrary functions $h(\cdot)$ of the coordinate multiplication and differentiation operators:

$$\mathcal{F}^a \left[h(u) f(u) \right] = h \left(\cos \alpha \, \mathcal{U}_a - \sin \alpha \, \mathcal{D}_a \right) f_a(u_a), \tag{4.173}$$

$$\mathcal{F}^a \left[h((i2\pi)^{-1} d/du) f(u) \right] = h \left(\sin \alpha \, \mathcal{U}_a + \cos \alpha \, \mathcal{D}_a \right) f_a(u_a), \tag{4.174}$$

where $h(u)$ is a function expandable in a Taylor series (Namias 1980a). In operator form the above can be written as

$$h(\mathcal{U}) = h \left(\cos \alpha \, \mathcal{U}_a - \sin \alpha \, \mathcal{D}_a \right), \tag{4.175}$$

$$h(\mathcal{D}) = h \left(\sin \alpha \, \mathcal{U}_a + \cos \alpha \, \mathcal{D}_a \right). \tag{4.176}$$

These properties can be used to derive a variety of interesting identities. For instance,

$$\mathcal{F}^a[h(u) f(u)] = h \left(\cos \alpha \, u_a - \sin \alpha \, (i2\pi)^{-1} \frac{d}{du_a} \right) \mathcal{F}^a[f(u)], \tag{4.177}$$

$$= f \left(\cos \alpha \, u_a - \sin \alpha \, (i2\pi)^{-1} \frac{d}{du_a} \right) \mathcal{F}^a[h(u)], \tag{4.178}$$

where the second equation follows by interchanging the roles of f and h (Namias 1980a).

Let us now recall equation 2.176, whose solutions are known to be the Hermite-Gaussian functions:

$$\frac{d^2 f(u)}{du^2} + 4\pi^2 \left(\frac{2l + 1}{2\pi} - u^2 \right) f(u) = 0. \tag{4.179}$$

We have noted that the ordinary Fourier transform of a solution of this equation is also a solution (equation 2.177). More generally, all fractional Fourier transforms of a solution are also solutions of this equation. This can be shown by taking the fractional Fourier transform of both sides of the above equation and observing that it is of exactly the same form. One way of performing this derivation is to directly use equations 4.161 and 4.162. Alternatively, let us write the above equation in operator form as

$$\left[\pi(\mathcal{U}^2 + \mathcal{D}^2) - (l + 1/2) \right] f(u) = 0. \tag{4.180}$$

Since $\mathcal{U}^2 + \mathcal{D}^2 = \mathcal{U}_a^2 + \mathcal{D}_a^2$ for all a (page 131), we can also write

$$\left[\pi(\mathcal{U}_a^2 + \mathcal{D}_a^2) - (l + 1/2) \right] f_a(u_a) = 0, \tag{4.181}$$

from which the desired result follows:

$$\frac{d^2 f_a(u_a)}{du_a^2} + 4\pi^2 \left(\frac{2l + 1}{2\pi} - u_a^2 \right) f_a(u_a) = 0. \tag{4.182}$$

Using equation 4.171 it is possible to derive the following commutators (Aytür and Ozaktas 1995, Ozaktas and Aytür 1995):

$$[\mathcal{U}_a, \mathcal{U}_{a'}] = \frac{i}{2\pi} \sin[\pi(a'-a)/2], \qquad (4.183)$$

$$[\mathcal{U}_a, \mathcal{D}_{a'}] = \frac{i}{2\pi} \cos[\pi(a'-a)/2], \qquad (4.184)$$

whose special cases $(a' - a) = 0$ and $(a' - a) = \pm 1$ the reader may wish to examine. Equation 4.171 allows us to derive uncertainty relations between two arbitrary domains a and a'. Referring to section 2.7, let

$$\sigma_a^2 \equiv \sigma_{\mathcal{U}_a}^2 = \left[\int (u_a - \eta_a)^2 |f_a(u_a)|^2 \, du_a \right] / \|f\|^2, \qquad (4.185)$$

$$\eta_a \equiv \eta_{\mathcal{U}_a} = \left[\int u_a |f_a(u_a)|^2 \, du_a \right] / \|f\|^2. \qquad (4.186)$$

Now, using equation 2.262 and the commutator for $[\mathcal{U}_a, \mathcal{U}_{a'}]$ we obtain the following uncertainty relation (Aytür and Ozaktas 1995, Ozaktas and Aytür 1995):

$$\sigma_a \sigma_{a'} \geq \frac{|\sin[\pi(a'-a)/2]|}{4\pi}, \qquad (4.187)$$

which is a special case of equation 3.163. Further discussion of uncertainty principles in the context of fractional Fourier transforms may be found in Mustard 1991, where a new family of uncertainty principles which are invariant under the fractional Fourier transform are developed. One of these is stronger than the conventional Heisenberg uncertainty relation (Mustard 1987b, c, 1991).

Finally, we also briefly mention some results related to the transformation of the second-order moments of the Wigner distribution (the reader may wish to review the discussion on page 107). The centralized second-order moment matrix (the moment of inertia tensor) given by equation 3.146 has two invariants under rotation of the Wigner distribution (or equivalently under fractional Fourier transformation of the signal). These two invariants are the trace and determinant of the matrix. The trace is the moment of inertia of the Wigner distribution. The square root of the determinant is a measure of the support area, much in the manner that the square root of the variance of a one-dimensional function is a measure of its support length. Mustard has noted that since these quantities are invariant under fractional Fourier transformation, they "state something intrinsic about the underlying signal, rather than something contingent on its particular representation," and that either of them "represents an intrinsic property of the underlying object represented by the given function" (Mustard 1991).

4.8 Phase shift and translation operators

Now we define generalizations of the phase shift operator $\mathcal{PH}_\xi = \exp(i2\pi\xi\mathcal{U})$ and the translation operator $\mathcal{SH}_\xi = \exp(i2\pi\xi\mathcal{D})$. In this section we will find the notation $\mathcal{PH}_a(\xi) \equiv \exp(i2\pi\xi\mathcal{U}_a)$ and $\mathcal{SH}_a(\xi) \equiv \exp(i2\pi\xi\mathcal{D}_a)$ more convenient because we are introducing the subscript a. When $a = 0$ these correspond to $\mathcal{PH}_0(\xi) \equiv \mathcal{PH}_\xi$ and

$\mathcal{SH}_0(\xi) \equiv \mathcal{SH}_\xi$ respectively. Recalling that the effect of $\mathcal{U}_a$ and $\mathcal{D}_a$ is to coordinate multiply and differentiate in the ath domain respectively, it is possible to show that

$$\mathcal{PH}_a(\xi) f_a(u_a) = e^{i2\pi\xi u_a} f_a(u_a), \tag{4.188}$$

$$\mathcal{SH}_a(\xi) f_a(u_a) = f_a(u_a + \xi). \tag{4.189}$$

The effect of $\mathcal{PH}_a(\xi)$ in the $(a + 1)$th domain is simply translation by ξ to the right, and the effect of $\mathcal{SH}_a(\xi)$ in the $(a - 1)$th domain is simply a multiplication by $\exp(-i2\pi\xi u_{a-1})$.

Having agreed on the notation to be used, it is now possible to deduce the following operator relations by using equation 4.36 or 4.171 in the definitions of $\mathcal{PH}_a(\xi)$ and $\mathcal{SH}_a(\xi)$ and employing Glauber's formula (equation 2.246):

$$\mathcal{PH}_a(\xi) = e^{i\pi\xi^2 \sin\alpha \cos\alpha} \mathcal{PH}_0(\xi\cos\alpha)\, \mathcal{SH}_0(\xi\sin\alpha)$$

$$= e^{-i\pi\xi^2 \sin\alpha \cos\alpha} \mathcal{SH}_0(\xi\sin\alpha)\, \mathcal{PH}_0(\xi\cos\alpha), \tag{4.190}$$

$$\mathcal{SH}_a(\xi) = e^{-i\pi\xi^2 \sin\alpha \cos\alpha} \mathcal{PH}_0(-\xi\sin\alpha)\, \mathcal{SH}_0(\xi\cos\alpha)$$

$$= e^{i\pi\xi^2 \sin\alpha \cos\alpha} \mathcal{SH}_0(\xi\cos\alpha)\, \mathcal{PH}_0(-\xi\sin\alpha). \tag{4.191}$$

These operator relations directly correspond to properties 3 and 4 in table 4.3. It is possible to write similar relations when the roles of the ath and zeroth domains are interchanged, or more generally between the ath and a'th domains by using equation 4.172 (Aytür and Ozaktas 1995, Ozaktas and Aytür 1995).

We see that the $\mathcal{PH}_a(\xi) = \exp(i2\pi\xi\mathcal{U}_a)$ operator, which simply results in a phase shift in the u_a domain, corresponds to a translation followed by a phase shift (or a phase shift followed by a translation) in the u domain. (The fact that translation and phase shifting do not commute accounts for the additional phase factor coming from Glauber's formula.) We see that the $\mathcal{SH}_a(\xi) = \exp(i2\pi\xi\mathcal{D}_a)$ operator, which simply results in a translation in the u_a domain, corresponds to a translation followed by a phase shift (or a phase shift followed by a translation) in the u domain. The amount of translation and phase shift is given by cosine and sine multipliers corresponding to the "projection" of the translation or phase shift between fractional domains. A phase shift (or translation) by an amount ξ in the ath domain corresponds to a phase shift (translation) by an amount $\xi\cos(\alpha - \alpha')$ in the α'th domain and a translation (phase shift) by an amount $\xi\sin(\alpha - \alpha')$. As an alternative explanation, consider figure 4.15 which illustrates a displacement from the origin $(0, 0)$ to the arbitrary point (u, μ) in the time-frequency plane. This displacement corresponds to a time shift of u and a frequency shift of μ. The same displacement with respect to the u_a and μ_a axes corresponds to a u_a shift of $u_a = u\cos\alpha + \mu\sin\alpha$ and a μ_a shift of $\mu_a = -u\sin\alpha + \mu\cos\alpha$. In general, the combined action of a u shift and a μ shift of the original time function $f(u)$ will correspond to a u_a shift and a μ_a shift of $f_a(u_a)$, as given by the presented results (D. Mustard, private communication 1999).

A complementary discussion of time-shift and frequency-shift operators and their fractional generalization may be found in Akay and Boudreaux-Bartels 1998a. Mustard (private communication 1999) has suggested that formulas such as equations 4.190 and 4.191 relating the phase shift and translation operators in different domains are so fundamental that they should be defining requirements of the fractional

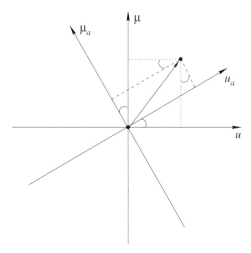

Figure 4.15. Time and frequency shifts with respect to different domains. The four indicated angles are all equal to α.

Fourier transform, in much the same way as the relations between the coordinate multiplication and differentiation operators were the defining requirements in definition D. Use of the phase shift and translation operators may have certain advantages stemming from the fact that they are bounded operators.

4.9   Fractional Fourier domains

In this section we consolidate and summarize several facts from which emerges the concept of *fractional Fourier domains* (Ozaktas and others 1994a, Ozaktas and Aytür 1995, Aytür and Ozaktas 1995). The reader carefully following this chapter has probably already grasped this notion.

Consistent with the conventional manner in which the time and frequency (or space and frequency, or position and momentum) axes are drawn orthogonal to each other in phase space, we speak of two representations which are related through a Fourier transform as being orthogonal to each other. The operator $\mathcal{U}_a$ is said to be orthogonal to the operator $\mathcal{D}_a = \mathcal{U}_{a+1}$ (equivalently, the operation of multiplying by u_a is said to be orthogonal to the operation $(i2\pi)^{-1}d/du_a$). Likewise, multiplication by a phase factor is said to be orthogonal to a corresponding translation, and so forth. In general, the domains (axes) associated with two representations that are related through an $(a' - a)$th order fractional Fourier transform make an angle $(a' - a)\pi/2$ with each other (figure 4.16). Coordinate multiplication or differentiation in one of these domains results in a combination of these two operations in the other domain, as most elegantly expressed by equation 4.172. Likewise, multiplication by a phase factor or a translation in one of these domains results in a combination of these two operations in the other domain, as given by equations 4.190 and 4.191 or their generalizations. The weighting factors appearing in these equations are cosines and sines of the angle between these two domains. The commutator and uncertainty

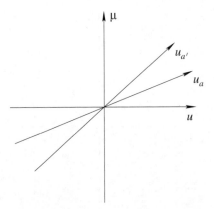

Figure 4.16. Fractional Fourier domains. The representation $f_{a'}(u_{a'})$ in the a'th domain is related to the representation $f_a(u_a)$ in the ath domain through an $(a' - a)$th order fractional Fourier transform.

relation between nonorthogonal domains are also interpreted in terms of this geometric picture.

As noted before, the zeroth and first domains are nothing but the ordinary time (or space) and frequency domains and the second and third domains correspond to the inverted time (or space) and frequency domains. We know that $\mathcal{F}^{a\pm 2} = \mathcal{F}^a \mathcal{P} = \mathcal{P} \mathcal{F}^a$, so that $u_{a\pm 2} = -u_a$ ($\mathcal{P}$ is the parity operator). The angle between the u_a domain and the $u_{a\pm 2} = -u_a$ domain is $\pm\pi$, as expected.

There is nothing special about the $a = 0$ representation which we commonly refer to as the time or space domain. This domain merely corresponds to the choice of origin of the parameter a. We recall from chapter 2 that a signal f has an existence independent of the particular representation in which it is explicitly written. Let us also recall that a particular representation in the form of a function $f(\cdot)$ is independent of the dummy variable used. We have most commonly used u to denote the time (or space) variable and have written the time (or space) domain representation of a signal f in the form $f(u)$. Likewise, although the representation of the signal f in the ath domain $f_a(\cdot)$ is independent of the dummy variable used (we have indeed simply written $f_a(u)$ throughout most of this chapter), in certain contexts it is found convenient to label this axis as u_a and write the ath domain representation as $f_a(u_a)$. This notation has already been used in this chapter whenever we felt the need to avoid confusion.

Figure 4.16 was first motivated by the Wigner rotation and Radon transform properties; see equations 4.131 and 4.146 and the associated discussions. The integral projection of the Wigner distribution onto the u_a axis (the ath fractional Fourier domain) is simply $|f_a(u)|^2$, the squared magnitude of the representation of the signal in the ath fractional Fourier domain. Recall that the integral projection onto the u axis ($a = 0$) gives the squared magnitude of the time-domain representation, and the integral projection onto the μ axis ($a = 1$) gives the squared magnitude of the frequency-domain representation of the signal. (The fact that many other time-frequency distributions also satisfy a rotation property, as given by equation 4.157, yields yet further support to these notions.)

We have already mentioned that the distribution of the energy of a signal in the time-frequency plane as given by its Wigner distribution should be considered as a geometrical entity, independent of the choice of the time and frequency axes. The introduction of these axes merely corresponds to choosing the origin of the parameter a. In certain applications it might be of interest to seek the most natural and intrinsic choice of the u and μ axes (or equivalently, the most natural origin for the order parameter a). These might not always be the original time or space axes, just as the "normal coordinates" of a physical system are not always the axes we use to set up the problem.

The operators $\mathcal{U}$ and $\mathcal{D}$ may be considered orthogonal basis (or coordinate) vectors spanning a kind of phase space. The operators $\mathcal{U}_a$ and $\mathcal{D}_a$ then correspond to the basis vectors of a new rotated coordinate system, and are related to the old coordinate system through equation 4.36 or 4.171. The expansion coefficient of $\mathcal{U}_a$ in the direction of $\mathcal{U}$ is given by $\cos a$, and so forth. (Formally, this can be found by evaluating the inner product of $\mathcal{U}_a = \cos a\, \mathcal{U} + \sin a\, \mathcal{D}$ with $\mathcal{U}$, taking the inner product of $\mathcal{U}$ with itself as unity and with $\mathcal{D}$ as zero. It should also be noted that the orthogonality of $\mathcal{U}$ and $\mathcal{D}$ is a geometric notion which should be interpreted in phase space. It does not imply that $\int [uf(u)]^* [(i2\pi)^{-1} df(u)/du]\, du = 0$, which is not true.)

As a natural extension, one may consider arbitrary nonorthogonal basis vectors as well. These are related to each other according to equation 3.162 and are associated with linear canonical transforms. In a similar spirit, it is also possible to express $\mathcal{U}_a$ in terms of *any* two operators $\mathcal{U}_{a'}$ and $\mathcal{U}_{a''}$, provided the latter two are not collinear, and express any function of an arbitrary number of operators $\Upsilon(\mathcal{U}_{b'}, \mathcal{U}_{b''}, \dots)$ in the form $\Upsilon'(\mathcal{U}_{a'}, \mathcal{U}_{a''})$. Furthermore, it is possible to define the space spanned by any two noncollinear operators $\mathcal{U}_a$ and $\mathcal{U}_{a'}$, and operations such as inner products and norms. These further extensions are not developed here.

4.10 Chirp bases and chirp transforms

Most of what appears in this section will already be evident to those well versed in the fundamentals of function spaces. We begin by rewriting the inverse fractional Fourier transform as

$$f(u) = \int f_a(u') K_{-a}(u, u')\, du'. \tag{4.192}$$

This can be interpreted as an expansion of $f(u)$ in terms of the basis functions $K_{-a}(u, u')$, interpreted as functions of u with index u'. Here $f_a(u')$, the representation of the abstract signal f in the ath domain, serves as the expansion coefficient. $K_{-a}(u, u')$ may be thought of as the time-domain (u-domain) representation of the abstract basis signal $(K_{-a})_{u'}$ so that the expansion can be written in abstract form as

$$f = \int f_a(u')(K_{-a})_{u'}\, du'. \tag{4.193}$$

If $a = 0$ the basis signals $(K_{-a})_{u'}$ reduce to $\delta_{u'}$, the impulse basis signals. If $a = 1$ the $(K_{-a})_{u'}$ reduce to $\mathrm{har}_{u'}$, the harmonic basis signals (page 20). Remember that the time-domain representation of a signal can be interpreted as the coefficients of the expansion of the signal in terms of the impulse basis, and the frequency-domain representation can be interpreted as the coefficients of the expansion of the signal in terms of the harmonic basis. Thus we see that, similarly, the ath domain representation can be interpreted as the coefficients of the expansion of the signal in terms of the *chirp basis* $(K_{-a})_{u'}$. There is a different chirp basis corresponding to each value of a. The fractional Fourier transformation relates the expansion coefficients (representations) with respect to these different basis sets, and thus corresponds to a change of basis.

One way to show that each of these chirp basis sets is an orthonormal set is to first note that $K_a(u, u')$ is nothing but the fractional Fourier transform of $\delta(u - u')$. (To see this, take the fractional Fourier transform of both sides of the identity $f(u) = \int f(u')\delta(u - u')\, du'$ to obtain $f_a(u) = \int f(u')\mathcal{F}^a[\delta(u - u')]\, du'$.) Abstracting away from the u representation we can write $(K_a)_{u'} = \mathcal{F}^a \delta_{u'}$. Then the orthonormality of $(K_a)_{u'}$ and $(K_a)_{v'}$ follows from the unitarity of $\mathcal{F}^a$ and the orthonormality of the impulse basis:

$$\langle (K_a)_{u'}, (K_a)_{v'} \rangle = \langle \mathcal{F}^a \delta_{u'}, \mathcal{F}^a \delta_{v'} \rangle = \langle \delta_{u'}, \delta_{v'} \rangle = \delta(u' - v'). \tag{4.194}$$

The explicit closure relation can also be found in a similar manner based on the closure relation for the impulse basis. In their most explicit forms, the closure and orthonormality relations are

$$\int K_a^*(u, u') K_a(u, v')\, du = \delta(u' - v'), \tag{4.195}$$

$$\int K_a^*(u, u') K_a(v, u')\, du' = \delta(u - v). \tag{4.196}$$

At this point the reader may also wish to look back to figure 4.11 and the associated discussion explaining how chirp functions are affected by fractional Fourier transformation. An expanded discussion of these issues may be found in Ozaktas and others 1994a.

The representation of a signal in the ath domain can be written as a superposition of delta functions in the same domain:

$$f_a(u_a) = \int f_a(u')\delta(u_a - u')\, du', \tag{4.197}$$

or as a superposition of harmonics in the domain orthogonal to that domain (which is the $(a + 1)$th domain):

$$f_a(u_a) = \int F_a(\mu_a)e^{i2\pi\mu_a u_a}\, d\mu_a \tag{4.198}$$

(where $F_a = f_{a+1}$, $\mu_a = u_{a+1}$, and $\mu = \mu_0 = u_1$), or more generally as a superposition of chirp functions in some other a'th domain:

$$f_a(u_a) = \int f_{a'}(u_{a'}) K_{a-a'}(u_a, u_{a'})\, du_{a'}. \tag{4.199}$$

Remember that the representation of a signal in the ath domain is the same as the ath fractional Fourier transform of the representation of the signal in the zeroth domain. If we know the representation of the signal in the a'th domain, we can find its representation in the ath domain by taking its $(a - a')$th fractional Fourier transform.

We summarize as follows:

1. Basis functions in the ath domain, be they delta functions or harmonics, are in general chirp functions in the a'th domain.
2. The representation of a signal in the ath domain can be obtained from its representation in the a'th domain by taking the inner products (projections) of the representation in the a'th domain with basis functions in the target ath domain.
3. This operation, having the form of a chirp transform, is equivalent to taking the $(a - a')$th fractional Fourier transform of the representation in the a'th domain.

4.11   Two-dimensional fractional Fourier transforms

The natural extension of the fractional Fourier transform to two dimensions is

$$f_{\mathbf{a}}(\mathbf{q}) = f_{a_u,a_v}(u,v) = \mathcal{F}^{\mathbf{a}}f(\mathbf{q}) = \mathcal{F}^{a_u,a_v}f(u,v)$$

$$= \iint K_{a_u,a_v}(u,v;u',v')f(u',v')\,du'\,dv', \qquad (4.200)$$

$$K_{a_u,a_v}(u,v;u',v') = K_{a_u}(u,u')K_{a_v}(v,v'),$$

and similar for higher dimensions. Here $\mathbf{q} = u\hat{\mathbf{u}} + v\hat{\mathbf{v}}$ and $\mathbf{a} = a_u\hat{\mathbf{u}} + a_v\hat{\mathbf{v}}$ where $\hat{\mathbf{u}}$ and $\hat{\mathbf{v}}$ are unit vectors in the u and v directions. $K_a(u,u')$ is the one-dimensional kernel defined in equation 4.4. Since most results generalize to two and higher dimensions in a straightforward manner, in this section we will only show how a number of important results look in two dimensions. A more comprehensive listing of properties may be found in Sahin, Ozaktas, and Mendlovic 1998. Notice that different transform orders a_u and a_v are allowed in the two dimensions, although some authors have defined higher-dimensional transforms with only a single order parameter (Karasik 1994). The effect of a one-dimensional fractional Fourier transform (say in the u direction) on a two-dimensional function is interpreted in the obvious manner by treating the other dimension (in this case v) as a parameter. Denoting such one-dimensional transforms as $\mathcal{F}^{a_u\hat{\mathbf{u}}}$ and $\mathcal{F}^{a_v\hat{\mathbf{v}}}$, it becomes possible to write

$$\mathcal{F}^{\mathbf{a}} = \mathcal{F}^{a_u\hat{\mathbf{u}}}\mathcal{F}^{a_v\hat{\mathbf{v}}} = \mathcal{F}^{a_v\hat{\mathbf{v}}}\mathcal{F}^{a_u\hat{\mathbf{u}}}, \qquad (4.201)$$

constituting a concise statement of the separability of the two-dimensional transform. Notice that the notation we have introduced makes these equations compatible with the index additivity property, so that it is possible to deduce identities such as $\mathcal{F}^{0.7\hat{\mathbf{u}}}\mathcal{F}^{0.5\hat{\mathbf{u}}-0.3\hat{\mathbf{v}}}\mathcal{F}^{0.2\hat{\mathbf{v}}}\mathcal{F}^{0.8\hat{\mathbf{u}}+0.1\hat{\mathbf{v}}} = \mathcal{F}^{2.0\hat{\mathbf{u}}} = \mathcal{P}_u$, where $\mathcal{P}_u$ is the parity operator in the u dimension. A more general nonseparable definition has also been proposed in Sahin, Kutay, and Ozaktas 1998.

Most of the results and properties presented for the one-dimensional case are easily generalized to two and higher dimensions by virtue of the separability of the transform, as embodied by equation 4.201. The eigenvalue equation for the two-dimensional transform may be written as

$$\mathcal{F}^{\mathbf{a}}\psi_{lm}(\mathbf{q}) = e^{-i(a_u l + a_v m)\pi/2}\psi_{lm}(\mathbf{q}), \qquad (4.202)$$

where $\psi_{lm}(\mathbf{q}) = \psi_l(u)\psi_m(v)$ are the two-dimensional Hermite-Gaussian functions. The Wigner rotation property takes the form

$$W_{f_{a_u,a_v}}(u,v;\mu,\nu)$$
$$= W_f(u\cos\alpha_u - \mu\sin\alpha_u, v\cos\alpha_v - \nu\sin\alpha_v; u\sin\alpha_u + \mu\cos\alpha_u, v\sin\alpha_v + \nu\cos\alpha_v).$$
$$(4.203)$$

Likewise, the integral projections of the Wigner distribution satisfy

$$\mathcal{RDN}_{\alpha_v}[\mathcal{RDN}_{\alpha_u}[W_f(u,v;\mu,\nu)]] = |f_{a_u,a_v}(u,v)|^2, \qquad (4.204)$$

where $\mathcal{RDN}_{\alpha_u}[W_f(u,v;\mu,\nu)]$ denotes the integral projection $\int W_f(u,v;\mu,\nu)\,d\mu_{a_u}$, where $\mu_{a_u} = -u\sin\alpha_u + \mu\cos\alpha_u$, and similarly for $\mathcal{RDN}_{\alpha_v}$. The double integral projection $\mathcal{RDN}_{\alpha_v}\mathcal{RND}_{\alpha_u}$ projects the Wigner distribution on the two-dimensional fractional Fourier domain defined by the u_{a_u} and v_{a_v} axes ($u_{a_u} = u\cos\alpha_u + \mu\sin\alpha_u$ and $v_{a_v} = v\cos\alpha_v + \nu\sin\alpha_v$). These properties can also be interpreted as special cases of equation 3.183 by using the four-dimensional rotation matrix

$$\begin{bmatrix} \cos\alpha_u & 0 & \sin\alpha_u & 0 \\ 0 & \cos\alpha_v & 0 & \sin\alpha_v \\ -\sin\alpha_u & 0 & \cos\alpha_u & 0 \\ 0 & -\sin\alpha_v & 0 & \cos\alpha_v \end{bmatrix}. \qquad (4.205)$$

This matrix replaces the common two-dimensional rotation matrix appearing in our discussion of the one-dimensional transform.

The solution of the two-dimensional differential equation

$$\left[-\frac{1}{4\pi}\left(\frac{\partial^2}{\partial u^2} + \frac{\partial^2}{\partial v^2}\right) + \pi(u^2 + v^2) - 1\right] f(u,v;a_u,a_v)$$
$$= i\frac{2}{\pi}\left(\frac{\partial f(u,v;a_u,a_v)}{\partial a_u} + \frac{\partial f(u,v;a_u,a_v)}{\partial a_v}\right) \qquad (4.206)$$

with initial condition $f(u,v;0,0) = f(u,v)$ is given by $f(u,v;a_u,a_v) = f_{a_u,a_v}(u,v)$. If the transform orders (a_u,a_v) are taken equal to a and the differential equation is rewritten for $f(u,v;a)$ with the two partial derivatives on the right-hand side replaced by $\partial f(u,v;a)/\partial a$, then the solution becomes $f(u,v;a) = f_{a,a}(u,v)$. The hyperdifferential form of the two-dimensional fractional Fourier transform operator is

$$\mathcal{F}^{\mathbf{a}} = e^{-i\pi(a_u\mathcal{H}_u + a_v\mathcal{H}_v)/2}, \qquad (4.207)$$

where $\mathcal{H}_u = \pi(\mathcal{D}^2 + \mathcal{U}^2) - \frac{1}{2}$ and likewise for $\mathcal{H}_v$.

Most of the transform pairs given in table 4.1 can be generalized easily if the function to be transformed is separable. Even when this is not the case, it is often relatively easy to generalize these properties if the orders in both dimensions are equal. One simply needs to judiciously employ the following result which can be demonstrated by

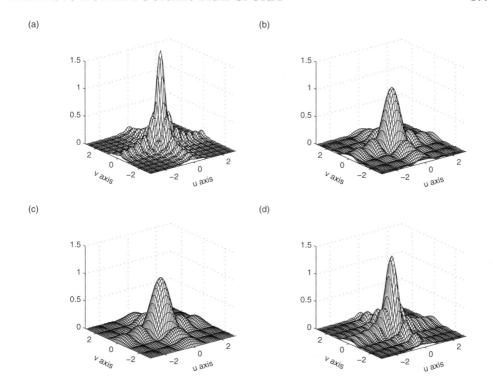

Figure 4.17. Magnitudes of the fractional Fourier transforms of the rectangle function: (a) $a_u = a_v = 0.3$, (b) $a_u = a_v = 0.7$, (c) $a_u = a_v = 1$, (d) $a_u = 0.3$, $a_v = 0.7$.

straightforward algebraic manipulations:

> If a two-dimensional function $f(u,v)$ is rotated by an angle ϕ in the u-v plane, then the fractional Fourier transform of the rotated function with identical orders $a_u = a = a_v$, can be obtained by rotating the fractional Fourier transform of $f(u,v)$.

In the most general case it becomes necessary to resort to complex Gaussian integrals to obtain the desired transform pair.

The magnitudes of various fractional Fourier transforms of the rectangle function $\text{rect}(u,v) = \text{rect}(u)\text{rect}(v)$ are given in figure 4.17. The same is repeated for the circle (or cylinder, or top hat) function $\text{rect}(\sqrt{u^2 + v^2})$ in figure 4.18.

Separability also allows the properties given in table 4.2 to be generalized in a fairly trivial manner (using equation 4.201). Most of the properties given in table 4.3 are likewise easily generalized if the function in question is also separable. If not, it is often still possible to obtain fairly straightforward generalizations so long as the operations involved are separable. Separability means that the operation $\mathcal{O}(p_u, p_v)$ with parameters (p_u, p_v) can be expressed as $\mathcal{O}(p_u, p_v) = \mathcal{O}_u(p_u)\mathcal{O}_v(p_v) = \mathcal{O}_v(p_v)\mathcal{O}_u(p_u)$. Most generalizations of the operations in table 4.3 will be separable (for instance, scaling $f(u,v)$ as $f(u/M_u, v/M_v)$ or coordinate multiplication in the form $u^{n_u} v^{n_v} f(u,v)$). When this is the case, we use equation 4.201 and the fact that

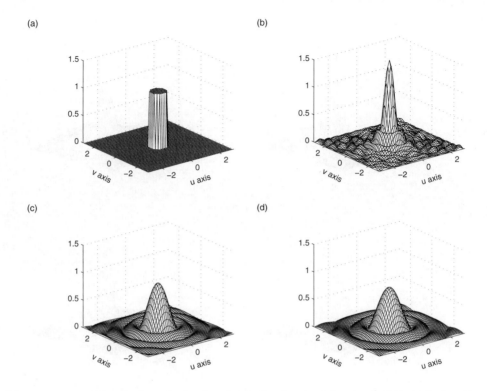

Figure 4.18. Magnitudes of the fractional Fourier transforms of the circle function:
(a) $a_u = a_v = 0$, (b) $a_u = a_v = 0.3$, (c) $a_u = a_v = 0.7$, (d) $a_u = a_v = 1$.

an operator acting on u and another acting on v commute, to write

$$\mathcal{F}^{\mathbf{a}}\mathcal{O}(p_u, p_v) = \mathcal{F}^{a_u\,\hat{\mathbf{u}}}\mathcal{F}^{a_v\,\hat{\mathbf{v}}}\mathcal{O}_u(p_u)\mathcal{O}_v(p_v) = \mathcal{F}^{a_u\,\hat{\mathbf{u}}}\mathcal{O}_u(p_u)\mathcal{F}^{a_v\,\hat{\mathbf{v}}}\mathcal{O}_v(p_v)$$
$$= \mathcal{O}'_u(p_u)\mathcal{F}^{a'_u\,\hat{\mathbf{u}}}\mathcal{O}'_v(p_v)\mathcal{F}^{a'_v\,\hat{\mathbf{v}}} = \mathcal{O}'_u(p_u)\mathcal{O}'_v(p_v)\mathcal{F}^{a'_u\,\hat{\mathbf{u}}}\mathcal{F}^{a'_v\,\hat{\mathbf{v}}} = \mathcal{O}'(p_u, p_v)\mathcal{F}^{\mathbf{a'}}, \quad (4.208)$$

where we have used one-dimensional commutation relations of the form $\mathcal{F}^{a_u\,\hat{\mathbf{u}}}\mathcal{O}_u(p_u) = \mathcal{O}'_u(p_u)\mathcal{F}^{a'_u\,\hat{\mathbf{u}}}$ deduced from table 4.3. We note again that the separability of $f(u,v)$ has not been assumed.

On the other hand, when it is the case that $\mathcal{O}(p_u, p_v)$ is not separable, but both the function $f(u,v)$ and the operation $\mathcal{O}(p_u, p_v)$ are rotationally symmetric, and the transform orders a_u, a_v are equal, then it is again possible to employ the result quoted on page 177 to obtain relatively simple results. An example of such an operation is to multiply $f(u,v)$ with $\sqrt{u^2 + v^2}$.

Speaking of rotational symmetry, we finally consider the fractional Hankel transform. Let $f(u,v)$ be a rotationally symmetric function so that it can expressed as a function of the radial coordinate $q = \sqrt{u^2 + v^2}$ in the form $f(q)$. Also let us set $a_u = a = a_v$. Then the result quoted on page 177 implies that the fractional transform $f_{a,a}(u,v)$ will also be rotationally symmetric and thus can also be written as $f_{a,a}(q)$. The transform mapping the one-dimensional function $f(q)$ to the one-dimensional function $f_{a,a}(q)$ is referred to as the ath order fractional Hankel transform. By using the coordinate transformation $u = q \cos \phi$, $v = q \sin \phi$ to go from rectangular coordinates (u, v) to polar coordinates (q, ϕ) and employing a number of trigonometric identities, it is possible to show that

$$f_{a,a}(q) = (1 - i \cot \alpha) \int_0^\infty e^{i\pi(\cot \alpha \, q^2 + \cot \alpha \, q'^2)} J_0(2\pi \csc \alpha \, qq') f(q') \, 2\pi q' \, dq', \quad (4.209)$$

where we have used the standard integral $2\pi J_0(w) = \int_0^{2\pi} \exp(-iw \cos \phi) \, d\phi$ for the zeroth-order Bessel function of the first kind, J_0, to evaluate the integral over ϕ from 0 to 2π. Naturally, for $a = 0$ the Hankel transform corresponds to the identity operation. When the order $a = \pm 1$ and $\alpha = a\pi/2 = \pm\pi/2$, equation 4.209 reduces to the ordinary Hankel transform defined by equation 2.297:

$$F(q) = \int_0^\infty J_0(2\pi qq') f(q') \, 2\pi q' \, dq', \quad (4.210)$$

$$f(q) = \int_0^\infty J_0(2\pi qq') F(q) \, 2\pi q' \, dq'. \quad (4.211)$$

Remembering the motivation of the fractional Hankel transform as the two-dimensional fractional Fourier transform of rotationally symmetric functions expressed in polar coordinates, it readily follows that index additivity holds. Consequently, the inverse of the ath order transform is simply the $-a$th transform. Since the forward and inverse ordinary Hankel transforms are identical, it follows that fractional Hankel transforms with even integer orders are equal to the identity operation, and those with negative integer orders are equal to those with the corresponding positive integer orders. In other words, the fractional Hankel transform is even and periodic in a with period 2.

Since applying the ordinary Hankel transform twice is equal to the identity operation, the ordinary Hankel transform has the two eigenvalues 1 and -1. The eigenfunctions are Gaussian functions multiplied by Laguerre polynomials, and thus may be referred to as Laguerre-Gaussian functions. If we write the eigenvalue associated with the lth order Laguerre-Gaussian function as $\exp(-il\pi)$, then one way of defining the ath order fractional Hankel transform is to assign the eigenvalues $\exp(-ial\pi)$ to the lth eigenfunction, in the same spirit as definition B of the fractional Fourier transform. This indeed leads to the definition of the fractional Hankel transform we have given above (Namias 1980b). The Laguerre-Gaussian series representations of the two-dimensional fractional Fourier transform and the fractional Hankel transform are also discussed in Yu and others 1998a, b. The associated scaling property is discussed in Sheppard and Larkin 1998.

The transform defined in equation 4.210 may be more specifically referred to as the Hankel transform of zero Bessel order, with the corresponding fractional transform referred to as the fractional Hankel transform of zero Bessel order. More general Hankel transforms are defined by replacing the zeroth-order Bessel function with Bessel functions of arbitrary orders. The order of the Bessel function involved should not be confused with the fractional order. These are discussed in Namias 1980b and with greater rigor and generality in Kerr 1991, 1992 and Karp 1995, to which the reader is referred for further details.

Finally, we note that the affine property given in equation 2.296 does not generalize to the fractional Fourier transform. That is, it is not possible to express the fractional

Fourier transform of $f(au + bv + e, cu + dv + f)$ as a similarly transformed version of the fractional Fourier transform of $f(u, v)$ of any order. The desire to define the transform in a more general manner such that it satisfies this property has motivated the definition of a nonseparable version of the two-dimensional fractional Fourier transform (Sahin, Kutay, and Ozaktas 1998). This nonseparable transform is more general than the separable transform and allows one to specify the two transform orders in arbitrary nonorthogonal directions, rather than being restricted to the u and v directions.

Although various results on two-dimensional transforms have been given in various sources, the first comprehensive reference solely devoted to the separable two-dimensional transform is Sahin, Ozaktas, and Mendlovic 1998.

4.12   Extensions and applications

4.12.1   Fractional Fourier transforms in braket notation

The reader familiar with the braket notation of quantum mechanics will have noted that much of what we have done can also be expressed in this powerful notation (Cohen-Tannoudji, Diu, and Laloë 1977). Although we have chosen not to employ this notation for the benefit of those not well versed with it, here we briefly show how some expressions would look in this notation (Aytür and Ozaktas 1995).

We will let $|u_a\rangle$ denote the basis vector whose explicit appearance in the ath domain is $\langle u'_a|u_a\rangle = \delta(u_a - u'_a)$. Then, the representation of $|f\rangle$ in the ath domain is written as

$$f_a(u_a) = \langle u_a|f\rangle = \mathcal{F}^a[f_0(u_0)](u_a) = \int K_a(u_a, u_0)f_0(u_0)\, du_0. \qquad (4.212)$$

$f_a(u_a)$ is simply the expansion coefficient when $|f\rangle$ is written as a superposition of the basis vectors $|u_a\rangle$:

$$|f\rangle = \int f_a(u_a)|u_a\rangle\, du_a. \qquad (4.213)$$

To find the representation of $|f\rangle$ in the a'th domain (with respect to the basis vectors $|u_{a'}\rangle$), we simply project onto this set of basis vectors:

$$f_{a'}(u_{a'}) = \langle u_{a'}|f\rangle = \int f_a(u_a)\langle u_{a'}|u_a\rangle\, du_a, \qquad (4.214)$$

from which we recognize $\langle u_{a'}|u_a\rangle$ as the fractional Fourier transform kernel $K_{(a'-a)}(u_{a'}, u_a)$. We may also write the orthonormality relation and the closure relation respectively as

$$\langle u_a|u'_a\rangle = \delta(u_a - u'_a), \qquad (4.215)$$

$$\int |u_a\rangle\langle u_a|\, du_a = \mathcal{I}. \qquad (4.216)$$

4.12.2 Complex-ordered fractional Fourier transforms

If we let $a = a_r + ia_i$ denote a complex number, with a_r and a_i denoting its real and imaginary parts, we would naturally expect

$$\mathcal{F}^a = \mathcal{F}^{a_r}\mathcal{F}^{ia_i}, \tag{4.217}$$

so that complex-ordered transforms may be separated into the real-ordered transforms which we discussed in this chapter, and imaginary-ordered transforms. The definition given in equation 4.4 may be assumed to remain valid for complex and imaginary values of a, although some care is required, especially with regard to the factors up front. Although we are unaware of a rigorous discussion of these issues in the context of the fractional Fourier transform, treatments of complex-parametered linear canonical transforms certainly have a bearing on complex-ordered fractional Fourier transforms (Wolf 1979). The class of transforms discussed in Hille and Phillips 1957: chapter 21, Beckner 1975, Weissler 1979, and Byun 1993 is also likely to be relevant to the study of imaginary- and complex-ordered transforms.

Complex-ordered transforms have been discussed in an optical context in Shih 1995a, Bernardo and Soares 1996, and Bernardo 1997.

4.12.3 Relation to wavelet transforms

Fractional Fourier transforms are closely related to a particular wavelet transform family. Rewriting equation 4.4 in the form

$$f_a(u_a) = \int K_a(u_a, u) f(u)\, du, \tag{4.218}$$

and making the change of variable $v = u_a \sec\alpha$, and denoting the left-hand side by $g(v)$, we obtain the following result (Ozaktas and others 1994a):

$$g(v) \equiv f_a\left(\frac{v}{\sec\alpha}\right) = C(\alpha)\, e^{-i\pi v^2 \sin^2\alpha} \int \exp\left[i\pi\left(\frac{v-u}{\tan^{1/2}\alpha}\right)^2\right] f(u)\, du, \tag{4.219}$$

where $C(\alpha)$ depends only on α. Taking $\tan^{1/2}\alpha$ as the scale parameter, the convolution represented by the above integral is a wavelet transform where the wavelet family is obtained from the quadratic-phase function $w(u) = e^{i\pi u^2}$. This class of wavelet transforms has been discussed in detail in Onural 1993, Onural, Kocatepe, and Ozaktas 1994, and Onural and Kocatepe 1995.

In chapter 10 we will discuss filtering of functions in different fractional Fourier domains. Thus, based on the above, these operations can also be interpreted as filtering in the corresponding wavelet transform domains.

4.12.4 Application to neural networks

Neural networks based on the fractional Fourier transform have been proposed in Lee and Szu 1994. This essentially involves taking a discrete form of the fractional Fourier transform kernel as a particular matrix of weights. Shin and others (1998) demonstrate that appropriate use of the fractional Fourier transform can improve both the learning convergence and the recall rate of the neural network. It remains to be seen whether further developments in this area will be forthcoming.

4.12.5   Chirplets and other approaches

Throughout this book, we mention many applications and properties of the fractional Fourier transform. Some of these previously appeared in varied contexts without explicitly employing the fractional transform. These were generalizations of existing methods or answered the need to handle signals with time-varying frequency content, such as signals which locally exhibited linearly varying instantaneous frequency. Of particular interest are works which deal with so-called *chirplets* (Mann and Haykin 1992, 1995, Mihovilović and Bracewell 1991, 1992, and Bultan 1999). What is common to these approaches is that they either deal with expansions into chirp bases or chirp transforms or other similar methods. Recently, there has also been increased interest in other approaches to handle similar problems, including those dealing with rotations and projections in the time-frequency plane; a random sampling of papers may include Ristic and Boashash 1993, Barbarossa 1995, Owechko 1998, Wang, Chan, and Chui 1998. Still other works have generalized wavelet transforms to obliquely defined cells. We believe that the fractional Fourier transform unifies all such approaches and is the purest, most elegant, and most powerful way to formulate them.

4.12.6   Other fractional operations and transforms

Fractional calculus, meaning fractional differentiation and integration (Engheta 1997), has long been an established area of study in mathematics, where the definition and study of fractional transforms is also sometimes encountered.

Recent interest in fractional transforms from a signal processing perspective has been inspired by work on the fractional Fourier transform. Of the many transforms studied, we might mention the fractional cosine, sine, and Hartley transforms (Lohmann and others 1996b, Pei and others 1998), the fractional Mellin transform (Akay and Boudreaux-Bartels 1998d), the fractional Hilbert transform (Lohmann, Mendlovic, and Zalevsky 1996), the fractional Radon transform (Zalevsky and Mendlovic 1996b), fractional wavelet transforms (Mendlovic and others 1997), the fractional Gabor transform (Zhang and others 1997), windowed fractional Fourier and windowed fractional Hartley transforms (Zhang and others 1998c), and the fractional wave packet transform (Huang and Suter 1998). The discrete rotational Gabor transform discussed in Akan and Chaparro 1996 is also essentially a fractional transform.

In principle it should not be difficult to propose fractional versions of "Fourier-like" transforms, transforms which are in essence some kind of frequency-domain representation. These include many of the unitary transforms commonly used in signal and image processing, such as Walsh, Hadamard, Haar, and so on. They do not seem to have received much attention yet. A satisfactory definition of the fractional Laplace and z-transforms would probably require greater care.

Zayed (1998b) discusses the analytic signal and Hilbert transform associated with the fractional Fourier transform. Just as the ordinary analytic signal is obtained by suppressing the negative frequencies of the ordinary Fourier transform of a signal, this fractional analytic signal is also obtained by suppressing the negative wing of the fractional Fourier transform.

The fractional Fourier transform is also related to a recently proposed extension of the common Laplace and Fourier transforms (Onural, Erden, and Ozaktas 1997).

An important mathematical device in classical and quantum mechanics is the *Legendre transformation* (Goldstein 1980), which plays an important role in phase-space approaches such as Hamiltonian mechanics and optics (Born and Wolf 1980). The fractional Legendre transformation (Alonso and Forbes 1995) is a generalization consistent with the fractional Fourier transform as defined in this book. Although not discussed in this book, it is an integral part of a complete theory of fractional Fourier transforms and phase-space representations.

A review of various fractional transforms emphasizing applications in optics is Lohmann, Mendlovic, and Zalevsky 1998.

4.13 Historical and bibliographical notes

The earliest work we are aware of containing what is essentially the fractional Fourier transform is Wiener 1929. Wiener sets out to find the kernel of the transformation whose eigenfunctions are the Hermite-Gaussian functions, but whose eigenvalues are of a more general form than those of the ordinary Fourier transform. As such, his premises are similar to those in our definition B. However, Wiener proceeds by using the relation given in table 2.7, property 3 to obtain a differential equation for the kernel, which he solves. (This is not unrelated to the approach we take in definition D.) He seems to have been motivated to develop the integral transform relationship so as to provide a basis for certain developments involving group theory and operator methods in quantum mechanics by Weyl (1927, 1930).

Condon 1937 seems to be the first work directly concerned with constructing the definition of the fractional Fourier transform, although the author does not use this term, and does not discuss any but its most elementary properties. Bargmann (1961) briefly discusses the transform in a broader context, and refers to Condon.

An important early work is Kober 1939, where the fractional Fourier and fractional Hankel transforms are obtained by a procedure essentially equivalent to our definition B. The general framework provided allows consideration of other transforms as well. Guinand (1956), citing Kober, discusses the relationship between fractional powers of unit matrices and fractional transformations. De Bruijn (1973), also referring to Kober, very briefly discusses the transform in a broader context.

Another early reference, cited in Khare 1974, is Patterson 1959; it gives a generalized transform in a physical context, which includes the fractional Fourier transform. A transform whose nth power is equal to the identity is mentioned in Antosik, Mikusiński, and Sikorski 1973:199.

Namias (1980a, b), apparently unaware of previous work, posed the fractional Fourier transform as the fractional power of the ordinary transform and presented several of its properties, its hyperdifferential form, and its relation to certain differential equations. Missing is the relationship of the transform to time-frequency representations. McBride and Kerr (1987) and Kerr (1988a, b, 1991, 1992) put the work of Namias on more solid ground.

Mustard seems to have developed several interesting results independently (1987a, b, c, 1989, 1991, 1996). He cites the work of Condon and Bargmann but not Kober or Namias. Mustard also cites the work of Taylor (1984) on pseudo-differential operators which involves the same group of integral operators. In particular, he discusses the relationship of the fractional Fourier transform to the Wigner distribution and new classes of uncertainty relationships that are invariant under the fractional Fourier transform.

It is practically impossible to trace all variants, relatives, or essentially equivalent forms of the fractional Fourier transform since it can be found under many different guises. For instance, the so-called oscillator semigroup (Howe 1988) is essentially the same as the family of fractional Fourier transforms. As phrased by Mustard, other "ghosts or shadows" of the fractional Fourier transform may be found in Miller 1968, Vilenkin 1965, Weissler 1979 and maybe other places referring to the oscillator or Hermite semigroup or the Heisenberg group or yet other entities.

Also, since fractional Fourier transforms are a special case of linear canonical transforms, all previous works on linear canonical transforms in some sense also cover fractional Fourier transforms; see the references in chapter 3 and in Wolf 1979. In some cases the fractional Fourier transform is not given any special attention and not recognized as such. In other cases the authors have commented on it as a one-parameter subclass with the Fourier transform as a special case, and sometimes even mentioned that it somehow interpolates between the identity and Fourier operations. However, they have mostly not recognized the fractional Fourier transform as the fractional power of the Fourier transform and have not recognized its many properties.

Given the nonuniqueness of the definition of the fractional Fourier transform, some authors single out the fractional Fourier transform defined in this book by making reference to some of the names mentioned above, such as Condon. However, this may not be appropriate given the complexity and possible incompleteness of the history.

The above history does not account for the many appearances of the expression for the fractional Fourier transform in varied contexts. In the course of their work, many researchers have come across a relation between two functions which is essentially a fractional Fourier transform, without recognizing it as such, or unaware of the many properties or the significance of the transform. Some examples of works where this occurs include Yurke, Schleich, and Walls 1990 and Ghatak and Thyagarajan 1980: equation 3.25.

We have already noted that the kernel of the fractional Fourier transform corresponds to the Green's function of the Schrödinger equation associated with the quantum-mechanical harmonic oscillator and propagation of light in quadratic graded-index media. Related equations appear in other contexts as well, such as in diffusive processes like heat transfer and Brownian motion, so the fractional Fourier transform may be relevant in these areas too. As ubiquitous as it is, the fractional Fourier transform is not always recognized as the fractional power of the Fourier transform, so its many properties are not exploited and the accompanying insights go unrealized.

Mendlovic and Ozaktas reinvented the transform in 1992, and Lohmann joined them with an alternative redefinition in terms of the Wigner distribution (Mendlovic and Ozaktas 1993a, b, Mendlovic, Ozaktas, and Lohmann 1993, Özaktaş and Mendlovic 1993, Ozaktas and Mendlovic 1993a, b, Ozaktas, Mendlovic, and Lohmann 1993, Lohmann 1993a, b, Ozaktas and others 1994a). These definitions were later demonstrated to be equivalent to each other and to earlier definitions (Mendlovic,

Ozaktas, and Lohmann 1994a). Almeida (1993, 1994) independently reintroduced the transform, initially referring to it as the "angular" Fourier transform. In a series of papers, Wood and Barry (1994a, b, c) dealt with the Radon transform of the Wigner distribution and its applications. Although they were not aware of the relationship with fractional Fourier transforms, later authors were quick to note the connection.

The publications in the early 1990s led to much activity in the signal processing and optics communities (see chapters 9, 10, 11 for further references), finally ending the earlier spatial and temporal isolation of authors, which had resulted in the transform being reinvented several times. A milestone was the publication of *Status Report on the Fractional Fourier Transform* 1995, which helped make the several people then working on the subject more visible to each other. A short and entertaining historical introduction to the transform is Larkin 1995a, b; however, this history is now known to be incomplete.

The fractional Fourier transform is not at all directly related to fractals or their applications in signal theory or optics (Lakhtakia and Caulfield 1992, Lakhtakia 1993, Uozumi and Asakura 1994), although some publications do deal with fractals and fractional Fourier transforms (Alieva 1996b, Alieva and Agullo-Lopez 1996a, 1998, Sheridan 1996).

A concise and self-contained introduction to the fractional Fourier transform and its applications to wave and beam propagation, optics, and signal processing is Ozaktas, Kutay, and Mendlovic 1998, 1999, and may be consulted by those needing a shorter reference.

5

Time-Order and Space-Order Representations

5.1 Introduction

This chapter is a brief exposition of time-order (or space-order) signal representations. Just like time-frequency and time-scale (or space-frequency and space-scale) signal representations, these representations constitute an alternative way of displaying the content of a signal, with the potential to reveal features which may not be evident upon examination of its other representations. Similar to time-frequency and time-scale representations, they are redundant representations in that the information contained in a one-dimensional signal is displayed in two dimensions.

We will consider two variations of the time-order representation, the rectangular time-order representation and the polar time-order representation. The rectangular time-order representation is simply $f_a(u)$ interpreted as a two-dimensional function, with u the horizontal coordinate and a the vertical coordinate. The polar time-order representation is simply $f_{2\alpha/\pi}(\rho)$ interpreted as a polar two-dimensional function where ρ is the radial coordinate and $\alpha = a\pi/2$ is the angular coordinate. Both representations are complex valued.

The rectangular space-order representation has been originally referred to as "the (u, a) chart," and the polar space-order representation as "the (ρ, α) chart" (Mendlovic and others 1995d). In image processing and optics, one often deals with two-dimensional signals. In that case, just as with space-frequency representations, the space-order representations are four-dimensional functions $f_{a_u, a_v}(u, v)$. In certain cases it may be beneficial to reduce the number of dimensions to three by choosing $a_u = a_v = a$.

5.2 The rectangular time-order representation

The rectangular time-order representation is simply $f_a(u)$ interpreted as a two-dimensional function, with u the horizontal coordinate and a the vertical coordinate. As such, it is a display of all the fractional Fourier transforms of $f(u)$ next to each other. In other words, the representations of the signal f in all fractional domains are displayed simultaneously. We have already seen such a display of the fractional Fourier transforms of the rectangle function in figure 4.5, which provides our first example of a rectangular time-order representation. Based on the relationship between optical propagation and the fractional Fourier transform discussed in chapter 9, the space-order representation of $f(u)$ also corresponds to a display of the evolution of a one-dimensional optical field through quadratic graded-index media, free space, or another quadratic-phase system (appropriately normalized in the two latter cases, as discussed in chapter 9). For instance, figure 4.5 corresponds to diffraction of light from a rectangular aperture, showing how the rectangular distribution of light evolves into a sinc function.

Formally, we will denote the rectangular time-order representation of a signal f by $T_f(u, a)$, so that we define

$$T_f(u, a) \equiv f_a(u). \tag{5.1}$$

Figure 5.1 illustrates the definition of the rectangular time-order representation.

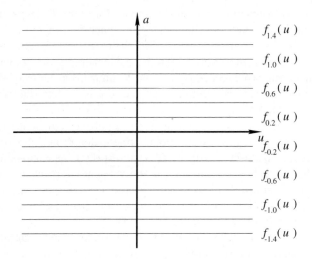

Figure 5.1.   The rectangular time-order representation (Ozaktas and Kutay 2000a).

Although we will not elaborate, it is evident that all of the properties of the fractional Fourier transform can be interpreted as properties of the time-order representation. In particular, the following simple identities are sometimes useful in dealing with the representations of the products and convolutions of functions:

$$T_{h(u)f(u)}(u, a) = T_{H(u)*F(u)}(u, a - 1), \tag{5.2}$$
$$T_{h(u)*f(u)}(u, a) = T_{H(u)F(u)}(u, a - 1). \tag{5.3}$$

Of course, the rectangular representation is periodic in a with period 4.

Figure 5.2 illustrates the rectangular time-order representations of the signal $\exp(-\pi 2u^2)$ and the signal $\mathrm{rect}(u)$.

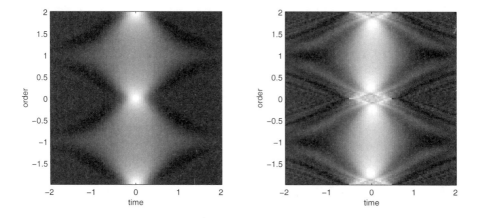

Figure 5.2. Magnitudes of the rectangular time-order representations of $\exp(-\pi 2u^2)$ (left) and $\text{rect}(u)$ (right) (Ozaktas and Kutay 2000a).

5.3 Optical implementation

This section can be skipped by readers with no interest in optics. Those interested should read it after reading section 9.4.3.

An optical system can be used to generate a two-dimensional display of the space-order representation of a one-dimensional signal. This amounts to using a two-dimensional optical system to simultaneously perform the fractional Fourier transform operation on a one-dimensional signal for many values of a at once. The key element of such a system is a fractional Fourier transforming scheme whose order can be adjusted by adjusting the focal lengths of the lenses, but in which the lens spacings are constant. Since the spacings are constant regardless of the order, multiple channels each consisting of this basic one-dimensional scheme can be lined up along the second dimension to realize a multiple-channel fractional Fourier transformer, with each channel realizing a fractional Fourier transform of different order.

A detailed description of such an optical system with experimental results may be found in Mendlovic and others 1996a. First consider the canonical configuration in figure 9.6, which consists of a lens followed by a section of free space followed by another lens. In this configuration, the focal lengths of the lenses and the length of the section of free space depend on the order. To eliminate order-dependent lens spacings, the section of free space was simulated by a Fourier transforming stage followed by an order-dependent lens followed by another Fourier transforming stage. The Fourier transforming stages employed consisted of a lens followed by a section of free space followed by another lens. When each group of adjacent lenses are replaced by a single equivalent lens, the final result is a configuration consisting of three lenses separated by two sections of free space, such that the order can be adjusted freely by choosing the focal lengths of the lenses only. The reader may either consult the referenced work or use matrix algebra to easily derive the required focal lengths for each order. Three lineups of one-dimensional lenses with gradually varying focal length were realized by using multifacet masks, consisting of narrow strips each of which is a one-dimensional Fresnel zone plate with appropriate focal length (to realize the fractional order associated with each channel). Experimental results may be found in the cited work.

5.4   The polar time-order representation

The polar time-order representation is simply $f_{2\alpha/\pi}(\rho)$ interpreted as a polar two-dimensional function where ρ is the radial coordinate and α is the angular coordinate. As such, it is a display of all the fractional Fourier transforms of $f(u)$ such that $f_a(\rho)$ lies along the radial line making angle $\alpha = a\pi/2$ with the horizontal u axis. As in the rectangular representation, all transforms are displayed simultaneously.

Formally, we will denote the polar time-order representation of a signal f by $T_f(\rho, \alpha)$, so that we define

$$T_f(\rho, \alpha) \equiv f_{2\alpha/\pi}(\rho). \tag{5.4}$$

Since $f_a(\rho)$ is periodic in a with period 4, $T_f(\rho, \alpha)$ is periodic in α with period 2π, as all polar functions must be. $T_f(\rho, \alpha)$ is defined for negative values of ρ as well. This does not pose any inconsistency since $f_{a\pm2}(\rho) = f_a(-\rho)$, from which it also follows that $T_f(\rho, \alpha) = T_f(-\rho, \alpha \pm \pi)$. Figure 5.3 illustrates the definition of the polar time-order representation.

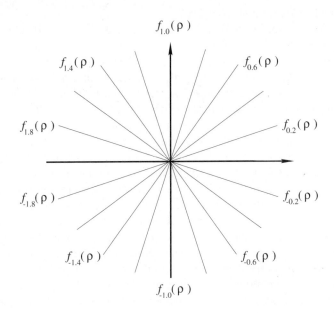

Figure 5.3.   The polar time-order representation (Ozaktas and Kutay 2000a).

Figure 5.4 illustrates the polar time-order representations of the signal $\exp(-\pi 2u^2)$ and the signal $\mathrm{rect}(u)$.

The polar time-order representation is directly related to the concept of fractional Fourier domains. Each fractional Fourier transform $f_a(\rho)$ "lives" in the ath domain, defined by the radial line making angle $\alpha = a\pi/2$ with the u axis. The rectangular

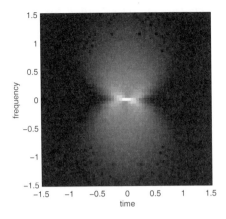 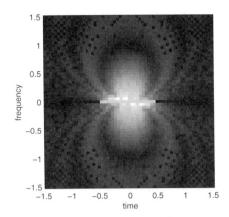

Figure 5.4. Magnitudes of the polar time-order representations of $\exp(-\pi 2u^2)$ (left) and rect(u) (right) (Ozaktas and Kutay 2000a).

time-order space is not a time-frequency space (a phase space) in the conventional sense, in that it is not defined by two Fourier-dual coordinates. However, the polar time-order space can be thought of as a time-frequency space (a phase space) since the horizontal and vertical axes indeed correspond to time and frequency. The slices of the polar representation (see section 2.10.1) are simply equal to the fractional Fourier transforms:

$$\mathcal{SLC}_\alpha[T_f(\rho, \alpha)](u) = f_a(u), \qquad \alpha = a\pi/2. \tag{5.5}$$

In particular, the slice at $\alpha = 0$ is the time-domain representation $f(u)$, and the slice at $\alpha = \pi/2$ is the frequency-domain representation $F(u)$. Other slices correspond to fractional transforms of other orders. We also know that the Radon transform of the Wigner distribution is given by

$$\mathcal{RDN}_\alpha[W_f(u, \mu)](\rho) = |f_{2\alpha/\pi}(\rho)|^2 = |T_f(\rho, \alpha)|^2. \tag{5.6}$$

Thus, the Radon transform of the Wigner distribution, interpreted as a polar function, corresponds to the absolute square of the time-order representation defined in this chapter. The relationship of time-order representations to the Wigner distribution and ambiguity function will be further discussed in section 5.5.

The origin of the polar time-order representation where $\rho = 0$ and the angle α is indefinite should be avoided since it does not contain any physically meaningful information. As with such polar displays, the resolution of the display is lower for smaller values of ρ. For instance, if we consider $T_f(\rho, \alpha)$ being displayed on a computer screen, we will have greater difficulty discerning features near the origin where a greater amount of information needs to be packed. Thus in practice, a certain central region of the display will not be able to represent $T_f(\rho, \alpha)$ accurately.

We now move on to discuss a number of properties of this representation. First, we note that obtaining the original function from the distribution is trivial:

$$f(u) = f_0(u) = T_f(u, 0). \tag{5.7}$$

Table 5.1. Properties of the polar time-order representation (Ozaktas and Kutay 2000a).

	$f(u)$	$T_f(\rho, \alpha)$
1.	$f(-u)$	$T_f(-\rho, \alpha)$
2.	$|M|^{-1} f(u/M)$	$\sqrt{\frac{1 - i \cot \alpha}{1 - iM^2 \cot \alpha}} \exp\left[i\pi\rho^2 \cot \alpha \left(1 - \frac{\cos^2 \alpha'}{\cos^2 \alpha}\right)\right] T_f\left(\frac{M\rho \sin \alpha'}{\sin \alpha}, \alpha'\right)$
3.	$f(u - \xi)$	$\exp(i\pi\xi^2 \sin \alpha \cos \alpha) \exp(-i2\pi\rho\xi \sin \alpha) T_f(\rho - \xi \cos \alpha, \alpha)$
4.	$\exp(i2\pi\xi u) f(u)$	$\exp(-i\pi\xi^2 \sin \alpha \cos \alpha) \exp(i2\pi\rho\xi \cos \alpha) T_f(\rho - \xi \sin \alpha, \alpha)$
5.	$u^n f(u)$	$[\cos \alpha \, \rho - \sin \alpha \, (i2\pi)^{-1} d/d\rho]^n T_f(\rho, \alpha)$
6.	$[(i2\pi)^{-1} d/du]^n f(u)$	$[\sin \alpha \, \rho + \cos \alpha \, (i2\pi)^{-1} d/d\rho]^n T_f(\rho, \alpha)$
7.	$f(u)/u$	$-i \csc \alpha \, \exp(i\pi\rho^2 \cot \alpha) \int_{-\infty}^{2\pi\rho} T_f(\rho', \alpha) \exp(-i\pi\rho'^2 \cot \alpha) \, d\rho'$
8.	$\int_\xi^u f(u') \, du'$	$\sec \alpha \, \exp(-i\pi\rho^2 \tan \alpha) \int_\xi^\rho T_f(\rho', \alpha) \exp(i\pi\rho'^2 \tan \alpha) \, d\rho'$
9.	$f^*(u)$	$T_f^*(\rho, -\alpha)$
10.	$f^*(-u)$	$T_f^*(-\rho, -\alpha)$
11.	$[f(u) + f(-u)]/2$	$[T_f(\rho, \alpha) + T_f(-\rho, \alpha)]/2$
12.	$[f(u) - f(-u)]/2$	$[T_f(\rho, \alpha) - T_f(-\rho, \alpha)]/2$

ξ and M are real but $M \neq 0, \pm\infty$. $\alpha' = \arctan(M^{-2} \tan \alpha)$ where α' is taken to be in the same quadrant as α. Property 7 is not valid when α is an integer multiple of π and property 8 is not valid when α is an odd integer multiple of $\pi/2$.

Obtaining the Fourier transform of the function or indeed any other fractional transform is likewise a direct consequence of the definition.

The time-order representation of the a'th fractional Fourier transform of a function is simply a rotated version of the time-order representation of the original function:

$$T_{f_{a'}}(\rho, \alpha) = T_f(\rho, \alpha + \alpha'), \tag{5.8}$$

where $\alpha' = a'\pi/2$. In particular, the time-order representation of the Fourier transform of a function is a $\pi/2$-rotated version of the original. Since the time-order representation is linear, the representation of any linear combination of functions is the same as the linear combination of their representations.

Various properties of the polar time-order representation follow immediately from table 4.3, and are presented in table 5.1. In addition to the properties given in the table, we also consider the effect of chirp multiplication (corresponding to a thin lens in optics) and chirp convolution (corresponding to Fresnel propagation in optics). Letting $g(u) = \exp(-i\pi q u^2) f(u)$, a moderate amount of algebra leads to

$$T_g(\rho, \alpha) = \sqrt{\frac{1 - i \cot \alpha}{1 - i \cot \alpha'}} \, e^{i\pi\rho^2 \cot \alpha \left(1 - \frac{\sin \alpha' \cos \alpha'}{\sin \alpha \cos \alpha}\right)} \, T_f\left(\frac{\rho \sin \alpha'}{\sin \alpha}, \alpha'\right), \tag{5.9}$$

which we see essentially amounts to a combination of rotation and radial scaling. Here $\alpha' = \text{arccot}(\cot \alpha - q)$ with the cotangent being inverted such that if $\alpha \in [0, \pi)$ then α' is also in $[0, \pi)$, and if $\alpha \in [-\pi, 0)$ then α' is also in $[-\pi, 0)$. The nature of this geometric transformation is illustrated in figure 5.5. Figure 5.5a shows a square and figure 5.5b shows the effect of ordinary vertical shearing on this square. Figure 5.5c, on the other hand, shows the effect of the geometric transformation inherent in equation 5.9. Figure 5.5d, e, f show the same for a circle. Overall, figure 5.5 comparatively illustrates the effect of chirp multiplication on the Wigner distribution (b, e) and the polar time-order representation (c, f).

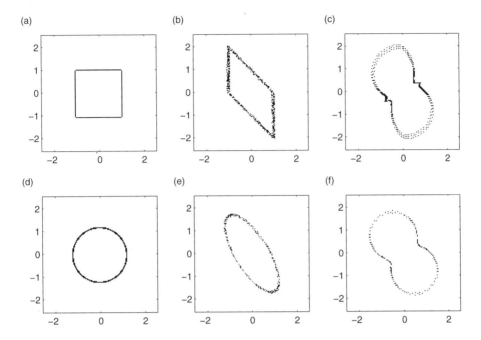

Figure 5.5. Effect of chirp multiplication on the Wigner distribution and the polar time-order representation ($q = 1$). After Mendlovic and others 1995d.

Now let $g(u) = [e^{-i\pi/4}\sqrt{1/r}\exp(i\pi u^2/r)] * f(u)$. Since chirp convolution in the time domain corresponds to chirp multiplication in the frequency domain as $G(\mu) = \exp(-i\pi r\mu^2)F(\mu)$, the above result can be combined with the fact that the Fourier transform corresponds to a $\pi/2$ rotation of the time-order representation to obtain

$$T_g(\rho,\alpha) = \sqrt{\frac{1 + i\tan\alpha}{1 - i\cot\alpha'}}\; e^{-i\pi\rho^2\tan\alpha\left(1+\frac{\sin\alpha'\cos\alpha'}{\sin\alpha\cos\alpha}\right)}\; T_f\left(\frac{-\rho\sin\alpha'}{\cos\alpha}, \alpha'\right), \qquad (5.10)$$

where $\alpha' = \operatorname{arccot}(-\tan\alpha - r)$. The cotangent is inverted such that if $\alpha \in [\pi/2, 3\pi/2)$ then α' is also in $[\pi/2, 3\pi/2)$, and if $\alpha \in [-\pi/2, \pi/2)$ then α' is also in $[-\pi/2, \pi/2)$.

It is possible to show that the chirp multiplication and convolution properties are indeed consistent under concatenation. For instance, let us assume that we multiply $g(u) = \exp(-i\pi qu^2)f(u)$ with another chirp to obtain $g'(u) = \exp(-i\pi q'u^2)g(u)$, so that $g'(u) = \exp[-i\pi(q + q')u^2]f(u)$. Now, using equation 5.9 we can find $T_g(\rho,\alpha)$ and using the same equation again we can find $T_{g'}(\rho,\alpha)$. Alternatively, we can use equation 5.9 once with $(q+q')$ to obtain $T_{g'}(\rho,\alpha)$ directly from $T_f(\rho,\alpha)$, as the reader can verify through straightforward algebra and trigonometry (Mendlovic and others 1995d).

The effect of an arbitrary linear canonical transform (quadratic-phase system) on the polar time-order representation can be found by decomposing the linear canonical transform into chirp multiplications and convolutions (equations 3.121 and 3.122).

5.5 Relationships with the Wigner distribution and the ambiguity function

We now return to the discussion of the relationship of time-order representations with other time-frequency representations, which we postponed after equation 5.6. Referring to the discussion beginning on page 162, it is easy to show as a consequence of the projection-slice theorem that

$$A_f(-\rho \sin \alpha, \rho \cos \alpha) = \mathcal{F}|f_{2\alpha/\pi}(\rho)|^2 = T_f(\rho, \alpha + \pi/2) * T_f^*(-\rho, \alpha + \pi/2). \quad (5.11)$$

$A_f(-\rho \sin \alpha, \rho \cos \alpha)$ can be interpreted as the slice of $A_f(\bar{u}, \bar{\mu})$ at an angle $\alpha + \pi/2$. Alternatively, $A_f(-\rho \sin \alpha, \rho \cos \alpha)$ can be interpreted as the slice of $A_f(-\bar{u}, \bar{\mu})$ at an angle α with respect to the $\bar{\mu}$ axis, with $\bar{\mu}$ taken as the horizontal axis and $\bar{u}$ taken as the vertical axis. After straightforward algebra, the above result can be cast in the slightly simpler form of

$$\mathcal{SLC}_\alpha[A_f(\bar{u}, \bar{\mu})](\rho) = A_f(\rho \cos \alpha, \rho \sin \alpha) = T_f(\rho, \alpha) * T_f^*(-\rho, \alpha)$$
$$= f_{2\alpha/\pi}(\rho) * f_{2\alpha/\pi}^*(-\rho) = f_{2\alpha/\pi}(\rho) \star f_{2\alpha/\pi}(\rho) = R_{f_{2\alpha/\pi} f_{2\alpha/\pi}}(\rho), \quad (5.12)$$

where $R_{f_{2\alpha/\pi} f_{2\alpha/\pi}}(\rho)$ denotes the deterministic autocorrelation of $f_{2\alpha/\pi}(\rho)$. We see that just as oblique projections of the Wigner distribution correspond to the squared magnitudes of the fractional Fourier transforms of the function, the oblique slices of the ambiguity function correspond to the autocorrelations of the fractional Fourier transforms of the function.

Having discussed the Radon transform (projections) of the Wigner distribution and the slices of the ambiguity function, we now turn our attention to the slices of the Wigner distribution and the Radon transform of the ambiguity function. To proceed, we first write the definition of the Wigner distribution for f_a:

$$W_{f_a}(u, \mu) = \int f_a(u + u'/2) f_a^*(u - u'/2) e^{-i2\pi \mu u'} \, du', \quad (5.13)$$

whose slice at the angle $\pi/2$ is easily obtained as

$$\mathcal{SLC}_{\pi/2}[W_{f_a}(u, \mu)](\rho) = W_{f_a}(0, \rho) = \int f_a(u'/2) f_a^*(-u'/2) e^{-i2\pi \rho u'} \, du'$$
$$= \mathcal{F}[f_a(u/2) f_a^*(-u/2)](\rho). \quad (5.14)$$

Now, we know that $W_{f_a}(u, \mu) = W_f(u \cos \alpha - \mu \sin \alpha, u \sin \alpha + \mu \cos \alpha)$, so that

$$W_{f_a}(0, \rho) = W_f(-\rho \sin \alpha, \rho \cos \alpha) = W_f[\rho \cos(\alpha + \pi/2), \rho \sin(\alpha + \pi/2)]$$
$$= \mathcal{SLC}_{\alpha + \pi/2}[W_f(u, \mu)](\rho). \quad (5.15)$$

Combining equations 5.14 and 5.15 finally leads us to the desired expression for the slices of the Wigner distribution:

$$\mathcal{SLC}_\alpha[W_f(u, \mu)](\rho) = \int f_{2\alpha/\pi - 1}(u'/2) f_{2\alpha/\pi - 1}^*(-u'/2) e^{-i2\pi \rho u'} \, du'$$
$$= 4 f_{2\alpha/\pi}(2\rho) * f_{2\alpha/\pi}^*(2\rho) = 4 T_f(2\rho, \alpha) * T_f^*(2\rho, \alpha). \quad (5.16)$$

The slice of the Wigner distribution at angle α is equal to the convolution of $2T_f(2\rho, \alpha) = 2f_{2\alpha/\pi}(2\rho)$ with its conjugate. Since $T_f(2\rho, \alpha)$ is a function of polar coordinates, its slice at an angle α is simply $T_f(2\rho, \alpha)$ itself. Thus the slice of the Wigner distribution at a certain angle is equal to the convolution of the slice of the time-order representation with its conjugate.

Now, application of the projection-slice theorem allows us to arrive at the following result for the Radon transform of the ambiguity function:

$$\mathcal{RDN}_\alpha[A_f(\bar{u}, \bar{\mu})](\rho) = f_{2\alpha/\pi}(\rho/2)f^*_{2\alpha/\pi}(-\rho/2) = T_f(\rho/2, \alpha)T^*_f(-\rho/2, \alpha). \quad (5.17)$$

The special case of $\alpha = 0$ yields

$$\mathcal{RDN}_0[A_f(\bar{u}, \bar{\mu})](\rho) = \int A_f(\rho, \bar{\mu})\, d\bar{\mu} = f(\rho/2)f^*(-\rho/2), \quad (5.18)$$

which could have also been derived directly from the definition of the ambiguity function.

Table 5.2 summarizes the Radon transforms and slices of the Wigner distribution and ambiguity function. We see that the results of these operations can be expressed in terms of the fractional Fourier transform and the polar time-order representation. For both the Wigner distribution and the ambiguity function, the Radon transform is of product form and the slice is of convolution form. The essential difference between the Wigner distribution and the ambiguity function lies in the scaling of ρ by 2 or $1/2$ on the right-hand side.

Table 5.2.   Radon transforms and slices of the Wigner distribution and the ambiguity function (Ozaktas and Kutay 2000a).

$\mathcal{RDN}_\alpha[W_f(u, \mu)](\rho) = f_{2\alpha/\pi}(\rho)f^*_{2\alpha/\pi}(\rho) = T_f(\rho, \alpha)T^*_f(\rho, \alpha)$
$\mathcal{RDN}_\alpha[A_f(\bar{u}, \bar{\mu})](\rho) = f_{2\alpha/\pi}(\rho/2)f^*_{2\alpha/\pi}(-\rho/2) = T_f(\rho/2, \alpha)T^*_f(-\rho/2, \alpha)$
$\mathcal{SLC}_\alpha[W_f(u, \mu)](\rho) = 2f_{2\alpha/\pi}(2\rho) * 2f^*_{2\alpha/\pi}(2\rho) = 2T_f(2\rho, \alpha) * 2T^*_f(2\rho, \alpha)$
$\mathcal{SLC}_\alpha[A_f(\bar{u}, \bar{\mu})](\rho) = f_{2\alpha/\pi}(\rho) * f^*_{2\alpha/\pi}(-\rho) = T_f(\rho, \alpha) * T^*_f(-\rho, \alpha)$

The upper row can also be expressed as $|f_{2\alpha/\pi}(\rho)|^2 = |T_f(\rho, \alpha)|^2$.

We already know the slice of $T_f(\rho, \alpha)$ to be simply given by $f_{2\alpha/\pi}(\rho)$, by definition. Now we embark on deriving the Radon transform of $T_f(\rho, \alpha)$. A polar-to-rectangular coordinate conversion allows us to write $T_f(\rho, \alpha)$ in rectangular coordinates as follows:

$$T_{f,\text{rect}}(u, \mu) \equiv T_f(\rho, \alpha) \qquad \rho = \sqrt{u^2 + \mu^2}, \quad \alpha = \arctan(\mu/u). \quad (5.19)$$

The Radon transform of $T_f(\rho, \alpha)$ at an angle ϕ is

$$\mathcal{RDN}_\phi[T_f(\rho, \alpha)](\varrho) = \int T_{f,\text{rect}}(\varrho\cos\phi - \mu'\sin\phi, \varrho\sin\phi + \mu'\cos\phi)\, d\mu', \quad (5.20)$$

leading to

$$\mathcal{RDN}_\phi[T_f(\rho, \alpha)](\varrho) = \int T_f\left[\sqrt{\varrho^2 + \mu'^2}, \arctan\left(\frac{\varrho\sin\phi + \mu'\cos\phi}{\varrho\cos\phi - \mu'\sin\phi}\right)\right] d\mu'.$$

$$(5.21)$$

Introducing the following change of integration variable from μ' to θ:

$$\mu' = \varrho \tan\theta, \qquad d\mu' = \varrho \sec^2\theta \, d\theta, \tag{5.22}$$

we can write the above integral as

$$\mathcal{RDN}_\phi[T_f(\rho,\alpha)](\varrho) = \int_{-\pi/2}^{\pi/2} T_f(\varrho \sec\theta, \phi+\theta) \, \varrho \sec^2\theta \, d\theta$$
$$= \int_{-\pi/2}^{\pi/2} f_{2(\phi+\theta)/\pi}(\varrho \sec\theta) \, \varrho \sec^2\theta \, d\theta, \tag{5.23}$$

which will be our final expression for the Radon transform of $T_f(\rho,\alpha)$.

We will denote the two-dimensional ordinary Fourier transform of $T_f(\rho,\alpha)$, expressed in rectangular coordinates, as $\tilde{T}_{f,\text{rect}}(\bar{\mu},\bar{u})$:

$$\tilde{T}_{f,\text{rect}}(\bar{\mu},\bar{u}) \equiv \iint T_{f,\text{rect}}(u,\mu)e^{-i2\pi(\bar{\mu}u+\bar{u}\mu)} \, du \, d\mu \tag{5.24}$$

The same Fourier transform relation can also be expressed in polar coordinates $\bar{\rho} = \sqrt{\bar{u}^2 + \bar{\mu}^2}$, $\bar{\alpha} = \arctan(\bar{\mu}/\bar{u})$ as

$$\tilde{T}_f(\bar{\rho},\bar{\alpha}) = \int_0^{2\pi} \int_0^\infty T_f(\rho,\alpha)e^{-i2\pi\bar{\rho}\rho\cos(\bar{\alpha}-\alpha)}\rho \, d\rho \, d\alpha \tag{5.25}$$

where $\tilde{T}_f(\bar{\rho},\bar{\alpha}) \equiv \tilde{T}_{f,\text{rect}}(\bar{\mu},\bar{u})$. The slice of $\tilde{T}_f(\bar{\rho},\bar{\alpha})$ at an angle ϕ is simply

$$\mathcal{SLC}_\phi[\tilde{T}_f(\bar{\rho},\bar{\alpha})](\varrho) = \tilde{T}_f(\varrho,\phi). \tag{5.26}$$

Now, application of the projection-slice theorem allows us to write

$$\mathcal{SLC}_\phi[\tilde{T}_f(\bar{\rho},\bar{\alpha})](\varrho) = \tilde{T}_f(\varrho,\phi) = \mathcal{F}\left[\mathcal{RDN}_\phi[T_f(\rho,\alpha)]\right](\varrho)$$
$$= \int \left[\int_{-\pi/2}^{\pi/2} T_f(\varrho'\sec\theta, \phi+\theta) \, \varrho' \sec^2\theta \, d\theta\right] e^{-i2\pi\varrho\varrho'} \, d\varrho'$$
$$= \frac{i}{2\pi} \int_{-\pi/2}^{\pi/2} f'_{2(\phi+\theta)/\pi+1}(\varrho\cos\theta) \, \sec\theta \, d\theta, \tag{5.27}$$

where $f'_{2(\phi+\theta)/\pi+1}(w)$ denotes the derivative $df_{2(\phi+\theta)/\pi+1}(w)/dw$.

Having obtained its slice, what remains is to write an expression for the Radon transform of $\tilde{T}_f(\bar{\rho},\bar{\alpha})$, which follows without much difficulty from the projection-slice theorem:

$$\mathcal{RDN}_\phi[\tilde{T}_f(\bar{\rho},\bar{\alpha})](\varrho) = \mathcal{F}^{-1}\mathcal{SLC}_{\phi+\pi}[T_f(\rho,\alpha)](\varrho) = \mathcal{F}^{-1}f_{2\phi/\pi+2}(\varrho)$$
$$= f_{2\phi/\pi+1}(\varrho) = T_f(\varrho, \phi+\pi/2). \tag{5.28}$$

Thus, we have now completed a set of four expressions for the Radon transforms and slices of the polar time-order representation $T_f(\rho, \alpha)$ and its two-dimensional Fourier transform $\tilde{T}_f(\bar{\rho}, \bar{a})$ (table 5.3). The slice of $T_f(\rho, \alpha)$ at a certain angle is simply equal to the fractional Fourier transform $f_a(\rho)$ by definition (with $\alpha = a\pi/2$). The Radon transform of $\tilde{T}_f(\bar{\rho}, \bar{a})$ at an angle ϕ is given by $f_{b+1}(\rho)$ or $T_f(\rho, \phi+\pi/2)$, a $\pi/2$ rotated version of $T_f(\rho, \alpha)$ (with $\phi = b\pi/2$). The remaining two relations are more complicated and are given by equations 5.23 and 5.27. We already know that the time-frequency representation whose projections are equal to $|f_a(u)|^2$ is the Wigner distribution. We now see that the time-frequency representation whose projections are equal to $f_a(u)$ is the two-dimensional Fourier transform of the polar time-order representation (within a rotation).

Table 5.3.   Radon transforms and slices of the polar time-order representation and its two-dimensional Fourier transform (Ozaktas and Kutay 2000a).

$$\mathcal{RDN}_\phi[T_f(\rho, \alpha)](\varrho) = \int_{-\pi/2}^{\pi/2} f_{2(\phi+\theta)/\pi}(\varrho \sec \theta)\, \varrho \sec^2 \theta\, d\theta$$
$$\mathcal{RDN}_\phi[\tilde{T}_f(\bar{\rho}, \bar{a})](\varrho) = f_{2\phi/\pi+1}(\varrho)$$
$$\mathcal{SLC}_\phi[T_f(\rho, \alpha)](\varrho) = f_{2\phi/\pi}(\varrho)$$
$$\mathcal{SLC}_\phi[\tilde{T}_f(\bar{\rho}, \bar{a})](\varrho) = \frac{i}{2\pi} \int_{-\pi/2}^{\pi/2} f'_{2(\phi+\theta)/\pi+1}(\varrho \cos \theta)\, \sec \theta\, d\theta$$

Looking back, we see that we have derived a total of eight expressions for the Radon transforms and slices of the Wigner distribution and its two-dimensional Fourier transform (the ambiguity function), and the Radon transforms and slices of the polar time-order representation and its two-dimensional Fourier transform (tables 5.2 and 5.3).

The polar time-order representation is a linear time-frequency representation, unlike the Wigner distribution which is a quadratic time-frequency representation. The reader may recall from chapter 3 that the Wigner distribution can be interpreted as giving the distribution of signal energy over time and frequency. As such it is an example of an *energetic* time-frequency representation. In contrast, the ambiguity function has qualities reminiscent of correlation, and it is an example of a *correlative* time-frequency representation. We also know that the Wigner distribution and ambiguity function constitute a two-dimensional Fourier transform pair. The polar time-order representation and its two-dimensional Fourier transform fall into neither category and do not belong to Cohen's class of shift-invariant representations. Their importance stems from the fact that the Radon transforms (integral projections) and slices of the Wigner distribution and the ambiguity function can be expressed in terms of products or convolutions of various scaled forms of the time-order representation and its two-dimensional Fourier transform.

5.6   Applications of time-order representations

Here we will discuss the application of time-order representations to the recognition of acoustic signals. First, the time-order representation of a reference signal $h(u)$ is computed and stored. Then we calculate the correlation of an incoming acoustic signal

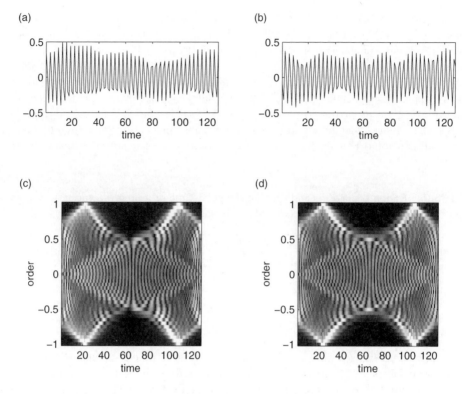

Figure 5.6.   (a) Reference signal $h(u)$; (b) a different signal $f(u)$; (c) $T_h(u, a)$ for the reference signal $h(u)$; (d) $T_f(u, a)$ for the signal $f(u)$.

$f(u)$ with the reference using either rectangular or polar time-order representations:

$$C_{\text{rect}}(u, a) = \iint T_f(u', a') T_h^*(u' - u, a' - a) \, du' \, da', \tag{5.29}$$

$$C_{\text{pol}}(u, \mu) = \iint T_{f,\text{rect}}(u', \mu') T_{h,\text{rect}}^*(u' - u, \mu' - \mu) \, du' \, d\mu', \tag{5.30}$$

where $T_{f,\text{rect}}(u', \mu')$ and $T_{h,\text{rect}}(u', \mu')$ are the polar time-order representations expressed in rectangular coordinates. Numerical evidence suggests that this approach leads to superior discrimination resulting from the additional dimension, as compared to direct ordinary time-domain correlation.

Figure 5.6a shows a 128 pixel (15.625 ms) segment from the middle of an acoustic signal sampled at a rate of 8192 Hz. This segment is taken as the reference signal. Figure 5.6c shows the rectangular time-order representation of this reference signal. Figure 5.7a presents the correlation given by equation 5.29 when the input signal is the same as the reference. Figure 5.7c presents the cross section of this correlation at $a = 0$. For comparison, the correlation between the reference signal and a second similar but different acoustic signal (shown in figure 5.6b) is presented in figure 5.7b. The cross section of this correlation, shown in figure 5.7d, exhibits a much smaller peak. Finally, in figure 5.7e we show the direct ordinary time-domain correlation of the reference signal with itself and in figure 5.7f we show the direct ordinary time-domain correlation of the reference signal with the second different signal. The peak obtained is much less distinct and highly oscillatory. Overall, it is clear that the discrimination which can be obtained from figure 5.7e and f is poorer than the discrimination which can be obtained from figure 5.7c and d.

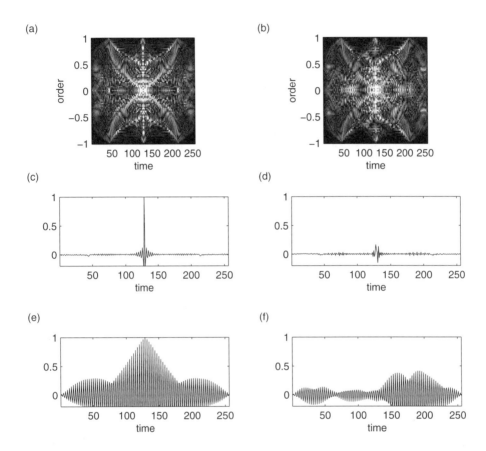

Figure 5.7. (a) Correlation given by equation 5.29 when the input signal is equal to the reference signal. (b) Correlation given by equation 5.29 when the input signal is equal to the second, different signal. (c) One-dimensional cross section of part a at $a = 0$. (d) One-dimensional cross section of part b at $a = 0$. (e) Ordinary time-domain correlation of the reference signal with itself. (f) Ordinary time-domain correlation of the reference signal with the second signal.

5.7 Other applications of the fractional Fourier transform in time- and space-frequency analysis

Here we briefly mention a number of other applications of the fractional Fourier transform in time-frequency analysis. Fonollosa and Nikias (1994) and Fonollosa (1996) have used the fractional Fourier transform to define new positive time-frequency distributions. Positive time-frequency distributions and so-called marginal properties are further studied in Sang, Williams, and O'Neill 1996. The fact that fractional Fourier transforms of different orders correspond to the Radon transforms of the Wigner distribution of a signal has led to the notion of generalized time-frequency distributions which have "generalized marginal" properties (Xia and others 1996). Indeed, the fractional Fourier transform plays an important role in the theory of time- or space-frequency (phase-space) representations, especially when rotations or projections of time-frequency representations are involved; a random sampling of papers may include Ristic and Boashash 1993, Barbarossa 1995, Wang, Chan, and Chui 1998.

5.8   Historical and bibliographical notes

Space-order representations were first defined in Mendlovic and others 1995d, where they were referred to as "the (u, a) chart" and "the (ρ, α) chart." Most of the results presented in sections 5.2 and 5.4 either appeared or were anticipated in this work. Section 5.5 is essentially based on Ozaktas and Kutay 2000a.

The possibility of what essentially amounts to some kind of discrete polar time-order representation was mentioned in Dickinson and Steiglitz 1982. However, the discrete fractional Fourier transform defined there is substantially different from that discussed in this book (see the discussion starting on page 139).

6

The Discrete Fractional
Fourier Transform

6.1 Introduction

In this chapter we will discuss a definition of the discrete fractional Fourier transform which generalizes the discrete Fourier transform (DFT) in the same sense that the continuous fractional Fourier transform generalizes the continuous ordinary Fourier transform. This definition is based on a particular set of eigenvectors of the DFT, which constitutes the discrete counterpart of the set of Hermite-Gaussian functions. It exactly satisfies all the desirable properties expected of a discrete counterpart of the fractional Fourier transform, such as unitarity, index additivity, and reduction to the DFT for unit order.

While we believe that the definition to be discussed is the most satisfactory among the different alternatives suggested so far, a definitive and widely accepted definition of the transform has not yet been established. A number of intrinsic difficulties explain why the development and establishment of a satisfactory definition of the discrete transform consistent with the continuous transform has been slow. The fractional Fourier transform corresponds to rotation of the Wigner distribution, so that its discrete counterpart should correspond to rotation of the discrete Wigner distribution. However, rotation of discrete lattices is inherently problematic as far as an exact self-consistent theory is concerned, for the reason that it is not obvious how the rotated lattice is to be interpolated to the original lattice in the "right" way. The issue of precisely how the discrete Wigner distribution should be defined has not been resolved in a totally satisfactory way either.

Apart from the definition of the discrete fractional Fourier transform, we will also discuss an algorithm which can compute the N samples of $f_a(u)$ in terms of the N samples of $f(u)$ in $O(N \log N)$ time. The reason for existence of such an algorithm will disappear when a fast fractional Fourier transform algorithm is developed for the discrete fractional Fourier transform. Unfortunately, such a fast algorithm which computes the exact discrete fractional Fourier transform is presently not known. However, the fact that the samples of $f_a(u)$ can be computed in terms of the samples of $f(u)$ in $O(N \log N)$ time very strongly suggests that such an algorithm can be found. Once such an algorithm is established, the algorithm computing the samples of $f_a(u)$ in terms of the samples of $f(u)$ would no longer be used, and the discrete fractional transform would be used to approximate the continuous fractional transform, in the same way that the DFT is used to approximate the ordinary Fourier transform.

While constituting an important gap in the theory, this has little or no impact in practice since the algorithm presented in section 6.7 can be used to approximate the continuous transform with the same degree of accuracy with which the discrete transform can approximate the continuous transform. The matrix mapping the samples of $f(u)$ to those of $f_a(u)$ is likewise a good approximation to the discrete fractional Fourier transform matrix, so that the $O(N \log N)$ algorithm we have for this matrix satisfactorily serves as a surrogate for the—as yet unavailable—fast algorithm for the discrete transform.

The definition of the discrete fractional Fourier transform discussed in this chapter is based on an idea by Pei and Yeh (1997) which was consolidated in Candan 1998 and Candan, Kutay, and Ozaktas 1999, 2000.

This definition has the following properties, which could plausibly be posed as requirements which a discrete definition of the transform must satisfy to begin with: (i) unitarity, (ii) index additivity, (iii) reduction to the DFT when the order is equal to unity, and (iv) approximation of the continuous transform in some satisfactory sense. An additional requirement may be stated as (v) continuity of the transform in the order, which is also satisfied by the present definition. The first two are essential properties of the continuous transform, and the third is necessary for the discrete fractional Fourier transform to constitute a generalization of the ordinary DFT. In order to achieve a theory exhibiting the internal consistency and analytical elegance we take for granted with the ordinary DFT, we wish these properties to be satisfied exactly. (The fast algorithm mapping the samples of $f(u)$ to those of $f_a(u)$ is rejected as a definition of the discrete transform, since it satisfies these properties merely to a good degree of approximation, but not exactly.) Beyond these, it would be desirable for the discrete transform to satisfy as many operational properties of the continuous transform as possible.

Any of the six definitions of the fractional Fourier transform presented in chapter 4 can be taken as a starting point for analogously defining its discrete version. The approach taken here is based on definition B, and is also related to definitions E and F. Since definition B involves the Hermite-Gaussian functions, the first task lying before us is to determine suitable discrete counterparts of the Hermite-Gaussian functions. Then it will be a relatively easy matter to define the discrete fractional Fourier transform in section 6.3.

6.2   Discrete Hermite-Gaussian functions

In this section we define the set of N-point discrete Hermite-Gaussian functions (or vectors) $\mathbf{hg}_{n/N}$. The subscript n/N means that this is the nth member of a set with a total of N members. When there is no need to be explicit, we will usually denote the members of this set of vectors more simply as $\mathbf{hg}_n$ and the elements of the nth member as hg_{nl}, $0 \le l \le N - 1$. When N is even, n takes on the values $n = 0, 1, \ldots, N - 2, N$, skipping $N - 1$. When N is odd, n takes on the values $n = 0, 1, \ldots, N - 1$. This peculiarity—inherited from the ordinary discrete Fourier transform—will be discussed later on. Tolerating this peculiarity, rather than defining things so as to cover it up, allows us to interpret n as the number of zero crossings of these discrete functions. These N-point functions can also be interpreted as periodic functions defined over

$-\infty < l < \infty$ with period N. This N-member set of N-point functions constitutes an orthonormal basis set spanning the space of all finite-energy N-point signals (or the space of finite-power signals having period N). The discrete Hermite-Gaussian functions are analogous to their continuous counterparts in many ways, and provide a satisfactory approximation to them. Readers willing to accept the existence of such a set of functions on faith may skip the remainder of this section.

Continuous Hermite-Gaussian functions were discussed in section 2.5.2 as the eigenfunctions of the ordinary Fourier transform. They constitute an orthonormal basis for the space of finite-energy signals. The nth Hermite-Gaussian function has the eigenvalue $\exp(-in\pi/2)$ and n zero crossings. The defining differential equation 2.176 for the Hermite-Gaussian functions $\psi_n(u)$ can be rewritten in eigenvalue equation form as follows:

$$\pi(\mathcal{D}^2 + \mathcal{U}^2)\psi_n(u) = (n + 1/2)\psi_n(u),$$
$$\mathcal{H}\psi_n(u) = \lambda_n\psi_n(u), \tag{6.1}$$

where we have used the fact that $\mathcal{D}$ and $\mathcal{U}$ can be replaced by $(i2\pi)^{-1}d/du$ and u in the time domain, and we let $\mathcal{H} = \pi(\mathcal{D}^2 + \mathcal{U}^2)$ and $\lambda_n = (n + 1/2)$. Some readers may recognize $\mathcal{H}$ as the Hamiltonian of the quantum-mechanical harmonic oscillator. The Hermite-Gaussian functions are the unique finite-energy eigensolutions of equation 6.1 (Birkhoff and Rota 1989:337). Since we know that coordinate multiplication in the time domain corresponds to differentiation in the Fourier domain, we can also write $\mathcal{H} = \pi(\mathcal{D}^2 + \mathcal{F}\mathcal{D}^2\mathcal{F}^{-1})$, a form which will be useful in what follows. (To keep the notation simple, in this chapter we are defining $\mathcal{H} = \pi(\mathcal{D}^2 + \mathcal{U}^2)$, despite the fact that in section 3.4.9 we defined $\mathcal{H}_{fF} = \pi(\mathcal{D}^2 + \mathcal{U}^2)$, and in definition F of chapter 4 we defined $\mathcal{H} = \pi(\mathcal{D}^2 + \mathcal{U}^2) - 1/2$. In any case, the eigenvectors of $\pi(\mathcal{D}^2 + \mathcal{U}^2)$ and $\pi(\mathcal{D}^2 + \mathcal{U}^2) - 1/2$ are the same and their eigenvalues differ only by $1/2$.)

We recall the following well-known theorem (page 34). If $\mathcal{A}$ and $\mathcal{B}$ commute, there exists a common eigensignal set between $\mathcal{A}$ and $\mathcal{B}$. Now, it is easy to see that $\mathcal{F}$ and $\mathcal{H}$ commute:

$$\mathcal{F}\mathcal{H}\mathcal{F}^{-1} = \pi\mathcal{F}\mathcal{D}^2\mathcal{F}^{-1} + \pi\mathcal{F}\mathcal{U}^2\mathcal{F}^{-1} = \pi\mathcal{U}^2 + \pi\mathcal{D}^2 = \mathcal{H}, \tag{6.2}$$

where we used $\mathcal{F}\mathcal{D}^2\mathcal{F}^{-1} = \mathcal{U}^2$ and $\mathcal{F}\mathcal{U}^2\mathcal{F}^{-1} = \mathcal{D}^2$. The $\mathcal{U}^2$ and $\mathcal{D}^2$ operators exchange places under Fourier transformation, and since $\mathcal{H}$ is symmetric with respect to $\mathcal{U}^2$ and $\mathcal{D}^2$, it remains the same under Fourier transformation. Thus it follows that the Hermite-Gaussian functions, which are the unique finite-energy eigenfunctions of $\mathcal{H}$, are also eigenfunctions of $\mathcal{F}$.

We will define the discrete Hermite-Gaussian functions as eigensolutions of a difference equation which is analogous to the defining differential equation 6.1 of the continuous Hermite-Gaussian functions. First, we define the finite difference operator $\mathcal{D}_h$ through

$$\frac{\mathcal{D}_h}{h}f(u) \equiv \frac{1}{i2\pi}\frac{f(u + h/2) - f(u - h/2)}{h}. \tag{6.3}$$

$\mathcal{D}_h$ is the discrete analog of $\mathcal{D}$, and $\mathcal{D}_h/h$ serves as a better and better approximation to $\mathcal{D}$ as $h \to 0$. Likewise, the second difference operator $\mathcal{D}_h^2$ can be defined as

$$\frac{\mathcal{D}_h^2}{h^2} f(u) \equiv \frac{1}{(i2\pi)^2} \frac{f(u+h) - 2f(u) + f(u-h)}{h^2}. \tag{6.4}$$

Using the hyperdifferential form of the translation operation $f(u + h) = \exp(i2\pi h\mathcal{D})f(u)$, we formally obtain

$$\frac{\mathcal{D}_h^2}{h^2} = \frac{1}{(i2\pi)^2} \frac{e^{i2\pi h\mathcal{D}} - 2 + e^{-i2\pi h\mathcal{D}}}{h^2} = \frac{1}{(i2\pi)^2} \frac{2\cos(2\pi h\mathcal{D}) - 2}{h^2} = \mathcal{D}^2 + O(h^2), \tag{6.5}$$

where $O(h^2)$ denotes terms in h^2 or higher powers of h. Next we consider the discrete analog $\mathcal{U}_h^2 \equiv \mathcal{F}(\mathcal{D}_h^2)\mathcal{F}^{-1}$ of $\mathcal{U}^2 = \mathcal{F}\mathcal{D}^2\mathcal{F}^{-1}$. Using equation 6.5 and the relation $\mathcal{F}\exp(i2\pi h\mathcal{D})\mathcal{F}^{-1} = \exp(-i2\pi h\mathcal{U})$, we obtain

$$\frac{\mathcal{U}_h^2}{h^2} \equiv \mathcal{F}\frac{\mathcal{D}_h^2}{h^2}\mathcal{F}^{-1} = \frac{1}{(i2\pi)^2} \frac{2\cos(2\pi h\mathcal{U}) - 2}{h^2} = \mathcal{U}^2 + O(h^2). \tag{6.6}$$

Thus the discrete approximation $\mathcal{H}_h/h^2$ of $\mathcal{H} = \pi(\mathcal{D}^2 + \mathcal{U}^2) = \pi(\mathcal{D}^2 + \mathcal{F}\mathcal{D}^2\mathcal{F}^{-1})$ can be obtained as

$$\frac{\mathcal{H}_h}{h^2} \equiv \pi\left(\frac{\mathcal{D}_h^2}{h^2} + \mathcal{F}\frac{\mathcal{D}_h^2}{h^2}\mathcal{F}^{-1}\right) = \frac{\pi}{(i2\pi)^2}\left(\frac{2\cos(2\pi h\mathcal{D}) - 2}{h^2} + \frac{2\cos(2\pi h\mathcal{U}) - 2}{h^2}\right)$$

$$= \pi(\mathcal{D}^2 + \mathcal{U}^2) + O(h^2) = \mathcal{H} + O(h^2), \tag{6.7}$$

from which we see that the finite difference analog of $\mathcal{H}$ we have defined is an $O(h^2)$ approximation of $\mathcal{H}$. Now, let us explicitly write the eigenvalue equation $(\mathcal{H}_h/h^2)f(u) = \pi(\mathcal{D}_h^2/h^2 + \mathcal{U}_h^2/h^2)f(u) = \lambda f(u)$ in the time domain by using equations 6.4 and 6.6:

$$\frac{\pi}{(i2\pi)^2 h^2}\{f(u+h) - 2f(u) + f(u-h) + 2[\cos(2\pi hu) - 1]f(u)\} = \lambda f(u). \tag{6.8}$$

Finally, we switch to discrete variables by letting $u = lh$ and obtain the finite-difference equation for $f_l \equiv f(lh)/h^2$ as

$$\frac{\pi}{(i2\pi)^2}\{f_{l+1} - 2f_l + f_{l-1} + 2[\cos(2\pi l/N) - 1]f_l\} = \lambda f_l, \tag{6.9}$$

where we have let $h^2 = 1/N$ so that $u = lh = l/\sqrt{N}$. Since the coefficients of the above equation are periodic in l with period N, it will have solutions f_l which are also periodic. This implies that the solutions will have N degrees of freedom and that it is sufficient to concentrate on a single period of f_l, say that defined by $0 \le l \le N - 1$. It also implies that there are only N distinct (not redundant) equations, which form a system of N equations in N unknowns of the form

$$\mathbf{Hf} = \lambda\mathbf{f}, \tag{6.10}$$

with $\mathbf{f} = [f_0 \ f_1 \ \dots \ f_{N-1}]^{\mathrm{T}}$ and

$$
\mathbf{H} = \frac{\pi}{(i2\pi)^2}
\begin{bmatrix}
-2 & 1 & 0 & \cdots & 0 & 1 \\
1 & 2\cos(\frac{2\pi}{N}) - 4 & 1 & \cdots & 0 & 0 \\
0 & 1 & 2\cos(\frac{2\pi}{N}2) - 4 & \cdots & 0 & 0 \\
\vdots & \vdots & \vdots & \ddots & \vdots & \vdots \\
1 & 0 & 0 & \cdots & 1 & 2\cos(\frac{2\pi}{N}(N-1)) - 4
\end{bmatrix}.
$$

$$(6.11)$$

This symmetric matrix $\mathbf{H}$ commutes with the DFT matrix $\mathbf{F}$, ensuring the existence of common eigenvectors. Furthermore, this common eigenvector set can be shown to be *unique* and *orthogonal*, as we will substantiate below. It is this eigenvector set $\{\mathbf{hg}_n\}$ with index n, which will be taken as the discrete counterpart of the continuous Hermite-Gaussian functions, since $\mathbf{H}$ has been defined to be the discrete counterpart of $\mathcal{H}$.

Just as $\mathcal{H}$ commutes with $\mathcal{F}$, ensuring a common set of eigensignals, it is also the case that $\mathbf{H}$ and $\mathbf{F}$ commute, ensuring a common set of eigenvectors. To prove that $\mathbf{H}$ and $\mathbf{F}$ indeed commute, we write $\mathbf{H} = \pi(\mathbf{D}^2 + \mathbf{U}^2)$, where $\mathbf{D}^2$ is the circulant matrix corresponding to the system whose impulse response is $h(l) = (-1/4\pi^2)[\delta(l+1) - 2\delta(l) + \delta(l-1)]$ (the second difference operation), and $\mathbf{U}^2$ is the diagonal matrix defined as $\mathbf{U}^2 = \mathbf{F}\mathbf{D}^2\mathbf{F}^{-1}$ (the discrete counterpart of $\mathcal{U}^2$). Since $h(l)$ is an even function, it also follows that $\mathbf{D}^2 = \mathbf{F}\mathbf{U}^2\mathbf{F}^{-1}$. Then the commutativity of $\mathbf{H}$ and $\mathbf{F}$ follows in the same manner as in the continuous case (equation 6.2).

Before continuing any further, we give the definitions of *even* and *odd* discrete functions. Let $\mathbf{f} = [f_0 \ f_1 \ \dots \ f_{N-1}]^{\mathrm{T}}$ be a vector of length N. This vector is even if it is equal to its coordinate-inverted version: $f_l = f_{N-l}$ for $l = 1, 2, \dots, N-1$. It is odd if it is equal to the negative of its coordinate-inverted version: $f_0 = 0$ and $f_l = -f_{N-l}$ for $l = 1, 2, \dots, N-1$. The parity (or coordinate-inversion) matrix $\mathbf{P}$ is defined such that if $\mathbf{g} = \mathbf{Pf}$, then $g_0 = f_0$ and $g_l = f_{N-l}$, $l = 1, 2, \dots, N-1$. Even vectors satisfy $\mathbf{Pf} = \mathbf{f}$ and odd vectors satisfy $\mathbf{Pf} = -\mathbf{f}$. The even and odd parts of an arbitrary $\mathbf{f}$ can be obtained as $\mathbf{f}_{\mathrm{ev}} \equiv (\mathbf{f} + \mathbf{Pf})/2$ and $\mathbf{f}_{\mathrm{od}} \equiv (\mathbf{f} - \mathbf{Pf})/2$. Finally, $\mathbf{F}^2 = \mathbf{P}$ and $\mathbf{P}^2$ is equal to the identity matrix.

Since $\mathbf{H}$ is real symmetric and thus also Hermitian, it possesses an orthogonal set of eigenvectors. And since it commutes with $\mathbf{F}$, there is a common eigenvector set between these two matrices. Furthermore, it can be shown that (i) this common eigenvector set is unique, and that (ii) this set can be ordered according to the number of zero crossings n of each member. (These will be demonstrated further below, after we complete our main argument.) The nth member of this ordered N-member set will be denoted by $\mathbf{hg}_{n/N}$ and will be defined as the nth N-point discrete Hermite-Gaussian function (or vector). This definition of the discrete Hermite-Gaussian functions is supported by the analogies listed in table 6.1. We also numerically compare the discrete Hermite-Gaussian functions $\mathbf{hg}_n$ with the corresponding continuous Hermite-Gaussian functions $\psi_n(u)$ in figure 6.1. We see that the discrete functions well reflect the nature of their continuous counterparts. Further numerical comparisons may be found in Candan 1998.

Before summarizing the main result, we must make a number of important comments. Until now we talked mostly about the eigenvectors of $\mathbf{H}$ and $\mathbf{F}$, but not

Table 6.1. Discrete Hermite-Gaussian functions versus continuous Hermite-Gaussian functions (Candan, Kutay, and Ozaktas 2000).

Continuous Hermite-Gaussian functions
1. Satisfy a generating differential equation
2. Are eigenfunctions of the continuous Fourier transform
3. Form an orthonormal basis in function space
4. Can be ordered by their number of zero crossings;
   the continuous Hermite-Gaussian with n zero crossings has eigenvalue $\exp(-in\pi/2)$ under the continuous Fourier transform

Discrete Hermite-Gaussian functions
1. Satisfy a generating difference equation
2. Are eigenvectors of the discrete Fourier transform
3. Form an orthonormal basis in $\mathbf{R}^N$
4. Can be ordered by their number of zero crossings;
   the discrete Hermite-Gaussian with n zero crossings has eigenvalue $\exp(-in\pi/2)$ under the discrete Fourier transform

---

so much about their eigenvalues. We had seen that the eigenvalue of $\mathcal{H}$ corresponding to the eigenfunction $\psi_n(u)$ is $n + 1/2$. The eigenvalue of $\mathbf{H}$ corresponding to the eigenvector $\mathbf{hg}_n$ is $\kappa_n + 1/2$, where the members of the sequence κ_n are close, but not exactly equal to n (Barker and others 2000). Turning our attention to the Fourier transform, we know that the eigenvalue of $\mathcal{F}$ corresponding to the eigenfunction $\psi_n(u)$ is $\exp(-in\pi/2)$. Extensive numerical evidence strongly suggests that the eigenvalue of $\mathbf{F}$ corresponding to the eigenvector $\mathbf{hg}_n$ is also $\exp(-in\pi/2)$. We already know that the eigenvalues of $\mathbf{F}$ are $\pm 1, \pm i$ and that the eigenvalues corresponding to the even eigenvectors are ± 1 and those corresponding to the odd eigenvectors are $\pm i$, but this is not sufficient to fully determine which eigenvalue corresponds to each $\mathbf{hg}_n$, the eigenvector with n zero crossings. A formal proof of the statement in table 6.1 that the discrete Hermite-Gaussian $\mathbf{hg}_n$ with n zero crossings has eigenvalue $\exp(-in\pi/2)$, is presently not available. Nevertheless, based on the strong numerical evidence, we will take it to be true that the eigenvalue of $\mathbf{F}$ corresponding to the eigenvector $\mathbf{hg}_n$ is $\exp(-in\pi/2)$. The reader may also note that given the close correspondence exhibited in figure 6.1, it seems very implausible indeed that the eigenvalues $\pm 1, \pm i$ correspond to the discrete eigenvectors $\mathbf{hg}_n$ differently than they correspond to the continuous eigenfunctions $\psi_n(u)$.

To summarize, in this section we have shown that the matrix $\mathbf{H}$ (obtained as the discrete analog of the operator $\mathcal{H}$) has a unique set of eigenvectors which can be put into one-to-one correspondence with the Hermite-Gaussian functions, and thus can serve as discrete analogs of these functions. In fact, here we define them as *the* discrete Hermite-Gaussian functions (or vectors). These eigenvectors are also eigenvectors of the DFT matrix $\mathbf{F}$ (although not the only possible set), and constitute the most crucial ingredient in our definition of the discrete fractional Fourier transform in section 6.3.

Early works on the eigenvalues and eigenvectors of the ordinary DFT matrix include McClellan and Parks 1972, McClellan 1973, Yarlagadda 1977, and Dickinson and

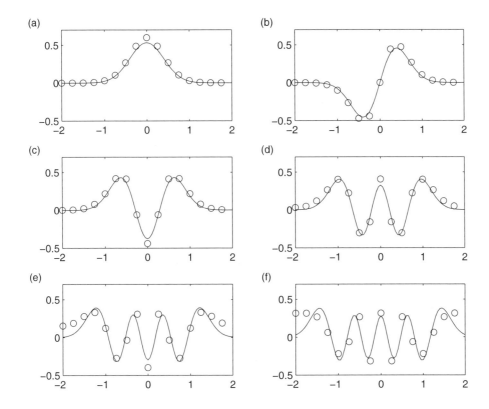

Figure 6.1. Comparison of $\mathbf{hg}_{n/N}$ with $\psi_n(u)$ for $N = 16$; we set $u = l/\sqrt{N}$ so that the continuous interval $[-2, 2]$ corresponds to the discrete period $-N/2 \le l \le N/2 - 1$: (a) $n = 0$, (b) $n = 1$, (c) $n = 2$, (d) $n = 4$, (e) $n = 6$, (f) $n = 8$ (Candan, Kutay, and Ozaktas 1999, 2000).

Steiglitz 1982. Some authors, including Dickinson and Steiglitz (1982), deal with the matrix obtained by adding a multiple of the identity matrix to $\mathbf{H}$, rather than $\mathbf{H}$ itself. The eigenvectors of both matrices are the same.

The rest of this section will deal with the uniqueness of the common eigenvector set and its ordering (indexing), which we promised to demonstrate. Readers willing to take these results for granted may directly skip to section 6.3.

We first embark on showing that the common eigenvector set of $\mathbf{H}$ and $\mathbf{F}$ is unique. It is known that eigenvectors of the DFT matrix are either even or odd vectors. (To see this, remember that since $\mathbf{F}^4$ is the identity matrix, the eigenvalues λ of $\mathbf{F}$ must satisfy $\lambda^4 = 1$ and thus $\lambda^2 = \pm 1$. Now, let $\mathbf{f}$ be an eigenvector so that $\mathbf{Ff} = \lambda\mathbf{f}$ and $\mathbf{F}^2\mathbf{f} = \lambda^2\mathbf{f} = \pm\mathbf{f}$. But we also know that $\mathbf{F}^2$ is the coordinate-inversion (parity) operator $\mathbf{P}$, from which the result follows.) Thus the common eigenvector set in question must also consist of even and odd vectors.

The eigenstructure of the matrix $\mathbf{H}$ given in equation 6.11 has been extensively studied (for instance, Wiegmann and Zabrodin 1995). It is sometimes referred to as *Harper's matrix*, with the associated eigenvalue equation being referred to as

Harper's equation. Some authors have associated it with certain Mathieu equations and Sturm-Liouville problems. For these the reader is referred to treatments of difference equations (McLachlan 1964, Hildebrand 1968, Wilkinson 1988, Kelley and Peterson 1991, Agarwal 1992). Harper's equation has also received considerable interest in the context of the Bloch electron problem in physics (Rammal and Bellissard 1990). A number of facts regarding this matrix will be invoked without proof, relying on the references cited. A more detailed exposition of these issues can be found in Candan 1998 and Candan, Kutay, and Ozaktas 1999, 2000.

It is known that when N is not a multiple of 4, all of the eigenvalues of $\mathbf{H}$ are distinct. Since $\mathbf{H}$ is a real symmetric matrix, it follows that all of its eigenvectors are orthogonal to each other, and thus the set of eigenvectors of $\mathbf{H}$ is unique (within multiplicative constants). Since we have already shown that $\mathbf{H}$ has a common eigenvector set with the DFT matrix, this unique set of eigenvectors of $\mathbf{H}$ must then also be a set of eigenvectors of the DFT matrix. The normalized version of this set of eigenvectors is what we define as the discrete version of the Hermite-Gaussian functions.

When N is a multiple of 4, the matrix $\mathbf{H}$ still has distinct eigenvalues with the exception of one eigenvalue which has the value of zero with degeneracy two. The eigenvectors corresponding to all eigenvalues except this one are orthogonal to each other. The two eigenvectors corresponding to the zero eigenvalue can be chosen to be orthogonal, again because $\mathbf{H}$ is a real symmetric matrix. There are many ways of choosing these two eigenvectors such that they are orthogonal; however, there is only one way to choose them such that one is an even vector and one is an odd vector. Since we are seeking the common set of eigenvectors between $\mathbf{H}$ and the DFT matrix, and since we know that all eigenvectors of the DFT matrix are either even or odd vectors, we have no choice but to choose the even and odd eigenvectors corresponding to the zero eigenvalue; other choices could not be eigenvectors of the DFT matrix. This requirement resolves the ambiguity associated with choosing the eigenvectors corresponding to the zero eigenvalue when N is a multiple of 4, and again uniquely determines the common set of eigenvectors of $\mathbf{H}$ and the DFT matrix.

What remains is to order (index) this set of eigenvectors so that they are in one-to-one correspondence with the continuous Hermite-Gaussian functions. One of the most distinguishing features of the Hermite-Gaussian functions is their zero crossings. The nth Hermite-Gaussian function has n zero crossings. Thus it seems natural to order the discrete eigenvectors in terms of their zero crossings as well. However, direct application of such a procedure is numerically problematic because some components of the eigenvectors can have very small values, making direct counting of the number of zero crossings difficult. Fortunately, it is known that ordering the eigenvectors by looking at the values of their eigenvalues with respect to $\mathbf{H}$ is equivalent to ordering them in terms of their zero crossings, an approach which is also numerically much more reliable. This procedure will be made more precise in the following paragraphs.

First, we must agree on what we mean by a zero crossing of a discrete vector. A vector $\mathbf{f}$ has a zero crossing at l if $f_l f_{l+1} < 0$. We treat f_l as a periodic sequence, counting the number of zero crossings in one period, say $[0, \ldots, N-1]$, also including the zero crossing at the endpoints of the period, such as when $f_{N-1} f_N = f_{N-1} f_0 < 0$.

To translate the described procedure into computational code, it is helpful to introduce some additional devices. We first introduce a matrix $\mathbf{V}$ which decomposes a given vector into its even and odd components. When we multiply a given N-point vector $\mathbf{f}$ with the matrix $\mathbf{V}$, we obtain a vector whose first $N/2 + 1$ or $(N + 1)/2$ components give the leftmost part of $\mathbf{f}_{\mathrm{ev}}$, and whose last $N/2 - 1$ or $(N - 1)/2$ components give the rightmost part of $-\mathbf{f}_{\mathrm{od}}$, depending on whether N is even or odd. For $N = 5$ we have

$$
\mathbf{V} = \frac{1}{\sqrt{2}} \begin{bmatrix}
\sqrt{2} & 0 & 0 & 0 & 0 \\
0 & 1 & 0 & 0 & 1 \\
0 & 0 & 1 & 1 & 0 \\
0 & 0 & 1 & -1 & 0 \\
0 & 1 & 0 & 0 & -1
\end{bmatrix},
\tag{6.12}
$$

and the vector $\mathbf{Vf}$ is

$$
\frac{1}{\sqrt{2}} [\sqrt{2} f_0, \; f_1 + f_{-1}, \; f_2 + f_{-2}, \; f_2 - f_{-2}, \; f_1 - f_{-1}]^{\mathrm{T}}
$$

$$
= \frac{1}{\sqrt{2}} [\sqrt{2} f_0, \; f_1 + f_4, \; f_2 + f_3, \; f_2 - f_3, \; f_1 - f_4]^{\mathrm{T}}, \quad (6.13)
$$

where we remember that the arguments are interpreted modulo N. The elements of $\mathbf{V}$ have been normalized so as to ensure $\mathbf{V} = \mathbf{V}^{\mathrm{T}} = \mathbf{V}^{-1}$, despite the fact that this slightly spoils the interpretation of the leftmost components as the even part of $\mathbf{f}$.

Now, consider the similarity transformation $\mathbf{V H V}^{-1}$, whose result turns out to be a block-diagonal matrix of the form

$$
\mathbf{V H V}^{-1} = \begin{bmatrix} \mathbf{Ev} & \mathbf{0} \\ \mathbf{0} & \mathbf{Od} \end{bmatrix},
\tag{6.14}
$$

where the matrices $\mathbf{Ev}$ and $\mathbf{Od}$ are symmetric tridiagonal matrices with the dimensions $N/2 + 1$ or $(N + 1)/2$ and $N/2 - 1$ or $(N - 1)/2$, depending on whether N is even or odd, respectively. ($\mathbf{Ev}$ and $\mathbf{Od}$ are *not* the matrices which produce the even and odd parts of a vector.) Since it is known that symmetric tridiagonal matrices have distinct eigenvalues (Wilkinson 1988), these two matrices have unique sets of orthogonal eigenvectors. It is also possible to show without much difficulty that when these eigenvectors are zero padded and multiplied with $\mathbf{V}$, what we obtain is precisely the unique orthogonal even-odd eigenvector set of $\mathbf{H}$. (We earlier discussed that although the eigenvectors of $\mathbf{H}$ are sometimes not unique, the requirement that they be even or odd vectors is sufficient to single out a unique set.) The zero padding is from the right for the even eigenvectors and from the left for the odd eigenvectors.

Our strategy is to obtain the eigenvectors of $\mathbf{H}$ by zero padding the eigenvectors of the symmetric tridiagonal matrices $\mathbf{Ev}$, $\mathbf{Od}$ and multiplying the results by $\mathbf{V}$, rather than by directly solving the eigenvalue equation for $\mathbf{H}$. This indirect procedure enables us to order the eigenvectors in terms of their numbers of zero crossings quite easily. An explicit expression for the eigenvectors of symmetric tridiagonal matrices is given in Wilkinson 1988:316. Combining this expression with the Sturm sequence theorem (Wilkinson 1988:300), one can show that the eigenvector of the $\mathbf{Ev}$ or $\mathbf{Od}$ matrix with the highest eigenvalue has no zero crossing, the eigenvector with the second highest eigenvalue has one zero crossing, and so on. Furthermore, one can show that the $\mathbf{Ev}$ and $\mathbf{Od}$ matrices have eigenvectors whose numbers of zero crossings range from 0 to $N/2$ or $(N - 1)/2$ and $(N - 4)/2$ or $(N - 3)/2$, depending on whether N is even

Table 6.2. Eigenvalue multiplicity of the DFT matrix (McClellan and Parks 1972).

N	1	$-i$	-1	i
$4j - 3$	j	$j - 1$	$j - 1$	$j - 1$
$4j - 2$	j	$j - 1$	j	$j - 1$
$4j - 1$	j	j	j	$j - 1$
$4j$	$j + 1$	j	j	$j - 1$

j is a positive integer. The entries give the number of eigenvectors with the indicated eigenvalue for the value of N shown in the leftmost column. Even when N is a multiple of 4, the multiplicity of the eigenvalues is not equal to $N/4$.

or odd, respectively. Zero padding and multiplying by **V** the eigenvector of **Ev** with j zero crossings, yields the even eigenvector of **H** with $2j$ zero crossings. Likewise, zero padding and multiplying by **V** the eigenvector of **Od** with j zero crossings, yields the odd eigenvector of **H** with $2j + 1$ zero crossings. This procedure not only enables us to accurately determine the number of zero crossings of each eigenvector, but also demonstrates that each of the eigenvectors of **H** has a different number of zero crossings, so that each vector can be assigned a unique index equal to its number of zero crossings.

In conclusion, we have presented a procedure for finding and ordering the common eigenvector set $\{\mathbf{hg}_n\}$ of the matrix **H** and the DFT matrix, such that the nth member of this eigenvector set has n zero crossings and is even or odd according to whether n is even or odd. The overall procedure is summarized in steps 1 to 5 of table 6.4.

6.3 The discrete fractional Fourier transform

Our definition of the discrete fractional Fourier transform will be in the spirit of definition B in chapter 4. We recall from equation 4.23 that the kernel of the continuous fractional Fourier transform has the following spectral expansion (singular-value decomposition):

$$K_a(u, u') = \sum_{n=0}^{\infty} \psi_n(u) e^{-ian\pi/2} \psi_n(u'). \tag{6.15}$$

where $\psi_n(u)$ is the nth Hermite-Gaussian function. This kernel maps a function $f(u)$ into its fractional Fourier transform $f_a(u) = \int K_a(u, u') f(u') \, du'$. The factor $\exp(-ian\pi/2)$ is the ath power of the nth eigenvalue $\exp(-in\pi/2)$ of the ordinary Fourier transform. When $a = 1$ we have $K_1(u, u') = \exp(-i2\pi uu')$, the kernel of the ordinary Fourier transform. We will define the discrete fractional Fourier transform matrix $\mathbf{F}^a$ as the discrete analog of equation 6.15:

$$F_{ll'}^a \equiv \sum_{\substack{n=0 \\ n \neq N-1+(N \bmod 2)}}^{N} hg_{nl} \, e^{-ian\pi/2} \, hg_{nl'}, \tag{6.16}$$

Table 6.3. Properties of the discrete fractional Fourier transform (Candan 1998, Candan, Kutay, and Ozaktas 2000).

$\mathbf{F}^a\mathbf{f} = \mathbf{f}_a$
1. $\mathbf{F}^a[\sum_j \alpha_j \mathbf{f}_j] = \sum_j \alpha_j[\mathbf{F}^a\mathbf{f}_j]$
2. $\mathbf{F}^{a_2}\mathbf{F}^{a_1}\mathbf{f} = \mathbf{F}^{a_1}\mathbf{F}^{a_2}\mathbf{f} = \mathbf{F}^{a_2+a_1}\mathbf{f} = \mathbf{f}_{a_1+a_2}$
3. $\mathbf{F}^1\mathbf{f} = \mathbf{F}\mathbf{f} = \text{DFT of } \mathbf{f}$
4. $\mathbf{F}^a(\mathbf{Pf}) = \mathbf{Pf}_a$
5. $\mathbf{F}^a\mathbf{f}^* = (\mathbf{f}_{-a})^*$
6. $\mathbf{F}^a\mathbf{f}_{\text{ev}} = (\mathbf{f}_a)_{\text{ev}}$
7. $\mathbf{F}^a\mathbf{f}_{\text{od}} = (\mathbf{f}_a)_{\text{od}}$
8. $\mathbf{f}^{\text{H}}\mathbf{g} = \mathbf{f}_a^{\text{H}}\mathbf{g}_a$

$\mathbf{f}_a = \mathbf{F}^a\mathbf{f}$ is the discrete fractional Fourier transform of $\mathbf{f}$. α_j are arbitrary complex constants. $\mathbf{P} = \mathbf{F}^2$ is the parity matrix. Other properties, such as those for shift and coordinate multiplication, have not yet been explicitly derived.

where hg_{nl} is the lth element of $\mathbf{hg}_n$ and $F_{ll'}^a$ is the ll'th element of $\mathbf{F}^a$, with $0 \le l \le N-1$, $0 \le l' \le N-1$. The discrete fractional Fourier transform of a vector $\mathbf{f}$ is simply $\mathbf{f}_a = \mathbf{F}^a\mathbf{f}$. In analogy with definition B of chapter 4, this amounts to defining the discrete transform through its eigenvalues $\exp(-ian\pi/2)$ and eigenvectors $\mathbf{hg}_n$. The peculiar range of the summation is due to the fact that there does not exist an eigenvector with $N-1$ or N zero crossings when N is even or odd respectively ($N \bmod 2 = 0$ when N is even and $= 1$ when N is odd). This skipping of an index is related to the similarly peculiar eigenvalue multiplicity of the ordinary DFT matrix, as a careful examination of table 6.2 will reveal. Several elementary properties of the discrete fractional Fourier transform are given in table 6.3.

We will make two passing comments. First, the reader might find it useful to interpret or express the right-hand side of the spectral expansion given in equation 6.16 in certain different ways. Equation 6.16 amounts to first expanding the function to be transformed in terms of discrete Hermite-Gaussian functions, multiplying each component with the respective eigenvalue, and reassembling the components back to obtain the transform (just as in definition B of chapter 4). The same spectral expansion can also be written in matrix form as

$$\mathbf{F}^a = \mathbf{HG}\,\mathbf{\Lambda}_a\,\mathbf{HG}^{\text{T}}, \qquad (6.17)$$

where $\mathbf{HG}$ is a matrix whose columns consist of the discrete Hermite-Gaussian vectors and $\mathbf{\Lambda}_a$ is the diagonal matrix of eigenvalues.

Second, the fact that there are N terms in the summation of equation 6.16 (as well as the fact that there are N Hermite-Gaussian vectors with N elements), is satisfying in the sense that, if a signal can be represented by N numbers as a vector (perhaps the samples of an underlying continuous signal), then it should be possible to represent it again with N coefficients when expanded in terms of Hermite-Gaussian vectors. (This is related to the physically motivated discussion found on page 244.)

We now show that the first three of the requirements discussed on page 202 are automatically satisfied by the discrete fractional Fourier transform defined in

Table 6.4.   Procedure for generating the N-point $\mathbf{F}^a$ matrix (Candan 1998, Candan, Kutay, and Ozaktas 1999, 2000).

1.	Generate the matrices $\mathbf{H}$ and $\mathbf{V}$
2.	Generate the matrices $\mathbf{Ev}$ and $\mathbf{Od}$ using equation 6.14
3.	Find the eigenvectors and eigenvalues of $\mathbf{Ev}$ and $\mathbf{Od}$
4.	Sort these eigenvectors in descending order of eigenvalues, calling them $\mathbf{ev}_0, \mathbf{ev}_1, \ldots$ and $\mathbf{od}_0, \mathbf{od}_1, \ldots$
5.	Define $\mathbf{hg}_{2j} = \mathbf{V}\,[\,\mathbf{ev}_j^{\mathrm{T}}\,|\,0\cdots 0\,]^{\mathrm{T}}$, $\mathbf{hg}_{2j+1} = \mathbf{V}\,[\,0\cdots 0\,|\,\mathbf{od}_j^{\mathrm{T}}\,]^{\mathrm{T}}$ and normalize them
6.	Define $\mathbf{F}^a$ according to equation 6.16

equation 6.16. Unitarity follows from the fact that the eigenvalues $\exp(-ian\pi/2)$ have unit magnitude (the matrix $\mathbf{HG}$ appearing in equation 6.17 is unitary since its columns are orthonormal, and the matrix $\mathbf{\Lambda}_a$ is unitary since the eigenvalues $\exp(-ian\pi/2)$ have unit magnitude). Index additivity can be easily demonstrated by multiplying the matrices $\mathbf{F}^{a_1}$ and $\mathbf{F}^{a_2}$ and using the orthonormality of the $\mathbf{hg}_n$. Reduction to the ordinary DFT when $a = 1$ follows from the fact that when $a = 1$ equation 6.16 reduces to the spectral expansion of the ordinary DFT matrix (the $\mathbf{hg}_n$ are eigenvectors of the DFT matrix with eigenvalue $\exp(-in\pi/2)$, see page 206).

We note that there are two ambiguities which arise in defining the fractional Fourier transform through a spectral expansion of the form of equation 6.16. The first concerns the eigenstructure of the DFT. Since the DFT matrix has only four distinct eigenvalues, these eigenvalues are degenerate and the eigenvector set is not unique. For this reason, it is necessary to specify the particular eigenvector set used in a definition of the form of equation 6.16. In the continuous case this ambiguity is resolved by choosing the Hermite-Gaussian functions as the eigenfunctions, or equivalently by choosing those eigenfunctions of the Fourier transform which are also eigenfunctions of $\mathcal{H}$ (the common eigenfunction set of the commuting operators $\mathcal{H}$ and $\mathcal{F}$). Since our aim was to obtain a definition of the discrete transform which is completely analogous to the continuous transform, we have resolved the corresponding ambiguity in the discrete case by choosing the common eigenvector set $\{\mathbf{hg}_n\}$ of the DFT matrix $\mathbf{F}$ and the matrix $\mathbf{H}$, which we defined to be the discrete counterparts of the Hermite-Gaussian functions.

The second ambiguity arises in taking the fractional power of the eigenvalues $\exp(-in\pi/2)$, since the fractional power operation is not single-valued. Again, we resolved this ambiguity by analogy with the continuous case, by taking the fractional power to be $\exp(-ian\pi/2)$. These two ambiguities, their resolution, and the distinct definitions of the fractional Fourier transform they lead to were discussed in detail in sections 4.2 and 4.3.

The overall procedure for obtaining the discrete fractional Fourier transform matrix is summarized in table 6.4. As a simple example, let us examine the fractional Fourier transforms of the discrete rectangle function. The discrete fractional Fourier transform can be used to approximately map the samples of the continuous rectangle function to the samples of its continuous fractional Fourier transform. Better and better

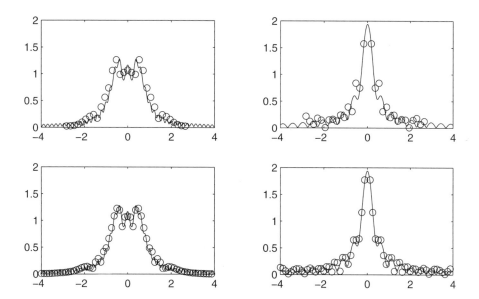

Figure 6.2. The solid curves show the magnitudes of the continuous fractional Fourier transforms of the rectangle function $\text{rect}(u/2)$. The circles show the values of the discrete transforms. The plots on the left are for $a = 0.25$ and the plots on the right are for $a = 0.75$. The upper plots are for $N = 32$ and the lower plots are for $N = 64$. We set $u = l/\sqrt{N}$ as in figure 6.1. (Candan 1998, Candan, Kutay, and Ozaktas 1999, 2000)

approximations can be achieved by increasing N. This is illustrated in figure 6.2 for two different values of a and two different values of N. The continuous transforms have been calculated using brute force numerical integration so as to be accurate within a tolerance of 0.001. The N samples are taken in the interval $[-\sqrt{N}/2, \sqrt{N}/2]$ around the origin with the sampling interval equal to $1/\sqrt{N}$, with samples at $-\sqrt{N}/2, -\sqrt{N}/2 + 1/\sqrt{N}, \ldots, \sqrt{N}/2 - 1/\sqrt{N}$. Extensive numerical experimentation indicates that the discrete fractional Fourier transform provides an approximation to the continuous fractional Fourier transform in the same sense and accuracy as the ordinary DFT provides an approximation to the ordinary continuous Fourier transform (Candan 1998). N should be chosen large enough that both the original function and the transform have negligible energy outside the interval $[-\sqrt{N}/2, \sqrt{N}/2]$. (A stronger condition which will ensure that the chosen value of N is sufficient for all values of a requires that the Wigner distribution of the function have negligible energy outside a disk of radius $\sqrt{N}/2$.)

6.4 Definition in hyperdifference form

Any of the six definitions of the continuous fractional Fourier transform can be used as a starting point to define the discrete fractional Fourier transform. What is desired is a self-consistent definition which satisfies counterparts of the more essential properties of the continuous fractional Fourier transform, and as many of the less essential ones as possible. In particular, it is desirable that it reduce to the ordinary discrete Fourier transform, as conventionally defined, when $a = 1$. (However, we cannot have absolute assurance that the conventional definition of the DFT is the purest and most appropriate definition possible, so that trying to maintain consistency with it, rather than redefining it as well, may turn out to be an impediment to progress.)

The definition already discussed in this chapter is modeled on definition B of chapter 4. Approaches based on definition A, employing sampling, interpolation, delta trains, periodic replications, and so forth, have so far not been satisfactory. Approaches based on definitions C and D are hindered by the fact that a proper understanding of discrete lattice rotations seems to be lacking (Richman and Parks 1997), and also because it is not totally certain what the purest and most appropriate definition of the discrete Wigner distribution should be. An approach based on definitions E and F (which are closely related to each other) will be discussed here. Unfortunately, this definition turns out to be slightly different than the definition presented in section 6.3. However, as N is increased, there is no difference between the two definitions as far as approximating the continuous transform is concerned.

Remembering that the matrix $\mathbf{H} = \pi(\mathbf{D}^2 + \mathbf{U}^2)$ is the discrete analog of $\mathcal{H} = \pi(\mathcal{D}^2 + \mathcal{U}^2)$, we can define the discrete fractional Fourier transform $\mathbf{F}^a_{\mathrm{def\,F}}$ in analogy with equation 4.63:

$$\mathbf{F}^a_{\mathrm{def\,F}} \equiv e^{-i(a\pi/2)[\pi(\mathbf{D}^2+\mathbf{U}^2)-1/2]}. \tag{6.18}$$

Functions of matrices are defined in the same manner as functions of operators (in term of their series expansions). They can be calculated by finding the eigenvalues and eigenvectors of the matrix and evaluating the function for the eigenvalues; see texts such as Strang 1988 for details. This procedure results in the following spectral expansion for the ll'th element of $\mathbf{F}^a_{\mathrm{def\,F}}$:

$$(F^a_{\mathrm{def\,F}})_{ll'} = \sum_{\substack{n=0 \\ n \neq N-1+(N\bmod 2)}}^{N} hg_{nl} \, e^{-i(a\pi/2)\kappa_n} \, hg_{nl'}, \tag{6.19}$$

which the reader should compare with equation 6.16. Here κ_n is the eigenvalue of $\pi(\mathbf{D}^2 + \mathbf{U}^2) - 1/2$ corresponding to the eigenvector $\mathbf{hg}_n$. The eigenvalues of $\pi(\mathcal{D}^2 + \mathcal{U}^2) - 1/2$, the continuous counterpart of $\pi(\mathbf{D}^2 + \mathbf{U}^2) - 1/2$, are simply n; this means definitions A and F are equivalent in the continuous case. However, the eigenvalues κ_n are close, but not equal to n, so the corresponding discrete definitions are not identical.

Which of equations 6.16 and 6.19 is a purer definition is not presently evident, although we favor equation 6.16 because it has simpler eigenvalues and reduces to the ordinary discrete Fourier transform, as conventionally defined, when $a = 1$. On the other hand, equation 6.19 may turn out to fit better into a framework involving rotations of the Wigner distribution on a discrete lattice.

Although we will not go into any details, it is worth pointing out that the discrete fractional Fourier transform is also related to the discrete analog of the quantum-mechanical harmonic oscillator (Atakishiev and Suslov 1991, Atakishiyev and others 1998). The differential equation appearing in definition E in chapter 4 is the Schrödinger equation for the harmonic oscillator. Just as the continuous fractional

Fourier transform can be interpreted as the time-evolution operator for this system, the discrete fractional Fourier transform can be interpreted as the evolution operator (and thus solution) of the discrete counterpart of the harmonic oscillator, which is essentially defined by the difference equations appearing in this chapter.

Further discussion of these issues, as well as plots of the eigenvalues of $\mathbf{H}$ may be found in Barker and others 2000.

6.5 Higher-order discrete analogs

The simplest and purest approach to defining the discrete version of a continuous transform is to employ first-order analogs. Nevertheless, it will be instructive to briefly discuss higher-order analogs of $\mathcal{H}$, and the definitions of the higher-order discrete Hermite-Gaussian functions and the discrete fractional Fourier transform which consequently emerge. These are not only of theoretical interest, but may also be useful when computational accuracy and efficiency are more important than analytical purity.

In section 6.2 we wrote a difference equation constituting a first-order approximation to the differential equation defining the Hermite-Gaussian functions, and took the solutions of this difference equation as the discrete Hermite-Gaussian functions. Here we will write higher-order difference equations which are better approximations to the differential equation in question. The error term in the first-order difference equation was $O(h^2)$. Here we will consider jth order equations whose error term is $O(h^{2j})$. The associated matrix will be denoted by $\mathbf{H}_j$ such that the matrix $\mathbf{H}$ appearing in the first-order case now becomes $\mathbf{H}_1 \equiv \mathbf{H}$.

We begin by defining the jth order difference analog $\mathcal{D}^2_{j,h}$ of the second derivative operator $\mathcal{D}^2$:

$$\frac{\mathcal{D}^2_{j,h}}{h^2} f(u) \equiv \frac{1}{(i2\pi)^2} \frac{1}{h^2} \sum_{k=1}^{j} (-1)^{k-1} \frac{2[(k-1)!]^2}{(2k)!} \left(\mathcal{D}^2_h\right)^k, \tag{6.20}$$

where $\mathcal{D}^2_h$ is defined by equation 6.4. That $\mathcal{D}^2_{j,h}$ is an $O(h^{2j})$ approximation to $\mathcal{D}^2$ can be demonstrated by straightforward series expansion (Candan 1998). When $j = 1$ we have $\mathcal{D}^2_{1,h} \equiv \mathcal{D}^2_h$, consistent with equation 6.4. As an example, let us consider the $O(h^4)$ approximation $\mathcal{D}^2_{2,h}$ of $\mathcal{D}^2$. Upon switching to discrete variables, as in passing from equation 6.8 to equation 6.9, we obtain the following result for $j = 2$:

$$\mathcal{D}^2_{2,h} f_l = \frac{1}{(i2\pi)^2} \left[-\frac{1}{12} f_{l+2} + \frac{4}{3} f_{l+1} - \frac{5}{2} f_l + \frac{4}{3} f_{l-1} - \frac{1}{12} f_{l-2} \right]. \tag{6.21}$$

Proceeding similarly as in the first-order case, we can obtain the jth order version $\mathbf{H}_j$ of the matrix $\mathbf{H}$ appearing in equation 6.11. It can also be shown without much difficulty that the eigenvectors of $\mathbf{H}_j$ are also eigenvectors of the DFT matrix $\mathbf{F}$. These eigenvectors then serve as the jth order discrete analogs of the Hermite-Gaussian functions. Although our discussion of the uniqueness and the ordering of the eigenvectors for $j = 1$ does not immediately generalize to $j > 1$, extensive numerical simulations strongly suggest that higher-order discrete Hermite-Gaussian functions and discrete fractional Fourier transforms do satisfy all of the requirements discussed on page 202. See Candan 1998 for further details.

6.6   Discussion

The definition of the discrete fractional Fourier transform presented in section 6.3 exactly satisfies the essential properties of the continuous transform. However, a simple closed-form expression for the discrete Hermite-Gaussian functions and the discrete fractional Fourier transform matrix is presently not known. Likewise, the discrete counterparts of the multitude of operational properties of the continuous Hermite-Gaussian functions, such as recurrence relations, differentiation properties, and so forth, and the discrete counterparts of the many interesting and useful properties of the continuous fractional Fourier transform, remain to be derived.

Presently, a fast $O(N \log N)$ algorithm for exactly computing the discrete fractional Fourier transform as defined in section 6.3 is not known. However, there already exists an algorithm which can compute the samples of the continuous fractional Fourier transform $f_a(u)$ in terms of the samples of the original function $f(u)$, in $O(N \log N)$ time. The approximation inherent in this algorithm (discussed in section 6.7) is the same as that inherent in approximating the continuous ordinary or fractional Fourier transform with the discrete ordinary or fractional Fourier transform. Thus, to the extent that we wish to use the discrete fractional Fourier transform for the purpose of approximately computing (simulating) the continuous transform, the lack of a fast $O(N \log N)$ algorithm for the discrete transform defined is of no consequence. However, the linear mapping realized by the algorithm of section 6.7 does not *exactly* correspond to the discrete transform defined in section 6.3. From a theoretical and conceptual standpoint, it would be desirable to have a fast algorithm which exactly computes the discrete fractional Fourier transform (just as the FFT exactly computes the DFT), and then to use the discrete transform to suitably approximate the continuous transform.

Although it is clear from numerical experimentation that the discrete fractional Fourier transform approaches the linear relation between the samples of a function and its fractional Fourier transform with increasing N, a rigorous statement of this remains to be established. (The same holds for the relationship between the discrete Hermite-Gaussian functions and the samples of the continuous Hermite-Gaussian functions.) The relationship between the ordinary continuous and discrete Fourier transforms is embodied in equation 3.64. A similar relationship for the fractional case is not known. Nevertheless, certain forms of convergence results have been presented in Barker 2000.

It is well understood that periodicity (or equivalently, finite extent) in either of the time or frequency domains implies discreteness in the other, and that discreteness in either domain implies periodicity (or finite extent) in the other. If periodicity and discreteness are simultaneously present in either domain, then they are simultaneously present in the other domain as well, implying a finite number of degrees of freedom N. This is the basis of the definition of the ordinary DFT. We have seen in chapter 4 that most dual properties of the ordinary Fourier transform had generalizations to the fractional case which satisfactorily interpolated them (for instance, the coordinate multiplication and differentiation or translation and phase shift properties in table 4.3). By analogy, given that periodicity and discreteness respectively imply discreteness and periodicity in the ordinary Fourier domain, we would expect them to imply some quality which is intermediate between periodicity and discreteness in a fractional Fourier domain. As the order parameter a approaches 0 or 1, the signal should approach a periodic or discrete one, and for intermediate values it must somehow exhibit a mixture of these qualities whose extent is determined by the value of a. This interpolation between periodicity and discreteness is presently not well understood.

In certain later chapters we will embark on a priori discrete formulations of certain signal processing problems. It will be assumed that multiplication of a signal vector with the fractional Fourier transform matrix $\mathbf{F}^a$ appearing in these formulations can be accomplished in $O(N \log N)$ time. Although we have just explained that an algorithm for exactly doing so is not available, the algorithm of section 6.7 can (and has) been used as a surrogate for such an algorithm and does provide a sufficient degree of approximation. The existence of such an algorithm increases confidence that a fast algorithm for exactly computing the discrete fractional Fourier transform will also emerge.

A number of avenues for further investigation have been discussed in Candan 1998. These include an approach based on sampling a doubly lowpass filtered version of the continuous transform kernel (or equivalently employing samples of lowpass filtered versions of the Hermite-Gaussian functions in the spectral expansion approach), and other approaches based on conventional sampling and periodic replication concepts and Poisson theorems. Another approach to defining the discrete fractional Fourier transform may involve the prolate spheroidal functions (Slepian and Pollak 1961). Yet another promising avenue seems to be the application of concepts from group theory: clearly defining the principles which take us from the group of orthogonal rotations to the group of continuous rotations, may allow deduction of the discrete fractional Fourier transform matrix from the ordinary DFT matrix.

Other open problems include derivation of further operational properties of the discrete Hermite-Gaussian functions and the discrete fractional Fourier transform, and the development of discrete versions of other continuous fractional transforms or fractional versions of the many discrete unitary transforms commonly employed in signal and image processing (Jain 1989).

One of the most interesting avenues for future research is to establish the relationship between the discrete fractional Fourier transform and the discrete Wigner distribution. Despite considerable work on the discrete Wigner distribution and discrete phase space (Segal 1963, Weil 1964, Claasen and Mecklenbräuker 1980b, 1983, Peyrin and Prost 1986, Galetti and de Toledo Piza 1988, Aldrovandi and Galetti 1990, Richman, Parks, and Shenoy 1998, and O'Neill, Flandrin, and Williams 1999), the purest and most appropriate way of defining the discrete Wigner distribution does not seem to have been agreed upon. We might expect study of the relationship of the Wigner distribution with the fractional Fourier transform to contribute to the establishment of the definition of the discrete Wigner distribution, leading to a consolidation of the theory of discrete time-frequency analysis.

These developments will likely be parallel to those in discrete or finite optics and quantum mechanics, which also heavily employ phase-space concepts (Santhanam and Tekumalla 1976, Balian and Itzykson 1986, Galetti and de Toledo Piza 1988, Aldrovandi and Galetti 1990, Atakishiev and Suslov 1991, Athanasiu and Floratos 1994, Leonhardt 1996, Atakishiyev and Wolf 1997, Atakishiyev, Chumakov, and Wolf 1998, Hakioğlu 1998, 1999, Luis and Peřina 1998). Discrete systems have started to attract an increasing amount of activity in physical contexts which have traditionally been treated in a continuous framework. Discrete physical systems should not be considered merely as contrived and artificial constructs of interest only to the pure mathematician. Neither are they merely vehicles for numerical simulation of underlying continuous systems. Whether real physical systems are inherently continuous or discrete is almost a philosophical question. Traditionally, most macroscopic physical systems, such as those arising in information optics, have been considered to be intrinsically continuous in nature. However, it can be argued that since all physical systems have finite extent and resolution, and thus a finite number of degrees of freedom or modes, it is physically more meaningful to deal with these finite degrees of freedom and how they are mapped from the input to the output, rather than with continuous functions. Thus it can be argued that discrete models give us a more meaningful description of reality than continuous models, especially from an information viewpoint.

6.7   Discrete computation of the fractional Fourier transform

Although the fractional Fourier transform is a linear integral transform, computing it with standard numerical integration techniques is not an efficient approach. The highly oscillatory nature of the kernel (especially when a is close to an even integer) would require a much larger number of samples than implied by the time- or space-bandwidth product of the signals to be transformed. This is a wasteful procedure, since most of the contributions to the integral from these samples would eventually cancel each other. It is possible to avoid this difficulty as follows. If $a \in [0.5, 1.5]$ or $a \in [2.5, 3.5]$, we evaluate the integral directly since the oscillations are not excessive. If $a \in [-0.5, 0.5]$ or $a \in [1.5, 2.5]$, we use $\mathcal{F}^a = \mathcal{F}\mathcal{F}^{a-1}$, noting that in this case the $(a-1)$th transform can be evaluated directly. Still, this approach would require $O(N^2)$ time to compute the whole transform, even if we need not oversample too much. Similar considerations apply to computation of other linear canonical transforms and in particular the Fresnel integral, which appears in diffraction theory (Mendlovic, Zalevsky, and Konforti 1997). Another approach to computing the fractional Fourier transform is to use the spectral expansion of the kernel. This amounts to first expanding the function into Hermite-Gaussian functions, multiplying with the respective eigenvalues, and then summing the components to obtain the transform. Although this method is quite robust, it also requires $O(N^2)$ time.

In this section we briefly discuss how the samples of the fractional Fourier transform $f_a(u)$ of a function $f(u)$ can be computed in terms of the samples of $f(u)$, in $O(N \log N)$ time. For a broader discussion of such approaches, the reader is referred to Ozaktas and others 1996.

The defining equation (equation 4.4) can be put in the form

$$f_a(u) = A_\alpha e^{i\pi \cot \alpha\, u^2} \int e^{-i2\pi \csc \alpha\, uu'} \left[e^{i\pi \cot \alpha\, u'^2} f(u') \right] du'. \tag{6.22}$$

We assume that the representations $f_a(u_a)$ of the signal f in all fractional Fourier domains are approximately confined to the interval $[-\Delta u/2, \Delta u/2]$ (that is, a sufficiently large percentage of the signal energy is confined to these intervals). This assumption corresponds to assuming that the Wigner distribution of $f(u)$ is approximately confined to a circle of diameter Δu (by virtue of equation 4.146). Again, this means that a sufficiently large percentage of the energy of the signal is contained in this circle. We can ensure that this assumption is valid for any signal by choosing

Δu sufficiently large. Under this assumption, and initially limiting the order a to the interval $0.5 \leq |a| \leq 1.5$, the modulated function $\exp(i\pi \cot \alpha \, u'^2)f(u')$ may be assumed to be approximately band-limited to $[-\Delta u, \Delta u]$ in the frequency domain (Ozaktas and others 1996). Thus $\exp(i\pi \cot \alpha \, u'^2)f(u')$ can be represented by the interpolation formula (section 3.3):

$$e^{i\pi \cot \alpha \, u'^2} f(u') = \sum_{l=-N}^{N-1} e^{i\pi \cot \alpha (l/2\Delta u)^2} f\left(\frac{l}{2\Delta u}\right) \operatorname{sinc}\left[2\Delta u \left(u' - \frac{l}{2\Delta u}\right)\right], \quad (6.23)$$

where $N \equiv (\Delta u)^2$. The summation ranges from $-N$ to $N-1$ since $f(u')$ is assumed to be zero outside $[-\Delta u/2, \Delta u/2]$. By using equation 6.23 and equation 6.22, and changing the order of integration and summation, we obtain

$$f_a(u) = A_\alpha e^{i\pi \cot \alpha \, u^2} \sum_{l=-N}^{N-1} e^{i\pi \cot \alpha (l/2\Delta u)^2} f\left(\frac{l}{2\Delta u}\right)$$
$$\times \int e^{-i2\pi \csc \alpha \, uu'} \operatorname{sinc}\left[2\Delta u \left(u' - \frac{l}{2\Delta u}\right)\right] du'. \quad (6.24)$$

By recognizing the integral as $(1/2\Delta u) \exp[-i2\pi \csc \alpha \, u \, (l/2\Delta u)]\operatorname{rect}(\csc \alpha \, u/2\Delta u)$, we obtain

$$f_a(u) = \frac{A_\alpha}{2\Delta u} \sum_{l=-N}^{N-1} e^{i\pi \cot \alpha \, u^2} e^{-i2\pi \csc \alpha \, u \, (l/2\Delta u)} e^{i\pi \cot \alpha (l/2\Delta u)^2} f\left(\frac{l}{2\Delta u}\right), \quad (6.25)$$

since $\operatorname{rect}(\csc \alpha \, u/2\Delta u) = 1$ in the interval $|u| \leq \Delta u/2$. Finally, the samples of $f_a(u)$ are obtained as follows ($k = -N$ to $N-1$):

$$f_a\left(\frac{k}{2\Delta u}\right) = \frac{A_\alpha}{2\Delta u} \sum_{l=-N}^{N-1} e^{i\pi\left[\cot \alpha (k/2\Delta u)^2 - 2\csc \alpha kl/(2\Delta u)^2 + \cot \alpha (l/2\Delta u)^2\right]} f\left(\frac{l}{2\Delta u}\right), \quad (6.26)$$

which is a finite summation allowing us to obtain the samples of the fractional Fourier transform $f_a(u)$ in terms of the samples of the original function $f(u)$. Direct computation of equation 6.26 would require $O(N^2)$ operations. A fast $O(N \log N)$ algorithm can be obtained by putting equation 6.26 into the following form:

$$f_a\left(\frac{k}{2\Delta u}\right) = \frac{A_\alpha}{2\Delta u} e^{i\pi(\cot \alpha - \csc \alpha)(k/2\Delta u)^2}$$
$$\times \sum_{l=-N}^{N-1} e^{i\pi \csc \alpha ((k-l)/2\Delta u)^2} e^{i\pi(\cot \alpha - \csc \alpha)(l/2\Delta u)^2} f\left(\frac{l}{2\Delta u}\right). \quad (6.27)$$

The summation is now recognizable as a convolution, which can be computed in $O(N \log N)$ time by using the fast Fourier transform (FFT). The result is then obtained by a final chirp multiplication. The overall procedure takes $O(N \log N)$ time.

We limited ourselves to $0.5 \leq |a| \leq 1.5$ in deriving the above algorithm. Using the index additivity property of the fractional Fourier transform, we can extend this range to all values of a easily. For the range $-0.5 \leq a \leq 0.5$ we can write

$$\mathcal{F}^a = \mathcal{F}^{a+1-1} = \mathcal{F}^{a+1} \mathcal{F}^{-1}. \tag{6.28}$$

Since $0.5 \leq (a+1) \leq 1.5$ we can use the above algorithm in conjunction with the ordinary inverse Fourier transform to compute $f_a(u)$; the overall procedure still takes $O(N \log N)$ time.

6.8 Historical and bibliographical notes

An earlier discussion of the discrete fractional Fourier transform matrix may be found in Ozaktas and Mendlovic 1993b, where it was shown how this matrix can be expressed in terms of discrete analogs of the Hermite-Gaussian functions and the eigenvalues of the fractional Fourier transform operation; that is, in the form of a spectral expansion. Samples of the continuous Hermite-Gaussian functions were considered as their approximate discrete analogs. Since the sampled Hermite-Gaussian functions are not exactly orthogonal, it was suggested that the set of orthonormal eigenvectors discussed in Dickinson and Steiglitz 1982 could be employed instead. Later, Pei and Yeh (1997) observed that this set of eigenvectors was a particularly suitable choice because they were not only orthonormal, but they also resembled the continuous Hermite-Gaussian functions well enough to constitute their discrete versions.

The development of the discrete fractional Fourier transform as it appears in this chapter is based on Candan 1998 and Candan, Kutay, and Ozaktas 1999, 2000, which were inspired by the observation by Pei and Yeh (1997) of the similarity of the eigenvectors discussed in Dickinson and Steiglitz 1982 to the samples of the continuous Hermite-Gaussian functions. Pei and Yeh went on to define the discrete transform in terms of these eigenvectors. They justified their claims by numerical observations and simulations and offered several insights. Candan (1998) provided an analytical development consolidating this definition.

Several other works dealing with the discrete fractional Fourier transform have appeared (Arıkan and others 1996, Deng, Caulfield, and Schamschula 1996, García, Mas, and Dorsch 1996, Ozaktas and others 1996, Atakishiyev and Wolf 1997, Deng and others 1997, Atakishiyev, Vicent, and Wolf 1999, Tucker, Ojeda-Castañeda, and Cathey 1999). However, these works do not satisfy at least one of the five requirements listed on page 202. Nevertheless, some do provide satisfactory approximations to the continuous transform and can be employed for this purpose. Of particular interest is the approach of Atakishiyev and Wolf (1997) and Atakishiyev, Vicent, and Wolf (1999), which results in explicit expressions for (a distinct but related definition of) the discrete Hermite-Gaussian functions and the discrete fractional Fourier transform matrix. Unfortunately, this definition does not reduce to the ordinary DFT when $a = 1$, although it is unitary and satisfies index additivity. (However, it does seem to reduce to another legitimate discrete version of the continuous ordinary Fourier transform.) Another work dealing with the definition of discrete versions of the Hermite-Gaussian functions is Grünbaum 1982. The definitions of the discrete fractional Fourier transform appearing in Dickinson and Steiglitz 1982, Bailey and Swarztrauber 1991, 1994, and Santhanam and McClellan 1996 correspond to completely distinct and unrelated definitions of the fractional Fourier transform. Given the important role periodicity plays in the study of discrete transforms, we might also mention Alieva and Barbé 1998a which deals with the fractional Fourier transforms of periodic signals.

Research on the discrete fractional Fourier transform and associated concepts is currently in a state of flux, so this chapter may soon become incomplete. The discrete fractional Hartley transform and its relationship to the discrete fractional Fourier transform is discussed in Pei and others 1998. Pei, Yeh, and Tseng (1999) discuss how the discrete fractional Fourier transform can be defined based on a discrete Hermite-Gaussian set obtained by orthogonalizing the sample vectors of the continuous Hermite-Gaussian functions. Pei and Yeh have also discussed the two-dimensional discrete fractional Fourier transform (1998a), the discrete fractional Hilbert transform (1998b), and the discrete fractional Hadamard transform (1999). Labunets and Labunets (1998) discuss fast algorithms for the fractional Fourier transform. Barker and others 2000 is a useful work providing a perspective complementary to ours.

7

Optical Signals and Systems

7.1 Introduction

In this book we will restrict our attention to the familiar class of centered optical systems. A typical example of such a system is shown in figure 7.1. It is composed of a number of spherical lenses and spatial filters, separated by sections of free space and centered about the optical axis (customarily chosen as the z axis). Other optical components and features that may appear in such systems include cylindrical and anamorphic lenses, mirrors, prisms, gratings, diffractive optical elements, sections of graded-index media, sections of homogeneous media with arbitrary refractive indices, and planar or spherical interfaces between such media. Axially or rotationally symmetric systems are those whose axially rotated versions are indistinguishable from themselves. A spherical lens is a rotationally symmetric component, whereas a cylindrical lens or prism is not. Centered systems consisting only of spherical lenses and sections of free space are axially symmetric.

The optical components of which optical systems are composed, can be viewed as elementary optical systems themselves (figure 7.2). Each optical component has its own input and output planes, and alters the distribution of light incident on its input plane in a certain way to produce the distribution of light at its output plane. If the effect of each optical component is known, the overall effect of the optical system can be found.

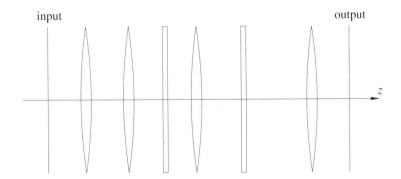

Figure 7.1. Centered optical system consisting of four convex lenses, two spatial filters (shown as thin slabs), and seven sections of free space.

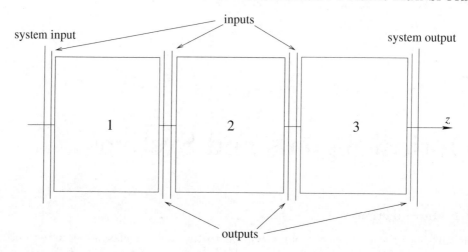

Figure 7.2. An optical system as a sequence of tandem optical components or subsystems. The output of each subsystem is the input of the next subsystem. The input (output) of the first (last) subsystem is the input (output) of the overall system.

7.2 Notation and conventions

Dimensionless variables and parameters were employed in the previous chapters for simplicity and purity (see section 2.1.2). In this and later chapters dealing with optical signals and systems, we will employ variables with real physical dimensions. We will exercise great care to ensure that the correspondence between the equations, results, and properties presented in both sets of chapters is self-evident, and that translating dimensionless and dimensional equations into each other is straightforward. This will allow the results of dimensionless chapters to be easily employed in dimensional chapters.

The space- and frequency-domain representations of a signal f have their dimensional and dimensionless forms related as follows:

$$\hat{f}(x) \equiv \frac{1}{\sqrt{s}} f(x/s) \equiv \frac{1}{\sqrt{s}} f(u), \tag{7.1}$$

$$\hat{F}(\sigma_x) \equiv \sqrt{s}\, F(s\sigma_x) \equiv \sqrt{s}\, F(\mu), \tag{7.2}$$

where $u \equiv x/s$, $\mu \equiv s\sigma_x$. The scale parameter $s > 0$ has the same dimension as x and $1/\sigma_x$, usually meters. The circumflex ˆ designates functions taking dimensional space or frequency arguments such as x, y, z, $\sigma_x, \sigma_y, \sigma_z$, regardless of domain or representation. As usual, lower case denotes the space-domain representation, and upper case denotes the frequency-domain representation of a signal. Although $f(u) = f(x/s)$ has a dimensionless argument, nothing prevents it from having a dimension itself (such as V, V/m, or V/m$\Omega^{1/2}$, the latter whose square is W/m^2).

$F(\mu) = \int f(u) \exp(-i2\pi\mu u)\, du$ is the Fourier transform of $f(u)$ as defined in chapter 2. With the above conventions, the reader may verify that $\hat{F}(\sigma_x)$ is the Fourier transform of $\hat{f}(x)$ defined as

$$\hat{F}(\sigma_x) = \int \hat{f}(x)e^{-i2\pi\sigma_x x}\,dx, \tag{7.3}$$

$$\hat{f}(x) = \int \hat{F}(\sigma_x)e^{i2\pi\sigma_x x}\,d\sigma_x, \tag{7.4}$$

and that

$$\int |\hat{f}(x)|^2\,dx = \int |f(u)|^2\,du = \int |F(\mu)|^2\,d\mu = \int |\hat{F}(\sigma_x)|^2\,d\sigma_x, \tag{7.5}$$

so that the signal energy is the same regardless of whether it is evaluated in dimensional or dimensionless coordinates. Since the defining relations for the Fourier transform look the same in dimensional and dimensionless coordinates, all properties of the Fourier transform will also look the same. We also note that the dimension of $\hat{F}(\sigma_x)$ is the dimension of $\hat{f}(x)$ multiplied by the dimension of x.

Now, let us consider the dimensionless and dimensional forms of linear system or transformation integrals where the input and output variables x' and x are of the same dimension:

$$g(u) = \int h(u, u')f(u')\,du', \tag{7.6}$$

$$\hat{g}(x) = \int \hat{h}(x, x')\hat{f}(x')\,dx', \tag{7.7}$$

where $\hat{h}(x, x') \equiv s^{-1}h(x/s, x'/s) \equiv s^{-1}h(u, u')$. Our conventions ensure that these equations are consistent and that if $h(u, u')$ is unitary, $\hat{h}(x, x')$ will also be unitary. (To see this, the reader may show that the orthonormality relation between the rows/columns of $h(u, u')$, given by $\int h(u, u'')h^*(u'', u')\,du'' = \delta(u - u')$, implies $\int \hat{h}(x, x'')\hat{h}^*(x'', x')\,dx'' = \delta(x - x')$.) Thus we can translate a dimensionless linear relation into a dimensional one simply by replacing both the input and output functions and the kernel with their dimensional counterparts. $\hat{f}(x)$ and $\hat{g}(x)$ usually have the same dimension so that $\hat{h}(x, x')$ has the inverse of the dimension of x.

The dimensionless and dimensional forms of the Wigner distribution can likewise be consistently written in a similar form:

$$W_f(u, \mu) = \int f(u + u'/2)f^*(u - u'/2)e^{-i2\pi\mu u'}\,du', \tag{7.8}$$

$$\hat{W}_{\hat{f}}(x, \sigma_x) = \int \hat{f}(x + x'/2)\hat{f}^*(x - x'/2)e^{-i2\pi\sigma_x x'}\,dx'. \tag{7.9}$$

These have the dimension of $|\hat{f}(x)|^2$ times the dimension of x, which is the same as the dimension of signal energy.

Delta functions require slight care. Remember that $\delta(x - x') = \delta(su - su') = s^{-1}\delta(u - u')$. Thus the sifting property also looks the same whether dimensionless or dimensional:

$$f(u) = \int \delta(u - u')f(u')\,du', \tag{7.10}$$

$$\hat{f}(x) = \int \delta(x - x')\hat{f}(x')\,dx'. \tag{7.11}$$

Let us consider a discrete basis expansion:

$$f(u) = \sum_l C_l \psi_l(u), \qquad C_l = \frac{\int \psi_l^*(u) f(u) \, du}{\int |\psi_l(u)|^2 \, du}, \tag{7.12}$$

$$\hat{f}(x) = \sum_l C_l \hat{\psi}_l(x), \qquad C_l = \frac{\int \hat{\psi}_l^*(x) \hat{f}(x) \, dx}{\int |\hat{\psi}_l(x)|^2 \, dx}, \tag{7.13}$$

where $\hat{\psi}_l(x) \equiv s^{-1/2} \psi_l(x/s)$. The dimension of $\hat{f}(x)$ and $\hat{\psi}_l(x)$ (or $f(u)$ and $\psi_l(u)$ for that matter) need not be the same; it is only necessary that the dimension of $\hat{\psi}_l(x)$ multiplied by that of C_l be equal to that of $\hat{f}(x)$. If the basis is normalized, the denominators in the above expressions would normally not be written, leaving what seems to be a dimensionally inconsistent expression. In such cases the reader should remember that there is an implicit value of unity in the denominator which carries a dimension. Such implicit unity values which actually carry a dimension appear in other contexts as well, a possibility the reader should remain aware of. For instance, a quadratic-phase optical signal expressed as $\exp[i\pi\sigma(x^2 + y^2)/z]$ might not seem to have a dimension, but it actually has the dimension of the scalar amplitude of light hidden in the value of unity in front of the exponential function.

In two dimensions the corresponding conventions are

$$\hat{f}(x,y) \equiv \frac{1}{s} f(x/s, y/s) \equiv \frac{1}{s} f(u,v), \tag{7.14}$$

$$\hat{F}(\sigma_x, \sigma_y) \equiv s F(s\sigma_x, s\sigma_y) \equiv s F(\mu, \nu), \tag{7.15}$$

$$\hat{h}(x,y;x',y') \equiv \frac{1}{s^2} h(x/s, y/s; x'/s, y'/s) \equiv \frac{1}{s^2} h(u,v;u',v'), \tag{7.16}$$

where $u \equiv x/s$, $v \equiv y/s$, $\mu \equiv s\sigma_x$, $\nu \equiv s\sigma_y$.

In general, all of the results and equations in dimensionless chapters will remain to hold true, for the simple reason that the derivations leading to them cannot "know" whether a dimension is attributed to a variable or not. We simply replace u, v and μ, ν with x, y and σ_x, σ_y, functions such as $f(u,v)$ and $F(\mu,\nu)$ by $\hat{f}(x,y)$ and $\hat{F}(\sigma_x, \sigma_y)$, and kernels such as $h(u,v;u',v')$ by $\hat{h}(x,y;x',y')$, as shown in the above basic relations. The circumflex is not used for physical parameter distributions which never appear in dimensionless contexts, such as the refractive index, which we simply write as $n(x,y,z)$ rather than $\hat{n}(x,y,z)$.

We will also replace certain dimensionless parameters with parameters of appropriate physical dimension. For instance, chirp multiplication takes a function $f(u)$ to $\exp(-i\pi q u^2) f(u)$. Physically, we will see that this models the action of a thin lens, and that the corresponding dimensional relation takes $\hat{f}(x)$ to $\exp(-i\pi x^2/\lambda f)\hat{f}(x)$, from which we see that the parameter q has been replaced with the physical parameter $1/\lambda f$ whose dimension is equal to the inverse square of the dimension of x. Two sets of parameters are important enough that we will use a special convention to stress that they are the dimensional counterparts of the corresponding dimensionless parameters: $\hat{A}$, $\hat{B}$, $\hat{C}$, $\hat{D}$ or $\hat{\alpha}$, $\hat{\beta}$, $\hat{\gamma}$ will denote the dimensional counterparts of the dimensionless parameters A, B, C, D or α, β, γ characterizing linear canonical transforms.

Finally, we mention some additional notation that will appear transiently at the beginning of the next section and then disappear. We use $\tilde{f}(x,t)$ to denote functions of both space and time and $\check{f}(x, f_o)$ to denote functions of space and temporal frequency f_o; $\check{f}(x, f_o)$ is the temporal Fourier transform of $\tilde{f}(x,t)$. The spatial Fourier transform of $\check{f}(x, f_o)$ is denoted by $\check{F}(\sigma_x, f_o)$. Finally, when we concentrate on monochromatic signals and the frequency dependence is dropped, we write $\hat{f}(x)$ or $\hat{F}(\sigma_x)$, which are then employed throughout the rest of the book.

The conventions we adopt are not new or unusual; they are implicitly employed in many texts without any special discussion and usually go unnoticed by readers, until they stumble upon a dimensionally inconsistent equation or obtain some dimensional paradox. We believe a conscious awareness of dimensions is important enough to warrant an explicit discussion.

7.3   Wave optics

Optical signals are most commonly represented by the complex amplitude or intensity of light as a function of space and/or time. We will usually deal with systems in which signals are represented by the amplitude of light as a function of the transverse spatial coordinates x and y over a given plane $z = $ constant. The distribution of light representing the signal propagates from left to right in the positive z direction, being operated on or transformed in the process. The distributions of light on the input and output planes in figure 7.1 represent the input and output of the optical system.

In this book we will mostly restrict our attention to optical systems consisting of linear and time-invariant components, and assume that the behavior of light can be adequately described by a scalar theory. Sections of free space or other homogeneous or inhomogeneous media will also be treated as components; in any event we will assume they are linear, isotropic, and nondispersive. We will also assume that we are dealing with systems and light sources for which we can assume that the light is *quasi-monochromatic* (effectively temporally coherent). We will, however, discuss both spatially coherent and spatially incoherent systems.

The output of such a system is related to its input by a relation of the form

$$\tilde{g}(\mathbf{r}, t) = \int_{\mathbf{r}} \int_t \tilde{h}(\mathbf{r}, \mathbf{r}', t - t') \tilde{f}(\mathbf{r}', t') \, dt' \, d\mathbf{r}', \tag{7.17}$$

where $\tilde{g}(\mathbf{r}, t)$ and $\tilde{f}(\mathbf{r}, t)$ represent the amplitude of light as functions of space and time over the output and input planes.

The wavelength (or center wavelength) of the light used will be denoted by λ, in the medium of propagation. The speed of light in vacuum will be denoted by c so that the frequency f_{oc} of a monochromatic optical wave of wavelength λ satisfies $c = f_{oc} n_A \lambda$, where n_A is the refractive index of some medium A. The wave number is defined as $\sigma \equiv 1/\lambda$ and is equal to the magnitude of the wave vector $\boldsymbol{\sigma}$. For a plane wave, $\boldsymbol{\sigma}$ points in the direction of propagation of the wave. $\mathbf{r}$ denotes the vector (x, y) or (x, y, z) and $\boldsymbol{\sigma}$ denotes the vector (σ_x, σ_y) or $(\sigma_x, \sigma_y, \sigma_z)$, depending on the context.

We will further restrict our attention to *first-order* centered systems. These are systems which have the general appearance of the system shown in figure 7.1 and for which a number of simplifying approximations can be employed. For the time being we satisfy ourselves by noting that they are precisely the same approximations employed in the theory of optical systems referred to as Fourier optics (Goodman 1996).

The intensity of a wave at a certain point is defined as the power per unit area at that point. Poynting's theorem (Ramo, Whinnery, and Van Duzer 1994) gives the intensity in terms of the electric field vector $\mathbf{E}$ as $|\mathbf{E}|^2/\eta$ where η is the intrinsic impedance of free space. We will assume that the scalar amplitudes we are working with have been normalized so that the intensity is given by (Saleh and Teich 1991:44)

$$\tilde{I}_{\tilde{f}}(\mathbf{r}, t) \equiv 2\overline{[\tilde{f}(\mathbf{r}, t)]^2}, \tag{7.18}$$

where the time average denoted by the overbar is taken over an interval much longer than the optical period but sufficiently shorter than the time over which the envelope of $\tilde{f}(\mathbf{r}, t)$ changes appreciably. In the monochromatic case, the real field $\tilde{f}(\mathbf{r}, t)$ can be written as

$$\tilde{f}(\mathbf{r}, t) = \hat{A}(\mathbf{r}) \cos[2\pi f_{oc} t + \hat{\varphi}(\mathbf{r})], \tag{7.19}$$

where $\hat{A}(\mathbf{r})$ and $\hat{\varphi}(\mathbf{r})$ are real-valued functions. The corresponding analytic signal $\tilde{f}_{as}(\mathbf{r}, t)$ is $\hat{A}(\mathbf{r}) \exp[-i2\pi f_{oc} t - i\hat{\varphi}(\mathbf{r})]$, and the complex amplitude which we will denote by $\hat{f}(\mathbf{r})$ is $\hat{A}(\mathbf{r}) \exp[-i\hat{\varphi}(\mathbf{r})]$.

7.3.1 The wave equation

The function $\tilde{f}(\mathbf{r}, t)$ describing the scalar amplitude distribution of light as a function of the position vector $\mathbf{r} = (x, y, z)$ and time t in a linear isotropic nondispersive medium with time-invariant refractive index distribution $n(\mathbf{r})$ satisfies the wave equation

$$\frac{\partial^2 \tilde{f}}{\partial x^2} + \frac{\partial^2 \tilde{f}}{\partial y^2} + \frac{\partial^2 \tilde{f}}{\partial z^2} - \frac{n^2(\mathbf{r})}{c^2} \frac{\partial^2 \tilde{f}}{\partial t^2} = 0. \tag{7.20}$$

We will assume that the variation of $n(\mathbf{r})$ is small over distances comparable to the wavelengths of light we deal with. Since the wave equation is linear, any linear superposition of solutions is also a solution. The wave equation can be solved uniquely if the distribution of $\tilde{f}$ over some surface is specified at some time. Furthermore, if new boundary conditions are specified as the linear superposition of some set of boundary conditions for which the solution is already known, the new solution can be written as the same linear superposition.

We take the temporal Fourier transform of both sides of the above equation, with the temporal Fourier transform of $\tilde{f}(\mathbf{r}, t)$ defined as

$$\check{f}(\mathbf{r}, f_o) = \int \tilde{f}(\mathbf{r}, t) e^{i2\pi f_o t} \, dt. \tag{7.21}$$

The temporal Fourier transform is defined with a positive sign in the exponent, in contrast to the spatial Fourier transform, which is defined in the conventional manner with a negative sign in the exponent. This is consistent with the interpretation of the spatio-temporal Fourier transform as the coefficient of expansion in terms of planes waves of the form $\exp[i2\pi(\boldsymbol{\sigma} \cdot \boldsymbol{r} - f_o t)]$ (Saleh and Teich 1991:925). The temporal Fourier transform of the wave equation is

$$\frac{\partial^2 \check{f}}{\partial x^2} + \frac{\partial^2 \check{f}}{\partial y^2} + \frac{\partial^2 \check{f}}{\partial z^2} + \frac{4\pi^2 n^2 f_o^2}{c^2} \check{f} = 0. \tag{7.22}$$

This equation is known as the Helmholtz equation. If we solve this equation for $\check{f}(\mathbf{r}, f_o)$ for all f_o, a temporal inverse Fourier transform operation will give us $\check{f}(\mathbf{r}, t)$. The analytic signal $\check{f}_{as}(\mathbf{r}, t)$ corresponding to $\check{f}(\mathbf{r}, t)$ is defined as the inverse Fourier transform of $[1 + \text{sgn}(f_o)]\check{f}(\mathbf{r}, f_o)$ and also satisfies the wave equation 7.20. In the event that we are dealing with monochromatic waves of specified frequency f_{oc}, we have $\check{f}(\mathbf{r}, f_o) = 0.5\,\hat{f}(\mathbf{r})\delta(f_o - f_{oc}) + 0.5\,\hat{f}^*(\mathbf{r})\delta(f_o + f_{oc})$. In this case the Fourier transform of the analytic signal is simply $\hat{f}(\mathbf{r})\delta(f_o - f_{oc})$ and the signal can be represented by the complex amplitude or phasor $\hat{f}(\mathbf{r})$. That $\hat{f}(\mathbf{r})$ also satisfies the Helmholtz equation can be shown most directly by substituting a monochromatic component of the form $\hat{f}(\mathbf{r})\exp(-i2\pi f_{oc}t)$ in equation 7.20 to obtain

$$\frac{\partial^2 \hat{f}}{\partial x^2} + \frac{\partial^2 \hat{f}}{\partial y^2} + \frac{\partial^2 \hat{f}}{\partial z^2} + \frac{4\pi^2 n^2 f_{oc}^2}{c^2}\hat{f} = 0. \tag{7.23}$$

In the monochromatic case, the intensity is simply related to $\hat{f}(\mathbf{r})$ as follows:

$$\hat{I}_{\hat{f}}(\mathbf{r}) = |\hat{f}(\mathbf{r})|^2, \tag{7.24}$$

and does not depend on time (Saleh and Teich 1991:46).

Two complete sets of solutions of equation 7.23 for a homogeneous medium $n(\mathbf{r}) = n = \text{constant}$, are the set of plane waves and the set of spherical waves respectively given as follows (Saleh and Teich 1991:47–48):

$$\hat{f}(x, y, z) = e^{i2\pi \boldsymbol{\sigma} \cdot \boldsymbol{r}} = e^{i2\pi(\sigma_x x + \sigma_y y + \sigma_z z)}, \qquad \boldsymbol{\sigma} \in \mathbf{R}^3, \tag{7.25}$$

$$\hat{f}(x, y, z) = \frac{e^{i2\pi \sigma r}}{i\lambda r}, \qquad \sigma \in \mathbf{R}, \tag{7.26}$$

where $r \equiv |\mathbf{r}|$ and $\sigma^2 \equiv |\boldsymbol{\sigma}|^2 = \sigma_x^2 + \sigma_y^2 + \sigma_z^2$. For each value of σ, there is a corresponding optical frequency $f_{oc} = \sigma c/n$. Other solutions can be expressed as linear superpositions of the members of either of these or yet other sets of solutions, as will be elaborated later.

Although the above solutions may be verified by direct substitution, it is instructive to note how they can be directly obtained. First take the three-dimensional spatial Fourier transform of equation 7.22, replacing the derivatives $\partial^2/\partial x^2$, $\partial^2/\partial y^2$, $\partial^2/\partial z^2$ with $(i2\pi\sigma_x)^2$, $(i2\pi\sigma_y)^2$, $(i2\pi\sigma_z)^2$. This results in an equation for the four-dimensional spatio-temporal Fourier transform of $\check{f}$:

$$\left[\sigma_x^2 + \sigma_y^2 + \sigma_z^2 - (nf_o/c)^2\right]\check{F}(\sigma_x, \sigma_y, \sigma_z, f_o) = 0, \tag{7.27}$$

where the spatial Fourier transform $\check{F}(\sigma_x, \sigma_y, \sigma_z, f_o)$ of $\check{f}(x, y, z, f_o)$ is defined as

$$\check{F}(\sigma_x, \sigma_y, \sigma_z, f_o) = \iiint \check{f}(x, y, z, f_o)e^{-i2\pi(\sigma_x x + \sigma_y y + \sigma_z z)}\, dx\, dy\, dz. \tag{7.28}$$

It immediately follows from equation 7.27 that $\check{F}(\sigma_x, \sigma_y, \sigma_z, f_o)$ can be nonzero only where $\sigma_x^2 + \sigma_y^2 + \sigma_z^2 - (nf_o/c)^2 = 0$. This implies delta-function-type solutions which can be inverse transformed to obtain the plane and spherical wave solutions presented, a task we leave to the reader. If the light is monochromatic with frequency f_{oc}, then $\sigma_x, \sigma_y, \sigma_z$ must satisfy $\sigma^2 = \sigma_x^2 + \sigma_y^2 + \sigma_z^2 = (nf_{oc}/c)^2$.

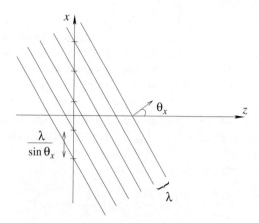

Figure 7.3. Wavefronts of a plane wave making angle θ_x with the z axis.

If we know the spatial variation of a plane wave at some plane $z = z_1$, we can easily determine its spatial variation at any other plane $z = z_2 = z_1 + d$. For instance, if the spatial variation of a wave with $\sigma = 10$ at the plane $z = 0$ is given by $K \exp[i2\pi(4x + 6y)]$ where K is some complex constant, we can deduce the complete three-dimensional distribution of the wave as

$$K e^{i2\pi(4x+6y+\sqrt{10^2-4^2-6^2}\,z)} \tag{7.29}$$

and thus determine the complex amplitude distribution at any other plane. For instance, at $z = 5.2$ we have $K \exp[i2\pi(4x + 6y + 6.9z)] = K \exp[i2\pi(4x + 6y + 36)] = K \exp[i2\pi(4x + 6y)] \exp[i2\pi(36)]$. We see that the form of the distribution remains unchanged but is affected by a phase factor of $\exp[i2\pi(36)]$. (In this and similar numerical discussions we assume the dimensions of the numerical factors are implied. For instance, $\sigma = 10$ has the dimensions of inverse length.)

Figure 7.3 shows the wavefronts of a plane wave making angle θ_x with the y-z plane (θ_x is the complement to $\pi/2$ of the angle made with the x axis). We wish to examine the variation of this wave as a function of x on the $z = 0$ plane. The period of the optical wave is λ. Along the x axis this translates into a period of $\lambda/\sin\theta_x$, which corresponds to the spatial frequency σ_x of the wave along the x direction. Similar considerations apply for the y direction. Thus the following relations hold:

$$\sin\theta_x = \lambda\sigma_x, \tag{7.30}$$

$$\sin\theta_y = \lambda\sigma_y. \tag{7.31}$$

The $z = 0$ profile of a plane wave $\exp[i2\pi(\sigma_x x + \sigma_y y + \sigma_z z)]$ is a two-dimensional spatial harmonic whose spatial frequencies are related to the direction of propagation of the plane wave as given by equations 7.30 and 7.31.

The spherical wave solution given in equation 7.26 is the solution corresponding to a point source $\delta(x, y, z)$ centered at the origin; it is the *Green's function* of the wave equation. For future reference, we also note the common *Fresnel approximation* of a spherical wave. With $r^2 = x^2 + y^2 + z^2$ and under the assumption that $z^2 \gg x^2 + y^2$, we obtain

$$\hat{f}(x,y,z) = \frac{e^{i2\pi\sigma r}}{i\lambda r} \approx \frac{e^{i2\pi\sigma z}}{i\lambda z} \exp\left[i2\pi\sigma\left(\frac{x^2+y^2}{2z}\right)\right], \tag{7.32}$$

where we replaced $r \approx z$ in the denominator but $r \approx z + (x^2 + y^2)/2z$ in the exponent because of the greater sensitivity of the imaginary exponent to small changes (Saleh and Teich 1991:49). Equation 7.32 is also known as a *parabolic wave*.

Sections 7.3.2, 7.3.3, and 7.3.4 will address from a different perspective the problem of obtaining the distribution of light on a plane $z = z_2 = z_1 + d$, given the distribution of light on a plane $z = z_1$. Readers willing to take the results for granted may skip to section 7.4.

7.3.2 Plane wave decomposition

First, let us assume that the distribution of light at $z = z_1$, which we refer to as the input $\hat{f}(x,y)$, is of the form

$$\hat{f}(x,y) = K e^{i2\pi(\sigma_x x + \sigma_y y)}, \tag{7.33}$$

where K is a complex constant and σ_x and σ_y are the spatial frequencies of this two-dimensional harmonic function. We recognize this as the profile of a plane wave with wave vector components σ_x, σ_y, and $\sigma_z = \sqrt{\sigma^2 - \sigma_x^2 - \sigma_y^2}$:

$$K e^{i2\pi(\sigma_x x + \sigma_y y)} \exp\left[i2\pi\sqrt{\sigma^2 - \sigma_x^2 - \sigma_y^2}\,(z - z_1)\right]. \tag{7.34}$$

The output $\hat{g}(x,y)$ observed at the plane $z = z_2 = z_1 + d$ is then given by

$$\hat{g}(x,y) = K e^{i2\pi(\sigma_x x + \sigma_y y)} \exp\left(i2\pi\sqrt{\sigma^2 - \sigma_x^2 - \sigma_y^2}\,d\right) = \hat{H}(\sigma_x, \sigma_y)\hat{f}(x,y), \tag{7.35}$$

where

$$\hat{H}(\sigma_x, \sigma_y) = \exp\left(i2\pi\sqrt{\sigma^2 - \sigma_x^2 - \sigma_y^2}\,d\right). \tag{7.36}$$

We see that harmonic functions are eigenfunctions of propagation over a section of free space, with eigenvalue $\hat{H}(\sigma_x, \sigma_y)$. Since harmonics are profiles of plane waves, it is sometimes also said that plane waves are eigenfunctions of propagation in free space. (It is also possible to pose the same in terms of temporal evolution by showing that if at any instant in time we observe a plane wave in space, we will observe a plane wave at all consecutive times, which more directly justifies referring to plane waves as eigenfunctions of propagation in free space.)

Let us return to the problem of relating the output to the input when the input is an arbitrary distribution of light, and not necessarily a two-dimensional harmonic. An arbitrary distribution of light at the plane z_1, denoted by $\hat{f}(x,y)$, can be written as a linear superposition of harmonics as follows:

$$\hat{f}(x,y) = \int\int \hat{F}(\sigma_x, \sigma_y)\, e^{i2\pi(\sigma_x x + \sigma_y y)}\, d\sigma_x\, d\sigma_y, \tag{7.37}$$

where $\hat{F}(\sigma_x, \sigma_y)$ is the Fourier transform of $\hat{f}(x, y)$. Since we know that a linear superposition of inputs will produce the same linear superposition of outputs, the distribution of light at the plane z_2 can be obtained easily as

$$\hat{g}(x, y) = \iint \hat{F}(\sigma_x, \sigma_y) e^{i2\pi \sqrt{\sigma^2 - \sigma_x^2 - \sigma_y^2} d} e^{i2\pi(\sigma_x x + \sigma_y y)} d\sigma_x d\sigma_y. \tag{7.38}$$

We see that the effect of free-space propagation in the Fourier domain is

$$\hat{G}(\sigma_x, \sigma_y) = e^{i2\pi \sqrt{\sigma^2 - \sigma_x^2 - \sigma_y^2} d} \hat{F}(\sigma_x, \sigma_y) = \hat{H}(\sigma_x, \sigma_y)\hat{F}(\sigma_x, \sigma_y), \tag{7.39}$$

where $\hat{H}(\sigma_x, \sigma_y)$ is given by equation 7.36. This result can be written in the space domain as a two-dimensional convolution

$$\hat{g}(x, y) = \hat{h}(x, y) * * \hat{f}(x, y), \tag{7.40}$$

where $\hat{h}(x, y)$ is the inverse Fourier transform of $\hat{H}(\sigma_x, \sigma_y)$. A simple analytical expression for $\hat{h}(x, y)$ is not known. However, the exponent of $\hat{H}(\sigma_x, \sigma_y)$ is commonly approximated as

$$\sqrt{\sigma^2 - \sigma_x^2 - \sigma_y^2} \approx \sigma - \frac{\sigma_x^2 + \sigma_y^2}{2\sigma}, \tag{7.41}$$

under the assumption that $\sigma^2 \approx \sigma_z^2 \gg (\sigma_x^2 + \sigma_y^2)$. Then we approximately obtain

$$\hat{H}(\sigma_x, \sigma_y) = e^{i2\pi\sigma d} \exp[-i\pi\lambda d(\sigma_x^2 + \sigma_y^2)]. \tag{7.42}$$

For a discussion of the validity of this approximation, known as the Fresnel approximation, see Goodman 1996 or Saleh and Teich 1991. With this approximation, the inverse transform $\hat{h}(x, y)$ becomes

$$\hat{h}(x, y) = h_0 \exp[i\pi(x^2 + y^2)/\lambda d], \tag{7.43}$$

$$h_0 = \frac{e^{i2\pi\sigma d}}{i\lambda d},$$

which is nothing but the parabolic approximation of the spherical wave given earlier as equation 7.32. The relation between $\hat{f}$ and $\hat{g}$ takes the form

$$\hat{g}(x, y) = \hat{h}(x, y) * * \hat{f}(x, y) = h_0 \iint e^{(i\pi/\lambda d)[(x-x')^2 + (y-y')^2]} \hat{f}(x', y') dx' dy', \tag{7.44}$$

an expression known as the Fresnel integral or Fresnel transform (see equation 2.29). It gives the amplitude distribution of light at the plane $z = z_2 = z_1 + d$ in terms of the amplitude at the plane $z = z_1$. It is the solution of the paraxial Helmholtz equation to be discussed in section 7.3.3, as can be shown by direct substitution. On the other hand, the exact form with the square root in the exponent is the solution of the exact Helmholtz equation 7.23. The fact that the system represented by the Fresnel integral is space-invariant, is consistent with the fact that the eigenfunctions of this system are harmonic functions (page 14).

Here we have not included the classic derivations through which the Fresnel integral is traditionally arrived at (Goodman 1996, Iizuka 1987, Yu 1983, Born and Wolf 1980). We only note that the kernel appearing in equation 7.44 is nothing but the Fresnel approximation of a spherical wave (see equation 7.32). Thus equation 7.44 is essentially an approximation of a weighted superposition of spherical waves. This interpretation is known as the Huygens-Fresnel principle. Each point in the input plane is considered to be a secondary source with amplitude $\hat{f}(x, y)$, which gives rise to a spherical wave. Superposing all of these spherical waves gives us the amplitude distribution $\hat{g}(x, y)$ at the output plane. The mathematical expression of the Huygens-Fresnel principle is known as the Rayleigh-Sommerfeld diffraction formula (Goodman 1996):

$$\hat{g}(x, y) = \frac{1}{i\lambda} \int\int \hat{f}(x', y') \frac{e^{i2\pi r/\lambda}}{r} \cos\theta \, dx' \, dy', \tag{7.45}$$

where $r = \sqrt{d^2 + (x - x')^2 + (y - y')^2}$ and θ is the angle between the line joining the input point (x', y') to the output point (x, y), and the z axis (so that $\cos\theta = d/r$). This integral is interpreted as a superposition of diverging spherical waves originating from "secondary sources" located at the input plane; $\cos\theta$ is an "obliquity factor." For an excellent discussion of the developments leading to this equation, see Goodman 1996. It is possible to arrive at equation 7.44 from equation 7.45 by employing the Fresnel approximation of the spherical wave given in equation 7.32.

7.3.3   The paraxial wave equation

Here we present a number of approaches closely related to those of section 7.3.2, but which nevertheless provide different perspectives. We take the Fourier transform of equation 7.23 with respect to x and y (or equivalently, we consider solutions of the form $\hat{f}(x, y, z) = \hat{F}(\sigma_x, \sigma_y, z) \exp[i2\pi(\sigma_x x + \sigma_y y)])$ to obtain

$$\frac{\partial^2 \hat{F}}{\partial z^2} = -4\pi^2(\sigma^2 - \sigma_x^2 - \sigma_y^2)\hat{F}. \tag{7.46}$$

A solution of this equation corresponding to propagation in the positive z direction is

$$\hat{F}(\sigma_x, \sigma_y, z) = e^{i2\pi\sqrt{\sigma^2 - \sigma_x^2 - \sigma_y^2}\, z} \, \hat{F}(\sigma_x, \sigma_y, 0). \tag{7.47}$$

We use the two-dimensional functions $\hat{f}(x, y)$ and $\hat{g}(x, y)$ to denote two-dimensional fields on the $z = z_1$ and $z = z_2$ planes respectively, and the three-dimensional function $\hat{f}(x, y, z)$ to denote three-dimensional fields. Thus we will write $\hat{F}(\sigma_x, \sigma_y, z_1) \equiv \hat{F}(\sigma_x, \sigma_y)$ and $\hat{F}(\sigma_x, \sigma_y, z_2) \equiv \hat{G}(\sigma_x, \sigma_y)$. Now, writing equation 7.47 once for z_1 and once for z_2 and eliminating $\hat{F}(\sigma_x, \sigma_y, 0)$, we obtain

$$\hat{G}(\sigma_x, \sigma_y) = e^{i2\pi\sqrt{\sigma^2 - \sigma_x^2 - \sigma_y^2}\, d} \, \hat{F}(\sigma_x, \sigma_y), \tag{7.48}$$

which is the same as equation 7.39, from which the same argument leads us to equation 7.44 ($d = z_2 - z_1$).

Another approach is as follows. Restricting ourselves to waves traveling in the positive z direction, the "square root" of equation 7.46 may be written as

$$\frac{\partial \hat{F}}{\partial z} = +i2\pi\sqrt{\sigma^2 - \sigma_x^2 - \sigma_y^2} \, \hat{F}. \tag{7.49}$$

(Although we are being far from rigorous here, our final result will nevertheless be correct.) Introducing what is essentially the Fresnel approximation at this point we obtain

$$\frac{\partial \hat{F}}{\partial z} = +i2\pi(\sigma - \sigma_x^2/2\sigma - \sigma_y^2/2\sigma)\hat{F}. \tag{7.50}$$

The solution to equation 7.50 is

$$\hat{F}(\sigma_x, \sigma_y, z) = e^{i2\pi(\sigma - \sigma_x^2/2\sigma - \sigma_y^2/2\sigma)z} \, \hat{F}(\sigma_x, \sigma_y, 0), \tag{7.51}$$

from which one can deduce equation 7.42.

It is also possible to work in the space domain. Starting directly with equation 7.23 and again formally taking the "square root" of the operators yields

$$\frac{\partial \hat{f}}{\partial z} = +i\sqrt{4\pi^2\sigma^2 + \frac{\partial^2}{\partial x^2} + \frac{\partial^2}{\partial y^2}} \, \hat{f}. \tag{7.52}$$

Introducing what is essentially the Fresnel approximation at this point, we obtain

$$\frac{\partial \hat{f}}{\partial z} = +i\left[2\pi\sigma + \frac{1}{4\pi\sigma}\left(\frac{\partial^2}{\partial x^2} + \frac{\partial^2}{\partial y^2}\right)\right]\hat{f}. \tag{7.53}$$

The Fresnel integral given in equation 7.44 is an exact solution of this differential equation. These discussions were inspired by Bastiaans 1979c.

A *paraxial wave* is one whose wave vectors make small angles with the optical axis. In other words, the wavefront normals are paraxial rays (Saleh and Teich 1991). Just as a temporally narrowband signal has harmonic components concentrated around a certain center frequency, a paraxial wave has plane-wave components whose wave vectors are concentrated around the optical axis. The major spatial variation is along the z axis, so we can write $\hat{f}(\mathbf{r})$ in the form $\hat{f}(\mathbf{r}) = \hat{A}(\mathbf{r})\exp(i2\pi\sigma z)$, where $\hat{A}(\mathbf{r})$ is a complex envelope. Paraxial waves can be interpreted as spatially narrowband modulated plane waves, just as temporally narrowband signals can be interpreted as temporally narrowband modulated harmonics. Substituting this form for $\hat{f}(\mathbf{r})$ in equation 7.53 we obtain

$$\frac{\partial^2 \hat{A}}{\partial x^2} + \frac{\partial^2 \hat{A}}{\partial y^2} + i4\pi\sigma\frac{\partial \hat{A}}{\partial z} = 0. \tag{7.54}$$

Equations such as 7.53 or 7.54 are known as paraxial wave equations or paraxial Helmholtz equations. A more conventional derivation of equation 7.54 is as follows (Saleh and Teich 1991). We substitute $\hat{f}(\mathbf{r}) = \hat{A}(\mathbf{r})\exp(i2\pi\sigma z)$ in equation 7.23, and employ $\partial\hat{A}/\partial z \ll 2\pi\sigma\hat{A}$ and $\partial^2\hat{A}/\partial z^2 \ll 4\pi^2\sigma^2\hat{A}$, which are mathematical statements of paraxiality, and are also referred to as the slowly varying envelope approximation (since $\hat{A}(\mathbf{r})$ varies slowly with $\mathbf{r}$). This derivation again results in equation 7.54.

The parabolic wave is an exact solution of equation 7.54 (Saleh and Teich 1991):

$$\hat{A}(\mathbf{r}) = \frac{1}{i\lambda z}e^{i2\pi\sigma(x^2+y^2)/2z}. \tag{7.55}$$

In preparation for section 7.3.4, we also write the following more general solution of equation 7.54 (Saleh and Teich 1991):

$$\hat{A}(\mathbf{r}) = \frac{1}{i\lambda\hat{q}(z)} e^{i2\pi\sigma(x^2+y^2)/2\hat{q}(z)}, \tag{7.56}$$

where $\hat{q}(z) = z - i\check{z}$, $\check{z} = $ constant. $\hat{q}$ is *not* a dimensional version of the parameter q appearing in chapter 2, but we nevertheless choose this notation to conform with convention.

7.3.4 Hermite-Gaussian beams

A more general solution of equation 7.54 is

$$\hat{A}(\mathbf{r}) \propto \frac{1}{i\lambda\hat{q}(z)} \exp[i\pi\sigma(x^2+y^2)/\hat{q}(z)], \tag{7.57}$$

where $\hat{q}(z) = z - i\check{z}$. Here $\check{z}$ is a constant which is referred to as the Rayleigh range. $\hat{q}(z)$ is referred to as the complex radius of curvature or simply as the $\hat{q}$-parameter. If we define the beam size $W(z)$ and the wavefront radius of curvature $R(z)$ through

$$\frac{1}{\hat{q}(z)} \equiv \frac{1}{R(z)} + i\frac{\lambda}{W^2(z)}, \tag{7.58}$$

it is possible to show that

$$\hat{A}(\mathbf{r}) = \frac{2^{1/2}}{W(z)} \exp\left[-\frac{\pi(x^2+y^2)}{W^2(z)}\right] \exp\left[i2\pi\sigma\frac{(x^2+y^2)}{2R(z)} - i\zeta(z)\right], \tag{7.59}$$

where

$$W(z) \equiv W_0\left[1 + (z/\check{z})^2\right]^{1/2}, \tag{7.60}$$

$$R(z) \equiv z\left[1 + (\check{z}/z)^2\right], \tag{7.61}$$

$$\zeta(z) \equiv \arctan(z/\check{z}), \tag{7.62}$$

and $W_0^2 \equiv W^2(0) \equiv \lambda\check{z}$. We have normalized $\hat{A}(\mathbf{r})$ so that it has unit energy. We are choosing to employ the parameter $W(z)$ which we refer to as the beam size, rather than the more commonly used $w(z)$ known as the beam radius. These two parameters are simply related by $W^2(z) = \pi w^2(z)$ so that $W_0^2 = \pi w_0^2$. The interpretations of these parameters are discussed in many texts such as Saleh and Teich 1991. The distribution of light represented by equation 7.59 is known as a *Gaussian beam*.

The paraxial approximation of a spherical wave originating at $x = y = z = 0$ can be written in the form $(i\lambda R)^{-1} \exp[i\pi(x^2+y^2)/\lambda R]$ where $R = z$ is the radius of curvature of the wavefronts at z. Comparing this with equation 7.57, we see that the Gaussian beam can be interpreted as a spherical wave with complex "radius" $\hat{q}$. When $W_0 = \infty$ the beam has infinite transverse extent and the complex radius $\hat{q}$ reduces to a real radius (equation 7.58). When $\hat{q}$ is complex the imaginary part manifests itself as the beam size:

$$e^{i\pi(x^2+y^2)/\lambda\hat{q}} = e^{i\pi(x^2+y^2)/\lambda R}e^{-\pi(x^2+y^2)/W^2}. \tag{7.63}$$

An even more general set of solutions of equation 7.54 are the Hermite-Gaussian beams. Unlike the set of plane waves, this is a discrete set with countably many members, enumerated by (l, m):

$$\hat{A}(\mathbf{r}) = \frac{1}{W(z)} \, \psi_l \left(\frac{x}{W(z)} \right) \, \psi_m \left(\frac{y}{W(z)} \right) \, \exp \left[i2\pi\sigma \frac{(x^2 + y^2)}{2R(z)} - i(l + m + 1)\zeta(z) \right],$$

(7.64)

where ψ_l, ψ_m are the Hermite-Gaussian functions defined in section 2.5.2. The $l = 0$, $m = 0$ beam is simply the Gaussian beam given in equation 7.59. The Hermite-Gaussian beams share the parabolic wavefronts of the Gaussian beam, but exhibit different intensity distributions. Further discussion of their physical characteristics can be found in texts such as Saleh and Teich 1991.

Hermite-Gaussian beams are a complete orthonormal set of solutions of the paraxial wave equation. Just like the set of plane waves, they can be used to construct arbitrary solutions. Let us assume that the amplitude distribution of light at the plane $z = 0$ is given by $\hat{f}(x, y)$. At $z = 0$, the Hermite-Gaussian beams are equal to $W_0^{-1} \psi_l(x/W_0) \, \psi_m(y/W_0)$. We first expand $\hat{f}(x, y)$ as

$$\hat{f}(x, y, 0) \equiv \hat{f}(x, y) = \sum_{l=0}^{\infty} \sum_{m=0}^{\infty} C_{lm} \frac{1}{W_0} \, \psi_l \left(\frac{x}{W_0} \right) \, \psi_m \left(\frac{y}{W_0} \right),$$

(7.65)

where

$$C_{lm} = \iint \frac{1}{W_0} \, \psi_l \left(\frac{x}{W_0} \right) \, \psi_m \left(\frac{y}{W_0} \right) \hat{f}(x, y) \, dx \, dy.$$

(7.66)

Now, it is possible to find the amplitude distribution at any z as

$$\hat{f}(x, y, z) \equiv \hat{g}(x, y) = \sum_{l=0}^{\infty} \sum_{m=0}^{\infty} C_{lm} \, e^{i2\pi\sigma z} \frac{1}{W(z)} \, \psi_l \left(\frac{x}{W(z)} \right) \, \psi_m \left(\frac{y}{W(z)} \right)$$

$$\times \exp \left[i2\pi\sigma \frac{(x^2 + y^2)}{2R(z)} - i(l + m + 1)\zeta(z) \right]. \quad (7.67)$$

This result relating the distribution of light at an arbitrary plane to that at $z = 0$ is equivalent to Fresnel's integral, although it is not straightforward to show analytically. That this is indeed the case will be apparent when we discuss the fractional Fourier transform formulation of optical propagation in chapter 9.

There exist many complete sets of solutions of the wave equation. Which set is preferred depends on the situation. Usually, it is best to work with the set that constitutes the eigenfunctions of the system through which light will pass. In free space, plane waves are the natural choice. Hermite-Gaussian functions, on the other hand, are eigenfunctions of spherical mirror resonators and periodic lens waveguides, and are thus useful for such systems. While Hermite-Gaussian functions are not strictly eigenfunctions of free space, they nevertheless retain their general form upon propagation through free space so that they can also be used with relative ease in this case as well. Gaussian beam propagation can also be formulated in cylindrical coordinates, in which case one obtain the so-called Laguerre-Gaussian beams instead of the Hermite-Gaussian beams.

Finally, we discuss how the parameters of Hermite-Gaussian beams change as a result of propagating through an optical system characterized by a linear canonical transform with matrix parameters $\hat{A}, \hat{B}, \hat{C}, \hat{D}$ (such optical systems will be discussed in detail in chapter 8). The $\hat{q}$-parameter of the output beam can be simply related to the $\hat{q}$-parameter of the input beam as follows:

$$(\lambda \hat{q}_{\text{out}}) = \frac{\hat{A}(\lambda \hat{q}_{\text{in}}) + \hat{B}}{\hat{C}(\lambda \hat{q}_{\text{in}}) + \hat{D}}, \tag{7.68}$$

a result whose similarity to equation 3.155 is worth pointing out. The relationship between the dimensional parameters $\hat{A}, \hat{B}, \hat{C}, \hat{D}$ and the dimensionless A, B, C, D will be discussed on page 269. The most straightforward way of deriving equation 7.68 is to take the linear canonical transform of a Gaussian beam.

Before proceeding any further, we must define a new parameter called the *accumulated Gouy phase shift* (Erden and Ozaktas 1997). The conventional Gouy phase shift $\zeta(z)$ defined in equation 7.62 is the on-axis phase of a Gaussian beam with respect to the beam waist in excess of the phase of a plane wave $\exp(i2\pi\sigma z)$. It is not independent from the beam size W and the wavefront radius of curvature R. Of greater interest from an input-output perspective is the phase shift accumulated by the beam as it passes through several lenses and sections of free space, with respect to a single reference point in the system. Thus, we define the accumulated Gouy phase shift $\tilde{\zeta}$ of a Gaussian beam passing through an optical system as the on-axis phase accumulated by the beam in excess of the factor $\exp(i2\pi\sigma z)$. (This latter factor is the on-axis phase that would be accumulated by a plane wave.) Mathematically,

$$-\tilde{\zeta} \equiv \angle[\hat{A}(0, 0, z_{\text{out}})] - \angle[\hat{A}(0, 0, z_{\text{in}})], \tag{7.69}$$

where $\hat{A}(0, 0, z_{\text{out}})$ and $\hat{A}(0, 0, z_{\text{in}})$ denote the on-axis values of the output and input Gaussian beams, and $\angle[\cdot]$ denotes the phase.

Let us now consider a Gaussian beam with parameters W_{in} and R_{in} input to an optical system characterized by the parameters $\hat{A}, \hat{B}, \hat{C}, \hat{D}$. Also, let the accumulated Gouy phase with respect to some reference point be given as $\tilde{\zeta}_{\text{in}}$. Denoting the corresponding output parameters by W_{out}, R_{out}, and $\tilde{\zeta}_{\text{out}}$, it is possible to obtain the following results (Erden and Ozaktas 1997):

$$W_{\text{out}}^2 = \left(\hat{A} + \frac{\hat{B}}{\lambda R_{\text{in}}}\right)^2 W_{\text{in}}^2 + \frac{\hat{B}^2}{W_{\text{in}}^2}, \tag{7.70}$$

$$\frac{1}{\lambda R_{\text{out}}} = \frac{\left(\hat{C} + \frac{\hat{D}}{\lambda R_{\text{in}}}\right)\left(\hat{A} + \frac{\hat{B}}{\lambda R_{\text{in}}}\right) + \frac{\hat{B}\hat{D}}{W_{\text{in}}^4}}{\left(\hat{A} + \frac{\hat{B}}{\lambda R_{\text{in}}}\right)^2 + \frac{\hat{B}^2}{W_{\text{in}}^4}}, \tag{7.71}$$

$$\tilde{\zeta}_{\text{out}} = \tilde{\zeta}_{\text{in}} + \arctan\left[\frac{\hat{B}}{\left(\hat{A} + \frac{\hat{B}}{\lambda R_{\text{in}}}\right) W_{\text{in}}^2}\right]. \tag{7.72}$$

The first two of these equations are a consequence of equation 7.68, whereas the last one is demonstrated in Erden and Ozaktas 1997. The narrowest part of the beam, known as the waist, is observed where the beam size is minimum. When the beam is incident to the system at its waist so that $R_{in} = \infty$ and $W_{in} = W_0$, and if we assume $\tilde{\zeta}_{in} = 0$, the above relations reduce to

$$\frac{1}{\lambda R_{out}} = \frac{\hat{B}/\hat{A}}{\hat{A}^2 W_0^4 + \hat{B}^2} + \frac{\hat{C}}{\hat{A}}, \tag{7.73}$$

$$W_{out}^2 = \hat{A}^2 W_0^2 + \hat{B}^2/W_0^2, \tag{7.74}$$

$$\tilde{\zeta}_{out} = \arctan\left(\frac{\hat{B}}{\hat{A}W_{in}^2}\right). \tag{7.75}$$

The accumulated Gouy phase shift is an independent parameter which complements the beam size and wavefront radius of curvature to constitute three parameters which uniquely characterize the beam with respect to a reference point in the system. This means that knowledge of these three parameters at any single plane in the system allows them to be calculated at any other plane in the system. Furthermore, measurement of these parameters allows one to uniquely recover the parameters characterizing the first-order system through which the beam propagates (Erden and Ozaktas 1997).

7.4   Wave-optical characterization of optical components

7.4.1   Sections of free space

Despite the fact that they often consist of no more than the stretch of space between two other components, it is common to look upon sections of free space as optical components in their own right. The input and output of this component are the light distributions on the planes $z = z_1$ and $z = z_2 = z_1 + d$ bounding the section of free space from the left and the right; d denotes the length of the section of free space. We have already determined the relation between the input and output in several ways in the preceding section. Here we consolidate the main results.

The output $\hat{g}(x, y)$ is related to the input $\hat{f}(x, y)$ through the Fresnel integral which is essentially a chirp convolution in the space domain and a chirp multiplication in the frequency domain:

$$\hat{g}(x, y) = \hat{h}(x, y) * *\hat{f}(x, y), \tag{7.76}$$

$$\hat{h}(x, y) = e^{i2\pi\sigma d} \frac{1}{i\lambda d} \exp\left[\frac{i\pi(x^2 + y^2)}{\lambda d}\right],$$

$$\hat{G}(\sigma_x, \sigma_y) = \hat{H}(\sigma_x, \sigma_y)\hat{F}(\sigma_x, \sigma_y), \tag{7.77}$$

$$\hat{H}(\sigma_x, \sigma_y) = e^{i2\pi\sigma d} \exp\left[-i\pi\lambda d(\sigma_x^2 + \sigma_y^2)\right].$$

Recall that $\hat{H}(\sigma_x, \sigma_y)$ is the two-dimensional Fourier transform of $\hat{h}(x, y)$ and the eigenvalue associated with the harmonic whose transverse spatial frequencies are (σ_x, σ_y).

We also write for reference the one-dimensional versions of these results:

$$\hat{g}(x) = \hat{h}(x) * \hat{f}(x), \tag{7.78}$$

$$\hat{h}(x) = e^{i2\pi\sigma d} e^{-i\pi/4} \sqrt{\frac{1}{\lambda d}} \exp\left[\frac{i\pi x^2}{\lambda d}\right],$$

$$\hat{G}(\sigma_x) = \hat{H}(\sigma_x)\hat{F}(\sigma_x), \tag{7.79}$$

$$\hat{H}(\sigma_x) = e^{i2\pi\sigma d} \exp\left[-i\pi\lambda d\sigma_x^2\right].$$

7.4.2 Thin lenses

A thin lens is a special kind of spatial filter that plays an important role in realizing optical systems. Ideally, thin lenses are phase-only multiplicative filters with transmittance function

$$\hat{h}(x, y) = \exp\left[-\frac{i\pi(x^2 + y^2)}{\lambda f}\right], \tag{7.80}$$

such that the amplitude distribution $\hat{g}(x, y)$ immediately after the lens is related to the distribution $\hat{f}(x, y)$ immediately before the lens through

$$\hat{g}(x, y) = \hat{h}(x, y)\hat{f}(x, y). \tag{7.81}$$

Thin lenses are assumed to have no thickness. The parameter f is referred to as the *focal length* of the lens. Lenses are referred to as positive or negative according to the sign of their focal length. The last equation can be written in the frequency domain as

$$\hat{G}(\sigma_x, \sigma_y) = \hat{H}(\sigma_x, \sigma_y) * *\hat{F}(\sigma_x, \sigma_y), \tag{7.82}$$

$$\hat{H}(\sigma_x, \sigma_y) = -i\lambda f \exp\left[i\pi\lambda f(\sigma_x^2 + \sigma_y^2)\right].$$

The one-dimensional versions of the above expressions are

$$\hat{g}(x) = \hat{h}(x)\hat{f}(x), \tag{7.83}$$

$$\hat{h}(x) = \exp\left[\frac{-i\pi x^2}{\lambda f}\right],$$

$$\hat{G}(\sigma_x) = \hat{H}(\sigma_x) * \hat{F}(\sigma_x), \tag{7.84}$$

$$\hat{H}(\sigma_x) = e^{-i\pi/4}\sqrt{\lambda f} \exp\left[i\pi\lambda f\sigma_x^2\right].$$

Lenses can be realized by grinding convex or concave spherical surfaces on both sides of a thin slab of glass with refractive index n_{gl}. The focal length is related to the radii of curvature of the surfaces by the following formula (Saleh and Teich 1991):

$$\frac{1}{f} = \left(\frac{n_{gl}}{n} - 1\right)\left(\frac{1}{R_{right}} - \frac{1}{R_{left}}\right). \tag{7.85}$$

R_{left} is the radius of the left surface and R_{right} is the radius of the right surface. The sign convention is such that surfaces which are convex towards the $+z$ direction are positive. Here n is the refractive index of the medium in which the lens is situated, which in most cases is air so that $n \approx 1$. We also note that λ is the wavelength in the same medium, and not in the glass.

Attenuation inside the lens and reflection from its surfaces are usually neglected, but sometimes it is desired to account for the finite size of the lens by defining a pupil function $\hat{p}(x,y)$, which is unity within the lens aperture and zero outside:

$$\hat{h}(x,y) = \hat{p}(x,y)e^{-i\pi(x^2+y^2)/\lambda f}. \tag{7.86}$$

The effects of attenuations, reflections, and aberrations are sometimes handled by absorbing them into the pupil function.

The above results are for spherical lenses which are rotationally symmetric around the z axis. More generally, the transmittance function may take the form

$$\hat{h}(x,y) = \exp\left[\frac{-i\pi}{\lambda}\left(\frac{x^2}{f_{xx}} + \frac{2xy}{f_{xy}} + \frac{y^2}{f_{yy}}\right)\right], \tag{7.87}$$

where f_{xx}, f_{xy}, f_{yy} are the three parameters characterizing the focal characteristics of the lens. Such lenses are often referred to as *anamorphic* lenses. If we express the above transmittance in terms of the rotated coordinates (x',y') (where the x' axis makes a positive angle $(1/2)\text{arccot}[f_{xy}(1/f_{yy} - 1/f_{xx})/2]$ with respect to the x axis), the cross term disappears:

$$\hat{h}(x',y') = \exp\left[\frac{-i\pi}{\lambda}\left(\frac{x'^2}{f_{x'x'}} + \frac{y'^2}{f_{y'y'}}\right)\right], \tag{7.88}$$

where $f_{x'x'}$ and $f_{y'y'}$ can be expressed in terms of f_{xx}, f_{xy}, f_{yy}. When $1/f_{x'x'} = 0$ (or $1/f_{y'y'} = 0$) we obtain a cylindrical lens with focal length $f_{y'y'}$ (or $f_{x'x'}$). Any anamorphic lens with given f_{xx}, f_{xy}, f_{yy} can be simulated by two orthogonally positioned cylindrical lenses with focal lengths $f_{x'x'}$ and $f_{y'y'}$.

7.4.3 Quadratic graded-index media

A quadratic graded-index medium is a medium characterized by a refractive index distribution of the form (Yariv 1989)

$$n^2(x,y) = n_0^2[1 - (x/\chi_x)^2 - (y/\chi_y)^2], \tag{7.89}$$

where χ_x, χ_y, and n_0 are the medium parameters. More generally these parameters may be functions of z, but we will not treat this case. We will further restrict ourselves to the special case $\chi_x = \chi_y = \chi$. The one-dimensional version of this index distribution is taken as

$$n^2(x) = n_0^2[1 - (x/\chi)^2]. \tag{7.90}$$

We start by substituting equation 7.89 in equation 7.23:

$$\frac{\partial^2 \hat{f}}{\partial x^2} + \frac{\partial^2 \hat{f}}{\partial y^2} + \frac{\partial^2 \hat{f}}{\partial z^2} + 4\pi^2\sigma^2[1 - (x^2 + y^2)/\chi^2]\hat{f} = 0, \tag{7.91}$$

where $\sigma = n_0 f_{oc}/c$. We seek positive-z traveling eigensolutions of the form $\hat{f}(x, y, z) = \hat{A}(x, y) \exp(i2\pi\sigma_z z)$. Now, it is possible to show that if the function $\hat{A}_x(x)$ satisfies

$$\frac{\partial^2 \hat{A}_x}{\partial x^2} + \frac{4\pi^2\sigma^2}{\chi^2}\left(\frac{\sigma_x^2\chi^2}{\sigma^2} - x^2\right)\hat{A}_x = 0, \tag{7.92}$$

and if the function $\hat{A}_y(y)$ satisfies an identical equation in y, then $\hat{A}(x, y) = \hat{A}_x(x)\hat{A}_y(y)$ is a solution of equation 7.91, provided $\sigma_x^2 + \sigma_y^2 + \sigma_z^2 = \sigma^2$.

Introducing the dimensionless variable $u = x/s$ where $s > 0$ is a scaling parameter here taken equal to $s^2 = \chi/\sigma$, the above equation reduces to

$$\frac{\partial^2 A_x}{\partial u^2} + 4\pi^2(s^2\sigma_x^2 - u^2)A_x = 0, \tag{7.93}$$

where $s^{-1/2}A_x(x/s) \equiv \hat{A}_x(x)$. Equation 7.93 is precisely the same as equation 2.176 and thus has the following discrete set of solutions for $l = 0, 1, \ldots$:

$$\hat{A}_x(x) = s^{-1/2}\psi_l(x/s), \qquad s^2\sigma_x^2 = (2l + 1)/2\pi. \tag{7.94}$$

A similar discussion for y leads to the following discrete set of solutions for $m = 0, 1, \ldots$:

$$\hat{A}_y(y) = s^{-1/2}\psi_m(y/s), \qquad s^2\sigma_y^2 = (2m + 1)/2\pi. \tag{7.95}$$

σ_x and σ_y can only assume the discrete values dictated by the above equations. For given l, m the value of σ_z is

$$\sigma_z = \sqrt{\sigma^2 - \sigma_x^2 - \sigma_y^2} = \sqrt{\sigma^2 - \frac{l + m + 1}{\pi s^2}}. \tag{7.96}$$

This can be expanded to first order as

$$\sigma_z \approx \sigma - \frac{l + m + 1}{2\pi s^2\sigma} = \sigma - \frac{l + m + 1}{2\pi\chi}. \tag{7.97}$$

Thus, each eigensolution (or eigenmode) $s^{-1}\psi_l(x/s)\psi_m(y/s)$ propagates through the graded-index medium with a propagation constant $\sigma - (l + m + 1)/2\pi\chi$. In the one-dimensional case, the corresponding result is

$$\sigma_z = \sqrt{\sigma^2 - \frac{l + 1/2}{\pi s^2}} \approx \sigma - \frac{l + 1/2}{2\pi\chi}. \tag{7.98}$$

The reader may also wish to note the similarity of the eigenmodes of quadratic graded-index media with Hermite-Gaussian beams in free space, with the identification of the beam size $W(z)$ with the scale parameter s appearing here.

Now, let us assume that an arbitrary distribution of light $\hat{f}(x, y)$ is incident on such a medium at $z = 0$. This distribution can be expanded in terms of the Hermite-Gaussian functions as

$$\hat{f}(x, y, 0) \equiv \hat{f}(x, y) = \sum_{l=0}^{\infty}\sum_{m=0}^{\infty} C_{lm}\frac{1}{s}\psi_l\left(\frac{x}{s}\right)\psi_m\left(\frac{y}{s}\right), \tag{7.99}$$

$$C_{lm} = \iint \frac{1}{s}\psi_l\left(\frac{x}{s}\right)\psi_m\left(\frac{y}{s}\right)\hat{f}(x, y)\,dx\,dy.$$

and the amplitude distribution at any z can be written as

$$\hat{f}(x,y,z) \equiv \hat{g}(x,y) = \sum_{l=0}^{\infty} \sum_{m=0}^{\infty} C_{lm} e^{i2\pi\sigma z} e^{-i(l+m+1)z/\chi} \frac{1}{s} \psi_l\left(\frac{x}{s}\right) \psi_m\left(\frac{y}{s}\right). \quad (7.100)$$

When $z = (\pi/2)\chi$ it is possible to show, using equation 2.184 and substituting for C_{lm}, that

$$\hat{g}(x,y) = e^{i2\pi\sigma z} \frac{e^{-i\pi/2}}{s^2} \iint \hat{f}(x',y') e^{-i2\pi(xx'+yy')/s^2} \, dx' \, dy', \quad (7.101)$$

which we recognize as being essentially a Fourier transform relation. Thus, propagation over a distance $z = (\pi/2)\chi$ in such a medium results in Fourier transformation. It also follows from the properties of repeated Fourier transformation that propagation over a distance $z = \pi\chi$ results in an inverted image, and that propagation over a distance $z = 2\pi\chi$ results in an erect image. Now, letting the parameter a denote the *fractional order*, it is possible to show by using equation 4.23 (or table 2.8, property 9) and by substituting for C_{lm}, that when $z = a(\pi/2)\chi$ we obtain

$$\hat{g}(x,y) = e^{i2\pi\sigma z} \frac{e^{-ia\pi/2}}{s^2} \iint K_a(x/s, x'/s) K_a(y/s, y'/s) \hat{f}(x',y') \, dx' \, dy', \quad (7.102)$$

where $K_a(u,u')$ is the kernel of the fractional Fourier transform (equation 4.4). We see that for arbitrary values of z, the effect of propagation in quadratic graded-index media can be interpreted as a fractional Fourier transform, a result we will further elaborate in chapter 9. An explicit integral transform relating $\hat{g}(x,y)$ to $\hat{f}(x,y)$ does appear in some treatments of graded-index media (Ghatak and Thyagarajan 1980: equation 3.25), although it has not been recognized as the fractional Fourier transform until Ozaktas and Mendlovic 1993a, b and Mendlovic and Ozaktas 1993a.

In the one-dimensional case, the corresponding input-output relation is

$$\hat{g}(x) = \sum_{l=0}^{\infty} C_l e^{i2\pi\sigma z} e^{-i(l+1/2)z/\chi} \frac{1}{\sqrt{s}} \psi_l\left(\frac{x}{s}\right). \quad (7.103)$$

When $z = (\pi/2)\chi$,

$$\hat{g}(x) = e^{i2\pi\sigma z} \frac{e^{-i\pi/4}}{s} \int \hat{f}(x') e^{-i2\pi xx'/s^2} \, dx', \quad (7.104)$$

and for arbitrary $z = a(\pi/2)\chi$

$$\hat{g}(x) = e^{i2\pi\sigma z} \frac{e^{-ia\pi/4}}{s} \int K_a(x/s, x'/s) \hat{f}(x') \, dx'. \quad (7.105)$$

We now turn our attention to equations 7.70 and 7.71 which we specialize for graded-index media. Assuming the waist of the beam coincides with the input plane $(1/R_{in} = 0)$ and borrowing the $\hat{A}, \hat{B}, \hat{C}, \hat{D}$ parameters from chapter 8 (equation 8.42) we obtain

$$\frac{1}{R_{out}} = \frac{-\left(1 - \frac{s^4}{W_{in}^4}\right) \frac{\lambda}{s^2} \sin(2d/\chi)}{\left(1 - \frac{s^4}{W_{in}^4}\right) \cos(2d/\chi) + \left(1 + \frac{s^4}{W_{in}^4}\right)}, \tag{7.106}$$

$$\frac{W_{out}^2}{W_{in}^2} = \frac{1}{2}\left(1 - \frac{s^4}{W_{in}^4}\right) \cos(2d/\chi) + \frac{1}{2}\left(1 + \frac{s^4}{W_{in}^4}\right). \tag{7.107}$$

We see that in general, the wavefront radius and beam size oscillate periodically with d. When the input beam size W_{in} "matches" the natural scale parameter $s = \sqrt{\chi/\sigma}$ of the medium ($W_{in} = s$), we obtain $1/R_{out} = 0$ and $W_{out} = W_{in}$ for all d.

In the rest of this section, which can be omitted without loss of continuity, we will discuss the number of degrees of freedom a graded-index medium of finite transverse extent Δx can support (Ozaktas and Mendlovic 1993b). Our analysis implicitly assumed that the medium is of infinite transverse extent and that equation 7.89 holds for all x, y. Of course, this is physically not possible and would result in negative values for the refractive index in equation 7.89. We note that apart from this abstraction, our analysis is fairly exact and does not even employ the slowly varying envelope approximation leading to the paraxial Helmholtz equation, but rather employs the more general Helmholtz equation 7.91. The major assumption behind equation 7.91 (apart from ignoring the vector nature of light) is that the refractive index changes little over distances of the order of a wavelength. This is valid provided $\chi \gg \lambda$, which, as we will see below, must always be satisfied anyway.

The fact that all physical systems are of finite extent has several implications. First, since the amplitude distribution of light and its Fourier transform are both (approximately) confined to finite intervals, a finite number of samples (degrees of freedom) are sufficient to represent both. Second, Hermite-Gaussian functions beyond a certain order will not be relevant because their energy content will mostly lie outside the finite extent of the medium.

Since it is difficult to manufacture large index variations, and more fundamentally since it is necessary that $n(x, y) \geq 1$, we must ensure that $(x^2 + y^2) \ll \chi^2$ if such a system is to be realizable. Thus the extent of the medium must satisfy $\Delta x \ll \chi$. We will now show that if this condition is satisfied, the following consequences hold (Ozaktas and Mendlovic 1993b):

1. The one-dimensional space-bandwidth product, or the number of degrees of freedom the medium can support, is $\approx \Delta x^2/s^2$.
2. The number of Hermite-Gaussian modes whose energies lie predominantly within the medium is $\approx \Delta x^2/s^2$.
3. The first-order approximation for σ_z (equation 7.97) is valid for these modes.

To show the first of the above, we note that both the original amplitude distribution of light and its Fourier transform will be confined to an extent of Δx. This spatial extent corresponds to an interval of length $\Delta x/s^2$ in units of spatial frequency (equation 7.101). Since $\Delta x/s^2$ represents the double-sided spatial bandwidth of the amplitude distribution, it follows that the Nyquist sampling interval is $s^2/\Delta x$. Therefore, $\Delta x/(s^2/\Delta x) = \Delta x^2/s^2$ samples are sufficient to fully represent the amplitude distribution of light propagating through the medium. This quantity is the space-bandwidth product of the medium or the number of degrees of freedom the medium can support. The two-dimensional space-bandwidth product is $(\Delta x^2/s^2)^2$.

To show the second claim, we refer back to page 40, where we stated that most of the energy of the lth Hermite-Gaussian function is concentrated in the interval $[-\sqrt{(l+1/2)/\pi}, \sqrt{(l+1/2)/\pi}]$. Examination of higher-order Hermite-Gaussian functions further reveals that most of the energy within this interval tends to be concentrated close to the endpoints of the interval, rather than around the origin. Thus, Hermite-Gaussian functions $\psi_l(x/s)$ whose energies mostly lie within the interval $[-\Delta x/2, \Delta x/2]$ are those which satisfy $\sqrt{(l+1/2)/\pi} < \Delta x/2s$, or whose orders l are (approximately) less than $\Delta x^2/s^2$. In other words, only these first $\Delta x^2/s^2$ modes are relevant and the summations can be truncated after this mode. In two dimensions the total number of relevant modes is $(\Delta x^2/s^2)^2$.

It is satisfying that the number of relevant modes is equal to the number of spatial degrees of freedom. Thus regardless of whether we prefer to represent the amplitude distribution of light in terms of its samples or in terms of the coefficients C_{lm} of its Hermite-Gaussian expansion, we need the same number of samples or coefficients.

Since the relevant modes have orders $l, m < \Delta x^2/s^2$, it follows that the first-order approximation of σ_z given by equation 7.97 is always accurate. The first-order expansion holds when $(l + m + 1)/(\pi s^2 \sigma^2) \ll 1$. For the relevant modes $(l+m+1)/(\pi s^2\sigma^2) < (\Delta x^2/s^2)/(\pi s^2\sigma^2) = \Delta x^2/\pi\chi^2$. Since we had assumed $\Delta x \ll \chi$ for a realizable medium, it is ensured that this quantity is always $\ll 1$.

That the number of relevant modes should be equal to the space-bandwidth product can also be seen as follows: If the space-bandwidth product is N, we can sample both sides of the expansion given in equation 7.99 at N points without loss of information. This gives us N linear equations in the unknown expansion coefficients C_{lm}. If the number of modes included in the expansion is less than N, the set of equations will be overdetermined, meaning that this number of modes is not sufficient to match the original function at the sample points. If the number of modes is more than N, the set of equations will be underdetermined, meaning that there are redundant modes. When the number of modes is equal to N, the expansion coefficients can be uniquely solved to match the original function at the sample points.

An interesting and important interpretation of a quadratic graded-index medium is as the limit of a large number of positive lenses interspersed between short sections of free space. As the number of lenses becomes larger and the sections of free space become shorter, such a lens system approaches a quadratic graded-index medium. This will be discussed in chapter 9.

A brief overview of graded-index media in a Fourier optics context with useful references is Gómez-Reino, Bao, and Pérez 1996.

7.4.4 Extensions

It is not difficult to also allow for homogeneous regions with refractive indices other than unity, spherical refracting surfaces between such homogeneous regions, and spherical mirrors (Saleh and Teich 1991). We will not explicitly discuss such components, since they can be handled by simple tricks. Homogeneous regions with refractive index $n \neq 1$ can be handled by working with normalized angles, spherical refractive surfaces are treated like lenses by inserting an infinitesimal section of free space on both sides, and spherical mirrors can be handled by folding the optical axis.

Sections of free space, thin lenses, and quadratic graded-index media belong to the class of *quadratic-phase systems*, which will be further discussed in chapter 8. An example of a common component which is not in this category is a prism (Saleh and Teich 1991). The effect of a prism is similar to the effect of tilting the optical axis and can be analyzed in the same manner. In certain instances, one may also want to be able to deal with transverse displacements of the optical axis. Prisms, as well as tilts and displacements of the optical axis, will not be considered in this book.

7.4.5   Spatial filters

Spatial filters are optical components whose output $\hat{g}(x, y)$ is equal to their input $\hat{f}(x, y)$ multiplied by a complex transmittance function $\hat{h}(x, y)$:

$$\hat{g}(x, y) = \hat{h}(x, y)\hat{f}(x, y). \tag{7.108}$$

If $\hat{h}$ is real and positive, the filter is referred to as a magnitude-only filter. If $|\hat{h}| = 1$ the filter is referred to as a phase-only filter. Unless the material exhibits gain, the filter function satisfies $|\hat{h}| \leq 1$.

Such filters can be realized by thin transmissive elements whose refractive index, attenuation coefficient, or thickness is a function of (x, y).

Thin plate with variable thickness: A thin plate of homogeneous refractive index n_{pl} and variable thickness $d(x, y)$ in a medium with refractive index n will exhibit complex transmittance

$$\hat{h}(x, y) = e^{i2\pi\sigma d_0} \exp[i2\pi\sigma(n_{\mathrm{pl}}/n - 1)d(x, y)], \tag{7.109}$$

where $\sigma = n f_{oc}/c$. This transmittance is defined between two planes separated by a distance d_0 between which the variable thickness plate is completely contained. This formula is valid in the paraxial approximation and when the thickness d_0 is sufficiently small (Saleh and Teich 1991). The constant factor $\exp(i2\pi\sigma d_0)$ is often dropped. The transmittance function for a thin lens is a special case of the above formula.

Thin plate with graded index: Consider a thin slab of uniform thickness d_0 but variable refractive index $n_0[1 + \rho_{\mathrm{pl}}(x, y)]$. In this case the transmittance function is (Saleh and Teich 1991)

$$\hat{h}(x, y) = e^{i2\pi\sigma d_0} \exp[i2\pi\sigma\rho_{\mathrm{pl}}(x, y)d_0], \tag{7.110}$$

where $\sigma = n_0 f_{oc}/c$. When $\rho_{\mathrm{pl}}(x, y) = -(x^2 + y^2)/2\chi^2$ the plate behaves like a thin lens. In this case the plate can also be interpreted as a thin section of quadratic graded-index media (since $n_0[1 + \rho_{\mathrm{pl}}(x, y)]$ is an approximation to equation 7.89).

Amplitude filters can be realized by using thin plates with variable attenuation coefficients. It is also possible to have spatial filters which simultaneously have variable attenuation, thickness, and refractive index, resulting in more general complex spatial filters. In this book we will assume that the complex transmittance function is specified and we will not involve ourselves with its physical realization.

Apertures are spatial filters whose transmittance function $\hat{h}(x,y)$ takes only two values: 1 or 0. They can be physically realized simply by cutting out the desired shape in an opaque (nontransparent) material. It is assumed that the distribution of light behind the opaque regions is zero, and that the distribution of light behind the transparent (cut out) regions is equal to the incident distribution of light. Although not strictly true, this is usually a sufficiently accurate approximation, known as the Kirchhoff approximation (Goodman 1996, Lohmann 1986).

7.4.6 Fourier-domain spatial filters

A Fourier-domain filter is an optical subsystem whose output $\hat{g}(x,y)$ is equal to its input $\hat{f}(x,y)$ convolved with $\hat{h}(x,y)$:

$$\hat{g}(x,y) = \hat{h}(x,y) * *\hat{f}(x,y), \tag{7.111}$$

$$\hat{G}(\sigma_x,\sigma_y) = \hat{H}(\sigma_x,\sigma_y)\hat{F}(\sigma_x,\sigma_y). \tag{7.112}$$

If $\hat{H}$ is real and positive, the filter is referred to as a magnitude-only filter. If $|\hat{H}| = 1$ the filter is referred to as a phase-only filter. Unless the material exhibits gain, the filter function satisfies $|\hat{H}| \leq 1$. Such a filter can be realized by sandwiching a thin spatial filter between a forward and inverse Fourier transform stage (figure 7.4).

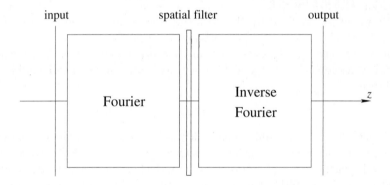

Figure 7.4.   A Fourier-domain filtering system consists of a Fourier transform followed by a spatial filter followed by an inverse Fourier transform.

Fourier transform stages: Although the Fourier transform can be realized optically in many different ways (as will be seen in chapter 9), the most common configurations are (i) a lens of focal length f sandwiched between two sections of free space of length $d = f$, (ii) a section of free space of length d sandwiched between two lenses of focal length $f = d$, and (iii) a section of quadratic graded-index media of length $(\pi/2)\chi$. Ignoring uninteresting constant phase factors which do not depend on (x,y), all of these transform an input $\hat{f}(x,y)$ into the Fourier transform $\hat{F}(\sigma_x,\sigma_y)$ with $\sigma_x = x/s^2, \sigma_y = y/s^2$:

$$\hat{F}(x/s^2, y/s^2) \propto \iint \hat{f}(x',y')e^{-i2\pi(xx'+yy')/s^2} \, dx' \, dy', \tag{7.113}$$

where s is given by $\sqrt{\lambda d} = \sqrt{\lambda f}$ for (i) and (ii), and $s = \sqrt{\chi/\sigma}$ for (iii). The first two cases can be easily demonstrated by using Fresnel's integral and the transmittance function of a thin lens (Goodman 1996). The third case is demonstrated by equation 7.101. It will be very easy to verify these results using matrix algebra after the matrix representation of such systems is introduced in chapter 8. It is also possible to rewrite equation 7.113 in the purer form

$$F(x/s, y/s) \propto \iint f(x'/s, y'/s) \, \exp\left[-i2\pi\left(\frac{x\,x'}{s\,s} + \frac{y\,y'}{s\,s}\right)\right] \frac{dx'}{s}\frac{dy'}{s}, \quad (7.114)$$

where $sF(s\sigma_x, s\sigma_y) = \hat{F}(\sigma_x, \sigma_y)$ and $s^{-1}f(x'/s, y'/s) = \hat{f}(x', y')$. This form can be directly translated to the dimensionless Fourier transform relation $F(\mu, \nu) = \iint f(u, v)\exp[-i2\pi(\mu u + \nu v)]\,du\,dv$ with $u = x'/s$, $\mu = x/s$ and similarly for v and ν.

Inverse Fourier transforms can be easily obtained by noting that the inverse Fourier transform is simply the forward Fourier transform followed by flipping the coordinate axes. Thus if we choose the output coordinates to be opposite in direction to the input coordinates, the same configuration will give us the inverse transform.

Using the Fourier and inverse Fourier transform as building blocks, it is possible to realize the desired Fourier-domain filtering system by employing a spatial filter of the form

$$\hat{H}(x/s^2, y/s^2) \quad (7.115)$$

between the Fourier and inverse Fourier blocks in figure 7.4. When this system is realized by using type (i) Fourier transform stages, it is commonly known as a $4f$ system. In dimensionless notation $H(s\sigma_x, s\sigma_y) = \hat{H}(\sigma_x, \sigma_y)$, the filter is given by $H(x/s, y/s)$.

7.4.7 General linear systems

It is also possible to optically realize arbitrary linear systems of the form

$$\hat{g}(x, y) = \iint \hat{h}(x, y; x', y')\hat{f}(x', y')\,dx'\,dy' \quad (7.116)$$

by using systems such as matrix-vector product architectures (Goodman 1996) or a class of systems which may be collectively referred to as multifacet architectures (Mendlovic and Ozaktas 1993c, Ozaktas, Brenner, and Lohmann 1993, Ozaktas and Mendlovic 1993c). However, the realization of such systems are usually considered inefficient or costly since optical components with space-bandwidth product $O(N^2)$ are needed to realize systems for signals with space-bandwidth product N.

7.4.8 Spherical reference surfaces

Although not a component in a physical sense, here we also discuss spherical reference surfaces for future reference. Usually we specify the amplitude distribution of light on a planar reference surface. Sometimes, however, it is more convenient to specify

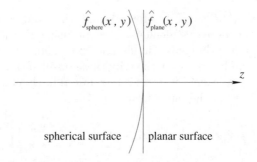

Figure 7.5. Spherical and planar reference surfaces. $R > 0$ as drawn.

the amplitude distribution on a spherical reference surface of radius R. Referring to figure 7.5, the relation between the distributions with respect to the planar and spherical surfaces is

$$\hat{f}_{\text{plane}}(x, y) = \hat{f}_{\text{sphere}}(x, y) \exp[i\pi(x^2 + y^2)/\lambda R]. \tag{7.117}$$

In the spherical case, the coordinates x, y are those perpendicularly dropping from the sphere onto the plane. The above relation is valid for small curvatures, so that the surface is "thin" in the same sense that a thin lens is thin. The reader can verify this relation easily by considering the expressions for plane waves and diverging/converging spherical waves.

7.4.9 Remarks

We have ignored attenuations and reflections from the components discussed, concentrating mostly on how they modify the phase of the incident optical wave. Furthermore, we have focused our attention to quadratic-phase systems for which higher-order dependences of the phase on x, y can be neglected. (Linear terms— corresponding to shifts and tilts—are also not included for simplicity, though most results can be easily generalized to include them as well.) We will see in chapter 8 that quadratic-phase systems mathematically correspond to linear canonical transforms, and they play a central role in the first-order study of optical systems.

7.5   Geometrical optics

In geometrical optics, light is represented by light *rays*, which are in general curvilinear paths along which light energy travels. A distribution of light can be represented by a bundle of rays whose trajectories (and sometimes intensities) are specified. Since we are considering centered optical systems, we will usually represent the distribution of light over a given plane $z = $ constant by specifying the points (x, y) and angles (θ_x, θ_y) at which the rays intersect this plane. θ_x is the angle the ray makes with the y-z plane, and θ_y is the angle the ray makes with the x-z plane. In other words, θ_x is the complement to $\pi/2$ of the angle the ray makes with the x axis, and θ_y is the complement to $\pi/2$ of the angle the ray makes with the y axis. The signs of the angles are the same as the signs of the slopes of the rays. In order to make the correspondence with wave optics more transparent, we will often choose to work with $\sigma_x \equiv \sin\theta_x/\lambda$ and $\sigma_y \equiv \sin\theta_y/\lambda$ instead of the angles themselves, where λ is the wavelength of light in the medium of propagation. From a purely geometrical-optical perspective, σ_x and σ_y may be interpreted merely as normalized angles. However, it is worth keeping in mind that σ_x, σ_y can also be interpreted as the transverse spatial frequencies of a plane wave propagating in the direction of the ray (figure 7.3 and equations 7.30 and 7.31).

Optical components are characterized by how they map a ray incident on their input plane to a ray exiting from their output plane. We will again restrict our attention to first-order centered systems for which (i) the paraxial (small-angle) approximation can be employed ($\theta_x, \theta_y \ll 1$), and (ii) in which any lenses are thin lenses. In the paraxial approximation, angles are taken equal to their sines/tangents and the slopes of the rays so that $\sigma_x \approx \theta_x/\lambda$, $\sigma_y \approx \theta_y/\lambda$. In a first-order system, the parameters (x, y) and (θ_x, θ_y) characterizing a ray at the output of a system are related to the corresponding parameters characterizing the ray at the input through a linear relation; higher-order dependences are neglected. These approximations are in direct correspondence with those made in section 7.3, which essentially amounted to neglecting terms beyond the quadratic *in the phase* of integral kernels.

The sections on geometrical optics proceed more or less in parallel with the sections on wave optics.

7.5.1 The ray equation

The ray equation governs the trajectory of rays $\mathbf{r}(s) \equiv (x(s), y(s), z(s))$ traveling in an inhomogeneous medium (Saleh and Teich 1991):

$$\frac{d}{ds}\left(n\frac{d\mathbf{r}}{ds}\right) = \nabla n, \qquad (7.118)$$

where ∇n is the gradient of the refractive index distribution $n(\mathbf{r})$. Each value of the parameter s corresponds to a point along the ray. Solution of the above equation subject to specified boundary conditions gives a set of trajectories representing a bundle of rays. The boundary conditions may take the form of the positions and directions of a bundle of rays crossing a given plane. (The parameter s used here in accordance with widespread convention should not be confused with the scale parameter introduced in section 7.2.)

In a homogeneous medium it is easy to show that solutions of the ray equation take the form of linear trajectories; the paths of light are straight lines. Two commonly encountered bundles are those that correspond to a plane wave and a spherical wave. In the former, the bundle consists of a set of rays which are all parallel to each other. In the latter, the rays emanate from a common origin and diverge outwards in all directions. More general bundles of rays correspond to more general waves.

If we know the angles (θ_x, θ_y) that the ray bundle corresponding to a plane wave makes with a particular plane $z = z_1$, then we know that the ray crossing this plane at (x, y) will cross a second plane $z = z_2 = z_1 + d$ at the point $(x + \theta_x d, y + \theta_y d)$, still making the same angles. We have assumed small angles so that $\tan\theta_x \approx \sin\theta_x \approx \theta_x$, and likewise for θ_y.

Turning our attention to the ray bundle corresponding to a spherical wave originating from the origin $(0,0,0)$, it is possible to show that the ray crossing a plane

$z = $ constant, at the point (x, y), will be making an angle of $(\theta_x, \theta_y) = (x/z, y/z)$ with that plane (again in the paraxial approximation).

Let us consider two homogeneous half-spaces which meet at a plane $z = $ constant, such that the medium to the left has a refractive index n_{left} and the medium to the right has an index n_{right}. Then the following relations hold for the angles characterizing the ray coming from the left and the ray leaving towards the right:

$$n_{\text{left}} \sin \theta_{x\,\text{left}} = n_{\text{right}} \sin \theta_{x\,\text{right}},$$
$$n_{\text{left}} \sin \theta_{y\,\text{left}} = n_{\text{right}} \sin \theta_{y\,\text{right}}. \tag{7.119}$$

These relationships, known as Snell's law, cover both the condition that the incoming and outgoing rays and the normal to the surface all lie in the same plane, and the condition relating the angles made with the normal. Snell's law takes a particularly simple form when stated in terms of the normalized angles σ_x, σ_y:

$$\sigma_{x\,\text{left}} = \sigma_{x\,\text{right}},$$
$$\sigma_{y\,\text{left}} = \sigma_{y\,\text{right}}, \tag{7.120}$$

which simply states that (σ_x, σ_y) is conserved at the boundary. In the paraxial approximation we obtain

$$n_{\text{left}} \theta_{x\,\text{left}} = n_{\text{right}} \theta_{x\,\text{right}},$$
$$n_{\text{left}} \theta_{y\,\text{left}} = n_{\text{right}} \theta_{y\,\text{right}}. \tag{7.121}$$

Although the law of refraction at refractive index discontinuities is strongly associated with the name of Snell (and sometimes Descartes), the law appeared in the work of Ibn Sahl some 650 years before Snell. For this reason it has been suggested that it be referred to as the Ibn Sahl law (Wolf and Krötzsch 1995).

7.5.2   Fermat's principle and the eikonal equation

An important concept in geometrical optics is the *optical path length* along a ray. The optical path length from point A to point B is defined as the line integral

$$\int_A^B n(\mathbf{r})\, ds, \tag{7.122}$$

where ds is the differential element along the path of integration (Saleh and Teich 1991). *Fermat's principle* is a statement of the laws of geometrical optics. According to this principle, among all the possible paths connecting the points A and B, only the paths whose optical path length variations with respect to small deviations are zero, correspond to actual light rays. By "small deviations" we mean perturbations of the path in question such that the perturbed path still lies in a narrow tubular neighborhood of the path. The "optical path length variation" is simply the difference between the optical path length of the perturbed path and the original path. Such paths whose variations are zero are known to be either paths with minimum or maximum optical path length, or paths for which the optical path length exhibits an "inflection." Since a minimum is the most commonly encountered case and since the optical path length is proportional to the time it takes light to travel along the path, this principle is also known as the *principle of least time* (Saleh and Teich 1991). Sometimes there are several paths for which the variations are zero. An important special case is that of point imaging when a whole bundle of rays emanating from point A all arrive at point B, in which case all the paths have the same optical path length.

The ray equation can be derived from Fermat's principle by using the calculus of variations (Marcuse 1982, Born and Wolf 1980). (The ray equation is the so-called *Euler equation* of the variational problem.) This variational principle can also be interpreted in terms of wave-optical concepts, providing significant insight on the relationship between rays and waves. The path which is the actual optical ray corresponds to the path along which the wave contributions constructively add up, whereas along other paths which are not actual rays the contributions destructively interfere. Constructive interference occurs where the phase varies slowly, particularly where the variation of the phase is zero. Since the optical path length is associated with the phase, this directly corresponds to the variation of the optical path length being zero. These notions find their mathematical expression in the stationary-phase integral (section 2.10.3). An excellent discussion of these issues may be found in Lohmann 1986. Further discussion of these relationships and Fermat's principle is beyond the scope of this book. For this we refer the reader to Lohmann 1986, Marcuse 1982, and Born and Wolf 1980.

Another important concept is the *eikonal*. The eikonal $\hat{S}(\mathbf{r})$ is a function of position such that (i) its equilevel surfaces are everywhere orthogonal to the optical rays, and (ii) the optical path lengths along all rays from one equilevel surface to another are equal. The rays lie along the gradient of $\hat{S}(\mathbf{r})$. An alternative statement of the laws of geometrical optics is the eikonal equation (Saleh and Teich 1991):

$$|\nabla \hat{S}(\mathbf{r})|^2 = n^2(\mathbf{r}). \tag{7.123}$$

The optical path length along a ray between points A and B is simply equal to the difference between the value of the eikonal at these points (Saleh and Teich 1991):

$$\int_A^B n(\mathbf{r})\, ds = \int_A^B |\nabla \hat{S}(\mathbf{r})|\, ds = \hat{S}(B) - \hat{S}(A). \tag{7.124}$$

The eikonal equation is equivalent to Fermat's principle, and the ray equation can be derived from the eikonal equation as well (Marcuse 1982, Born and Wolf 1980).

The equilevel surfaces of the eikonal are sometimes referred to as *geometrical wavefronts*. They are often close approximations to the physical wavefronts, but deviate to a greater extent in those regions where geometrical optics does not provide a satisfactory description of the behavior of light (such as in those regions where light is tightly focused).

It is possible to substitute an expression of the form $\hat{f}(\mathbf{r}) = \hat{A}(\mathbf{r}) \exp[i2\pi(f_{oc}/c)\hat{S}(\mathbf{r})]$ in the Helmholtz equation 7.23 and show that in the limit $f_{oc} \to \infty$ one obtains the eikonal equation. This supports the association between the phase and the optical path length. This classic derivation, which the reader may find in many texts such as Saleh and Teich 1991 and Marcuse 1982, is often used to motivate the fact that geometrical optics is a limiting case of wave optics, which holds when the wavelength is small.

7.5.3 Hamilton's equations

We now briefly discuss the Hamiltonian formulation of geometrical optics (Marcuse 1982). We first define the Hamiltonian $\hat{H}$ of a system with refractive index distribution $n(x, y, z)$ as

$$\hat{H}(x, y, \sigma_x, \sigma_y; z) \equiv -\sqrt{n^2(x, y, z)f_{oc}^2/c^2 - \sigma_x^2 - \sigma_y^2}. \qquad (7.125)$$

As before, $\sigma_x \equiv \sin\theta_x/\lambda$ and $\sigma_y \equiv \sin\theta_y/\lambda$, where $\lambda = c/n(x, y, z)f_{oc}$. We can now write the celebrated Hamilton equations:

$$\frac{dx}{dz} = \frac{\partial\hat{H}}{\partial\sigma_x}, \qquad \frac{dy}{dz} = \frac{\partial\hat{H}}{\partial\sigma_y}, \qquad (7.126)$$

$$\frac{d\sigma_x}{dz} = -\frac{\partial\hat{H}}{\partial x}, \qquad \frac{d\sigma_y}{dz} = -\frac{\partial\hat{H}}{\partial y}. \qquad (7.127)$$

It is possible to show that Hamilton's equations are equivalent to the ray equation (Marcuse 1982, Goldstein 1980), so that they also constitute an alternative statement of the laws of geometrical optics. Here x, y, σ_x, σ_y are treated as functions of z and give us the position and angles of a ray propagating through the system in the positive z direction. Noting that $dx/ds = \sin\theta_x$ and $dy/ds = \sin\theta_y$, where $ds = \sqrt{dx^2 + dy^2 + dz^2}$ is the differential element along the ray, it is possible to write σ_x and σ_y in terms of the derivatives of x and y with respect to z as follows:

$$\sigma_x = \frac{\sin\theta_x}{\lambda} = \frac{f_{oc}}{c}n(x, y, z)\sin\theta_x = \frac{f_{oc}n(x, y, z)}{c}\frac{dx}{ds}$$

$$= \frac{f_{oc}}{c}\frac{n(x, y, z)dx/dz}{\sqrt{1 + (dx/dz)^2 + (dy/dz)^2}},$$

$$\sigma_y = \frac{\sin\theta_y}{\lambda} = \frac{f_{oc}}{c}n(x, y, z)\sin\theta_y = \frac{f_{oc}n(x, y, z)}{c}\frac{dy}{ds}$$

$$= \frac{f_{oc}}{c}\frac{n(x, y, z)dy/dz}{\sqrt{1 + (dx/dz)^2 + (dy/dz)^2}}, \qquad (7.128)$$

These equations allow us to express the Hamiltonian in terms of x, y and their derivatives rather than x, y and σ_x, σ_y, a form which is sometimes preferred.

The total derivative of the Hamiltonian with respect to z can be written as

$$\frac{d\hat{H}}{dz} = \frac{\partial\hat{H}}{\partial x}\frac{dx}{dz} + \frac{\partial\hat{H}}{\partial y}\frac{dy}{dz} + \frac{\partial\hat{H}}{\partial\sigma_x}\frac{d\sigma_x}{dz} + \frac{\partial\hat{H}}{\partial\sigma_y}\frac{d\sigma_y}{dz} + \frac{\partial\hat{H}}{\partial z}. \qquad (7.129)$$

By using Hamilton's equations, it immediately follows that the total derivative $d\hat{H}/dz$ is equal to $\partial\hat{H}/\partial z$ since the other terms cancel out. If $n(x, y, z)$ does not depend on z then $d\hat{H}/dz = \partial\hat{H}/\partial z = 0$, which means that the Hamiltonian remains the same along a ray $(x(z), y(z), \sigma_x(z), \sigma_y(z))$.

The results obtained by solving Hamilton's equations will be the same as those obtained by solving the ray equation. As instructive as it is, the analogy between the Hamiltonian formulation in optics and the Hamiltonian formulation in mechanics will not be pursued in this book. We refer interested readers to Goldstein 1980, Sekiguchi and Wolf 1987, and the other advanced books on mathematical optics that are referred to at the end of chapter 8. However, we note that point particles are to the wave functions of quantum mechanics as rays are to optical waves. Also, the important Hamilton-Jacobi partial differential equation associated with the Hamiltonian given in equation 7.125 is nothing but the eikonal equation 7.123 (Marcuse 1982).

In the paraxial approximation we take $\sin\theta_x = \theta_x$, $\sin\theta_y = \theta_y$ and $dx/ds = dx/dz$, $dy/ds = dy/dz$ so that

$$
\sigma_x = \frac{\theta_x}{\lambda} = \frac{f_{oc}}{c} n(x,y,z)\theta_x = \frac{f_{oc}}{c} n(x,y,z)\frac{dx}{dz},
$$

$$
\sigma_y = \frac{\theta_y}{\lambda} = \frac{f_{oc}}{c} n(x,y,z)\theta_y = \frac{f_{oc}}{c} n(x,y,z)\frac{dy}{dz}. \tag{7.130}
$$

In this case the Hamiltonian simplifies to

$$
\hat{H}(x,y,\sigma_x,\sigma_y;z) = \frac{(\sigma_x^2 + \sigma_y^2)}{2n f_{oc}/c} - n f_{oc}/c \approx \frac{(\sigma_x^2 + \sigma_y^2)}{2n_0 f_{oc}/c} + \Delta n f_{oc}/c - n_0 f_{oc}/c, \tag{7.131}
$$

where we have let $n = n_0 - \Delta n$ with the assumption that $\Delta n \ll n_0$ and we have replaced $n \approx n_0$ in the denominator (Marcuse 1982). The constant term $-n_0 f_{oc}/c$ may be dropped since it will disappear in Hamilton's equations anyway.

Although we have not provided a derivation of Hamilton's equations from the ray equation, we mention a crude but simple and instructive derivation which allows us to see their consistency in the paraxial case. Referring ahead to equation 7.136 and using equation 7.130, we can obtain $d\sigma_x/dz \approx (f_{oc}/c)\partial n/\partial x$, which in turn is equal to $-\partial\hat{H}/\partial x$ evaluated from equation 7.131. The other equation $\partial\hat{H}/\partial\sigma_x = dx/dz$ is also easily seen to carry the same information as equation 7.130. (The equations for y are of course identical.)

Finally, we show that in their paraxial form, Hamilton's equations have a particularly simple interpretation in terms of Snell's law. (The rest of this section can be omitted if desired.) First, consider the equation $dx/dz = \partial\hat{H}/\partial\sigma_x$. The right-hand side is readily evaluated from the paraxial Hamiltonian as $\sigma_x/(n f_{oc}/c) = \sigma_x\lambda = \theta_x$ so that we are left with $dx/dz = \theta_x$: the rate of change of x is simply given by θ_x, which in the paraxial approximation also equals the slope of the ray. Second, let us consider the equation $d\sigma_x/dz = -\partial\hat{H}/\partial x$. We will show that this equation corresponds to Snell's law. The right-hand side evaluates to $(f_{oc}/c)\partial n/\partial x = -(f_{oc}/c)\partial\Delta n/\partial x$. Thus the incremental change in σ_x can be written as

$$
\sigma_x(z + dz) = \sigma_x(z) + (f_{oc}/c)\frac{\partial n(x,z)}{\partial x} dz. \tag{7.132}
$$

Using $\sigma_x = \theta_x/\lambda \approx n_0\theta_x(f_{oc}/c)$ to change to angles, and noting that the axial and transverse increments in the ray position dz and dx are related by $dx = \theta_x dz$, we get an angular deflection given by

$$
d\theta_x \equiv \theta_x(x + dx) - \theta_x(x) = \frac{1}{\theta_x}\frac{dn}{n_0}, \tag{7.133}
$$

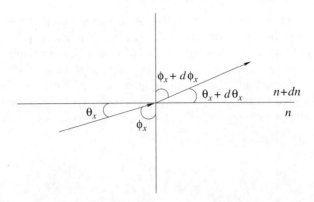

Figure 7.6. Deflection of a ray by an infinitesimal refractive index change.

where $dn = (\partial n/\partial x)dx$. Now let us write Snell's law for an interface parallel to the optical axis such that the refractive index on one side is n and that on the other side is $n + dn$:

$$\sin[\phi_x(x + dx)]\,(n + dn) = \sin[\phi_x(x)]\,n, \qquad (7.134)$$

where the angle $\phi_x(x)$ complements the angle θ_x to $\pi/2$ and is thus a large angle (figure 7.6). Writing $\phi_x(x + dx) = \phi_x(x) + d\phi_x$ and expanding the sine of a sum, it follows that equation 7.134 implies an angular deflection of

$$d\phi_x \equiv \phi_x(x + dx) - \phi_x(x) = -\tan[\phi_x(x)]\,\frac{dn}{n} \approx -\tan[\phi_x(x)]\,\frac{dn}{n_0}. \qquad (7.135)$$

Since the angular deflections $d\theta_x$ and $-d\phi_x$ are equal, and since $\tan\phi_x = \cot\theta_x = 1/\tan\theta_x \approx 1/\theta_x$, it follows that the second Hamilton equation corresponds to Snell's law. We also note that the derivative of the refractive index with respect to z does not come into play since for paraxial angles, the deflective force of changes with respect to x is dominant.

Chapter 3 of Marcuse 1982 is particularly recommended for a very accessible discussion of these and related topics. Further references are given at the end of chapter 8.

7.6 Geometrical-optical characterization of optical components

7.6.1 Sections of free space

If we restrict ourselves to paraxial rays which make small angles with the z axis, then $ds \approx dz$ and we can write the ray equation as follows (Saleh and Teich 1991):

$$\frac{d}{dz}\left(n\frac{dx}{dz}\right) \approx \frac{\partial n}{\partial x}, \qquad \frac{d}{dz}\left(n\frac{dy}{dz}\right) \approx \frac{\partial n}{\partial y}. \qquad (7.136)$$

In a homogeneous medium where n is constant we have

$$\frac{d^2 x}{dz^2} = 0, \qquad \frac{d^2 y}{dz^2} = 0, \qquad (7.137)$$

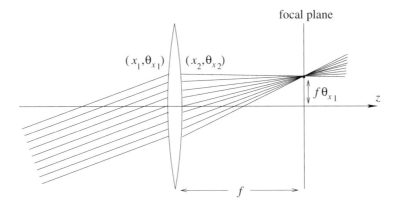

Figure 7.7. In the paraxial approximation, parallel rays incident on a thin lens intersect at a single point lying on the focal plane.

which imply that $x(z)$ and $y(z)$ both increase linearly with z, corresponding to the fact that the rays are straight lines. If a ray intercepts the plane $z = z_1$ at (x_1, y_1) making angles $(\theta_{x1}, \theta_{y1})$ with the y-z and x-z planes respectively, then this ray will intercept the plane $z = z_2 = z_1 + d$ at (x_2, y_2) making angles $(\theta_{x2}, \theta_{y2})$, where

$$x_2 = x_1 + \theta_{x1}d, \qquad y_2 = y_1 + \theta_{y1}d, \tag{7.138}$$
$$\theta_{x2} = \theta_{x1}, \qquad \theta_{y2} = \theta_{y1}. \tag{7.139}$$

These equations can also be easily derived from the paraxial form of Hamilton's equations.

7.6.2 Thin lenses

Thin spherical lenses are characterized by their focal length f (equation 7.85), which is the distance from the lens along the optical axis at which an incident bundle of parallel rays intersect at a point (figure 7.7). The thickness of a thin lens is assumed to be sufficiently small so that the point (x_2, y_2) from which a ray leaves the lens is equal to the point (x_1, y_1) at which the ray is incident on the lens. Employing Snell's law twice, it is possible to show—under the paraxial approximation—that for a thin lens the angles $(\theta_{x2}, \theta_{y2})$ of a ray leaving the lens are related to the angles $(\theta_{x1}, \theta_{y1})$ of the ray incident on the lens through relations which depend only on the focal length of the lens (and not on the separate surface curvatures). Taken together, these results provide the geometrical-optical characterization of a thin lens in the paraxial approximation:

$$x_2 = x_1, \qquad y_2 = y_1, \tag{7.140}$$
$$\theta_{x2} = \theta_{x1} - \frac{x_1}{f}, \qquad \theta_{y2} = \theta_{y1} - \frac{y_1}{f}. \tag{7.141}$$

7.6.3   Quadratic graded-index media

The refractive index distribution of a quadratic graded-index medium was given in equation 7.89. Assuming $\chi_x = \chi_y = \chi$, substituting in equation 7.136, and noting that the refractive index does not depend on z, we obtain

$$\frac{d^2x}{dz^2} = \frac{1}{n^2}\frac{-xn_0^2}{\chi^2}, \qquad \frac{d^2y}{dz^2} = \frac{1}{n^2}\frac{-yn_0^2}{\chi^2}. \tag{7.142}$$

Usually χ is large so that the refractive index distribution $n(x,y)$ appearing in the denominators can be replaced by n_0 within the extent of the medium (Saleh and Teich 1991), leading to

$$\frac{d^2x}{dz^2} = -\frac{x}{\chi^2}, \qquad \frac{d^2y}{dz^2} = -\frac{y}{\chi^2}. \tag{7.143}$$

These equations can also be derived from the paraxial form of Hamilton's equations with the Hamiltonian given in equation 7.131. For instance, we can use $dx/dz = \partial\hat{H}/\partial\sigma_x = c\sigma_x/n_0 f_{oc}$ and $d\sigma_x/dz = -\partial\hat{H}/\partial x = (f_{oc}/c)\partial n/\partial x$, leading to $n_0 d^2x/dz^2 = \partial n/\partial x$. Now, using equation 7.89 we again obtain equation 7.143.

Equation 7.143 has oscillatory harmonic solutions $x(z)$ and $y(z)$. The angles of the ray are given by the derivatives of the position of the ray: $\theta_x(z) = dx(z)/dz$, $\theta_y(z) = dy(z)/dz$. Making use of this fact, it is possible to show that if a ray is incident onto a quadratic graded-index medium at the plane $z = z_1$ at (x_1, y_1) making angles $(\theta_{x1}, \theta_{x2})$, then at the plane $z = z_2 = z_1 + d$, the position $(x_2, y_2) = (x(z_2), y(z_2))$ and angles $(\theta_{x2}, \theta_{y2}) = (\theta_x(z_2), \theta_y(z_2))$ will be given by

$$x_2 = x_1\cos(d/\chi) + \theta_{x1}\chi\sin(d/\chi), \tag{7.144}$$

$$y_2 = y_1\cos(d/\chi) + \theta_{y1}\chi\sin(d/\chi), \tag{7.145}$$

$$\theta_{x2} = -\frac{x_1}{\chi}\sin(d/\chi) + \theta_{x1}\cos(d/\chi), \tag{7.146}$$

$$\theta_{y2} = -\frac{y_1}{\chi}\sin(d/\chi) + \theta_{y1}\cos(d/\chi). \tag{7.147}$$

The period of oscillation is $2\pi\chi$.

7.6.4   Extensions

Homogeneous regions with refractive indices other than unity, spherical refracting surfaces between such homogeneous regions, and spherical mirrors are easily handled with Snell's law and the law of reflection (Saleh and Teich 1991). Again we will not explicitly discuss such components, since they can be handled by simple tricks. Homogeneous regions with refractive index $n \neq 1$ can be handled by working with $\sigma_x = \theta_x/\lambda$ and $\sigma_y = \theta_y/\lambda$ instead of the angles themselves, where λ is the wavelength in the homogeneous medium. Spherical refractive surfaces can be treated like lenses by inserting an infinitesimal section of free space on both sides, and spherical mirrors can be handled by folding the optical axis.

All of the components we have treated so far were characterized by linear relations between the output position and angles and input position and angles:

$$x_2 = \hat{A}_x x_1 + \hat{B}_x \frac{\theta_{x1}}{\lambda}, \qquad y_2 = \hat{A}_y y_1 + \hat{B}_y \frac{\theta_{y1}}{\lambda}, \qquad (7.148)$$

$$\frac{\theta_{x2}}{\lambda} = \hat{C}_x x_1 + \hat{D}_x \frac{\theta_{x1}}{\lambda}, \qquad \frac{\theta_{y2}}{\lambda} = \hat{C}_y y_1 + \hat{D}_y \frac{\theta_{y1}}{\lambda}, \qquad (7.149)$$

where the parameters $\hat{A}, \hat{B}, \hat{C}, \hat{D}$ with subscripts x, y are constants. An example of a common component which cannot be characterized by these relations is a prism (Saleh and Teich 1991), which would require the addition of constant terms to the right-hand sides of the above equations. The effect of a prism is similar to the effect of tilting the optical axis and can be analyzed in the same manner. In certain instances, one might also want to be able to deal with transverse displacements of the optical axis, which also require the addition of constant terms on the right-hand sides of the above equations. Prisms, as well as tilts and displacements of the optical axis, will not be considered in this book.

7.6.5   Spatial filters

The most common spatial filters discussed in terms of geometrical optics are opaque apertures. Rays are completely blocked when they intercept the nontransparent parts of an aperture but pass unhindered through the transparent parts.

Systems involving more general spatial filters are usually analyzed with wave optics; we will exclusively do so in this book. Nevertheless, it is possible to assign an amplitude or intensity to each ray in a bundle and thus represent quite general distributions of light. Attenuating or amplifying filters will then modify this amplitude or intensity. Phase filters (including thin plates with variable thickness or graded index) have the effect of bending the rays by an angle determined by the local partial derivatives of the phase function. However, such approaches are less frequently employed.

7.6.6   Fourier-domain spatial filters

Similar comments apply to Fourier-domain spatial filters, which are likewise usually analyzed with wave optics. However, it is instructive to examine the behavior of rays in the three Fourier transform stages discussed on page 247 (figure 7.8). It is seen in all parts of the figure that parallel rays converge to a point on the Fourier plane, corresponding to the fact that the Fourier transform of a harmonic function is a delta function. Rays emanating from a point on the input plane appear in the Fourier plane as a bundle of parallel rays, whose angle with respect to the optical axis is determined by the position of the point. The position x_F and angle θ_{xF} of a ray at the Fourier plane are related to those at the input plane through the relations:

$$x_F = s^2 \sigma_{x\,\text{in}}, \qquad (7.150)$$

$$\sigma_{xF} = -\frac{x_{\text{in}}}{s^2}, \qquad (7.151)$$

where $\sigma_x = \theta_x / \lambda$. This equation is the geometrical-optical analog of equation 7.113. It is an instructive exercise to verify the above relations using simple geometry or

(i)

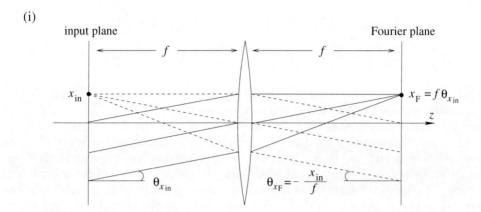

(ii)

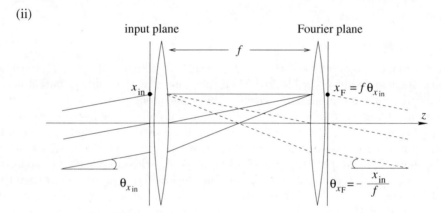

(iii)

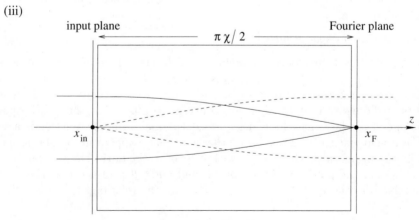

Figure 7.8. Fourier transform stages introduced on page 247; the lenses have focal length f: (i) and (ii) $s = \sqrt{\lambda f}$, (iii) $s = \sqrt{\chi/\sigma}$.

the input-output relations for each component. The intuition gained by carefully examining figure 7.8 from both wave-optical and geometrical-optical perspectives goes a long way towards understanding the relationship between these two descriptions of light.

7.6.7 General linear systems

Once again such systems will be exclusively dealt with in wave-optical terms.

7.6.8 Spherical reference surfaces

In certain cases it may be desirable to specify the intercepts and angles of rays with respect to spherical reference surfaces. It is possible to write relations between the ray parameters with respect to the planar and spherical reference surfaces shown in figure 7.5. However, these are not presented since we will not be making use of such results.

7.6.9 Remarks

We have ignored attenuations and reflections from the components discussed, concentrating only on how they bend the rays (which corresponds to modification of the phase of the wave field). Furthermore, we are focusing our attention on systems for which the ray position and angles at the output are linearly related to those at the input, assuming that higher-order dependences can be neglected. This corresponds to neglecting higher than quadratic terms in the phase in wave optics. (Constant terms— corresponding to shifts and tilts—are also not included for simplicity, though most results can be easily generalized to include them as well.) We will see in chapter 8 that systems represented by equations 7.148 and 7.149 play a central role in the first-order study of optical systems.

7.7 Partially coherent light

In sections 7.3 and 7.4 we assumed that the distribution of light at any given plane could be represented by deterministic functions $\hat{f}(x, y)$. In some cases, either due to the intrinsic random nature of light sources or the random nature of the media which they travel through, it is more appropriate to represent light as random processes.

In this book we generally assume that the light is temporally stationary and ergodic, and also *quasi-monochromatic*, which means it can effectively be assumed to be *temporally coherent*. The temporal power spectral density of such light at any given point in space is confined to a narrow band of temporal frequencies. (This assumption requires that the path length differences in the optical system are much smaller than the *coherence length* of the light.) For definitions and further discussion of these concepts, we refer the reader to Goodman 1985 and Saleh and Teich 1991.

Light which is quasi-monochromatic may nevertheless exhibit different degrees of *spatial coherence*. Light with partial spatial coherence is characterized by its autocorrelation $\hat{R}_{\hat{f}\hat{f}}$, better known in optics as its *mutual intensity*, defined by

$$\hat{R}_{\hat{f}\hat{f}}(x_1, y_1; x_2, y_2) \equiv \langle \hat{f}(x_1, y_1)\hat{f}^*(x_2, y_2) \rangle, \tag{7.152}$$

where the angle brackets denote an ensemble average. When $x_1 = x_2 = x$ and $y_1 = y_2 = y$, we obtain the average intensity

$$\hat{R}_{\hat{f}\hat{f}}(x, y; x, y) = \langle |\hat{f}(x, y)|^2 \rangle = \hat{I}_{\hat{f}}(x, y). \tag{7.153}$$

When the light is *spatially coherent*, the average is redundant and $\hat{R}_{\hat{f}\hat{f}}(x_1, y_1; x_2, y_2) = \hat{f}(x_1, y_1)\hat{f}^*(x_2, y_2)$. This corresponds to the deterministic case, where we usually simply work with the amplitude $\hat{f}(x, y)$ rather than the mutual intensity. Light may be treated as spatially coherent if the function $\hat{R}_{\hat{f}\hat{f}}(x_1, y_1; x_2, y_2)$ does not become small for the range of values x_1, y_1, x_2, y_2 can take within the aperture of the optical system. When the light is *spatially incoherent*, $\hat{R}_{\hat{f}\hat{f}}(x_1, y_1; x_2, y_2)$ behaves like the delta function $\delta(x_1 - x_2, y_1 - y_2)$. This corresponds to the case where distinct spatial points, even very close ones, do not exhibit any correlation with each other. Light may be treated as spatially incoherent if the spatial extent over which the value of $\hat{R}_{\hat{f}\hat{f}}(x_1, y_1; x_2, y_2)$ is substantially greater than zero is smaller than the resolution of the optical system.

The definitions and results presented in section 2.8 on random processes directly apply and can be used, for instance, to determine the mutual intensity $\hat{R}_{\hat{g}\hat{g}}$ at the output of an optical system characterized by the kernel $\hat{h}(x, y; x', y')$, in terms of the mutual intensity $\hat{R}_{\hat{f}\hat{f}}$ at its input:

$$\hat{R}_{\hat{g}\hat{g}}(x_1, y_1; x_2, y_2) = \int\int\int\int \hat{R}_{\hat{f}\hat{f}}(x_1', y_1'; x_2', y_2')$$
$$\times \hat{h}(x_1, y_1; x_1', y_1')\hat{h}^*(x_2, y_2; x_2', y_2') \, dx_1' \, dy_1' \, dx_2' \, dy_2'. \tag{7.154}$$

If the system is a thin spatial filter with $\hat{h}(x, y; x', y') = \hat{h}(x, y)\delta(x - x', y - y')$, then $R_{\hat{g}\hat{g}}(x_1, y_1; x_2, y_2) = \hat{h}(x_1, y_1)\hat{h}^*(x_2, y_2)R_{\hat{f}\hat{f}}(x_1, y_1; x_2, y_2)$, and $\hat{I}_{\hat{g}}(x, y) = |\hat{h}(x, y)|^2 \hat{I}_{\hat{f}}(x, y)$. When the light is spatially coherent, equation 7.154 simply reduces to a duplicated form of the coherent relation

$$\hat{g}(x, y) = \int\int \hat{h}(x, y; x', y')\hat{f}(x', y') \, dx' \, dy'. \tag{7.155}$$

On the other hand, when the light is spatially incoherent, equation 7.154 leads to the following relation in terms of the intensities:

$$\hat{I}_{\hat{g}}(x, y) = \kappa \int\int |\hat{h}(x, y; x', y')|^2 \hat{I}_{\hat{f}}(x', y') \, dx' \, dy', \tag{7.156}$$

where κ is a constant (Goodman 1985:206, Saleh and Teich 1991:368).

7.8 Fourier optical systems

The purpose of this chapter has been to present the basic tools needed to analyze *Fourier optical systems*. In this book, this term refers to centered optical systems consisting of arbitrary concatenations of sections of free space in the Fresnel approximation, thin spherical lenses, sections of quadratic graded-index media, and thin spatial filters. Although more general systems can also be analyzed in terms of Fourier transforms and linear systems theory, such systems will not be dealt with in this book. We will see in chapter 8 that centered systems consisting of arbitrary concatenations of sections of free space, thin lenses, and sections of graded-index media constitute the class of *first-order optical systems* or *quadratic-phase systems* (Bastiaans 1979a). Thus Fourier optical systems consist of thin spatial filters sandwiched between any number of first-order optical systems.

The key results that will be most frequently used are that describing the effect of a section of free space (equation 7.76), that describing the effect of a thin lens (equation 7.80), that describing the effect of quadratic graded-index media (equation 7.100), and that describing the effect of a thin spatial filter (equation 7.108). (The relation describing graded-index media will be seen in chapter 9 to reduce to a simple fractional Fourier transform relation.) Using these results consecutively, it is possible to analyze any first-order or Fourier optical system and obtain either the overall input-output relation or the distribution of light at any desired plane.

Although we will not present the explicit derivations, here we will list the overall input-output relations for a number of important elementary systems. (The derivations are much simplified by the matrix formalism to be discussed in chapter 8.)

First, we assume that a spatial filter with amplitude transmittance $\hat{f}(x, y)$ is situated at $z = 0$ and illuminated with a unit-amplitude plane wave. Then it is possible to show that the amplitude distribution $\hat{g}(x, y)$ at the plane $z = d$ when $d \to \infty$ is given by (Saleh and Teich 1991)

$$\hat{g}(x, y) = \frac{e^{i2\pi\sigma d}}{i\lambda d} e^{i\pi(x^2 + y^2)/\lambda d} \hat{F}\left(\frac{x}{\lambda d}, \frac{y}{\lambda d}\right), \qquad (7.157)$$

where $\hat{F}$ represents the Fourier transform of $\hat{f}$.

Another important system is the so-called $2f$ system, which consists of a lens of focal length f sandwiched between two sections of free space of length f each. This system was discussed in section 7.4.6, where it was shown that it acts as a Fourier transformer. Yet another important system is the $4f$ system, also discussed in section 7.4.6. This system can be used to convolve an input light distribution with a desired function, or in other words to implement a desired Fourier-domain filter. The necessary filters can be realized holographically, by using computer-aided and/or lithographic techniques, or by using a spatial light modulator. These methods are beyond the scope of this book.

The possibility of realizing desired convolutions and spatial filters has led to a vast and diverse array of applications often referred to as analog optical information processing (or signal processing). Very broadly speaking, these applications can be roughly classified into two categories: those which are applications of convolution, and those which are applications of correlation. Mathematically, correlation is closely related to convolution; however, their interpretations are very different. The former category includes applications such as beam shaping, image enhancement, and Wiener filtering for image restoration and noise removal. The latter category includes matched filtering and more advanced approaches in pattern recognition.

The single-lens imaging system is a very important system in optics which consists of an "object" situated a distance d_o to the left of a lens with focal length f, with the "image" observed a distance d_i to the right, such that

$$\frac{1}{f} = \frac{1}{d_o} + \frac{1}{d_i}, \tag{7.158}$$

an equation known as the imaging condition. If the object (input) complex amplitude distribution is $\hat{f}(x,y)$, the image (output) complex amplitude distribution is

$$\propto e^{-i\pi(x^2+y^2)/\lambda M f} \, \hat{f}(x/M, y/M), \tag{7.159}$$

where $M \equiv -d_i/d_o$ is the *magnification* of the imaging system. This result is valid for an infinite lens whose transmittance function is given by equation 7.80. In reality, the lens will have a finite aperture. If the aperture is characterized by an aperture function $\hat{p}(x,y)$ which is 1 where the aperture is transparent and 0 where it is opaque, such a lens can be modeled by the transmittance function $\hat{p}(x,y) \exp[-i\pi(x^2+y^2)/\lambda f]$. Using this new transmittance it is possible to reanalyze the single-lens coherent imaging system (as well as the $2f$ or $4f$ systems discussed above) to find the correct input-output relation. It is stated in many sources that the overall effect of a finite lens aperture is to result in a space-invariant system whose transfer function is proportional to $\hat{p}(-\lambda d_i \sigma_x, -\lambda d_i \sigma_y)$. However, this widely stated result is not correct and the overall coherent imaging system is not space-invariant (Rhodes 1998). The lens aperture cannot be a Fourier-domain filter since the Fourier transform is in general not observed at the lens but elsewhere in the system. We will see in chapter 9 that greater insight into this system can be gained by using fractional Fourier transforms. We also note that when incoherent light is used, the same system does become space-invariant and can be modeled by a point spread function or a Fourier-domain transfer function.

Real imaging systems may consist of several lenses with different aperture sizes. An approximate but useful way of analyzing such systems is to determine the limiting aperture of the system (known as the aperture stop) and find its image with respect to the output plane (known as the exit pupil), using geometrical optics techniques (see section 8.4.3). Then, to determine the resolution of the system, we can instead analyze the problem of a converging spherical wave incident on the exit pupil. If the exit pupil were infinite in extent, this converging spherical wave would be focused to a point at the image plane, corresponding to perfect imaging with no blur. The finiteness of the exit pupil leads to a finite spot size instead, determining the resolution of the system. If the exit pupil is a distance d to the left of the image plane and is represented by the function $\hat{p}(x,y)$, then at the image plane the distribution of light is proportional to

$$e^{i\pi(x^2+y^2)/\lambda d} \, \hat{P}\left(\frac{x}{\lambda d}, \frac{y}{\lambda d}\right). \tag{7.160}$$

Thus, instead of a point, we observe a distribution of light given by the Fourier transform of the exit pupil function. If one considers a simple circular or rectangular aperture of diameter D, then it is easy to show that this distribution has a width of about $\lambda d/D$. This quantity is a measure of the resolution of the imaging system. Image features closer than $\lambda d/D$ cannot be distinctly resolved from each other. In certain cases the distance d corresponds to the focal length f of the overall imaging system or compound lens, so that the size of the smallest resolvable feature is written as $f_{\#}\lambda$, where $f_{\#} \equiv f/D$ is referred to as the "f-number." Smaller f-numbers mean better resolution. The fact that the size of the smallest resolvable feature is inversely proportional to the size of the aperture is, of course, a consequence of the uncertainty relation discussed in section 2.7.

7.9 Further reading

Our coverage of elementary optics in this chapter has been brief and skewed towards the needs of later chapters. Classic introductory books on general optics include Saleh and Teich 1991, Möller 1988, Klein and Furtak 1986, Hecht, Zajac, and Guardino 1997, and Jenkins and White 1976. More advanced books include Born and Wolf 1980, Stavroudis 1972, Solimeno, Crosignani, and Di Porto 1986, and Mandel and Wolf 1995.

Classic introductory texts on Fourier optics and optical information processing include Goodman 1996, Lohmann 1986, Papoulis 1968, Iizuka 1987, Reynolds and others 1989, Yu 1983, Gaskill 1978, and Cathey 1974. More advanced treatments include VanderLugt 1992, Yu and Jutamulia 1992, Javidi and Horner 1994, Boone 1997, and *Optical Pattern Recognition* 1998. For a discussion of sampling theory in the context of optics, see Gori 1993.

Although this book does not deal with quantum optics, the fact that the mathematical techniques discussed have found many applications in these areas warrants inclusion of a few references: Walls and Milburn 1994, Mandel and Wolf 1995, and Yamamoto and İmamoğlu 1999.

8

Phase-Space Optics

8.1 Wave-optical and geometrical-optical phase spaces

In chapter 2 we discussed the basic concepts used in the study of signals and systems, and in chapter 7 we discussed optical signals and systems. In chapter 3 we discussed time- or space-frequency representations and linear canonical transforms, and in this chapter we discuss the use of these concepts in optics. Our main interest will be in the spatial distribution of light so that we will usually deal with space-frequency representations rather than time-frequency representations. Despite the fact that optical signals are most commonly two-dimensional, leading to four-dimensional space-frequency representations, in this chapter we will mostly discuss one-dimensional optical signals leading to two-dimensional space-frequency representations, since they are much easier to visualize. *Phase space* is the space in which space-frequency representations exist, and is also referred to as the space-frequency plane for one-dimensional signals and systems. One of the two dimensions of phase space is usually a spatial coordinate, whereas the other dimension may be either spatial frequency, the angle or sine of the angle or slope of a ray, or a quantity corresponding to momentum.

We will discuss both wave-optical and geometrical-optical phase-space representations. Let us consider an optical signal corresponding to the amplitude distribution of light at a given plane. Although other alternatives are also possible, we will employ the Wigner distribution $\hat{W}_{\hat{f}}(x, \sigma_x)$ of the optical signal $\hat{f}(x)$ as its phase-space representation (equation 3.16):

$$\hat{W}_{\hat{f}}(x, \sigma_x) = \int \hat{f}(x + x'/2)\hat{f}^*(x - x'/2)e^{-i2\pi\sigma_x x'}\, dx'. \tag{8.1}$$

The properties given in equations 3.17, 3.18, and 3.19 remain valid, with the appropriate replacement of dimensional variables. When we integrate the Wigner distribution over all spatial frequencies, we obtain the intensity distribution

$$\hat{I}_{\hat{f}}(x) = |\hat{f}(x)|^2 = \int \hat{W}_{\hat{f}}(x, \sigma_x)\, d\sigma_x. \tag{8.2}$$

When we integrate the Wigner distribution over space, we obtain the spectral distribution of power

$$|\hat{F}(\sigma_x)|^2 = \int \hat{W}_{\hat{f}}(x, \sigma_x)\, dx. \tag{8.3}$$

When we integrate over both x and σ_x, we obtain the total power of the optical signal:

$$\iint \hat{W}_{\hat{f}}(x, \sigma_x)\, dx\, d\sigma_x = \text{signal power.} \tag{8.4}$$

Notice that the optical signal power (which might be expressed in watts) is the quantity corresponding to the mathematical concept of "energy" defined in chapter 2. This is because our main interest is in the distribution of this total power in space and spatial frequency. Temporally, the light is assumed to be quasi-monochromatic, so it is appropriate to consider the energy in a unit time interval (which is the power), rather than the total energy (the energy from $t = -\infty$ to $+\infty$ would be infinite). If we were dealing with signals of finite duration rather than quasi-monochromatic signals, then we could consider the spatio-temporal Wigner distribution whose integral over all variables could be interpreted as the actual energy in joules (Mendlovic and Zalevsky 1997).

Roughly speaking, the Wigner distribution of an optical signal gives us the distribution of optical power over space and spatial frequency. In other words, it provides us information about the local frequency content of the optical wave at a certain location.

The Wigner distributions of several elementary signals were given in table 3.1. We see that the Wigner distribution of a plane wave $\exp(i2\pi\sigma_{x0}x)$ (which makes an angle $\arcsin(\lambda\sigma_{x0})$ with the z axis), is given by $\delta(\sigma_x - \sigma_{x0})$. A point source $\delta(x - x_0)$ has a Wigner distribution given by $\delta(x - x_0)$. Likewise, the remaining entries of the table can be interpreted respectively as a parabolic wave, a Gaussian beam, and light coming through a simple aperture.

We will not further repeat the content of chapter 3, but the reader should bear in mind that the results and discussions presented there are directly applicable to optical signals and systems. We will also not discuss space-frequency representations other than the Wigner distribution in an optical context. Use of the ambiguity function in optics is discussed in Papoulis 1974. A discussion of wavelets from an optics perspective is given in Li and Sheng 1998. We also recall that the Fresnel integral can be cast in the form of a wavelet transform; see page 68 and Onural 1993.

Whereas the coordinates of wave-optical phase space are space and spatial frequency, the coordinates of geometrical-optical phase space will be chosen as the position and angle of the rays. In some contexts the sine of the angle or the slope of the ray is considered instead; however, in the paraxial approximation where all angles are assumed to be small, all of these are equivalent. The reader should just bear in mind that the angle can also be interpreted as the slope of the ray or the derivative of the function $x(z)$ describing the trajectory of the ray.

As usual, the position will be denoted by x. As for the angle, we will choose to work with the normalized angle $\sigma_x \equiv \sin\theta_x/\lambda \approx \theta_x/\lambda$ introduced on page 248. Although we will interpret σ_x primarily as an angle in geometrical-optical contexts, we are using the same notation we use for spatial frequency, since σ_x directly corresponds to the spatial frequency of the spatial harmonic at $z = 0$ associated with a plane wave whose wave vector is parallel to the ray with angle θ_x. This association allows us to look at wave-optical and geometrical-optical phase spaces in a unified manner. It is also possible to interpret θ_x/λ as a quantity corresponding to momentum, which allows an analogy with the phase space of mechanics whose coordinates consist of spatial variables and momenta associated with these spatial variables.

We recall from chapter 7 that in geometrical optics a distribution of light is represented by a bundle of rays. In general, there will be a continuum of rays crossing a given plane at different points x making different normalized angles $\sigma_x = \theta_x/\lambda$. Each such ray will be represented in phase space by the point (x, σ_x). It is also possible to assign weights to each ray and thus to each point in phase space. These weights might be represented as a function of (x, σ_x) defined over the whole x-σ_x plane. If there is no ray at position x making angle σ_x, the value of this function is zero at that point (x, σ_x). We will further discuss the relationship between the two phase spaces throughout this chapter.

Figure 8.1 shows how several different ray bundles look in phase space, where we have assumed all rays to have the same weight. More general ray bundles will cover more general regions. The usual case is for these regions to consist of one or more well-connected regions, rather than a scattering of isolated points. As these bundles of rays pass through optical systems, the region and its boundary will be distorted such that any ray on the boundary will continue to remain on the boundary. The closest wave-optical analog of such a boundary is a contour of the Wigner distribution which contains "most" of the signal energy (say 99%).

Here we also briefly point out the relationship of the phase-space quantities discussed above to *radiometric* or *photometric* quantities (Born and Wolf 1980). Consider an extended light source illuminating a certain region of space. The power (energy per unit time) coming from an element of area δA centered at a point P on this source, and falling within an element of solid angle $\delta \Omega$ centered around the direction of Θ, is given by $B(P, \Theta) \cos \theta \, \delta A \, \delta \Omega$. Here θ is the angle between the surface normal at P and the direction of Θ, and $B(P, \Theta)$ is referred to as the *photometric brightness*. Here the term "photometric" implies that we are referring to physical quantities measured in watts, and so on, rather than measures of visual sensation. The $\cos \theta$ term is an obliquity factor corresponding to the fact that the projection of the area element in the direction Θ is given by $\cos \theta \delta A$. We observe that as a function of position and angle, the brightness is also a kind of phase-space density, analogous to the Wigner distribution. When we integrate the brightness over the area of the source as $\int B(P, \Theta) \cos \theta \, dA$, we obtain the *photometric intensity* as a function of the direction of Θ. Integration of this quantity over all relevant angles yields the total power. When we integrate the brightness throughout the solid angle as $\int B(P, \Theta) \cos \theta \, d\Omega$, we obtain the *photometric illumination* as a function of the point P. Integration of this quantity over the source area yields the total power. The integrals of $B(P, \Theta)$ over one of the variables are analogous to the projections of the Wigner distribution. Integration over both variables gives us the total power. Further discussion of these quantities may be found in Born and Wolf 1980:181–183. Note that the field of radiometry and photometry displays a wealth of terminology which might be potentially confusing. The terminology used above (Born and Wolf 1980) is not consistent with that used in some other sources and elsewhere in this book. When reference is made to visual sensation, as opposed to true physical magnitude, photometric power, brightness, intensity, and illumination are replaced by luminous energy, luminance, luminous intensity (candle power), and illumination, respectively. Yet another set of terms—in the same respective order—are power, radiance, intensity, and irradiance. A brief introduction to the relationship of radiometric quantities to the Wigner distribution may be found in Bastiaans 1997.

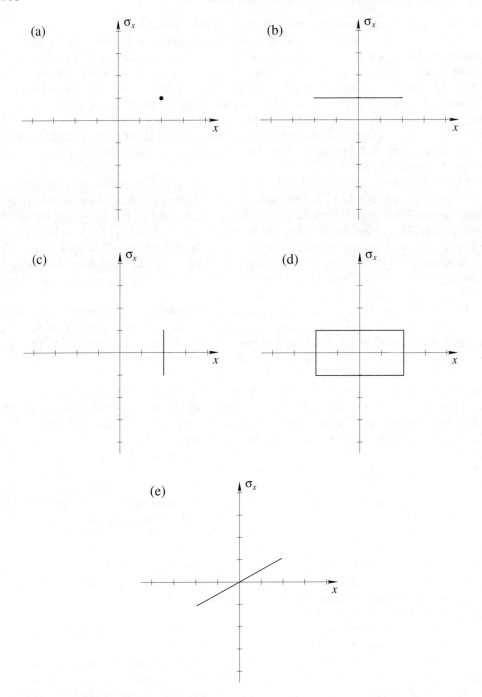

Figure 8.1. Representation of various ray bundles in phase space: (a) a single ray, (b) a bundle of parallel rays, (c) a bundle of rays emanating from the same point, (d) a bundle of rays uniformly distributed over different positions and angles, (e) a bundle of rays corresponding to a spherical wave emanating from a point to the left.

What is important for our purpose is to realize that conventional radiometric and photometric quantities are essentially analogous to the Wigner distribution and its integrals (projections or marginals).

Two-dimensional phase space is defined by the variables $(x, y, \sigma_x, \sigma_y)$. Since working with these four variables makes the notation more cumbersome and requires dealing with four-dimensional functions, we will mostly restrict ourselves to the one-dimensional case.

8.2 Quadratic-phase systems and linear canonical transforms

Sections of free space in the Fresnel approximation, thin lenses, and sections of quadratic graded-index media, as well as arbitrary combinations of these belong to the class of *quadratic-phase systems* (Bastiaans 1979a), which are mathematically the same as the class of linear canonical transforms we discussed in chapter 3. The output of a one-dimensional quadratic-phase system is related to its input through

$$\hat{g}(x) = \int \hat{h}(x, x') \hat{f}(x') \, dx', \qquad (8.5)$$

$$\hat{h}(x, x') = \sqrt{\hat{\beta}} \, e^{-i\pi/4} \exp\left[i\pi(\hat{\alpha}x^2 - 2\hat{\beta}xx' + \hat{\gamma}x'^2)\right],$$

where $\hat{\alpha}$, $\hat{\beta}$, $\hat{\gamma}$ are the three independent parameters of the system (equation 3.66). We have chosen the normalization such that the system is unitary. In two dimensions

$$\hat{g}(x, y) = \iint \hat{h}(x, y; x', y') \hat{f}(x', y') \, dx' \, dy', \qquad (8.6)$$

$$\hat{h}(x, y; x', y') = -i\hat{\beta} \exp\left[i\pi\left(\hat{\alpha}(x^2 + y^2) - 2\hat{\beta}(xx' + yy') + \hat{\gamma}(x'^2 + y'^2)\right)\right].$$

It is also possible to consider the case where the parameters are different for the two dimensions (Bastiaans 1979a, Sahin, Ozaktas, and Mendlovic 1998).

Again, s denotes the implicit scale parameter relating the dimensionless variables u, v to the variables x, y with dimensions of length: $x \equiv su$ and $y \equiv sv$. As in chapter 7, $s\hat{f}(x, y) \equiv f(x/s, y/s) \equiv f(u, v)$ and $s^{-1}\hat{F}(\sigma_x, \sigma_y) \equiv F(s\sigma_x, s\sigma_y) \equiv F(\mu, \nu)$. We are also introducing the definitions $\hat{\alpha} \equiv \alpha/s^2$, $\hat{\beta} \equiv \beta/s^2$, $\hat{\gamma} \equiv \gamma/s^2$ and $\hat{A} \equiv A$, $\hat{B} \equiv s^2 B$, $\hat{C} \equiv C/s^2$, $\hat{D} \equiv D$. These relate the dimensionless parameters of chapters 2, 3, and 4 to the dimensional parameters employed in this chapter and chapter 7. These sets of parameters are again related to each other through equation 3.74:

$$\begin{bmatrix} \hat{A} & \hat{B} \\ \hat{C} & \hat{D} \end{bmatrix} = \begin{bmatrix} \hat{\gamma}/\hat{\beta} & 1/\hat{\beta} \\ -\hat{\beta} + \hat{\alpha}\hat{\gamma}/\hat{\beta} & \hat{\alpha}/\hat{\beta} \end{bmatrix} = \begin{bmatrix} \hat{\alpha}/\hat{\beta} & -1/\hat{\beta} \\ \hat{\beta} - \hat{\alpha}\hat{\gamma}/\hat{\beta} & \hat{\gamma}/\hat{\beta} \end{bmatrix}^{-1}, \qquad (8.7)$$

with $\hat{A}\hat{D} - \hat{B}\hat{C} = 1$. Once again, it is the case that the matrix associated with the system obtained by concatenating two systems, is given by the product of the matrices of the two systems. The kernel in equation 8.5 can be rewritten in terms of $\hat{A}$, $\hat{B}$, $\hat{C}$, $\hat{D}$ as

$$\hat{h}(x, x') = \sqrt{\frac{1}{\hat{B}}} \, e^{-i\pi/4} \exp\left[\frac{i\pi}{\hat{B}}\left(\hat{D}x^2 - 2xx' + \hat{A}x'^2\right)\right]. \qquad (8.8)$$

This formula is found in the literature in a variety of contexts and has been referred to by various names; for instance, "quadratic-phase systems" in Bastiaans 1979a, "linear canonical transforms" in Wolf 1979, "generalized Huygens integral" in Siegman 1986, "generalized Fresnel transform" in James and Agarwal 1996, "special affine Fourier transforms" in Abe and Sheridan 1994a, b.

The effect of a quadratic-phase system on the Wigner distribution of an optical signal follows from the discussion of section 3.4.2. The Wigner distribution of the output signal $\hat{W}_{\hat{g}}$ is related to that of the input signal by

$$\hat{W}_{\hat{g}}(x, \sigma_x) = \hat{W}_{\hat{f}}(\hat{D}x - \hat{B}\sigma_x, -\hat{C}x + \hat{A}\sigma_x). \tag{8.9}$$

The kernel relating these Wigner distributions is given by

$$\hat{W}_{\hat{g}}(x, \sigma_x) = \int\!\!\int \hat{K}_{\hat{h}}(x, \sigma_x; x', \sigma'_x)\, \hat{W}_{\hat{f}}(x', \sigma'_x)\, dx'\, d\sigma'_x, \tag{8.10}$$

$$\hat{K}_{\hat{h}}(x, \sigma_x; x', \sigma'_x) = \delta(x' - \hat{D}x + \hat{B}\sigma_x)\, \delta(\sigma'_x + \hat{C}x - \hat{A}\sigma_x).$$

The kernel $\hat{K}_{\hat{h}}(x, \sigma_x; x', \sigma'_x)$ associated with optical components which are not linear canonical transforms (such as a spatial filter) can be determined from equation 3.44.

As in chapter 3, we are restricting ourselves to systems with real $\hat{A}$, $\hat{B}$, $\hat{C}$, $\hat{D}$. Allowing complex parameters makes it possible to deal with attenuating apertures of Gaussian profile, and propagation in certain media exhibiting attenuation (or gain). Most of the results presented here are also valid for such systems with complex parameters (Siegman 1986).

When dealing with two-dimensional systems, the above matrices become four-dimensional matrices. The most compact approach is usually to define a two-dimensional space vector and a two-dimensional spatial frequency vector. In terms of these, the four-dimensional matrices become 2×2 block matrices of 2×2 matrices, and most results bear a direct formal similarity to the one-dimensional results. Two-dimensional linear canonical transforms and their matrices were briefly discussed in section 3.5. The reader is referred to Bastiaans 1979a for a discussion of two-dimensional systems in an optical context. We also note that when dealing with axially symmetric systems, one-dimensional analysis is sufficient for most purposes.

8.3 Optical components

We now individually discuss the optical components introduced in section 7.4. Our discussion will be uniform and comprehensive, running in parallel to the wave-optical and geometrical-optical discussion of the same components in chapter 7. The results presented are one-dimensional versions of the results presented in that chapter, so they are not rederived.

For each optical component, we will first state which mathematical operation the optical component corresponds to. These operations were already studied in detail in chapters 2 and 3 (see tables 2.2 and 3.7). We will then state the forward and inverse kernels $\hat{h}(x, x')$ and $\hat{h}^{-1}(x, x')$ of the component, and show how the output amplitude distribution $\hat{g}(x)$ is related to the input distribution $\hat{f}(x)$. We recall that these quantities are related by

$$\hat{g}(x) = \int \hat{h}(x, x') \hat{f}(x') \, dx'. \tag{8.11}$$

We will also state how the Fourier transform of the output $\hat{G}(\sigma_x)$ is related to the Fourier transform $\hat{F}(\sigma_x)$ of the input. Comparing the kernels with the general form of the kernel for quadratic-phase systems

$$\hat{h}(x, x') = e^{-i\pi/4} \sqrt{\frac{1}{\hat{B}}} \, \exp\left[\frac{i\pi}{\hat{B}} \left(\hat{D}x^2 - 2xx' + \hat{A}x'^2\right)\right], \tag{8.12}$$

will reveal that most of the components under consideration are quadratic-phase systems with different matrix parameters, and enable us to identify their $\hat{A}$, $\hat{B}$, $\hat{C}$, $\hat{D}$ parameters.

We will then state the kernel $\hat{K}_{\hat{h}}(x, \sigma_x; x', \sigma'_x)$ relating the Wigner distribution of the output $\hat{W}_{\hat{g}}(x, \sigma_x)$ to the Wigner distribution of the input $\hat{W}_{\hat{f}}(x, \sigma_x)$:

$$\hat{W}_{\hat{g}}(x, \sigma_x) = \iint \hat{K}_{\hat{h}}(x, \sigma_x; x', \sigma'_x) \, \hat{W}_{\hat{f}}(x', \sigma'_x) \, dx' \, d\sigma'_x, \tag{8.13}$$

as well as an expression directly relating $\hat{W}_{\hat{g}}(x, \sigma_x)$ to $\hat{W}_{\hat{f}}(x, \sigma_x)$. When the system in question is a quadratic-phase system, these relations take the form

$$\hat{K}_{\hat{h}}(x, \sigma_x; x', \sigma'_x) = \delta(x' - \hat{D}x + \hat{B}\sigma_x) \, \delta(\sigma'_x + \hat{C}x - \hat{A}\sigma_x), \tag{8.14}$$

$$\hat{W}_{\hat{g}}(x, \sigma_x) = \hat{W}_{\hat{f}}(\hat{D}x - \hat{B}\sigma_x, -\hat{C}x + \hat{A}\sigma_x). \tag{8.15}$$

The parameters $\hat{A}$, $\hat{B}$, $\hat{C}$, $\hat{D}$ of a component, deduced by comparison with equation 8.12, are also always consistent with the above two equations.

In sections 7.5 and 7.6 we characterized rays by noting their position x and angle θ_x (or normalized angle $\sigma_x = \theta_x/\lambda$) at a given plane. We derived linear relations between the output position and normalized angle (x_2, σ_{x2}) and the input position and normalized angle (x_1, σ_{x1}) for each component. We now observe that all of these linear relations can be written as a simple matrix equation of the form

$$\begin{bmatrix} x_2 \\ \sigma_{x2} \end{bmatrix} = \begin{bmatrix} \hat{A} & \hat{B} \\ \hat{C} & \hat{D} \end{bmatrix} \begin{bmatrix} x_1 \\ \sigma_{x1} \end{bmatrix}, \tag{8.16}$$

with the same parameters $\hat{A}$, $\hat{B}$, $\hat{C}$, $\hat{D}$ deduced from wave-optical considerations.

We emphasize that the $\hat{A}$, $\hat{B}$, $\hat{C}$, $\hat{D}$ parameters of quadratic-phase components can be determined in two ways. We can obtain the kernel $\hat{h}(x, x')$ characterizing the component based on wave optics, and compare it with equation 8.12 to identify $\hat{A}$, $\hat{B}$, $\hat{C}$, $\hat{D}$. Alternatively, we can write the relation between the input and output ray vectors in matrix form based on geometrical optics, and identify these four parameters from this relation. The fact that both approaches yield the same result reflects an underlying correspondence between the approximations employed in the wave-optical and geometrical-optical approaches. Wave optics and geometrical optics become operationally equivalent under these conditions, despite the fact they are quite different descriptions of the behavior of light.

We will also provide a figure for each component, illustrating how a rectangle representing either the support of the Wigner distribution or a bundle of rays is changed by this component. These figures illustrate the effect of the component in phase space. (The reader should compare them with figures 3.3 and 3.6.)

We will then note the eigenfunctions of the component, and also present the associated hyperdifferential form, as well as a differential equation description of the component. However, we will not discuss them in great detail in this chapter, referring the reader to section 3.4.9 for further elaboration of the mathematical relationships between these concepts.

Bastiaans has been a major contributor to the use of Wigner distributions and quadratic-phase systems in optics (1978, 1979a, b, c, d, 1989, 1997).

8.3.1   Sections of free space

The physical process of propagation through a section of free space of length d in the Fresnel approximation, mathematically corresponds to *chirp convolution*. Positive values of d correspond to forward propagation and negative values of d correspond to backward propagation. The associated kernel and its inverse are given by

$$\hat{h}(x, x') = e^{-i\pi/4} \sqrt{\frac{1}{\lambda d}} \, e^{i\pi(x-x')^2/\lambda d}, \tag{8.17}$$

$$\hat{h}^{-1}(x, x') = e^{i\pi/4} \frac{1}{\sqrt{\lambda d}} \, e^{-i\pi(x-x')^2/\lambda d}, \tag{8.18}$$

where the trivial phase factors $\exp(\pm i2\pi\sigma d)$ have been dropped. The output can be expressed in terms of the input as

$$\hat{g}(x) = e^{-i\pi/4} \sqrt{\frac{1}{\lambda d}} \int e^{i\pi(x-x')^2/\lambda d} \hat{f}(x') \, dx' = \left[e^{-i\pi/4} \sqrt{\frac{1}{\lambda d}} \, e^{i\pi x^2/\lambda d} \right] * f(x). \tag{8.19}$$

In the Fourier domain,

$$\hat{G}(\sigma_x) = e^{-i\pi\lambda d\sigma_x^2} \, \hat{F}(\sigma_x). \tag{8.20}$$

The kernel relating the Wigner distribution of the output to that of the input, and the Wigner distribution of the output are given by

$$\hat{K}_{\hat{h}}(x, \sigma_x; x', \sigma_x') = \delta(x - \lambda d\sigma_x - x') \, \delta(\sigma_x - \sigma_x'), \tag{8.21}$$

$$\hat{W}_{\hat{g}}(x, \sigma_x) = \hat{W}_{\hat{f}}(x - \lambda d\sigma_x, \sigma_x). \tag{8.22}$$

Upon comparison with equations 8.12 and 8.15, equations 8.17 and 8.22 imply

$$\begin{bmatrix} \hat{A} & \hat{B} \\ \hat{C} & \hat{D} \end{bmatrix} = \begin{bmatrix} 1 & \lambda d \\ 0 & 1 \end{bmatrix} = \begin{bmatrix} 1 & -\lambda d \\ 0 & 1 \end{bmatrix}^{-1}. \tag{8.23}$$

The system inverse of a section of free space of length d is a section of free space of length $-d$, which can be interpreted as backward propagation. The equations relating the position and normalized angle of the output ray to those of the input ray can be written in the form

$$\begin{bmatrix} x_2 \\ \sigma_{x2} \end{bmatrix} = \begin{bmatrix} 1 & \lambda d \\ 0 & 1 \end{bmatrix} \begin{bmatrix} x_1 \\ \sigma_{x1} \end{bmatrix} = \begin{bmatrix} \hat{A} & \hat{B} \\ \hat{C} & \hat{D} \end{bmatrix} \begin{bmatrix} x_1 \\ \sigma_{x1} \end{bmatrix}. \tag{8.24}$$

Figure 8.2 shows how a rectangle representing either the support of the Wigner distribution or a bundle of rays is transformed through this operation.

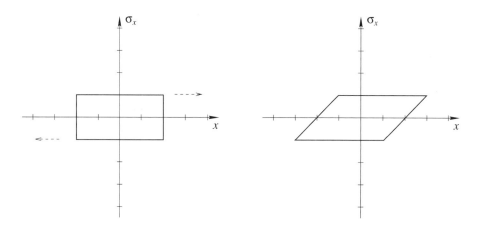

Figure 8.2.   The effect of free-space propagation in phase space is horizontal shearing (in the space dimension).

The eigenfunctions of propagation through free space are of the form $\hat{f}(x) = \exp(i2\pi\sigma_{x0}x)$, where σ_{x0} is any real number. Physically, this corresponds to a spatial harmonic in the transverse plane, and a plane wave throughout space. The effect of chirp convolution on such functions is simply to multiply them with a unit-magnitude complex constant (a phase factor).

The hyperdifferential form of the free-space propagation operator is given by

$$\exp\left(i\frac{\lambda d}{4\pi}\frac{d^2}{dx^2}\right). \tag{8.25}$$

Since the operator corresponding to free-space propagation is a function of d/dx only, all operators of d/dx will remain invariant under free-space propagation. In other words, they commute with free-space propagation. For instance, if we propagate the derivative of an input field $\hat{f}(x)$, we will obtain the derivative of the original output $\hat{g}(x)$.

Let us now denote the amplitude distribution of light as a function of both x and z by $\hat{f}(x, z)$. Then, if we solve the differential equation

$$-\frac{1}{4\pi}\frac{\partial^2 \hat{f}(x, z)}{\partial x^2} = i\frac{\partial \hat{f}(x, z)}{\partial(\lambda z)} \tag{8.26}$$

subject to the boundary condition $\hat{f}(x, 0) = \hat{f}(x)$, the solution $\hat{f}(x, z)$ will be related to $\hat{f}(x)$ through the Fresnel integral, with z replacing the distance of propagation d.

8.3.2 Thin lenses

The physical process of passing through a thin lens of focal length f in the paraxial approximation, mathematically corresponds to *chirp multiplication*. Positive values of f correspond to positive lenses and negative values of f correspond to negative lenses. The associated kernel and its inverse are given by

$$\hat{h}(x, x') = e^{-i\pi x^2/\lambda f}\, \delta(x - x'),\tag{8.27}$$

$$\hat{h}^{-1}(x, x') = e^{i\pi x^2/\lambda f}\, \delta(x - x').\tag{8.28}$$

The output can be expressed in terms of the input as

$$\hat{g}(x) = e^{-i\pi x^2/\lambda f}\, \hat{f}(x).\tag{8.29}$$

In the Fourier domain,

$$\hat{G}(\sigma_x) = e^{-i\pi/4}\sqrt{\lambda f}\int e^{i\pi\lambda f(\sigma_x - \sigma_x')^2}\hat{F}(\sigma_x')\, d\sigma_x' = \left[e^{-i\pi/4}\sqrt{\lambda f}\, e^{i\pi\lambda f\sigma_x^2}\right] * \hat{F}(\sigma_x).\tag{8.30}$$

The kernel relating the Wigner distribution of the output to that of the input, and the Wigner distribution of the output are given by

$$\hat{K}_{\hat{h}}(x, \sigma_x; x', \sigma_x') = \delta(\sigma_x + x/\lambda f - \sigma_x')\delta(x - x'),\tag{8.31}$$

$$\hat{W}_{\hat{g}}(x, \sigma_x) = \hat{W}_{\hat{f}}(x, \sigma_x + x/\lambda f).\tag{8.32}$$

Upon comparison with equations 8.12 and 8.15, equations 8.27 and 8.32 imply

$$\begin{bmatrix} \hat{A} & \hat{B} \\ \hat{C} & \hat{D} \end{bmatrix} = \begin{bmatrix} 1 & 0 \\ -1/\lambda f & 1 \end{bmatrix} = \begin{bmatrix} 1 & 0 \\ 1/\lambda f & 1 \end{bmatrix}^{-1}.\tag{8.33}$$

The system inverse of a thin lens of focal length f is a thin lens of focal length $-f$: the inverse of a positive lens is a negative lens and vice versa. The equations relating the position and normalized angle of the output ray to those of the input ray can be written in the form

$$\begin{bmatrix} x_2 \\ \sigma_{x2} \end{bmatrix} = \begin{bmatrix} 1 & 0 \\ -1/\lambda f & 1 \end{bmatrix}\begin{bmatrix} x_1 \\ \sigma_{x1} \end{bmatrix} = \begin{bmatrix} \hat{A} & \hat{B} \\ \hat{C} & \hat{D} \end{bmatrix}\begin{bmatrix} x_1 \\ \sigma_{x1} \end{bmatrix}.\tag{8.34}$$

Figure 8.3 shows how a rectangle representing either the support of the Wigner distribution or a bundle of rays is transformed through this operation.

The eigenfunctions of passing through a thin lens are of the form $\hat{f}(x) = \delta(x - x_0)$, where x_0 is any real number. Physically, this corresponds to a point source. The effect of chirp multiplication on such functions is simply to multiply them with a unit-magnitude complex constant (a phase factor).

The hyperdifferential form of the thin lens operator is given by

$$\exp\left(-i\frac{\pi}{\lambda f}x^2\right).\tag{8.35}$$

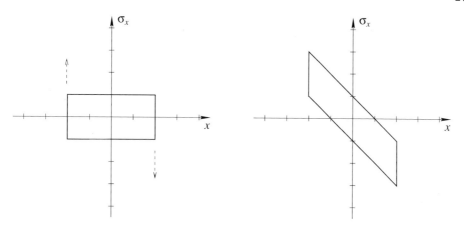

Figure 8.3. The effect of a thin lens in phase space is vertical shearing (in the frequency dimension).

Since the operator corresponding to a thin lens is a function of x only, all operators of x will remain invariant under passage through a thin lens. In other words, they commute with the thin lens. For instance, if $x\hat{f}(x)$ is incident on the lens, at the output we will obtain $x\hat{g}(x)$.

The associated differential equation does not have a meaningful physical interpretation and is thus not presented here.

8.3.3 Quadratic graded-index media

The physical process of propagation through a section of quadratic graded-index medium with parameter χ and of length d mathematically corresponds to *fractional Fourier transformation*. Positive values of d correspond to forward propagation and negative values of d correspond to backward propagation. The associated kernel and its inverse are given by

$$\hat{h}(x,x') = \begin{cases} \dfrac{e^{-id/2\chi}}{\sqrt{\lambda\chi}}A_\alpha\,e^{\frac{i\pi}{\lambda\chi}\left(\cot\alpha\,x^2-2\csc\alpha\,xx'+\cot\alpha\,x'^2\right)} & d \neq j\pi\chi \\ e^{-id/2\chi}\,\delta(x-x') & d = 2j\pi\chi \\ e^{-id/2\chi}\,\delta(x+x') & d = (2j\pm1)\pi\chi \end{cases}$$

$$(8.36)$$

$$\hat{h}^{-1}(x,x') = \begin{cases} \dfrac{e^{id/2\chi}}{\sqrt{\lambda\chi}}A_\alpha^*\,e^{-\frac{i\pi}{\lambda\chi}\left(\cot\alpha\,x^2-2\csc\alpha\,xx'+\cot\alpha\,x'^2\right)} & d \neq j\pi\chi \\ e^{id/2\chi}\,\delta(x-x') & d = 2j\pi\chi \\ e^{id/2\chi}\,\delta(x+x') & d = (2j\pm1)\pi\chi \end{cases}$$

$$(8.37)$$

where j is an integer, $\alpha = d/\chi$, and $A_\alpha = \sqrt{1-i\cot\alpha}$. (The fractional Fourier order a is related to α by $\alpha = a\pi/2$.) In the above equations, the trivial phase factors $\exp(\pm i2\pi\sigma d)$ have been dropped. This kernel can be derived from equation 7.103 by employing the last property in table 2.8.

The output can be expressed in terms of the input as

$$\hat{g}(x) = e^{-id/2\chi}(\lambda\chi)^{-1/4} f_a(x/\sqrt{\lambda\chi}) = e^{-id/2\chi}\hat{f}_a(x) \tag{8.38}$$

for $d \neq j\pi\chi$. Here $f_a(u)$ denotes the ath order fractional Fourier transform of $f(u)$, where $(\lambda\chi)^{-1/4} f(x/\sqrt{\lambda\chi}) = \hat{f}(x)$. In the Fourier domain,

$$\hat{G}(\sigma_x) = e^{-id/2\chi}(\lambda\chi)^{1/4} F_a(\sqrt{\lambda\chi}\,\sigma_x) = e^{-id/2\chi}\hat{F}_a(\sigma_x). \tag{8.39}$$

Here $F_a(u)$ denotes the ath order fractional Fourier transform of $F(u)$, where $(\lambda\chi)^{1/4} F(\sqrt{\lambda\chi}\,\sigma_x) = \hat{F}(\sigma_x)$. The kernel relating the Wigner distribution of the output to that of the input, and the Wigner distribution of the output are given by

$$\hat{K}_{\hat{h}}(x,\sigma_x;x',\sigma'_x) = \delta(x\cos\alpha - \sigma_x\lambda\chi\sin\alpha - x')\,\delta(x\sin\alpha/\lambda\chi + \sigma_x\cos\alpha - \sigma'_x), \tag{8.40}$$

$$\hat{W}_{\hat{g}}(x,\sigma_x) = \hat{W}_{\hat{f}}(x\cos\alpha - \sigma_x\lambda\chi\sin\alpha, x\sin\alpha/\lambda\chi + \sigma_x\cos\alpha). \tag{8.41}$$

Upon comparison with equations 8.12 and 8.15, equations 8.36 and 8.41 imply

$$\begin{bmatrix} \hat{A} & \hat{B} \\ \hat{C} & \hat{D} \end{bmatrix} = \begin{bmatrix} \cos\alpha & \lambda\chi\sin\alpha \\ -\sin\alpha/\lambda\chi & \cos\alpha \end{bmatrix} = \begin{bmatrix} \cos\alpha & -\lambda\chi\sin\alpha \\ \sin\alpha/\lambda\chi & \cos\alpha \end{bmatrix}^{-1}. \tag{8.42}$$

The equations relating the position and normalized angle of the output ray to those of the input ray can be written in the form

$$\begin{bmatrix} x_2 \\ \sigma_{x2} \end{bmatrix} = \begin{bmatrix} \cos\alpha & \lambda\chi\sin\alpha \\ -\sin\alpha/\lambda\chi & \cos\alpha \end{bmatrix} \begin{bmatrix} x_1 \\ \sigma_{x1} \end{bmatrix} = \begin{bmatrix} \hat{A} & \hat{B} \\ \hat{C} & \hat{D} \end{bmatrix} \begin{bmatrix} x_1 \\ \sigma_{x1} \end{bmatrix}. \tag{8.43}$$

Figure 8.4 shows how a rectangle representing either the support of the Wigner distribution or a bundle of rays is transformed through this operation.

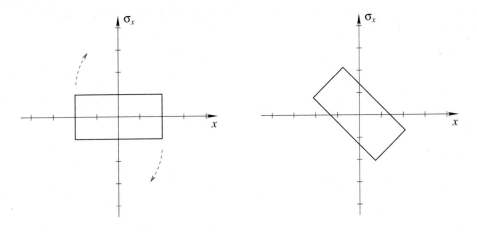

Figure 8.4. The effect of propagation through graded-index media in phase space is rotation.

The eigenfunctions of propagation through graded-index media are of the form $\hat{f}(x) = (\lambda\chi)^{-1/4}\psi_l(x/\sqrt{\lambda\chi})$, where $\psi_l(\cdot)$ is the lth order Hermite-Gaussian function. The effect of fractional Fourier transformation on such functions is simply to multiply them with a unit-magnitude complex constant (a phase factor).

The hyperdifferential form of the quadratic graded-index media propagation operator is given by

$$\exp\left[-i\alpha\left(-\frac{\lambda\chi}{4\pi}\frac{d^2}{dx^2} + \pi\frac{x^2}{\lambda\chi}\right)\right]. \tag{8.44}$$

Since the operator corresponding to propagation in this medium is a function of $\lambda\chi(i2\pi)^{-2}d^2/dx^2 + x^2/\lambda\chi$ only, all operators of $\lambda\chi(i2\pi)^{-2}d^2/dx^2 + x^2/\lambda\chi$ will remain invariant under propagation in this medium. In other words, they will commute with propagation in this medium.

Let us now denote the amplitude distribution of light as a function of both x and z by $\hat{f}(x, z)$. Then, if we solve the differential equation

$$-\frac{\lambda\chi}{4\pi}\frac{\partial^2 \hat{f}(x, z)}{\partial x^2} + \pi\frac{x^2}{\lambda\chi}\hat{f}(x, z) = i\chi\frac{\partial \hat{f}(x, z)}{\partial z} \tag{8.45}$$

subject to the boundary condition $\hat{f}(x, 0) = \hat{f}(x)$, the solution $\hat{f}(x, z)$ will be related to $\hat{f}(x)$ through a fractional Fourier transformation, with z replacing the distance of propagation d.

Quadratic graded-index media also exhibiting a quadratically varying attenuation coefficient are discussed in Siegman 1986. We only note that propagation in such media can be related to complex-ordered fractional Fourier transforms.

8.3.4   Extensions

Although the components discussed in section 7.4.4 will not be of direct interest to us in this book, we briefly mention that they also have simple effects in phase space. As mentioned in section 7.4.4, homogeneous regions with refractive index $n \neq 1$ are transparently handled by normalizing spatial frequencies and angles by n, and spherical refractive surfaces are handled in the same manner as lenses. Prisms as well as tilts and displacements of the optical axis simply correspond to shifts in phase space. A prism or a tilt of the optical axis corresponds to a shift in the σ_x dimension of phase space, and a displacement of the optical axis corresponds to a shift in the x dimension.

8.3.5   Spatial filters

The physical process of spatial filtering with a spatial filter $\hat{h}(x)$ mathematically corresponds to *multiplicative filtering*. The associated kernel and its inverse are given by

$$\hat{h}(x, x') = \hat{h}(x)\delta(x - x'), \tag{8.46}$$

$$\hat{h}^{-1}(x, x') = [1/\hat{h}(x)]\delta(x - x'). \tag{8.47}$$

The output can be expressed in terms of the input as

$$\hat{g}(x) = \hat{h}(x)\hat{f}(x).$$

(8.48)

In the Fourier domain,

$$\hat{G}(\sigma_x) = \int \hat{H}(\sigma_x - \sigma'_x)\hat{F}(\sigma'_x)\,d\sigma'_x = \hat{H}(\sigma_x) * \hat{F}(\sigma_x).$$

(8.49)

The kernel relating the Wigner distribution of the output to that of the input, and the Wigner distribution of the output are given by

$$\hat{K}_{\hat{h}}(x, \sigma_x; x', \sigma'_x) = \hat{W}_{\hat{h}}(x, \sigma_x - \sigma'_x)\,\delta(x - x'),$$

(8.50)

$$\hat{W}_{\hat{g}}(x, \sigma_x) = \int \hat{W}_{\hat{h}}(x, \sigma_x - \sigma'_x)\,\hat{W}_{\hat{f}}(x, \sigma'_x)\,d\sigma'_x.$$

(8.51)

The effect of a spatial filter in phase space is to convolve the Wigner distribution along the frequency dimension with the Wigner distribution of the filter function.

The eigenfunctions of spatial filters are of the form $\hat{f}(x) = \delta(x - x_0)$, where x_0 is any real number. Physically, this corresponds to a point source. The effect of spatial filters on such functions is simply to multiply them with a complex constant.

8.3.6   Fourier-domain spatial filters

The physical process of Fourier-domain spatial filtering mathematically corresponds to *convolutive filtering*. The associated kernel and its inverse are given by

$$\hat{h}(x, x') = \hat{h}(x - x'),$$

(8.52)

$$\hat{h}^{-1}(x, x') = \hat{h}^{-1}(x - x').$$

(8.53)

The output can be expressed in terms of the input as

$$\hat{g}(x) = \int \hat{h}(x - x')\hat{f}(x')\,dx' = \hat{h}(x) * \hat{f}(x).$$

(8.54)

In the Fourier domain,

$$\hat{G}(\sigma_x) = \hat{H}(\sigma_x)\hat{F}(\sigma_x).$$

(8.55)

The kernel relating the Wigner distribution of the output to that of the input, and the Wigner distribution of the output are given by

$$\hat{K}_{\hat{h}}(x, \sigma_x; x', \sigma'_x) = \hat{W}_{\hat{h}}(x - x', \sigma_x)\,\delta(\sigma_x - \sigma'_x),$$

(8.56)

$$\hat{W}_{\hat{g}}(x, \sigma_x) = \int \hat{W}_{\hat{h}}(x - x', \sigma_x)\,\hat{W}_{\hat{f}}(x', \sigma_x)\,dx'.$$

(8.57)

The effect of a Fourier-domain spatial filter in phase space is to convolve the Wigner distribution along the space dimension with the Wigner distribution of the filter function.

The eigenfunctions of Fourier-domain spatial filters are of the form $\hat{f}(x) = \exp(i2\pi\sigma_{x0}x)$, where σ_{x0} is any real number. Physically, this corresponds to a spatial harmonic in the transverse plane, and a plane wave throughout space. The effect of a Fourier-domain spatial filter on such functions is simply to multiply them with a complex constant.

Fourier transform stages: We also discuss the effect in phase space of the Fourier transform stages presented on page 246. The following applies to all three systems, provided the appropriate value for the parameter s is inserted.

The associated kernel and its inverse are given by

$$\hat{h}(x, x') \propto e^{-i2\pi xx'/s^2}, \tag{8.58}$$

$$\hat{h}^{-1}(x, x') \propto e^{i2\pi xx'/s^2}. \tag{8.59}$$

The output can be expressed in terms of the input as

$$\hat{g}(x) \propto \int e^{-i2\pi xx'/s^2} \hat{f}(x')\, dx' = \hat{F}(x/s^2). \tag{8.60}$$

In the Fourier domain,

$$\hat{G}(\sigma_x) \propto \int e^{-i2\pi s^2 \sigma_x \sigma_x'} \hat{F}(\sigma_x')\, d\sigma_x' = \hat{f}(-s^2\sigma_x). \tag{8.61}$$

The kernel relating the Wigner distribution of the output to that of the input, and the Wigner distribution of the output are given by

$$\hat{K}_{\hat{h}}(x, \sigma_x; x', \sigma_x') = \delta(s^2\sigma_x + x')\, \delta(x/s^2 - \sigma_x'), \tag{8.62}$$

$$\hat{W}_{\hat{g}}(x, \sigma_x) = \hat{W}_{\hat{f}}(-s^2\sigma_x, x/s^2). \tag{8.63}$$

Upon comparison with equations 8.12 and 8.15, equations 8.58 and 8.63 imply

$$\begin{bmatrix} \hat{A} & \hat{B} \\ \hat{C} & \hat{D} \end{bmatrix} = \begin{bmatrix} 0 & s^2 \\ -1/s^2 & 0 \end{bmatrix} = \begin{bmatrix} 0 & -s^2 \\ 1/s^2 & 0 \end{bmatrix}^{-1}. \tag{8.64}$$

The equations relating the position and normalized angle of the output ray to those of the input ray can be written in the form

$$\begin{bmatrix} x_2 \\ \sigma_{x2} \end{bmatrix} = \begin{bmatrix} 0 & s^2 \\ -1/s^2 & 0 \end{bmatrix} \begin{bmatrix} x_1 \\ \sigma_{x1} \end{bmatrix} = \begin{bmatrix} \hat{A} & \hat{B} \\ \hat{C} & \hat{D} \end{bmatrix} \begin{bmatrix} x_1 \\ \sigma_{x1} \end{bmatrix}. \tag{8.65}$$

Figure 8.5 shows how a rectangle representing either the support of the Wigner distribution or a bundle of rays is transformed through this operation.

The eigenfunctions of propagation through these Fourier transform stages are of the form $\hat{f}(x) = s^{-1/2}\psi_l(x/s)$, where $\psi_l(\cdot)$ is the lth order Hermite-Gaussian function. The effect of Fourier transformation on such functions is simply to multiply them with a unit-magnitude complex constant (a phase factor).

The hyperdifferential form of the Fourier transform operator is given by

$$\propto \exp\left[-i(\pi/2)\left(-\frac{s^2}{4\pi}\frac{d^2}{dx^2} + \pi\frac{x^2}{s^2}\right)\right]. \tag{8.66}$$

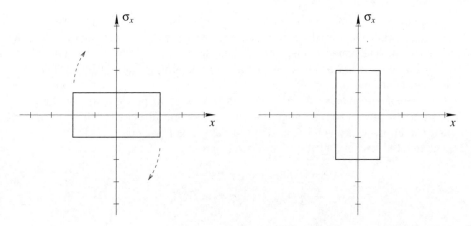

Figure 8.5. The effect of Fourier transform stages in phase space.

Since the operator corresponding to Fourier transformation is a function of $s^2(i2\pi)^{-2}d^2/dx^2 + x^2/s^2$ only, all operators of $s^2(i2\pi)^{-2}d^2/dx^2 + x^2/s^2$ will remain invariant under Fourier transformation. In other words, they will commute with the Fourier transform.

8.3.7 General linear systems

The effect of an arbitrary linear system

$$\hat{g}(x) = \int \hat{h}(x, x')\hat{f}(x')\,dx' \tag{8.67}$$

can be described in phase space by the kernel $\hat{K}_{\hat{h}}(x, \sigma_x; x', \sigma'_x)$ as follows:

$$\hat{W}_{\hat{g}}(x, \sigma_x) = \iint \hat{K}_{\hat{h}}(x, \sigma_x; x', \sigma'_x)\,\hat{W}_{\hat{f}}(x', \sigma'_x)\,dx'\,d\sigma'_x, \tag{8.68}$$

$$\hat{K}_{\hat{h}}(x, \sigma_x; x', \sigma'_x) = \iint \hat{h}(x + x''/2, x' + x'''/2)\,\hat{h}^*(x - x''/2, x' - x'''/2)$$
$$\times\, e^{-i2\pi x''\sigma_x + i2\pi x'''\sigma'_x}\,dx''\,dx''',$$

which is simply the dimensional form of equation 3.44.

8.3.8 Spherical reference surfaces

Since we have seen in our discussion of spherical reference surfaces in section 7.4.8 that the effect of employing such reference surfaces is mathematically equivalent to passage through a lens, their effect in phase space also corresponds to vertical shearing (along the frequency dimension).

8.3.9   Discussion

At this point we have completed a rather comprehensive and unified description of so-called *first-order optical systems*. This is simply another name for *quadratic-phase systems*, which are mathematically equivalent to the linear canonical transforms discussed in chapter 3. Free-space propagation in the Fresnel approximation, transmission through thin lenses, and propagation through quadratic graded-index media, and their arbitrary combinations fall into this class. The wealth of properties and results derived in chapter 3 for linear canonical transforms can be directly applied to these optical components and systems composed of them. For instance, we can readily translate the uncertainty relation given in equation 3.163 to an optical context:

$$\Delta x' \Delta x \geq |\hat{B}|/4\pi, \tag{8.69}$$

where x' denotes the coordinate of the input plane and x denotes the coordinate of the output plane. For instance, specializing to free-space propagation over a distance d, we have $\hat{B} = \lambda d$ so that the product of the spreads of the input and output light distributions cannot be less than $\lambda d/4\pi$.

When we combine first-order optical components and systems with arbitrary spatial filters (which are not first-order components), we obtain a class of systems which we will refer to as *Fourier optical systems.*

We have discussed common optical components that will be of interest to us from both wave optics and geometrical optics perspectives, and also discussed the effects of these components in phase space, be it the wave-optical phase space of Wigner distributions, or the geometrical-optical phase space of ray bundles. We saw that the relationship of the wave and geometrical optics perspectives is particularly transparent in phase space: the support of the Wigner distribution and the region representing the ray bundle are transformed in the same manner.

At this point we have a rather broad array of tools for characterizing and analyzing first-order optical systems: their forward and inverse kernels, their input-output relations in the space and Fourier domains, their effect on the Wigner distribution, the ray matrices (which are nothing but the matrices characterizing these systems as linear canonical transforms), the parallelogram-type geometric distortions which they effect in phase space, and their eigenfunctions, Hamiltonians $\mathcal{H}$, hyperdifferential forms, and associated differential equations. Most of these were already discussed in a purely mathematical and signal theoretical context in chapter 3. Here we have reexpressed them in an optical context and showed their relation to wave-optical and geometrical-optical concepts. The reader who has grasped these different forms in which common optical components and systems can be characterized, how to translate between them, and how to concatenate different components, can be said to have acquired a rather broad and comprehensive understanding of the theory of first-order and Fourier optical systems. The importance of studying the parallels between the results presented in chapters 3 and 7 and the present chapter cannot be overstressed.

The 2×2 matrices characterizing first-order optical components and systems constitute one of the most efficient ways of manipulating and concatenating them. As already discussed, these matrices can be interpreted in many different ways. The following three 2×1 vectors (which are equivalent in the paraxial approximation) transform with the $\hat{A}\hat{B}\hat{C}\hat{D}$ matrix:

$$\left[\begin{array}{c} x \\ \sigma_x \end{array}\right] = \left[\begin{array}{c} x \\ \theta_x/\lambda \end{array}\right] = \left[\begin{array}{c} x \\ dx(z)/d(\lambda z) \end{array}\right]. \qquad (8.70)$$

Here σ_x can be interpreted either as spatial frequency or normalized angle θ_x/λ. Of course, under the paraxial approximation, θ_x is equal to $\sin\theta_x$ and $\tan\theta_x$ and thus also the slope and derivative of the ray. To this list we could also add the 2×1 vector formed by replacing the second component by *momentum*, which is the second phase-space variable most commonly encountered in classical and quantum mechanics. We also recall from chapter 3 that the coordinate multiplication and differentiation operators $[\mathcal{U}\ \mathcal{D}]^\mathrm{T}$ also transform according to the (dimensionless) $ABCD$ matrix. (We refrain from defining dimensional versions of these operators to avoid further proliferation of notation.) The intimate relation between the differentiation operator and spatial frequency is immediate if we recall that the effect of $\mathcal{D}$ in the frequency domain is merely to multiply by μ. Alternatively, if we consider a signal of the form $\exp[i2\pi\hat{\phi}(x)]$, we see that its derivative at a certain point is simply given by the signal multiplied by the instantaneous frequency $d\hat{\phi}(x)/dx$ at that point.

The unity and underlying equivalence of these 2×1 vectors should already be evident from the discussions of chapters 2, 3, 7, and the present one. Any of these pairs of quantities constituting the two variables of phase space (for one-dimensional systems) are said to be *conjugate variables*. Although the first member of each pair, corresponding to time or space or the generic coordinate variable is often the "original" variable in which physical relations are first expressed, this priority disappears in phase space, and there is complete symmetry between the two conjugate variables. We also find it useful to refer to these variables as being *orthogonal* to each other. We further know from chapter 4 that there is also complete symmetry between variables associated with all fractional Fourier domains (variables which are not orthogonal to each other).

We will have more to say on the transformation of 2×1 vectors in section 8.7.

8.4   Imaging and Fourier transformation

In this section we will discuss imaging and Fourier transforming systems in the context of first-order optics (Ozaktas and Erden 1997), and also present a number of general theorems for determining image and Fourier transform planes.

8.4.1   Imaging systems

We begin with the historically prior case of imaging. The most general transform kernel, allowing for the possibility of a residual quadratic-phase term, may be expressed as

$$\hat{h}(x, x') = K e^{i\pi x^2/\lambda R} \sqrt{M}\, \delta(x - Mx'). \qquad (8.71)$$

The system is energy preserving when the magnitude of the constant K is unity. This kernel maps a function $\hat{f}(x)$ to $K\exp(i\pi x^2/\lambda R)|M|^{-1/2}\hat{f}(x/M)$. M is referred to as the magnification and R is the radius of the spherical surface on which the perfect image is observed. When $R = \infty$ the quadratic-phase term disappears and the perfect image is observed on a planar surface. The above kernel is a special case of equation 8.5 with $\hat{\alpha}, \hat{\beta}, \hat{\gamma} \to \infty$ such that $\hat{\gamma}/\hat{\beta} = M$, $\hat{\alpha} - \hat{\beta}^2/\hat{\gamma} = 1/\lambda R$. The matrix associated with this kernel is given by

$$\begin{bmatrix} M & 0 \\ M/\lambda R & 1/M \end{bmatrix}, \tag{8.72}$$

which the reader should compare with equation 3.115, from which it is seen that this matrix corresponds to perfect imaging (scaling) followed by chirp multiplication. Now, interpreting the above as a ray matrix, the following well-recognized conditions may be easily deduced:

1. If and only if a given quadratic-phase system is an imaging system, a ray emanating from the axial point at the input plane ($x_1 = 0$) is mapped onto the axial point at the output plane ($x_2 = 0$) (figure 8.6a). Furthermore, all rays emanating from an off-axis point at the input plane will be mapped to a common point at the output plane.
2. If a ray parallel to the optical axis at the input plane ($\theta_{x1} = 0$) is also parallel to the optical axis at the output plane ($\theta_{x2} = 0$), the imaging is perfect with no residual phase curvature ($R = \infty$), and the magnification M is given by the ratio of the distance of the output ray to the optical axis, to the distance of the input ray to the optical axis: $M = x_2/x_1$ (figure 8.6a).
3. If a ray parallel to the optical axis at the input plane is not parallel to the optical axis at the output plane, the magnification is found in the same way as in perfect imaging, and the radius of curvature of the residual quadratic-phase factor is simply that defined by the slope of the ray: $R = x_2/\theta_{x2}$ (figure 8.6b).

As an example, we consider the classical single-lens imaging system which consists of a section of free space of length d_1 followed by a lens with focal length f followed by another section of free space of length d_2, such that $1/f = 1/d_1 + 1/d_2$. The matrix associated with the overall system can be found easily by multiplying the matrices of its three components:

$$\begin{bmatrix} -d_2/d_1 & 0 \\ -1/\lambda f & -d_1/d_2 \end{bmatrix}, \tag{8.73}$$

from which we see that $M = -d_2/d_1$ and $R = fd_2/d_1$. We see that the single-lens imaging system does not provide a perfect image on a planar surface but results in an additional phase factor. This also means that parallel rays entering the system do not emerge parallel to each other.

Introduction of spherical reference surfaces at the input and/or output does not alter the basic imaging condition, but simply changes the overall phase factor appearing in equation 8.71. (This is because the product of lower triangular matrices is always lower triangular.) If we choose these input and output spherical reference surfaces in a particular manner, it is possible to eliminate the phase factor in the kernel relating the input and output with respect to these reference surfaces. To determine the radii R_1 and R_2 of the input and output spherical reference surfaces which eliminate the phase factor, we consider the following identity:

$$\begin{bmatrix} M & 0 \\ M/\lambda R & 1/M \end{bmatrix} = \begin{bmatrix} 1 & 0 \\ 1/\lambda R_2 & 1 \end{bmatrix}\begin{bmatrix} M & 0 \\ 0 & 1/M \end{bmatrix}\begin{bmatrix} 1 & 0 \\ -1/\lambda R_1 & 1 \end{bmatrix} \tag{8.74}$$

(a)

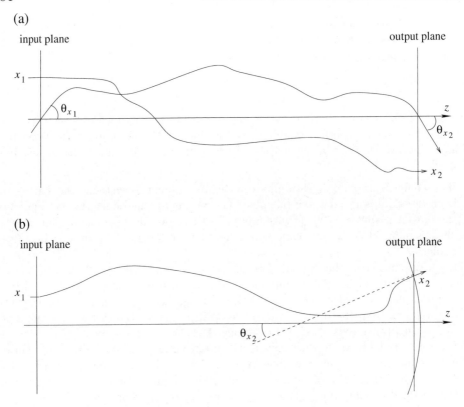

Figure 8.6. Rays in an imaging system (Ozaktas and Erden 1997).

where R_1 and R_2 satisfy $M/R = M/R_2 - 1/MR_1$. Here the rightmost matrix takes us from a planar reference surface to a spherical reference surface with radius R_1 and the matrix immediately to the right of the equal sign takes us from a spherical reference surface with radius R_2 to a planar reference surface. A particular choice of R_1 and R_2 which satisfies this relation is $R_1 = (R/M)(1 - 1/M)$ and $R_2 = R(1 - 1/M)$. Figure 8.7 shows these reference surfaces, between which perfect imaging with no additional phase factor is obtained.

The reader may also wish to similarly analyze the telescopic imaging system which consists of a section of free space of length f_1, followed by a lens of focal length f_1, a section of free space of length $f_1 + f_2$, a lens of focal length f_2, and finally a section of free space of length f_2. This system has magnification $M = -f_2/f_1$. The image is perfect without any residual phase factor.

8.4.2 Fourier transforming systems

We now turn our attention to Fourier transformation. The most general transform kernel allowing for the possibility of a residual quadratic-phase term may be expressed as

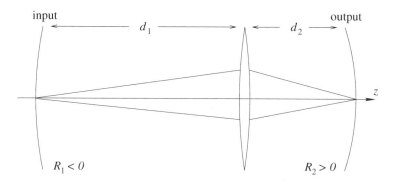

Figure 8.7. Single-lens imaging system with spherical reference surfaces. The figure is drawn for a positive lens ($f > 0$) and $d_1, d_2 > 0$ for which $R_1 < 0$ and $R_2 > 0$.

$$\hat{h}(x, x') = K e^{i\pi x^2 / \lambda R} \sqrt{\frac{1}{s^2 M}} e^{-i2\pi xx' / s^2 M}. \tag{8.75}$$

The system is energy preserving when the magnitude of the constant K is unity. This kernel maps a function $\hat{f}(x)$ into $K \exp(i\pi x^2 / \lambda R) \sqrt{1/s^2 M} \, \hat{F}(x/s^2 M)$, where $F(\mu)$ is the Fourier transform of $f(u)$. Here s is the implicit scale factor discussed in section 7.2 relating functions which take dimensionless and dimensional arguments: $s^{1/2} \hat{f}(x) \equiv f(x/s) \equiv f(u)$ and $s^{-1/2} \hat{F}(\sigma_x) \equiv F(s\sigma_x) \equiv F(\mu)$. M is the magnification associated with the Fourier transformation and R is the radius of the spherical surface on which the perfect Fourier transform is observed. When $R = \infty$ the quadratic-phase term disappears and the perfect Fourier transform is observed on a planar surface. The above kernel is a special case of equation 8.5 with $\hat{\gamma} = 0$, $\hat{\alpha} = 1/\lambda R$, and $\hat{\beta} = 1/s^2 M$. The matrix associated with the Fourier transform kernel is given by

$$\begin{bmatrix} 0 & s^2 M \\ -1/s^2 M & s^2 M / \lambda R \end{bmatrix}, \tag{8.76}$$

which the reader should compare with equation 3.120, from which it is seen that this matrix corresponds to perfect Fourier transformation (with a certain magnification) followed by chirp multiplication. Now, interpreting the above as a ray matrix, the following well-recognized conditions may be easily deduced:

1. If and only if a given quadratic-phase system is a Fourier transforming system, a ray parallel to the optical axis at the input plane ($\theta_{x1} = 0$) will pass through the axial point at the output plane ($x_2 = 0$) (figure 8.8a). Furthermore, all parallel rays making the same angle with the optical axis at the input plane will be mapped to a common point at the output plane.
2. If a ray emanating from the axial point at the input plane ($x_1 = 0$) emerges parallel to the optical axis at the output plane ($\theta_{x2} = 0$), the Fourier transformation is perfect with no residual phase curvature ($R = \infty$), and the magnification M is given by the ratio of the distance of the output ray to the optical axis, to the angle of the input ray: $M = x_2 \lambda / \theta_{x1} s^2$ (figure 8.8a).

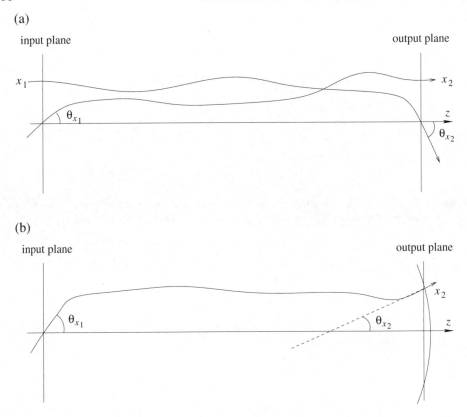

Figure 8.8. Rays in a Fourier transforming system (Ozaktas and Erden 1997).

3. If a ray emanating from the axial point at the input plane does not emerge parallel
    to the optical axis at the output plane, the scale factor is found in the same way
    as in perfect Fourier transformation, and the radius of curvature of the residual
    quadratic-phase factor is simply that defined by the slope of the ray: $R = x_2/\theta_{x2}$
    (figure 8.8b).

Overall, we see that Fourier transforming systems map ray positions into ray angles,
and ray angles into ray positions. This is to be expected due to the close association
between ray angles and spatial frequencies.

   We have already discussed three examples of Fourier transforming systems on
pages 246 and 279. All of these systems provide perfect Fourier transforms without any
residual phase. A system with residual phase can be corrected by either working with
spherical reference surfaces or appending thin lenses at the input or output planes.

   We conclude with a result which states that any quadratic-phase system can
be interpreted as a magnified Fourier transform by choosing appropriate spherical
reference surfaces. This fact directly follows from the following matrix identity:

$$\begin{bmatrix} \hat{A} & \hat{B} \\ \hat{C} & \hat{D} \end{bmatrix} = \begin{bmatrix} 1 & 0 \\ 1/\lambda R_2 & 1 \end{bmatrix} \begin{bmatrix} 0 & M' \\ -1/M' & 0 \end{bmatrix} \begin{bmatrix} 1 & 0 \\ -1/\lambda R_1 & 1 \end{bmatrix}, \tag{8.77}$$

where $R_1 = -\hat{B}/\hat{A}\lambda$, $R_2 = \hat{B}/\hat{D}\lambda$, and $M' = \hat{B}$. Alternatively, any quadratic-phase system can be realized by appending lenses with focal lengths $f_1 = R_1$ and $f_2 = -R_2$ to the input and output planes of a magnified Fourier transforming system.

8.4.3   General theorems for image and Fourier transform planes

We now consider a generic first-order optical system of the general form shown in figure 8.9, consisting of several thin lenses separated by arbitrary distances, an object $\hat{f}(x)$, and a point light source. Although not shown in the figure, the system may also contain sections of quadratic graded-index media or more generally any system which can be characterized by a linear $\hat{A}\hat{B}\hat{C}\hat{D}$ relation. That part of the system between the point source and object serves to illuminate the object, and that part to the right of the object forms a sequence of images and Fourier transforms. Typically, we expect to first observe a Fourier transform, followed by an inverted image, followed by an inverted Fourier transform, followed by an erect image, and so on. (In chapter 9 we will see that in between the object, Fourier, and image planes, we observe the fractional Fourier transforms of the object.)

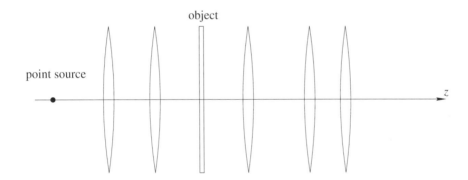

Figure 8.9.   Generic first-order optical system.

The planes at which the images are observed are quite easily determined by finding those planes at which the system matrix from the object to that plane is of the form of equation 8.72. The magnifications and residual phase factors are also readily obtained from this procedure. (The image planes can also be less systematically found by consecutive application of the equation $1/f = 1/d_1 + 1/d_2$.)

Likewise, the planes at which the Fourier transforms are observed are quite easily determined by finding those planes at which the system matrix from the object to that plane is of the form of equation 8.76. The magnifications and residual phase factors are again readily obtained from this procedure. It turns out that *the planes at which the Fourier transforms are observed are precisely the planes at which the images of the point source would have been observed had there been no object. Furthermore, the phase factor preceding the Fourier transform is the same as that which would have been observed at the transform plane had the object been replaced with a point source.* Thus the location of the Fourier transform does not depend on the location of the object,

object

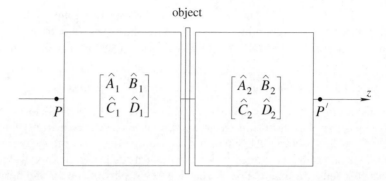

Figure 8.10. Derivation of general Fourier transforming condition. P represents the point source and P' one of its images had there been no object transparency.

although the residual phase factor does. This result encompasses and generalizes the many Fourier transforming configurations given in texts on Fourier optics. We now present a derivation of this very general result which is stated in several sources without proof (VanderLugt 1992, Goodman 1996).

This result is intuitively plausible in that having no object is the same as having an object whose transmittance is unity, so that the Fourier transform plane being the plane where the point source is imaged is consistent with the fact that the Fourier transform of unity is a delta function.

To derive this result, we refer to figure 8.10. Here we have lumped all components lying between the point source and object transparency into a single first-order system characterized by $\hat{A}_1$, $\hat{B}_1$, $\hat{C}_1$, $\hat{D}_1$, and we have lumped all components lying between the object and a point to which the source is imaged into another first-order system characterized by $\hat{A}_2$, $\hat{B}_2$, $\hat{C}_2$, $\hat{D}_2$. If P' is an image of P when there is no object, the following necessarily holds (equation 8.72)

$$\begin{bmatrix} M & 0 \\ M/\lambda R & 1/M \end{bmatrix} = \begin{bmatrix} \hat{A}_2 & \hat{B}_2 \\ \hat{C}_2 & \hat{D}_2 \end{bmatrix} \begin{bmatrix} \hat{A}_1 & \hat{B}_1 \\ \hat{C}_1 & \hat{D}_1 \end{bmatrix} \qquad (8.78)$$

for some M and R. Since the upper right element of an imaging matrix is zero, all rays emanating from point P, regardless of their angle, arrive at P'. The result we are seeking to prove states that the Fourier transform of the object is observed at P' regardless of how the left-hand side is decomposed into the two matrices appearing on the right-hand side. The left part of the system characterized by $\hat{A}_1, \hat{B}_1, \hat{C}_1, \hat{D}_1$ merely illuminates the object transparency with a spherical wave. This is fully equivalent to putting a thin lens of appropriate focal length f immediately before the object and illuminating it with a plane wave. Since the positions of the object transparency and thin lens can be interchanged without effect, this in turn is equivalent to plane wave illumination of the object transparency followed by the thin lens followed by the system characterized by $\hat{A}_2, \hat{B}_2, \hat{C}_2, \hat{D}_2$. Thus if we can show that the thin lens followed by the system characterized by $\hat{A}_2, \hat{B}_2, \hat{C}_2, \hat{D}_2$ is a Fourier transforming system, we will have demonstrated the claimed result. This amounts to showing that the matrix

$$
\begin{bmatrix} \hat{A}_2 & \hat{B}_2 \\ \hat{C}_2 & \hat{D}_2 \end{bmatrix} \begin{bmatrix} 1 & 0 \\ -1/\lambda f & 1 \end{bmatrix} \tag{8.79}
$$

is in the form of equation 8.76. In order to show this, it remains to determine the focal length f of the thin lens which illuminates the object with the same spherical wave which would have been generated by the point source followed by the system characterized by $\hat{A}_1, \hat{B}_1, \hat{C}_1, \hat{D}_1$. Thinking in terms of rays, the point source is represented by a bundle of rays of the general form $[0 \; \sigma_x]^T$, and the plane wave illuminating the thin lens is represented by a bundle of rays of the general form $[x \; 0]^T$. Thus, equating the bundle of rays resulting from a point source followed by the system $\hat{A}_1, \hat{B}_1, \hat{C}_1, \hat{D}_1$, to a plane wave followed by the lens of focal length f, we obtain

$$
\begin{bmatrix} \hat{A}_1 & \hat{B}_1 \\ \hat{C}_1 & \hat{D}_1 \end{bmatrix} \begin{bmatrix} 0 \\ \sigma_x \end{bmatrix} = \begin{bmatrix} 1 & 0 \\ -1/\lambda f & 1 \end{bmatrix} \begin{bmatrix} x \\ 0 \end{bmatrix}, \tag{8.80}
$$

from which we get $f = -\hat{B}_1/\lambda\hat{D}_1$. This lens, when illuminated by a plane wave, produces a spherical wave whose radius of curvature is $\hat{B}_1/\lambda\hat{D}_1$. Now we can evaluate the matrix product in equation 8.79 using this value of f:

$$
\begin{bmatrix} \hat{A}_2 + \hat{B}_2\hat{D}_1/\hat{B}_1 & \hat{B}_2 \\ \hat{C}_2 + \hat{D}_1\hat{D}_2/\hat{B}_1 & \hat{D}_2 \end{bmatrix}. \tag{8.81}
$$

Using the condition $\hat{A}_2\hat{B}_1 + \hat{B}_2\hat{D}_1 = 0$ which follows from equation 8.78, the above equation reduces to

$$
\begin{bmatrix} 0 & \hat{B}_2 \\ -1/\hat{B}_2 & \hat{D}_2 \end{bmatrix}, \tag{8.82}
$$

which upon comparison with equation 8.76 is seen to be a Fourier transform matrix with magnification $M = \hat{B}_2$ and residual phase radius of curvature $R = \hat{B}_2/\lambda\hat{D}_2$ (here we take $s = 1$ for simplicity). Thus we have shown that the Fourier transform is observed at locations where the point source is imaged and does not depend on the location of the object. The magnification and curvature, however, do depend on where the object is. We now also show that the residual phase factor observed is the same phase factor as would be observed using a point source located in place of the object. Again representing a point source by $[0 \; \sigma_x]^T$ and passing it through the system $\hat{A}_2, \hat{B}_2, \hat{C}_2, \hat{D}_2$, we obtain

$$
\begin{bmatrix} \hat{A}_2 & \hat{B}_2 \\ \hat{C}_2 & \hat{D}_2 \end{bmatrix} \begin{bmatrix} 0 \\ \sigma_x \end{bmatrix} = \begin{bmatrix} \hat{B}_2\sigma_x \\ \hat{D}_2\sigma_x \end{bmatrix}. \tag{8.83}
$$

The right-hand side corresponds to a bundle of rays representing a spherical wave with radius of curvature $\hat{B}_2/\lambda\hat{D}_2$, which is the same as the radius of curvature of the residual phase factor accompanying the Fourier transform.

Naturally, it is also possible to think of the above derivation in terms of direct illumination of an object by a diverging or converging spherical wave, rather than a point source followed by illumination optics. (Such illumination can of course be obtained by a plane wave followed by a thin lens.) In this case we would say that the Fourier transform is observed at the plane where the illuminating spherical wave would have come to focus, had the object transparency been removed. We finally note that the results also hold for virtual sources and when the point source is at $-\infty$ (corresponding to plane wave illumination).

We now turn our attention to another closely related result. But first we must very briefly introduce the *aperture stop*, and the *entrance pupil* and *exit pupil* of an optical system, referring the reader to textbooks such as Jenkins and White 1976 and Born and Wolf 1980 for a more complete explanation. These concepts are most easily understood in terms of a centered rotationally symmetric optical system consisting of many sections of free space, lenses, and apertures. The aperture stop is the limiting aperture in such an optical system. It is the aperture which limits the solid angle of the fan or cone of rays (the ray bundle) which can pass through the system. Increasing the diameter of any aperture other than the aperture stop will not increase the solid angle of the cone of rays which can pass through the system. Increasing the diameter of the aperture stop, on the other hand, will let a greater number of rays pass. (However, if the diameter is increased too much, some other aperture may start limiting the solid angle and become the aperture stop.) The exit pupil and the entrance pupil are the images of the aperture stop with respect to the image and object sides. That is, the image of the stop formed by everything to the right of the stop is the exit pupil, and the image of the stop formed by everything to the left of the stop is the entrance pupil. Alternatively, the exit pupil is where the stop seems to be when we look from the image side and the entrance pupil is where the stop seems to be when we look from the object side. Ray matrices can be used to determine their position and size. Since we cannot offer a more complete discussion, an example should help make these concepts more concrete. Figure 8.11 shows an imaging system consisting of two lenses and an aperture stop located between the lenses (the lenses themselves are assumed to have large apertures). The entrance and exit pupils are shown.

Now, consider an imaging system which under ideal conditions (no apertures) would map an on-axis delta function object to an on-axis delta function image. In reality there will always be a limiting aperture (the aperture stop) determining the spatial frequency response of the system. If the object is an on-axis delta function, the Fourier transform of the aperture stop will be observed at the output plane. To see this, we again refer to figure 8.10, this time interpreting P as an on-axis point object, P' as its ideal on-axis image, and the object in that figure as an aperture stop. It now follows from the analysis associated with figure 8.10 that we observe the Fourier transform of the aperture stop at the image plane.

We can now state the result we wish to demonstrate: *The amplitude distribution observed at the image plane will be the same as that which would have been observed had the exit pupil been illuminated with a spherical wave converging towards the ideal point image through free space. This observed amplitude distribution is a scaled version of the Fourier transform of the exit pupil.* The latter part of this statement is consistent with the result stated in the previous paragraph because the exit pupil is simply an image of the aperture stop so that their Fourier transforms are simply scaled versions of each other.

We now proceed to demonstrate the stated result. The left part of the system

(a)

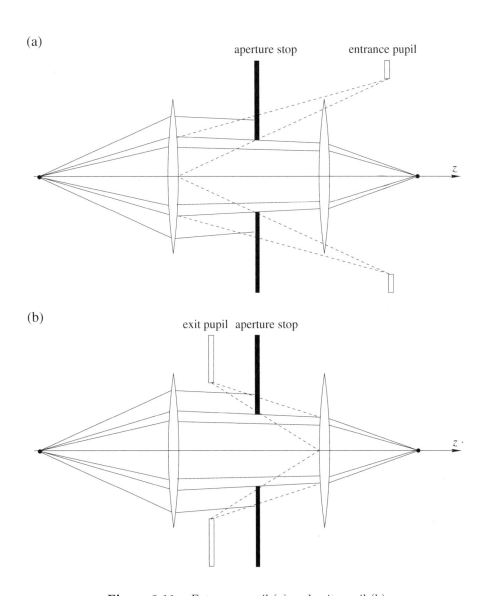

Figure 8.11. Entrance pupil (a) and exit pupil (b).

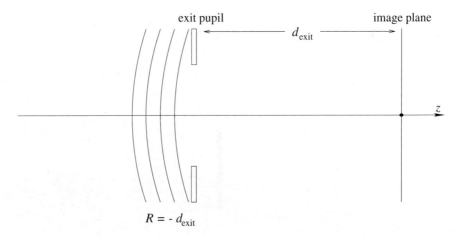

Figure 8.12. The exit pupil illuminated by a converging spherical wave.

characterized by $\hat{A}_1, \hat{B}_1, \hat{C}_1, \hat{D}_1$ merely serves to illuminate the aperture stop with a spherical wave of radius $\hat{B}_1/\lambda\hat{D}_1$. Thus, the result stated in the preceding paragraph is equivalent to the following claim. The second part of the system consisting of the aperture stop illuminated by a spherical wave followed by the system characterized by $\hat{A}_2, \hat{B}_2, \hat{C}_2, \hat{D}_2$, produces the same amplitude distribution as the exit pupil illuminated by a spherical wave of radius $-d_{\text{exit}}$ followed by a section of free space of length d_{exit} (figure 8.12). Here d_{exit} is the distance of the exit pupil from the image plane. First, we write the equation which enables us to determine the size and location of the exit pupil:

$$\left[\begin{array}{cc} 1 & -\lambda d_{\text{exit}} \\ 0 & 1 \end{array} \right] \left[\begin{array}{cc} \hat{A}_2 & \hat{B}_2 \\ \hat{C}_2 & \hat{D}_2 \end{array} \right] = \left[\begin{array}{cc} M_{\text{exit}} & 0 \\ M_{\text{exit}}/\lambda R_{\text{exit}} & 1/M_{\text{exit}} \end{array} \right]. \qquad (8.84)$$

The left-hand side of this equation consists of the matrix associated with that part of the system falling to the right of the aperture stop, followed by free-space propagation over a distance $-d_{\text{exit}}$. (Here the minus sign reflects the fact that d_{exit} is measured towards the left, so that if $d_{\text{exit}} > 0$ the location of the exit pupil is to the left of the image plane.) The right-hand side of this equation has the form of equation 8.72 and reflects the fact that the exit pupil is the image of the aperture stop as seen from the image side. By solving this equation we can determine the magnification M_{exit} between the exit pupil and the aperture stop as $1/\hat{D}_2$, and the distance of the exit pupil towards the left of the image plane as $d_{\text{exit}} = \hat{B}_2/\lambda\hat{D}_2$. Now we can demonstrate the claimed equivalence result through the following identity which the reader should be able to verify:

$$\left[\begin{array}{cc} \hat{A}_2 & \hat{B}_2 \\ \hat{C}_2 & \hat{D}_2 \end{array} \right] \left[\begin{array}{cc} 1 & 0 \\ \hat{D}_1/\hat{B}_1 & 1 \end{array} \right] = \left[\begin{array}{cc} 1 & \lambda d_{\text{exit}} \\ 0 & 1 \end{array} \right] \left[\begin{array}{cc} 1 & 0 \\ -1/\lambda d_{\text{exit}} & 1 \end{array} \right] \left[\begin{array}{cc} M_{\text{exit}} & 0 \\ 0 & 1/M_{\text{exit}} \end{array} \right].$$

$$(8.85)$$

The second term on the left-hand side corresponds to illumination of the aperture stop by a spherical wave with radius $\hat{B}_1/\lambda\hat{D}_1$, which is equal to $-\hat{B}_2/\lambda\hat{A}_2$ by virtue

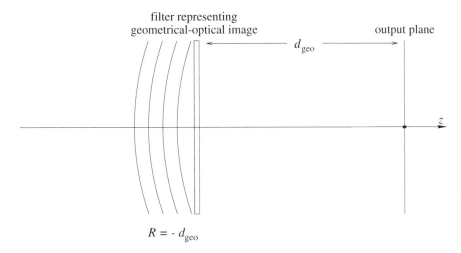

Figure 8.13. The input of the system is the geometrical-optical image of the original input object as seen from the output side, and is situated at the same location where this image is observed. It is illuminated by a spherical wave converging towards the axis at the output plane.

of the imaging condition 8.78. The last term on the right-hand side reflects the fact that the exit pupil is a magnified image of the aperture stop. The central term on the right-hand side represents a spherical wave converging towards the on-axis image point, and the first term corresponds to free-space propagation over a distance d_{exit}. Thus the two sides of the above identity represent the two systems whose equivalence we were seeking to prove.

The above discussion has been carried out for an on-axis object and image point. If it can be assumed that the images corresponding to off-axis object points are simply shifted versions of the Fourier transform of the exit pupil, we are then led to a simple space-invariant model for the whole system with the impulse response being given by the Fourier transform of the exit pupil. Unfortunately, the system is often not space-invariant so that this is not possible. See the discussion following equation 7.159.

We will conclude this section by considering the demonstrated result from a different perspective. Let us assume $\hat{f}(x)$ (or $\hat{f}(x,y)$) is the input of an optical system characterized by $\hat{A}_2, \hat{B}_2, \hat{C}_2, \hat{D}_2$. We assume $\hat{f}(x)$ is illuminated with a spherical wave which, in the absence of $\hat{f}(x)$, would have been focused to the axis at the output plane. The output of the system is then the same as that resulting from the system shown in figure 8.13. Here the input is the geometrical-optical image of the original input object $\hat{f}(x)$ as seen from the output side, illuminated with a spherical wave converging towards the axis at the output plane. This result is simply a more general form of the previous result with an arbitrary function replacing the aperture stop, and thus the matrix algebra behind this result is the same. The significance of this result is that it allows a separation of geometrical-optical effects from diffraction effects. By replacing the original $\hat{f}(x)$ by its geometrical-optical image, and the system characterized by $\hat{A}_2, \hat{B}_2, \hat{C}_2, \hat{D}_2$ by free space, we separate out the geometric effect, leaving us with the diffraction effect in the pure form represented by figure 8.13.

A complementary discussion of some of these issues may be found in Goodman 1996.

8.5   Decompositions and duality in optics

In section 3.4.4 we discussed how general linear canonical transforms (quadratic-phase systems) could be expressed in terms of each other or decomposed into simpler component systems. These decompositions can be interpreted in terms of matrices, abstract operators, operations such as chirp multiplications and convolutions, integral transforms, and geometric distortions. Now we are in a position to see that they can also be interpreted in terms of optical components such as sections of free space, thin lenses, quadratic graded-index media, and imaging and Fourier transforming systems. Such decompositions of optical systems have many applications. Two particularly important decompositions are those showing how any quadratic-phase system can be realized by using only two sections of free space and a thin lens (equation 3.121) or two thin lenses and a section of free space (equation 3.122).

As an example, consider the decompositions

$$
\begin{bmatrix} 0 & \lambda f \\ -1/\lambda f & 0 \end{bmatrix} = \begin{bmatrix} 1 & 0 \\ -1/\lambda f & 1 \end{bmatrix} \begin{bmatrix} 1 & \lambda d \\ 0 & 1 \end{bmatrix} \begin{bmatrix} 1 & 0 \\ -1/\lambda f & 1 \end{bmatrix}
$$
$$
= \begin{bmatrix} 1 & \lambda d \\ 0 & 1 \end{bmatrix} \begin{bmatrix} 1 & 0 \\ -1/\lambda f & 1 \end{bmatrix} \begin{bmatrix} 1 & \lambda d \\ 0 & 1 \end{bmatrix}, \tag{8.86}
$$

with $f = d$, which are special cases of equations 3.122 and 3.121. These decompositions show how an optical Fourier transform can be realized either as a lens followed by free space followed by another lens, or free space followed by a lens followed by free space. These are nothing but the Fourier transform stages (ii) and (i) we introduced on page 246. These decompositions can also be cast in abstract operator notation if desired (equations 3.130 and 3.129). We will later see in chapter 9 that optical fractional Fourier transforms can also be realized by similar decompositions but with $f \neq d$.

As another example, let us consider the simulation of a section of free space of negative length $-d$, where $d > 0$. Using

$$
\begin{bmatrix} 1 & -\lambda d \\ 0 & 1 \end{bmatrix} = \begin{bmatrix} 0 & -Ms^2 \\ 1/Ms^2 & 0 \end{bmatrix} \begin{bmatrix} 1 & 0 \\ -1/\lambda f & 1 \end{bmatrix} \begin{bmatrix} 0 & Ms^2 \\ -1/Ms^2 & 0 \end{bmatrix}, \tag{8.87}
$$

with the condition $\lambda^2 f d = -s^4 M^2$, we see that a negative section of free space can be simulated with two Fourier transform stages and a negative lens. This configuration can also be useful for simulating an inconveniently long section of free space. Alternatively, a lens of given focal length we happen not to have at hand, can be simulated by using a section of free space of appropriate length and two Fourier transform stages. A zoom lens can likewise be simulated by employing a variable section of free space.

In an optical context, equation 3.138 (or 3.139) means that any quadratic-phase system can be interpreted as a magnified fractional (or ordinary) Fourier transform by choosing appropriate spherical reference surfaces. An alternative interpretation is that any quadratic-phase system can be implemented by appending lenses at the input and output surfaces of a fractional (or ordinary) Fourier transformer.

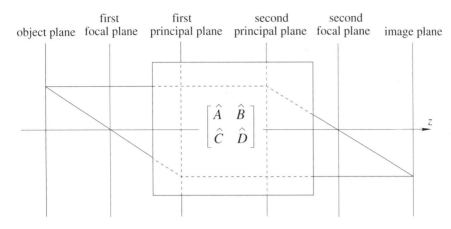

Figure 8.14.   Principal and focal planes.

More generally, equation 3.137 means that any quadratic-phase system can be interpreted as a magnified version of another simply by choosing appropriate spherical reference surfaces. A more general treatment of the synthesis of quadratic-phase systems, also covering anamorphic two-dimensional systems, is Sahin, Ozaktas, and Mendlovic 1998.

The decompositions under consideration can also be used to discuss a number of traditional concepts in geometrical optics. Considering an imaging system, the second focal plane is defined as the plane where a ray coming from the left parallel to the optical axis intersects the optical axis (figure 8.14). Likewise, the first focal plane is the plane where a ray coming from the right parallel to the optical axis intersects the optical axis. The second principal plane is that plane where the ray coming from the left seems to bend towards the second focal point. In reality, when we have a complex optical system consisting of several components, the ray will actually undergo several bends before finally heading for the second focal point. Thus the figure does not show the actual trajectory of the ray, but a fictitious trajectory corresponding to the total refractive effect being concentrated at a single plane, which we refer to as the second principal plane. The first principal plane is likewise defined for a ray coming from the right. The first and second focal lengths f_1 and f_2 are measured from the principal planes to the focal planes and are equal in magnitude when the refractive indices of the object and image spaces are equal. A third pair of planes called the first and second nodal planes are also often introduced. However, we will not separately define them since they coincide with the principal planes when the refractive indices of the object and image spaces are equal. The rationale behind these definitions is that they enable us to treat a complex optical system consisting of many components as if it were a thin lens. The first principal plane plays the role of the input plane of the thin lens and the second principal plane plays the role of the output plane of the thin lens. The light distribution or bundle of rays incident on the first principal plane emerges from the second principal plane just as if these planes were the input and output faces of a thin lens. In other words, the matrix relating the ray vector at the second principal plane to that at the first principal plane is of lower triangular form. With reference to

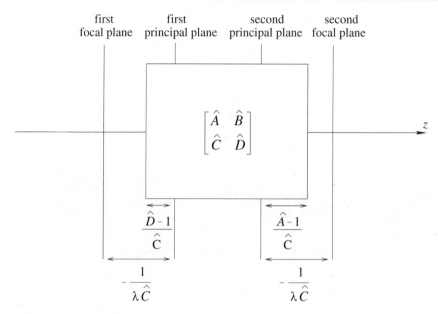

Figure 8.15. Determination of principal and focal planes. The figure is drawn for the case $\hat{C} < 0$, $(1 - \hat{A})/\hat{C} < 0$ and $(1 - \hat{D})/\hat{C} < 0$.

the optical system characterized by $\hat{A}$, $\hat{B}$, $\hat{C}$, $\hat{D}$ shown in figure 8.15, let us consider the following decomposition:

$$\begin{bmatrix} 1 & 0 \\ \hat{C} & 1 \end{bmatrix} = \begin{bmatrix} 1 & (1 - \hat{A})/\hat{C} \\ 0 & 1 \end{bmatrix} \begin{bmatrix} \hat{A} & \hat{B} \\ \hat{C} & \hat{D} \end{bmatrix} \begin{bmatrix} 1 & (1 - \hat{D})/\hat{C} \\ 0 & 1 \end{bmatrix}. \tag{8.88}$$

The left-hand side of this equation is the matrix from the first principal plane to the second principal plane, and indeed corresponds to a thin lens with focal length $f = -1/\lambda\hat{C}$. The two matrices surrounding the original system matrix on the right-hand side, represent sections of free space and thus tell us the locations of the principal planes with respect to the actual physical input and output planes of the original system.

Here we also mention the concept of *dual* optical systems (Lohmann 1954, Papoulis 1968). The concept of duality was introduced in equations 3.141 and 3.142. In an optical context, a thin lens of focal length f is the dual of a section of free space of length d provided $s^4 = \lambda^2 fd$, where s is some implicit scale factor with the dimension of length. This means that the action of one is the same as the action of the other under a Fourier transform with this implicit scale factor s:

$$\begin{bmatrix} 1 & 0 \\ -1/\lambda f & 1 \end{bmatrix} = \begin{bmatrix} 0 & -s^2 \\ 1/s^2 & 0 \end{bmatrix} \begin{bmatrix} 1 & \lambda d \\ 0 & 1 \end{bmatrix} \begin{bmatrix} 0 & s^2 \\ -1/s^2 & 0 \end{bmatrix}. \tag{8.89}$$

The dual of a section of quadratic graded-index medium is again a section of quadratic graded-index medium. The dual of a system consisting of several components is easily found by replacing each component by its dual, with the value of s fixed. The wave fields $\hat{f}(x)$ are replaced by their Fourier transforms $\hat{F}(x/s^2)$. Rays are replaced by dual rays: if a ray is represented by the vector $[x \ \sigma_x]^\mathrm{T}$, the dual of this ray is represented by $[s^2\sigma_x \ x/s^2]^\mathrm{T}$. If an optical system maps a given input $\hat{f}(x)$ into an output $\hat{g}(x)$, then the dual optical system will map the input $\hat{F}(x/s^2)$ into $\hat{G}(x/s^2)$. Likewise, if an optical system maps an input ray into an output ray, the dual optical system will map the dual of the original input ray into the dual of the original output ray.

In closing this section, we mention a closely related approach to the study of first-order optical systems known as *operator optics* or *operator algebra*. A brief exposition may be found in Goodman 1996, or the reader may wish to consult some of the original references, of which we provide a selection: Butterweck 1977, 1981, Shamir 1979, Nazarathy and Shamir 1980, 1981, 1982a, b, Stoler 1981, Nazarathy, Hardy, and Shamir 1982, 1986, Ojeda-Castañeda and Noyola-Isgleas 1988, Ruiz and Rabal 1997. This approach has also been generalized to deal with higher-order systems. A useful introductory exposition to the use of operators and the applications of linear canonical transforms and group theoretical methods in optics may be found in Seger and Lenz 1992.

The operators employed in such approaches are simply dimensional forms of the operators $\mathcal{Q}_q$, $\mathcal{R}_r$, $\mathcal{F}$, $\mathcal{M}$ and the like which appear in this book, and are used to represent optical components such as thin lenses and sections of free space. Once a list of commutation relations and formulas are tabulated, one can easily manipulate concatenations of operators representing components, or decompose complex systems into elementary components. In this book we prefer to use matrices to represent these operators since the ease with which these matrices are multiplied allows us to dispense with a pretabulated list of commutation relations and formulas. Incidentally, this is also the reason why Baker-Campbell-Hausdorff type relations have not been employed in this book; these relations are not transparent and directly verifiable (see equation 2.245 and the accompanying discussion). Nevertheless, the operator method has its place as an elegant and useful formulation of Fourier optics and interested readers are especially recommended to study the works of Nazarathy and Shamir cited above.

8.6   Relations between wave and geometrical optics

In this section we will present several results which will serve to further highlight the parallels between wave optics and geometrical optics perspectives for first-order systems.

8.6.1   *Phase of the system kernel and Hamilton's point characteristic*

This section assumes slight familiarity with Hamilton's characteristic functions (Born and Wolf 1980).

First, we begin by showing how the phase of the kernel transforming the wave fields can be related to geometrical-optical concepts. Let the kernel of a general linear system be expressed as

$$\hat{h}(x, x') = |\hat{h}(x, x')| e^{i2\pi \hat{V}(x, x')}, \tag{8.90}$$

where the magnitude and phase are both assumed to be slowly varying. Approximating the phase as

$$\hat{V}(x, x') \approx \hat{V}(0,0) + \frac{\partial \hat{V}}{\partial x} x + \frac{\partial \hat{V}}{\partial x'} x', \tag{8.91}$$

and taking the magnitudes out of the integral in equation 8.68, we can explicitly calculate the kernel transforming the Wigner distributions as (Bastiaans 1979b)

$$\hat{K}_{\hat{h}}(x, \sigma_x; x', \sigma'_x) = |\hat{h}(x, x')|^2 \, \delta[\sigma_x - \partial \hat{V}(x, x')/\partial x] \, \delta[\sigma'_x + \partial \hat{V}(x, x')/\partial x']. \tag{8.92}$$

These delta functions imply a pointwise mapping of the values of the Wigner distribution in phase space. Which phase-space point is mapped to which phase-space point is governed by the relations

$$\sigma_x = \frac{\partial \hat{V}(x, x')}{\partial x}, \qquad \sigma'_x = -\frac{\partial \hat{V}(x, x')}{\partial x'}. \tag{8.93}$$

Recalling that σ_x can be interpreted as a normalized ray angle, the above equations for $\hat{V}(x, x')$ can be recognized as those satisfied by *Hamilton's point characteristic* (Bastiaans 1979b). The point characteristic gives the optical path length from the input point x' to the output point x, and is known to satisfy equations 8.93 (Born and Wolf 1980). (To be precise, we note that the quantity $\hat{V}(x, x')$ appearing here is actually Hamilton's point characteristic normalized by the wavelength of light in free space.) Thus it may be concluded that the wave-optical phase $\hat{V}(x, x')$ corresponds to the geometrical-optical entity known as Hamilton's point characteristic. The reader may also wish to relate this discussion to the eikonal (page 251) and Hamilton's equations (page 252).

   Similar derivations are also possible for the other three representations of the system kernel: $\hat{h}_{\text{spac}\to\text{freq}}(\sigma_x, x')$, $\hat{h}_{\text{freq}\to\text{spac}}(x, \sigma'_x)$, and $\hat{h}_{\text{freq}\to\text{freq}}(\sigma_x, \sigma'_x)$. (For instance, $\hat{h}_{\text{spac}\to\text{freq}}(\sigma_x, x')$ relates the Fourier transform of the output to the input: $\hat{G}(\sigma_x) = \int \hat{h}_{\text{spac}\to\text{freq}}(\sigma_x, x') \hat{f}(x') \, dx'$.) Bastiaans (1979b) shows that these likewise correspond to Hamilton's angle and mixed characteristics.

   We will now derive Hamilton's point characteristic for a first-order optical system in terms of the $\hat{A}\hat{B}\hat{C}\hat{D}$ parameters. Since this discussion is more readily interpreted from a geometrical optics perspective, we will use $[x_1 \ \sigma_{x1}]^T$ and $[x_2 \ \sigma_{x2}]^T$, rather than primed quantities, to denote the input and output ray vectors respectively. First-order systems can be characterized by a linear relation between the input and output ray vectors:

$$x_2 = \hat{A}x_1 + \hat{B}\sigma_{x1},$$
$$\sigma_{x2} = \hat{C}x_1 + \hat{D}\sigma_{x1}, \tag{8.94}$$

which can be solved to yield

$$\sigma_{x1} = \frac{1}{\hat{B}}\left(-\hat{A}x_1 + x_2\right),$$
$$\sigma_{x2} = \frac{1}{\hat{B}}\left(-(\hat{A}\hat{D} - \hat{B}\hat{C})x_1 + \hat{D}x_2\right). \tag{8.95}$$

Now, using these together with equations 8.93 which are known to be satisfied by the point characteristic, we can write

$$\frac{\partial \hat{V}(x_2, x_1)}{\partial x_1} = \frac{1}{\hat{B}}\left(\hat{A}x_1 - x_2\right),$$

$$\frac{\partial \hat{V}(x_2, x_1)}{\partial x_2} = \frac{1}{\hat{B}}\left(-(\hat{A}\hat{D} - \hat{B}\hat{C})x_1 + \hat{D}x_2\right). \tag{8.96}$$

Integrating these with respect to x_1 and x_2 respectively, we obtain

$$\hat{V}(x_2, x_1) = \frac{1}{\hat{B}}\left(\hat{A}x_1^2/2 - x_2x_1\right) + \text{term independent of } x_1,$$

$$\hat{V}(x_2, x_1) = \frac{1}{\hat{B}}\left(\hat{D}x_2^2/2 - (\hat{A}\hat{D} - \hat{B}\hat{C})x_2x_1\right) + \text{term independent of } x_2. \tag{8.97}$$

Consistency of these equations requires $\hat{A}\hat{D} - \hat{B}\hat{C} = 1$ and

$$\hat{V}(x_2, x_1) = \frac{1}{2\hat{B}}\left(\hat{D}x_2^2 - 2x_2x_1 + \hat{A}x_1^2\right) + \text{term independent of } x_1 \text{ and } x_2. \tag{8.98}$$

Thus we have shown that the point characteristic of an optical system characterized by a linear $\hat{A}\hat{B}\hat{C}\hat{D}$ relation is of the above quadratic form. Referring back to equation 8.8, we see that $2\pi\hat{V}(x_2, x_1)$ is identical to the quadratic form appearing as the phase of that equation. This does not surprise us since the derivation starting with equation 8.90 showed that the phase of the transform kernel corresponds to Hamilton's point characteristic. Also note that the quadratic form of $\hat{V}$ is consistent with the approximation made in equation 8.91: neglecting second- and higher-order derivatives of $\hat{V}$ amounts to assuming it can be expressed as a quadratic function.

To consolidate, first-order systems are characterized by linear relations between input and output ray vectors, or equivalently, linear distortions in phase space: The output ray intercepts and angles are related linearly to the input ray intercepts and angles, and the effect in phase space is a parallelogram-type distortion. These linear relations are a consequence of the paraxial and other simplifying approximations. In terms of wave kernels, first-order systems are characterized by quadratic terms in the phase. It is assumed that higher-order terms can be neglected.

We conclude this section by pointing out the work of Alonso and Forbes (1995), which discusses a generalization of Hamilton's formalism for geometrical optics.

8.6.2   Transport equations for the Wigner distribution

We now turn our attention to *transport equations* for the Wigner distribution (Bastiaans 1979c, d, 1997), which are differential equations describing how the Wigner distribution $\hat{W}_{\hat{f}_z(x)}(x, \sigma_x)$ of the transverse field $\hat{f}_z(x)$ evolves with z. We first repeat equation 7.52 for a single transverse dimension:

$$\frac{\partial \hat{f}_z(x)}{\partial z} = i\sqrt{4\pi^2\sigma^2 + \frac{\partial^2}{\partial x^2}}\,\hat{f}_z(x), \tag{8.99}$$

which can be rearranged to read

$$\frac{\partial \hat{f}_z(x)}{\partial z} = i2\pi \sqrt{\sigma^2 - \left(\frac{1}{i2\pi}\frac{\partial}{\partial x}\right)^2} \, \hat{f}_z(x), \tag{8.100}$$

which in turn can be expressed in terms of the Hamiltonian given in equation 7.125 with $\sigma = 1/\lambda = n(x,z)f_{\text{oc}}/c$ as

$$\frac{-1}{i2\pi}\frac{\partial \hat{f}_z(x)}{\partial z} = \hat{H}(x, (i2\pi)^{-1}\partial/\partial x; z) \, \hat{f}_z(x), \tag{8.101}$$

where $\hat{H}(x, (i2\pi)^{-1}\partial/\partial x; z)$ is a differential operator. We are not surprised to see $(i2\pi)^{-1}\partial/\partial x$ in place of σ_x, since we know that the operator version of σ_x, corresponding to multiplication by σ_x in the frequency domain, corresponds to a derivative with respect to x in the space domain. Now, if $\hat{f}_z(x)$ evolves according to an equation of this general form, it is possible to show that $\hat{W}_{\hat{f}_z(x)}(x, \sigma_x)$ evolves according to the following equation (Bastiaans 1997):

$$\frac{\partial \hat{W}_{\hat{f}_z(x)}(x, \sigma_x)}{\partial z} = \frac{\partial \hat{H}}{\partial x}\frac{\partial \hat{W}_{\hat{f}_z(x)}}{\partial \sigma_x} - \frac{\partial \hat{H}}{\partial \sigma_x}\frac{\partial \hat{W}_{\hat{f}_z(x)}}{\partial x}. \tag{8.102}$$

(This equation can also be put in the form $\partial \hat{W}_{\hat{f}_z(x)}/\partial z = [\cdots]\hat{W}_{\hat{f}_z(x)}$ if desired.) Here $\hat{f}_z(x)$ is the amplitude distribution of light at a plane intersecting the optical axis at z, and $\hat{W}_{\hat{f}_z(x)}$ is the Wigner distribution of $\hat{f}_z(x)$. The solution of this equation tells us how the Wigner distribution of the transverse light distribution propagates along the optical axis z. (We already know how the Wigner distribution propagates as a function of z in a quadratic-phase system, such as free space or quadratic graded-index media. It simply undergoes shearing, rotation, or a more general linear distortion. The present formulation is more general and is valid for arbitrary weakly inhomogeneous refractive index distributions.)

Let us consider the total derivative of $\hat{W}_{\hat{f}_z(x)}(x, \sigma_x)$ with respect to z:

$$\frac{d\hat{W}_{\hat{f}_z(x)}}{dz} = \frac{\partial \hat{W}_{\hat{f}_z(x)}}{\partial x}\frac{dx}{dz} + \frac{\partial \hat{W}_{\hat{f}_z(x)}}{\partial \sigma_x}\frac{d\sigma_x}{dz} + \frac{\partial \hat{W}_{\hat{f}_z(x)}}{\partial z}. \tag{8.103}$$

The partial derivative of $\hat{W}_{\hat{f}_z(x)}$ with respect to z tells us how the value of $\hat{W}_{\hat{f}_z(x)}$ changes with increasing z for a given (x, σ_x). The total derivative, on the other hand, tells us the total change in $\hat{W}_{\hat{f}_z(x)}$ when $x(z)$ and $\sigma_x(z)$ are also varying with z. If $x(z)$ and $\sigma_x(z)$ are governed by Hamilton's equations (which are equivalent to the ray equation), the total derivative will tell us the change in $\hat{W}_{\hat{f}_z(x)}$ along an optical ray. Now, using Hamilton's equations $dx/dz = \partial \hat{H}/\partial \sigma_x$ and $d\sigma_x/dz = -\partial \hat{H}/\partial x$ and equation 8.102, we obtain

$$\frac{d\hat{W}_{\hat{f}_z(x)}}{dz} = 0. \tag{8.104}$$

The value of the Wigner distribution remains constant along a ray. (The same result can also be written in the form $d\hat{W}_{\hat{f}_z(x)}/ds = 0$, where ds is a differential element along a ray which satisfies the ray equation.) Referring the reader to Bastiaans 1979c, d for details and a precise statement of the conditions under which this result holds, and Bastiaans 1997 for a complementary discussion, here we focus on the interpretation of this result. For instance, let us consider a particular phase-space point (x_1, σ_{x1}) at the input plane, which corresponds to a ray emanating from the point x_1 with normalized angle σ_{x1}. The value of the Wigner distribution along all points along the ray will be the same. Alternatively, if we want to determine the value of the Wigner distribution at a certain phase-space point (x_2, σ_{x2}) at the output plane corresponding to a ray incident at the point x_2 with normalized angle σ_{x2}, all we need to do is to trace back this ray to the input plane and determine the value of the Wigner distribution at the phase-space point the ray originated from (Bastiaans 1979c). In other words, solving equation 8.102 is equivalent to solving the ray equation, since if we know how the rays propagate, we can also determine the Wigner distribution at any plane.

The kernel $\hat{K}_{\hat{h}}(x, \sigma_x; x', \sigma'_x)$ can be interpreted as the response of the system to a distribution of light whose Wigner distribution is $\delta(x - x')\,\delta(\sigma_x - \sigma'_x)$, which can be thought of as representing a single ray (Bastiaans 1978). There does not exist a distribution of light with such a Wigner distribution. However, such a fictitious Wigner distribution represents the same idealization and approximation as the concept of a ray of light. For reasons that should by now be obvious, Bastiaans refers to $\hat{K}_{\hat{h}}$ as the *ray spread function*, as opposed to the *point spread function* $\hat{h}(x, x')$.

As another viewpoint, we may consider the value of the Wigner distribution at the phase-space point (x, σ_x) as representing the intensity or weight of the ray passing through the point x with normalized angle σ_x. In a lossless system, each ray belonging to a bundle preserves its intensity or weight as it propagates through an inhomogeneous medium. Then the Wigner distribution function represents the intensities or weights associated with the whole bundle of rays. As light propagates through such a medium, the values of the Wigner distribution at given phase-space points are mapped in one-to-one fashion to other phase-space points, precisely according to the mapping of the ray vectors. In other words, the effect of propagation through this medium on the Wigner distribution is a distortion which perfectly parallels the distortion of the bundle of rays representing the same light distribution. We already knew this to be the case for quadratic-phase systems (recall that the ray matrix was found identical to the matrix governing the distortion of the Wigner distribution as it passes through a quadratic-phase system). We now see that this result is more generally true, and have further confirmation for the analogy between the support of the Wigner distribution and the region representing the bundle of rays.

It is also possible to relate our discussion to the local spatial spectrum of a light distribution $\hat{f}(x)$. We have already discussed the global spatial spectrum of $\hat{f}(x)$ in chapter 7, seeing that an arbitrary wave field can be expressed as the superposition of plane waves with different amplitudes and directions. In the local spatial spectrum, we concentrate on a certain small region and examine the spatial spectrum within this region. Mathematically, this is precisely what space-frequency representations such as the windowed Fourier transform or the Wigner distribution provide us. The value of the Wigner distribution at a certain phase-space point (x, σ_x) tells us the strength of the localized plane wave component at point x traveling along the direction defined by the normalized angle σ_x. It is such a localized plane wave component passing through a given point at a certain angle with which we can associate a ray (perpendicular to the wavefronts of the plane wave component).

8.6.3 Discussion

When dealing with first-order systems, equivalent results are obtained regardless of whether one uses wave optics or geometrical optics, so that a unified mathematical treatment is possible. The $\hat{A}\hat{B}\hat{C}\hat{D}$ matrix may be interpreted either as the ray matrix, or the matrix characterizing the linear canonical transform describing how the wave fields are transformed. Speaking of the equivalence of geometrical optics and wave optics requires care, since geometrical optics is normally considered to be an approximation to wave optics. There are two points to be made in this regard: (i) we are concerned with a certain restricted class of systems, namely first-order systems characterized by only three parameters; and (ii) even for these systems, the equivalence in question does not imply that the geometrical optics viewpoint will properly predict the wave fields everywhere (for instance, near a focal point). However, given the limited number of parameters of the systems under consideration, the ray matrix properly summarizes the operational characteristics of the kernel governing the propagation of the wave fields, so that this kernel and the output wave field can be easily obtained when desired.

The detailed discussion of quadratic graded-index media on page 348 will provide further illustration of the correspondence between the wave-optical and geometrical-optical pictures. That discussion will also reveal that geometrical optics is to wave optics what classical mechanics is to quantum mechanics, in the sense that geometrical optics and classical mechanics represent formally similar approximations of wave optics and quantum mechanics, respectively.

8.7 Quadratic-exponential signals

In this section we will discuss two special families of signals whose dependence on the spatial coordinate x is quadratic in the exponent. Apart from a constant multiplicative factor, quadratic-phase *systems* are characterized by three parameters (one of the four matrix parameters is redundant because of the unit-determinant condition). A single ray is characterized by two real parameters, the position x and normalized angle σ_x, which we write as a 2×1 vector $[x \ \sigma_x]^{\mathrm{T}}$. The quadratic-exponential signals we will discuss are likewise characterized by only two parameters.

This section may be skipped without loss of continuity.

8.7.1 Ray-like signals

In the previous section we noted that a distribution of light whose Wigner distribution is $\delta(x - x_0)\,\delta(\sigma_x - \sigma_{x0})$ could be associated with a ray whose position is x_0 and whose normalized angle is σ_{x0}. We also commented that no actual distribution of light has such a Wigner distribution. A signal whose Wigner distribution comes as close as physically possible to $\delta(x - x_0)\,\delta(\sigma_x - \sigma_{x0})$ is

$$\hat{w}_{x_0,\sigma_{x0}}(x) = 2^{1/4}\Delta_x^{-1/2}\,e^{-\pi(x-x_0)^2/\Delta_x^2}\,e^{i2\pi\sigma_{x0}x}. \tag{8.105}$$

The Wigner distribution of this signal can be expressed as (Bastiaans 1978, 1989)

$$2 \exp \left[-2\pi \left(\frac{(x - x_0)^2}{\Delta_x^2} + \frac{(\sigma_x - \sigma_{x0})^2}{\Delta_{\sigma_x}^2} \right) \right], \qquad (8.106)$$

where $\Delta_{\sigma_x} = 1/\Delta_x$. This is a distribution of light which is spatially concentrated within a region of extent Δ_x around the point x_0, and spectrally concentrated within a region of extent Δ_{σ_x} around the frequency σ_{x0}. This distribution of light is as close as we can physically get to a ray passing through x_0 with normalized angle σ_{x0}.

Such a signal $\hat{w}_{x_0,\sigma_{x0}}(x)$ is characterized by two parameters corresponding to position and frequency, just as a ray is characterized by two parameters corresponding to position and normalized angle, and of course we already know that normalized angle corresponds to frequency. Thus both a ray and $\hat{w}_{x_0,\sigma_{x0}}(x)$ can be characterized by a 2×1 vector $[x_0 \ \sigma_{x0}]^\mathrm{T}$ which is transformed by the $\hat{A}\hat{B}\hat{C}\hat{D}$ matrix upon passage through a quadratic-phase system. When $\hat{w}_{x_0,\sigma_{x0}}(x)$ is input to a quadratic-phase system, the output is another similar signal characterized by the vector $[(\hat{A}x_0 + \hat{B}\sigma_{x0}), (\hat{C}x_0 + \hat{D}\sigma_{x0})]^\mathrm{T}$, which tells us where the output signal is centered in the space-frequency plane. We are not dealing with the shape parameters of $\hat{w}_{x_0,\sigma_{x0}}(x)$, which are also transformed in passing through the system; we are concentrating only on its center (x_0, σ_{x0}).

A ray and the signal $\hat{w}_{x_0,\sigma_{x0}}(x)$ can both be seen as elementary signals characterized by two real parameters. Just as more general distributions of light can be represented in geometrical optics as a bundle of rays, they can also be expressed in wave optics as a linear superposition of signals of the form $\hat{w}_{x_0,\sigma_{x0}}(x)$. If we let $x_0 = l\,\delta x$ and $\sigma_{x0} = m\,\delta\sigma_x$, we immediately recognize these signals as Gaussian basis functions appearing in a Gabor expansion (equation 3.7), so that the linear superposition in question is nothing but a Gabor expansion. Thinking of a distribution of light as a bundle of rays corresponds to thinking of it in terms of its Gabor expansion. The Gabor coefficient may be associated with the weight of either the ray in question or a "plane wave" localized in both space and spatial frequency, with its wavefronts perpendicular to the ray.

In section 8.6.2 we discussed how to associate a bundle of weighted rays with an arbitrary wave field $\hat{f}(x)$. We simply find the Wigner distribution of the wave field and assign $\hat{W}_{\hat{f}}(x, \sigma_x)$ as a weight to the ray $[x \ \sigma_x]^\mathrm{T}$. We now consider an alternative approach. Let us write $\hat{f}(x) = |\hat{f}(x)| \exp[i2\pi\hat{\phi}(x)]$ and concentrate on a localized part of this signal by multiplying it with a window function $\propto \exp[-\pi(x - x_0)^2/\Delta_x^2]$ centered around x_0. Assuming the magnitude and phase are slowly varying, we obtain approximately

$$\propto e^{-\pi(x-x_0)^2/\Delta_x^2} |\hat{f}(x_0)| e^{i2\pi(d\hat{\phi}(x_0)/dx)x}. \qquad (8.107)$$

We immediately recognize this to be of the same form as the signal $\hat{w}_{x_0,\sigma_{x0}}(x)$ in equation 8.105, with $\sigma_{x0} = d\hat{\phi}(x_0)/dx$. What we have done is to associate a localized part of $\hat{f}(x)$ with $\hat{w}_{x_0,\sigma_{x0}}(x)$, which in turn we had associated with the ray $[x_0 \ \sigma_{x0}]^\mathrm{T}$. What we have found is that the normalized angle σ_{x0} corresponds to the instantaneous spatial frequency $d\hat{\phi}(x_0)/dx$. This derivative can also be recognized as the transverse component of the gradient of the phase function. (The gradient itself is perpendicular to the wavefronts and points along the direction of the ray.) Finally, let us recall from table 3.2 the following theorem:

$$\int \sigma_x \hat{W}_{\hat{f}}(x, \sigma_x) \, d\sigma_x \propto \frac{d\hat{\phi}(x)}{dx}, \tag{8.108}$$

from which we conclude that the derivative of the phase for a certain value of x also corresponds to the average frequency of the signal at that value of x, where the average is taken in the space-frequency plane by weighting with the Wigner distribution. This result is more general than interpreting the derivative of the phase as an instantaneous spatial frequency, which has a meaningful physical interpretation only when a narrow band of frequencies dominates at each spatial position.

8.7.2 Complex Gaussian signals

We now turn our attention to another family of quadratic-exponential functions which may be referred to as complex Gaussian signals. They are also characterized by two parameters (equation 3.148):

$$\hat{f}(x) = (2/W^2)^{1/4} \, e^{i\pi x^2 / \lambda \hat{q}}, \tag{8.109}$$

where $\hat{q}$ is the *complex radius of curvature* corresponding to two real parameters, and $W^2 > 0$. These signals were already discussed in a purely mathematical context in section 3.4.6. As in section 7.3.4, we will denote the real and imaginary parts of $\hat{q}$ in the form $1/\hat{q} = 1/R + i\lambda/W^2$. (Signals with a term linear in x in the exponent are also complex Gaussian signals from a physical viewpoint, but are not considered here for simplicity.) In passing, we note that such signals can also be represented by what is known as the *complex ray representation* (Arnaud 1973).

Now, it follows from the discussion of section 3.4.6 that when such a signal passes through a quadratic-phase system, it remains a complex Gaussian signal, and the complex radius of curvature $\hat{q}'$ at the output is related to $\hat{q}$ through

$$(\lambda \hat{q}') = \frac{\hat{A}(\lambda \hat{q}) + \hat{B}}{\hat{C}(\lambda \hat{q}) + \hat{D}}. \tag{8.110}$$

The Wigner distribution of a complex Gaussian signal can be written as follows (Bastiaans 1979a, 1989):

$$\hat{W}_{\hat{f}}(x, \sigma_x) = 2 \exp\left\{ -2\pi \left[\frac{x^2}{W^2} + W^2 \left(\sigma_x - \frac{x}{\lambda R} \right)^2 \right] \right\}$$

$$= 2 \exp\left\{ -2\pi \left[\left(\frac{1}{W^2} + \frac{W^2}{\lambda^2 R^2} \right) x^2 - 2 \left(\frac{W^2}{\lambda R} \right) x\sigma_x + (W^2) \sigma_x^2 \right] \right\} \tag{8.111}$$

and it retains this form when the signal passes through a quadratic-phase system. The values of R' and W' at the output can be easily found by using equation 8.110. When $1/W^2 = 0$ we expect to recover the simple chirp function $\propto \exp(i\pi x^2/\lambda R)$. However, since the chirp function does not have unit energy, this does not follow as a special case of equation 8.109 by letting $1/W^2 \to 0$. Nevertheless, the result

$(\lambda R') = [\hat{A}(\lambda R) + \hat{B}]/[\hat{C}(\lambda R) + \hat{D}]$ still holds. We know that the Wigner distribution of the chirp function is given by $\hat{W}_{\hat{f}}(x, \sigma_x) = \delta(\sigma_x - x/\lambda R)$, which is a line delta concentrated along $\sigma_x = x/\lambda R$. When such a function passes through a quadratic-phase system, its Wigner distribution becomes $\delta(\sigma_x - x/\lambda R')$, where R' is given by the linear fractional transformation formula above.

We now consider the moments of the Wigner distribution of complex Gaussian signals, which were mathematically defined in section 3.4.5. Since constant or linear terms were excluded from the exponent of our signal, the first-order moments $\overline{x}$ and $\overline{\sigma_x}$ will be zero. The second-order moments are given by the following expressions (Bastiaans 1989):

$$\overline{(x - \overline{x})^2} = \overline{x^2} = \iint x^2 \hat{W}_{\hat{f}}(x, \sigma_x)\, dx\, d\sigma_x = W^2/4\pi,$$

$$\overline{(\sigma_x - \overline{\sigma_x})^2} = \overline{\sigma_x^2} = \iint \sigma_x^2 \hat{W}_{\hat{f}}(x, \sigma_x)\, dx\, d\sigma_x = (1/W^2 + W^2/\lambda^2 R^2)/4\pi,$$

$$\overline{(x - \overline{x})(\sigma_x - \overline{\sigma_x})} = \overline{x\sigma_x} = \iint x\sigma_x \hat{W}_{\hat{f}}(x, \sigma_x)\, dx\, d\sigma_x = W^2/4\pi\lambda R. \qquad (8.112)$$

Upon passage through a quadratic-phase system, these moments transform according to equation 3.147, which can be cast in the form of expressions relating R', W' to R, W (which the reader may show are consistent with similar expressions obtained directly from equation 8.110). The reader may also wish to relate the present considerations to the discussion on page 109 by noting that

$$\frac{\overline{x\sigma_x}}{\overline{x^2}} + i\frac{\sqrt{\overline{x^2}\,\overline{\sigma_x^2} - \overline{x\sigma_x}^2}}{\overline{x^2}} \qquad (8.113)$$

evaluates to $1/\lambda R + i/W^2 = 1/\lambda\hat{q}$. We also note that had we considered the more general case where the first-order moments are not zero, they would have been transformed by equation 3.145.

More generally, equations 3.145 and 3.147, or the equivalent linear fractional transformation given in equation 3.155, hold for the first- and second-order moments of any signal. Thus, in situations where it is not necessary to fully characterize a signal by its functional representation $\hat{f}(x)$, and it is sufficient to characterize it by its moments, the moments of the output can be easily related to the moments of the input through simple matrix relations. This amounts to approximating the signal by the closest-fitting complex Gaussian signal. Of course, complex Gaussian signals are fully characterized by their first and second moments, so that if we have a complex Gaussian signal itself, we will have complete knowledge of the output once we find its moments.

The second-order moments of the Wigner distribution have been discussed in an optical context in Bastiaans 1979a, 1989, 1991b. Higher-order moments have been discussed in Dragoman 1994.

8.8   Optical invariants

Some of the more fundamental statements in physics take the form of conservation laws. Certain quantities or properties have the quality of remaining invariant as a function of time, after propagation through space, or upon passage through certain classes of systems. One such quantity is of course energy, which within our framework is given by the square of the norm of a signal. We know that the energy of a signal remains unchanged upon passage through a system if the system is unitary, which physically means that it does not contain attenuating or amplifying components. Quadratic-phase systems are unitary and thus conserve energy. On the other hand, a multiplicative filter $\hat{h}(x)$ is usually not, unless $|\hat{h}(x)| = 1$ for all x.

In this section we will discuss another invariant which is of central importance in optics, often called the *optical invariant*. It takes many forms, has many different interpretations, and is referred to by many different names. We have already seen one manifestation of it: the area of support of the Wigner distribution or the ray bundle corresponding to a distribution of light.

As preparation, we recall from page 252 that for an inhomogeneous medium of one transverse dimension with refractive index distribution $n(x, z)$, the Hamiltonian $\hat{H}$ and Hamilton's equations are given by

$$\hat{H}(x, \sigma_x; z) = -\sqrt{n^2(x, z)f_{oc}^2/c^2 - \sigma_x^2}, \qquad (8.114)$$

$$\frac{dx}{dz} = \frac{\partial \hat{H}}{\partial \sigma_x}, \qquad \frac{d\sigma_x}{dz} = -\frac{\partial \hat{H}}{\partial x}. \qquad (8.115)$$

Hamilton's equations can be written compactly as a single vector equation of the following form (Goldstein 1980):

$$\begin{bmatrix} dx/dz \\ d\sigma_x/dz \end{bmatrix} = \begin{bmatrix} 0 & 1 \\ -1 & 0 \end{bmatrix} \begin{bmatrix} \partial \hat{H}/\partial x \\ \partial \hat{H}/\partial \sigma_x \end{bmatrix}. \qquad (8.116)$$

The 2×2 matrix appearing in this expression is conventionally denoted by the symbol **J**. Under the paraxial approximation the Hamiltonian takes the form

$$\hat{H}(x, \sigma_x; z) = \frac{\sigma_x^2}{2n(x, z)f_{oc}/c} - n(x, z)f_{oc}/c. \qquad (8.117)$$

Hamilton's equations are equivalent to the ray equation and thus embody the laws of geometrical optics. We also recall the discussion starting on page 253 which showed that (at least in the paraxial case) Hamilton's equations corresponded to Snell's law and the continuity of light rays. In this section we will derive many consequences of Hamilton's equations.

Our discussion of optical invariants is intended to be instructive, rather than complete and general. Our derivations are not always rigorous or complete, nor are they set up for the most general case possible.

8.8.1 *Invariance of density and area in phase space*

One form the optical invariant takes is the density or area in phase space. The invariance of phase-space density and area in optics is an instance of a more general theorem appearing in many branches of physics called *Liouville's theorem* (Arnaud 1976, Marcuse 1982).

Let us consider a bundle of rays occupying an arbitrary region in phase space. We will let ρ denote the density of the rays in phase space. As z increases, each ray will move in phase space according to the functions $x(z)$, $\sigma_x(z)$. Thus, the "velocity" of the phase-space point corresponding to a ray is given by $\mathbf{v} = [dx(z)/dz \; d\sigma_x(z)/dz]^{\mathrm{T}}$. We now write the continuity equation

$$\frac{\partial \rho}{\partial z} + \nabla \cdot (\rho \mathbf{v}) = 0, \tag{8.118}$$

which is essentially a statement of the fact that rays do not disappear and are not created. (That this equation holds can be seen by considering the equation $(\partial/\partial z) \iint \rho \, dx \, d\sigma_x = -\int \rho \mathbf{v} \cdot \mathbf{dl}$. The double integral on the left-hand side is taken over some region and simply gives the total number of rays in that region. $\mathbf{dl}$ is a vector line element normal to the boundary of the region. The integral on the right-hand side is taken over the boundary and tells us the number of rays exiting the region. The divergence theorem states that the right-hand side is equal to $-\iint \nabla \cdot (\rho \mathbf{v}) \, dx \, d\sigma_x$, from which the continuity equation follows upon noting that these equations hold for any arbitrary region.) The continuity equation can be explicitly written as

$$\frac{\partial \rho}{\partial z} + \frac{\partial}{\partial x}\left(\rho \frac{dx(z)}{dz}\right) + \frac{\partial}{\partial \sigma_x}\left(\rho \frac{d\sigma_x(z)}{dz}\right) = 0. \tag{8.119}$$

Now, using Hamilton's equations, expanding the product rule for differentiation, and again using Hamilton's equations, we obtain

$$\frac{\partial \rho}{\partial z} + \frac{\partial \rho}{\partial x}\frac{\partial x}{\partial z} + \frac{\partial \rho}{\partial \sigma_x}\frac{\partial \sigma_x}{\partial z} = 0, \tag{8.120}$$

whose left-hand side we recognize as the total derivative of ρ and thus conclude that

$$\frac{d\rho}{dz} = 0, \tag{8.121}$$

which means that the density of rays (or phase-space points) remains invariant upon propagation through a medium with arbitrary refractive index. As we move along the rays with increasing z, the boundary of the region enclosing the ray bundle changes shape. However, since the density of rays remains constant, the fact that the total number of rays is constant also implies that the area of this region remains invariant. Mathematically,

$$\text{number of rays} = \text{density of rays} \times \text{phase-space area}, \tag{8.122}$$

so that invariance of the density implies invariance of the phase-space area. The area occupied by a bundle of rays in phase space is one of the most common forms of the optical invariant. (The reader may wish to imagine a fluid spread on a table. If the surface density of the fluid remains constant, the area occupied by it will also remain constant no matter how the region changes.) The above discussion is based on Marcuse 1982, where it is also shown that the invariance of density and area also holds when the refractive index distribution is discontinuous.

Let us write the phase-space coordinates of the rays $[x \; \sigma_x]^{\mathrm{T}}$ at an arbitrary plane z as a function of their respective phase-space coordinates $[x_0 \; \sigma_{x0}]^{\mathrm{T}}$ at some other plane z_0 as follows:

$$x = x(x_0, \sigma_{x0}),$$
$$\sigma_x = \sigma_x(x_0, \sigma_{x0}). \tag{8.123}$$

These equations tell us the phase-space location of a ray at the plane z in terms of its phase-space location at the plane z_0. Now, consider the area integrals $\iint dx_0 \, d\sigma_{x0}$ and $\iint dx \, d\sigma_x$ taken over the respective regions enclosing the ray bundle at the planes z_0 and z. If we make a substitution of variables using equations 8.123 in the second of these integrals, this integral will be identical to the first integral except that it will be multiplied by the Jacobian of the variable transformation in question. Since we have seen that the area is invariant, the two integrals must be equal, and it follows that the Jacobian associated with the transformation must be unity.

(We recall that the area elements in such a variable transformation are related by $dx \, d\sigma_x = |\mathbf{M}| \, dx_0 \, d\sigma_{x0}$, where the Jacobian matrix $\mathbf{M}$ is given by

$$\mathbf{M} = \left[\begin{array}{cc} \partial x/\partial x_0 & \partial x/\partial \sigma_{x0} \\ \partial \sigma_x/\partial x_0 & \partial \sigma_x/\partial \sigma_{x0} \end{array} \right], \tag{8.124}$$

and $|\mathbf{M}|$ is the determinant of the matrix, referred to as the Jacobian of the transformation.)

The results just derived are generalizations of results we already know to hold for quadratic-phase systems: The area of the region enclosing a ray bundle is invariant and the determinant $\hat{A}\hat{D} - \hat{B}\hat{C}$ of the linear transformation matrix (the $\hat{A}\hat{B}\hat{C}\hat{D}$ matrix) is equal to unity.

The reader can also show, in a manner similar to the way we showed equation 8.121, that $\nabla \cdot \mathbf{v} = 0$: the divergence of the velocity vector is zero. The total flux of the phase-space "fluid" out of the boundary enclosing the ray bundle is zero. In other words, the phase-space fluid behaves like an incompressible fluid (Lanczos 1970).

Before we close this section, we make note of the physical interpretations of the total derivative and the partial derivative of an entity $\hat{K}(x, \sigma_x; z)$ with respect to z. The total derivative $d\hat{K}/dz$ is the rate of change of $\hat{K}$ as we follow a ray $[x(z) \; \sigma_x(z)]^{\mathrm{T}}$, whereas the partial derivative $\partial \hat{K}/\partial z$ is the explicit rate of change with respect to z for fixed x and σ_x. The two derivatives are of course related by

$$\frac{d\hat{K}}{dz} = \frac{\partial \hat{K}}{\partial x}\frac{dx}{dz} + \frac{\partial \hat{K}}{\partial \sigma_x}\frac{d\sigma_x}{dz} + \frac{\partial \hat{K}}{\partial z}. \tag{8.125}$$

If $d\hat{K}/dz = 0$ the entity $\hat{K}$ remains invariant along a ray. We have already seen that the phase-space density ρ is such an invariant entity (equation 8.121). We have also seen that the value of the Wigner distribution remains constant along a ray (equation 8.104), consistent with our interpretation of the Wigner distribution as the phase-space density in wave optics.

8.8.2   The symplectic condition and canonical transformations

If a ray has the phase-space coordinates $[x_0 \; \sigma_{x0}]^{\mathrm{T}}$ at z_0, the phase-space coordinates $[x \; \sigma_x]^{\mathrm{T}}$ at an arbitrary value of z will be given by transformation expressions having the form of equation 8.123 (Goldstein 1980):

$$x = x(x_0, \sigma_{x0}; z),$$
$$\sigma_x = \sigma_x(x_0, \sigma_{x0}; z), \tag{8.126}$$

which are solutions $x(z)$, $\sigma_x(z)$ of equation 8.116 (or the ray equation) corresponding to the "initial conditions" $x(z_0) = x_0$ and $\sigma_x(z_0) = \sigma_{x0}$. The Jacobian matrix $\mathbf{M}(z)$ of the transformation given in equation 8.126 is

$$\mathbf{M}(z) = \begin{bmatrix} \partial x/\partial x_0 & \partial x/\partial \sigma_{x0} \\ \partial \sigma_x/\partial x_0 & \partial \sigma_x/\partial \sigma_{x0} \end{bmatrix}. \tag{8.127}$$

We will now show that this Jacobian matrix satisfies the following relation for all z:

$$\mathbf{J} = \mathbf{M}^{\mathrm{T}}\mathbf{J}\mathbf{M}. \tag{8.128}$$

This relation, known as the *symplectic condition*, has already appeared as equation 3.179, where we mentioned that matrices $\mathbf{M}$ satisfying such a relation are referred to as *symplectic matrices* (Goldstein 1980). The *symplectic form* which they preserve is nothing but phase-space area or density, as will be further discussed in section 8.8.3.

To demonstrate equation 8.128, we first note that it is a matter of simple matrix algebra to show that this equation is equivalent to the unit-determinant condition $|\mathbf{M}| = 1$:

$$\frac{\partial x}{\partial x_0}\frac{\partial \sigma_x}{\partial \sigma_{x0}} - \frac{\partial x}{\partial \sigma_{x0}}\frac{\partial \sigma_x}{\partial x_0} = 1. \tag{8.129}$$

But we have already seen in the previous section that this determinant, the Jacobian of the transformation, is equal to unity. Thus equation 8.128 also holds. We conclude that for systems with only one transverse dimension, *the unit-Jacobian condition and phase-space area conservation are fully equivalent to the symplectic condition*.

The unit-Jacobian condition and thus equation 8.128 can also be derived directly from Hamilton's equations. First, note from equation 8.126 that $|\mathbf{M}|$ is a function of $(x_0, \sigma_{x0}; z)$. Second, note that when $z = z_0$ the Jacobian matrix is the identity matrix and its determinant is unity. Thus if we show that $d|\mathbf{M}(z)|/dz = 0$ for all $(x_0, \sigma_{x0}; z)$, we will have shown that $|\mathbf{M}(z)| = 1$ along all rays for all z. Concentrating on a particular ray, (x_0, σ_{x0}) is merely a label for that ray, so there is no distinction between the total and partial derivatives of x, σ_x, and $|\mathbf{M}|$ with respect to z. Now we explicitly take the derivative of $|\mathbf{M}|$ with respect to z:

$$\frac{\partial |\mathbf{M}|}{\partial z} = \frac{\partial^2 x}{\partial z \partial x_0}\frac{\partial \sigma_x}{\partial \sigma_{x0}} + \frac{\partial x}{\partial x_0}\frac{\partial^2 \sigma_x}{\partial z \partial \sigma_{x0}} - \frac{\partial^2 x}{\partial z \partial \sigma_{x0}}\frac{\partial \sigma_x}{\partial x_0} - \frac{\partial x}{\partial \sigma_{x0}}\frac{\partial^2 \sigma_x}{\partial z \partial x_0}. \tag{8.130}$$

By exchanging the order of the mixed partial derivatives and using Hamilton's equations we obtain

$$\frac{\partial |\mathbf{M}|}{\partial z} = \frac{\partial^2 \hat{H}}{\partial x_0 \partial \sigma_x}\frac{\partial \sigma_x}{\partial \sigma_{x0}} - \frac{\partial x}{\partial x_0}\frac{\partial^2 \hat{H}}{\partial \sigma_{x0} \partial x} - \frac{\partial^2 \hat{H}}{\partial \sigma_{x0} \partial \sigma_x}\frac{\partial \sigma_x}{\partial x_0} + \frac{\partial x}{\partial \sigma_{x0}}\frac{\partial^2 \hat{H}}{\partial x_0 \partial x}. \tag{8.131}$$

The factor $\partial^2 \hat{H}/\partial x_0 \partial \sigma_x$ appearing in the first term is equal to

$$\frac{\partial^2 \hat{H}}{\partial x_0 \partial \sigma_x} = \frac{\partial}{\partial x_0}\frac{\partial \hat{H}}{\partial \sigma_x} = \left(\frac{\partial}{\partial x}\frac{\partial \hat{H}}{\partial \sigma_x}\right)\frac{\partial x}{\partial x_0} + \left(\frac{\partial}{\partial \sigma_x}\frac{\partial \hat{H}}{\partial \sigma_x}\right)\frac{\partial \sigma_x}{\partial x_0}. \tag{8.132}$$

It is possible to write similar chain rules for the corresponding factors appearing in the three remaining terms. Upon substitution, all terms cancel and we are left with zero, completing the derivation. A similar derivation is given in Arnaud 1976.

The reader may rightly think that equation 8.128 is just a fancy way of writing the unit-Jacobian condition. One advantage of equation 8.128 is that it remains true in two and higher dimensions, with the zeros and ones in the matrix $\mathbf{J}$ being replaced by higher-dimensional zero and identity matrices. In this case the symplectic condition is equivalent to a set of conditions involving the blocks of $\mathbf{M}$ which is more complicated than the unit-determinant condition (Bastiaans 1979a, Folland 1989).

What we have shown is that a transformation of the form of equation 8.126 which is a solution of Hamilton's equations or the ray equation, and thus which corresponds to the propagation of actual light rays (rather than being some arbitrary transformation), satisfies the symplectic condition.

Equation 8.126 was introduced as a relation which gives us the output phase-space coordinates of rays in terms of their input phase-space coordinates. Mathematically, this equation describes a change of coordinates in phase space from (x_0, σ_{x0}) to (x, σ_x). Thus, the propagation of light rays from one plane to another can be viewed as a coordinate transformation in phase space. The coordinates of the ray at the output plane give the location of the ray in the input plane expressed with respect to the new coordinate system.

Transformations of the form of equation 8.126 satisfying the symplectic condition are called *canonical transformations* (Goldstein 1980). Since transformations corresponding to the propagation of actual light rays satisfy the symplectic condition for all z, the equations describing the propagation of light rays take the form of continuously unfolding canonical transformations. We will see in chapter 9 that in a broad class of optical systems this canonical transformation can be interpreted as the fractional Fourier transform. Also recall that linear canonical transforms (quadratic-phase systems) corresponded to linear transformations of the form $x = \hat{A}x_0 + \hat{B}\sigma_{x0}$, $\sigma_x = \hat{C}x_0 + \hat{D}\sigma_{x0}$ in phase space, constituting a special case of equation 8.126.

We saw in chapter 3 that linear canonical transforms (as well as the matrices characterizing them) constitute a group. More generally, the canonical transformations discussed here (as well as their Jacobian matrices satisfying the symplectic condition) also constitute a group (Goldstein 1980).

8.8.3   The Lagrange invariant

We motivated the concept of symplectic forms in section 3.4.11. Most optical invariants, in their various forms, are essentially symplectic forms, and the optical systems we are considering are symplectic systems preserving these forms. Certain optical invariants which are explicit symplectic forms are collectively referred to as the *Lagrange invariant*.

Again, we consider a ray whose phase-space coordinates at z are given by $[x(x_0, \sigma_{x0}; z) \ \sigma_x(x_0, \sigma_{x0}; z)]^{\mathrm{T}}$ where $[x_0 \ \sigma_{x0}]^{\mathrm{T}}$ are the coordinates at z_0. Now, consider the following symplectic form:

$$j_{\mathrm{L}} \equiv [\partial x/\partial x_0 \ \partial x/\partial \sigma_{x0}] \begin{bmatrix} 0 & 1 \\ -1 & 0 \end{bmatrix} \begin{bmatrix} \partial \sigma_x/\partial x_0 \\ \partial \sigma_x/\partial \sigma_{x0} \end{bmatrix}, \tag{8.133}$$

which is one form of the Lagrange invariant. If the symplectic condition (equation 8.128) holds, then the symplectic form j_L will have the same value for all z. In section 8.8.7 we will see that the symplectic condition ensures the invariance of a more general class of entities known as *Poisson brackets*. (The form given in equation 8.133 will then be seen to be the Poisson bracket $[x, \sigma_x]_{x_0, \sigma_{x0}}$.) In the particular case of equation 8.133, however, invariance is trivial since the right-hand side evaluates to

$$j_L = \frac{\partial x}{\partial x_0} \frac{\partial \sigma_x}{\partial \sigma_{x0}} - \frac{\partial x}{\partial \sigma_{x0}} \frac{\partial \sigma_x}{\partial x_0}, \tag{8.134}$$

which is equal to the Jacobian determinant $|\mathbf{M}|$. The fact that the symplectic condition ensures the invariance of j_L becomes a trivial statement since the symplectic condition is equivalent to the unit-determinant condition to begin with. (Nevertheless, closing the circle in this way does serve as a consistency check.)

It remains to give a physical interpretation of the entity j_L whose invariance we have just demonstrated. In general, rays emanating from lines of constant x_0 and σ_{x0} in the phase plane at z_0 will appear as curved lines in the phase plane at z, and the differential elements dx_0 and $d\sigma_{x0}$ will be mapped to differentials which are not orthogonal to each other. dx and $d\sigma_x$ are related to dx_0 and $d\sigma_{x0}$ by

$$\begin{bmatrix} dx \\ d\sigma_x \end{bmatrix} = \begin{bmatrix} \partial x/\partial x_0 & \partial x/\partial \sigma_{x0} \\ \partial \sigma_x/\partial x_0 & \partial \sigma_x/\partial \sigma_{x0} \end{bmatrix} \begin{bmatrix} dx_0 \\ d\sigma_{x0} \end{bmatrix} = \mathbf{M} \begin{bmatrix} dx_0 \\ d\sigma_{x0} \end{bmatrix}, \tag{8.135}$$

where $\mathbf{M}$ is the Jacobian matrix. The area of the parallelogram area element defined by the images of dx_0 and $d\sigma_{x0}$ can be calculated by taking the "cross product" of $[\partial x/\partial x_0 \ \partial \sigma_x/\partial x_0]^T dx_0$ and $[\partial x/\partial \sigma_{x0} \ \partial \sigma_x/\partial \sigma_{x0}]^T d\sigma_{x0}$ (corresponding to $d\sigma_{x0} = 0$ and $dx_0 = 0$ respectively), which evaluates to $j_L dx_0 d\sigma_{x0} = |\mathbf{M}| dx_0 d\sigma_{x0}$, a result familiar from calculus. Since $j_L = |\mathbf{M}| = 1$, the area of the differential phase-area element corresponding to a differential bundle of rays remains constant for all values of z.

Now we consider another form of the Lagrange invariant (Bastiaans 1979a). Considering two distinct rays $[x_1 \ \sigma_{x1}]^T$ and $[x_2 \ \sigma_{x2}]^T$ at a given plane, we define their symplectic form as

$$[x_1 \ \sigma_{x1}] \begin{bmatrix} 0 & 1 \\ -1 & 0 \end{bmatrix} \begin{bmatrix} x_2 \\ \sigma_{x2} \end{bmatrix} = x_1 \sigma_{x2} - x_2 \sigma_{x1}. \tag{8.136}$$

This symplectic form corresponds to the (signed) area of the parallelogram defined by the two ray vectors; see section 3.4.11 and Guillemin and Sternberg 1984. We have already discussed the invariance of this form in section 3.4.11. Note that here $[x_1 \ \sigma_{x1}]^T$ and $[x_2 \ \sigma_{x2}]^T$ are *not* the positions and angles of the same ray at the input and output, but rather the positions and angles of two distinct rays at the same plane, consistent with section 3.4.11.

8.8.4   The Smith-Helmholtz invariant and Abbe's sine condition

An important special case of the Lagrange invariant will here be referred to as the Smith-Helmholtz invariant. The invariants we discussed until now remained unchanged throughout the optical system (for all z). The Smith-Helmholtz invariant, on the other hand, has the same value at the object plane and any number of image planes. These planes are connected to each other through a matrix of the form

$$\begin{bmatrix} M & 0 \\ M/\lambda R & 1/M \end{bmatrix}. \tag{8.137}$$

Again, considering the two rays $[x_1 \ \sigma_{x1}]^T$ and $[x_2 \ \sigma_{x2}]^T$ of the previous section (Bastiaans 1979a), let us take the second ray to be a ray passing through the origin at the object plane ($x_2 = 0$). If the system is an imaging system, this ray will also pass through the origin at the image plane. Thus the second term of the Lagrange invariant $x_1\sigma_{x2} - x_2\sigma_{x1}$ will be zero at both the object and image planes, implying the invariance of the first term $x_1\sigma_{x2}$. This result can also be deduced directly from the above matrix. We simply consider two rays, the first crossing the object plane at x_1 and the other crossing the object plane at the origin making angle σ_{x2}. It immediately follows from the matrix that the first ray will cross the image plane at Mx_1 and the second ray will cross it at the origin making angle σ_{x2}/M. Thus the product $(Mx_1)(\sigma_{x2}/M) = x_1\sigma_{x2}$ is the same at both object and image planes. This is directly related to the fact that if the ray bundle (or Wigner distribution) occupies a rectangular region in phase space, the effect of imaging is to simply change the aspect ratio of this rectangle without changing its area.

Here we see the optical invariant in its barest and most transparent form. Nothing more is involved than the fact that spatial and angular magnifications are inverses of each other (as implied by the unit-determinant condition), so that the product of spatial and angular features remains invariant. Remembering that angles correspond to spatial frequencies, we can also discuss the same in the language of wave optics. A perfect imaging system maps an object $\hat{f}(x)$ into the image $\propto \hat{f}(x/M)$. If the spatial extent of the object is Δx, that of the image is $M\Delta x$. We will let $\hat{F}(\sigma_x)$ denote the Fourier transform of $\hat{f}(x)$ with bandwidth $\Delta\sigma_x$. The Fourier transform of the image is $\propto \hat{F}(M\sigma_x)$ and is of extent $\Delta\sigma_x/M$. We see that as spatial extent is magnified, frequency extent (bandwidth) is demagnified by the same amount. The phase-space area $(M\Delta x)(\Delta\sigma_x/M)$ of the image is equal to that of the object $\Delta x\Delta\sigma_x$. Now, let us assume that both the function and its Fourier transform are centered around the origin, so that the largest nonnegligible frequency component is $\sigma_{x\max} = \Delta\sigma_x/2$ and the largest value of x for which the function is not negligible is $x_{\max} = \Delta x/2$. By recalling the association between ray angles θ_x and spatial frequencies σ_x given by $\sin\theta_x = \lambda\sigma_x$, we find that the largest ray angle $\theta_{x\max}$ satisfies $\sin\theta_{x\max} = \lambda\Delta\sigma_x/2$. From this we conclude that $x_{\max}\sin\theta_{x\max}$ is an invariant between object and image planes. In the paraxial approximation, this reduces to the invariance of $x_{\max}\theta_{x\max}$ discussed in the previous paragraph.

The invariance of the product of a transverse extent and the sine of an angle between object and image planes is known as *Abbe's sine condition*. Thus the Smith-Helmholtz invariant is the paraxial form of Abbe's sine condition. Another way of thinking about this invariant is as follows (Born and Wolf 1980). We consider a first-order expansion of Hamilton's point characteristic $\hat{V}(x, x')$ around the origins of the object (input) and image (output) planes:

$$\hat{V}(x, x') \approx \hat{V}(0,0) + \frac{\partial\hat{V}}{\partial x}x + \frac{\partial\hat{V}}{\partial x'}x' = \hat{V}(0,0) + \sigma_x x - \sigma'_x x',$$

$$\hat{V}(x, x') - \hat{V}(0,0) \approx \sigma_x x - \sigma'_x x', \tag{8.138}$$

where we have used equation 8.93. Here x' and x are points close to the axis, and since the derivatives are evaluated at the origin, σ'_x and σ_x are normalized angles of rays passing through the origin at the object and image planes. If $\sigma'_x = 0$ we have a ray coinciding with the optical axis so that also $\sigma_x = 0$, implying that the right-hand side and thus the left-hand side of the last displayed equation are equal to zero. But since the left-hand side of the displayed equation is independent of the normalized angles σ'_x and σ_x, it follows that it is always equal to zero and therefore $\sigma_x x = \sigma'_x x'$.

We also mention a much less well known special case of the Lagrange invariant which holds not between object and image planes, but between planes connected by a perfect Fourier transform matrix (equation 8.76 with $R \to \infty$):

$$\begin{bmatrix} 0 & s^2 M \\ -1/s^2 M & 0 \end{bmatrix}. \tag{8.139}$$

It is easy to see that $|x\sigma_x|$ is an invariant since the same product formed at the Fourier plane is $|s^2 M \sigma_x (-x/s^2 M)| = |x\sigma_x|$.

8.8.5 The constant brightness theorem

We saw on page 267 that the *brightness* represented the power per area per solid angle. Thus, expressed in suitable units, the brightness can be thought to represent power per unit phase-space area (the power density in phase space). Assuming the brightness is uniform over the region of interest in phase space, we can therefore write

$$\text{total power} = \text{brightness} \times \text{phase-space area}. \tag{8.140}$$

Now, since the phase-space area is invariant, the brightness will also be invariant in a lossless and gainless system. This is the constant brightness theorem. A more traditional approach may be found in Born and Wolf 1980:188–189. The above relation is closely related to equation 8.122, with brightness corresponding to ray density and total power corresponding to the total number of rays.

More generally, when the brightness is not uniform, $B(P, \Theta)$ essentially corresponds to the Wigner distribution $\hat{W}_{\hat{f}}(x, y, \sigma_x, \sigma_y)$. In one-dimensional notation, the generalization of equation 8.140 may be written as the familiar phase-space integral

$$\text{total power} = \iint \hat{W}_{\hat{f}}(x, \sigma_x) \, dx \, d\sigma_x. \tag{8.141}$$

The many results derived throughout our discussion of optical invariants are essentially consequences of Hamilton's equations and therefore the basic laws governing the propagation of light. The law of constant brightness is also known to be a consequence of the second law of thermodynamics; if it could be violated, one could heat up a hotter body with a colder one (Boyd 1983). This provides further insight into the physics underlying various forms of the optical invariant.

8.8.6 The unit-determinant condition for inhomogeneous media

In this section we present an alternative derivation of the unit-determinant condition for inhomogeneous media, which may be more appealing to those who prefer to think in physical rather than abstract terms. We consider a weakly inhomogeneous medium with refractive index distribution $n(x, z)$. With the paraxial Hamiltonian given in equation 7.131, Hamilton's equations take the form

$$\frac{dx}{dz} = \frac{\sigma_x}{n f_{oc}/c}, \qquad \frac{d\sigma_x}{dz} = (f_{oc}/c)\frac{\partial n}{\partial x}, \tag{8.142}$$

leading to

$$x(z + dz) = x(z) + \frac{1}{n f_{oc}/c} dz\, \sigma_x,$$

$$\sigma_x(z + dz) = \sigma_x(z) + (f_{oc}/c)\, dz\, \frac{\partial n}{\partial x}. \tag{8.143}$$

A linear matrix relation will result only if $n(x, z)$ can be approximated by a quadratic function of the form $K_2(z)x^2/2 + K_0(z)$, in which case

$$\begin{bmatrix} x(z + dz) \\ \sigma_x(z + dz) \end{bmatrix} = \begin{bmatrix} 1 & dz(n f_{oc}/c)^{-1} \\ dz(f_{oc}/c)K_2(z) & 1 \end{bmatrix}\begin{bmatrix} x(z) \\ \sigma_x(z) \end{bmatrix}. \tag{8.144}$$

More generally, the Jacobian matrix will replace the above matrix, corresponding to the lower left term being replaced by $dz(f_{oc}/c)\partial^2 n/\partial x^2$. In either case the determinant of the matrix is of the form $1 - [\cdots]dz^2$. Since the deviation of this determinant from unity is only of second order, the determinant of the system between any two planes will be unity. For instance, if the distance between two planes is given by L, then the overall determinant will be the product of L/dz terms of the form $1 - [\cdots]dz^2$, where the bracketed factor varies from term to term. Although we do not provide the details, readers may convince themselves that this product is unity by considering the case where the bracketed factor is some constant, say ι. Then, by noting the standard limit $\lim_{w\to\infty}(1 - c/w)^w = \exp(-c)$, we find that $\lim_{dz\to 0}(1 - \iota dz^2)^{L/dz} = \exp(-dz\,\iota L) = 1$.

8.8.7 Poisson brackets

The *Poisson bracket* $[\hat{K}, \hat{L}]_{x,\sigma_x}$ of $\hat{K}(x, \sigma_x)$ and $\hat{L}(x, \sigma_x)$ with respect to (x, σ_x) is defined as follows (Goldstein 1980):

$$[\hat{K}, \hat{L}]_{x,\sigma_x} \equiv \frac{\partial \hat{K}}{\partial x}\frac{\partial \hat{L}}{\partial \sigma_x} - \frac{\partial \hat{K}}{\partial \sigma_x}\frac{\partial \hat{L}}{\partial x} = [\partial\hat{K}/\partial x \;\; \partial\hat{K}/\partial\sigma_x]\begin{bmatrix} 0 & 1 \\ -1 & 0 \end{bmatrix}\begin{bmatrix} \partial\hat{L}/\partial x \\ \partial\hat{L}/\partial\sigma_x \end{bmatrix} \tag{8.145}$$

The matrix Poisson bracket of two vectors $[\hat{K}_1 \;\; \hat{K}_2]^T$ and $[\hat{L}_1 \;\; \hat{L}_2]^T$ is defined as

$$\left[[\hat{K}_1 \;\; \hat{K}_2]^T, [\hat{L}_1 \;\; \hat{L}_2]^T\right]_{x,\sigma_x} \equiv \begin{bmatrix} \left[\hat{K}_1, \hat{L}_1\right]_{x,\sigma_x} & \left[\hat{K}_1, \hat{L}_2\right]_{x,\sigma_x} \\ \left[\hat{K}_2, \hat{L}_1\right]_{x,\sigma_x} & \left[\hat{K}_2, \hat{L}_2\right]_{x,\sigma_x} \end{bmatrix}. \tag{8.146}$$

Using this definition, a number of special cases and properties follow immediately:

$$\left[[x_0 \;\; \sigma_{x0}]^T, [x_0 \;\; \sigma_{x0}]^T\right]_{x_0,\sigma_{x0}} = \mathbf{J}, \tag{8.147}$$

$$\left[[x \;\; \sigma_x]^T, [x \;\; \sigma_x]^T\right]_{x,\sigma_x} = \mathbf{J}, \tag{8.148}$$

$$\left[[x \;\; \sigma_x]^T, [x \;\; \sigma_x]^T\right]_{x_0,\sigma_{x0}} = \mathbf{MJM}^T, \tag{8.149}$$

where $\mathbf{J}$ is the 2×2 matrix appearing in equation 8.145 and $\mathbf{M}$ is the Jacobian matrix given in equation 8.124. The first two equations are merely two instances of the same equation. Now, if $[x \ \sigma_x]^{\mathrm{T}}$ is related to $[x_0 \ \sigma_{x0}]^{\mathrm{T}}$ through an actual physical ray transformation of the form of equation 8.126, then we know that $\mathbf{M}$ is symplectic and the right-hand side of the third equation becomes equal to $\mathbf{J}$. (This can be demonstrated in a manner similar to equation 8.128. In one dimension, the conditions $\mathbf{MJM}^{\mathrm{T}} = \mathbf{J}$ and $\mathbf{M}^{\mathrm{T}}\mathbf{JM} = \mathbf{J}$ are equivalent to each other and the unit-determinant condition.) Thus the right-hand sides of the second and third equations become identical, and we conclude that the Poisson bracket on the left-hand side is invariant under canonical transformations of the form of equation 8.126 (Goldstein 1980). In other words, it has the same value regardless of which variables it is evaluated with respect to (provided these are related through canonical transformations). The reader may have already noted that Poisson brackets are *symplectic forms* (section 3.4.11).

We now show the invariance of all Poisson brackets (Goldstein 1980). First we write the following chain rule:

$$\begin{bmatrix} \partial \hat{L}/\partial x_0 \\ \partial \hat{L}/\partial \sigma_{x0} \end{bmatrix} = \mathbf{M}^{\mathrm{T}} \begin{bmatrix} \partial \hat{L}/\partial x \\ \partial \hat{L}/\partial \sigma_x \end{bmatrix},$$

$$\begin{bmatrix} \partial \hat{L}/\partial x_0 & \partial \hat{L}/\partial \sigma_{x0} \end{bmatrix} = \begin{bmatrix} \partial \hat{L}/\partial x & \partial \hat{L}/\partial \sigma_x \end{bmatrix} \mathbf{M}, \tag{8.150}$$

which of course also hold for $\hat{K}$. Now we write the Poisson bracket $[\hat{K}, \hat{L}]_{x_0, \sigma_{x0}}$ as

$$[\partial \hat{K}/\partial x_0 \ \partial \hat{K}/\partial \sigma_{x0}] \begin{bmatrix} 0 & 1 \\ -1 & 0 \end{bmatrix} \begin{bmatrix} \partial \hat{L}/\partial x_0 \\ \partial \hat{L}/\partial \sigma_{x0} \end{bmatrix}$$

$$= [\partial \hat{K}/\partial x \ \partial \hat{K}/\partial \sigma_x] \mathbf{M} \begin{bmatrix} 0 & 1 \\ -1 & 0 \end{bmatrix} \mathbf{M}^{\mathrm{T}} \begin{bmatrix} \partial \hat{L}/\partial x \\ \partial \hat{L}/\partial \sigma_x \end{bmatrix}. \tag{8.151}$$

The condition $\mathbf{MJM}^{\mathrm{T}} = \mathbf{J}$ leads us to conclude that $[\hat{K}, \hat{L}]_{x_0, \sigma_{x0}} = [\hat{K}, \hat{L}]_{x, \sigma_x}$. Poisson brackets are invariant under canonical transformations, which physically means they are invariants of ray propagation. Because of this invariance, the subscripts of the brackets are sometimes dropped. The reader may check that as binary operators, Poisson brackets are not associative. They obey a nonassociative algebra, known as a *Lie algebra*. The vector cross product and the commutator of two operators can also be used to construct similar Lie algebras. (Goldstein 1980)

We now consider the total derivative of some entity $\hat{K}(x, \sigma_x; z)$ with respect to z:

$$\frac{d\hat{K}}{dz} = \frac{\partial \hat{K}}{\partial x} \frac{dx}{dz} + \frac{\partial \hat{K}}{\partial \sigma_x} \frac{d\sigma_x}{dz} + \frac{\partial \hat{K}}{\partial z} = \frac{\partial \hat{K}}{\partial x} \frac{\partial \hat{H}}{\partial \sigma_x} - \frac{\partial \hat{K}}{\partial \sigma_x} \frac{\partial \hat{H}}{\partial x} + \frac{\partial \hat{K}}{\partial z} = [\hat{K}, \hat{H}] + \frac{\partial \hat{K}}{\partial z}.$$
$$\tag{8.152}$$

This is an equation for the evolution of the entity $\hat{K}$ in terms of Poisson brackets (Goldstein 1980). If we take $\hat{K} = x$ or $\hat{K} = \sigma_x$ as special cases, we obtain $dx/dz = [x, \hat{H}]$ or $d\sigma_x/dz = [\sigma_x, \hat{H}]$, which the reader can show to be equivalent to Hamilton's equations. If we take $\hat{K} = \hat{H}$, we recover the familiar result $d\hat{H}/dz = \partial \hat{H}/\partial z$. If $\hat{K}$ is an invariant ($d\hat{K}/dz = 0$), then we must have $-[\hat{K}, \hat{H}] = [\hat{H}, \hat{K}] = \partial \hat{K}/\partial z$. An example of the latter case has already been encountered in equation 8.102, whose right-hand side we can now recognize to be $[\hat{H}, \hat{W}_{\hat{f}_z(x)}]$. (Goldstein 1980)

When $\partial \hat{K}/\partial z = 0$ it is possible to write a formal series solution for the equation $d\hat{K}/dz = [\hat{K}, \hat{H}]$, and by recognizing the similarity of this expansion to that of an exponential function, express the series in "hyperbracket" form (Goldstein 1980). Although we do not pursue this any further, we mention that it closely corresponds to the hyperdifferential form for operators discussed in chapter 3.

8.8.8 The number of degrees of freedom

In chapter 3 we discussed at length the concept of the number of degrees of freedom of a set of signals, seeing that it is given by the area of the space-frequency support of the set of signals. In this section we have seen that the area of the space-frequency support is an invariant upon passage through a broad class of optical systems. If we consider members of the set of signals in question to be inputs to such an optical system, the set of signals consisting of the outputs will also have the same number of degrees of freedom. Thus the number of degrees of freedom is another form of the optical invariant. Such optical systems preserve the information content of signals passing through them. This is also consistent with the reciprocity of wave and ray propagation.

In a typical optical system, the number of degrees of freedom of the set of signals which can pass through that system can be determined by examining the *space-frequency aperture* of the system. Although this phase-space aperture may in general have different shapes (Lohmann and others 1996a), it is commonly assumed to be of rectangular form with the spatial extent determined by a spatial aperture in the object or image plane, and the frequency extent determined by an aperture in a Fourier plane. If these apertures are of length Δx_{sp} and Δx_{fr} respectively, and the scale factor relating these planes to each other is s, then the number of degrees of freedom is given by $\Delta x_{sp} \Delta x_{fr}/s^2$ (see page 88). The number of degrees of freedom and information-carrying capacity of refractive and diffractive lenses, as well as how these scale with various parameters have been discussed in Ozaktas and Urey 1993 and Ozaktas, Urey, and Lohmann 1994.

The concept of the number of degrees of freedom or the space-bandwidth product has received considerable attention in the area of information optics in a way which is not reflected by our limited discussion. For a sampling of historical works, see for instance, Toraldo di Francia 1955, 1969, Gabor 1961, Walther 1967, Winthrop 1971, Frieden 1971, Gori and Guattari 1971, 1973, Bendinelli and others 1974, Lohmann 1986, and the further references in these works. More recent works are diverse and too numerous to list here. We will note van Dekker and van den Bos 1997, Mendlovic and Lohmann 1997, Mendlovic, Lohmann, and Zalevsky 1997, and Miller 1998. The first is a review of past and present approaches to optical resolution. The second and third deal with the adaptation of the space-frequency aperture of an optical system so as to maximally benefit from the available space-frequency area, even when its shape does not match the space-frequency support of the signals to be processed. Miller examines the (functional) singular-value decomposition for propagation between two volumes. This amounts to finding pairs of orthonormal functions $\hat{\psi}_l(x')$ (at the input) and $\hat{\phi}_l(x)$ (at the output) which couple into each other with strength g_l (the singular value). (This means that $\hat{\psi}_l$ at the input is mapped to $g_l \hat{\phi}_l$ at the output.) These pairs of functions constitute independent spatial "channels" of information transfer, or "communication modes" between the input and the output. The "strength" of each channel, given by the singular value, tells us how reliable that channel is. Channels with very small strengths will couple the signal weakly and perhaps below the noise floor and are thus not reliable. Miller shows (at least when the medium of propagation is free space) that the sum of the squared connection strengths is a constant, so that even if there are an infinite number of channels, only a finite number of them can have substantial strength and thus be considered reliable: there is effectively a finite number N of channels or spatial degrees of freedom. In our context, N corresponds to the phase-space area or space-bandwidth product. In many typical physical systems, the eigenvalues (or singular values), when listed in decreasing order, are seen to have values comparable to each other (and often to unity) until we come close to the Nth eigenvalue. Then the eigenvalues decrease sharply and become close to zero as we pass the Nth eigenvalue.

These considerations are closely related to the problem of solving so-called *Fredholm integral equations of the first kind*:

$$\hat{g}(x) = \int \hat{h}(x, x') \hat{f}(x') \, dx', \tag{8.153}$$

where $\hat{g}(x)$ is measured with a finite degree of accuracy, $\hat{h}(x, x')$ is assumed known, and we seek to find $\hat{f}(x')$. Such problems are known as *inverse problems* and they are considered to be *ill-posed* in the sense that even if the error in measuring $\hat{g}(x)$ is small, the error in the estimate of $\hat{f}(x')$ may be large (*Image Recovery: Theory and Applications* 1987). This is a consequence of the fact that if we look at the discrete spectral expansion of some $\hat{h}(x, x')$ corresponding to a real physical system, we see that the number of terms for which the eigenvalues are substantial are finite, and thus the system effectively filters out a portion of the information inherent in $\hat{f}(x')$. The system does not pass all of the information inherent in the function $\hat{f}(x')$, but rather only partial information represented by a finite number of degrees of freedom. The remaining information is lost. In most systems the lost information corresponds to the high-frequency content of $\hat{f}(x')$.

8.9 Partially coherent light

The study of partially coherent light is an area in which the phase-space picture and the Wigner distribution have found great use. We have already discussed in chapter 7 that partially coherent light is characterized as a random process, and we have already defined the Wigner distribution of a random process in chapter 3. Thus, we already know how to obtain the Wigner distribution of partially coherent light and characterize it in phase space.

Unfortunately, we will not have space in this book to further develop this subject, and must satisfy ourselves by referring the reader to a short list of references: Bastiaans 1981b, 1986a, b, 1991a, Friberg 1981, 1986. Bastiaans 1997 is a particularly useful starting point.

8.10   Further reading

The subject matter of this chapter naturally leads the way to a deeper study of mathematical optics. Similar to the study of advanced classical and quantum mechanics, the study of advanced mathematical optics is heavily based on phase-space concepts. In fact, there is a rather close mathematical analogy between mechanics and optics, which the reader may have already noted. For instance, see Lanczos 1970 or Goldstein 1980. The calculus of variations plays an important part in the development of variational principles such as that of Fermat. Elementary expositions may be found in Hildebrand 1965, Born and Wolf 1980, and Goldstein 1980. Lanczos 1970 is an insightful text and Caratheodory 1965 is a translation of a classic work on the subject.

Relatively advanced texts containing discussions on general geometrical optics and mathematical optics, including the Hamiltonian formulation, include Synge 1937, 1954, Born and Wolf 1980, Stavroudis 1972, Arnaud 1976, Marcuse 1982, and Solimeno, Crosignani, and Di Porto 1986. Luneburg 1964, Kline and Kay 1965, and Deschamps 1972 are noted for the development of geometrical optics based on the foundation of Maxwell's equations. Luneburg's work also contains extensions of several of the concepts discussed in this chapter. Pegis 1961, Arnaud 1973, and Sekiguchi and Wolf 1987 specifically concentrate on Hamiltonian methods. Wolf and Krötzsch 1995 discusses discontinuities and refraction in the Hamiltonian framework.

A sampling of works dealing with Lie methods might include Dragt and Finn 1976, Dragt 1982, Wolf 1986a, Ferraro and Caelli 1988, and the edited volumes *Lie Methods in Optics* 1986, 1989. As for symplectic and group-theoretical techniques, we might mention the mathematical text by Folland (1989) and the physics or optics oriented works Stavroudis 1972: chapters 13, 14, 15, Guillemin and Sternberg 1981, 1984: chapter 1, Bacry and Cadilhac 1981, Navarro-Saad and Wolf 1986a, Perelomov 1986, Wolf 1986b, and Kauderer 1990.

An additional sampling of relatively recent works on phase-space optics more from a geometrical optics perspective includes Tanaka 1986, Navarro-Saad and Wolf 1986b, Wright and Garrison 1987, Wolf 1991, 1993, Wolf and Kurmyshev 1993, and Campbell 1994.

Bastiaans has been a major contributor to the use of phase-space techniques from a Fourier optics perspective (1978, 1979a, b, c, d, 1980, 1981a, b, 1982a, b, 1986a, b, 1989, 1991a, b, 1994). This body of work deals with the use of the Wigner distribution, the Gabor expansion and other space-frequency representations as well as linear canonical transforms in optics. A very readable exposition is Bastiaans 1997, which is particularly recommended as a complement to this chapter. A number of other relatively early works which have been found instructive are Papoulis 1974, Bartelt, Brenner, and Lohmann 1980, Bamler and Glünder 1983, and Easton, Ticknor, and Barrett 1984. An earlier work bringing together partial coherence, radiometry, and the Wigner distribution is Walther 1968. The Wigner distribution of polychromatic wave fields is discussed in Wolf 1996.

A review of the Wigner distribution in optics is Dragoman 1997. A recent collection of works on the subject is *Wigner Distributions and Phase Space in Optics* 2000.

In this book we have totally excluded the discussion of quantum optics in phase space. For this the reader is referred to Gardiner 1991, Schleich, Mayr, and Krähmer 1999, and Carmichael 1999.

9

The Fractional Fourier Transform in Optics

9.1 Applications of the transform to wave and beam propagation

A considerable amount of work has been done on the application of the fractional Fourier transform to wave and beam propagation problems, mostly in an optical context. The presentation of this chapter will also be phrased in the notation and terminology of optics. Nevertheless, the reader should have no difficulty translating the results to other propagation, diffraction, and scattering phenomena which are mathematically equivalent or similar.

Whenever we can express the result of an optical problem (such as Fraunhofer diffraction) in terms of the Fourier transform, we tend to think of this as a simple and elegant result. This is justified by the fact that the Fourier transform has many simple and useful properties which make it attractive to work with. In an optical system involving many lenses separated by arbitrary distances (figure 9.1), Fourier transforms and images occur at certain privileged planes, whose locations can be determined using the results presented in section 8.4.3. In such a system, one typically observes first a Fourier transform, then an inverted image, then an inverted Fourier transform, then an erect image, then another Fourier transform, and so on. Often all we know about what happens in between these planes is that the amplitude distribution is given by complicated nested integrals. We will see in this chapter that the distribution of light at intermediate planes can be expressed in terms of the fractional Fourier transform (which we know also has several useful properties and operational formulas). Thus the fractional Fourier transform completes in a very natural way the study of optical systems often called "Fourier optics."

Fourier optical systems can be analyzed using geometrical optics, Fresnel integrals (spherical wave expansions), plane wave expansions, Hermite-Gaussian beam expansions, and as we will see, fractional Fourier transforms. The several approaches prove useful in different situations and provide different viewpoints which complement each other. The fractional Fourier transform approach is appealing in that it describes the continuous evolution of a wave as it propagates through a system.

The propagation of light along the $+z$ direction can be viewed as a process of continual fractional Fourier transformation. As light propagates, its distribution evolves through fractional transforms of increasing orders. The order $a(z)$ of the fractional transform observed at z is a continuous monotonic increasing function of z.

input object

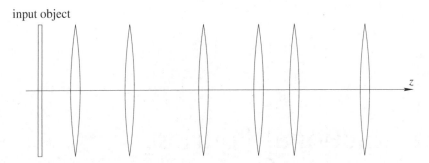

Figure 9.1. Optical system consisting of many lenses.

(The fractional transform at z is observed on a spherical reference surface intersecting the optical axis at that location.) One of the central results of diffraction theory is that the far-field diffraction pattern is the Fourier transform of the diffracting object. At closer distances we observe the fractional Fourier transforms of the diffracting object. The effect of a thin lens on such a propagating waveform is merely to bend the spherical reference surface into another with different radius, so that we can continue tracking the evolution of the wave in terms of fractional Fourier transforms starting from this new reference surface. Thus we can use the transform to model the propagation of waves through optical systems consisting of an arbitrary number of thin lenses sandwiched between arbitrary sections of free space.

Scale parameters and dimensions: In this chapter we will see that the output of a fairly broad class of optical systems can be expressed as the fractional Fourier transform of the input. This is a generalization of the well-known fact that in certain special planes one observes the ordinary Fourier transform. However, when we are dealing with fractional Fourier transforms, the choice of scale and dimensions must always be noted, as this will have an effect on the fractional order observed at a given plane in the system. We recall from chapter 4 that the ath fractional Fourier transform of $f(u/M)$ is not a scaled version of $f_a(u)$ (table 4.3, property 2), but a scaled version of a fractional Fourier transform of $f(u)$ of another order:

$$|M|\sqrt{\frac{1 - i\cot\alpha}{1 - iM^2\cot\alpha}}\ \exp\left[i\pi u^2\cot\alpha\left(1 - \frac{\cos^2\alpha'}{\cos^2\alpha}\right)\right]\ f_{a'}\left(\frac{Mu\sin\alpha'}{\sin\alpha}\right), \qquad (9.1)$$

$$\alpha' = \arctan(M^{-2}\tan\alpha).$$

Now consider two different scale parameters s_1 and $s_2 = Ms_1$ used to translate the actual physical function $\hat{f}(x)$ to a function with dimensionless argument as $\hat{f}(x) \equiv s_1^{-1/2}f_1(x/s_1)$ and $\hat{f}(x) \equiv s_2^{-1/2}f_2(x/s_2)$ (section 7.2). Since the two dimensionless functions are related by $f_1(u) = M^{-1/2}f_2(u/M)$, the ath order transform of $f_1(u)$ will not be a scaled version of the ath order transform of $f_2(u)$. This means that the fractional Fourier transform of $\hat{f}(x)$ with x measured in meters is not simply a scaled version of the transform of $\hat{f}(x)$ with x measured in centimeters, although the two original functions are simply scaled versions of each other. This fact is not a cause for alarm, however. There will be no inconsistency or paradox if we fix the choice of units/scale by specifying the parameter s for each plane of the system, and then write all fractional Fourier transform relations between functions which take dimensionless arguments. (We will often choose s to be the same for all planes, although we will also consider the case where it is specified differently for different planes.) For instance, we will frequently deal with systems whose outputs are related to their inputs through a fractional Fourier transform

$$g(u) = (\mathcal{F}^a f)(u) = f_a(u) \tag{9.2}$$

in dimensionless arguments. To find the physical output $\hat{g}(x)$ when the physical input is $\hat{f}(x)$, we first translate $\hat{f}(x)$ to the dimensionless form $f(u)$ by using $\hat{f}(x) \equiv s^{-1/2} f(x/s) = s^{-1/2} f(u)$, take the fractional Fourier transform of $f(u)$ to obtain $g(u) = f_a(u)$, and then translate $g(u)$ to $\hat{g}(x)$ by using $\hat{g}(x) \equiv s^{-1/2} g(x/s) = s^{-1/2} g(u)$:

$$\hat{g}(x) = s^{-1/2} (\mathcal{F}^a f)(x/s) \equiv s^{-1/2} f_a(x/s) \equiv \hat{f}_a(x). \tag{9.3}$$

Note that $\hat{f}_a(x)$ is the dimensional form of $f_a(u)$, and it is *not* the ath order fractional Fourier transform of $\hat{f}(x)$. We will never fall in error as long as we never talk about the fractional Fourier transform of a function with dimensional arguments.

We stress that far from being unusual, this practice is necessary in dealing with transformations between spaces with different or undefined dimensions. The space variable is measured in meters, and the spatial frequency variable in inverse meters, but we do not know with what units we must measure the independent variable in a fractional Fourier domain. This is not a peculiarity of the fractional Fourier transform; the same applies to any distance in the space-frequency (or phase-space) plane, for which a metric can be defined only after passing to dimensionless variables. This issue is usually totally avoided in purely mathematical texts where all variables are implicitly dimensionless. And with the ordinary Fourier transform, one is able to get away without special attention since the scaling theorem ensures that the Fourier transform of a scaled function is simply a scaled version of the Fourier transform of the original function.

9.2 Overview

In this section we will review some of the essential tools and results from previous chapters and give an extended self-contained overview of this chapter. The following sections of this chapter will elaborate or extend the concepts discussed here.

As we have seen in earlier chapters, optical systems involving an arbitrary sequence of thin lenses separated by arbitrary sections of free space (under the Fresnel approximation), as well as arbitrary sections of quadratic graded-index media, belong to the class of *quadratic-phase systems*, which are mathematically equivalent to linear canonical transforms:

$$\hat{g}(x) = \int \hat{h}(x, x') \hat{f}(x') \, dx', \tag{9.4}$$

$$\hat{h}(x, x') = \sqrt{\hat{\beta}} \, e^{-i\pi/4} \exp\left[i\pi(\hat{\alpha} x^2 - 2\hat{\beta} x x' + \hat{\gamma} x'^2) \right],$$

where $\hat{\alpha} = \alpha/s^2$, $\hat{\beta} = \beta/s^2$, $\hat{\gamma} = \gamma/s^2$ are real parameters, and we are using the notational conventions introduced at the beginning of chapter 7. We also recall that the broader class of systems consisting of arbitrary thin filters sandwiched between arbitrary quadratic-phase systems were referred to as *Fourier optical systems*. For the same reasons as in chapter 8, we will mostly work with one-dimensional notation, despite the fact that most optical systems are two dimensional. The results of this chapter are also easily generalized to include systems containing spherical refracting surfaces and mirrors, since these essentially behave like lenses under the same approximations.

We repeat for convenience the kernels $\hat{h}(x, x')$ associated with a thin lens with focal length f, free-space propagation over a distance d, and a quadratic graded-index medium of length d exhibiting a refractive index profile $n^2(x) = n_0^2[1 - (x/\chi)^2]$:

$$\hat{h}_{\text{lens}}(x, x') = \delta(x - x') \, \exp\left[\frac{-i\pi x^2}{\lambda f}\right], \tag{9.5}$$

$$\hat{h}_{\text{space}}(x, x') = e^{i2\pi\sigma d} e^{-i\pi/4} \sqrt{\frac{1}{\lambda d}} \, \exp\left[\frac{i\pi(x - x')^2}{\lambda d}\right], \tag{9.6}$$

$$\hat{h}_{\text{grin}}(x, x') = e^{i2\pi\sigma d} \frac{e^{-id/2\chi}}{\sqrt{\lambda\chi}} A_\alpha \exp\left[\frac{i\pi}{\lambda\chi}\left(\cot\alpha \, x^2 - 2\csc\alpha \, xx' + \cot\alpha \, x'^2\right)\right], \tag{9.7}$$

where $\sigma = 1/\lambda$, $\alpha = d/\chi$, and $A_\alpha = \sqrt{1 - i\cot\alpha}$ (equation 8.36). These kernels are special cases of the kernel given in equation 9.4. Furthermore, arbitrary concatenations of kernels of this form will always result in a kernel which is again of the same form.

We saw in chapters 3 and 8 that apart from a constant factor which has no effect on the resulting spatial distribution, a member of the class of quadratic-phase systems is completely characterized by the three parameters $\hat{\alpha}$, $\hat{\beta}$, $\hat{\gamma}$ or equivalently by the matrix

$$\begin{bmatrix} \hat{A} & \hat{B} \\ \hat{C} & \hat{D} \end{bmatrix} \equiv \begin{bmatrix} \hat{\gamma}/\hat{\beta} & 1/\hat{\beta} \\ -\hat{\beta} + \hat{\alpha}\hat{\gamma}/\hat{\beta} & \hat{\alpha}/\hat{\beta} \end{bmatrix} \tag{9.8}$$

with $\hat{A}\hat{D} - \hat{B}\hat{C} = 1$, such that if several systems each characterized by such a matrix are cascaded, the matrix characterizing the overall system can be found by multiplying the individual matrices. In chapter 8 we saw that the matrix defined above also corresponds to the well-known ray matrix employed in geometrical optics, relating the input and output ray intercepts x and normalized angles σ_x through the relation

$$\begin{bmatrix} x_2 \\ \sigma_{x2} \end{bmatrix} = \begin{bmatrix} \hat{A} & \hat{B} \\ \hat{C} & \hat{D} \end{bmatrix} \begin{bmatrix} x_1 \\ \sigma_{x1} \end{bmatrix}. \tag{9.9}$$

The matrices corresponding to a thin lens, a section of free space, and a quadratic graded-index medium are also repeated for convenience:

$$\hat{\mathbf{M}}_{\text{lens}} = \begin{bmatrix} 1 & 0 \\ -1/\lambda f & 1 \end{bmatrix}, \tag{9.10}$$

$$\hat{\mathbf{M}}_{\text{space}} = \begin{bmatrix} 1 & \lambda d \\ 0 & 1 \end{bmatrix}, \tag{9.11}$$

$$\hat{\mathbf{M}}_{\text{grin}} = \begin{bmatrix} \cos[(d/d_0)\pi/2] & (\lambda\chi)\sin[(d/d_0)\pi/2] \\ -(\lambda\chi)^{-1}\sin[(d/d_0)\pi/2] & \cos[(d/d_0)\pi/2] \end{bmatrix}, \tag{9.12}$$

where $d_0 \equiv \chi\pi/2$. The effect of these systems in phase space was extensively discussed in chapter 8.

9.2.1   Quadratic-phase systems as fractional Fourier transforms

We know from chapters 3 and 4 that the one-parameter class of fractional Fourier transforms is a subclass of the class of three-parameter quadratic-phase systems (linear canonical transforms). If we allow an additional magnification parameter M and a phase radius of curvature parameter R, the family of magnified fractional Fourier transforms with phase curvature will also have three parameters and can be put in one-to-one correspondence with the family of quadratic-phase systems. The kernel of this three-parameter transform may be written as

$$\hat{g}(x) = \int \hat{h}(x, x') \hat{f}(x') \, dx', \tag{9.13}$$

$$\hat{h}(x, x') = K e^{i\pi x^2/\lambda R} \sqrt{\frac{1}{s^2 M}} \, A_{a\pi/2}$$

$$\times \exp\left[\frac{i\pi}{s^2}\left(\frac{x^2}{M^2}\cot(a\pi/2) - 2\frac{xx'}{M}\csc(a\pi/2) + x'^2 \cot(a\pi/2)\right)\right],$$

which has the general form of a quadratic-phase system. (The pure mathematical form of the fractional Fourier transform given by equation 4.4 is recovered by setting $x/s = u$, $x'/s = u'$, $M = 1$, $R = \infty$, $K = 1$.) The system is energy preserving when the magnitude of the constant K is unity. This kernel maps a function $\hat{f}(x) = s^{-1/2} f(x/s)$ into $K \exp(i\pi x^2/\lambda R) \sqrt{1/sM} \, f_a(x/sM)$, where $f_a(u)$ is the ath order fractional Fourier transform of $f(u)$. We will avoid using the angular order $\alpha = a\pi/2$ as much as possible in contexts where it may be confused with the linear canonical transform parameter α. Here $M > 0$ is the magnification associated with the transform and R is the radius of the spherical surface on which the perfect fractional Fourier transform is observed. When $R = \infty$ the quadratic phase factor disappears and the perfect fractional Fourier transform is observed on a planar surface.

The above family of kernels is in one-to-one correspondence with the family of kernels given in equation 9.4. The parameters $\hat{\alpha}$, $\hat{\beta}$, and $\hat{\gamma}$ are recognized to be related to the parameters $a\pi/2$, M, and R through the relations

$$\hat{\alpha} = \frac{\cot(a\pi/2)}{s^2 M^2} + \frac{1}{\lambda R},$$

$$\hat{\beta} = \frac{\csc(a\pi/2)}{s^2 M},$$

$$\hat{\gamma} = \frac{\cot(a\pi/2)}{s^2}. \tag{9.14}$$

Alternatively, the matrix parameters are related to $a\pi/2$, M, and R through the following relations (equation 9.8):

$$\hat{A} = M \cos(a\pi/2),$$

$$\hat{B} = s^2 M \sin(a\pi/2),$$

$$\hat{C} = -\frac{\sin(a\pi/2)}{s^2 M} + \frac{M\cos(a\pi/2)}{\lambda R},$$

$$\hat{D} = \frac{\cos(a\pi/2)}{M} + \frac{s^2 M \sin(a\pi/2)}{\lambda R}, \tag{9.15}$$

which can be inverted to yield

$$\tan(a\pi/2) = \frac{1}{s^2}\frac{\hat{B}}{\hat{A}}, \tag{9.16}$$

$$M = \sqrt{\hat{A}^2 + (\hat{B}/s^2)^2}, \tag{9.17}$$

$$\frac{1}{\lambda R} = \frac{1}{s^4}\frac{\hat{B}/\hat{A}}{\hat{A}^2 + (\hat{B}/s^2)^2} + \frac{\hat{C}}{\hat{A}}. \tag{9.18}$$

The above result essentially means that any quadratic-phase system can be interpreted as a magnified fractional Fourier transform, perhaps with a residual phase curvature. Thus the relatively large class of optical systems which can be modeled as quadratic-phase systems, can also be interpreted as fractional Fourier transformers. In section 9.8 we will further see that all Fourier optical systems, which consist of spatial filters sandwiched between quadratic-phase systems, can be interpreted as consecutive spatial filtering operations in fractional Fourier domains.

We see that an optical system consisting of an arbitrary concatenation of sections of free space, thin lenses, and sections of quadratic graded-index media can be characterized either as a linear canonical transform with parameters $\hat{\alpha}$, $\hat{\beta}$, $\hat{\gamma}$ (or $\hat{A}$, $\hat{B}$, $\hat{C}$, $\hat{D}$), or as a fractional Fourier transform of order a, magnified by M, and observed on a spherical surface with radius R. Although both interpretations are legitimate and may be beneficial in different circumstances, we admit a preference for the fractional Fourier transform description because the three parameters are simpler to interpret. M and R refer simply to the magnification of the observed transform and the radius of the surface on which it is observed. The order a begins from 0 at the input of the system, and then monotonically increases as a function of z. Fractional Fourier transforms of increasing orders describe the evolution of light as it propagates through the system. The reader will be able to see this graphically through a numerical example in section 9.2.4. But first we will consider two elementary examples: propagation in quadratic graded-index media and diffraction in free space, and then treat the more general case of arbitrary compositions of thin lenses and sections of free space.

9.2.2   Quadratic graded-index media

Quadratic graded-index media have a natural and direct relationship with the fractional Fourier transform. Light is simply fractional Fourier transformed as it propagates through such media. Comparing the matrix given in equation 9.12 with equation 9.15, we immediately conclude that propagation through a section of graded-index media results in a fractional Fourier transform of order $a = d/d_0$, provided the scale parameter s is chosen such that $s^2 = \lambda\chi$. Agreeing on this choice of s, there is no magnification ($M = 1$) and no residual phase curvature ($R = \infty$). Readers who have studied chapters 3 and 8 will know that since the associated matrix is a rotation matrix, as light propagates through quadratic graded-index media its Wigner distribution rotates.

Quadratic graded-index media realize fractional Fourier transforms in their purest and simplest form. If the distribution of light at the input plane is given by $f(x/s)$, then the distribution of light at the output plane is proportional to $f_a(x/s)$, where the transform order a increases linearly with distance of propagation. A quadratic graded-index medium of length d_0 takes a Fourier transform, while one of $2d_0$ results in an inverted image and one of $4d_0$ in an erect image. To obtain a fractional Fourier transformer of fractional order a, we simply cut a piece whose length is ad_0.

The same result can be arrived at from scratch by starting from the Helmholtz equation for quadratic graded-index media, finding its modes (which are the Hermite-Gaussian functions), and constructing arbitrary solutions as linear superpositions of these modes (section 7.4.3). Quadratic graded-index media will be further discussed in section 9.5.

Other areas where situations analogous to optical guiding in graded-index media occur, and thus which may potentially be formulated in terms of the fractional Fourier transform, include propagation of ion beams and magnetospheric physics (Bracewell 1999).

9.2.3 Fresnel diffraction

Although propagation in quadratic graded-index media corresponds to the fractional Fourier transform in its purest form, it is of interest to discuss the more basic problem of diffraction from a planar screen with complex amplitude transmittance $\hat{t}(x)$. The complex amplitude distribution $\hat{g}(x)$ of light in a diffraction plane at distance d is given by the Fresnel integral (equation 9.6):

$$\hat{g}(x) = \int \hat{h}_{\text{space}}(x, x')\hat{t}(x')\, dx',$$
(9.19)

$$\hat{h}_{\text{space}}(x, x') = e^{i2\pi\sigma d} e^{-i\pi/4} \sqrt{\frac{1}{\lambda d}} \, \exp\left[\frac{i\pi (x - x')^2}{\lambda d}\right],$$

assuming illumination of the screen by a unit plane wave. Now, it is possible to cast this integral in the form of the integral in equation 9.13 by identifying

$$\tan(a\pi/2) = \frac{\lambda d}{s^2},$$
(9.20)

$$M = \sqrt{1 + \left(\frac{\lambda d}{s^2}\right)^2},$$
(9.21)

$$\lambda R = \frac{s^4 + \lambda^2 d^2}{\lambda d}.$$
(9.22)

These results can also be arrived at by comparing equation 9.15 with equation 9.11, or by specializing equations 9.16, 9.17, and 9.18 to $\hat{A} = \hat{D} = 1$, $\hat{B} = \lambda d$, $\hat{C} = 0$.

We conclude that at a distance d from the diffracting object or aperture, we observe the ath order fractional Fourier transform of the object on a spherical reference surface with radius R. The transform is magnified by M. As d is increased from 0 to ∞, the order a of the fractional transform increases according to $a = (2/\pi)\arctan(\lambda d/s^2)$ from 0 to 1 (figure 9.2). We see that the diffraction of light can be viewed as a process of continual fractional Fourier transformation. As light propagates, its distribution evolves through fractional transforms of increasing order. Letting $d \to \infty$, we obtain $a = 1$, $M = \lambda d/s^2 \propto d$, and $R = d$, which we readily associate with the Fraunhofer diffraction pattern (the Fourier transform of the diffracting screen). In this limit, the magnification and radius of curvature are both proportional to the distance d.

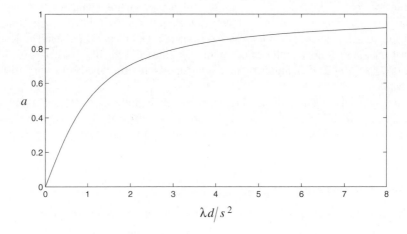

Figure 9.2.   Plot of $a = (2/\pi)\arctan(\lambda d/s^2)$ as a function of $\lambda d/s^2$.

More generally, there exists a fractional Fourier transform relation between the amplitude distribution of light on two spherical reference surfaces of given radii and separation. It is possible to determine the order and scale parameters associated with this fractional transform given the radii and separation of the surfaces. Alternatively, given the desired order and scale parameters, it is possible to determine the necessary radii and separation. These results will be presented in section 9.3.

The propagation of light in free space, as described by the Fresnel integral, can also be interpreted as a certain wavelet transform (Onural 1993, Onural, Kocatepe, and Ozaktas 1994, Onural and Kocatepe 1995). The relationship between the fractional Fourier transform and this wavelet transform was discussed in section 4.12.3, so that Onural's results are easily related to those of the present chapter.

In this context we also mention the work of Alonso and Forbes (1997), who take a fractional transformation approach in developing a new representation for the scalar wave propagator, which remains valid in the presence of caustics and has other advantages.

9.2.4   Multi-lens systems

We now return to our discussion of the general case. We have seen that every quadratic-phase system can be interpreted as a fractional Fourier transforming system, hence the fractional Fourier transform can describe all systems consisting of an arbitrary number of lenses separated by arbitrary distances (whereas imaging and Fourier transforming systems are only special cases).

As discussed earlier, the kernel in equation 9.4 can be cast in the form of the kernel in equation 9.13 by identifying a, M, and R according to equations 9.16, 9.17, and 9.18. For instance, given an optical system characterized by the focal lengths and separations of the lenses, we can find the matrix parameters $\hat{A}(z)$, $\hat{B}(z)$, $\hat{C}(z)$, $\hat{D}(z)$ characterizing the system from the input (taken as $z = 0$) up to any plane $z = $ constant, and then use the equations in question to find $a(z)$, $M(z)$, and $R(z)$.

A concrete example will be useful. Figure 9.3a shows a system consisting of several thin lenses drawn as vertical chain-dotted lines, whose focal lengths have been indicated in meters right above them (also see figure 9.1). The vertical axis is in arbitrary units. The input plane is taken as $z = 0$. The output plane is variable, ranging from $z = 0$ to $z = 2$ m. Let $\hat{A}(z)$, $\hat{B}(z)$, $\hat{C}(z)$, $\hat{D}(z)$ denote the matrix parameters of the section of the system occupying the interval $[0, z]$, which can be readily calculated using the matrices for lenses and sections of free space and the concatenation property. Also let $[x(z)\ \sigma_x(z)]^{\mathrm{T}}$ denote the ray vector at z. Then,

$$\left[\begin{array}{c} x(z) \\ \sigma_x(z) \end{array} \right] = \left[\begin{array}{cc} \hat{A}(z) & \hat{B}(z) \\ \hat{C}(z) & \hat{D}(z) \end{array} \right] \left[\begin{array}{c} x(0) \\ \sigma_x(0) \end{array} \right]. \tag{9.23}$$

Two rays have been drawn through the system in figure 9.3a. We further let $a(z)$, $M(z)$, $R(z)$ represent the order, magnification, and phase radius of curvature of the fractional Fourier transform observed at z, which can be determined by using equations 9.16, 9.17, and 9.18. These three functions of z are plotted in figure 9.3 (Ozaktas and Erden 1997). Letting j denote an arbitrary integer, when $a = 4j$ we observe an erect image, when $a = 4j+2$ we observe an inverted image, when $a = 4j+1$ we observe the ordinary Fourier transform, and when $a = 4j - 1$ we observe an inverted Fourier transform (which is the same as an inverse Fourier transform). The reader should study the behavior of the two rays in conjunction with the graphs in figure 9.3. At $z = 0.4$ m we obtain an ordinary Fourier transform ($a = 1$) as a result of the conventional $2f$ system occupying the interval $[0, 0.4]$. An inverted image ($a = 2$) is observed at $z \approx 0.65$ m, where we see that $M < 1$ and $R > 0$, as confirmed by an examination of the rays. (The ray represented by the solid line crosses the $z = 0.65$ m plane at a negative value (implying an inverted image) smaller than unity in magnitude (implying $M < 1$), with a slope indicating divergence (implying $R > 0$).) An inverted Fourier transform ($a = 3$) is observed at $z \approx 1.2$ m, almost coincident with the lens at that location. An erect image ($a = 4$) is observed at $z \approx 1.4$ m, immediately after the lens at that location. The phase curvature $1/R$ of this image has a very small negative value and the magnification M is slightly smaller than 1. Before leaving our example, we remind the reader that the curves in figure 9.3 are not numerically obtained; they are described by piecewise analytical expressions. The imaging systems discussed in Bernardo and Soares (1994b) provide additional useful examples which the reader may wish to study in a similar manner. (Ozaktas and Erden 1997)

It is instructive to compare multi-lens systems with quadratic graded-index media. Figure 9.4a shows the trajectory of two rays through such a medium and the other parts of figure 9.4 show how a, M, and R vary as functions of z. We see that the order a increases linearly, the magnification parameter M is constant, and $1/R = 0$. The rays follow sinusoidal trajectories. Propagation in such media allows the relationship

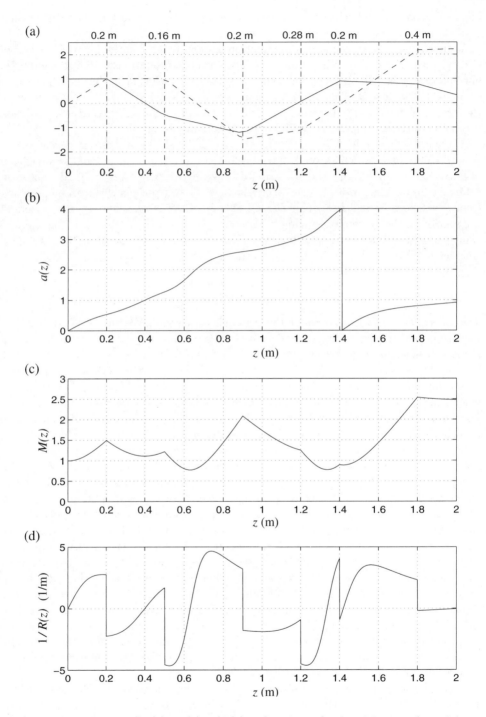

Figure 9.3. Evolution of $a(z)$, $M(z)$, $1/R(z)$ as functions of z: $\lambda = 0.5\,\mu$m and $s = 0.3$ mm (Ozaktas and Erden 1997, Erden 1997).

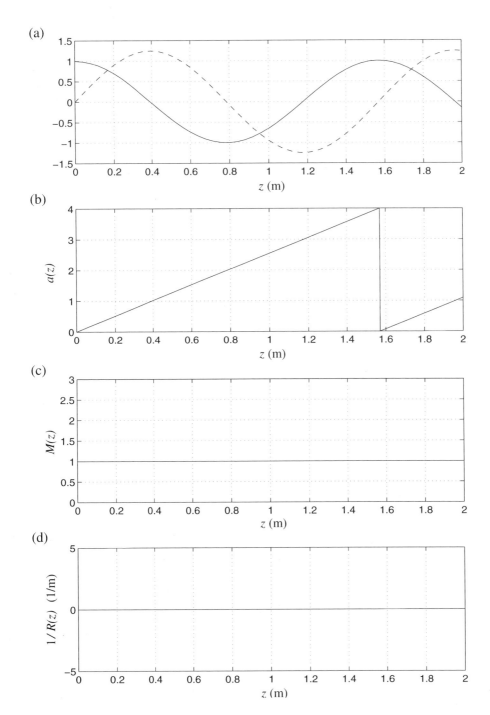

Figure 9.4. Figure 9.3 repeated for a quadratic graded-index medium with $\chi = 0.25$ m, $n_0 = 1.4$, and $s = 0.3$ mm (Ozaktas and Erden 1997, Erden 1997).

between the behavior of rays and the fractional transform order a to be seen in its purest form, and should make it easier to comprehend the more complicated case of multi-lens systems. It is important to note that this behavior is obtained only when the scale parameter s is chosen to have its "natural" value of $s^2 = \lambda\chi$. For other values of s, the magnification is not constant, as will be discussed in section 9.5.4. We also note that we deliberately ensured that the value of s in both the multi-lens system and the graded-index medium are identical so that the plots are comparable to each other.

We conclude with a brief comment on Fourier optical systems, which consist of an arbitrary number of thin filters sandwiched between arbitrary quadratic-phase systems. It readily follows from the results of this section that any Fourier optical system can be modeled as filters sandwiched between fractional Fourier transform stages, or as repeated filtering in consecutive fractional Fourier domains, as will be further discussed in section 9.8 (Ozaktas and Mendlovic 1996a).

9.2.5   Optical implementation of the fractional Fourier transform

Until now our interest was in analyzing or interpreting optical systems in terms of the fractional Fourier transform. Now we consider the problem of designing systems for optically realizing the fractional Fourier transform, perhaps as a component in an optical signal processing system. In general, one can obtain a fractional Fourier transformer with desired order, magnification, and phase curvature by using equation 9.15 to find the matrix elements $\hat{A}$, $\hat{B}$, $\hat{C}$, $\hat{D}$ that result in the desired transform. Then it remains to use an appropriate decomposition (for example, equation 3.121 or 3.122) for this matrix in order to realize the system with lenses and sections of free space.

Rather than pursuing this general approach, here we will mention three particularly simple systems which map $\hat{f}(x) = s^{-1/2}f(x/s)$ into $\hat{g}(x) \propto f_a(x/s)$. These correspond to and generalize the three Fourier transforming stages discussed in chapters 7 and 8, and reduce to them as special cases when $a = 1$.

Conceptually the simplest is to use a section of quadratic graded-index media of length $d = \chi(a\pi/2) = ad_0$ with $s^2 = \lambda\chi$ (section 9.2.2).

In practice, systems consisting of bulk lenses may be preferred. Two such systems were first presented by Lohmann (1993a, b). The first system consists of a section of free space of length d followed by a lens of focal length f followed by a second section of free space of length d. To obtain an ath order fractional Fourier transform with scale parameter s, we must choose d and f according to

$$d = \frac{s^2}{\lambda} \tan(a\pi/4), \tag{9.24}$$

$$f = \frac{s^2}{\lambda} \csc(a\pi/2). \tag{9.25}$$

The second system consists of a lens of focal length f followed by a section of free space of length d followed by a second lens of focal length f. This time d and f must be chosen according to

$$d = \frac{s^2}{\lambda} \sin(a\pi/2), \tag{9.26}$$

$$f = \frac{s^2}{\lambda} \cot(a\pi/4). \tag{9.27}$$

These systems will be discussed again in section 9.4.3.

9.2.6 Hermite-Gaussian expansion approach

We have already seen that the propagation of light can be viewed as a process of continual fractional Fourier transformation. Here we will discuss the same process, but this time in terms of Hermite-Gaussian beam expansions, rather than Fresnel integrals or plane wave expansions. We will see that the order of the fractional Fourier transform is proportional to the Gouy phase shift accumulated with propagation.

Considering a single transverse dimension x so that the amplitude distribution of light is given by the function $\hat{f}_z(x)$, let $\hat{f}_0(x)$ denote the distribution at the plane $z = 0$. We can expand this function in terms of the Hermite-Gaussian functions:

$$\hat{f}_0(x) = \sum_{l=0}^{\infty} C_l \frac{1}{s^{1/2}} \psi_l\left(\frac{x}{s}\right), \tag{9.28}$$

$$C_l = \int \hat{f}_0(x) \frac{1}{s^{1/2}} \psi_l\left(\frac{x}{s}\right) dx.$$

We can interpret the function $s^{-1/2}\psi_l(x/s)$ as the amplitude distribution of a one-dimensional lth order Hermite-Gaussian beam with beam size $W_0 = s$ at its waist. Then it becomes an easy matter to write the amplitude distribution $\hat{f}_z(x)$ at an arbitrary plane, since we know how each of the Hermite-Gaussian components propagates (section 7.3.4):

$$\hat{f}_z(x) = \sum_{l=0}^{\infty} C_l \frac{1}{W^{1/2}(z)} \psi_l\left(\frac{x}{W(z)}\right) \exp\left[i2\pi\sigma z + \frac{i2\pi\sigma x^2}{2R(z)} - i(l + 1/2)\zeta(z)\right], \tag{9.29}$$

where $W(z) = W_0[1 + (z/\check{z})^2]^{1/2}$ is the beam size. The Rayleigh range $\check{z}$ is related to W_0 by the relation $W_0^2 = \lambda\check{z}$. $R(z) = z[1 + (\check{z}/z)^2]$ is the radius of curvature of the wavefronts and $\zeta(z) = \arctan(z/\check{z})$ is the Gouy phase shift.

Using equation 4.23, equation 9.29 can be written in a very simple form in terms of the fractional Fourier transform. Let us switch to functions with dimensionless arguments: $\hat{f}_z(x) = s^{-1/2}f_z(x/s)$. Then the amplitude distribution at any plane is given by

$$\hat{f}_z(x) = \frac{1}{W^{1/2}(z)} e^{i2\pi\sigma z} e^{-i\zeta(z)/2} e^{i2\pi\sigma x^2/2R(z)} \left(\mathcal{F}^{a(z)} f_0(u)\right)\left(\frac{x}{W(z)}\right), \tag{9.30}$$

where

$$a(z) = \frac{2}{\pi}\zeta(z). \tag{9.31}$$

In equation 9.30 the fractional Fourier transform is taken with respect to u, and $f_0(u) = s^{1/2}\hat{f}_0(su)$. Rewriting

$$\alpha(z) = \frac{\pi}{2}a(z) = \zeta(z), \tag{9.32}$$

we see that the "angular order" a of the fractional Fourier transform in question is simply equal to the Gouy phase shift accumulated in propagating from 0 to z. As $z \to \infty$ we see that $\zeta(z) \to \pi/2$ and $a(z) \to 1$, corresponding to the ordinary Fourier transform. This is precisely the same limit discussed in section 9.2.3.

This result can be generalized for propagation between two spherical references surfaces with arbitrary radii; see section 9.6 and Ozaktas and Mendlovic 1994. Let the radius of the surface at $z = z_1$ be denoted by R_1 and that of the surface at $z = z_2$ be denoted by R_2. The radius is positive if the surface is convex to the right. Then there exists a fractional Fourier transform between these two surfaces whose order is given by

$$a = \frac{\zeta(z_2) - \zeta(z_1)}{\pi/2}. \tag{9.33}$$

It is well known that if there holds a certain relation between R_1, R_2, and $z_2 - z_1$, one obtains an ordinary Fourier transform relation between two spherical surfaces. We see that for other values of these parameters we obtain a fractional Fourier transform relation. Given any two spherical surfaces, what we need to do to find the order a of the fractional Fourier transform between them, is to find the Rayleigh range and waist location of a Gaussian beam that would "fit" into these surfaces, and then calculate a from equation 9.33.

We may also think of a complex amplitude distribution "riding" on a Gaussian beam wavefront. The spatial dependence of the propagating wavefront is like a carrier defining spherical surfaces, on top of which the complex amplitude distribution rides, being fractional Fourier transformed in the process.

Since laser resonators commonly consist of two spherical mirrors, it becomes possible to characterize such resonators in terms of a fractional order parameter, again obtained from equation 9.33. The well-known stability (or confinement) condition for spherical mirror resonators can be stated in a particularly simple form in terms of a: as long as a is real, we have a stable resonator. (In our discussion we have implicitly assumed that a is real, which means that we have implicitly assumed stable resonators.) Unstable resonators are described by values of a which are not real. The stability condition is further discussed in section 9.6 and Ozaktas and Mendlovic 1994.

The Hermite-Gaussian approach can also be used to discuss multi-lens systems such as those shown in figure 9.1 and figure 9.3a. When dealing with such systems, one must consider the accumulated Gouy phase shift with respect to the input plane at $z = 0$, rather than the conventional Gouy phase shift with respect to the last waist of the beam; see page 237 and Erden and Ozaktas 1997. Then the fractional order parameter $\alpha(z) = (\pi/2)a(z)$ shown in figure 9.3 can be associated with the accumulated Gouy phase. Furthermore, readers familiar with the propagation of Gaussian beams may have noted the similarity between the behavior of the functions $R(z)$ and $M(z)$ shown in figure 9.3, and the behavior of the wavefront radius and beam size of a Gaussian beam propagating through the system. In section 9.7.4 we discuss how the expressions for the parameters of a Gaussian beam are related to the expressions for $\alpha(z) = (\pi/2)a(z)$, $M(z)$, and $R(z)$ in equations 9.16, 9.17, and 9.18. The main result can be stated as follows. Let the output of an arbitrary system consisting of lenses and sections of free space be interpreted as a fractional Fourier transform of the input of order $\alpha(z)$, with magnification factor $M(z)$, observed on a spherical surface of radius $R(z)$. Let a Gaussian beam whose waist is located at $z = 0$ with waist size W_{G0} exhibit an accumulated Gouy phase shift $\tilde{\zeta}_G(z)$, beam size $W_G(z)$, and wavefront radius $R_G(z)$ at the output of the same system. If the scale parameter s appearing in equations 9.13 and 9.15 is equal to W_{G0}, then $\alpha(z) = \tilde{\zeta}_G(z)$, $M(z) = W_G(z)/W_{G0}$, and $R(z) = R_G(z)$. (Ozaktas and Erden 1997)

9.3 General fractional Fourier transform relations in free space

In this section we show that there exists a fractional Fourier transform relation between the (appropriately scaled) amplitude distributions of light on two spherical surfaces of given radii and separation. This basic result provides an alternative statement of the law of propagation of light, and allows us to pose the fractional Fourier transform as a tool for analyzing and describing a rather general class of optical systems. It also enables us to arrive at a very general class of fractional Fourier transforming systems with variable input and output scale parameters, and to state the necessary and sufficient conditions for a fractional Fourier transform in full generality.

Again, we will be customarily speaking of "spherical surfaces," despite the fact that for one-dimensional systems it would be more accurate to speak of circular (or cylindrical) surfaces.

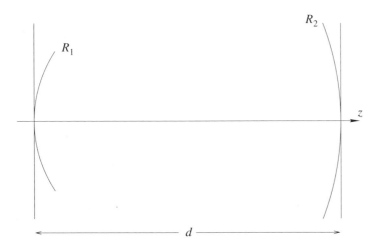

Figure 9.5. Planar and spherical reference surfaces. The figure has been drawn such that $R_1 < 0$ and $R_2 > 0$. The distance d is always taken to be positive. (Ozaktas and Mendlovic 1995a)

9.3.1 The fractional Fourier transform and Fresnel's integral

We refer to figure 9.5. The complex amplitude distribution with respect to the first and second spherical reference surfaces will be denoted by $\hat{f}_{sr1}(x')$ and $\hat{f}_{sr2}(x)$ respectively. The distribution with respect to the planar surfaces tangent to the spherical surfaces on the optical axis will be denoted by $\hat{f}_1(x')$ and $\hat{f}_2(x)$. If the radii of the spherical surfaces are denoted by R_1 and R_2, we have

$$\hat{f}_2(x) = e^{i\pi x^2/\lambda R_2} \hat{f}_{sr2}(x), \tag{9.34}$$

$$\hat{f}_1(x') = e^{i\pi x'^2/\lambda R_1} \hat{f}_{sr1}(x'). \tag{9.35}$$

Assuming propagation from left to right, $\hat{f}_2(x)$ is related to $\hat{f}_1(x')$ by a Fresnel integral:

$$\hat{f}_2(x) = \frac{e^{i2\pi\sigma d}}{\sqrt{i\lambda d}} \int e^{i\pi(x-x')^2/\lambda d} \hat{f}_1(x')\,dx'. \tag{9.36}$$

Combining the last three equations, we can obtain a relation between $\hat{f}_{sr1}(x')$ and $\hat{f}_{sr2}(x)$. To enable comparison of this relation with equation 4.4, we introduce the dimensionless variables $u' \equiv x'/s_1$ and $u \equiv x/s_2$, where s_1 and s_2 are real-valued scale parameters with dimensions of length. Also introducing the dimensionless functions $f_{sr1}(u') \equiv s_1^{1/2}\hat{f}_{sr1}(u's_1)$ and $f_{sr2}(u) \equiv s_2^{1/2}\hat{f}_{sr2}(us_2)$, we obtain

$$f_{sr2}(u) = \frac{e^{i2\pi\sigma d}\, s_1^{1/2} s_2^{1/2}}{\sqrt{i\lambda d}} \int \exp\left[\frac{i\pi}{\lambda d}\left(g_2 s_2^2 u^2 - 2s_2 s_1 uu' + g_1 s_1^2 u'^2\right)\right] f_{sr1}(u')\,du', \tag{9.37}$$

where we make the definitions

$$g_1 \equiv 1 + d/R_1, \qquad g_2 \equiv 1 - d/R_2. \tag{9.38}$$

Now, comparing this result with the definition of the fractional Fourier transform (equation 4.4), we conclude that $f_{sr2}(u)$ will be proportional to the fractional Fourier transform of $f_{sr1}(u')$; that is,

$$f_{sr2}(u) = \left[\frac{e^{i2\pi\sigma d}\, s_1^{1/2} s_2^{1/2}\, e^{i[\pi\,\mathrm{sgn}(\alpha)/4 - \alpha/2]}\, |\sin\alpha|^{1/2}}{\sqrt{i\lambda d}}\right] (\mathcal{F}^a f_{sr1})(u), \tag{9.39}$$

if and only if

$$g_2 \frac{s_2^2}{\lambda d} = \cot\alpha, \tag{9.40}$$

$$g_1 \frac{s_1^2}{\lambda d} = \cot\alpha, \tag{9.41}$$

$$\frac{s_1 s_2}{\lambda d} = \csc\alpha. \tag{9.42}$$

These three equations are the necessary and sufficient conditions for equation 9.39 to hold. Speaking of $f_{sr1}(u')$ as the input and $f_{sr2}(u)$ as the output of our system, in the following sections we discuss the consequences of these equations from three perspectives:

1. *Analysis.* Given the radii R_1, R_2 and separation d of the spherical surfaces, what is the order a of the fractional Fourier transform relation existing between the amplitude distributions on these surfaces, and what are the scale parameters s_1, s_2 associated with the input and output?

2. *Synthesis (or design).* Given that we want to design an ath order fractional Fourier transformer with scale parameters s_1, s_2 specified for the input and output, how should we choose the radii R_1, R_2 and separation d of the spherical surfaces?

3. *Propagation.* Given the radius R_1 and scale parameter s_1 associated with the input, what are the radius R_2 and scale parameter s_2 associated with the output situated a distance d to the right, as well as the order a of the resulting fractional Fourier transform?

For greater generality, in the following sections we allow s_1, s_2 to take negative values as well, contrary to our previous practice of requiring them to be strictly positive.

9.3.2 Analysis

Problem: Given R_1, R_2, and d; find s_1, s_2, and a (or equivalently $\alpha = a\pi/2$). That is, we are given the reference surfaces, and wish to find the order and scale parameters of the resulting transform.

Equations 9.40 and 9.41 imply

$$g_1 s_1^2 = g_2 s_2^2. \tag{9.43}$$

Now, using the identity $\cot^2 \alpha + 1 = \csc^2 \alpha$ and equations 9.40 to 9.43, we obtain

$$s_2^4 = (\lambda d)^2 (g_2/g_1 - g_2^2)^{-1}, \tag{9.44}$$
$$s_1^4 = (\lambda d)^2 (g_1/g_2 - g_1^2)^{-1}. \tag{9.45}$$

Note that equation 9.43 implies $g_1 g_2 \geq 0$ and equations 9.44 and 9.45 imply $(g_2/g_1 - g_2^2) \geq 0$ and $(g_1/g_2 - g_1^2) \geq 0$. These conditions can be summarized in the form

$$0 \leq g_1 g_2 \leq 1. \tag{9.46}$$

If a fractional transform relation is to hold, R_1, R_2, and d must be specified such that this condition holds. (If figure 9.5 is interpreted as a spherical mirror resonator, this equation is the stability or confinement condition of the resonator; see section 9.6.)

We see that given R_1, R_2, and d; $|s_2|$ and $|s_1|$ are uniquely determined by equations 9.44 and 9.45. Then equation 9.40 or 9.41 enables us to determine α according to

$$\tan \alpha = \pm \left(\frac{1}{g_1 g_2} - 1 \right)^{1/2}, \tag{9.47}$$

where the $\pm$ is determined according to the common sign of g_1 and g_2. The ambiguity in the inverse tangent function is resolved by examining the sign of $\csc \alpha$. Equation 9.42 tells us that if we choose the signs of s_1 and s_2 such that $s_1 s_2 \geq 0$, then $\csc \alpha \geq 0$ so that α lies in $[0, \pi]$. On the other hand, if $s_1 s_2 \leq 0$, then $\csc \alpha \leq 0$ so α lies in $[-\pi, 0]$. These results are consistent with parity properties of the fractional Fourier transform.

Result: A fractional Fourier transform relation exists between two spherical surfaces of radii R_1, R_2 and separation d if and only if $0 \leq g_1 g_2 \leq 1$. Granted that this condition is satisfied, $|s_1|$ and $|s_2|$ are determined by equations 9.44 and 9.45, and α is determined within $\pm\pi$ by equation 9.47. The quadrant of α is determined by our choice of the signs of s_1 and s_2, or vice versa.

9.3.3 Synthesis

Problem: Given s_1, s_2, and a (or equivalently $\alpha = a\pi/2$); find R_1, R_2, and d. That is, we wish to design a fractional Fourier transforming system with specific order and scale parameters.

Equation 9.42 implies that the sign of $s_1 s_2$ must be the same as the sign of $\csc\alpha$, which is the same as the sign of α. Granted that s_1, s_2, and α have been specified consistent with this requirement, equation 9.42 determines d:

$$d = \frac{s_1 s_2}{\lambda}\sin\alpha. \tag{9.48}$$

Then, equations 9.40 and 9.41 give g_1 and g_2, which in turn give R_1 and R_2:

$$1 + d/R_1 \equiv g_1 = \frac{s_2}{s_1}\cos\alpha, \tag{9.49}$$

$$1 - d/R_2 \equiv g_2 = \frac{s_1}{s_2}\cos\alpha. \tag{9.50}$$

Note that g_1 and g_2 as given by these equations will always satisfy $g_1 g_2 = \cos^2\alpha$ and thus $0 \le g_1 g_2 \le 1$.

Result: A fractional Fourier transform relation of order a between two spherical surfaces with the input and output scaled by s_1 and s_2 respectively, can be obtained if and only if $\operatorname{sgn}(s_1 s_2) = \operatorname{sgn}(\sin\alpha)$. Granted that this condition is satisfied, we must choose R_1, R_2, and d according to equations 9.48, 9.49, and 9.50.

9.3.4 Propagation

Problem: Given s_1, R_1, and d; find a (or α), s_2, and R_2. That is, given the radius of the spherical reference surface and the scale parameter on the input side, find them at a distance d to the right, as well as the order of the resulting transform.

Since R_1 and d are given, g_1 is also known. Using equations 9.40, 9.41, and 9.42 it is possible to obtain

$$\tan\alpha = \frac{\lambda d}{g_1 s_1^2}, \tag{9.51}$$

$$s_2^2 = g_1^2 s_1^2 + \frac{(\lambda d)^2}{s_1^2}, \tag{9.52}$$

$$1 - d/R_2 \equiv g_2 = \frac{g_1 s_1^4}{g_1^2 s_1^4 + (\lambda d)^2}. \tag{9.53}$$

We are free to choose the sign of s_2, which together with the specified sign of s_1 determines the sign of $\csc\alpha$. This determines the quadrant of α in equation 9.51. If we assume that s_1, s_2 are both positive, α lies in the interval $[0, \pi]$. It now becomes the case that α, as given by equation 9.51, is a continuous monotonic increasing function of d. Let us consider the distribution of light as it propagates over increasing distances d to the right. For larger values of d, we will observe fractional transforms of the initial distribution of larger orders α; the amplitude distribution of light is continuously fractional Fourier transformed as it propagates.

Result: Given a distribution of light on a spherical reference surface of radius R_1 and scale parameter s_1, we can observe its fractional Fourier transform on another spherical reference surface a distance d to the right. The radius R_2 and scale parameter s_2 for this surface, and the order of the fractional transform are given by equations 9.51, 9.52, and 9.53, where the quadrant of α is determined by our choice of the sign of s_2.

9.3.5   *Discussion*

We have seen that there exists a fractional Fourier transform relation between two spherical reference surfaces of given radii and separation. It is possible to determine the order and scale parameters associated with this fractional transform, given the radii and separation of the surfaces. Alternatively, given the desired order and scale parameters, it is possible to determine the necessary radii and separation.

Is the fact that the fractional Fourier transforms are in general observed on spherical surfaces (rather than planar surfaces) a concession? No. Remember that the Fourier transforms and images occuring in multi-lens systems are also in general observed on spherical surfaces.

The propagation result (as well as the analysis and synthesis results) should stand up to a transitivity test. To understand what this means, consider three spherical reference surfaces. Now, the results presented can be applied between any two of these three surfaces. We can apply our results for propagation from the first surface to the second, and also from the second to the third. If we eliminate the variables associated with the second surface from these results, we will obtain a relation between the first and third surfaces. Transitivity means that this relation is the same as the relation we would have obtained by directly applying our results for propagation from the first surface to the third. This must necessarily be true if our results are to be consistent. The algebra involved in explicitly verifying that this is indeed the case is quite lengthy and is not presented (Pellat-Finet and Bonnet 1994).

9.4   Illustrative applications

We now present several illustrations of the results presented in section 9.3. We show how the Fresnel diffraction integral can be expressed in terms of a fractional Fourier transform. We then discuss how axially centered systems composed of an arbitrary number of lenses separated by arbitrary distances can be analyzed using fractional Fourier transforms. As an instructive example we will consider the classical single-lens imaging configuration. We also specify the general conditions under which an arbitrary system is a fractional Fourier transformer.

For convenience we return to the practice of restricting s_1 and s_2 to positive values. This implies $\csc \alpha > 0$ so that α lies in the interval $[0, \pi]$. Using equation 9.42, equation 9.39 can now be written in the simpler form

$$f_2(u) = \left[e^{i2\pi\sigma d} e^{-ia\pi/4} \right] \, (\mathcal{F}^a f_{\mathrm{sr}1})\,(u). \tag{9.54}$$

The phase factor $\exp(i2\pi\sigma d)$ is associated with propagation over the distance d. The phase factor $\exp(-ia/2)$ corresponds to the Gouy phase shift (see equation 9.30). Apart from these factors, $f_{\mathrm{sr}2}(u)$ is simply the fractional Fourier transform of $f_{\mathrm{sr}1}(u)$.

9.4.1 Fresnel diffraction as fractional Fourier transformation

We now revisit the problem of Fresnel diffraction within the above general framework. Let us assume that a plane wave of unit amplitude illuminates a planar screen with complex amplitude transmittance $\hat{t}(x)$. We can handle this case by letting $R_1 \to \infty$ in figure 9.5. We then have $\hat{f}_{sr1}(x') = \hat{f}_1(x') = \hat{t}(x')$. Since $g_1 = 1$ in this case, equation 9.41 implies $\cot \alpha \geq 0$ so that α lies in the interval $[0, \pi/2]$.

We also assume that the scale parameter s_1 associated with the input plane is specified freely. Then equation 9.41 gives us the order of the fractional transform observed at distance d from the screen. The scale s_2 of the transform at this distance can be found from equation 9.42. Finally, equation 9.40 determines g_2 and hence R_2. Thus the observed field $\hat{f}_2(x)$ at a distance $d > 0$ is given by

$$\hat{f}_2(x) = \left[e^{i2\pi\sigma d} e^{-ia\pi/4} s_2^{-1/2} \right] e^{i\pi x^2/\lambda R_2} \left(\mathcal{F}^a t \right)(x/s_2), \tag{9.55}$$

with

$$\frac{a\pi}{2} = \alpha = \arctan\left(\frac{\lambda d}{s_1^2} \right), \tag{9.56}$$

$$s_2 = s_1 \sqrt{1 + \frac{(\lambda d)^2}{s_1^4}}, \tag{9.57}$$

$$R_2 = d \left[1 + \frac{s_1^4}{(\lambda d)^2} \right], \tag{9.58}$$

where $t(u') = s_1^{1/2} \hat{t}(u' s_1)$. The spherical phase factor $\exp(i\pi x^2/\lambda R_2)$ can be eliminated by observing the output on a spherical reference surface with radius R_2; that is, by observing $\hat{f}_{sr2}(x)$ instead of $\hat{f}_2(x)$. (Of course, the various phase factors would have no effect if we were observing the intensity only.)

Once again we see that the Fresnel diffraction integral can be formulated as a fractional Fourier transform. As d is increased from 0 to ∞, the order a of the fractional transform increases according to equation 9.56 from 0 to 1 (figure 9.2). Letting $d \to \infty$, we get the intuitively appealing $a = 1$, $s_2 = \lambda d/s_1$, and $R_2 = d$, corresponding to the familiar Fraunhofer diffraction pattern (the Fourier transform of the diffracting screen).

9.4.2 The symmetric case

Referring to figure 9.5, let us consider the special case $-R_1 = R_2 \equiv R$ and $g_1 = g_2 \equiv g \equiv 1 - d/R$. Equations 9.40 and 9.41 then imply $s_1 = s_2$, which we denote by s. The condition $0 \leq g_1 g_2 \leq 1$ becomes $0 \leq g^2 \leq 1$, or more simply $|g| \leq 1$. This implies $0 \leq d/R \leq 2$ and hence $R \geq d/2$.

Equations 9.40 and 9.42 now become

$$g \frac{s^2}{\lambda d} = \cot \alpha, \tag{9.59}$$

$$\frac{s^2}{\lambda d} = \csc \alpha, \tag{9.60}$$

from which it also follows that

$$g = \cos \alpha. \tag{9.61}$$

Given R and d such that $|g| \leq 1$, equation 9.61 immediately determines α. Then either of equations 9.59 or 9.60 determines s. Alternatively, given α and s, we can use the same equations to find d and g, and hence R.

9.4.3 Fractional Fourier transform between planar surfaces

Since there exists a fractional Fourier transform relation between the two spherical surfaces shown in figure 9.5, by using lenses to compensate the spherical phase factors at both surfaces, we can obtain a fractional Fourier transform between two planar surfaces. We just choose lenses with focal lengths $f_1 = -R_1$ and $f_2 = R_2$. Thus, using our synthesis result, it is possible to design a fractional Fourier transformer of any given order a and any desired input and output scale parameters s_1 and s_2.

Let us restrict ourselves to the symmetric case for which $R \geq d/2 > 0$. Then, with two positive lenses of focal length $f = R > 0$, we obtain Lohmann's type II fractional Fourier transforming system (figure 9.6a) (Lohmann 1993b).

Alternatively, let us consider an asymmetric pair of spherical surfaces with $R_1 \to \infty$ and $R_2 = R$. Let us follow this by a thin lens of focal length $f = R/2$, whose effect is to map the amplitude distribution on the spherical surface of radius R onto a spherical surface of radius $-R$. Now, let us follow the lens with a second asymmetric pair of spherical surfaces with $R_1 = -R$ and $R_2 \to \infty$. The overall system consists of a stretch of free space followed by a lens followed by another stretch of free space. This is Lohmann's type I fractional Fourier transforming system (figure 9.6b).

We will refer to these realizations of the fractional Fourier transform as canonical realizations type II and type I, since they are applications of the canonical decompositions given in equations 3.122 and 3.121 to the fractional Fourier transform. They are summarized below within the framework of this chapter. Although we do not explicitly derive the results presented below, the reader should be able to do so without difficulty. An alternative discussion of these systems is to be found in Lohmann's original paper (Lohmann 1993b).

Canonical fractional Fourier transforming configurations have planar input and output reference surfaces separated by a distance ℓ and with scale parameter s. The focal lengths of the lenses are denoted by f. The output $\hat{g}$ of such a system is related to the input $\hat{f}$ by the relation

$$\hat{g}(x) = e^{i2\pi\sigma\ell} e^{-ia\pi/4} s^{-1/2} \left(\mathcal{F}^a f\right)(x/s), \tag{9.62}$$

where $\hat{g}(x) = s^{-1/2} g(x/s)$ and $\hat{f}(x') = s^{-1/2} f(x'/s)$. The output is essentially the ath order fractional Fourier transform of the input.

Canonical type II: A type II system consists of a lens of focal length f followed by a section of free space of length d followed by a second lens of focal length f (figure 9.6a). Thus its total length is $\ell = d$. If the order of the fractional transform $a\pi/2 = \alpha$ and the scale s are specified, then the separation of the lenses d and their focal length f

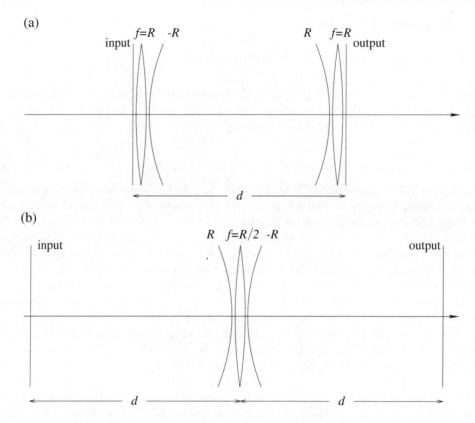

Figure 9.6. Lohmann's canonical realizations: (a) type II and (b) type I. The separations and focal lengths are given by equations 9.63, 9.64 in part a and equations 9.67, 9.68 in part b. (Ozaktas and Mendlovic 1995a)

must be chosen according to

$$d = \frac{s^2}{\lambda} \sin \alpha, \tag{9.63}$$

$$f = \frac{s^2}{\lambda} \cot(\alpha/2). \tag{9.64}$$

Alternatively, provided $f \geq d/2$, we can find α and s in terms of d and f by using

$$\alpha = \arccos(1 - d/f), \tag{9.65}$$

$$s^4 = \frac{\lambda^2 df}{2 - d/f}. \tag{9.66}$$

Canonical type I: A type I system consists of a section of free space of length d followed by a lens of focal length f followed by a second section of free space of length d (figure 9.6b). Thus its total length is $\ell = 2d$. If the order of the fractional transform $a\pi/2 = \alpha$ and the scale s are specified, then the separation of the lenses d and their focal length f must be chosen according to

$$d = \frac{s^2}{\lambda} \tan(\alpha/2), \tag{9.67}$$

$$f = \frac{s^2}{\lambda} \csc \alpha. \tag{9.68}$$

Alternatively, we can find α and s in terms of d and f by using

$$\alpha = \arccos(1 - d/f), \tag{9.69}$$
$$s^4 = \lambda^2 df(2 - d/f). \tag{9.70}$$

The canonical fractional Fourier transforming systems are the simplest symmetric configurations yielding a fractional Fourier transform between planar reference surfaces.

The above systems work well for the range $0 \le a \le 1$. The given equations are also valid for $1 \le a < 2$, but the system becomes rather impractical as a approaches 2. To realize fractional Fourier transforms with such values of a, it is best to first realize an ordinary Fourier transform and follow it with a fractional Fourier transform of order $a - 1$, which will now lie between 0 and 1. Likewise, in order to realize transforms of order in the range $0 \le 2 < 4$, it is best to first realize a parity operation $\mathcal{F}^2$, which leaves us with a transform of order $a - 2$. In most cases the parity operation can be realized by merely arranging inverted coordinates, or it can be physically realized by an inverting imaging system.

9.4.4 Classical single-lens imaging

Here we will derive the (well-known) single-lens imaging equations starting from an imaging condition stated in terms of fractional Fourier transforms.

Consider the classical single-lens imaging configuration where the object is located a distance $d_o > 0$ to the left of the lens and the image is located a distance $d_i > 0$ to the right of the lens, which has focal length $f > 0$ (figure 9.7). We can view this system as performing two consecutive fractional Fourier transform operations (from the object to the lens, and from the lens to the image), provided the radius R_- of the spherical reference surface just before the lens and the radius R_+ of the spherical reference surface just after the lens are related by

$$\frac{1}{R_+} = \frac{1}{R_-} - \frac{1}{f}. \tag{9.71}$$

This equation states that the amplitude distribution of light on the reference surface with radius R_- is mapped by the lens onto a reference surface with radius R_+. If we are to look into the combined effect of the two consecutive fractional transforms, we should choose the scale factors s_- and s_+ immediately before and immediately after the lens so that

$$s_+ = s_-. \tag{9.72}$$

Letting $\alpha_o \equiv a_o \pi/2$ and $\alpha_i \equiv a_i \pi/2$ denote the fractional orders from the object to the lens and from the lens to the image respectively, we can write the imaging condition

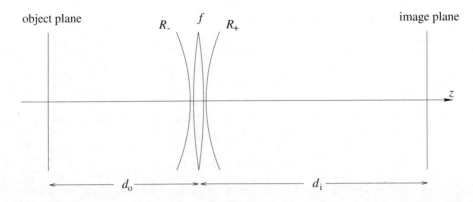

Figure 9.7.   Single-lens imaging system: $R_- > 0$ and $R_+ < 0$ (Ozaktas and Mendlovic 1995a).

(for an inverted image) as $a_o + a_i = \pm 2$ or equivalently

$$\alpha_o + \alpha_i = \pm\pi. \tag{9.73}$$

This condition follows from the fact that $\mathcal{F}^{\pm 2}[f(u)] = f(-u)$ for any $f(u)$.

Now, with the definitions $g_- \equiv 1 - d_o/R_-$ and $g_+ \equiv 1 + d_i/R_+$ we have the following results (see equations 9.40 and 9.41):

$$\cot \alpha_o = g_- \frac{s_-^2}{\lambda d_o}, \tag{9.74}$$

$$\cot \alpha_i = g_+ \frac{s_+^2}{\lambda d_i}. \tag{9.75}$$

It is now possible to show that equation 9.73 implies the well-known imaging condition

$$\frac{1}{f} = \frac{1}{d_o} + \frac{1}{d_i}. \tag{9.76}$$

Furthermore, letting s_o and s_i denote the scale factors associated with the object and image, equation 9.42 lets us write

$$\frac{s_o s_-}{\lambda d_o} = \csc \alpha_o, \tag{9.77}$$

$$\frac{s_+ s_i}{\lambda d_i} = \csc \alpha_i. \tag{9.78}$$

Now, since equation 9.73 implies $\csc \alpha_o = \csc \alpha_i$, we obtain the magnification s_i/s_o of the scale factors as

$$\frac{s_i}{s_o} = \frac{d_i}{d_o}, \tag{9.79}$$

again a familiar result.

9.4.5 Multi-lens systems as consecutive fractional Fourier transforms

More complicated systems involving several lenses separated by arbitrary distances can be analyzed in a similar manner. For simplicity let us restrict ourselves to positive lenses. Let the distance separating the input (or object) from the first lens be d_0, the distance separating the first and second lens d_1, and the distance separating the ith and $(i+1)$th lens d_i. The focal length of the ith lens will be denoted by f_i. In general, the subscripts $i-$ and $i+$ will be used to denote quantities immediately to the left and immediately to the right of the ith lens. α_i will denote the order of the fractional transform associated with propagation from lens i to lens $i+1$.

Assume that the input is specified with respect to a particular spherical reference surface of radius R_{0+}, and that the input scale parameter is denoted by s_{0+}. Just before the first lens in the system, we will observe the fractional Fourier transform of the input of order α_0, where α_0 can be calculated from $\cot \alpha_0 = g_{0+}s_{0+}^2/\lambda d_0$ (equation 9.41), where $g_{0+} \equiv 1 + d_0/R_{0+}$. The scale of this transform will be $s_{1-} = (\lambda d_0/s_{0+}) \csc \alpha_0$ (equation 9.42) and the transform will be observed on a spherical reference surface of radius R_{1-}, which we can calculate from $g_{1-} \equiv 1 - d_0/R_{1-}$ and $g_{1-} = (\lambda d_0/s_{1-}^2) \cot \alpha_0$ (equation 9.40).

Now, we can make our way through the lens using $s_{1+} = s_{1-}$ and $1/R_{1+} = 1/R_{1-} - 1/f_1$ (equations 9.71 and 9.72). Then, just before the second lens, on a spherical reference surface of radius R_{2-}, we will observe the fractional Fourier transform of the input of order $\alpha_0 + \alpha_1$, and so on.

Let us assume that all $d_0, d_1, \ldots, f_1, f_2, \ldots$ as well as s_{0+} and R_{0+} are specified. Then, we can work our way through the system iteratively using the following equations for $i = 0, 1, 2, \ldots$:

$$g_{i+} = 1 + d_i/R_{i+}, \tag{9.80}$$

$$\cot \alpha_i = \frac{g_{i+}s_{i+}^2}{\lambda d_i}, \tag{9.81}$$

$$s_{(i+1)-} = \frac{\lambda d_i}{s_{i+}} \csc \alpha_i, \tag{9.82}$$

$$s_{(i+1)+} = s_{(i+1)-}, \tag{9.83}$$

$$g_{(i+1)-} = \frac{\lambda d_i}{s_{(i+1)-}^2} \cot \alpha_i, \tag{9.84}$$

$$R_{(i+1)-} = \frac{d_i}{1 - g_{(i+1)-}}, \tag{9.85}$$

$$R_{(i+1)+} = \frac{f_{i+1} R_{(i+1)-}}{f_{i+1} - R_{(i+1)-}}. \tag{9.86}$$

Here s_{i+} and R_{i+} may be considered to be the state variables of the iteration. The cumulative order of the transform just before the ith lens is given by $\alpha_{\text{cum}\,i-1} = \alpha_0 + \alpha_1 + \cdots + \alpha_{i-1}$. This procedure allows us to find the orders of the transforms at the lenses, but of course it is also possible to calculate the order of the transform observed at any intermediate location (using the propagation result).

Thus we again see that in an optical system involving many lenses separated by arbitrary distances, the amplitude distribution is continually fractional Fourier transformed as it propagates through the system. The order $\alpha(z)$ of the fractional transform observed at the distance z along the optical axis is a monotonically increasing function. The transforms are observed on spherical reference surfaces. Wherever the order of the transform $\alpha(z)$ is equal to $(4j + 1)\pi/2$ for any integer j, we observe the Fourier transform of the input. Wherever the order is equal to $(4j + 2)\pi/2$, we observe an inverted image, and so on.

Of course, the piecewise approach taken here is simply another way of arriving at the identical results obtained in section 9.2.4, where we assumed the $\hat{A}\hat{B}\hat{C}\hat{D}$ matrix parameters to be specified as functions of z.

9.4.6 General fractional Fourier transform relations for quadratic-phase systems

Finally, we consider the class of quadratic-phase systems and state the general conditions a member of this class must satisfy so that its output can be interpreted as the fractional Fourier transform of its input. We also show that a fractional Fourier transformer can be made out of any quadratic-phase system by appending lenses of appropriate focal length at the input and output planes.

In essence, this amounts to a generalization of section 9.3 from free-space propagation to arbitrary quadratic-phase systems. A given quadratic-phase system (equation 9.4) can be interpreted as a fractional Fourier transform (equation 4.4) if there exist input and output scale parameters s_1, s_2 such that

$$\hat{\alpha}s_2^2 = \alpha = \cot(a\pi/2) = \hat{\gamma}s_1^2,$$
$$\hat{\beta}s_1s_2 = \beta = \csc(a\pi/2), \tag{9.87}$$

for some a. (Recall that when the same scale parameter s is used in both the input and output planes, we have $\hat{\alpha} = \alpha/s^2$, $\hat{\beta} = \beta/s^2$, $\hat{\gamma} = \gamma/s^2$. Here, however, since we are allowing different scale parameters in the output and input planes defined through $u = x/s_2$ and $u' = x'/s_1$, we have $\hat{\alpha} = \alpha/s_2^2$, $\hat{\beta} = \beta/s_1s_2$, $\hat{\gamma} = \gamma/s_1^2$.) It can be shown that such scale parameters can be found if and only if $0 \leq \hat{\alpha}\hat{\gamma} \leq \hat{\beta}^2$. Granted that this condition is satisfied, the necessary scale parameters and the order of the resulting transform are given by

$$s_1^4 = (\hat{\beta}^2\hat{\gamma}/\hat{\alpha} - \hat{\gamma}^2)^{-1},$$
$$s_2^4 = (\hat{\beta}^2\hat{\alpha}/\hat{\gamma} - \hat{\alpha}^2)^{-1},$$
$$\tan(a\pi/2) = \pm(\hat{\beta}^2/\hat{\alpha}\hat{\gamma} - 1)^{1/2}, \tag{9.88}$$

where the sign of $\tan(a\pi/2)$ is the same as the identical signs of $\hat{\alpha}$ and $\hat{\gamma}$, and a lies in the interval $[-2, 0]$ or $[0, 2]$ according to whether $\hat{\beta}s_1s_2 \leq 0$ or $\hat{\beta}s_1s_2 \geq 0$. For example, if $s_1s_2 \geq 0$, $\hat{\alpha} \geq 0$, $\hat{\beta} \geq 0$, then a lies in the interval $[0, 1]$.

If the condition $0 \leq \hat{\alpha}\hat{\gamma} \leq \hat{\beta}^2$ is not satisfied, it is still possible to observe a fractional Fourier transform between spherical (rather than planar) reference surfaces at the input and output. That is, it is possible to take *any* quadratic-phase system, and make a fractional Fourier transformer out of it by appending lenses to the input and output. (This result is essentially the same as that already discussed in section 9.2.1 from a somewhat different perspective.) To show this, note that the phase of the kernel between the new spherical reference surfaces will be modified according to $\hat{\alpha}_{new} = \hat{\alpha} - 1/\lambda R_2$, $\hat{\beta}_{new} = \hat{\beta}$, $\hat{\gamma}_{new} = \hat{\gamma} + 1/\lambda R_1$, where R_1 and R_2 are the radii of the input and output spherical reference surfaces. We can always ensure $0 \le \hat{\alpha}_{new}\hat{\gamma}_{new} \le \hat{\beta}_{new}^2$ by appropriate choice of R_1 and R_2.

It is instructive to repeat the same discussion in terms of matrices. The dimensional matrix elements are related to the dimensionless elements through $\hat{A} = \hat{\gamma}/\hat{\beta} = As_2/s_1$, $\hat{B} = 1/\hat{\beta} = Bs_1s_2$, $\hat{C} = -\hat{\beta} + \hat{\alpha}\hat{\gamma}/\hat{\beta} = C/s_1s_2$, $\hat{D} = \hat{\alpha}/\hat{\beta} = Ds_1/s_2$, which reduce to the relations appearing on page 269 when $s_1 = s_2$. An arbitrary quadratic-phase system can be interpreted as a fractional Fourier transformer if there exist scale parameters s_1 and s_2 such that the dimensionless matrix is equal to the matrix $[\cos(a\pi/2)\ \sin(a\pi/2);\ -\sin(a\pi/2)\ \cos(a\pi/2)]$ for some a. The necessary and sufficient condition for this can be expressed as $0 \le \hat{A}\hat{D} \le 1$ (or equivalently $-1 \le \hat{B}\hat{C} \le 0$). Granted that this is satisfied, the necessary scale parameters and the order of the resulting transform are given by

$$s_1^4 = \hat{B}^2(\hat{A}/\hat{D} - \hat{A}^2)^{-1},$$
$$s_2^4 = \hat{B}^2(\hat{D}/\hat{A} - \hat{D}^2)^{-1},$$
$$\cos(a\pi/2) = \pm(\hat{A}\hat{D})^{1/2}, \qquad (9.89)$$

where $\text{sgn}[\cos(a\pi/2)] = \text{sgn}(\hat{A}s_1/s_2)$ and a lies in the interval $[-2, 0]$ or $[0, 2]$ according to whether $\hat{B}/s_1s_2 \le 0$ or $\hat{B}/s_1s_2 \ge 0$. For example, if $s_1s_2 \ge 0$, $\hat{A} \ge 0$, $\hat{B} \ge 0$, then a lies in the interval $[0, 1]$.

If the condition $0 \le \hat{A}\hat{D} \le 1$ is not satisfied, it is still possible to make a fractional Fourier transformer out of this system by appending lenses at its input and output planes. That is, for *any* quadratic-phase system, it is possible to find spherical input and output reference surfaces between which a fractional Fourier transform relation exists. To show this, we multiply the matrix from the left and right by the matrix of a thin lens:

$$\begin{bmatrix} \hat{A}_{new} & \hat{B}_{new} \\ \hat{C}_{new} & \hat{D}_{new} \end{bmatrix} = \begin{bmatrix} 1 & 0 \\ -1/\lambda f_2 & 1 \end{bmatrix} \begin{bmatrix} \hat{A} & \hat{B} \\ \hat{C} & \hat{D} \end{bmatrix} \begin{bmatrix} 1 & 0 \\ -1/\lambda f_1 & 1 \end{bmatrix}, \qquad (9.90)$$

where f_1, f_2 are the focal lengths of the lenses. Multiplying the matrices we find $\hat{A}_{new} = \hat{A} - \hat{B}/\lambda f_1$ and $\hat{D}_{new} = \hat{D} - \hat{B}/\lambda f_2$. It is clearly possible to satisfy the condition $0 \le \hat{A}_{new}\hat{D}_{new} \le 1$ by appropriate choice of f_1 and f_2. Thus the same results are obtained whether we examine the phases of the kernels or the elements of the corresponding matrices.

As a by-product of the above discussion, we state for easy reference the matrix associated with an optical fractional Fourier transformer with input and output scale parameters s_1 and s_2:

$$\begin{bmatrix} (s_1/s_2)\cos(a\pi/2) & (s_1s_2)\sin(a\pi/2) \\ (-1/s_1s_2)\sin(a\pi/2) & (s_2/s_1)\cos(a\pi/2) \end{bmatrix}. \qquad (9.91)$$

The system represented by this matrix maps an input function $\hat{f}(x) = s_1^{-1/2}f(x/s_1)$ into an output function $\hat{g}(x) = s_2^{-1/2}f_a(x/s_2)$, where $f_a(u)$ is the ath order fractional

Fourier transform of $f(u)$. Of course, this matrix reduces to that given in equation 9.15 when $s_1 = s_2$, $R = \infty$, and $M = 1$.

It is perhaps worth explicitly contrasting the present approach of using two different scale parameters s_1 and s_2 for the input and output, with the approach of section 9.2.1 where we used only one scale parameter s but also a magnification parameter M. In section 9.2.1 we essentially concentrated on the "propagation" approach, fixing the scale parameter at the input of the system as $s = s_1$, and let the varying output scale parameter be denoted by $Ms = s_2$. The use of two different scale parameters provides an alternative perspective and is more in line with the emphasis on symmetry between input and output in section 9.3 and the present section.

To summarize, we asked ourselves whether a given quadratic-phase system can be interpreted as a fractional Fourier transformer by choosing the input and output scale parameters appropriately. We found the condition under which this is possible, and the required scale parameters. We then showed that when this condition does not hold, we can make it hold either by choosing to work with spherical reference surfaces, or by appending lenses at the input and output planes.

Let us look back at equation 3.137, which tells us how any quadratic-phase system can be interpreted as any other by appropriate scaling and appropriate choice of spherical reference surfaces (or by appending appropriate lenses). The considerations of this section correspond to the case where one of these systems is taken to be a general quadratic-phase system, and the other as a fractional Fourier transformer. In section 9.3 we considered the more specialized case where one of the systems was taken to be a section of free space and the other was a fractional Fourier transformer. (The dual special case where we take a lens instead of a section of free space is analogous in the formal mathematical sense, but less interesting from a physical and practical perspective.)

Although we do not further elaborate, it is straightforward to extend the analysis, synthesis, and propagation problems of section 9.3 from sections of free space to arbitrary quadratic-phase systems. For instance, the radii R_1 and R_2 of the spherical reference surfaces may be specified along with the quadratic-phase system, and we can inquire whether a fractional Fourier transform exists between these surfaces, and solve for s_1, s_2, a (analysis). Or, given desired values of s_1, s_2, a, we can choose the radii and parameters of the quadratic-phase system in some convenient way to achieve these values (synthesis). Alternatively, we may be given the parameters of the quadratic-phase system along with s_1, R_1, and we may find s_2, R_2, and a (propagation). On the other hand, if the order a is specified along with the quadratic-phase parameters, it does not become possible to specify both s_1 and s_2 independently, and one will be determined in terms of the other.

We conclude this section with two final remarks. First, it is worth being aware of the converse interpretations of our results. For instance, if a quadratic-phase system can be interpreted as a fractional Fourier transformer, this also means that the fractional Fourier transformer can be used to simulate or realize the quadratic-phase system. Second, in those cases where we are free to specify the order a of the fractional Fourier transform, choosing $a = 1$ leads to specialized results telling whether and how arbitrary quadratic-phase systems can be interpreted as a scaled ordinary Fourier transform between spherical surfaces.

9.5 Fractional Fourier transformation in quadratic graded-index media

As stated earlier, propagation in quadratic graded-index media corresponds to the fractional Fourier transform in its purest form. In fact, it was consideration of quadratic graded-index media which allowed definition of the fractional Fourier transform in an optical context (Mendlovic and Ozaktas 1993a, b, Ozaktas and Mendlovic 1993a, b), a definition which remarkably turned out to be identical to previous mathematical definitions. The reasoning was that since Fourier transforms and images are observed periodically in such a medium, and since the system is uniform in the z direction, the distributions of light at intermediate planes should correspond to fractional Fourier transforms.

In systems consisting of lenses separated by sections of free space, the transform is observed on curved surfaces and with variable scale. In contrast, in quadratic graded-index media, the transform is observed with constant scale and without residual curvature. Furthermore, the transform order increases linearly with distance of propagation. This is what is meant by observing the fractional Fourier transform in its "purest" form.

9.5.1 Propagation in quadratic-index media as fractional Fourier transformation

The results presented in previous chapters put us in a position to immediately state the fractional Fourier transforming property of quadratic graded-index media. We let the amplitude distribution of light at a certain plane perpendicular to the optical axis be denoted by $\hat{f}(x,y)$, and the amplitude distribution at another such plane situated a distance d further to the right be denoted by $\hat{g}(x,y)$. We have seen in equations 7.102 and 7.105 that they are related by

$$
\hat{g}(x,y) = e^{i2\pi\sigma d}\frac{e^{-ia\pi/2}}{s^2}\iint K_a(x/s,x'/s)K_a(y/s,y'/s)\hat{f}(x',y')\,dx'\,dy'
$$

$$
= e^{i2\pi\sigma d}\frac{e^{-ia\pi/2}}{s}f_a(x/s,y/s), \tag{9.92}
$$

in two dimensions, and

$$
\hat{g}(x) = e^{i2\pi\sigma d}\frac{e^{-ia\pi/4}}{s}\int K_a(x/s,x'/s)\hat{f}(x')\,dx'
$$

$$
= e^{i2\pi\sigma d}\frac{e^{-ia\pi/4}}{s^{1/2}}f_a(x/s), \tag{9.93}
$$

in one dimension, where $a\pi/2 = d/\chi$. Here $f_a(u,v)$ is the ath order fractional Fourier transform of $f(u,v) = s\hat{f}(su,sv)$ and $f_a(u)$ is the ath order fractional Fourier transform of $f(u) = s^{1/2}\hat{f}(su)$.

For the benefit of readers who read chapter 7 less carefully, here we outline the derivation leading to the above results in one dimension (Ozaktas and others 1994a). We begin by expanding the incident light distribution $\hat{f}(x)$ in terms of the

Hermite-Gaussian functions:

$$\hat{f}(x) = \sum_{l=0}^{\infty} C_l \frac{1}{s^{1/2}} \psi_l \left(\frac{x}{s}\right),$$ (9.94)

$$C_l = \int \frac{1}{s^{1/2}} \psi_l \left(\frac{x'}{s}\right) \hat{f}(x') \, dx'.$$

Now, it is known that the Hermite-Gaussian functions $s^{-1/2}\psi_l(x/s)$ are eigenmodes of propagation through quadratic graded-index media with $s^2 = \chi/\sigma$. That is, the lth order Hermite-Gaussian function $s^{-1/2}\psi_l(x/s)$ is mapped to $\exp(i2\pi\sigma_z z) \, s^{-1/2}\psi_l(x/s)$ where $2\pi\sigma_z \approx 2\pi\sigma - (l+1/2)/\chi$. Thus, upon propagation over a distance d, we obtain

$$\hat{g}(x) = e^{i2\pi\sigma d} e^{-id/2\chi} \sum_{l=0}^{\infty} C_l \, e^{-idl/\chi} \frac{1}{s^{1/2}} \psi_l \left(\frac{x}{s}\right).$$ (9.95)

Substituting for C_l leads us to

$$\hat{g}(x) = e^{i2\pi\sigma d} e^{-id/2\chi} \int \left[\sum_{l=0}^{\infty} e^{-idl/\chi} \frac{1}{s} \psi_l \left(\frac{x}{s}\right) \psi_l \left(\frac{x'}{s}\right)\right] \hat{f}(x') \, dx'.$$ (9.96)

Now, it remains to recognize the term in square brackets as a scaled fractional Fourier transform kernel (equation 4.23 or table 2.8, property 9). In essence, the result follows from the facts that (i) both quadratic graded-index media and the fractional Fourier transform have the same eigenfunctions, and (ii) the dependences of the eigenvalues or propagation constants on the index l are also identical.

We might summarize by saying that quadratic graded-index media are fractional Fourier transforming systems with input and output scale parameter $s_1 = s_2 = s = \sqrt{\chi/\sigma} = \sqrt{\lambda\chi}$, such that the order of the transform is related to the distance of propagation through $\alpha = a\pi/2 = d/\chi$. This result can also be deduced by comparing the matrix given in equation 8.42 to that in equation 9.15, which also implies $\alpha = d/\chi$, $M = 1$, $R = \infty$, and $s^2 = \lambda\chi$.

9.5.2 Analogy with the simple harmonic oscillator

Hamilton's equations were discussed in section 7.5.3 and the paraxial Hamiltonian was given in equation 7.131. Assuming the refractive index distribution for a quadratic graded-index medium can be approximated as $n(x) = n_0\sqrt{1 - (x/\chi)^2} \approx n_0(1 - x^2/2\chi^2)$, we can write Hamilton's equations for this medium as

$$\frac{dx}{dz} = \frac{1}{n_0 f_{oc}/c} \sigma_x, \qquad \frac{d\sigma_x}{dz} = -\frac{n_0 f_{oc}/c}{\chi^2} x.$$ (9.97)

With increasing z, the solution $[x(z) \ \sigma_x(z)]^{\mathsf{T}}$ of these equations traces an ellipse in phase space, just like a classical harmonic oscillator with one degree of freedom. Indeed, combining the two equations we obtain

$$\frac{d^2 x}{dz^2} + \frac{1}{\chi^2} x = 0,$$ (9.98)

which is the classical harmonic oscillator differential equation. Using the scale parameter $s^2 = \chi c/n_0 f_{oc} = \lambda\chi = \chi/\sigma$, it is possible to write the same equations in dimensionless form:

$$\frac{du}{dz} = \frac{\mu}{\chi}, \qquad \frac{d\mu}{dz} = -\frac{u}{\chi}, \tag{9.99}$$

in which the case the ellipse is transformed to a circle and the bundle of rays rotates in phase space, making one complete rotation for every $2\pi\chi$ of distance traveled along the optical axis z.

Rays are to geometrical optics what particles are to classical mechanics. Optical wave fields and the optical wave equation fulfill the same role in wave optics as wave functions (probability amplitudes) and the Schrödinger equation in quantum mechanics. Geometrical optics is the same kind of approximation to wave optics that classical mechanics is to quantum mechanics. We have just seen that Hamilton's ray equations for quadratic graded-index media are identical to the equations for a classical harmonic oscillator. Now we will see that the wave equation for the same medium is the same as the Schrödinger equation for a quantum-mechanical harmonic oscillator. Just as the ray bundle of geometrical optics rotates in geometrical-optical phase space and the ensemble of particles of classical mechanics rotates in classical phase space, the Wigner distribution of the wave field of wave optics rotates in wave-optical phase space and the Wigner distribution of the wave function of quantum mechanics rotates in quantum-mechanical phase space.

The Schrödinger equation is given by (Cohen-Tannoudji, Diu, and Laloë 1977)

$$i\hbar\frac{\partial \tilde{f}(x,t)}{\partial t} = \mathcal{H}\tilde{f}(x,t), \tag{9.100}$$

where $\tilde{f}(x,t)$ is the wave function. The quantum-mechanical Hamiltonian $\mathcal{H}$ for the harmonic oscillator is obtained from the classical Hamiltonian $p_x^2/2m + kx^2/2 = p_x^2/2m + m\omega^2 x^2/2$ (kinetic energy plus potential energy) by replacing the classical momentum p_x by the operator $-i\hbar\partial/\partial x$, where $\hbar$ is Planck's constant. Here m may be thought of as the mass of a particle and k as a spring constant so that $\omega = \sqrt{k/m}$ is the natural frequency of the oscillator. Schrödinger's equation now becomes

$$i\hbar\frac{\partial \tilde{f}(x,t)}{\partial t} = \left[-\frac{\hbar^2}{2m}\frac{\partial^2}{\partial x^2} + \frac{m\omega^2 x^2}{2}\right]\tilde{f}(x,t). \tag{9.101}$$

It is evident that by defining appropriate scale parameters, this equation can be put into the dimensionless form

$$\left[-\frac{1}{4\pi}\frac{\partial^2}{\partial u^2} + \pi u^2\right] f_a(u) = i\frac{2}{\pi}\frac{\partial f_a(u)}{\partial a}, \tag{9.102}$$

where u is a dimensionless variable corresponding to x, and a corresponds to t.

We now show that propagation in quadratic graded-index media is also described by equation 9.102. We begin with the Helmholtz equation given as equation 7.91. In our original derivation following this equation, we had assumed a solution of the form $\hat{f}(x,y,z) = \hat{A}(x,y)\exp(i2\pi\sigma_z z)$, where the envelope $\hat{A}(x,y)$ does not depend on z, and obtained the eigenmodes and associated propagation constants. To obtain an equation similar to equation 9.102, we will instead assume a solution of the more general form $\hat{f}(x, y, z) = \hat{A}(x, y, z) \exp(i2\pi\sigma_z z)$ in two dimensions or $\hat{f}(x, z) = \hat{A}(x, z) \exp(i2\pi\sigma_z z)$ in one dimension, where the envelope depends also on z. Now we substitute this form in the one-dimensional version of equation 7.91 (that is, without the $\partial^2/\partial y^2$ and y^2 terms). Next we follow the "conventional derivation" for obtaining paraxial wave equations referred to just after equation 7.54, by using the mathematical statements of paraxiality given there. This results in

$$-\frac{1}{4\pi}\frac{\partial^2 \hat{A}}{\partial x^2} + \frac{\pi\sigma^2 x^2}{\chi^2}\,\hat{A} = i\sigma\frac{\partial \hat{A}}{\partial z}, \qquad (9.103)$$

where $\sigma = n_0 f_{oc}/c$. This is the paraxial wave equation for propagation in quadratic graded-index media. Now, let us employ the scale parameter $s^2 = \chi/\sigma$ defined earlier to introduce the dimensionless variable $u = x/s$. Furthermore, let us remember that the angular order parameter α was related to the distance of propagation through $a\pi/2 = \alpha = z/\chi$. Then it is an easy matter to show that the above equation reduces precisely to equation 9.102 with the identification $f_a(u) = (\chi/\sigma)\hat{A}(su, \chi a\pi/2)$.

Thus we have shown that the paraxial wave equation for quadratic graded-index media is the same as the Schrödinger equation for the harmonic oscillator; both take the form of equation 9.102. The role played by time in the Schrödinger equation is played by the coordinate along the direction of propagation in the optical case. This equation is nothing but equation 3.171 with $p = \alpha = a\pi/2$ and $\mathcal{H} = \pi(\mathcal{D}^2 + \mathcal{U}^2)$ (the Hamiltonian corresponding to the fractional Fourier transform). It is also almost the same as equation 4.51, save the constant term $-1/2$ appearing in the square brackets in that equation. (This inconsequential discrepancy is the same as that discussed on page 125.) Of course, we already know from definition E of chapter 4 that the solution of this equation is expressed in terms of the fractional Fourier transform.

In conclusion, we have seen that the geometrical-optical picture of propagation in quadratic graded-index media is analogous to a classical harmonic oscillator, and the wave-optical picture is analogous to a quantum-mechanical harmonic oscillator. The differential equation defining the fractional Fourier transform physically corresponds to simple harmonic oscillation. The kernel of the transform is the Green's function of the simple harmonic oscillator. The harmonic oscillator has a central place in mechanics not only because it represents the simplest and most basic vibrational system, but also because systems disturbed from equilibrium, no matter what the underlying forces are, can often be satisfactorily modeled in terms of harmonic oscillations.

9.5.3   Quadratic graded-index media as the limit of multi-lens systems

Unlike with quadratic graded-index media, in the case of optical systems consisting of sections of free space and thin lenses, we saw that the fractional Fourier transform is observed with a scale parameter and residual phase curvature. Furthermore, the transform order does not increase linearly with the distance along the optical axis (although it is still a monotonic function). Here we will compare the fractional Fourier transforming action of quadratic graded-index media and multi-lens systems and see how the former can be viewed as a limiting case of the latter.

The ordinary optical Fourier transform operation is commonly realized through the

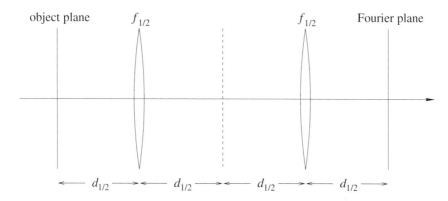

Figure 9.8. Optical Fourier transformer with two lenses: $d_{1/2} = (s^2/\lambda)\tan(\pi/8)$ and
$$f_{1/2} = (s^2/\lambda)\csc(\pi/4).$$

joint action of diffraction and focusing. Sections of free space result in light being diffracted and spread. Positive lenses focus and recollect the light. These operations are duals or Fourier conjugates of each other since mathematically one corresponds to chirp convolution and the other corresponds to chirp multiplication. In conventional optical Fourier transforming configurations, the act of focusing is fully concentrated at the lenses and segregated from the act of diffraction.

As a thought experiment, let us consider the optical Fourier transforming system shown in figure 9.8, in which the act of focusing is distributed over two lenses (Mendlovic and Ozaktas 1993a). The separation and focal lengths of the lenses have been chosen such that the overall system realizes the ordinary Fourier transform. The fact that the two subsystems are identical to each other suggests that each is a 0.5th order fractional Fourier transformer. Although we could not say on the face of it that each subsystem indeed realizes the 0.5th fractional Fourier transform, the reader may easily verify that this is the case by looking back at equations 9.67 and 9.68 for the type I canonical system (also see section 9.4.3).

Let us now extend this thought experiment and consider optical subsystems which, when cascaded q times, produce the ordinary Fourier transform. Such subsystems can be expected to realize the $1/q$th order fractional Fourier transform. Again, such subsystems can be realized by using equations 9.67 and 9.68. Furthermore, by cascading p of these subsystems, we can realize fractional Fourier transformers of any rational order p/q.

The larger the value of q, the more evenly distributed and less segregated the act of focusing is from the act of diffraction. In the limit $q \to \infty$, the actions of focusing and diffraction would be infinitesimally and uniformly interspersed. Of course, bulk optical systems with even moderately large q would be impractical. However, a physical embodiment of the limit $q \to \infty$ does exist, and is nothing but a quadratic graded-index medium. Such a medium can be thought to consist of infinitesimal layers in which focusing and diffraction take place simultaneously.

Before we further discuss this limit, it will be instructive to compare a bulk $2f$ system and a graded-index medium in phase space. Figure 9.9a shows how the support of the Wigner distribution or ray bundle is rotated after three shearing steps in a bulk $2f$

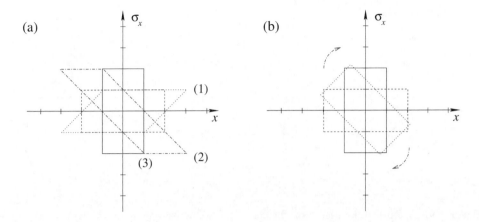

Figure 9.9. The actions in phase space of (a) $2f$ system, (b) quadratic graded-index
medium (Ozaktas and Mendlovic 1993a).

system consisting of a section of free space followed by a lens followed by another
section of free space. Figure 9.9b shows that in the case of a quadratic graded-index
medium, the action in phase space is a smooth rotation (which can be thought of as
alternating infinitesimal vertical and horizontal shears).

We now proceed to show that a quadratic graded-index medium can be seen as
the limit of a larger and larger number of weaker and weaker lenses (Ozaktas and
Mendlovic 1995a). We recall that the refractive index distribution in such a medium
has the following dependence on x:

$$n^2(x) = n_0^2[1 - (x/\chi)^2],\qquad(9.104)$$

where n_0 and χ are the medium parameters. We have seen that a section of length
d_{grin} of such a medium acts as a fractional Fourier transformer with order $\alpha = a\pi/2 =$
d_{grin}/χ and scale parameter $s^2 = \chi/\sigma_{\mathrm{grin}} = \chi c/n_0 f_{\mathrm{oc}}$.

We will consider type I canonical fractional Fourier transforming configurations,
characterized by equations 9.67 and 9.68 (the discussion is similar for type II
configurations). Assume that we cascade p type I systems each of length $2d_{\mathrm{I}}$ so that
the overall length of the bulk optical system is $d_{\mathrm{bulk}} = p2d_{\mathrm{I}}$. Now, keeping this overall
length as well as the scale parameter s fixed, we let $p \to \infty$ and $2d_{\mathrm{I}} \to 0$. Physically,
what we obtain is a large number of closely spaced lenses of weak focal power. In the
limit, the average refractive index distribution of this assembly will have a quadratic
dependence on x, similar to equation 9.104, if we set the fractional Fourier order and
scale parameter equal for both systems.

To see this, let us consider optical paths parallel to the optical axis. Using small-
angle approximations of equations 9.67 and 9.68, it is possible to show that the phase
collected along such a path through the infinite cascade of infinitesimal type I systems
is

$$\simeq -\pi\alpha \left(\frac{x}{s}\right)^2,\qquad(9.105)$$

where we have dropped the uninteresting constant term $2\pi\sigma d_{\text{bulk}}$. The collected phase is given by $2\pi f_{\text{oc}}/c$ times the optical path length (the integrated optical density). Turning our attention to quadratic-index media, equation 9.104 and $\alpha = a\pi/2 = d_{\text{grin}}/\chi$, $s^2 = \chi/\sigma_{\text{grin}}$ allow us to write the collected phase in exactly the same form as above, again by dropping an uninteresting constant phase term. This means that if we assume functional equivalence of the two systems (identical values of α and s), we can deduce their physical equivalence (identical collected phase, which is proportional to optical density).

In both systems the fractional Fourier order is linear in the length of the system and is given by $(s^2\sigma_{\text{grin}})^{-1}d_{\text{grin}}$ or $(s^2\sigma)^{-1}d_{\text{bulk}}$. The first of these is easily derived from $\alpha = d_{\text{grin}}/\chi$ and $s^2 = \chi/\sigma_{\text{grin}}$. To derive the second, we use $d_{\text{I}} = (s^2/\lambda)\tan\alpha_{\text{I}}/2$ and $f_{\text{I}} = (s^2/\lambda)\csc\alpha_{\text{I}}$ in the small-angle approximation to get $s^2 = \sqrt{2f_{\text{I}}d_{\text{I}}}\,\lambda$ and $\alpha = (\lambda/s^2)d_{\text{bulk}}$. To avoid confusion, it must be remembered that for the bulk system $\sigma = 1/\lambda$ corresponds to the wavelength of light in free space (whatever "free space" is: vacuum, air, or glass), and that for the graded-index medium σ_{grin} is the inverse of the wavelength in a medium with refractive index n_0. Because of this, although the collected phase for the two systems is the same, the lengths d_{grin} and d_{bulk} differ by a constant factor.

We finally mention another curious fact. If we fix the fractional Fourier order of the overall system, the overall length of the bulk system d_{bulk} is a decreasing function of p, attaining its minimum value when $p \to \infty$, when the system approaches the index distribution of a graded-index medium.

Application of a similar limiting process as discussed here may be found in Reardon and Chipman 1990.

9.5.4 Gaussian beams through quadratic graded-index media

We will conclude by repeating the discussion associated with equations 7.106 and 7.107 in the present context. Let us assume that the origin $z = 0$ of the optical axis is taken to coincide with the waist of a Gaussian beam: $\hat{q}(0) = -iW_0^2/\lambda = -i\check{z}$. Now, using equation 7.68 and the matrix for quadratic graded-index media given in equation 8.42, we obtain

$$R(z) = \frac{(\check{z}^2 - \chi^2)\cos(2z/\chi) + \check{z}^2 + \chi^2}{(\chi - \check{z}^2/\chi)\sin(2z/\chi)},$$

$$\frac{W^2(z)}{\lambda} = \frac{(\check{z}^2 - \chi^2)\cos(2z/\chi) + \check{z}^2 + \chi^2}{2\check{z}}. \tag{9.106}$$

$W^2(z)/\lambda$ oscillates between $\check{z}$ and $\chi^2/\check{z}$. In the special case $\check{z} = \chi$, we have

$$R(z) = \infty,$$

$$W^2(z) = W_0^2 = \lambda\check{z} = \text{constant}. \tag{9.107}$$

The beam profile does not change as it propagates. This choice of W_0 for the Gaussian beam $\propto \exp[-\pi x^2/W^2(z)] = \exp[-\pi x^2/W_0^2]$ corresponds to the value of s which makes the Gaussian beam $\propto \exp(-\pi x^2/s^2)$ an eigenfunction of quadratic graded-index media (section 7.4.3).

9.6   Hermite-Gaussian expansion approach

Here we expand on the discussion of section 9.2.6. We have seen in section 9.3 that the complex amplitude distributions of light on two spherical reference surfaces are related by a fractional Fourier transform. Here we provide an alternative derivation of this result in terms of Hermite-Gaussian beam expansions. It will follow that the order of the fractional Fourier transform between the reference surfaces is proportional to the Gouy phase shift between the two surfaces. This result provides new insight into wave propagation and spherical mirror resonators, as well as the possibility to exploit the fractional Fourier transform as a mathematical tool in analyzing such systems.

9.6.1   The fractional Fourier order and the Gouy phase shift

Let $\hat{f}_0(x, y)$ denote the complex amplitude distribution at the plane $z = 0$, of light propagating in the positive z direction. We are interested in the amplitude distribution at other planes $z \neq 0$. The most common approach is to use harmonic expansion or the Fresnel integral, as we have done earlier. Instead, here we will expand $\hat{f}_0(x, y)$ in terms of scaled Hermite-Gaussian functions, which also constitute a complete orthonormal set. We recall from section 2.5.2 that the normalized lth order Hermite-Gaussian function is given by

$$\psi_l(u) = \frac{2^{1/4}}{\sqrt{2^l l!}} \, H_l(\sqrt{2\pi}\, u) \, e^{-\pi u^2}, \qquad (9.108)$$

where $H_l(u)$ is the lth order Hermite polynomial. We can expand $\hat{f}_0(x, y)$ in terms of these functions as follows:

$$\hat{f}_0(x, y) = \sum_{l=0}^{\infty} \sum_{m=0}^{\infty} C_{lm} \frac{1}{s} \psi_l\left(\frac{x}{s}\right) \psi_m\left(\frac{y}{s}\right) \qquad (9.109)$$

$$C_{lm} = \iint \hat{f}_0(x, y) \frac{1}{s} \psi_l\left(\frac{x}{s}\right) \psi_m\left(\frac{y}{s}\right) dx\, dy.$$

Since $\int |\psi(x/s)|^2 \, dx = s$ for any $s > 0$, the function $s^{-1}\psi_l(x/s)\psi_m(y/s)$ is also normalized to unit signal energy.

We can interpret the function $s^{-1}\psi_l(x/s)\psi_m(y/s)$ as the amplitude distribution at $z = 0$ of a two-dimensional Hermite-Gaussian beam of order (l, m) with beam size $W_0 = s$ at its waist. Then it becomes an easy matter to write the amplitude distribution $\hat{f}_z(x, y)$ at an arbitrary plane z, since we know how each of the Hermite-Gaussian components propagates (section 7.3.4):

$$\hat{f}_z(x, y) = \sum_{l=0}^{\infty} \sum_{m=0}^{\infty} C_{lm} \frac{1}{W(z)} \psi_l\left(\frac{x}{W(z)}\right) \psi_m\left(\frac{y}{W(z)}\right)$$

$$\times \exp\left[i2\pi\sigma z + \frac{i2\pi\sigma(x^2 + y^2)}{2R(z)} - i(l + m + 1)\zeta(z)\right]. \qquad (9.110)$$

In this equation $W(z) = W_0[1 + (z/\breve{z})^2]^{1/2}$ is the beam size and $s = W_0 = W(0)$. We recall that the Rayleigh range $\breve{z}$ is related to W_0 by the relation $W_0^2 = \lambda \breve{z}$.

$R(z) = z[1 + (\check{z}/z)^2]$ is the radius of curvature of the wavefronts and

$$\zeta(z) = \arctan(z/\check{z}) \tag{9.111}$$

is the Gouy phase shift, as defined previously.

Now, using equation 4.23, equation 9.110 can be written in a very simple form in terms of the fractional Fourier transform as follows:

$$\hat{f}_z(x,y) = \frac{1}{W(z)} e^{i2\pi\sigma z} e^{-i\zeta(z)} e^{i2\pi\sigma(x^2+y^2)/2R(z)} \left(\mathcal{F}^{a(z)} f_0(u,v)\right) \left(\frac{x}{W(z)}, \frac{y}{W(z)}\right), \tag{9.112}$$

where

$$a(z) = \frac{\zeta(z)}{\pi/2}. \tag{9.113}$$

Here $\mathcal{F}^{a(z)} f_0(u,v)$ is the $a(z)$th order two-dimensional fractional Fourier transform of the function $f_0(u,v) = f_0(x/s, y/s) = s\hat{f}_0(x,y)$. Before discussing the interpretation of this result, let us put it to a simple test. Letting $z \to \infty$, we see that the resulting intensity pattern is simply the squared magnitude of the ordinary Fourier transform of $\hat{f}_0(x,y)$, consistent with what we know of Fraunhofer diffraction (also see sections 9.2.3 and 9.4.1).

Let us now consider any two planes at $z_1 \neq 0$ and $z_2 \neq 0$ such that $z_1 < z_2$, and relate the amplitude distributions $\hat{f}_{z_1}(x,y)$ and $\hat{f}_{z_2}(x,y)$ on these two planes. Equation 9.112 holds for both, regardless of the signs of z_1 and z_2. Let us introduce the scaled coordinates $u_2 = x/W(z_2)$, $v_2 = y/W(z_2)$, $u_1 = x/W(z_1)$, $v_1 = y/W(z_1)$ on the spherical surfaces with radii $R_2 = R(z_2)$ and $R_1 = R(z_1)$ (see figure 9.10). Working with these spherical reference surfaces enables us to eliminate the quadratic phase factor in equation 9.112. Then, in terms of dimensionless coordinates, the complex amplitude distributions of light on the two spherical surfaces shown in figure 9.10 can be related in this particularly simple form:

$$f_{z_2}(u_2, v_2) = [\mathcal{F}^{a_{z_{21}}} f_{z_1}(u_1, v_1)](u_2, v_2), \tag{9.114}$$

where

$$a_{z_{21}} \equiv a(z_2) - a(z_1) = \frac{\zeta(z_2) - \zeta(z_1)}{\pi/2}, \tag{9.115}$$

and we have dropped the uninteresting phase factor $\exp[i2\pi\sigma(z_2-z_1)-i(\zeta(z_2)-\zeta(z_1))]$ from the right-hand side of equation 9.114. Equation 9.114 is derived by first writing $f_{z_2}(u_2, v_2) \propto [\mathcal{F}^{a(z_2)} f_0(u,v)](u_2, v_2)$ from equation 9.112 and a similar equation for z_1, and then combining them.

The equalities $a(z_1) = 2\zeta(z_1)/\pi$ and $a(z_2) = 2\zeta(z_2)/\pi$ are merely special cases of equation 9.113. At $z = z_1$ we have the $a(z_1)$th fractional transform of the distribution at $z = 0$, which implies that at $z = 0$ we have the $-a(z_1)$th fractional transform of the distribution at $z = z_1$. At $z = z_2$ we have the $a(z_2)$th fractional transform of the distribution at $z = 0$. Thus at $z = z_2$ we have the $[a(z_2) - a(z_1)]$th transform of the distribution at z_1, as given in equation 9.115.

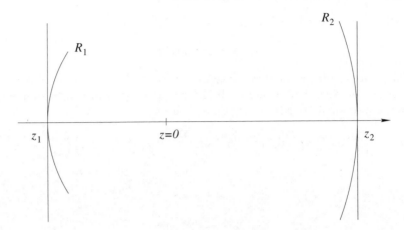

Figure 9.10. The complex amplitude distribution on the second spherical surface is the fractional Fourier transform of that on the first spherical surface. $f_{z_1}(u_1, v_1)$ and $f_{z_2}(u_2, v_2)$ denote the scaled complex amplitude distributions on the spherical surfaces 1 and 2. In the figure $z_1 < 0$ and $z_2 > 0$ but the results remain valid when both surfaces are on the same side of the $z = 0$ plane. (Ozaktas and Mendlovic 1994)

It is well known that if a certain relation between R_1, R_2 and $z_2 - z_1$ holds, one obtains an ordinary Fourier transform relation between two spherical surfaces. What we have shown is that for other values of these parameters, we obtain a fractional Fourier transform relation. Given any two surfaces as in figure 9.10, all we need to do to find the order a of the fractional Fourier transform relation between them is to find the Rayleigh range and waist location of a Gaussian beam that would "fit" into these surfaces, and then calculate a from equation 9.115.

The complex amplitude distribution with respect to any given reference sphere can be mapped harmlessly onto another reference sphere by using a lens of appropriate focal length. Conversely, the effect of an ideal thin lens can be interpreted merely as a change of the spherical reference surface used, with no change in the amplitude. Thus, any system consisting of lenses and sections of free space can be analyzed within the present framework, as consecutive fractional Fourier transforms. We have arrived at the same results discussed in sections 9.3 and 9.4, coming via a different path and from a different perspective.

9.6.2 Spherical mirror resonators and stability

Hermite-Gaussian beams are solutions of the wave equation in free space, but their profiles are not strictly eigenfunctions of free-space propapation. However, they are eigenfunctions of periodic lens systems and spherical mirror resonators. Thus, it will be instructive to relate the above results to spherical mirror resonators. Let us now interpret figure 9.10 as a resonator with the spherical surfaces interpreted as mirrors. Assume that the complex amplitude distribution of light at, say, the waist plane is known. After one round trip, we will observe at the same plane the $2a_{z_{21}}$th fractional Fourier transform of the initial distribution, where $a_{z_{21}}$ is given by equation 9.115.

(This is because the mirrors precisely reverse the quadratic phase factor so that we get twice the effect upon completing a round trip.) The system kernel corresponding to full round-trip propagation (from $z = 0$ to $z = 0$) in a spherical mirror resonator can be written as

$$\hat{h}(x, x') = e^{i2\pi\sigma 2(z_2 - z_1)} e^{-ia_{z_{21}}\pi} s^{-1} K_{2a_{z_{21}}}(x/s, x'/s). \tag{9.116}$$

In general, the $2a_{z_{21}}$th fractional Fourier transform is not of the same functional form as the initial distribution. If the initial distribution is to be a mode of the resonator, it must preserve its functional form after a round trip. That is, it must be an eigenfunction of the fractional Fourier transform operation. But we know very well that eigenfunctions of the fractional Fourier transform are the Hermite-Gaussian functions, which are also well known to be the modes of spherical mirror resonators.

It is possible to express $a_{z_{21}}$ (as given by equation 9.115) in terms of the radii R_1, R_2 and spacing $d = z_2 - z_1$ of the resonator mirrors by first calculating the Rayleigh range and waist location of a Gaussian beam that fits the resonator, and then calculating the Gouy phase shift. When the origin of the optical axis ($z = 0$) is chosen to coincide with the waist, the Rayleigh range and mirror locations are given as follows (Saleh and Teich 1991:332):

$$z_1 = \frac{d(R_2 - d)}{R_1 - R_2 + 2d}, \qquad z_2 = \frac{d(R_1 + d)}{R_1 - R_2 + 2d}, \tag{9.117}$$

$$\check{z}^2 = \frac{d(R_1 + d)(R_2 - d)(R_1 - R_2 + d)}{(R_1 - R_2 + 2d)^2}, \tag{9.118}$$

which determine the waist location $z = 0$ relative to the mirror locations. If the stability condition $0 \le (1 + d/R_1)(1 - d/R_2) \le 1$ is satisfied, it can be shown that $\check{z}^2 > 0$ so that $\check{z}$ is real (Saleh and Teich 1991:333). Having obtained z_1, z_2, and $\check{z}$, we can now obtain an expression for $\tan[\zeta(z_2) - \zeta(z_1)]$ by using the identity

$$\tan(\xi_2 - \xi_1) = \frac{\tan \xi_2 - \tan \xi_1}{1 + \tan \xi_2 \tan \xi_1}. \tag{9.119}$$

The result is

$$\zeta(z_2) - \zeta(z_1) = \arctan\left[\pm\sqrt{\frac{d|R_1 - R_2 + d|}{|R_1 + d|\,|R_2 - d|}} \right]. \tag{9.120}$$

Both $\zeta(z_1)$ and $\zeta(z_2)$ are limited to $[-\pi/2, \pi/2]$ and since $z_2 > z_1$, we have $\zeta(z_2) > \zeta(z_1)$. Thus $\zeta(z_2) - \zeta(z_1)$ can take values between 0 and π and the arctangent should be evaluated accordingly. With $\check{z}^2 > 0$ the $\pm$ is resolved according to the sign of $(R_1 - R_2 + d)/(R_1 - R_2 + 2d)$. Let us consider a symmetric resonator with $-R_1 = R_2 \equiv R > 0$. Then $-z_1 = d/2 = z_2$, $\check{z} = (d/2)(2R/d - 1)^{1/2}$, the stability condition is $0 \le d/R \le 2$, and we obtain

$$a\frac{\pi}{2} = \zeta(z_2) - \zeta(z_1) = \arctan\left[\frac{(2R/d - 1)^{1/2}}{R/d - 1} \right]. \tag{9.121}$$

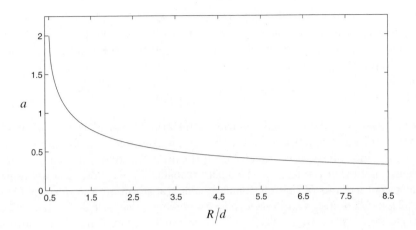

Figure 9.11.   The fractional order a as a function of R/d for a symmetric resonator.

(From now on we simply write a instead of $a_{z_{21}}$.) $a\pi/2$ monotonically decreases from π to 0 as R/d increases from $1/2$ to ∞ (figure 9.11). For $R/d < 1/2$ the resonator is unstable.

An important special case is the symmetrical confocal resonator in which the radius of the mirrors equals their spacing. In this case, half a round trip in the resonator corresponds to the ordinary Fourier transform ($a = 1$). Lipson and Lipson (1981) have discussed "quasi-confocal resonators," in which the beam profile repeats itself not after one round trip, but after several round trips. Such systems are easily analyzed within the framework of fractional Fourier transforms (Mendlovic, Ozaktas, and Lohmann 1994b). For instance, if the resonator in question has $a = 2/3$ and thus $2a = 4/3$, after three round trips the beam profile will repeat itself.

We now turn our attention to the stability (or confinement) condition for spherical mirror resonators (Siegman 1986, Saleh and Teich 1991). We will show that the following conditions are equivalent:

1. Resonator stability condition: $0 \leq (1 + d/R_1)(1 - d/R_2) \leq 1$.
2. $\check{z}^2 > 0$ or $\check{z}$ is real.
3. $0 \leq \cos^2(a\pi/2) \leq 1$ or a is real.

Stable resonators are characterized by real fractional Fourier orders. Since we discuss only real-ordered fractional Fourier transforms in this book, we have implicitly assumed that a is real throughout our discussion and thus we have implicitly assumed that the resonators we are dealing with are stable. Unstable resonators are described by values of a which are not real. Although we do not discuss such resonators here, references related to the interpretation of complex-ordered transforms in optics will be given later in this chapter.

That the resonator stability condition is equivalent to $\check{z}^2 > 0$ can be shown from equation 9.118 (Saleh and Teich 1991:333), demonstrating the equivalence of condition 1 to condition 2. That condition 1 is algebraically equivalent to condition 3 can be shown as follows. We substitute $d = z_2 - z_1$, $R_1 = R(z_1) = z_1[1 + (\check{z}/z_1)^2]$,

$R_2 = R(z_2) = z_2[1 + (\check{z}/z_2)^2]$ in the stability condition $0 \leq (1 + d/R_1)(1 - d/R_2) \leq 1$ and eliminate $\check{z}/z_1$, $\check{z}/z_2$ using $\tan \zeta(z_1) = z_1/\check{z}$, $\tan \zeta(z_2) = z_2/\check{z}$. Then we eliminate z_1/z_2 using $z_2/z_1 = \tan \zeta(z_2)/\tan \zeta(z_1)$. Now, purely trigonometric manipulations yield

$$0 \leq \cos^2[\zeta(z_2) - \zeta(z_1)] \leq 1,$$
$$0 \leq \cos^2(a\pi/2) \leq 1,$$
$$-1 \leq \cos(a\pi/2) \leq 1. \tag{9.122}$$

Finally, we comment on the equivalence between spherical mirror resonators and periodic lens systems. To obtain an equivalent periodic lens system, we "unfold" the resonator and replace the mirrors by thin lenses with focal lengths $-1/f_1 = 2/R_1$ and $1/f_2 = 2/R_2$ to obtain one period of the system. Since the type I and type II canonical configurations discussed in section 9.4.3 can constitute building blocks of periodic lens systems, equations 9.67 and 9.68 for the type I system and equations 9.63 and 9.64 for the type II system can be used to show the consistency of the results obtained in both cases. Hermite-Gaussian functions are eigenfunctions of both systems in that they retain their form after one full round trip in the resonator, or over one full period of the periodic lens system. Although all Hermite-Gaussian functions "fit" the same resonator since they share the same wavefronts, they accumulate different amounts of phase after one round trip; they have different eigenvalues. These different eigenvalues precisely correspond to those of the fractional Fourier transform, as was shown mathematically at the beginning of this section.

9.7 First-order optical systems

In this section we more fully develop the fractional Fourier transform formulation of propagation through arbitrary first-order optical systems (quadratic-phase systems). Some of the main results of this section were highlighted in sections 9.2.1 and 9.2.4. The reader may also wish to quickly review section 8.4, where we discussed the important special cases of imaging and Fourier transforming and how to determine the planes where images and Fourier transforms are observed (up to a residual quadratic phase factor). In this section we show that the distributions of light at intermediate planes can be interpreted as fractional Fourier transforms (again up to a quadratic phase factor).

9.7.1 Quadratic-phase systems as fractional Fourier transforms

We again consider centered optical systems composed of an arbitrary number of lenses separated by arbitrary distances (figures 9.1 and 9.3a), which we know to belong to the class of quadratic-phase systems. We know from chapters 3 and 4 that the one-parameter class of fractional Fourier transforms is a subclass of the class of three-parameter quadratic-phase systems (linear canonical transforms). If we allow an additional magnification parameter M and a phase radius of curvature parameter R, the family of magnified fractional Fourier transforms with phase curvature will also have three parameters and can be put in one-to-one correspondence with the family of quadratic-phase systems. We have written the kernel of this three-parameter transform in equation 9.13:

$$\hat{g}(x) = \int \hat{h}(x, x') \hat{f}(x') \, dx', \tag{9.123}$$

$$\hat{h}(x, x') = K e^{i\pi x^2/\lambda R} \sqrt{\frac{1}{s^2 M}} A_{a\pi/2}$$

$$\times \exp\left[\frac{i\pi}{s^2}\left(\frac{x^2}{M^2}\cot(a\pi/2) - 2\frac{xx'}{M}\csc(a\pi/2) + x'^2\cot(a\pi/2)\right)\right].$$

This kernel maps $\hat{f}(x) = s^{-1/2} f(x/s)$ into $K \exp(i\pi x^2/\lambda R) \sqrt{1/sM} f_a(x/sM)$, where $f_a(u)$ is the ath order fractional Fourier transform of $f(u)$.

To identify the one-to-one correspondence between the family of magnified fractional Fourier transforms with phase curvature and the family of quadratic-phase systems, we interpret equation 9.123 as a quadratic-phase system and use equation 9.8 to write the associated matrix:

$$\begin{bmatrix} \hat{A} & \hat{B} \\ \hat{C} & \hat{D} \end{bmatrix} = \begin{bmatrix} M\cos\alpha & s^2 M \sin\alpha \\ -\sin\alpha/s^2 M + M\cos\alpha/\lambda R & \cos\alpha/M + s^2 M \sin\alpha/\lambda R \end{bmatrix}, \tag{9.124}$$

where $\alpha = a\pi/2$. Inverting the above we obtain

$$\tan\alpha = \frac{1}{s^2}\frac{\hat{B}}{\hat{A}}, \tag{9.125}$$

$$M = \sqrt{\hat{A}^2 + (\hat{B}/s^2)^2}, \tag{9.126}$$

$$\frac{1}{\lambda R} = \frac{1}{s^4}\frac{\hat{B}/\hat{A}}{\hat{A}^2 + (\hat{B}/s^2)^2} + \frac{\hat{C}}{\hat{A}}. \tag{9.127}$$

What we have shown is that any quadratic-phase system can be interpreted as a magnified fractional Fourier transform, perhaps with a residual phase curvature. The order $\alpha = a\pi/2$ of the transform, the magnification M, and the radius of curvature R of the residual phase factor are given by the above expressions in terms of the parameters of the quadratic-phase system. This is illustrated in figure 9.3. It can also be interpreted in terms of the following matrix equation:

$$\begin{bmatrix} \hat{A} & \hat{B} \\ \hat{C} & \hat{D} \end{bmatrix} = \begin{bmatrix} 1 & 0 \\ 1/\lambda R & 1 \end{bmatrix}\begin{bmatrix} M & 0 \\ 0 & 1/M \end{bmatrix}\begin{bmatrix} \cos\alpha & s^2\sin\alpha \\ -s^{-2}\sin\alpha & \cos\alpha \end{bmatrix} \tag{9.128}$$

Multiplying out the right-hand side gives the matrix in equation 9.124.

For a given quadratic-phase system to be an imaging or Fourier transforming system, certain conditions have to be satisfied. However, any quadratic-phase system can always be interpreted as a fractional Fourier transforming system; the fractional Fourier transform is general enough to describe all quadratic-phase systems. An optical system consisting of an arbitrary concatenation of sections of free space, thin lenses, and sections of quadratic graded-index media can be characterized either as a quadratic-phase system (a linear canonical transform) with parameters $\hat{A}$, $\hat{B}$, $\hat{C}$, $\hat{D}$, or as a fractional Fourier transform of order a, magnified by M, and observed on a spherical surface with radius R. Both interpretations are legitimate and may be beneficial in different circumstances. In the fractional Fourier transform description, the three parameters are particularly simple to interpret. M and R refer simply to the magnification of the observed transform and the radius of the surface on which it is observed. The order a begins from 0 at the input of the system, and then monotonically increases as a function of z. Fractional Fourier transforms of increasing order describe the evolution of light as it propagates through the system. This was graphically illustrated through a numerical example in section 9.2.4.

We conclude this section with two remarks. First, we note that R, as given by equation 9.127, never becomes zero. To see this, observe that for finite sections of free space and lenses with focal lengths different than zero, $\hat{A}$, $\hat{B}$, $\hat{C}$, $\hat{D}$ are always finite. Furthermore, since $\hat{A}\hat{D} - \hat{B}\hat{C} = 1$, $\hat{B}$ and $\hat{A}$ cannot be zero at the same time. When $\hat{A} \neq 0$, $1/R$ is obviously finite. When $\hat{A} = 0$ we have $\hat{C} = -1/\hat{B}$, from which $1/R = 0$ so R is again not zero.

Second, consistency of equations 9.125 to 9.127 requires that they satisfy a transitivity property. Let us consider two quadratic-phase systems concatenated so that the overall system matrix is the product of the individual system matrices. Using the matrix of the first system, we can find the order, magnification, and radius of curvature of the fractional Fourier transform observed at the output of the first system, which is also the input of the second system. Then we can repeat this procedure to find the parameters of the fractional Fourier transform observed at the output of the second system. Alternatively, we can find these parameters directly in one step by using the matrix of the overall system. Transitivity means that the results found by the one-step procedure are always the same as those found by the two-step procedure. In order to check that this is indeed the case, we need more general forms of equations 9.125 to 9.127 valid for the case where the input distribution is defined with respect to a spherical reference surface with radius R_0:

$$\tan \alpha = \frac{1}{s^2} \frac{\hat{B}}{\left(\hat{A} + \frac{\hat{B}}{\lambda R_0}\right)}, \tag{9.129}$$

$$M^2 = \left(\hat{A} + \frac{\hat{B}}{\lambda R_0}\right)^2 + \left(\frac{\hat{B}}{s^2}\right)^2, \tag{9.130}$$

$$\frac{1}{\lambda R} = \frac{1}{s^4} \frac{\hat{B}/\hat{A}}{\left(\hat{A} + \frac{\hat{B}}{\lambda R_0}\right)^2 + \left(\frac{\hat{B}}{s^2}\right)^2} + \frac{\hat{C}}{\hat{A}} + \frac{1}{\lambda R_0} \frac{\left(\hat{A} + \frac{\hat{B}}{\lambda R_0}\right)/\hat{A}}{\left(\hat{A} + \frac{\hat{B}}{\lambda R_0}\right)^2 + \left(\frac{\hat{B}}{s^2}\right)^2}. \tag{9.131}$$

These equations indeed allow transitivity to be demonstrated. The order corresponding to the overall system is the sum of the orders corresponding to the individual systems, and the scale factor corresponding to the overall system is the product of the scale factors corresponding to the individual systems, provided we match the radius of curvature at the input of the second system to that emerging from the first system.

9.7.2 Geometrical-optical determination of fractional Fourier transform parameters

In a system composed of an arbitrary sequence of lenses separated by arbitrary distances (figure 9.1), one can determine the output amplitude distribution of light in terms of the input amplitude distribution by employing wave-optical methods. However, upon first inspection, it is often easier and more intuitive to ascertain the function and properties of such a system by tracing—on paper or in our imagination— a few rays or geometrical wavefronts through the system. Determining the location, magnification, and phase curvature associated with images and Fourier transforms by tracing only two rays is a common skill in which much experience and intuition have been invested (section 8.4). It is easier to visually grasp the nature of an optical system by tracing a few rays, rather than by evaluating nested integrals obtained by concatenating kernels of the components. For historical, psychological, and pedagogical reasons, geometrical-optical methods play an important role in our understanding of the first-order properties of optical systems. One example of the usefulness of geometrical optics for the analysis of coherent imaging systems has been given by Rhodes (1994).

In this section we show how the transform order, magnification, and phase curvature can be determined in a similar manner for the fractional Fourier transform. Our purpose is to develop the understanding and skill necessary to recognize fractional Fourier transforms and their parameters by visually examining ray traces. This should help readers be able to visualize or intuitively grasp the formation of the fractional Fourier transform in a similar way as they can visualize imaging or Fourier transforming operations. Of course, there is the intrinsic obstacle arising from the fact that the fractional Fourier transform is a more general and complicated transform, so that we cannot expect its visualization to be as easy as its special cases.

Although geometrical optics is only a limiting case of wave optics, it is often possible to extract a significant amount of information regarding a system by geometrical-optical analysis alone. This is not surprising since the $\hat{A}\hat{B}\hat{C}\hat{D}$ matrix describes the relationship between input and output wave fields as well as input and output rays. Examination of the behavior of a few rays will allow us to determine the $\hat{A}\hat{B}\hat{C}\hat{D}$ matrix or the parameters $\hat{\alpha}$, $\hat{\beta}$, $\hat{\gamma}$, which completely characterize the kernel appearing in equation 9.4, which in turn completely determines the input-output relationship of the system (also see section 8.6.3).

We first consider a ray parallel to the optical axis at the input plane ($\theta_{x1} = 0$) (figure 9.12b). Our first measurement M_{ray} is given by the ratio of the distance of the output ray to the optical axis, to the distance of the input ray to the optical axis ($M_{\mathrm{ray}} = x_2/x_1$). Our second measurement is the radius of curvature R_{ray} defined by the output ray ($R_{\mathrm{ray}} = x_2/\theta_{x2}$). Next we consider a ray emanating from the axial point at the input plane ($x_1 = 0$) (figure 9.12a) and measure the radius of curvature R'_{ray} defined by the output ray ($R'_{\mathrm{ray}} = x_2/\theta_{x2}$). A fourth measurement defined by $M'_{\mathrm{ray}} = x_2\lambda/s^2\theta_{x1}$, while redundant, will simplify our expressions. By employing the unit-determinant condition $\hat{A}\hat{D} - \hat{B}\hat{C} = 1$, M'_{ray} can be shown to be related to the other measurements M_{ray}, R_{ray}, R'_{ray} by the equation $s^2 M'_{\mathrm{ray}} M_{\mathrm{ray}}(R'_{\mathrm{ray}}{}^{-1} - R_{\mathrm{ray}}{}^{-1}) = \lambda$. Now, equations 9.9 and 9.124 allow us to write the equations

$$M \cos\alpha = M_{\mathrm{ray}}, \tag{9.132}$$

$$M \sin\alpha = M'_{\mathrm{ray}}, \tag{9.133}$$

$$-\frac{\tan\alpha}{s^2 M^2} + \frac{1}{\lambda R} = \frac{1}{\lambda R_{\mathrm{ray}}}, \tag{9.134}$$

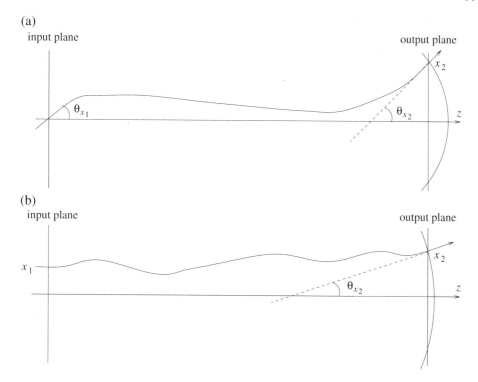

Figure 9.12.   Rays in a fractional Fourier transforming system (Ozaktas and Erden 1997, Erden 1997).

$$\frac{\cot\alpha}{s^2 M^2} + \frac{1}{\lambda R} = \frac{1}{\lambda R'_{\text{ray}}}.$$
(9.135)

Again, any one of these equations, for instance the one involving M'_{ray}, can be derived from the other three using the unit-determinant condition and is thus redundant. By using the first three equations and since $M > 0$ we obtain the unique solution

$$\tan\alpha = \frac{M'_{\text{ray}}}{M_{\text{ray}}},$$
(9.136)

$$M = \sqrt{(M_{\text{ray}})^2 + (M'_{\text{ray}})^2},$$
(9.137)

$$\frac{1}{\lambda R} = \frac{1}{s^2}\frac{M'_{\text{ray}}/M_{\text{ray}}}{(M_{\text{ray}})^2 + (M'_{\text{ray}})^2} + \frac{1}{\lambda R_{\text{ray}}}.$$
(9.138)

The quadrant of α is determined uniquely by the signs of M_{ray} and M'_{ray}. The result can be summarized as:

1. Any quadratic-phase system can be interpreted as a magnified fractional Fourier transform with a residual quadratic phase factor.
2. The transform order, magnification factor, and radius of the quadratic phase factor can be determined by using equations 9.136 through 9.138, based on three measurements obtained by sketching only two rays.

The key equations are equations 9.132 and 9.133. We see that the "projection" of the magnification factor M on the $\alpha = 0$ axis (which corresponds to ordinary imaging), given by $M \cos \alpha$, corresponds to M_{ray}, the ratio of the distances of intercept of the output and input rays (figure 9.12b). We also see that the "projection" of this magnification on the $\alpha = \pi/2$ axis (which corresponds to ordinary Fourier transforming), given by $M \sin \alpha$, corresponds to M'_{ray}, the ratio of the distance of intercept of the output ray to the angle of the input ray (figure 9.12a). If the magnification M "lies" totally along the $\alpha = 0$ axis, we have ordinary imaging and M is interpreted as the magnification of the image. If M "lies" totally along the $\alpha = \pi/2$ axis, we have ordinary Fourier transformation and M is interpreted as the magnification factor associated with the Fourier transform. In the general case, a combination of both effects is observed. This becomes more evident when we write the matrix appearing in equation 9.124 for the case $M = 1$ and $R = \infty$:

$$\begin{bmatrix} \cos \alpha & s^2 \sin \alpha \\ -s^{-2} \sin \alpha & \cos \alpha \end{bmatrix}. \tag{9.139}$$

In ordinary imaging, output distances from the axis are proportional to input distances from the axis, and output inclinations are proportional to input inclinations. In ordinary Fourier transformation, output distances from the axis are proportional to input inclinations, and output inclinations are proportional to input distances from the axis. In fractional Fourier transformation, both effects are combined such that the trigonometric weighting factors can be interpreted as projections in phase space.

Starting from this foundation, it takes only a little practice to start being able to recognize fractional Fourier transforms by sketching two rays through a given optical system. The reader wishing to master this skill should start by considering cases which deviate slightly from ordinary imaging and Fourier transformation. With practice, it is possible to recognize from the inclinations and intercepts of the rays the quadrant of α, the sense of curvature of the residual phase factor, and the rough magnitude of the magnification factor of fractional Fourier transforms, just as we are able to recognize them for ordinary images and Fourier transforms (section 8.4).

We will conclude this section with a number of additional comments. First we note that the matrix parameters are related to the measurement parameters through

$$\hat{A} = M_{\text{ray}},$$
$$\hat{A}/\hat{C} = \lambda R_{\text{ray}},$$
$$\hat{B}/\hat{D} = \lambda R'_{\text{ray}},$$
$$\hat{B} = s^2 M'_{\text{ray}}. \tag{9.140}$$

Using the unit-determinant condition and equation 9.138, it is possible to show the following interesting results:

$$\frac{(M_{\text{ray}})^2 + (M'_{\text{ray}})^2}{R} = \frac{(M_{\text{ray}})^2}{R_{\text{ray}}} + \frac{(M'_{\text{ray}})^2}{R'_{\text{ray}}}, \tag{9.141}$$

$$\frac{1}{R} = \frac{\cos^2 \alpha}{R_{\text{ray}}} + \frac{\sin^2 \alpha}{R'_{\text{ray}}}. \tag{9.142}$$

In deriving equations 9.136 to 9.138 we chose two rays, one with arbitrary intercept x_1 and one with arbitrary inclination θ_{x1}. But the results of course do not depend on the choice of x_1 and θ_{x1} as these will always cancel out in the final result. We can also think of these rays as probes through which we calculate the $\hat{A}\hat{B}\hat{C}\hat{D}$ matrix elements by using equation 9.140 (it is always possible to determine the matrix of a system by using two probe rays: one with zero intercept and arbitrary inclination and the other with zero inclination and arbitrary intercept). Then, equations 9.125 to 9.127 can be used to find α, M, and R, a procedure that will give the same results as equations 9.136 to 9.138.

9.7.3 *Differential equations for the fractional Fourier transform parameters*

Here we determine the differential equations governing the propagation of the order, magnification, and radius of curvature. The major purpose of this exercise is to reveal that these parameters have a reality of their own, and that in a sense they "propagate" through the system.

In examining figure 9.3, the reader might have noticed that there seem to be certain relations among the functions $(\pi/2)a(z) = \alpha(z)$, $M(z)$, and $R(z)$. We now derive these relations and use them to verify several observations regarding figure 9.3. Let x_0 and θ_{x0} denote the intercept and angle of a ray at $z = 0$. The intercept and angle of this ray at an arbitrary value of z are given by

$$
\begin{bmatrix} x(z) \\ \theta_x(z)/\lambda \end{bmatrix} = \begin{bmatrix} \hat{A}(z) & \hat{B}(z) \\ \hat{C}(z) & \hat{D}(z) \end{bmatrix} \begin{bmatrix} x_0 \\ \theta_{x0}/\lambda \end{bmatrix},
\tag{9.143}
$$

where $\hat{A}(z)$, $\hat{B}(z)$, $\hat{C}(z)$, $\hat{D}(z)$ are the matrix elements for the system lying between 0 and z. We repeat equations 9.125 to 9.127 for convenience:

$$
\tan\alpha(z) = \frac{1}{s^2}\frac{\hat{B}(z)}{\hat{A}(z)},
\tag{9.144}
$$

$$
M(z) = \sqrt{\hat{A}^2(z) + (\hat{B}(z)/s^2)^2},
\tag{9.145}
$$

$$
\frac{1}{\lambda R(z)} = \frac{1}{s^4}\frac{\hat{B}(z)/\hat{A}(z)}{\hat{A}^2(z) + (\hat{B}(z)/s^2)^2} + \frac{\hat{C}(z)}{\hat{A}(z)}.
\tag{9.146}
$$

To make further progress, we will examine free-space propagation, lenses, and quadratic graded-index media separately. Derivations of some results are postponed to the end of this section.

Free-space propagation

Let $z_1 < z$ denote the position of the last lens when we look back from the point z in a system consisting of lenses and sections of free space. Let $\hat{A}_1$, $\hat{B}_1$, $\hat{C}_1$, $\hat{D}_1$ denote the matrix elements of the system up to z_1. Then

$$
\begin{bmatrix} \hat{A}(z) & \hat{B}(z) \\ \hat{C}(z) & \hat{D}(z) \end{bmatrix} = \begin{bmatrix} 1 & \lambda(z - z_1) \\ 0 & 1 \end{bmatrix} \begin{bmatrix} \hat{A}_1 & \hat{B}_1 \\ \hat{C}_1 & \hat{D}_1 \end{bmatrix},
\tag{9.147}
$$

from which we can show that $d\hat{A}(z)/dz = \lambda\hat{C}_1$, $d\hat{B}(z)/dz = \lambda\hat{D}_1$ and $\hat{C}(z) = \hat{C}_1$, $\hat{D}(z) = \hat{D}_1$. Using these, it is possible to derive the following results from equations 9.144 to 9.146:

$$\frac{d\alpha(z)}{dz} = \frac{\lambda}{s^2}\frac{1}{M^2(z)}, \tag{9.148}$$

$$\frac{dM(z)}{dz} = \frac{M(z)}{R(z)}, \tag{9.149}$$

$$\frac{d}{dz}\left(\frac{1}{R(z)}\right) = \frac{\lambda^2}{s^4}\frac{1}{M^4(z)} - \frac{1}{R^2(z)}, \tag{9.150}$$

from which one can further derive several additional results:

$$\frac{d^2\alpha(z)}{dz^2} = \frac{\lambda}{s^2}\frac{1}{M^2(z)}\frac{-2}{R(z)} = -\frac{2}{R(z)}\frac{d\alpha(z)}{dz}, \tag{9.151}$$

$$\frac{d^2M(z)}{dz^2} = \frac{M(z)}{R^2(z)}\left(1 - \frac{dR(z)}{dz}\right) = \frac{\lambda^2}{s^4}\frac{1}{M^3(z)}. \tag{9.152}$$

In interpreting equations 9.148 to 9.150, let us first note the following: (i) $\hat{A}$, $\hat{B}$, $\hat{C}$, $\hat{D}$ are always finite for finite z if no lens has a focal length of zero. (ii) It is never the case that $M = 0$, since this would require $\hat{A}$ and $\hat{B}$ to be zero at the same time, which is not possible since we always have $\hat{A}\hat{D} - \hat{B}\hat{C} = 1$. (iii) M is always finite since $\hat{A}$ and $\hat{B}$ are also finite. (iv) It is never the case that $R = 0$, as we discussed on page 361.

We can now deduce the following conclusions for the free-space regions (the regions between any two lenses) of the system in figure 9.3a: $M(z)$ is an increasing function when $1/R(z) > 0$ and a decreasing function when $1/R(z) < 0$, since the signs of $dM(z)/dz$ and $1/R(z)$ are the same. Since $d^2M(z)/dz^2$ is always a finite number greater than zero, $M(z)$ is always concave up. As a consequence, $M(z)$ exhibits smooth minimums where $dM(z)/dz$ and $1/R(z)$ cross zero continuously (as at $z = 0.4$ m but not as at $z = 0.2$ m). Since $d\alpha(z)/dz$ always has a finite positive value, $\alpha(z)$ is a monotonic increasing function. Since $d^2\alpha(z)/dz^2$ is zero when $1/R(z)$ crosses zero continuously, the inflection points of $\alpha(z)$ occur where $1/R(z) = 0$. (Thus the points where $1/R(z) = 0$, $dM(z)/dz = 0$, and $d^2\alpha(z)/dz = 0$ all coincide.)

Again, starting from equations 9.148, 9.149, and 9.150, it is also possible to obtain

$$\frac{d^2\alpha'(z)}{dz^2} - \frac{3}{2\alpha'(z)}\left(\frac{d\alpha'(z)}{dz}\right)^2 + 2\alpha'^3(z) = 0, \tag{9.153}$$

$$\frac{d^2M(z)}{dz^2} - \frac{\lambda^2}{s^4}\frac{1}{M^3(z)} = 0, \tag{9.154}$$

$$\frac{d^2}{dz^2}\left(\frac{1}{R(z)}\right) + 6\left(\frac{1}{R(z)}\right)\frac{d}{dz}\left(\frac{1}{R(z)}\right) + 4\left(\frac{1}{R(z)}\right)^3 = 0, \tag{9.155}$$

where $\alpha'(z) \equiv d\alpha(z)/dz$. These may be interpreted as nonlinear wave equations for $\alpha'(z)$, $M(z)$, and $1/R(z)$ in free space. The solutions of these equations are given by equations 9.144 to 9.146, with $\hat{A}(z)$, $\hat{B}(z)$, $\hat{C}(z)$ being given by equation 9.147. The appearance of $\alpha'(z)$ rather than $\alpha(z)$ in the above second-order differential equation may be attributed to the fact that $\alpha(z)$ can be redefined freely by an additive constant.

Lenses

Let z_l denote the position of a particular lens in a system consisting of lenses and sections of free space. Let $\hat{A}_{1-}$, $\hat{B}_{1-}$, $\hat{C}_{1-}$, $\hat{D}_{1-}$ denote the matrix coefficients of the system just up to the lens at z_l. Then the matrix coefficients just after the lens may be obtained as

$$\left[\begin{array}{cc} \hat{A}_{1+} & \hat{B}_{1+} \\ \hat{C}_{1+} & \hat{D}_{1+} \end{array} \right] = \left[\begin{array}{cc} 1 & 0 \\ -1/\lambda f & 1 \end{array} \right] \left[\begin{array}{cc} \hat{A}_{1-} & \hat{B}_{1-} \\ \hat{C}_{1-} & \hat{D}_{1-} \end{array} \right], \tag{9.156}$$

from which we can show that $\hat{C}_{1+} - \hat{C}_{1-} = -\hat{A}_{1-}/\lambda f$, $\hat{D}_{1+} - \hat{D}_{1-} = -\hat{B}_{1-}/\lambda f$ and $\hat{A}_{1+} = \hat{A}_{1-}$, $\hat{B}_{1+} = \hat{B}_{1-}$. In the case of propagation through free space, $\hat{A}(z)$, $\hat{B}(z)$, $\hat{C}(z)$, $\hat{D}(z)$ are continuous functions so that $\alpha(z)$, $M(z)$, $R(z)$ are also continuous functions. In passing through a lens, we observe discontinuities in $\hat{C}(z)$ and $\hat{D}(z)$. Upon examination of equations 9.144 to 9.146, we see that $\alpha(z)$ and $M(z)$ will still be continuous, but $1/R(z)$ will make a discontinuous jump by the amount $(\hat{C}_{1+} - \hat{C}_{1-})/\hat{A}_{1-} = -1/\lambda f$:

$$\alpha_{1+} = \alpha_{1-}, \tag{9.157}$$

$$M_{1+} = M_{1-}, \tag{9.158}$$

$$\frac{1}{R_{1+}} = \frac{1}{R_{1-}} - \frac{1}{f}. \tag{9.159}$$

Equation 9.149 implies that these drops in $1/R(z)$ at the positive lenses will be matched by a discontinuous drop in $dM(z)/dz$, as we indeed observe in figure 9.3. Since $\alpha'(z) \propto M^{-2}(z)$ we can also add to the above "boundary conditions"

$$\alpha'_{1+} = \alpha'_{1-}, \tag{9.160}$$

and further show that the derivatives just before and just after the lens are related by

$$\left. \frac{d\alpha'(z)}{dz} \right|_{1+} = \left. \frac{d\alpha'(z)}{dz} \right|_{1-} + \frac{2\alpha'_{1-}}{f}, \tag{9.161}$$

$$\left. \frac{dM(z)}{dz} \right|_{1+} = \left. \frac{dM(z)}{dz} \right|_{1-} - \frac{M_{1-}}{f}, \tag{9.162}$$

$$\left[\frac{d}{dz} \left(\frac{1}{R(z)} \right) \right]_{1+} = \left[\frac{d}{dz} \left(\frac{1}{R(z)} \right) \right]_{1-} + \frac{1}{f} \left(\frac{2}{R_{1-}} - \frac{1}{f} \right). \tag{9.163}$$

These boundary conditions, together with the nonlinear wave equations presented previously, completely determine the behavior of $\alpha'(z)$, $M(z)$, and $1/R(z)$ for all values of z.

Quadratic graded-index media

For completeness, we show that equations 9.148 and 9.149 are also valid for quadratic graded-index media. With similar conventions as in the case of sections of free space, we can write

$$\begin{bmatrix} \hat{A}(z) & \hat{B}(z) \\ \hat{C}(z) & \hat{D}(z) \end{bmatrix} = \begin{bmatrix} \cos[(z-z_1)/\chi] & (\lambda\chi)\sin[(z-z_1)/\chi] \\ -(\lambda\chi)^{-1}\sin[(z-z_1)/\chi] & \cos[(z-z_1)/\chi] \end{bmatrix} \begin{bmatrix} \hat{A}_1 & \hat{B}_1 \\ \hat{C}_1 & \hat{D}_1 \end{bmatrix},$$

(9.164)

from which we can derive equations 9.148 and 9.149. In the above $\lambda = c/f_{oc}n_0$.

Derivations of some results

We will end this section by sketching the derivations of some of the results stated above. Readers may skip them if they wish.

First, we will consider the derivation of the first-order differential equation for $1/R(z)$ (equation 9.150). This will be accomplished indirectly by first finding $d^2M(z)/dz^2$. The first derivative of $M(z)$ is

$$\frac{dM(z)}{dz} = \frac{\lambda\hat{A}(z)\hat{C}(z) + (\lambda/s^4)\hat{B}(z)\hat{D}(z)}{M(z)},$$

(9.165)

which we write as

$$M(z)\frac{dM(z)}{dz} = \lambda\hat{A}(z)\hat{C}(z) + \frac{\lambda}{s^4}\hat{B}(z)\hat{D}(z).$$

(9.166)

By taking the derivative of both sides of equation 9.166 and using $\hat{A}(z)\hat{D}(z) - \hat{B}(z)\hat{C}(z) = 1$, somewhat lengthy but straightforward algebra leads to

$$\frac{d^2M(z)}{dz^2} = \frac{\lambda^2}{s^4}\frac{1}{M^3(z)}.$$

(9.167)

Now, using $dM(z)/dz = M(z)/R(z)$ and taking the derivative of both sides, we obtain

$$\frac{d^2M(z)}{dz^2} = \frac{M(z)}{R^2(z)}\left(1 - \frac{dR(z)}{dz}\right).$$

(9.168)

Comparing equation 9.167 and equation 9.168,

$$\frac{dR(z)}{dz} = 1 - \frac{\lambda^2}{s^4}\frac{R^2(z)}{M^4(z)},$$

(9.169)

or

$$\frac{d}{dz}\left(\frac{1}{R(z)}\right) = \frac{\lambda^2}{s^4}\frac{1}{M^4(z)} - \frac{1}{R^2(z)}.$$

(9.170)

We now proceed to the differential equations for $\alpha'(z)$, $M(z)$, and $1/R(z)$. Equation 9.154 for $d^2M(z)/dz^2$ has already been derived. We consider $(d^2/dz^2)[1/R(z)]$. Taking the derivative of both sides of equation 9.170 and using the expressions for $dM(z)/dz$ and $dR(z)/dz$ allows us to derive equation 9.155. We also know that

$$\frac{d^2\alpha(z)}{dz^2} = -\frac{2}{R(z)}\frac{d\alpha(z)}{dz}.$$

(9.171)

We see from this equation that it is not possible to obtain a second-order differential equation of $\alpha(z)$ alone. Taking the derivative of both sides of equation 9.171 and using the expressions for $d\alpha(z)/dz$ and $dR(z)/dz$ we arrive at equation 9.153.

We finally note that the approach taken for free-space propagation can be made somewhat more general. Rather than restricting ourselves to free space, if we make the more general assumptions $d\hat{A}(z)/dz = \lambda\hat{C}(z)$ and $d\hat{B}(z)/dz = \lambda\hat{D}(z)$, without assuming $\hat{C}(z)$ and $\hat{D}(z)$ to be constant, then we can show that equations 9.148 and 9.149 still hold, but equation 9.150 does not since we also used $d\hat{C}(z)/dz = 0$ and $d\hat{D}(z)/dz = 0$ in its derivation. The following will hold in place of equation 9.150:

$$\frac{d}{dz}\left(\frac{1}{R(z)}\right) = \frac{1}{M(z)}\frac{d^2 M(z)}{dz^2} - \frac{1}{R^2(z)}, \tag{9.172}$$

where

$$\frac{d^2 M(z)}{dz^2} = \frac{\lambda\hat{A}(z)d\hat{C}(z)/dz + (\lambda/s^4)\hat{B}(z)d\hat{D}(z)/dz}{M(z)} + \frac{\lambda^2}{s^4}\frac{1}{M^3(z)}. \tag{9.173}$$

Since the assumptions $d\hat{A}(z)/dz = \lambda\hat{C}(z)$ and $d\hat{B}(z)/dz = \lambda\hat{D}(z)$ hold for quadratic graded-index media, we can specialize the above equation as

$$\frac{d^2 M(z)}{dz^2} = \frac{-(1/\chi^2)\cos^2[(z - z_1)/\chi] - (\lambda^2/s^4)\sin^2[(z - z_1)/\chi]}{M(z)} + \frac{\lambda^2}{s^4}\frac{1}{M^3(z)}. \tag{9.174}$$

When $s^2 = \lambda\chi$ we have $M(z) = 1$, $dM(z)/dz = 0$, and $d^2 M(z)/dz^2 = 0$, as expected.

9.7.4 Fractional Fourier transform parameters and Gaussian beam parameters

Readers familiar with the propagation of Gaussian beams might have already noticed the similarity between the behavior of $\alpha(z) = (\pi/2)a(z)$, $M(z)$, and $R(z)$ (figure 9.3), and the common parameters of Gaussian beams, namely the Gouy phase shift $\tilde{\zeta}_G(z)$, the beam size $W_G(z)$, and the wavefront radius of curvature $R_G(z)$. (In this section we use the subscript G to distinguish the parameters of Gaussian beams.) Our present purpose is to discuss the precise relationship between these two sets of parameters.

We emphasize that $\tilde{\zeta}_G(z)$ denotes the accumulated Gouy phase shift with respect to the input plane at $z = 0$, rather than the conventional Gouy phase shift with respect to the last waist of the beam. Conventionally, the Gouy phase shift is given by $\arctan[(z - z_{lw})/\check{z}]$, where z_{lw} denotes the location of the last waist of the beam, and $\check{z}$ is the Rayleigh range of the beam. We define the accumulated Gouy phase shift of a Gaussian beam passing through an optical system as the phase accumulated by the beam in excess of the phase accumulated by a plane wave passing through the same system (page 237).

At a given location z, a Gaussian beam can be characterized by the complex parameter $\hat{q}_G(z)$ defined as

$$\frac{1}{\hat{q}_G(z)} = \frac{1}{R_G(z)} + i\frac{\lambda}{W_G^2(z)}, \tag{9.175}$$

where $R_G(z)$ is the wavefront radius and $W_G(z)$ is the beam size at z (section 7.3.4). If a Gaussian beam with complex parameter $\hat{q}_G(0)$ is incident at $z = 0$ on a system characterized by the matrix elements $\hat{A}(z)$, $\hat{B}(z)$, $\hat{C}(z)$, $\hat{D}(z)$, the complex parameter $\hat{q}_G(z)$ of the Gaussian beam at z is given by

$$\lambda \hat{q}_G(z) = \frac{\hat{A}(z)\lambda \hat{q}_G(0) + \hat{B}(z)}{\hat{C}(z)\lambda \hat{q}_G(0) + \hat{D}(z)}. \tag{9.176}$$

Let us consider a Gaussian beam whose waist (its narrowest part where the wavefront radius $R_G = \infty$) is located at $z = 0$ of the system shown in figure 9.3a. Let us denote the waist size by W_{G0}, which is related to the Rayleigh range $\check{z}$ through $W_{G0} = (\lambda \check{z})^{1/2}$. With $1/R_G(0) = 0$ and $W_G(0) = W_{G0}$, we have $1/\hat{q}_G(0) = i\lambda/W_{G0}^2$. Upon substitution in equation 9.176 and using equation 9.175, we can find expressions for $\hat{q}_G(z)$, $W_G(z)$, $R_G(z)$ at any location z in figure 9.3a, in terms of $\hat{A}(z)$, $\hat{B}(z)$, $\hat{C}(z)$, $\hat{D}(z)$. These expressions, along with a similar expression for the accumulated Gouy phase shift $\tilde{\zeta}_G(z)$, have already been presented as equations 7.73, 7.74, and 7.75:

$$\frac{1}{\lambda R_G(z)} = \frac{\hat{B}(z)/\hat{A}(z)}{\hat{A}^2(z)(W_{G0})^4 + \hat{B}^2(z)} + \frac{\hat{C}(z)}{\hat{A}(z)}, \tag{9.177}$$

$$W_G^2(z) = \hat{A}^2(z)(W_{G0})^2 + \hat{B}^2(z)/(W_{G0})^2, \tag{9.178}$$

$$\tan \tilde{\zeta}_G(z) = \frac{\hat{B}(z)}{\hat{A}(z)(W_{G0})^2}. \tag{9.179}$$

We are inquiring how the expressions for $\tilde{\zeta}_G(z)$, $W_G(z)$, and $R_G(z)$ are related to the expressions for $\alpha(z)$, $M(z)$, and $R(z)$ given in equations 9.144 to 9.146. Upon comparison of these equations with those appearing above, the result can be stated as follows:

> Let the output of an arbitrary multi-lens system be interpreted as a fractional Fourier transform of the input of order $\alpha(z)$, with magnification $M(z)$, observed on a spherical surface of radius $R(z)$.
>
> Let a Gaussian beam whose waist is located at $z = 0$ with waist size W_{G0} exhibit an accumulated Gouy phase shift $\tilde{\zeta}_G(z)$, beam size $W_G(z)$, and wavefront radius $R_G(z)$ at the output of the same system.
>
> If the scale parameter s appearing in equations 9.123 and 9.124 is equal to W_{G0}, then $\alpha(z) = \tilde{\zeta}_G(z)$, $M(z) = W_G(z)/W_{G0}$, and $R(z) = R_G(z)$.

The reader may recall that the relationship between $\tilde{\zeta}_G(z)$ and $\alpha(z)$ was also discussed in section 9.6, where we arrived at it following a different path.

For completeness, we also present the relationships between the two sets of parameters when $s \neq W_{G0}$, We use equation 9.124 to write the right-hand sides of equations 9.177, 9.178, and 9.179 in terms of $\alpha(z)$, $M(z)$, and $R(z)$:

$$\frac{1}{\lambda R_G} = \frac{\cos \alpha \sin \alpha [s^4 - (W_{G0})^4]}{M^2 s^2 [(W_{G0})^4 \cos^2 \alpha + s^4 \sin^2 \alpha]} + \frac{1}{\lambda R}, \tag{9.180}$$

$$\left(\frac{W_G}{W_{G0}}\right)^2 = M^2 \left[1 + \sin^2 \alpha \left(\frac{s^4}{(W_{G0})^4} - 1\right)\right], \tag{9.181}$$

$$\tan \tilde{\zeta}_G = \frac{s^2}{(W_{G0})^2} \tan \alpha, \tag{9.182}$$

where we have suppressed the dependence on z for simplicity. When $s \neq W_{G0}$, we find that $M(z)$ and $W_G(z)/W_{G0}$, $R(z)$ and $R_G(z)$, and $\alpha(z)$ and $\tilde{\zeta}_G(z)$ will be exactly equal at image planes ($\sin \alpha = 0$), but exhibit only a general resemblance elsewhere.

Originally, we posed $\tilde{\zeta}_G(z)$, $W_G(z)$, and $R_G(z)$ as parameters characterizing the state of the beam at a certain value of z, and $\alpha(z)$, $M(z)$, and $R(z)$ as parameters characterizing the optical system occupying the interval $[0, z]$. However, the parameters $\tilde{\zeta}_G(z)$, $W_G(z)$, and $R_G(z)$ also characterize the system occupying the interval $[0, z]$, since we can recover any other set of parameters characterizing the system from them. Likewise, the parameters $\alpha(z)$, $M(z)$, and $R(z)$, since they are functions of z, may be thought of as entities which "propagate" through the system. This picture is strengthened by the wave equations 9.153, 9.154, and 9.155 for these entities.

Similar comments hold for the parameters M_{ray}, M'_{ray}, R_{ray}, and R'_{ray} and the parameters $\alpha(z)$, $M(z)$, and $R(z)$ (equations 9.136 to 9.138). M_{ray}, M'_{ray}, R_{ray}, and R'_{ray} are related to the intercepts and angles characterizing the rays at a certain value of z, whereas $\alpha(z)$, $M(z)$, and $R(z)$ characterize the optical system occupying the interval $[0, z]$. However, the parameters M_{ray}, M'_{ray}, R_{ray}, and R'_{ray}, or the ray intercepts and angles themselves, also fully characterize the system, since the system matrix can be recovered from them. On the other hand, $\alpha(z)$, $M(z)$, and $R(z)$ also characterize—somewhat indirectly—the rays propagating through the system, since the behavior of the rays is related to the evolution of these functions.

We conclude this section with some additional remarks. If we launch a Gaussian beam into the system of figure 9.3 at its waist, the locations of subsequent waists will not in general coincide with the image planes, unless these are perfect image planes with no residual phase factor ($1/R(z) = 0$). In general, $1/R(z)$ and $1/R_G(z)$ are not zero at the image planes so that waists are not observed at these locations.

Given $W_G(z)$ and $R_G(z)$ of a Gaussian beam at a certain value of z, we can always find its waist location and waist size and thus its conventional Gouy phase shift. Unlike the accumulated Gouy phase shift, the conventional Gouy phase shift is not an independent parameter. Say we are using a Gaussian beam as a probe to determine the matrix elements of some system. By observing $\tilde{\zeta}_G(z)$, $W_G(z)$, and $R_G(z)$ of the probe Gaussian beam incident on the system at its waist, we can deduce the matrix elements by employing equation 9.124 and the results discussed in this section. The same is not possible with the conventional Gouy phase shift which is not an independent parameter.

9.7.5 Discussion

The characterization of optical systems is directly related to the characterization of optical rays, beams, or fields. If we know the parameters characterizing a particular system, we can also trace the evolution of the parameters of rays, beams, or fields propagating through the system. Likewise, if we know how the parameters of a number of rays, beams, or fields evolve through a restricted class of systems, we may be able to determine the parameters of the system.

The fractional Fourier transform allows us to characterize both quadratic-phase systems and the rays, beams, or fields passing through them. Quadratic-phase systems, which consist of lenses and sections of free space as well as quadratic graded-index media, can be characterized by their physical parameters (focal lengths and separations of lenses, and so on), the parameters $\hat{\alpha}, \hat{\beta}, \hat{\gamma}$ appearing in equation 9.4, or by the matrix appearing in equation 9.8. The major result of this section was to show that such systems can be interpreted as fractional Fourier transformers and thus can also be characterized by the parameters α, M, and R.

Optical rays are characterized by their intercepts and angles. Gaussian beams are characterized by their beam sizes and wavefront radii. We have closely examined the relationships between rays and beams propagating through a system, the matrix parameters $\hat{A}(z)$, $\hat{B}(z)$, $\hat{C}(z)$, $\hat{D}(z)$ corresponding to the system occupying the interval $[0, z]$, and the fractional Fourier transform parameters $\alpha(z)$, $M(z)$, $R(z)$. The evolution of these functions of z are related to the evolution of the ray or beam parameters as functions of z. Moreover, $\alpha(z)$, $M(z)$, and $R(z)$ were shown to satisfy wave equations. These observations tempt us to think of $\alpha(z)$, $M(z)$, and $R(z)$ as the parameters of some kind of disturbance propagating through the system. This notion was made concrete by showing that these parameters are actually equal to the parameters of a Gaussian beam propagating through the system when the scale parameter is chosen equal to the waist size of the Gaussian beam.

Many physical phenomena in areas other than optics are also described by quadratic-phase systems (linear canonical transforms) or one of its subclasses. As we have seen in this section, it is possible to interpret any quadratic-phase systems as a magnified fractional Fourier transform with residual phase factor in a physically meaningful way. Such meaningful interpretations may be possible in other areas as well.

In this section we dealt with systems composed of lenses separated by sections of free space. A more general class of systems is obtained by inserting thin spatial filters between such systems. Inserting filters at fractional Fourier transform planes provides the basis for various signal processing operations to be performed. This class of systems is discussed in the next section. Signal processing applications are discussed in chapters 10 and 11.

9.8   Fourier optical systems

So far we have concentrated on first-order optical systems, mathematically characterized as quadratic-phase systems (linear canonical transforms). Now we turn our attention to Fourier optical systems, which consist of an arbitrary number of thin spatial filters sandwiched between arbitrary quadratic-phase systems. They include optical systems composed of an arbitrary number of lenses and filters separated by arbitrary distances, under the standard approximations of Fourier optics (figure 9.13a). Sections of quadratic graded-index media may also appear in such systems.

Here we will show that every Fourier optical system is equivalent to, and can be modeled as, both of the following:

1. Consecutive filtering operations in several fractional Fourier domains (figure 9.13c). Each fractional Fourier transform stage transforms from one fractional domain to another, where a multiplicative filter is applied. More precisely, *every Fourier optical system is equivalent to a sequence of appropriately chosen multiplicative filters inserted between fractional Fourier transform stages with appropriately chosen orders.*
2. Consecutive filtering operations alternately in the space and frequency domains

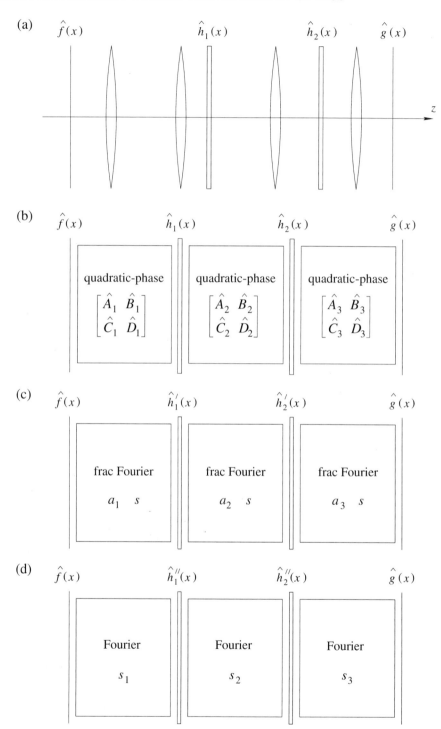

Figure 9.13. Fourier optical systems (Ozaktas and Mendlovic 1996a).

(figure 9.13d). Each time a Fourier transform is applied we alternate between the space and frequency domains, where multiplicative filters are applied. More precisely, *every Fourier optical system is equivalent to a sequence of appropriately chosen multiplicative filters inserted between appropriately scaled Fourier transform stages.*

These equivalences provide considerable conceptual simplification and insight regarding such systems, and should also facilitate their design and analysis. (Ozaktas and Mendlovic 1996a)

Figure 9.13a shows a Fourier optical system with input $\hat{f}(x)$ and output $\hat{g}(x)$, consisting of several lenses, filters, and sections of free space. The transmittance functions of the filters are indicated directly above them. Figure 9.13b shows the system modeled as a sequence of multiplicative filters sandwiched between quadratic-phase systems, each of which is characterized by its matrix parameters $\hat{A}_j, \hat{B}_j, \hat{C}_j, \hat{D}_j$. Figure 9.13c shows the system modeled as a sequence of multiplicative filters sandwiched between fractional Fourier transform stages, each of which is characterized by its order a_j. The scale parameter s is the same for all stages. Finally, figure 9.13d shows the system modeled as a sequence of multiplicative filters sandwiched between ordinary Fourier transform stages, each of which is characterized by its scale parameter s_j.

We now prove the results stated above, also showing how to obtain the appropriate multiplicative filters, transform orders, and scale parameters. We first recall that any quadratic-phase system can be expressed as the concatenation of a lens followed by a fractional Fourier transformer followed by another lens, as expressed by the following matrix equation:

$$\begin{bmatrix} \hat{A} & \hat{B} \\ \hat{C} & \hat{D} \end{bmatrix} = \begin{bmatrix} 1 & 0 \\ -1/\lambda f_r & 1 \end{bmatrix} \begin{bmatrix} \cos\alpha & s^2\sin\alpha \\ -s^{-2}\sin\alpha & \cos\alpha \end{bmatrix} \begin{bmatrix} 1 & 0 \\ -1/\lambda f_l & 1 \end{bmatrix}. \quad (9.183)$$

f_l and f_r are the focal lengths of the lenses on the left and right of the fractional Fourier transformer. The fractional Fourier transform matrix appearing in the middle of the right-hand side had appeared before in equation 9.139.

We solve the above equation for f_l, f_r, and α in terms of $\hat{A}$, $\hat{B}$, $\hat{C}$, $\hat{D}$, and s. The result is

$$\sin\alpha = \frac{\hat{B}}{s^2}, \qquad \alpha \in [-\pi/2, \pi/2],$$

$$\lambda f_l = \frac{\hat{B}}{\sqrt{1 - \hat{B}^2/s^4} - \hat{A}}, \qquad \lambda f_r = \frac{\hat{B}}{\sqrt{1 - \hat{B}^2/s^4} - \hat{D}}. \quad (9.184)$$

We observe that we are not fully free in choosing s; we are subject to the constraint $s^2 \geq |\hat{B}|$.

Each of the quadratic-phase systems in figure 9.13b can be replaced by a lens followed by a fractional Fourier transformer followed by a lens, with the help of these formulas. Now, absorbing the transmittance functions of the lenses in those of the adjacent filters by defining

$$\hat{h}'_j(x) = e^{-i\pi x^2/\lambda f_{l(j+1)}} \, \hat{h}_j(x) \, e^{-i\pi x^2/\lambda f_{r_j}}, \quad (9.185)$$

we arrive at the configuration of figure 9.13c, which is the first result we sought to show.

Returning to equation 9.183, let us now set $a = 1$ ($\alpha = \pi/2$) corresponding to the ordinary Fourier transform. This time we can solve for f_l, f_r, and s in terms of $\hat{A}$, $\hat{B}$, $\hat{C}$, $\hat{D}$ as follows:

$$s = \sqrt{\hat{B}}, \qquad \lambda f_l = \frac{-\hat{B}}{\hat{A}}, \qquad \lambda f_r = \frac{-\hat{B}}{\hat{D}}. \qquad (9.186)$$

If $\hat{B} < 0$ the above solution will not be valid, but a complementary solution can be obtained in the same manner by setting $a = -1$ (corresponding to an inverse Fourier transform) instead of $a = 1$. This time each of the quadratic-phase systems in figure 9.13b can be replaced by a lens followed by a scaled Fourier transform followed by a lens. By absorbing the transmittance functions of the lenses in those of adjacent filters as in equation 9.185, the second result is also proved. It is important to note that similar results can be found by setting a to any other specific value as well. The ordinary Fourier transform is no more privileged than fractional transforms of other orders.

We have seen in this chapter that fractional Fourier transforms can be realized optically by using canonical configurations type I or type II or sections of quadratic graded-index media. Thus our first result implies that every Fourier optical system can be realized (or simulated) by sandwiching spatial filters between canonical fractional Fourier transform configurations or appropriate length sections of graded-index media. Such realizations would constitute physical embodiments of figure 9.13c. We can also conclude that since the fractional Fourier transform can be digitally computed in the order of $N \log N$ time (chapter 6), a system with M filters can be digitally simulated in the order of $MN \log N$ time.

In chapter 10 we refer to filtering systems having the form of figure 9.13c as *repeated filtering systems*. In contrast to ordinary Fourier domain filtering systems, these systems employ a multitude of filters in several consecutive fractional Fourier domains. Whereas the use of a single ordinary Fourier domain filter allows us to realize only space-invariant (convolution-type) systems, repeated filtering systems represent a more general class of linear systems including many space-variant operations which are useful for a variety of applications such as the elimination of nonstationary noise and restoration of signals under space-variant distortion models (chapter 10).

Our first result means that the analysis or design of Fourier optical filtering systems (figure 9.13a) is equivalent and thus can be reduced to that of repeated filtering systems. In other words, every Fourier optical filtering system acts on the input in a way that is equivalent to applying multiplicative filters in consecutive fractional Fourier domains. All results, methods of analysis or design, and algorithms developed for repeated filtering systems (chapter 10) are thus also applicable to Fourier optical systems.

We showed how figure 9.13b can be reduced to both figure 9.13c and figure 9.13d. For completeness, we also mention how figure 9.13c can be reduced to figure 9.13d. This follows immediately from the fact that since a fractional Fourier transformer is a special kind of quadratic-phase system, it can also be decomposed into a lens followed by an ordinary Fourier transformer followed by a lens. We see that repeated filtering systems employing fractional Fourier transforms (figure 9.13c) can be reduced to repeated filtering systems employing only the ordinary Fourier transform (figure 9.13d). Applying multiplicative filters alternately in the space and frequency domains allows us to do everything that can be done by applying filters in fractional domains. The following generalization can also be demonstrated similarly. Applying multiplicative filters alternately in *any* two given domains (provided their orders do not differ by an integer multiple of 2) allows us to do everything that can be done by any of the configurations mentioned here. This generalization will be elaborated further below.

The fact that repeated filtering in fractional domains can be reduced to repeated filtering in ordinary space and frequency domains does not compromise the conceptual and practical utility of the fractional Fourier transform. The fractional transform may be conceptually indispensable in devising an algorithm or designing an effective filter, even if the system is then reduced to one which does not employ fractional Fourier transforms. Furthermore, one would not necessarily engage in such a reduction, since computation of the fractional transform—both optically and digitally—is not more difficult or costly than computation of the ordinary transform. The computation of ordinary and fractional transforms can both be reduced to each other. The implementation of figure 9.13c is not more difficult than that of figure 9.13d. Examples of the utility of the fractional Fourier transform in a wide variety of signal and image processing applications will be seen in chapters 10 and 11.

From a practical viewpoint, the implementation of the necessary filters may be much easier in certain domains, as compared to others. For instance, in chirp elimination (chapter 10), the filters necessary in fractional domains are simple apertures or knife edges, whereas in the ordinary space and frequency domains they would have to be complex and highly oscillatory functions. Furthermore, it may be easier to minimize deviations from the standard approximations of Fourier optics in one configuration as opposed to another. In conclusion, the equivalence results shown in this section should be used to increase the number of alternative physical realizations which are nominally equivalent, not to reduce them to one. These alternative realizations provide additional degrees of freedom which may allow us to deal effectively with certain practical and technical constraints such as the need to use catalog optics, the need to limit sensitivity to parameter deviations, limitations on the realizability of filters, and so forth.

As promised, we now further elaborate the fact that there is nothing special about the ordinary Fourier domain as far as our second result is concerned. We claim that applying multiplicative filters alternately in *any* two given domains (provided their orders do not differ by an integer multiple of 2) is sufficient to realize any Fourier optical system. From equation 9.183 we can find the scale parameter and focal lengths necessary to replace a quadratic-phase system with a fractional Fourier transformer of order α:

$$s^2 = \frac{\hat{B}}{\sin \alpha}, \qquad \lambda f_{\mathrm{l}} = \frac{\hat{B}}{\cos \alpha - \hat{A}}, \qquad \lambda f_{\mathrm{r}} = \frac{\hat{B}}{\cos \alpha - \hat{D}}. \qquad (9.187)$$

The sign of $\sin \alpha$ is constrained by the sign of $\hat{B}$. Note that consecutive domains whose orders differ by ± 2 are essentially the same domain within a sign reversal. Now, given any two domains α_1 and α_2 such that their difference is not an integer multiple of

π, it is always possible to alternate between the two by employing either a positive transform or a negative transform of order $\alpha = \pm|\alpha_2 - \alpha_1|$, as necessitated by the sign of $\sin \alpha$.

We conclude by briefly outlining an alternative formulation of our first result. We saw in section 9.4.6 that any quadratic-phase system can be interpreted as a fractional Fourier transform by choosing appropriate input and output scale parameters and spherical reference surfaces. Since there are sufficient degrees of freedom, we are allowed to fix the scale parameter and radius of the reference surface on the input side. Once this is done, the order of the transform, and the scale parameter and radius of the reference surface on the output side are determined in terms of $\hat{A}$, $\hat{B}$, $\hat{C}$, $\hat{D}$. By introducing the constraint that the input scale parameter and radius for each stage be set equal to the corresponding quantities at the output of the previous stage, it is possible to reduce a system of the form depicted in figure 9.13b to a system of the form depicted in figure 9.13c such that $\hat{h}_j(x) = \hat{h}'_j(x)$. When the radii of the spherical reference surfaces on both sides of the filter are the same, the amplitude distribution on the right is found simply by multiplying the amplitude distribution on the left by $\hat{h}_j(x)$. Unlike in the previous approach, here the scale parameter is different for each stage, but chirps need not be absorbed in the filters.

9.9 Locations of fractional Fourier transform planes

Here we comment on a number of (sometimes apparently paradoxical) issues related to the locations of fractional Fourier transform planes. Although our discussion is more general, it will be useful to think of multi-lens systems of the kind shown in figure 9.3a. Remember from section 8.4.3 that the Fourier transform of the object is always observed at the images of the source, and does not depend on the position of the object or the scale parameter. However, the location of the ath order fractional Fourier transform depends on both.

If we move the object towards the right, the continuum of fractional transforms are compressed towards the right, such that the first-order Fourier transform is always observed at the same position. This might seem paradoxical in the following sense. Let us assume the object is moved to the location where the 0.3rd transform was observed. Then, from index additivity, we would expect the subsystem from this location to the ordinary Fourier plane to correspond to a 0.7th order transform. Thus, with the object at this new location, we would expect to observe the 0.7th transform at the location where we originally observed the ordinary Fourier transform, and to observe the ordinary Fourier transform further to the right. But this cannot be and is not true. The ordinary Fourier transform is always observed at the same plane where the source is imaged. To resolve the apparent paradox, we remember that the 0.3rd transform is observed on a spherical reference surface. If we specify the same original object with respect to this spherical reference surface, then we would indeed observe its 0.7th transform at the original ordinary Fourier plane. However, if we specify the same original object with respect to a planar reference surface at the same location, we will observe the 1st transform at the original ordinary Fourier plane. Readers who wish may explicitly work this out using the example of figure 9.3, remembering that the fractional Fourier transform of a function multiplied by a quadratic phase factor is related to the fractional Fourier transform of the original function of a different order.

The location of the ath order fractional Fourier transform also depends on the choice of s. If we change s, the ordinary Fourier and image planes remain unchanged but the locations of fractional transform planes of other orders are compressed towards or decompressed away from these fixed planes. This is no paradox either, since we know that the fractional Fourier transform of a function with a different scale is related to the fractional Fourier transform of the original function of a different order.

The importance of properly interpreting the scale in optical fractional Fourier transforming systems, and the design of systems with desired scale has been discussed in Jiang 1995, Hua, Liu, and Li 1997b, Liu and others 1997b, Lohmann and others 1998, and Wang and Zhou 1998.

9.10 Wave field reconstruction, phase retrieval, and phase-space tomography

One of the important applications of the fractional Fourier transform is the technique known as *phase-space tomography*, which is used for spatial and temporal phase retrieval and reconstruction of both fully and partially coherent optical fields, as well as quantum-mechanical wave functions (Beck and others 1993, Smithey and others 1993, Raymer, Beck, and McAlister 1994a, b, James and Agarwal 1995, McAlister and others 1995, Tu and Tamura 1997, 1998). Earlier works which contain related ideas include Vogel and Risken 1989, Yurke, Schleich, and Walls 1990, and Vogel and Schleich 1991.

Before briefly outlining this approach, it is worth recalling a few facts about the Radon transform and the projection-slice theorem (section 2.10.1). This theorem stated that a two-dimensional function can be reconstructed from its Radon transform, which is the complete set of integral projections at different angles. Numerical methods for performing such reconstructions have been extensively studied because of their applications in medical imaging (Barrett 1984, Bracewell 1995). In phase-space tomography, the two-dimensional function to be reconstructed is the Wigner distribution of the signal. But we saw in equation 4.146 that the integral projections of the Wigner distribution at different angles are equal to the squared magnitudes of the fractional Fourier transforms of the original signal. Thus, knowledge of the magnitude or intensity of the fractional Fourier transforms for all orders is equivalent to knowledge of the projections of the Wigner distribution for all angles, which in turn is sufficient to fully reconstruct the Wigner distribution and (within a constant phase factor) the signal itself.

The application of the method takes a diversity of forms. However, in most cases the approach is based on generating a sufficiently large number of different-ordered fractional Fourier transforms of the signal and measuring their intensities. When dealing with spatial optical signals, this may be achieved by using any of the fractional Fourier transforming configurations discussed in this chapter. When dealing with temporal optical signals, this may be achieved by using temporal chirp multipliers and convolvers to realize temporal equivalents of these configurations (see section 9.11.1) (Beck and others 1993). When we are dealing with partially coherent fields, what we recover is the Wigner distribution of the random wave field. We know from equation 3.50 that the Wigner distribution of a random process is the Fourier transform of its ensemble-averaged autocorrelation function, known in optics as the mutual intensity function. Thus in the partially coherent case, phase-space tomography provides us with the mutual intensity of the optical wave field. Analogous considerations apply in the case of quantum-mechanical measurements, with intensities being replaced by probabilities and complex wave fields by complex wave functions.

Phase-space tomography is essentially a method of phase retrieval, since the complex amplitude of the signal is determined from intensity measurements. We will take a closer look at some of the issues by considering spatial optical signals. Let $\hat{f}(x)$ or $\hat{f}(x, y)$ represent the complex amplitude at $z = 0$ of an unknown signal which we desire to recover. Now, let this signal propagate either in free space or through an arbitrary quadratic-phase system. We have seen in this chapter that in either case the complex amplitude distribution at any plane perpendicular to the optical axis can be expressed as a magnified fractional Fourier transform, perhaps with a residual quadratic phase factor. The order of the transform is a monotonically increasing function of z. It therefore follows that knowing the intensity $\hat{I}_{\hat{f}}(x, z)$ or $\hat{I}_{\hat{f}}(x, y, z)$ for all values of z is equivalent to knowing the intensity of the fractional Fourier transforms for all orders. (The quadratic phase factor disappears when we look at the intensity, and the magnification is easily accounted for.) The finite range and sampling required for practical measurements will depend on the particular quadratic-phase system and the desired resolution.

First, let us concentrate on the one-dimensional coherent case. Knowing the intensity $\hat{I}_{\hat{f}}(x, z)$ for all values of z amounts to knowing $|\hat{f}_a(x_a)|^2$ for all values of a and thus enables us to recover $\hat{W}_{\hat{f}}(x, \sigma_x)$ and $\hat{f}(x)$. However, it is important to note that this is a very redundant procedure. A two-dimensional real measurement is being used to recover a one-dimensional complex measurement, whereas we would expect two one-dimensional real measurements to suffice. Indeed, intensity measurements at any two fractional Fourier transform planes is usually considered sufficient to recover $\hat{f}(x)$, if the orders of the transforms are not too close to each other (Cong, Chen, and Gu 1998b, c, d, Dong and others 1997). This is nothing but a generalization of the celebrated Gerchberg-Saxton iterative algorithm which has been employed to recover the complex signal from the intensity of itself and its Fourier transform (*Image Recovery: Theory and Applications* 1987). Nevertheless, phase retrieval from two intensity measurements can be problematic in certain circumstances, and there is certainly much more work that remains to be done in this area. Better results and greater noise immunity can be expected when more than two intensity measurements are available (Ivanov, Sivokon, and Vorontsov 1992). The multiple intensity measurement problem is of course equivalent to the problem of recovering a complex function from the magnitudes of several of its fractional Fourier transforms, a problem deserving further attention. As the number of intensity measurements approaches the number of degrees of freedom of the signals, we arrive at the phase-space tomography approach.

Similar comments apply to the two-dimensional coherent case, where the phase-space tomography approach is again redundant, and a relatively small number of intensity measurements, if not only two, is sufficient in most cases. In fact, much work has been done on the problem of recovery from only one intensity measurement supplemented with additional constraints (*Image Recovery: Theory and Applications* 1987).

We now turn our attention to the partially coherent case. Again, let us first consider the one-dimensional case, where the wave field is characterized by its mutual intensity $\hat{R}_{\hat{f}\hat{f}}(x_1, x_2)$. Knowledge of $\hat{I}_{\hat{f}}(x, z)$ for all values of z is the same as knowledge of the magnitudes of the fractional Fourier transforms for all values of a. This allows tomographic reconstruction of the Wigner distribution, from which we can obtain the mutual intensity. This time the approach is not redundant since we are recovering a two-dimensional function from a two-dimensional measurement.

Moving on to the two-dimensional case, at first we might expect to recover the mutual intensity $\hat{R}_{\hat{f}\hat{f}}(x_1, y_1; x_2, y_2)$ from knowledge of the intensity $\hat{I}_{\hat{f}}(x, y, z)$. However, normally we cannot expect to recover a four-dimensional function from a three-dimensional measurement. Indeed, full knowledge of the three-dimensional intensity distribution of a partially coherent wave field does not uniquely specify the wave field. In other words, two-dimensional partially coherent wave fields have more degrees of freedom than can be captured in their full three-dimensional intensity distributions, in contrast to the one-dimensional case (Nugent 1992 as corrected by Hazak 1992, Gori, Santarsiero, and Guattari 1993).

Let us consider the same ideas in terms of Wigner distributions and fractional Fourier transforms, recalling that knowledge of the Wigner distribution is equivalent to knowledge of the mutual intensity, and that knowledge of the fractional Fourier transforms is equivalent to knowledge of the intensities. In the one-dimensional case, knowledge of the intensities for all values of z means knowledge of the fractional Fourier transforms for all values of a and thus knowledge of the projections of the Wigner distribution for all angles. This is sufficient to recover the Wigner distribution and thus the mutual intensity. In two dimensions, however, knowledge of the intensities for all values of z means knowledge of the two-dimensional fractional Fourier transforms *with the same orders $a_x = a_y$ in both dimensions x and y.* In other words, as light propagates, only a single-parameter family of fractional transforms are generated. The situation would not change if we employed anisotropic media or anamorphic optical components. In this case the transform orders would be different for the two dimensions x and y, but we would still be measuring the fractional Fourier transforms of two-dimensional order (a_x, a_y) along a curve in a_x-a_y space. On the other hand, in order to recover the Wigner distribution by using the projection-slice theorem, it is necessary to know the intensity of the fractional Fourier transforms for a two-dimensional set of orders (a_x, a_y), varied independently so as to cover the whole region $[-1, 1] \times [-1, 1]$.

It is not possible to obtain such a four-dimensional measurement $|\hat{f}_{a_x, a_y}(x, y)|^2$ by propagating the light through a single setup. However, by employing several setups, each of whose parameters are adjusted appropriately, it is possible to obtain intensity measurements for a sufficient number of distinct pairs of orders (a_x, a_y), and use the projection-slice theorem to obtain the four-dimensional Wigner distribution and mutual intensity. General optical systems for generating fractional Fourier transforms with independently specified orders a_x and a_y are discussed in detail in Sahin, Ozaktas, and Mendlovic 1998. This approach has been developed in Raymer, Beck, and McAlister 1994a, which is probably the first work to deal with phase-space tomographic reconstruction of classical optical fields.

The fractional Fourier transform also appears in related issues arising in the context of quantum state characterization (Alieva and Barbé 1998b).

9.11 Extensions and applications

In this section we briefly mention several extensions and applications which we do not have space to discuss in greater detail, providing references for those interested.

9.11.1 Temporal optical implementation of the transform

The analogy between spatial and temporal optical systems has been discussed by many authors. For instance, see Papoulis 1968, 1994, Kolner and Nazarathy 1989, and Lohmann and Mendlovic 1992c. Since any quadratic-phase system can be realized in terms of chirp multiplications and convolutions, all it takes to realize the temporal equivalent of any spatial quadratic-phase system is the ability to temporally realize chirp multiplications and convolutions. These elementary operations have been realized by using electro-optic modulators, various kinds of dispersive devices, and other means.

Based on such concepts Lohmann and Mendlovic (1994b) have discussed the temporal optical implementation of the fractional Fourier transform. Dragoman and Dragoman (1998) discuss the extension of these approaches to a variety of other transforms. Ozaktas and Nuss (1996) describe an approach for realizing general space-variant linear systems, which can also be used to realize temporal optical fractional Fourier transforms.

9.11.2 Digital optical implementation of the transform

In this chapter we have exclusively limited our attention to analog optical systems. Digital optical implementation of the fractional Fourier transform is possible within the framework described in Ozaktas and Miller 1996.

9.11.3 Optical implementation of two-dimensional transforms

Although most of this chapter employed one-dimensional notation for simplicity, all of the results can be easily generalized to two-dimensional systems, especially when the transform order is the same in both dimensions. The more general case where the transform is still separable but the transform orders are different in the two dimensions was first discussed in Sahin, Ozaktas, and Mendlovic 1995 and Mendlovic and others 1995b. Design equations for dynamically adjustable fractional Fourier transformers with different orders in the two dimensions have been given in Erden and others 1997. A general and comprehensive approach to the design of optical fractional Fourier transforming systems, as well as general quadratic-phase systems, is found in Sahin 1996 and Sahin, Ozaktas, and Mendlovic 1998. In these works several alternative systems giving as much control as possible over the various parameters (such as the transform orders, scale, and system complexity) are presented. The optical and digital implementation of the nonseparable two-dimensional fractional Fourier transform is discussed in Sahin 1996 and Sahin, Kutay, and Ozaktas 1998.

9.11.4   Optical interpretation and implementation of complex-ordered transforms

Complex-ordered transforms have received specific attention in an optical context in Bernardo and Soares 1994a, 1996, Shih 1995a, Bernardo 1997, and Hua, Liu, and Li 1997e. There seems to be much that is unexplored in this area.

9.11.5   Incoherent optical implementation of the transform

Incoherent optical systems are known to be advantageous under certain circumstances. Since incoherent optical systems can deal only with positive real quantities represented as intensities, it is necessary to introduce a scheme to encode complex quantities, or to introduce and work with a real version of the fractional Fourier transform. The latter approach is taken in Mendlovic and others 1995c, where a real fractional Fourier transform is introduced, and an incoherent optical implementation based on a shearing interferometer is proposed. When equal path lengths are chosen for both branches of the interferometer, an ordinary real cosine transform is obtained. By changing the lengths of the branches so that they are no longer equal, it is possible to obtain fractional transforms of different orders, subject to a certain residual phase factor which is often insignificant.

9.11.6   Applications to systems with partially coherent light

Let us rewrite the one-dimensional version of equation 7.154 relating the output mutual intensity $\hat{R}_{\hat{g}\hat{g}}(x_1, x_2)$ to the input mutual intensity $\hat{R}_{\hat{f}\hat{f}}(x_1, x_2)$:

$$\hat{R}_{\hat{g}\hat{g}}(x_1, x_2) = \iint \hat{R}_{\hat{f}\hat{f}}(x_1', x_2')\hat{h}(x_1, x_1')\hat{h}^*(x_2, x_2') \, dx_1' \, dx_2'. \tag{9.188}$$

Here $\hat{h}(x_1, x_1')$ is the kernel characterizing the system. We have seen in this chapter that for very general classes of optical systems, this kernel can be cast in the form of a fractional Fourier transform kernel by choosing appropriate scale factors and/or spherical reference surfaces. It follows that for such systems, the output mutual intensity is essentially the double fractional Fourier transform of the input mutual intensity (Erden, Ozaktas, and Mendlovic 1996a). In the one-dimensional case we have essentially a two-dimensional transform, and in the two-dimensional case we have essentially a four-dimensional transform. We say "essentially" because the conjugate appearing on the second appearance of $\hat{h}$ will flip a sign associated with the second variable, an action with no more than minor bookkeeping consequences.

We conclude that the fractional Fourier transform is just as useful in dealing with partially coherent light as it is in dealing with fully coherent light. Just as it describes the propagation of the complex amplitude of fully coherent light, it describes the propagation of the mutual intensity of partially coherent light. The propagation of second-order statistical properties of a wave is further discussed in Lohmann, Mendlovic, and Shabtay 1999.

In section 10.4.5 we mention the use of fractional Fourier transform based filtering configurations for the synthesis of signals with desired second-order statistics, with direct application to the synthesis of optical wave fields with desired mutual intensities.

Further applications of the fractional Fourier transform to the study of partially coherent light and systems employing partially coherent light include Yoshimura and Iwai 1997, 1998, Simon and Mukunda 1998, and Tu and Tamura 1998.

9.11.7 Other applications of the transform in optics

An important application of the transform is in the area of optical beam shaping (Zalevsky, Mendlovic, and Dorsch 1996, Cong, Chen, and Gu 1998a, Cong and Chen 1999, Zhang and others 1998b, 1999b), where a desired beam profile is obtained by employing various kinds of (possibly phase-only) filters in fractional Fourier planes. The synthesis of beams with desired mutual intensities has also been considered (Erden, Ozaktas, and Mendlovic 1996b, Zalevsky, Mendlovic, and Ozaktas 2000).

Dragoman (1996) has discussed applications of the fractional Fourier transform to beam characterization. The relationship between the moments of an optical beam and the transform are discussed in Dragoman and Dragoman 1997b. The transform has also been used in the study of shape-invariant propagation of optical beams (Simon and Mukunda 1998).

The use of the fractional Fourier transform in lens design problems has been suggested by Dorsch and Lohmann (1995). Image quality parameters such as the Strehl ratio and the optical transfer function have been analyzed in terms of the transform in Granieri, Sicre, and Furlan 1998. The transform has been used in theoretical studies of the human eye (Pons and others 1999). The fractional Fourier transform of fractal functions and applications to the propagation of fractal fields has been the subject of papers by Alieva (1996b) and Alieva and Agullo-Lopez (1996a). The application of the transform to holography is discussed in Wolf and Rivera 1997. Here we also mention that the fractional Fourier transform has also found use in the modeling of atomic force microscopy (Dragoman and Dragoman 1997a).

Abe and Sheridan (1997) show how to optically estimate the transform order of an optical fractional Fourier transforming system whose parameters are not known. Since a broad class of optical systems can be interpreted as fractional Fourier transformers, this provides a means for estimation of the physical parameters of such optical systems or media. Abe and Sheridan (1995a) have also dealt with characterization of deviations or imperfections of optical systems, including fractional Fourier transforming systems.

9.11.8 Practical considerations for implementing the transform

The first actual optical implementation of the fractional Fourier transform was reported in Bitran and others 1995. In practical and experimental situations it is desirable to have a scheme or setup which allows one to realize fractional Fourier transforms of different orders easily. Such systems have been proposed by Lohmann (1995). A modular lens system for optically realizing fractional Fourier transforms of many different orders has been proposed by Dorsch (1995). An optical fractional Fourier transformer whose order can be varied dynamically in real time has been proposed in Erden and others 1997. Other works dealing with the practical aspects of the optical implementation of the transform, including ways to realize transforms with variable parameters, include Mendlovic and others 1996a, Andrés and others 1997, Granieri and others 1997, Kong and Lü 1997, and Lohmann and others 1998. Jiang (1995), Mendlovic and others (1996a), and Zhang and others (1998a, 1999a) discuss how several one-dimensional fractional Fourier transforms of different orders can be realized simultaneously with a two-dimensional optical system. The planar integrated-optical implementation of the transform has been demonstrated in Song and others 1997a, b and the fractional Fourier transforming properties of a hemispherical-rod microlens have been discussed in Dragoman and others 1998. Correction of aberrations of fractional Fourier transforming systems has been dealt with in Wolf and Krötzsch 1999.

In the rest of this section we will discuss an issue which is of particular relevance for experimental implementation. Although the canonical configurations of section 9.4.3 represent the simplest realizations of the fractional Fourier transform using lenses and sections of free space, they have one drawback. Referring to equations 9.67 and 9.68, we see that both d and f are determined once the scale s and the order a are specified. Since we would usually wish to keep the scale constant for different orders, this means that we have to use a different lens for each order. What would be desirable from a practical viewpoint is to be able to always use the same lenses and be able to realize transforms of different orders with the same scale, by simply adjusting the separations of the lenses and the input and output planes. The problem can be posed as follows. Find a sequence of fixed lenses and variable sections of free space such that the product of their matrices is equal to

$$\begin{bmatrix} (s_2/s_1)\cos(a\pi/2) & (s_1 s_2)\sin(a\pi/2) \\ (-1/s_1 s_2)\sin(a\pi/2) & (s_1/s_2)\cos(a\pi/2) \end{bmatrix}, \tag{9.189}$$

where s_1, s_2 are fixed and a is variable. Many solutions to this problem can be found and employed. Here we will mention an approach which, while not necessarily the most efficient, is conceptually simple (Lohmann 1995). Concentrating on the case $s_2 = s_1 = s$, we can simply employ canonical configuration type I with the variable lens simulated by sandwiching an adjustable section of free space between two ordinary Fourier transform stages (a section of free space between two Fourier transform stages behaves like a lens).

9.11.9   Other fractional operations and effects in optics

Fractional operations and effects have always been an attractive, elegant, and fruitful area of study. We have already discussed other fractional operations and transforms in section 4.12.6, to which the reader should refer for predominantly mathematics or signal processing oriented references. Here we discuss works dealing with fractional operations, transforms, and effects in optics and wave propagation. An extensive review is Lohmann, Mendlovic, and Zalevsky 1998.

The fractional derivative has found applications in optics; for instance, see Lancis and others 1997. The applications of fractional calculus (meaning fractional differentiation and integration) to electromagnetic theory were discussed by Engheta (1995, 1996a, b, c, 1997, 1998).

The optical implementation of the two-dimensional fractional Hilbert transform was discussed in Lohmann, Tepichín, and Ramírez 1997, and the fractional Gabor transform was discussed in Zhang and others 1997.

Shamir and Cohen (1995) discuss the "roots" of optical systems characterized as quadratic-phase systems, and show how they can be optically realized. (For example, the fifth root of a system is a system which, when repeated five times, is equivalent to the original system.) This issue is more carefully discussed in Simon and Wolf 2000.

A fractional effect in optics which deserves special mention is what has been referred to as the *fractional Talbot effect* (Testorf and Ojeda-Castañeda 1996). In the Talbot effect, also known as the self-imaging phenomenon (Patorski 1989), self-images of an input object are observed at the planes $z = jd_0$ for integer j, where d_0 is a characteristic distance. In Lohmann and Mendlovic 1992a it was shown that j could also assume certain rational values. In Lohmann and Mendlovic 1992b an example of an optical fractional Fourier transform was given without the authors noticing. Here a self-imaging configuration containing an odd number of identical stages is discussed. Noting the fact that self-imaging corresponds to a transform of order $a = 4$, each of the identical stages could be considered as a fractional Fourier transformer with order $a = 4/j$, where j is an odd integer. (Ozaktas and Mendlovic 1993a, Mendlovic and Ozaktas 1993a)

Granieri, Trabocchi, and Sicre (1995) and Bernardo (1997) discuss the relationships between the fractional Fourier transform and the self-imaging phenomenon. Other recent works on self-imaging, self-transforms, and self-transform objects include Arrizón, Ibarra, and Ojeda-Castañeda 1996 and Hua and Liu 1997.

9.12 Historical and bibliographical notes

The first discussion of the optical fractional Fourier transform in the context of quadratic graded-index media appeared in Mendlovic and Ozaktas 1993a, b, Ozaktas and Mendlovic 1993a, b, and Özaktaş and Mendlovic 1993. A simpler derivation of the fractional Fourier transforming property of quadratic graded-index media was later given in Ozaktas and others 1994a. Lohmann (1993a, b) discussed optical fractional Fourier transforming configurations composed of lenses and sections of free space. The relationship between these approaches was discussed and their equivalence demonstrated in Ozaktas, Mendlovic, and Lohmann 1993 and Mendlovic, Ozaktas, and Lohmann 1993, 1994a. Further discussion of the fractional Fourier transforming property of quadratic graded-index media and some of its applications may be found in Alieva and Agulló-López 1995. A more recent work on this subject is Zalevsky and Mendlovic 1997.

The Hermite-Gaussian expansion approach to propagation in free space, discussed in section 9.6, first appeared in Ozaktas and Mendlovic 1994. The relationship of the fractional Fourier transform to Fresnel diffraction, and fractional Fourier transform relations between spherical reference surfaces, as they appear in sections 9.3, 9.4, and 9.6, were first given in Ozaktas and Mendlovic 1994, 1995a. Closely related results were independently arrived at by Pellat-Finet and Bonnet (1994) and Pellat-Finet (1994, 1995). The close relationship between Fresnel diffraction and the fractional Fourier transform was also noted in Alieva and others 1994 and Gori, Santarsiero, and Bagini 1994. Abe and Sheridan 1995b and Alieva and others 1995 are comments on Alieva and others 1994. The discussion of first-order optical systems in section 9.7 was first presented in Ozaktas and Erden 1997 and the discussion of Fourier optical systems in section 9.8 first appeared in Ozaktas and Mendlovic 1996a.

Other works dealing with the relationship between the fractional Fourier transform and Fresnel propagation, various aspects of its optical implementation, and its applications to quadratic-phase and Fourier optical systems include Agarwal and Simon 1994, Bernardo and Soares 1994a, b, Granieri, Trabocchi, and Sicre 1995, Liu, Wu, and Li 1995, Liu and others 1995, Alieva and Agullo-Lopez 1996b, Bernardo 1996, Deng and others 1996, Hua, Liu, and Li 1997a, b, d, Jiang, Lu, and Zhao 1997, Lü and Kong 1997, Lü, Kong, and Zhang 1997, Pellat-Finet and Torres 1997, Dragoman 1998, Jagoszewski 1998a, b, Liu, Wu, and Fan 1998, and Sheppard 1998.

An early review of the fractional Fourier transform and its applications in optics is Ozaktas 1994. Later reviews and tutorials include Ozaktas and Mendlovic 1995b, 1996b, Mendlovic 1999, Mendlovic and Ozaktas 1999, and Ozaktas, Kutay, and Mendlovic 1998, 1999. Section 9.2 is based on Ozaktas, Kutay, and Mendlovic 1998, 1999. An entertaining and instructive elementary introduction to the Fourier and fractional Fourier transforms in an optical context is Lohmann 1994.

Although recent activity on the fractional Fourier transform in optics stems from publications made in the early 1990s, the relationship of the transform to optics was previously suggested in a few earlier isolated and neglected works. Ludwig (1988) considered a section of free space followed by a lens followed by another section of free space and showed that it produced the fractional Fourier transform mathematically defined by Condon (1937). He also noted that the transform indeed reduces to the special cases of imaging and Fourier transformation when the separations and focal lengths are chosen appropriately. Another neglected yet insightful work is that by Khare (1974), who pointed out potential applications of the transform to image recovery from intensity measurements in multiple planes (see section 9.10). Unfortunately, these very short works did not develop the insights they contained, nor were they followed up by other researchers or the authors themselves, and they were rediscovered only very recently. Khare cites the works of Wiener (1929), Condon (1937), and Patterson (1959), and also a few other works which seem to have employed related concepts in physical contexts. It seems that the physical insights regarding the relevance of such transforms to wave and beam propagation or diffraction can be traced back to Patterson 1959. Another work which is relevant in this context is Osipov 1992 (S. M. Sitnik, private communication 1999).

Of course, since fractional Fourier transforms are a special case of linear canonical transforms, the works of Bastiaans and others dating back to the late 1970s are also of great relevance (see chapter 8). However, in these works the fractional Fourier transform is not recognized or named as such and is not given any special attention, and the many results and applications developed in the 1990s and partly covered in this book do not appear.

10

Applications of the Fractional Fourier Transform to Filtering, Estimation, and Signal Recovery

10.1 Introduction

In chapter 4 we saw that the fractional Fourier transform leads to a generalization of time (or space) and frequency domains, which are central concepts in signal analysis and processing, as well as other areas. The continuum of *fractional Fourier domains* correspond to oblique axes in the time-frequency plane, with the ordinary time and frequency domains being two special cases. The intimate relationship of the fractional Fourier transform to time-frequency representations, as well as the central importance of the Fourier transform, suggest that the fractional Fourier transform should have many applications in signal analysis and processing.

In this chapter we will introduce the concept of *filtering in fractional Fourier domains*, and show how flexible general filtering configurations or "filter circuits" can be realized based on this concept. Within this framework one can flexibly and efficiently realize general shift-variant linear filters for a variety of applications. The focus of this chapter will be applications of the recovery or synthesis type. In signal recovery we typically have an observation g which is related to f through some system, and it is desired to estimate f. The observation g can be measured with some finite accuracy determined by noise or other errors. Signal restoration problems are signal recovery problems where g is a distorted, noisy, or otherwise degraded version of f. Signal reconstruction problems are signal recovery problems where g is some—perhaps quite complicated—mapping of f to another space such that g has no direct resemblance to f (the most common examples are reconstruction from projections problems). In signal synthesis a desired output signal g is specified, and we must choose the input of the system f so that g is observed at the output. All of these inverse problems are mathematically similar. In each case the problem is to estimate f from knowledge of g, also using any available prior knowledge regarding the nature of f and/or the nature and statistics of the measurement errors or noise, or the specified tolerance.

In many areas of science and engineering there are a number of analytically solvable basic problems that are of central importance. For instance, in quantum mechanics the harmonic oscillator and hydrogen atom are such problems. In signal processing, *optimal Wiener filtering* and *matched filtering* are two basic problems of such central importance. The core of this chapter focuses on optimal Wiener filtering in fractional Fourier domains. Chapter 11, on the other hand, deals with matched filtering with the fractional Fourier transform and discusses applications to pattern recognition. We believe that these two chapters represent only a fraction of the possible applications of the fractional transform in signal analysis and processing.

This chapter also contains a number of other basic results on signal analysis and processing, and a discussion of the concept of multiplexing in fractional domains. Of particular interest is the fractional Fourier domain decomposition (FFDD), in which a matrix is decomposed into fractional Fourier domain filtering operations. A procedure called pruning, analogous to truncation of the singular-value decomposition (SVD), underlies a number of applications of this decomposition.

Sections 10.2, 10.3, and 10.4 discuss single-domain, multistage, multichannel, and generalized filtering configurations and their applications. Section 10.5 discusses the elementary operations of convolution, multiplication, compaction, and filtering in fractional Fourier domains. Section 10.6 provides the derivation of the optimal single-domain filter and can be considered a continuation of section 10.2. Section 10.7 discusses how to determine the optimal filters in multichannel and multistage configurations and can be considered a continuation of section 10.3. Section 10.8 discusses the fractional Fourier domain decomposition, which is related to multichannel filtering. Section 10.9 discusses repeated filtering in the ordinary time and frequency domains, which is related to multistage filtering. Finally, section 10.10 discusses multiplexing in fractional Fourier domains.

10.2   Optimal Wiener filtering in fractional Fourier domains

In a large class of signal processing applications, signals we wish to recover are degraded by a known distortion, blur, and/or by noise, and the problem is to reduce or eliminate these degradations. The solution of such problems depends on the observation model and the objectives, as well as the prior knowledge available about the desired signal, degradation process, and noise. A commonly used observation model is

$$g(u) = \int h_{\mathrm{d}}(u, u') f(u')\, du' + n(u), \tag{10.1}$$

where $h_{\mathrm{d}}(u, u')$ is the kernel of the linear system that distorts or blurs the desired signal $f(u)$, and $n(u)$ is an additive noise term. The problem is to find an estimation operator represented by the kernel $h(u, u')$, such that the estimated signal

$$f_{\mathrm{est}}(u) = \int h(u, u') g(u')\, du' \tag{10.2}$$

optimizes some criterion of fidelity. Despite its limitations as a criterion of fidelity, one of the most commonly used objectives is to minimize the mean square error σ_{err}^2 defined as

$$\sigma_{\mathrm{err}}^2 = \left\langle \int |f_{\mathrm{est}}(u) - f(u)|^2\, du \right\rangle, \tag{10.3}$$

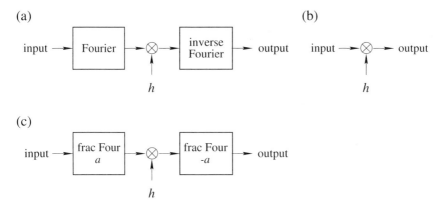

Figure 10.1. (a) Filtering in the frequency domain; (b) filtering in the time (or space) domain; (c) filtering in the ath order fractional Fourier domain (Erden 1997, Kutay and others 1998a, b, Erden, Kutay, and Ozaktas 1999a).

where the angle brackets denote an ensemble average. The estimation or recovery operator minimizing σ_{err}^2 is known as the optimal Wiener filter. The kernel $h(u, u')$ of this optimal filter satisfies the following relation (Lewis 1986):

$$R_{fg}(u, u') = \int h(u, u'') R_{gg}(u'', u') \, du'' \qquad \text{for all } u, u', \tag{10.4}$$

where $R_{fg}(u, u')$ is the statistical cross-correlation of $f(u)$ and $g(u)$, and $R_{gg}(u, u')$ is the statistical autocorrelation of $g(u)$. Needless to say, similar considerations apply to synthesis problems.

When $h_{\text{d}}(u, u')$ represents a time-invariant system and the input and noise processes are stationary, the optimal Wiener filter turns out to be time-invariant, and thus can be expressed as a convolution and implemented effectively with a multiplicative filter in the ordinary Fourier domain by using the fast Fourier transform algorithm (figure 10.1a). In this case the mean square error is not defined according to equation 10.3, which would be infinite, but as $\sigma_{\text{err}}^2 = \langle |f_{\text{est}}(u) - f(u)|^2 \rangle$. However, equation 10.4 remains valid, leading to the following frequency-domain filter $H(\mu)$ in the case of zero-mean noise which is independent from the signal:

$$H(\mu) = \frac{H_{\text{d}}^*(\mu) S_{ff}(\mu)}{|H_{\text{d}}(\mu)|^2 S_{ff}(\mu) + S_{nn}(\mu)}, \tag{10.5}$$

where $S_{ff}(\mu)$ and $S_{nn}(\mu)$ are the power spectral densities of $f(u)$ and $n(u)$ respectively.

In the more general case where $h_{\text{d}}(u, u')$ represents a system which is not necessarily time-invariant, or for nonstationary processes, the optimal Wiener filter will not be time-invariant and cannot be expressed as a multiplicative filter in the ordinary Fourier domain (a convolution in the time domain). In this case there is no fast algorithm for obtaining $f_{\text{est}}(u)$ as given by equation 10.2. We can still seek the frequency-domain multiplicative filter leading to the smallest error, but in general this will not provide a satisfactory result.

The dual of filtering in the ordinary frequency domain is filtering in the time (or space) domain (figure 10.1b). This operation simply corresponds to multiplying the original function with a filter or mask function. Filtering in the ordinary time (space) or frequency domains can be generalized to filtering in the ath order fractional Fourier domain (figure 10.1c). For $a = 1$ this reduces to the ordinary multiplicative Fourier domain filter and for $a = 0$ it reduces to time (space) domain multiplicative filtering.

To understand the basic motivation for filtering in fractional Fourier domains, consider figure 10.2, where the Wigner distributions of a desired signal and an undesired noise term are superimposed. We observe that the signal and noise overlap in both the 0th and 1st domains, but they do not overlap in the 0.5th domain (consider the projections onto the $u_0 = u$, $u_1 = \mu$, and $u_{0.5}$ axes). Although it is not possible to eliminate the noise in the time or frequency domains, we can eliminate it easily by using a simple amplitude mask in the 0.5th domain.

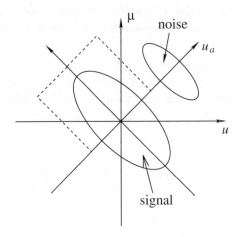

Figure 10.2.   Filtering in a fractional Fourier domain as observed in the time- or space-frequency plane. $a = 0.5$ as drawn. (Ozaktas and others 1994a)

Inspired by this example, we now formulate the problem of obtaining an estimate $f_{one}(u)$ of $f(u)$ by using the ath order fractional Fourier domain filtering configuration shown in figure 10.1c, rather than restricting ourselves to the configurations shown in figure 10.1a and b. We first take the ath order fractional Fourier transform of the observed signal $g(u)$, then multiply the transformed signal with a multiplicative filter function denoted by $h(u)$, and finally take the inverse ath order fractional Fourier transform of the resulting signal to obtain our estimate $f_{one}(u)$:

$$f_{one}(u) = \left[\mathcal{F}^{-a} \, \Lambda_h \, \mathcal{F}^a \right] g(u) = \mathcal{T}_{one} \, g(u), \tag{10.6}$$

where $\mathcal{F}^a$ is the ath order fractional Fourier transform operator, Λ_h denotes the operator corresponding to multiplication by the filter function $h(u)$, and $\mathcal{T}_{one}$ is the operator representing the overall filtering configuration. Since the fractional Fourier transform has efficient digital and optical implementations, the cost of fractional Fourier domain filtering is approximately the same as the cost of ordinary Fourier domain filtering. The remaining problem is to find the optimal multiplicative filter function $h(u)$ that minimizes the mean square error defined in equation 10.3 for the filtering configuration represented by equation 10.6. For a given transform order a, the optimal filter function $h(u)$ can be found analytically using the orthogonality principle or the calculus of variations:

$$h(u_a) = \frac{\iint K_a(u_a, u)K_{-a}(u_a, u')R_{fg}(u, u')\, du'\, du}{\iint K_a(u_a, u)K_{-a}(u_a, u')R_{gg}(u, u')\, du'\, du},\qquad(10.7)$$

where the statistical cross-correlation and autocorrelation functions $R_{fg}(u, u')$ and $R_{gg}(u, u')$ can be obtained from the functions $R_{ff}(u, u')$ and $R_{nn}(u, u')$, which are assumed to be known. Equation 10.7 will be derived in section 10.6. The corresponding mean square error can be calculated from equation 10.3 for different values of a, and the value of a resulting in the smallest error can be determined. The optimal fractional Fourier domain filter will result in an error which is smaller or equal than the optimal ordinary Fourier domain filter, since ordinary Fourier domain filtering is a special case corresponding to $a = 1$. In some cases the value of a resulting in the smallest error may turn out to be $a = 1$, but in general it will turn out to be some other value of a.

One particular case where single-domain fractional Fourier filtering is especially advantageous is when the distortion or noise is of a chirped nature. Such situations are encountered in many real-life applications. In particular, two-dimensional fractional Fourier domain filtering finds many applications in optical systems. This is because the types of distortion for which fractional Fourier domain filtering achieves greatest benefits arise naturally in optical systems. For instance, a major problem in reconstruction from holograms is the elimination of twin-image noise. Since this noise is essentially a modulated chirp signal, it can be dealt with by using fractional Fourier domain filtering. Other applications arise in holography where different chirp rates involved in in-line holograms can be used for extraction of three-dimensional object location information. The separation of these chirps directly yields location information (Onural and Özgen 1992). Yet another example is the correction of the effects of point or line defects found on lenses or filters in optical systems, which appear at the output plane in the form of chirp artifacts. Another application arises in synthetic aperture radar which employs chirps as transmitted pulses, so that the measurements are related to the terrain reflectivity function through a chirp convolution. This process results in chirp-type disturbances caused by moving objects on the terrain, which should be removed if high-resolution imaging is to be achieved. As a final example, we mention broadband interference excision in spread-spectrum communication systems (Akay and Boudreaux-Bartels 1998e).

More generally, simulations show that filtering in a single fractional domain works better for certain kinds of distortions and signal and noise statistics in comparison to others. The presence of time-varying distortions or blurs and nonstationary signals suggest that single-domain fractional Fourier filtering may be of use, but it does not guarantee significant improvements in every case. Nevertheless, applications of single-domain filtering are certainly not limited to chirp-type signals. The method is especially effective with such signals because they can be very easily eliminated or separated by notch filtering in a single fractional Fourier domain. Other types of signals can also be successfully dealt with as long as it is possible to sufficiently separate the desired and undesired parts with a single line in the time-frequency plane,

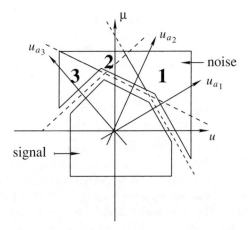

Figure 10.3. Filtering in multiple fractional Fourier domains as observed in the time- or space-frequency plane (Erden 1997, Erden, Kutay, and Ozaktas 1999a, Kutay 1999).

as motivated in figure 10.2.

Both one-dimensional and two-dimensional examples of the improvements obtained by using single-domain fractional Fourier filtering will be given in section 10.4. What is most important is that, whether implemented digitally or optically, the cost of single-domain filtering is about the same, irrespective of the order a. This includes the case $a = 1$ corresponding to the ordinary Fourier transform. Thus any improvements that come with the flexibility provided by the parameter a come without additional cost. In certain cases, however, the amount of improvement that can be achieved in this manner without additional cost may not be sufficient. In the next section we turn our attention to systems whose cost is greater than ordinary Fourier domain filtering, but which nevertheless provide good value in terms of performance for cost.

Single-domain filtering will be revisited in sections 10.5 and 10.6. A further generalization of the concept of filtering in fractional Fourier domains is filtering in linear canonical transform domains (Barshan, Kutay, and Ozaktas 1997). Here the linear canonical transform of the input is multiplied with a suitable filter function and then inverse transformed to return to the original domain.

10.3 Multistage, multichannel, and generalized filtering configurations

10.3.1 Introduction

We begin by considering figure 10.3, where the Wigner distributions of a desired signal and an undesired noise term are superimposed. This figure is similar to figure 10.2, but in this case we cannot find a single fractional Fourier domain where the noise can be separated from the signal. If we consider the projections of these Wigner distributions, we see that the signal and noise overlap in all fractional Fourier domains. Thus, we employ the following strategy. We first transform to the domain represented by u_{a_1} and eliminate that part of the noise marked by **1** by using a simple amplitude mask. Then we transform to the domain represented by u_{a_2} and eliminate that part of the noise

(a)

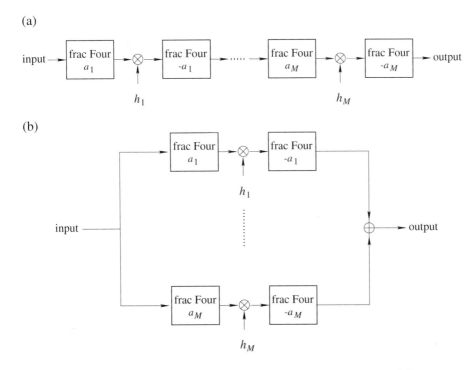

(b)

Figure 10.4. (a) Multistage filtering in fractional Fourier domains; (b) multichannel filtering in fractional Fourier domains (Kutay and others 1998a, b).

marked by **2**. Finally, we transform to the domain represented by u_{a_3} and eliminate the rest of the noise marked by **3**.

Motivated by this example, we introduce the concepts of multistage (repeated or serial) and multichannel (parallel) filtering in fractional Fourier domains. These systems consist of M single-domain fractional Fourier filtering stages in series or in parallel (figure 10.4). $M = 1$ corresponds to single-domain filtering in both cases. In the multistage system shown in figure 10.4a, the input is first transformed into the a_1th domain where it is multiplied by a filter $h_1(u)$. The result is then transformed back into the original domain and the same process is repeated M times consecutively. This amounts to sequentially visiting the domains $a_1, a_2, a_3, \ldots$, and applying a filter in each. On the other hand, the multichannel system consists of M single-domain blocks in parallel (figure 10.4b). For each channel k, the input is transformed to the a_kth domain, multiplied with a filter $h_k(u)$, and then transformed back. If these configurations are used to obtain an estimate $f_{\text{ser}}(u)$ or $f_{\text{par}}(u)$ of $f(u)$ in terms of $g(u)$, we have

$$f_{\text{ser}}(u) = \left[\mathcal{F}^{-a_M} \Lambda_{h_M} \cdots \mathcal{F}^{a_2 - a_1} \Lambda_{h_1} \mathcal{F}^{a_1} \right] g(u) = \mathcal{T}_{\text{ser}} \, g(u), \qquad (10.8)$$

$$f_{\text{par}}(u) = \left[\sum_{k=1}^{M} \mathcal{F}^{-a_k} \Lambda_{h_k} \mathcal{F}^{a_k} \right] g(u) = \mathcal{T}_{\text{par}} \, g(u), \qquad (10.9)$$

where $\mathcal{F}^{a_k}$ represents the a_kth order fractional Fourier transform operator, Λ_{h_k}

denotes the operator corresponding to multiplication by the filter function $h_k(u)$, and $\mathcal{T}_{\mathrm{ser}}$, $\mathcal{T}_{\mathrm{par}}$ the operators representing the overall filtering configurations. Both of these equations reduce to equation 10.6 for $M = 1$. The filter functions and transform orders should of course be chosen so as to optimize some criterion of fidelity or to minimize some measure of error, such as σ_{err}^2, as we will discuss more carefully below.

In the above paragraph, we posed the multistage and multichannel configurations as filtering systems for optimal signal estimation. They can also be used for cost-efficient synthesis of desired linear systems, transforms, or mappings, such as geometric distortion compensators, optical beam shapers and synthesizers, as well as linear recovery operators. In this case, given a general linear system $\mathcal{H}$—characterized by the kernel $h(u, u')$—which we wish to implement, we try to find the optimal orders a_k and filters $h_k(u)$, such that the overall linear operator $\mathcal{T}_{\mathrm{one}}$, $\mathcal{T}_{\mathrm{ser}}$, or $\mathcal{T}_{\mathrm{par}}$ (as given by equation 10.6, 10.8, or 10.9), is as close as possible to $\mathcal{H}$, according to some specified criterion (such as minimum Frobenius norm of the difference of the kernels).

It is helpful to clearly distinguish the two different ways in which these configurations can be used in a given application:

1. Starting with a signal recovery or synthesis problem, we determine the optimal linear estimation operator $\mathcal{H}$ using any models and methods considered appropriate, for instance by using equation 10.4. Or there may simply be a linear system $\mathcal{H}$ with kernel $h(u, u')$ we wish to implement for some other purpose. Then we seek the transform orders a_k and filters $h_k(u)$ such that $\mathcal{T}_{\mathrm{one}}$, $\mathcal{T}_{\mathrm{ser}}$, or $\mathcal{T}_{\mathrm{par}}$ is as close as possible to $\mathcal{H}$ according to some specified criterion.
2. We take equation 10.6, 10.8, or 10.9 as a constraint on the form of the linear estimation operator to be employed. Given a specific optimization criterion, such as minimum mean square error, we find the optimal values of a_k and $h_k(u)$ subject to this constraint.

Both approaches will be illustrated in section 10.4. The problem of determining the optimal orders and filters will be discussed in section 10.7.1.

10.3.2 Cost-performance trade-off

Multistage and multichannel filtering systems are a subclass of the class of general linear systems for which the output is related to the input through equation 10.2. In general, linear systems have N^2 degrees of freedom, where N is the time-bandwidth (or space-bandwidth) product of the signals. Obtaining $f_{\mathrm{est}}(u)$ in terms of $g(u)$ normally takes $O(N^2)$ time on a serial computer, and its optical implementation requires components whose space-bandwidth products are N^2, unless the system kernel $h(u, u')$ has some special structure which can be exploited. Shift-invariant (time- or space-invariant) systems are also a subclass of general linear systems whose system kernels $h(u, u')$ can always be expressed in the form $h(u, u') = h(u - u')$. They are a very restricted subclass with only N degrees of freedom, but can be implemented in $O(N \log N)$ time on a serial computer and with optical components whose space-bandwidth products are N.

We may think of shift-invariant systems and general linear systems as representing two extremes in a cost-performance trade-off. Shift-invariant systems exhibit low cost and low performance, whereas general linear systems exhibit high cost and high performance. Sometimes use of shift-invariant systems may be inadequate, but at the same time use of general linear systems may be overkill and prohibitively costly. Multistage and multichannel fractional Fourier domain filtering configurations interpolate between these two extremes, offering greater flexibility in trading off between cost and performance. In a given application, this flexibility may allow us to realize a system which is acceptable in terms of both cost and performance.

Both multistage and multichannel filtering configurations have at most $MN + M$ degrees of freedom. Their digital implementation will take $O(MN \log N)$ time since the fractional Fourier transform can be implemented in $O(N \log N)$ time (chapter 6). Optical implementation will require an M-stage or M-channel optical system, each with space-bandwidth product N (chapter 9). These configurations interpolate between general linear systems and shift-invariant systems in terms of both cost and flexibility. If we choose M to be small, cost and flexibility are both low; $M = 1$ corresponds to single-domain filtering. If we choose M larger, cost and flexibility are both higher; as M approaches N, the number of degrees of freedom approaches that of a general linear system. The increase in flexibility as M increases will often translate into a reduction of the estimation error σ_{err}^2, since we have more degrees of freedom to optimize over. Thus the trade-off between cost and flexibility as we vary M is also a trade-off between cost and accuracy.

There are two ways to look at the greater flexibility that comes with increasing M. First, larger values of M allow us to realize a broader class of linear systems since there are more degrees of freedom available. For a given value of M, we can realize a certain subset of all linear systems exactly or to some specified degree of accuracy. As M increases, the subset in question becomes larger and larger. Second, and perhaps more useful, is to consider the problem of approximating a given linear system. For a given value of M, we can approximate this system with a certain degree of accuracy (or error). For instance, a shift-invariant system can be realized with zero error with $M = 1$. In general, there will be a finite error for each value of M. As M is increased, the error will usually decrease (but never increase). Thus, in the context of a particular application or problem, we can seek the minimum value of M which results in the desired error, or the minimum error that can be achieved for a given value of M. This amounts to seeking the best performance for given cost, or least cost for given performance. Such cost-performance combinations are referred to as *Pareto optimal* cost-performance combinations. The locus of such Pareto optimal points constitutes the cost-performance trade-off curve. Given the relative emphasis we place on cost and performance, we can choose the most desirable point on this trade-off curve. The filtering configurations being discussed have been found to be particularly useful for relatively small values of M, on the low-cost low-flexibility side of the trade-off curve (section 10.4).

The cost-accuracy trade-off is illustrated in figure 10.5, where we have plotted both the cost and the error as functions of the number of filters M for a hypothetical application. The two plots show how the cost increases and the error decreases as we increase M. Eliminating M from these two graphs leads us to a graph of error versus cost.

Naturally, the value of M required to attain a given accuracy or error will be smaller for systems with kernels exhibiting greater regularity, or other more subtle forms of intrinsic structure. In such cases, direct implementation of the linear system

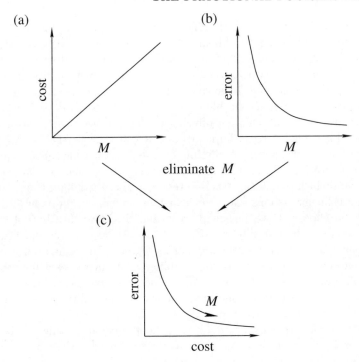

Figure 10.5. (a) Cost versus M, (b) error versus M, (c) error versus cost (Kutay 1999).

in $O(N^2)$ time or with $O(N^2)$ space-bandwidth product is clearly inefficient. The inherent structure can be exploited on a case-by-case basis through ingenuity or invention; most efficient algorithms and optical implementations are obtained in this manner. In contrast, use of multistage and multichannel filtering configurations provides a systematic way of obtaining an efficient implementation which does not require ingenuity on a case-by-case basis. This approach would be especially useful when the regularity or structure of the matrix is not simple or immediately apparent. A distinct circumstance in which this approach may be beneficial is when it is sufficient to implement the linear system with limited accuracy, and it is desirable to use the lowest-cost system compatible with that accuracy. This may be the case when some other component of the overall system limits the accuracy to a lower value anyway, or simply when the application itself demands limited accuracy.

Multistage and multichannel filtering configurations offer improvements in both digital and optical systems. We expect them to be especially beneficial in optical systems due to a number of reasons. First, the monetary cost of optical components, such as lenses, is a strong function of their space-bandwidth product. Thus, direct implementation of linear systems using matrix-vector product or multifacet architectures may be totally infeasible (Mendlovic and Ozaktas 1993c, Ozaktas, Brenner, and Lohmann 1993, Ozaktas and Mendlovic 1993c). For instance, processing a 256×256 image would require optical components with a very large space-bandwidth product of 256^4. Furthermore, the intrinsic amplitude or intensity level accuracy of analog optical systems is usually limited to quite modest dynamic ranges. Given this limited accuracy, it is pointless to try to implement the desired linear system using an expensive scheme which could in principle accommodate much greater accuracies (such as a conventional matrix-vector product architecture which maps out an exact matrix-vector product). The configurations presented allow one to approximate the desired linear system to a degree of accuracy which just matches the intrinsic accuracy of analog optical systems, in a manner which reduces the cost as much as possible. When it is the case that we employ analog optical systems to image digital optical signals (as in free-space optical interconnection systems), even lower accuracies are tolerable. In this case it is possible to make do with even smaller numbers of stages, since due to the digital nature of the signals, larger analog errors can be tolerated while still maintaining an acceptable eye pattern. For these reasons, the proposed scheme would be especially useful in optical systems, including analog signal processing systems, matrix processors, numerical processors, algebraic processors, and optical interconnection systems.

As M increases, the optical implementation of a multistage filtering configuration would begin to approach what we might refer to as a volume spatial filtering system. For instance, if the fractional Fourier transform stages were realized by sections of graded-index media, the system would consist of spatial filters alternating with short sections of graded-index media. We cannot expect a system constructed from discrete components to have a very large number of stages, due to practical limitations. However, if a technology for imprinting desired transmittance distributions in chosen profiles of, say, a quadratic-index fiber is developed, then the concept of multistage filtering would provide the necessary framework for designing such distributed volume spatial filtering systems, with an exciting array of potential applications. The limiting factor in such systems would most likely be the axial accuracy and resolution which would determine the minimum physical distance between two consecutive filters, and thus the minimum difference between the orders of two consecutive domains.

Although less explored, the discussed filtering architectures may also be expected to result in—possibly parallel and pipelined—efficient VLSI implementations of linear systems, transforms, and matrix operations.

10.3.3   Extensions and generalizations

An immediate generalization is to combine the serial and parallel filtering configurations in an arbitrary manner to obtain *generalized filtering configurations* or *generalized filter circuits* (figure 10.6).

The multistage and multichannel configurations may also be based on other transforms with fast or efficient implementations, such as linear canonical transforms (Barshan, Kutay, and Ozaktas 1997). Concentrating on equation 10.9, for instance, the essential idea is to approximate a general linear operator as a linear combination of operators with fast algorithms or space-bandwidth efficient implementations. If an acceptable approximation can be found with a value of M which is not too large, the cost can be significantly reduced.

It is straightforward to generalize the presented equations and results to the two-dimensional case. It would also be of interest to extend the described approaches to the fractional Hankel transform. This would be useful when dealing with optical systems in which there exists circular symmetry, but in which the noise and distortion may

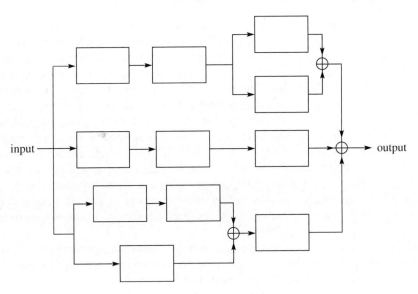

Figure 10.6. Generalized filter circuits; each block is of the form $\mathcal{F}^{-a_k}\Lambda_{h_k}\mathcal{F}^{a_k}$ (Kutay and others 1998b).

have radial dependences.

We conclude this section by presenting a discrete formulation of multistage and multichannel filtering. Assuming discrete signals of length N, we let h_{kl} denote the lth sample of the kth filter $h_k(u)$, and $\Lambda_{\mathbf{h}_k}$ denote the diagonal matrix whose diagonal elements are equal to the elements of the vector $\mathbf{h}_k = [h_{k1}\ h_{k2}\ \dots\ h_{kN}]^{\mathrm{T}}$. Then the discrete counterparts of equations 10.6, 10.8, and 10.9 are

$$\mathbf{f}_{\mathrm{one}} = \left[\mathbf{F}^{-a}\Lambda_{\mathbf{h}}\mathbf{F}^{a}\right]\mathbf{g} = \mathbf{T}_{\mathrm{one}}\mathbf{g}, \tag{10.10}$$

$$\mathbf{f}_{\mathrm{ser}} = \left[\mathbf{F}^{-a_M}\Lambda_{\mathbf{h}_M}\cdots\mathbf{F}^{a_2-a_1}\Lambda_{\mathbf{h}_1}\mathbf{F}^{a_1}\right]\mathbf{g} = \mathbf{T}_{\mathrm{ser}}\mathbf{g}, \tag{10.11}$$

$$\mathbf{f}_{\mathrm{par}} = \left[\sum_{k=1}^{M}\mathbf{F}^{-a_k}\Lambda_{\mathbf{h}_k}\mathbf{F}^{a_k}\right]\mathbf{g} = \mathbf{T}_{\mathrm{par}}\mathbf{g}, \tag{10.12}$$

where $\mathbf{F}^{a_k}$ represents the a_kth order discrete fractional Fourier transform matrix.

A number of further results and extensions will be discussed in sections 10.7, 10.8, and 10.9.

10.4   Applications of fractional Fourier domain filtering

In this section we will discuss several example applications of the filtering configurations introduced in the preceding sections. Most of the examples in this section are taken or adapted from Ozaktas and others 1994b, Kutay and others 1995, 1997, Erden 1997, Erden and Ozaktas 1998, Kutay and Ozaktas 1998, Kutay and others 1998a, b, Ozaktas, Erden, and Kutay 1998, Erden, Kutay, and Ozaktas 1999a, and Kutay 1999, where greater detail and further examples may be found.

This section can be omitted without loss of continuity.

10.4.1 Elementary signal separation examples

We begin by considering simple numerical examples (Ozaktas, Barshan, and Mendlovic 1994, Ozaktas and others 1995). Consider the signal $\exp[-\pi(u-4)^2]$ corrupted additively by $\exp(-i\pi u^2)\mathrm{rect}(u/16)$. The magnitude of their sum is displayed in figure 10.7a. The signal and noise overlap in the frequency domain as well. In figure 10.7b we show the 0.5th fractional Fourier transform of the signal plus noise. We observe that the signal and noise are well separated in this domain. The chirp distortion is transformed into a peaked function which does not exhibit significant overlap with the transform of the signal, so that it can be blocked out by a simple mask (figure 10.7c). Inverse transforming to the original domain, we obtain the desired signal cleansed of the chirp noise (figure 10.7d).

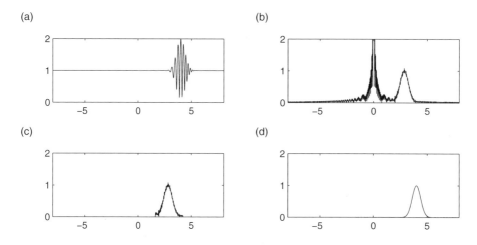

Figure 10.7. Elementary example 1: magnitudes are shown (Ozaktas, Barshan, and Mendlovic 1994, Ozaktas and others 1994b, 1995, 1996).

As an example of a situation requiring consecutive filtering in more than one domain (as in figure 10.3), we now consider a case in which the noise is also real. The signal $\exp(-\pi u^2)$ is corrupted additively by $\cos[2\pi(u^2/2 - 4u)]\mathrm{rect}(u/8)$, as shown in figure 10.8a. The 0.5th transform is shown in figure 10.8b. One of the complex-exponential components of the cosine chirp has been isolated in this domain and can be masked away, but the other still distorts the transform of the Gaussian. After masking out the isolated peaked function appearing at the right of figure 10.8b, we take the -1st transform (which is just an inverse Fourier transform) to arrive at the -0.5th domain (figure 10.8c). Here the other chirp component is isolated and can be blocked out by another simple mask. Finally, we take the 0.5th transform to come back to our home domain (figure 10.8d), where we have recovered our Gaussian signal, with a small error.

The last example has shown the need to visit two fractional Fourier domains to separate a real noise term. Although we do not provide the details, the symmetry exhibited by real signals in the frequency domain allows the same task to be accomplished in a single domain. The analytic signal associated with a real signal eliminates this redundancy, exhibiting a Wigner distribution confined to positive frequencies. Thus if we work with analytic signals, the filtering operation above can be realized by visiting only the 0.5th domain.

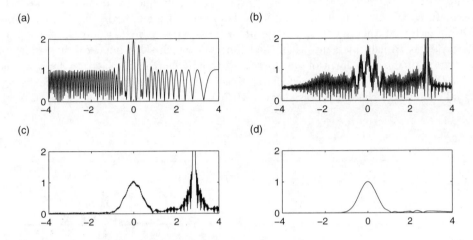

Figure 10.8. Elementary example 2: magnitudes are shown (Ozaktas, Barshan, and Mendlovic 1994, Ozaktas and others 1994b, 1995).

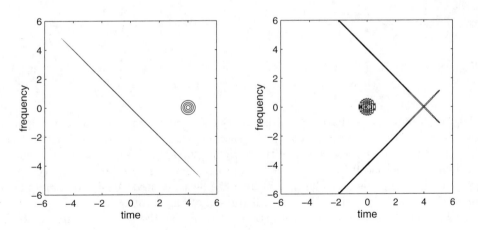

Figure 10.9. Wigner distributions for elementary examples 1 (left) and 2 (right). The separate Wigner distributions of the signal and noise are superimposed.

The examples above involve chirp corruptions which are particularly easy to separate in fractional Fourier domains (just as pure harmonic corruption is particularly easy to separate in the ordinary Fourier domain). However, it is possible to filter out much more general types of noise as well, as will be seen in later examples. Also, there is

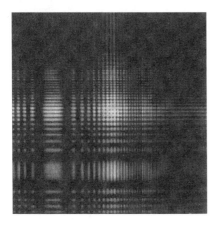 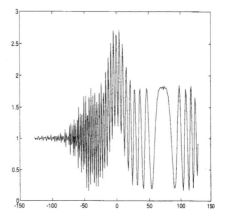

Figure 10.10.   Corrupted signal (Dorsch and others 1994).

nothing special about our choice of Gaussian signals, nor about the ±0.5th domains; it just turns out that these are the domains of choice for the examples considered above. What led us to transform to these particular domains and what gave us the confidence that doing so will get rid of the corruption becomes very transparent if we draw everything in the time-frequency plane. The Wigner distributions of the signal and noise for both elementary examples are shown in figure 10.9, from which the separation strategy clearly emerges.

10.4.2   Optical signal separation

Here we discuss a signal separation example similar to those in section 10.4.1 and present experimental results obtained from optical implementations of fractional Fourier domain filtering (Dorsch and others 1994). This section can be omitted without loss of continuity.

Optical implementation of the fractional Fourier transform is as easy as the optical implementation of the ordinary Fourier transform and can be achieved by using any of the setups discussed in chapter 9. Chirp or chirp-like signals occur widely in optical systems. They are not well localized in either the space or the spatial frequency domain, so it is not possible to eliminate them by filtering in the ordinary space or frequency domain.

The example we will present essentially involves a Gaussian signal corrupted by chirp noise:

$$g(u) = e^{-\pi u^2} + \cos[2\pi(u^2/2 - 4u)] + \text{constant}$$

$$= e^{-\pi u^2} + \frac{1}{2}e^{i2\pi(u^2/2-4u)} + \frac{1}{2}e^{-i2\pi(u^2/2-4u)} + \text{constant}. \qquad (10.13)$$

A bias term has been added to eliminate negative values. Figure 10.10 shows a simulated image and cross section of the corrupted signal. One of the chirps corresponds to a delta function in the $a = +0.5$th domain and the other corresponds to

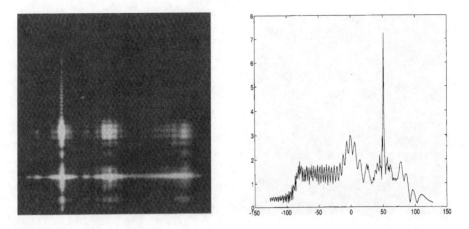

Figure 10.11. Corrupted signal in the $a = 0.5$ domain (Dorsch and others 1994).

a delta function in the $a = -0.5$th domain. Therefore we need to apply filters in both of these domains. The first step is to employ an optical fractional Fourier transformer of order $a = 0.5$ to arrive at the 0.5th fractional Fourier domain. Figure 10.11 shows the experimental image and simulated cross section of the corrupted signal in this domain. Here it is a simple matter to mask out the chirp component which has been transformed into a delta peak. Now we transform to the -0.5th domain by taking an ordinary inverse Fourier transform ($a = -1$), which is equivalent to a forward transform ($a = 1$) with an axis flip. Figure 10.12 shows the experimental image in this domain. Here the second chirp component is masked out. A final fractional Fourier transform of order $a = 0.5$ brings us back to the original space domain. Figure 10.13 shows the experimental image and simulated cross section of the final result, where the corruption has been eliminated.

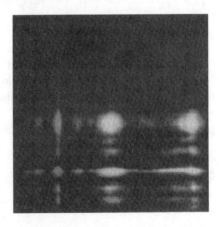

Figure 10.12. Partially processed signal in the $a = -0.5$ domain (Dorsch and others 1994).

 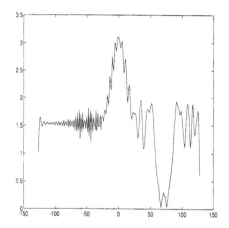

Figure 10.13.   Restored signal (Dorsch and others 1994).

10.4.3   System and transform synthesis

Here we consider the problem of synthesizing the Hadamard transform (Erden and Ozaktas 1998), one of the standard unitary transforms employed in image processing (Jain 1989). Letting the subscript F denote a Frobenius norm (the square root of the sum of the squares of the elements), we define the normalized synthesis error as

$$\varepsilon^2 = \frac{\|\mathbf{T} - \mathbf{H}\|_F^2}{\|\mathbf{H}\|_F^2},\tag{10.14}$$

where $\mathbf{H}$ is the Hadamard transform matrix and $\mathbf{T}$ is either of $\mathbf{T}_{\text{one}}$, $\mathbf{T}_{\text{ser}}$, or $\mathbf{T}_{\text{par}}$. In Erden and Ozaktas 1998 and Kutay 1999 it was shown that the 128-point Hadamard transform can be synthesized with 48% error using single-domain filtering, with 4% error using five-stage filtering, and with 8% error using seven-channel filtering. The multistage configuration offers smaller errors with a fewer number of stages. The orders were chosen according to $a_k = (k-1)/M, k = 1, 2, \ldots, M$ where M is the number of filters, with no attempt being made to optimize them. Optimizing the orders could in principle lead to even smaller errors. The block diagrams of the filter configurations can be interpreted either as a digital fast algorithm, or an efficient optical implementation of the Hadamard transform. Needless to say, the same procedure can be applied to a wide variety of transforms of interest, for which fast digital algorithms or efficient optical implementations may not be known.

10.4.4   Signal recovery and restoration

We begin by presenting several examples of single-domain filtering, which have been chosen mainly for their instructional value (Kutay 1995, Kutay and others 1995, 1997, Kutay and Ozaktas 1998). These examples are based on the observation model given in equation 10.1.

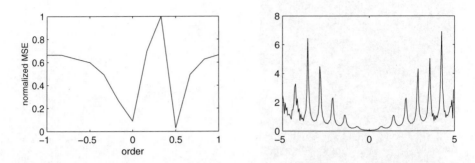

Figure 10.14. Normalized MSE for different values of the transform order a (left); magnitude of the optimal filter function $h(u_{0.5})$ in the $a_{\text{opt}} = 0.5$ domain (right) (Kutay 1995, Kutay and others 1997).

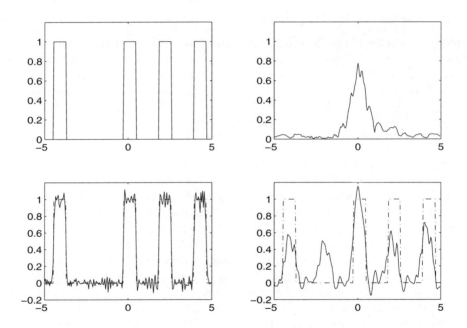

Figure 10.15. A realization of the input random process $f(u)$ (upper left). The corresponding output process $g(u)$ (upper right). The estimate $f_{\text{est}}(u)$ obtained by filtering in the $a = 0.5$ domain (lower left). The estimate obtained by filtering in the $a = 1$ domain (lower right). The original $f(u)$ is also superimposed in the lower parts of the figure for comparison. (Kutay 1995, Kutay and others 1997)

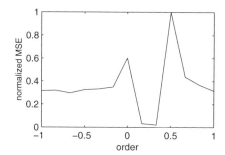 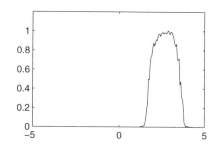

Figure 10.16. Normalized MSE for different values of the transform order a (left); magnitude of the optimal filter function $h(u_{0.33})$ in the $a_{opt} = 0.33$ domain (right) (Kutay 1995, Kutay and others 1997).

First, we consider a degradation model corresponding to a time-varying bandpass filter whose center frequency changes linearly with time: $h_d(u, u') = \exp[-i2\pi u (u - u')]\text{sinc}(u - u')$. One way of characterizing such a system is through its *time-varying frequency response*:

$$H_d(\mu, u) = \int_{-\infty}^{\infty} h_d(u, u') e^{-i2\pi\mu(u-u')} \, du' = \text{rect}(\mu + u). \tag{10.15}$$

The input process $f(u)$ is a sequence of rectangular pulses whose amplitudes take the value of 1 or 0 with equal probability. There is no noise process ($n(u) = 0$). For each transform order a, the optimal filter function $h(u_a)$ can be calculated from equation 10.7. The normalized mean square error (MSE) is plotted for different values of a in the left part of figure 10.14. (The normalization is with respect to the maximum value of the MSE.) The minimum MSE is obtained when $a = a_{opt} = 0.5$. The magnitude of the optimal filter function in the 0.5th domain is shown in the right part of the same figure. Figure 10.15 shows a realization of the input random process $f(u)$ (upper left) and the corresponding output process $g(u)$ (upper right). The input process is totally unrecognizable at the output. The estimate obtained by filtering $g(u)$ in the optimal 0.5th domain is shown in the lower left part of the same figure. For comparison, the lower right part of the figure shows the optimal estimate obtained by filtering in the ordinary Fourier domain ($a = 1$). Filtering in the optimal fractional domain is seen to be significantly better than filtering in the ordinary Fourier domain.

We now turn our attention to a noise removal problem. The input process $f(u)$ is a shifted Gaussian function which is deterministic except for a random amplitude factor and a random shift. That is, $f(u) = A \exp[-\pi(u - u_0)^2]$ where A and u_0 are random variables uniformly distributed on the intervals $[0, 3]$ and $[1, 3]$, respectively. The noise $n(u)$ is a finite-duration bandpass process modulated with a chirp function $\exp(-i\,1.73\,\pi u^2)$ so that its center frequency changes linearly with time. We assume there is no distortion or blur so that $h_d(u, u') = \delta(u - u')$. Again, for each transform order a, the optimal filter function $h(u_a)$ can be calculated from equation 10.7. The resulting normalized MSEs for different values of a, as well as the optimal filter function for the value of $a = a_{opt} = 0.33$ resulting in the smallest MSE, are plotted in figure 10.16. Realizations of the input and output processes, together with the

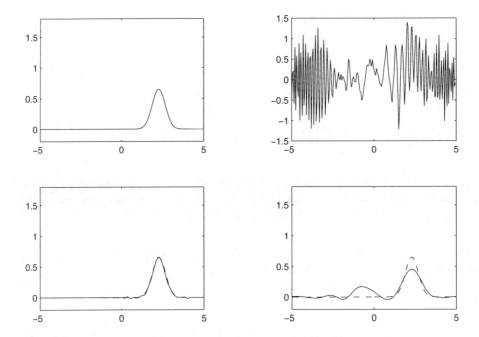

Figure 10.17. A realization of the input random process $f(u)$ (upper left). The corresponding output process $g(u)$ (upper right). The estimate $f_{\text{est}}(u)$ obtained by filtering in the $a = 0.33$ domain (lower left). The estimate obtained by filtering in the $a = 1$ domain (lower right). The original $f(u)$ is also superimposed in the lower parts of the figure for comparison. After Kutay 1995, Kutay and others 1997.

estimates obtained by filtering in the optimal fractional Fourier domain ($a = 0.33$) and the ordinary Fourier domain ($a = 1$), are shown in figure 10.17.

So far we have given one example in which $n(u) = 0$ and one example in which $h_{\text{d}}(u, u') = \delta(u - u')$. The following example illustrates the more general case where distortions or blurs and noise are simultaneously present. We consider two cases: (i) the optimal values of a for distortion or blur alone and noise alone are the same or nearly the same; (ii) they are significantly different.

We assume the same input process as in the last example, that the noise $n(u)$ is a finite-duration bandpass process modulated with a chirp function $\exp(-i\pi u^2)$, and that $h_{\text{d}}(u, u')$ is given by equation 10.15. The normalized MSEs for different values of a, as well as the optimal filter function for the value of $a = a_{\text{opt}} = 0.5$ resulting in the smallest MSE, are plotted in figure 10.18. The MSE obtained in the 0.5th domain is significantly less than that obtained in the ordinary Fourier domain because the optimal values of a for the distortion/blur and noise considered separately coincide. There is no trade-off involved in choosing the optimal domain. A sample realization and optimal estimate are shown in figure 10.19. Figure 10.20 shows the Wigner distributions of $f(u)$ and $g(u)$.

Now we consider the case where the optimal values of a for distortion/blur and noise considered separately are different and thus impose conflicting requirements in

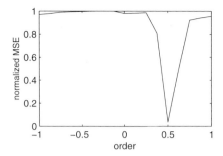 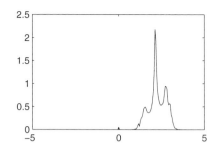

Figure 10.18. Normalized MSE for different values of the transform order a (left); magnitude of the optimal filter function $h(u_{0.5})$ in the $a_{opt} = 0.5$ domain (right) (Kutay 1995, Kutay and others 1997).

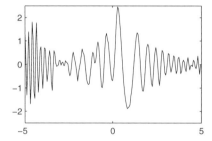 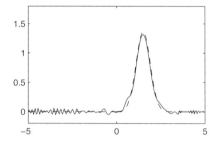

Figure 10.19. A realization of the output random process $g(u)$ (left); the estimate $f_{est}(u)$ obtained by filtering in the $a = 0.5$ domain with the original $f(u)$ superimposed for comparison (right). After Kutay 1995, Kutay and others 1997.

choosing the filtering domain. We assume everything to be the same as in the previous example; however, the chirp function modulating the noise process is assumed to be $\exp(-i\,1.73\,\pi u^2)$, rather than $\exp(-i\pi u^2)$. The normalized MSEs for different values of a, as well as the optimal filter function for the value of $a = a_{opt} = 0.5$ resulting in the smallest MSE, are plotted in figure 10.21. The minimum value of the MSE is much larger than obtained in previous examples. Figure 10.22 show a sample realization of $g(u)$ (upper left) and a sample estimate which is far from being satisfactory (upper right).

One way of improving upon the present result is to pursue a heuristic two-step filtering procedure. Based on the results of the previous examples, we might first go to the $a = 0.5$th domain and eliminate the distortion/blur by using the optimal filter function obtained when $n(u) = 0$. We can then go to the $a = 0.33$rd domain and eliminate the noise by using the optimal filter function obtained when $h_d(u, u') = \delta(u - u')$. The resulting estimate, shown in the lower right part of figure 10.22, is much better than that obtained by filtering in a single domain. The two orders in the above procedure have been chosen heuristically by looking at the optimal order resulting from a consideration of each effect alone. Figure 10.23 shows this two-step filtering process in

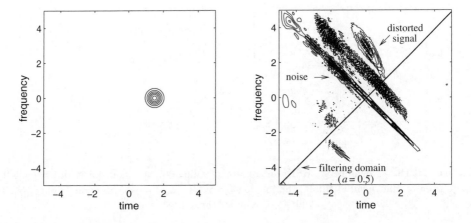

Figure 10.20. Wigner distribution of $f(u)$ (left) and $g(u)$ (right). After Kutay and others 1997.

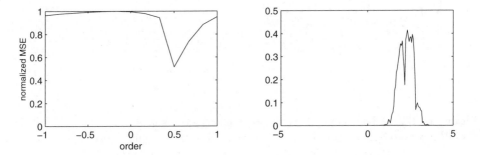

Figure 10.21. Normalized MSE for different values of the transform order a (left); magnitude of the optimal filter function $h(u_{0.5})$ in the $a_{\text{opt}} = 0.5$ domain (right) (Kutay 1995, Kutay and others 1997).

the time-frequency plane. This example was meant to show why single-domain filtering is sometimes inadequate and how multistage filtering greatly increases flexibility. In practice it would normally be preferable to directly formulate the problem within the general framework discussed in section 10.3, rather than engaging in such heuristic procedures.

We now consider a number of examples involving two-dimensional signals. The upper left part of figure 10.24 shows the original image $f(u, v)$. The upper right part shows this image corrupted by two chirp waveforms with random magnitudes such that the signal-to-noise ratio (SNR) is approximately unity. (The SNR is defined as signal energy divided by noise energy.) The optimal estimate, obtained with the fractional orders $a_u = 0.4$ and $a_v = -0.6$, is shown in the lower left part of the figure. The normalized MSE for this estimate is about 0.3% (in this and the following two-dimensional examples normalization is with respect to the energy of the original image). For comparison, the lower right part of the same figure shows the optimal

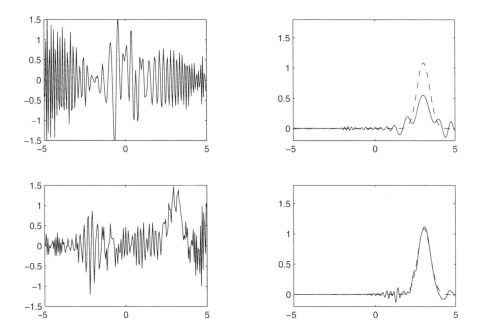

Figure 10.22. A realization of the output random process $g(u)$ (upper left). The estimate $f_{\text{est}}(u)$ obtained by filtering in the $a = 0.5$ domain with the original $f(u)$ superimposed for comparison (upper right). Successive filtering of $g(u)$ in two different domains: the result of eliminating distortion/blur in the $a = 0.5$ domain (lower left); the result of eliminating noise in the $a = 0.33$ domain (lower right). After Kutay 1995, Kutay and others 1997.

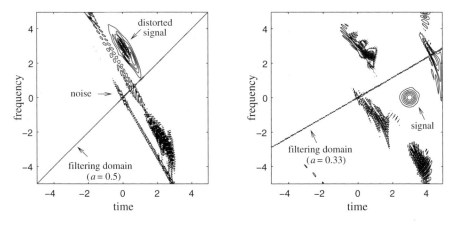

Figure 10.23. The left part shows the Wigner distribution of $g(u)$; the right part shows the Wigner distribution of the intermediate result in which the distortion/blur has been eliminated (the lower left part of figure 10.22) (Kutay and others 1997).

Figure 10.24. Original image (upper left); corrupted image with SNR ≈ 1 (upper right); estimate obtained by filtering in the optimal fractional Fourier domain (lower left); estimate obtained by filtering in the ordinary Fourier domain (lower right) (Kutay and Ozaktas 1998).

estimate obtained in the ordinary Fourier domain ($a_u = a_v = 1$), which is much less satisfactory with MSE equal to 3.5%.

We now repeat the same example with a signal-to-noise ratio approximately equal to 0.1. The results are presented in figure 10.25. The MSE obtained by filtering in the optimal fractional Fourier domain is 0.6%, whereas that obtained by filtering in the ordinary Fourier domain is 10%. Comparison of the lower parts of the figures reveals that the relative benefit of using fractional Fourier domain filtering with respect to ordinary Fourier domain filtering is greater for this lower value of signal-to-noise ratio.

In the following two examples, we consider images degraded by space-varying blurs and additive white Gaussian noise. These examples illustrate the results obtained when the blurs and noise are not of a chirped nature.

We first consider the following space-varying blur model:

$$h_{\mathrm{d}}(u, v; u', v') = \frac{1}{\iota_1 u + \iota_0} \, \mathrm{rect}\left(\frac{u - u'}{\iota_1 u + \iota_0} - \frac{1}{2}\right) \delta(v - v'), \qquad (10.16)$$

Figure 10.25. Original image (upper left); corrupted image with SNR ≈ 0.1 (upper right); estimate obtained by filtering in the optimal fractional Fourier domain (lower left); estimate obtained by filtering in the ordinary Fourier domain (lower right) (Kutay and Ozaktas 1998).

where ι_1 and ι_0 are parameters of the model. When $\iota_1 = 0$ this reduces to a space-invariant blur but when $\iota_1 \neq 0$ the width of the rectangle function depends on u and the blur is space-variant. We assume $\iota_1 = 0.01$ and $\iota_0 = 0.3$ and that additive white Gaussian noise with SNR $= 4$ is present. The upper left part of figure 10.26 shows the original image $f(u,v)$ and the upper right part shows the degraded image $g(u,v)$. The optimal estimate $f_{\text{est}}(u,v)$, obtained with the orders $a_u = 0.7$ and $a_v = 0.8$, is shown in the lower left part of the figure. The corresponding MSE is 0.97%. For comparison, the lower right part shows the optimal estimate obtained in the ordinary Fourier domain ($a_u = a_v = 1$), which is less satisfactory with MSE equal to 3.82%.

As our last single-domain filtering example, we consider images degraded by space-varying atmospheric turbulence (Goodman 1985). We assume a blur model of the general form

$$h_{\text{d}}(u,v;u',v') = \exp\left\{-\pi\kappa^2(u,v)\left[(u-u')^2 + (v-v')^2\right]\right\}, \tag{10.17}$$

where $\kappa(u,v)$ is a function of u and v. If $\kappa(u,v)$ could be assumed constant, the blur

Figure 10.26.   Original image (upper left); blurred and noisy image—rectangular blur function (upper right); estimate obtained by filtering in the optimal fractional Fourier domain (lower left); estimate obtained by filtering in the ordinary Fourier domain (lower right) (Kutay and Ozaktas 1998).

would become space-invariant. We will write $\kappa(u, v) = \kappa_0 + \Delta\kappa(u, v)$, where $\Delta\kappa(u, v)$ represents the fluctuations of κ around κ_0. In our example it has been assumed that $\kappa_0 = 0.1$ and $\Delta\kappa(u, v)$ is a function slowly fluctuating around 0 with standard deviation 0.005. Additive white Gaussian noise with SNR = 2 is assumed present. The results are shown in figure 10.27. The optimal domain turns out to be characterized by the orders $a_u = 0.4$ and $a_v = 0.7$ with a corresponding MSE of 2.1%. The MSE obtained by filtering in the ordinary Fourier domain is 5.2%.

The presented examples show that single-domain fractional Fourier filtering allows a significant reduction of the restoration error for chirp-like degradations and at least a substantial reduction for certain other types of common degradations, including space-varying blurs. We now turn our attention to examples in which the use of multiple stages or channels results in substantially greater benefits than filtering in a single domain.

Figure 10.27. Original image (upper left); blurred and noisy image—atmospheric turbulence (upper right); estimate obtained by filtering in the optimal fractional Fourier domain (lower left); estimate obtained by filtering in the ordinary Fourier domain (lower right) (Kutay and Ozaktas 1998).

We begin by reconsidering a variation of the atmospheric turbulence example just discussed, where a one-dimensional version of equation 10.17 was assumed as the blur model, and $\Delta\kappa$ has been modeled by a piecewise constant function whose values range between -0.005 and $+0.005$ (Erden 1997, Erden, Kutay, and Ozaktas 1999a, Kutay 1999). Initially, let us consider the first of the two different approaches distinguished on page 394. We obtain the kernel of the optimal general linear estimator $\mathcal{H}$ from equation 10.4 and determine the single-domain, multistage, or multichannel configuration which most closely approximates it. The resulting normalized mean square difference between $\mathbf{H}$ and $\mathbf{T}$ (discretized versions of $\mathcal{H}$ and $\mathcal{T}$) has been found to be 11% with $M = 4$ and $< 1\%$ with $M = 8$ for the multistage case, and 12% with $M = 4$ and $< 1.5\%$ with $M = 8$ for the multichannel case (Kutay 1999). Now, let us consider the second of the two approaches on page 394. This time, the resulting normalized MSE between the actual signal and its estimate is $< 1\%$ with $M = 4$ for

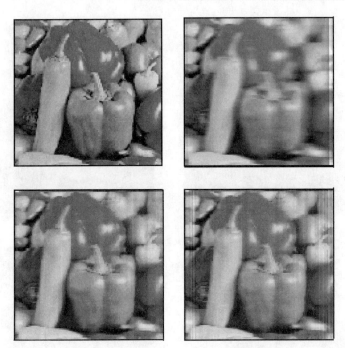

Figure 10.28.   Original image (upper left); blurred image—one-dimensional Gaussian point spread function (upper right); estimate obtained by multistage filtering ($M = 5$) (lower left); estimate obtained by multichannel filtering ($M = 5$) (lower right). Ordinary Fourier domain filtering gives very poor results (not shown). (Kutay and others 1998c, Kutay 1999)

both the multistage and multichannel configurations. (In this example the orders were chosen according to $a_k = k/M, k = 1, 2, \ldots, M$.)

Further results were presented in Kutay 1999 and Kutay and others 1999, where a similar blur model was employed. The authors considered restoration of images blurred by a space-variant one-dimensional Gaussian point spread function, whose effect we will represent by the operator $\mathcal{H}_d$ characterized by the kernel $h_d(u, v; u', v') = \exp[-\pi(u - u')^2/\omega(u)]\,\delta(v - v')$ with $\omega(u) = \omega_0 + \Delta\omega(u)$. The constant ω_0 has been taken equal to 20 and $\Delta\omega(u)$ has been generated by lowpass filtering white Gaussian noise such that its standard deviation is equal to 1. Again, we will consider both of the two different approaches distinguished on page 394. First, assume that we have no prior information about the original signal or the underlying random processes. In this circumstance, one approach would be to try to determine the recovery operator $\mathcal{H}$ as the inverse of $\mathcal{H}_d$, and then synthesize it in the form of $\mathcal{T}_{\mathrm{one}}$, $\mathcal{T}_{\mathrm{ser}}$, or $\mathcal{T}_{\mathrm{par}}$, as described in approach 1 on page 394. A more direct procedure is to solve the problem of minimizing $\varepsilon^2 = \|\mathcal{T}\mathcal{H}_d - \mathcal{I}\|_{\mathrm{F}}^2$, as described in approach 2 on page 394. Here $\mathcal{T}$ is $\mathcal{T}_{\mathrm{one}}$, $\mathcal{T}_{\mathrm{ser}}$, or $\mathcal{T}_{\mathrm{par}}$, and the subscript F denotes the Frobenius norm. The normalized squared error is 42% when a single-domain filter is used. With the multistage configuration, the resulting error is 24% with $M = 3$ filters, 10% with 4 filters, and 3% with 6 filters. With the multichannel configuration, the error is 23%, 12%, and 4% with 3, 4, and 6 filters respectively ($a_k = k/M, k = 1, 2, \ldots, M$). Now, let us assume that

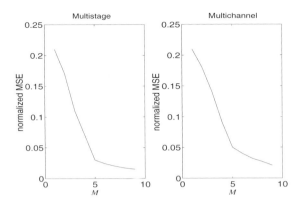

Figure 10.29. Minimum MSE as a function of the number of filters M (Kutay 1999).

the correlation functions associated with the underlying random processes are known. Then we can solve the problem of minimizing σ_{err}^2 (equation 10.3) subject to the constraint that the recovery operator $\mathcal{H}$ (corresponding to the kernel $h(u, u')$) has the form of $\mathcal{T}_{\text{one}}$, $\mathcal{T}_{\text{ser}}$, or $\mathcal{T}_{\text{par}}$. The results of this procedure are illustrated in figure 10.28. The corresponding normalized errors have been found to be 21%, 3%, and 5% for the single-domain, multistage ($M = 5$), and multichannel ($M = 5$) configurations respectively. Figure 10.29 provides plots of the minimum MSE as a function of the number of filters M for both multistage and multichannel configurations. These are actual examples of figure 10.5b. Had we used the most general linear optimal Wiener filter (equation 10.4), the resulting error would have been 1%. We see not only that significant improvements are possible with respect to single-domain ($M = 1$) filtering, but also that it is possible to approach the performance of the optimal general linear estimator by using only a moderate number of filters. The minimum errors obtained with the multistage and multichannel configurations are similar in this example, but this need not always be so. Furthermore, multistage may be preferable for certain values of M and multichannel for other values of M.

As a simple example of the use of filter circuits more general than multistage and multichannel configurations, we consider using the filter circuit shown in figure 10.30 for the same restoration problem. The operator representing the overall circuit configuration is given by

$$\mathcal{T}_{\text{cir}} = \sum_{k=1}^{2} \mathcal{F}^{-a_{2k}} \Lambda_{h_{2k}} \mathcal{F}^{a_{2k}-a_{2k-1}} \Lambda_{h_{2k-1}} \mathcal{F}^{a_{2k-1}}. \tag{10.18}$$

The optimal filter functions have been determined by modifying the optimization algorithms used for the multistage and multichannel cases (Kutay 1999), resulting in a normalized error of 2%. In this example the fractional domains have been chosen as $a_1 = 0.25$, $a_2 = 0.5$ for the upper branch and $a_3 = 0.75$, $a_4 = 1$ for the lower branch. No attempt has been made to optimize these orders or the structure (circuit diagram) of the filter circuit. How an appropriate filter structure should be chosen for a given application is a problem which remains largely unexplored. What is known is that

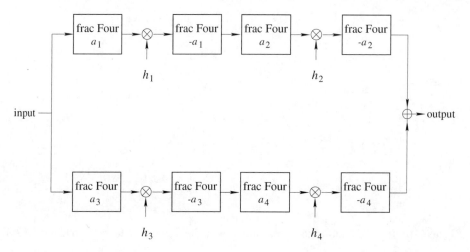

Figure 10.30. Filter circuit consisting of two branches of two consecutive filters each (Kutay 1999).

the results strongly depend on the structure chosen. The present example illustrates a case where the error is lower than the error obtained with both the multistage and multichannel configurations with five filters, representing an improvement in both performance and cost. In some simulations, however, less satisfactory results were obtained by using a filter circuit with a total of M filters, as compared to multichannel or multistage configurations with the same number of filters.

We now consider two one-dimensional examples involving chirp signals. First, we look at the problem of recovering a signal corrupted with several additive chirps (figure 10.31) (Kutay and others 1998b, Kutay 1999). In this example it is assumed that the Q chirps have uniformly distributed random amplitudes and time shifts, that their chirp rates are clustered around P (initially unknown) values, and that the correlation functions of the signal and the chirps are known. It is now possible to first determine the optimal recovery operator from equation 10.4 and then seek the best multistage or multichannel approximation to it. For $Q = 9$ and $P = 3$ the multistage configuration with $M = 3$ results in a normalized mean square signal recovery error of 2% and the multichannel configuration with $M = 3$ results in a normalized mean square signal recovery error of $< 1\%$.

In our second example involving chirp signals, we consider the problem of recovering a signal consisting of multiple chirp-like components buried in white Gaussian noise with a signal-to-noise ratio of 0.1 (figure 10.32) (Kutay and others 1998a, c, 1999, Kutay 1999). It is assumed that the signal consists of six chirps with uniformly distributed random amplitudes and time shifts, and that the chirp rates are known with a $\pm 5\%$ accuracy. Employing the first of the two approaches discussed on page 394, the optimal general linear recovery operator $\mathbf{H}$ for this problem (obtained from equation 10.4) can be approximated with a normalized mean square Frobenius error of 5.2% by using an $M = 6$ multichannel configuration, leading to a normalized mean square signal recovery error of 4.7%. With the second of the two approaches, the same number of filters results in a normalized mean square signal recovery error of 2.6%.

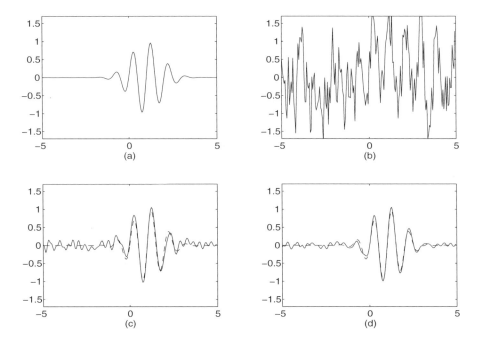

Figure 10.31. Original signal (upper left); corrupted signal (upper right); estimate obtained by multistage filtering ($M = 3$) (lower left); estimate obtained by multichannel filtering ($M = 3$) (lower right). The original signal is also superimposed in the lower parts of the figure for comparison. (Kutay 1999)

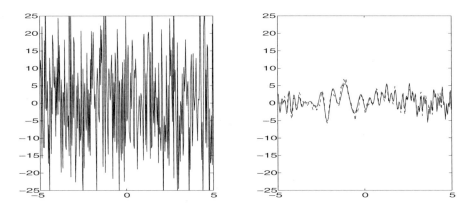

Figure 10.32. Multicomponent signal buried in noise (left); estimate obtained by multichannel filtering ($M = 6$) (right). The original signal is also superimposed in the right part for comparison. (Kutay 1999)

The results obtained in the above examples represent significant performance improvements with respect to single-domain filtering, but are much less costly to implement than general linear filtering. The use of multistage or multichannel configurations or generalized filter circuits has the potential to result in more desirable cost-performance combinations than possible with ordinary Fourier domain filters or general linear systems.

A broader class of signal recovery and restoration problems have been systematically treated in Kutay 1999, including the design of prefiltering and postfiltering configurations to compensate for system or channel degradations, and the use of the windowed fractional Fourier transform discussed in section 11.6.

10.4.5 Signal synthesis

Although space does not permit us to include further illustrative applications, we briefly mention applications to the problem of synthesizing random signals with desired second-order characteristics. This application was first studied in an optical context in Erden, Ozaktas, and Mendlovic 1996b and later in Zalevsky, Mendlovic, and Ozaktas 2000. These works deal with the problem of determining optimal fractional Fourier domain filters to synthesize signals with desired autocorrelation functions, which are known in optics as mutual intensity functions. A much broader class of such problems are studied in detail in Kutay 1999.

10.4.6 Free-space optical interconnection architectures

We now briefly turn our attention away from signal processing, to a problem which is central to the construction of free-space optically interconnected digital computing systems. We consider the problem of realizing one-to-one permutation-type interconnection patterns between N input and N output channels. Some common approaches as well as various difficulties in realizing arbitrary permutation patterns have been discussed in Mendlovic and Ozaktas 1993c and Ozaktas and Mendlovic 1993c; the latter concludes that multistage architectures based on regular patterns (such as the so-called perfect shuffle or Banyan) are most favorable. Here we discuss the use of fractional Fourier domain filtering configurations to realize such systems. This not only provides a systematic way of designing such systems, but the implementation may be more convenient and less costly since the present approach is based on the use of conventional spatial filters rather than micro-optical elements.

Any one-to-one interconnection architecture between N input and N output channels is characterized by an $N \times N$ permutation matrix. In such a matrix every row and column has only one nonzero element which is equal to one. The interconnection architecture can be designed by synthesizing its permutation matrix in the form of a fractional Fourier filtering configuration. As an example, consider the reverse perfect shuffle architecture shown in figure 10.33. This interconnection pattern can be synthesized exactly by using six filters in the multistage configuration, and with an error of 1% by using six filters in the multichannel configuration ($a_k = k/6, k = 1, 2, \ldots, 6$). A large number of other interconnection patterns which do not exhibit any obvious regularity have also been considered. In all cases they could be realized with a moderate number of filters.

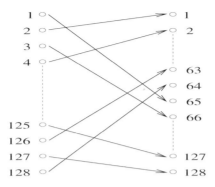

Figure 10.33. Reverse perfect shuffle interconnection architecture (Erden 1997, Erden and Ozaktas 1998).

Conventional multistage permutation networks can realize arbitrary permutations in $O(\log N)$ stages. Extensive numerical experimentation indicates that arbitrary permutations can also be realized with fractional Fourier filtering configurations in a similar number of stages. Unlike conventional approaches, however, use of fractional Fourier filtering configurations can allow a further reduction in the number of stages. This is because such configurations allow us to trade off cost and error, and the digital nature of the systems under consideration allows relatively large analog errors to be tolerated while still maintaining an acceptable digital error. While it may be possible to come up with comparable optical setups through ingenuity and invention, the present approach provides a systematic way of designing such systems with the ability to fine-tune cost and error in order to obtain the most desirable system overall.

10.5 Convolution and filtering in fractional Fourier domains

The previous sections aimed to provide an overview and present illustrative applications. The rest of this chapter more carefully develops several elementary concepts from scratch.

10.5.1 Convolution and multiplication in fractional Fourier domains

Consider two abstract signals h and f. When we speak of the convolution of these two signals in the ath fractional Fourier domain, we will mean that their ath order fractional Fourier domain representations $h_a(u_a)$ and $f_a(u_a)$ are convolved to give the corresponding representation of some new signal g:

$$g_a(u_a) = h_a(u_a) * f_a(u_a). \tag{10.19}$$

Likewise, when we speak of multiplying two signals in the ath fractional Fourier domain, we will mean

$$g_a(u_a) = h_a(u_a)f_a(u_a). \tag{10.20}$$

Of course, convolution (or multiplication) in the $a = 0$th domain is ordinary convolution (or multiplication) and convolution (or multiplication) in the $a = 1$st domain is ordinary multiplication (or convolution). More generally, convolution (or multiplication) in the ath domain is multiplication (or convolution) in the $(a \pm 1)$th domain (which is orthogonal to the ath domain), and convolution (or multiplication) in the ath domain is again convolution (or multiplication) in the $(a \pm 2)$th domain (the sign-flipped version of the ath domain). Convolution or multiplication in an arbitrary ath domain is an operation "interpolating" between the ordinary convolution and multiplication operations. (Ozaktas and others 1994a)

In many works, including some by the present authors, the concepts of fractional convolution and fractional correlation have been developed somewhat differently. (To avoid any possible confusion, in the above we were careful not to talk about "fractional convolution," but only "convolution in fractional domains.") In these works, fractional convolution has been defined as what happens in the u domain when we multiply the fractional Fourier transforms of functions. The ath order "fractional convolution" $g(u)$ of $h(u)$ with $f(u)$ is defined through

$$g_a(u_a) = h_a(u_a)f_a(u_a).$$ (10.21)

When $a = 1$ we have $G(\mu) = H(\mu)F(\mu)$, corresponding to ordinary convolution and revealing the motivation behind this definition. The two approaches are in essence not much different if we remember that "taking the ath order fractional convolution of two functions" is the same as "multiplying them in the ath order fractional Fourier domain."

Mustard (1990, 1998) defines the fractional convolution operator $*_a$ essentially according to equation 10.21. Then, $*_0$ corresponds to ordinary multiplication, $*_1$ corresponds to ordinary convolution, and $*_a = *_{(a\pm2)}$. The binary operator $*_a$ is associative, provided the same order appears in all instances. Furthermore, it can be shown that

$$\mathcal{F}^a(h *_{a'} f) = h_a *_{(a'-a)} f_a.$$ (10.22)

This constitutes a generalization of the convolution property of the ordinary Fourier transform. For instance, when $a' = 0$ and $a' = 1$, we have $\mathcal{F}^a(hf) = h_a *_{-a} f_a$ and $\mathcal{F}^a(h * f) = h_a *_{(1-a)} f_a$, respectively. With $a = 1$, these reduce to the familiar $\mathcal{F}(hf) = H * F$ and $\mathcal{F}(h * f) = HF$. Mustard (1998) derives an explicit relation for the fractional convolution of two functions, which is similar to equation 4.123. He also discusses the generalized "unit distribution," denoted $1_a(u)$, which for $a = 0$ reduces to the ordinary unit function 1 and for $a = 1$ reduces to the ordinary Dirac delta function $\delta(u)$. In general, $1_a(u) \equiv \mathcal{F}^{-a}[1](u)$. The function $1_a(u)$ is the unit for the $*_a$ operation. An alternative approach to fractional convolution and correlation may be found in Akay and Boudreaux-Bartels 1997.

Recall properties 14 and 15 from table 3.2, which tell us how the Wigner distribution is affected when we convolve or multiply a function $f(u)$ with another function $h(u)$. Property 14 tells us that if two functions are convolved in the u domain, their Wigner distributions are convolved along the u direction. Property 15 tells us that if two functions are convolved in the μ domain, their Wigner distributions are convolved along the μ direction. Now, if two functions are convolved in an arbitrary fractional

Fourier domain u_a, as a necessary consequence of the rotational symmetry of Wigner space (or the arbitrariness of choosing the origin of a), it follows that their Wigner distributions will be convolved along the u_a direction (Ozaktas and others 1994a, Mustard 1998). In this context, the reader may recall that shifting the ath domain representation $f_a(u_a)$ of a signal by a certain amount corresponds to shifting its Wigner distribution along the u_a direction by the same amount (page 165).

10.5.2   Compaction in fractional Fourier domains

A signal will be said to be compact in the ath domain if $f_a(u_a)$ is zero outside some finite interval around the origin of the u_a axis. Compaction of a signal f in the ath domain can be realized by multiplying $f_a(u_a)$ with the window

$$\text{rect}(u_a/\Delta u_a) \qquad (10.23)$$

in that domain. Compaction in the $a = 1$st domain corresponds to ordinary bandpass filtering. Compaction in the $a = 0$th domain corresponds to ordinary rectangular windowing.

Let us examine the effect of compaction on the Wigner distribution. Since compaction involves multiplication with a rectangle function, this implies convolution of the Wigner distribution of f with the Wigner distribution of the rectangle function, in the direction orthogonal to u_a, namely along the $u_{a+1} = \mu_a$ direction. The Wigner distribution of the rectangle function was given in table 3.1. From property 11 in table 3.2, the Wigner distribution is nonzero only along the corridor defined by the rectangle function. This means that compaction of a signal to a certain interval in the ath domain will also result in compaction of the Wigner distribution to a corresponding corridor orthogonal to the u_a axis.

Convolving the Wigner distribution of f with that of the rectangle function in the $\mu_a = u_{a+1}$ direction will result in a broadening of the Wigner distribution of f in the μ_a direction comparable to the extent of the Wigner distribution of the rectangle function in the μ_a direction (figure 10.34). This extent is simply $\sim 1/\Delta u_a$ so that compaction to an interval of extent $\sim \Delta u_a$ in any domain u_a necessarily results in a spread in the orthogonal domain u_{a+1} by the amount $\sim 1/\Delta u_a$. It also results in a spread $\sim |\sin[(a' - a)\pi/2]|/\Delta u_a$ in any other domain $u_{a'}$. Of course, this is a consequence of the generalized uncertainty relation discussed on page 169.

Because of the abrupt transitions of the rectangle function in the u_a domain, the spread of the Wigner distribution in the orthogonal domain u_{a+1} will be somewhat larger than fundamentally necessary. By using window functions with smoother transitions instead of the rectangle function, it is possible to reduce this spread to the fundamental minimum dictated by the uncertainty relation.

If a signal is compact in a given domain a, it is not compact in any other domain (Xia 1996). This is a generalization of the fact that a signal cannot be both time-limited and band-limited. It is not difficult to see this in terms of Wigner distributions. A signal which is compact in a certain domain will have a Wigner distribution which is confined to a time-frequency corridor perpendicular to that domain. Now, multiplying $f_a(u_a)$ with a rectangle function of appropriate width will not have any affect on the signal. Such a multiplication corresponds to convolution of the Wigner distribution along the u_{a+1} direction with the Wigner distribution of the rectangle function. This,

(a)   (b)

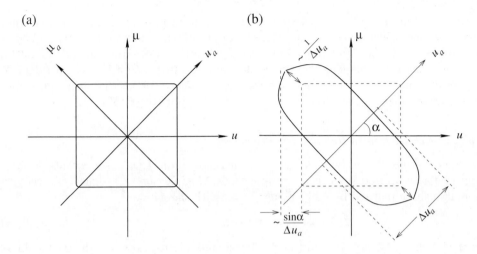

Figure 10.34. (a) Wigner distribution of a signal; (b) compaction in the ath domain
(Ozaktas and others 1994a).

of course, does not change the Wigner distribution, but demonstrates that the Wigner
distribution has tails along the u_{a+1} direction which never become zero. Thus, the
Wigner distribution cannot be confined to a corridor in any but the initial domain
a, and the signal cannot be compact in any other domain (except those whose order
differs from a by an integer multiple of 2).

Several sampling theorems for signals band-limited in the ath domain are given by
Xia (1996).

10.5.3 Filtering in fractional Fourier domains

Having understood the effect of convolution and compaction in fractional domains, we
now consider filtering of signals to separate undesired terms. Multiplicative filtering
in the ordinary Fourier domain limits us to shift-invariant operations; those that can
be expressed in the form of ordinary convolution. (In the discrete case this restricts
the system matrix to be of Toeplitz or circulant form.) Multiplicative filtering in the
ordinary time or space domain, on the other hand, limits us to ordinary windowing
operations. (In the discrete case this restricts the system matrix to be diagonal.)

Filtering in a fractional Fourier domain amounts to first finding the representation
$f_a(u_a)$ of the signal in the ath domain, and multiplying it by a filter function $h_a(u_a)$ to
obtain the ath domain representation of the filtered signal. Here we denote the filter
function by $h_a(u_a)$ rather than $h(u_a)$, so that filtering in the ath domain becomes
equivalent to multiplication in the ath domain (equation 10.20). For instance, if we
are given a time-domain function $f(u)$, we first find $f_a(u_a)$, multiply it with the filter
function $h_a(u_a)$ to obtain the filtered function $g_a(u_a) = h_a(u_a)f_a(u_a)$ in the ath
domain, which may then be transformed back to the time domain:

$$g(u) = \mathcal{F}^{-a}\left\{h_a(u_a)\mathcal{F}^a[f(u)]\right\} = \mathcal{F}^{-a}\Lambda_{h_a}\mathcal{F}^a f(u), \qquad (10.24)$$

where Λ_{h_a} denotes the operator corresponding to multiplication with the function $h_a(u_a)$ in the ath domain. The same equation can also be written in abstract form as $g = \mathcal{T}_{\text{one}} f$ with $\mathcal{T}_{\text{one}} = \mathcal{F}^{-a} \Lambda_{h_a} \mathcal{F}^a$. In the discrete case, the system matrix $\mathbf{T}_{\text{one}}$ will be of the form $\mathbf{T}_{\text{one}} = \mathbf{F}^{-a} \Lambda_{\mathbf{h}_a} \mathbf{F}^a$, where $\mathbf{F}^a$ is the discrete fractional Fourier transform matrix and $\Lambda_{\mathbf{h}_a}$ is a diagonal matrix. For any given value of a, $\mathbf{T}_{\text{one}}$ has N degrees of freedom, the same as a circulant matrix (N is the length of the signal vectors). However, the general form of $\mathbf{T}_{\text{one}}$ is different for different values of a, providing additional flexibility. Matrices of the form $\mathbf{F}^{-a} \Lambda_{\mathbf{h}_a} \mathbf{F}^a$, where $\Lambda_{\mathbf{h}_a}$ is an arbitrary diagonal matrix, may be referred to as *fractionally circulant* with order a.

As we have discussed in previous sections, a common use of filtering is to separate a desired signal from additive noise which corrupts it. First, consider the case where the ordinary Fourier transforms of the signal and noise do not overlap in the ordinary Fourier domain, although they might overlap in the ordinary time (or space) domain. The noise is easy to eliminate in the Fourier domain by using a window or mask with a value of unity throughout the extent of the signal, and zero elsewhere. We can similarly consider the case where the signals overlap in the Fourier domain but not in the time (or space) domain. Again, it is an easy matter to separate the signal from noise by employing a simple window or mask. Now, reconsider the situation depicted in figure 10.2 where the Wigner distributions of the signal and noise have been superimposed. Considering the projections of these Wigner distributions on the u and μ axes, we see that they overlap in both the $a = 0$th and $a = 1$st domains, although they do not overlap in the time-frequency plane as a whole. Although it is not possible to mask away the undesired noise in either the 0th or 1st domains, the noise is easily eliminated by the use of a simple window or mask in the ath domain, since the projections of the Wigner distributions do not overlap in this domain. This operation can be implemented just as easily and efficiently as ordinary Fourier domain filtering, since both the optical (chapter 9) and digital (chapter 6) implementations of the fractional Fourier transform are as easy and efficient as those of the ordinary Fourier transform.

It will be instructive to take a closer look at the effect of using windows or masks as noise separation filters. Multiplying a signal with a window or mask in a certain domain essentially amounts to compaction in that domain. As discussed in section 10.5.2, this will lead to spreading of the Wigner distribution along the domain orthogonal to the one in which the window is applied. In certain situations this may result in merging of the Wigner distributions of signal and noise. Let us consider figure 10.35, where we wish to cleanse a signal of space-bandwidth product $N = \Delta u \Delta \mu \gg 1$ from noise which is separated from the signal by the distances shown. If we use a rectangular mask of the form $\text{rect}(u/\Delta u)$ to eliminate the noise, this will result in broadening of the Wigner distribution of the signal by an amount $\sim 1/\Delta u$ in the μ direction. Since we do not want the signal to mix up with the noise in this process, we require $\delta \mu > 1/\Delta u$. This is a reasonable requirement, being a direct consequence of the uncertainty relation. It amounts to requiring that the area of the hatched region in the figure satisfies $\delta \mu \Delta u > 1$. We cannot hope to work with greater resolution in the time-frequency plane; a buffer region of unity area must lie between the signal and noise if we are to be able to separate them.

When the desired signal and noise have no significant overlap in the time-frequency plane, near-perfect recovery of the signal is in principle possible. More generally, the

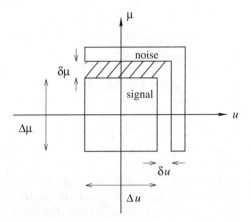

Figure 10.35. Limits to noise separation imposed by the uncertainty relation (Ozaktas and others 1994a).

signal and noise may have partial overlap so that perfect recovery is not possible and we seek to separate the signal from the noise as much as possible. In such situations it is possible to formulate the problem as a minimum mean square error estimation problem, as discussed in sections 10.2, 10.6, and 10.7.

10.6   Derivation of the optimal fractional Fourier domain filter

In this section we more fully treat the problem of optimal filtering in fractional Fourier domains, which was previously discussed in section 10.2 (Kutay 1995, Kutay and others 1995, 1997). As we have already noted, filtering in fractional Fourier domains can reduce the estimation error without increasing the computational cost with respect to ordinary Fourier domain filtering, since the cost of implementing the fractional Fourier transform is similar to that of implementing the ordinary Fourier transform.

Although our discussion will be in one-dimensional notation, the extension to two dimensions is straightforward. Allowing different orders along the u and v directions gives greater flexibility and in general will result in better performance. In this book we limit our attention to the separable fractional Fourier transform, in which the orders cannot be specified along arbitrary directions, but only along the orthogonal u and v axes. Allowing the transform orders to be specified along arbitrary directions results in still greater flexibility and performance, as demonstrated in Sahin 1996 and Sahin, Kutay, and Ozaktas 1998. Here one optimizes over and chooses not only the two orders, but also the directions along which they are specified. An example from optics illustrates the utility of this added flexibility. Line defects on optical components produce chirp-like distortions. Since the angle between the defects will in general not be 90°, allowing the transform orders to be specified along arbitrary directions will result in greater improvements.

10.6.1 Continuous time

Let an observed signal $g(u)$ be related to a signal $f(u)$ we wish to recover as follows:

$$g(u) = \int h_\mathrm{d}(u, u') f(u') \, du' + n(u), \qquad (10.25)$$

We assume that $h_\mathrm{d}(u, u')$, as well as the correlation functions $R_{ff}(u, u') = \langle f(u) f^*(u') \rangle$ and $R_{nn}(u, u') = \langle n(u) n^*(u') \rangle$ of the random signal $f(u)$ and noise $n(u)$ are known. We further assume that $n(u)$ is independent of $f(u)$ and that it has zero mean for all time: $\langle n(u) \rangle = 0$. Under these assumptions we can also find the cross-correlation function $R_{fg}(u, u') = \langle f(u) g^*(u') \rangle$ and the autocorrelation function $R_{gg}(u, u') = \langle g(u) g^*(u') \rangle$ by using equation 10.25.

In equation 10.2 we considered the most general linear estimator characterized by the kernel $h(u, u')$. Setting ourselves the objective of minimizing the mean square error

$$\sigma^2_\mathrm{err} = \left\langle \int |f_\mathrm{est}(u) - f(u)|^2 \, du \right\rangle \qquad (10.26)$$

between the actual signal $f(u)$ and our estimate $f_\mathrm{est}(u)$, we saw that the optimal kernel satisfies equation 10.4. Here we restrict ourselves to the class of linear estimators which correspond to filtering in a single fractional Fourier domain, in which case the estimate can be expressed as

$$f_\mathrm{est}(u) = \mathcal{F}^{-a} \Lambda_h \mathcal{F}^a \, g(u), \qquad (10.27)$$

where Λ_h denotes the operator corresponding to multiplication by the filter function $h(u)$. According to this equation, we first take the ath order transform of the observed signal $g(u)$, then multiply with the filter $h(u)$, and finally take the inverse ath order transform. For $a = 1$ this procedure corresponds to filtering in the ordinary Fourier domain. Shortly, we will solve for the optimal filter $h(u)$ minimizing the mean square error given in equation 10.26.

The class of fractional Fourier domain filters is a subclass of the class of all linear filters, so that the optimal filter we find will not be the most optimal among all linear filters. However, the class of fractional Fourier domain filters is a much broader class than the class of ordinary Fourier domain filters, so that in general the optimal filter we find will result in much smaller error than can be obtained with an ordinary Fourier domain filter. This reduction in error comes at no additional cost because the fractional Fourier transform can be implemented with the same cost as the ordinary Fourier transform. Use of the most optimal general linear filter can result in even smaller errors, but is much more costly to implement.

Since the fractional Fourier transform is unitary, σ^2_err can also be expressed in the ath domain:

$$\sigma^2_\mathrm{err} = \left\langle \int |f_{\mathrm{est}\,a}(u_a) - f_a(u_a)|^2 \, du_a \right\rangle, \qquad (10.28)$$

where $f_{\mathrm{est}\,a}(u_a) = h(u_a) g_a(u_a)$. Assuming the value of a is fixed for the moment, we are confronted with a calculus of variations problem where the *functional* σ^2_err is to be minimized with respect to the function $h(u_a)$. Referring the reader to Hildebrand 1965 for an introduction to the calculus of variations, here we present a brief derivation of the solution. We substitute $h(u_a) \to h(u_a) + \epsilon\,\delta h(u_a)$, where $\epsilon = \epsilon_r + i\epsilon_i$ is a complex scalar parameter and $\delta h(u_a)$ is an arbitrary perturbation term. For a given $\delta h(u_a)$, we can consider σ_{err}^2 to be a function of ϵ. The partial derivatives of $\sigma_{\text{err}}^2(\epsilon)$ are given by

$$\frac{\partial\,\sigma_{\text{err}}^2(\epsilon)}{\partial\,\epsilon_r} = -2\left\langle \int \Re\left[\kappa^*(u_a,\epsilon)\,\tilde{\kappa}(u_a,\epsilon)\right]du_a \right\rangle,$$

$$\frac{\partial\,\sigma_{\text{err}}^2(\epsilon)}{\partial\,\epsilon_i} = 2\left\langle \int \Im\left[\kappa^*(u_a,\epsilon)\,\tilde{\kappa}(u_a,\epsilon)\right]du_a \right\rangle. \tag{10.29}$$

where $\Re[\cdot]$ and $\Im[\cdot]$ denote the real and imaginary parts of a complex entity and

$$\kappa(u_a,\epsilon) = f_a(u_a) - f_{\text{est}\,a}(u_a,\epsilon), \qquad \tilde{\kappa}(u_a,\epsilon) = \frac{\partial\,f_{\text{est}\,a}(u_a,\epsilon)}{\partial\,\epsilon_r}. \tag{10.30}$$

Now, setting the partial derivatives at $\epsilon = 0$ to zero, we obtain

$$\left\langle \int \kappa^*(u_a,\epsilon)\,\tilde{\kappa}(u_a,\epsilon)\,du_a \right\rangle\Bigg|_{\epsilon=0} = 0, \tag{10.31}$$

which evaluates to

$$\left\langle \int \left[f_a(u_a) - f_{\text{est}\,a}(u_a,0)\right]^* \delta h(u_a) g_a(u_a)\,du_a \right\rangle = 0. \tag{10.32}$$

Since the last equation is true for all $\delta h(u_a)$ we can write

$$\left\langle \left[f_a(u_a) - f_{\text{est}\,a}(u_a,0)\right]^* g_a(u_a) \right\rangle = 0. \tag{10.33}$$

Some readers may recognize this result as the well-known orthogonality condition. Since $f_{\text{est}\,a}^*(u_a,0)$ can be expressed in terms of $h(u_a)$, by taking the complex conjugate of the above equation we obtain an expression for the optimal filter function:

$$h(u_a) = \frac{R_{f_a g_a}(u_a,u_a)}{R_{g_a g_a}(u_a,u_a)}, \tag{10.34}$$

where the above correlation functions are related to the correlation functions $R_{fg}(u,u')$ and $R_{gg}(u,u')$ through

$$R_{f_a g_a}(u_a,u_a) = \iint K_a(u_a,u)K_{-a}(u_a,u')R_{fg}(u,u')\,du'\,du,$$

$$R_{g_a g_a}(u_a,u_a) = \iint K_a(u_a,u)K_{-a}(u_a,u')R_{gg}(u,u')\,du'\,du. \tag{10.35}$$

Thus the optimal filter function is

$$h(u_a) = \frac{\iint K_a(u_a,u)K_{-a}(u_a,u')R_{fg}(u,u')\,du'\,du}{\iint K_a(u_a,u)K_{-a}(u_a,u')R_{gg}(u,u')\,du'\,du}, \tag{10.36}$$

as previously presented in equation 10.7.

The last equation provides us the optimal filter function in the ath fractional domain. To find the optimal value of a, we first insert the expression for the optimal filter into the expression for σ_{err}^2 and thus obtain an expression for the minimum error as a function of a:

$$\sigma_{\text{err}}^2(a) = \int \{ R_{f_a f_a}(u_a, u_a) - 2\Re\left[h^*(u_a) R_{f_a g_a}(u_a, u_a)\right]$$
$$+ h(u_a) h^*(u_a) R_{g_a g_a}(u_a, u_a) \} \, du_a, \quad (10.37)$$

where $h(u_a)$ is given by equation 10.36. Now, the optimal domain can be found by finding the minimizer of $\sigma_{\text{err}}^2(a)$. Usually this minimizer cannot be found analytically. It can be found by simply calculating $\sigma_{\text{err}}^2(a)$ for different values of a and choosing that value which results in the smallest error. Other more sophisticated and efficient minimization routines can also be employed.

To summarize the procedure, given $h_{\text{d}}(u, u')$ and the autocorrelations of the input f and noise n, we can find the cross-correlation between the input f and output g and the autocorrelation of the output g. These allow us to calculate the optimal filter function in the ath domain by using equation 10.36. The optimal value of a can be subsequently determined.

10.6.2 Discrete time

Here we formulate the problem of optimal filtering in fractional Fourier domains for discrete-time signals and systems, in complete analogy with the preceding section. Let an observed signal $\mathbf{g}$ be related to a signal $\mathbf{f}$ we wish to recover as follows:

$$\mathbf{g} = \mathbf{H}_{\text{d}} \mathbf{f} + \mathbf{n}, \quad (10.38)$$

where $\mathbf{g}, \mathbf{f}, \mathbf{n}$ are column vectors with N elements and $\mathbf{H}_{\text{d}}$ is an $N \times N$ matrix. We assume that $\mathbf{H}_{\text{d}}$, as well as the correlation matrices $\mathbf{R}_{\mathbf{ff}} = \langle \mathbf{f}\,\mathbf{f}^{\text{H}} \rangle$ and $\mathbf{R}_{\mathbf{nn}} = \langle \mathbf{n}\,\mathbf{n}^{\text{H}} \rangle$ of the random signal $\mathbf{f}$ and noise $\mathbf{n}$ are known. We further assume that $\mathbf{n}$ is independent of $\mathbf{f}$ and that all of its elements have zero mean. Under these assumptions we can also find the cross-correlation matrix $\mathbf{R}_{\mathbf{fg}} = \langle \mathbf{f}\,\mathbf{g}^{\text{H}} \rangle$ and the autocorrelation matrix $\mathbf{R}_{\mathbf{gg}} = \langle \mathbf{g}\,\mathbf{g}^{\text{H}} \rangle$ by using equation 10.38.

If we express the most general linear estimate in the form $\mathbf{f}_{\text{est}} = \mathbf{H}\mathbf{g}$ in analogy with equation 10.2, then the optimal general linear estimator $\mathbf{H}$ minimizing the mean square error

$$\sigma_{\text{err}}^2 = \frac{1}{N} \left\langle \|\mathbf{f}_{\text{est}} - \mathbf{f}\|^2 \right\rangle \quad (10.39)$$

between the actual signal $\mathbf{f}$ and our estimate $\mathbf{f}_{\text{est}}$, can be shown to satisfy $\mathbf{R}_{\mathbf{fg}} = \mathbf{H}\mathbf{R}_{\mathbf{gg}}$. Here we restrict ourselves to the class of linear estimators which correspond to filtering in a single fractional Fourier domain, in which case the estimate can be expressed as

$$\mathbf{f}_{\text{est}} = \mathbf{F}^{-a} \mathbf{\Lambda}_{\mathbf{h}} \mathbf{F}^a \mathbf{g}, \quad (10.40)$$

where $\mathbf{F}^a$ is the ath order discrete fractional Fourier transform matrix, $\mathbf{F}^{-a} = (\mathbf{F}^a)^{\text{H}}$, and $\mathbf{\Lambda}_{\mathbf{h}}$ is a diagonal matrix whose diagonal consists of the elements of the vector

h. When $a = 1$ the matrix $\mathbf{F}^a$ reduces to the DFT matrix and the overall operation corresponds to ordinary Fourier domain filtering.

Since the fractional Fourier transform is unitary, σ_{err}^2 can also be expressed in the ath domain:

$$\sigma_{\text{err}}^2 = \frac{1}{N} \left\langle \|\mathbf{f}_{\text{est}\,a} - \mathbf{f}_a\|^2 \right\rangle, \qquad (10.41)$$

where $\mathbf{f}_a = \mathbf{F}^a \mathbf{f}$ and $\mathbf{f}_{\text{est}\,a} = \mathbf{F}^a \mathbf{f}_{\text{est}} = \boldsymbol{\Lambda}_{\mathbf{h}} \mathbf{F}^a \mathbf{g} = \boldsymbol{\Lambda}_{\mathbf{h}} \mathbf{g}_a$. Assuming the value of a is fixed for the moment, we are confronted with the problem of minimizing σ_{err}^2 with respect to the complex vector $\mathbf{h}$. Let us write σ_{err}^2 explicitly in the form

$$\sigma_{\text{err}}^2 = \left\langle \frac{1}{N} \sum_{l=1}^{N} |f_{al} - h_l g_{al}|^2 \right\rangle, \qquad (10.42)$$

where f_{al}, g_{al}, and $h_l = h_{lr} + i h_{li}$ are the elements of the vectors $\mathbf{f}_a$, $\mathbf{g}_a$, and $\mathbf{h}$. The partial derivatives can be evaluated as

$$\frac{\partial \sigma_{\text{err}}^2}{\partial h_{lr}} = \frac{1}{N} \left\langle -2\Re[f_{al} g_{al}^*] + 2h_{lr} |g_{al}|^2 \right\rangle,$$

$$\frac{\partial \sigma_{\text{err}}^2}{\partial h_{li}} = \frac{1}{N} \left\langle -2\Im[f_{al} g_{al}^*] + 2h_{li} |g_{al}|^2 \right\rangle. \qquad (10.43)$$

Setting these to zero we obtain the elements of the optimal filter vector

$$h_l = \frac{\langle f_{al} g_{al}^* \rangle}{\langle |g_{al}|^2 \rangle}. \qquad (10.44)$$

Notice that the numerator is simply the (l, l)th element of $\mathbf{R}_{\mathbf{f}_a \mathbf{g}_a}$ and the denominator is simply the (l, l)th element of $\mathbf{R}_{\mathbf{g}_a \mathbf{g}_a}$, where

$$\mathbf{R}_{\mathbf{f}_a \mathbf{g}_a} = \langle \mathbf{f}_a \mathbf{g}_a^{\mathrm{H}} \rangle = \mathbf{F}^a \mathbf{R}_{\mathbf{ff}} \mathbf{H}_{\mathrm{d}}^{\mathrm{H}} \mathbf{F}^{-a},$$

$$\mathbf{R}_{\mathbf{g}_a \mathbf{g}_a} = \langle \mathbf{g}_a \mathbf{g}_a^{\mathrm{H}} \rangle = \mathbf{F}^a \left(\mathbf{H}_{\mathrm{d}} \mathbf{R}_{\mathbf{ff}} \mathbf{H}_{\mathrm{d}}^{\mathrm{H}} + \mathbf{R}_{\mathbf{nn}} \right) \mathbf{F}^{-a}. \qquad (10.45)$$

Equation 10.44 is fully analogous to equation 10.36.

10.7   Optimization and cost analysis of multistage and multichannel filtering configurations

Here we discuss the optimization and cost analysis of multichannel and multistage filtering configurations. Since we will concentrate on discrete-time signals and systems in this section, it is worth rephrasing some of our earlier comments in a discrete context. As discussed on page 394, one way of using fractional Fourier domain filtering configurations is to assume that there is some linear system matrix $\mathbf{H}$ we wish to implement, and then to seek the transform orders a_k and/or filter vectors $\mathbf{h}_k$ so that the overall filtering matrix $\mathbf{T}_{\text{one}}$, $\mathbf{T}_{\text{ser}}$, or $\mathbf{T}_{\text{par}}$ (equations 10.10 to 10.12) is as close as possible to $\mathbf{H}$. Naturally, the number of stages or channels required to approximate $\mathbf{H}$ to a given precision will be smaller for matrices exhibiting greater regularity or other more subtle forms of intrinsic structure. In such cases, directly multiplying the input signal vector with $\mathbf{H}$, which would take $O(N^2)$ time, would be clearly inefficient. The regularity or structure of a given matrix can be exploited on a case-by-case basis through ingenuity or invention; most sparse matrix algorithms and fast transform algorithms are obtained in this manner. The use of multistage and multichannel configurations provides a systematic way of obtaining an efficient implementation which is applicable even when the regularity or structure of the matrix is not readily discernible.

Further insight regarding the multichannel case will be gained when we discuss the fractional Fourier domain decomposition in section 10.8. Further insight regarding the multistage case will be gained when we discuss repeated time- and frequency-domain filtering in section 10.9.

10.7.1 Determination of the optimal filters

In section 10.6 we solved the problem of determining the optimal filter for the single-domain case. Now we look at the multichannel and multistage configurations.

In the multichannel case, the problem of determining the optimal filters can be exactly solved since $\mathbf{T}_{\text{par}}$ depends linearly on the elements of the filter vectors $\mathbf{h}_k$. Denoting the lth element of the kth filter by h_{kl}, we have

$$\mathbf{T}_{\text{par}} = \sum_{k=1}^{M} \mathbf{F}^{-a_k} \mathbf{\Lambda}_{\mathbf{h}_k} \mathbf{F}^{a_k} = \sum_{k=1}^{M} \sum_{l=1}^{N} h_{kl} \mathbf{w}_l^{-a_k} \left(\mathbf{w}_l^{-a_k}\right)^{\mathrm{H}}$$

$$= \sum_{k=1}^{M} \sum_{l=1}^{N} h_{kl} \mathbf{P}_{kl}, \tag{10.46}$$

where $\mathbf{w}_l^{-a_k}$ denotes the lth column of the matrix $\mathbf{F}^{-a_k}$. Here we have also defined the matrices $\mathbf{P}_{kl}$, indexed by kl, as $\mathbf{P}_{kl} = \mathbf{w}_l^{-a_k} \left(\mathbf{w}_l^{-a_k}\right)^{\mathrm{H}}$. These matrices constitute a family of "basis matrices" which are used to construct the matrix $\mathbf{T}_{\text{par}}$.

Now, say that we wish to approximate a given matrix $\mathbf{H}$ with $\mathbf{T}_{\text{par}}$ such that $\|\mathbf{T}_{\text{par}} - \mathbf{H}\|_{\mathrm{F}}^2$ is as small as possible. Here the subscript F denotes the Frobenius norm, the square root of the sum of the squares of the elements of the matrix. The optimization variables are the NM filter coefficients h_{kl} (N coefficients in each of M filters). This problem can be posed as a standard least-squares optimization problem. To see this, it is necessary to first "vectorize" the above equations. Let $\underline{\mathbf{T}}_{\text{par}}$ denote the $N^2 \times 1$ vector obtained by stacking the columns of $\mathbf{T}_{\text{par}}$ on top of each other, let $\underline{\mathbf{H}}$ be defined similarly in terms of $\mathbf{H}$, and let $\underline{\mathbf{P}}_j$ denote the $N^2 \times 1$ vector obtained by stacking the columns of $\mathbf{P}_{kl}$ on top of each other, where $j = (k-1)N + l$. Finally, let $\underline{\mathbf{h}}$ denote the $MN \times 1$ vector obtained by stacking the M filters $\mathbf{h}_1, \mathbf{h}_2, \ldots, \mathbf{h}_M$ on top of each other. With these conventions we obtain

$$\underline{\mathbf{T}}_{\text{par}} = \left[\underline{\mathbf{P}}_1 \, \underline{\mathbf{P}}_2 \, \cdots \, \underline{\mathbf{P}}_j \, \cdots \, \underline{\mathbf{P}}_{MN}\right] \underline{\mathbf{h}} \equiv \tilde{\mathbf{P}} \underline{\mathbf{h}}, \tag{10.47}$$

where the $N^2 \times MN$ matrix $\tilde{\mathbf{P}}$ is defined by the second equality. Once $\underline{\mathbf{T}}_{\text{par}}$ is written in the above form, our problem reduces to the standard one of minimizing $\|\tilde{\mathbf{P}}\underline{\mathbf{h}} - \underline{\mathbf{H}}\|^2$ over the filter coefficients $\underline{\mathbf{h}}$. The minimizer of this objective is known to satisfy the so-called normal equations (Strang 1988):

$$\tilde{\mathbf{P}}^H\underline{\mathbf{H}} = \tilde{\mathbf{P}}^H\tilde{\mathbf{P}}\underline{\mathbf{h}}. \tag{10.48}$$

In practice, when the dimensions are large, a variety of iterative or other techniques may be preferred to find the minimizer, rather than solving these equations.

In the case of multistage configurations, $\mathbf{T}_{\text{ser}}$ depends nonlinearly on the elements of the filter vectors $\mathbf{h}_k$ and the resulting nonlinear optimization problem is much more difficult. Nevertheless, an iterative approach has been successfully applied to this problem (Erden 1997, Erden and Ozaktas 1998, Erden, Kutay, and Ozaktas 1999a). A modification of these approaches has been used for the determination of the optimal filters in a generalized filter circuit (Kutay 1999).

There remains much that is unexplored regarding the optimization of fractional Fourier filtering configurations. For instance, it would be useful to know the minimum number of filters needed to approximate the given matrix $\mathbf{H}$ with some specified accuracy. It would also be useful to relate the minimum number of filters to some measure of the structural redundancy of the matrix, or to give a general characterization of the $\mathbf{T}$ matrices which can be realized with a given number of filters.

Although the optimal choice of orders remains a subject for future research, it seems that the number of filters is more important than the precise choice of orders, provided the orders are not very close to each other. Thus in the absence of any special indication, it would be natural to choose them uniformly spaced.

10.7.2   Rectangular system matrices

On page 398 we implicitly assumed that the matrices $\mathbf{T}_{\text{one}}$, $\mathbf{T}_{\text{ser}}$, and $\mathbf{T}_{\text{par}}$ are square $N \times N$ matrices. Here we assume that we are trying to realize a linear system relation of the form $\mathbf{g} = \mathbf{Hf}$, where $\mathbf{H}$, $\mathbf{f}$, and $\mathbf{g}$ have the dimensions $N_{\text{out}} \times N_{\text{in}}$, $N_{\text{in}} \times 1$ and $N_{\text{out}} \times 1$, respectively. In what follows we will indicate the dimensions of matrices and vectors as subscripts in order to keep things straight. The generalizations of equations 10.10 and 10.12 to this case become

$$(\mathbf{T}_{\text{one}})_{N_{\text{out}} \times N_{\text{in}}} = \mathbf{F}_{N_{\text{out}}}^{-a} \, (\mathbf{\Lambda_h})_{N_{\text{out}} \times N_{\text{in}}} \, \mathbf{F}_{N_{\text{in}}}^{a}, \tag{10.49}$$

$$(\mathbf{T}_{\text{par}})_{N_{\text{out}} \times N_{\text{in}}} = \sum_{k=1}^{M} \mathbf{F}_{N_{\text{out}}}^{-a_k} \, (\mathbf{\Lambda_{h_k}})_{N_{\text{out}} \times N_{\text{in}}} \, \mathbf{F}_{N_{\text{in}}}^{a_k}, \tag{10.50}$$

where $\mathbf{F}_N^a$ is the $N \times N$ discrete fractional Fourier transform matrix, and $(\mathbf{\Lambda_{h_k}})_{N_{\text{out}} \times N_{\text{in}}}$ is a rectangular diagonal matrix whose $N' = \min(N_{\text{in}}, N_{\text{out}})$ diagonal elements are the elements of the $N' \times 1$ filter vector $\mathbf{h}_k$. The diagonal of the rectangular matrix is anchored to its upper left corner.

There is considerable flexibility in generalizing equation 10.11 to rectangular matrices. The only necessary condition is to match the dimensions at the first and last stages. The dimensions of the intermediate stages may be freely chosen. Normally one would choose the dimensions of the intermediate stages to lie between N_{in} and N_{out}. One approach is to choose $\mathbf{\Lambda_{h_1}}$ to be $N_{\text{int}} \times N_{\text{in}}$ and $\mathbf{\Lambda_{h_M}}$ to be $N_{\text{out}} \times N_{\text{int}}$ and to choose the dimensions of $\mathbf{\Lambda_{h_k}}$ for intermediate values of k to be $N_{\text{int}} \times N_{\text{int}}$ where

N_{int} lies between N_{out} and N_{in}. Alternatively, we may choose to taper the dimensions gently from N_{in} to N_{out}. The best way to choose the dimensions of intermediate stages is presently not known.

10.7.3 Cost analysis

A simplified discussion of the cost of implementing multistage and multichannel filtering configurations was presented in section 10.3.2. We saw that if a given linear system can be satisfactorily approximated with such a configuration using a relatively small number of filters, the implementation cost can be significantly reduced. Here we will present a more detailed analysis of the cost of implementing such systems.

Let the input of some general linear system be represented by N_{in} samples and the output by N_{out} samples. Digital implementation of general linear systems takes $O(N_{\text{out}}N_{\text{in}})$ time (the time to multiply the system matrix with the input vector). Direct optical implementations of general linear systems using matrix-vector product or multifacet architectures require an optical system whose space-bandwidth product is $O(N_{\text{out}}N_{\text{in}})$ (Mendlovic and Ozaktas 1993c, Ozaktas, Brenner, and Lohmann 1993, Ozaktas and Mendlovic 1993c). On the other hand, the digital implementation of shift-invariant systems takes $O(N_{\text{in}} \log N_{\text{in}} + N' + N_{\text{out}} \log N_{\text{out}}) \sim O(N \log N)$ time by using the fast Fourier transform, where $N \equiv \max(N_{\text{out}}, N_{\text{in}})$ and $N' \equiv \min(N_{\text{out}}, N_{\text{in}})$. Optical implementation of shift-invariant systems requires a pair of optical Fourier transformers whose space-bandwidth products are $O(N_{\text{in}})$ and $O(N_{\text{out}})$. The cost of filtering in a single fractional Fourier domain is the same as that of implementing shift-invariant systems (which correspond to filtering in the ordinary Fourier domain).

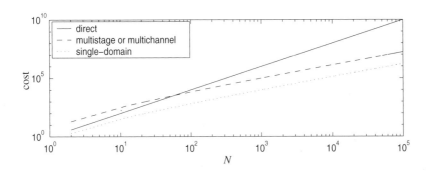

Figure 10.36. Cost of directly implementing a linear system compared with the cost of implementing multistage, multichannel, and single-domain filtering configurations ($N_{\text{out}} = N_{\text{in}} = N$, $M = 10$) (Kutay and others 1998c, 2000).

We now turn our attention to multistage and multichannel filtering configurations. These configurations consist of M single-domain filters. Thus the multichannel configuration can be digitally implemented in

$$O\left(M(N_{\text{in}} \log N_{\text{in}} + N' + N_{\text{out}} \log N_{\text{out}})\right) \sim O(MN \log N) \qquad (10.51)$$

time. Likewise, the multistage configuration can be digitally implemented in

$$O\left(N_{\text{in}}\log N_{\text{in}} + \sum_{k=1}^{M}[\min(N_{k-1}, N_k) + N_k \log N_k]\right) \sim O\left(\sum_{k=0}^{M} N_k \log N_k\right) \quad (10.52)$$

time. Normally, the dimensions N_k of the intermediate stages would lie between $N_0 \equiv N_{\text{in}}$ and $N_M \equiv N_{\text{out}}$. Therefore, the last expression is also $\sim O(MN \log N)$. We now consider the costs of optical implementation. The multichannel configuration requires M pairs of fractional Fourier transformers whose space-bandwidth products are $O(N_{\text{in}})$ and $O(N_{\text{out}})$. The multistage configuration requires $M+1$ fractional Fourier transformers whose space-bandwidth products are $O(N)$.

Figure 10.36 compares the time cost of directly implementing a linear system with that of implementing multistage or multichannel configurations with a moderate number of filters.

The above results which assume the use of a serial digital computer would be modified for parallel computers, but the reduction in cost obtained with multistage and multichannel configurations would remain. These results have also been generalized to the case where the knowledge of the rank of the linear system matrix is exploited to eliminate redundant computations (Kutay and others 2000).

10.8   The fractional Fourier domain decomposition

In this section we discuss the fractional Fourier domain decomposition (FFDD), and a procedure called *pruning*, which is analogous to truncation of the singular-value decomposition. We will see that a pruned FFDD is nothing but a multichannel filtering configuration.

10.8.1   *Introduction and definition*

The singular-value decomposition (SVD) plays a fundamental role in signal and system analysis, representation, and processing. The SVD of an arbitrary $N_{\text{out}} \times N_{\text{in}}$ complex matrix $\mathbf{H}$ is

$$\mathbf{H}_{N_{\text{out}} \times N_{\text{in}}} = \mathbf{U}_{N_{\text{out}} \times N_{\text{out}}} \, \mathbf{\Sigma}_{N_{\text{out}} \times N_{\text{in}}} \, \mathbf{V}^{\text{H}}_{N_{\text{in}} \times N_{\text{in}}}, \quad (10.53)$$

where $\mathbf{U}$ and $\mathbf{V}$ are unitary matrices whose columns are the eigenvectors of $\mathbf{HH}^{\text{H}}$ and $\mathbf{H}^{\text{H}}\mathbf{H}$ respectively. As usual, the superscript H denotes Hermitian transpose. $\mathbf{\Sigma}$ is a diagonal matrix whose elements λ_k (the singular values) are the nonnegative square roots of the eigenvalues of $\mathbf{HH}^{\text{H}}$ and $\mathbf{H}^{\text{H}}\mathbf{H}$. The number of strictly positive singular values is equal to the rank R of $\mathbf{H}$. The SVD can also be written in the form of an outer product (or spectral) expansion

$$\mathbf{H} = \sum_{k=1}^{R} \lambda_k \mathbf{u}_k \mathbf{v}_k^{\text{H}} \quad (10.54)$$

where $\mathbf{u}_k$ and $\mathbf{v}_k$ are the columns of $\mathbf{U}$ and $\mathbf{V}$. It is common to assume that the λ_k are ordered in decreasing value. An understanding of the fundamental properties of the

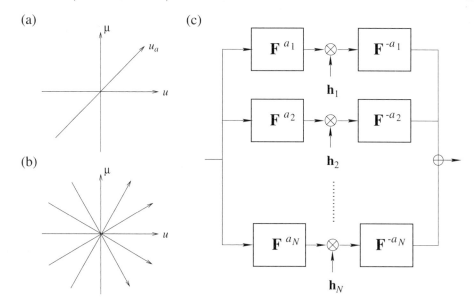

Figure 10.37. (a) The ath fractional Fourier domain u_a; the $a = 0$ and $a = 1$ domains are the ordinary time and frequency domains u and μ. (b) N equally spaced fractional Fourier domains. (c) Block diagram of the FFDD. (Kutay and others 1998a, b, 1999)

SVD (Strang 1988, Jain 1989) provides a context in which to understand the FFDD. While the FFDD may not match the SVD's central importance, it is of fundamental importance in its own right as an alternative which offers complementary insight and understanding.

We will denote the N-point ath order fractional Fourier transform matrix as $\mathbf{F}_N^a$. The columns of the inverse transform matrix $\mathbf{F}_N^{-a}$ constitute an orthonormal basis for the ath domain, just as the columns of the identity matrix constitute a basis for the time domain and the columns of the ordinary inverse DFT matrix constitute a basis for the frequency domain.

Now, let $\mathbf{H}$ be a complex $N_{\text{out}} \times N_{\text{in}}$ matrix and $\{a_1, a_2, \dots, a_N\}$ a set of $N = \max(N_{\text{out}}, N_{\text{in}})$ distinct real numbers such that $-1 < a_1 < a_2 < \cdots < a_N \leq 1$. For instance, we may take the a_k uniformly spaced in this interval. The corresponding fractional Fourier domains are illustrated in figure 10.37b. We define the FFDD of $\mathbf{H}$ as

$$\mathbf{H}_{N_{\text{out}} \times N_{\text{in}}} = \sum_{k=1}^{N} \mathbf{F}_{N_{\text{out}}}^{-a_k} \, (\boldsymbol{\Lambda}_{\mathbf{h}_k})_{N_{\text{out}} \times N_{\text{in}}} \, \left(\mathbf{F}_{N_{\text{in}}}^{-a_k}\right)^{\mathrm{H}}, \tag{10.55}$$

where the $\boldsymbol{\Lambda}_{\mathbf{h}_k}$ are $N_{\text{out}} \times N_{\text{in}}$ diagonal matrices with $N' = \min(N_{\text{out}}, N_{\text{in}})$ complex elements. Starting from the upper left corner, the lth diagonal element of $\boldsymbol{\Lambda}_{\mathbf{h}_k}$ is denoted as h_{kl}, $l = 1, 2, \dots, N'$ (the lth element of the column vector $\mathbf{h}_k$). When $\mathbf{H}$ is Hermitian (skew Hermitian), $\mathbf{h}_k$ is real (imaginary). We also recall that $\left(\mathbf{F}_{N_{\text{in}}}^{-a_k}\right)^{\mathrm{H}} = \mathbf{F}_{N_{\text{in}}}^{a_k}$. The FFDD always exists and is unique, as will be discussed further below.

Comparing and contrasting the FFDD with the SVD will help gain insight into the FFDD. If we compare one term on the right-hand side of equation 10.55 with the right-hand side of equation 10.53, we see that they are similar in that they both consist of three terms of corresponding dimensionality, the first and third being unitary matrices and the second being a diagonal matrix. But whereas the columns of $\mathbf{U}$ and $\mathbf{V}$ constitute orthonormal bases specific to $\mathbf{H}$, the columns of $\mathbf{F}_{N_{\text{out}}}^{-a_k}$ and $\mathbf{F}_{N_{\text{in}}}^{-a_k}$ constitute orthonormal bases for the a_kth fractional Fourier domain. Customization of the decomposition is achieved through the coefficients h_{kl} (and perhaps also the orders a_k).

When $\mathbf{H}$ is a square matrix of dimension N, the FFDD takes the simpler form

$$\mathbf{H} = \sum_{k=1}^{N} \mathbf{F}^{-a_k} \, \mathbf{\Lambda}_{\mathbf{h}_k} \, (\mathbf{F}^{-a_k})^{\mathrm{H}}, \tag{10.56}$$

where all matrices are $N \times N$. The continuous counterpart of the FFDD is similar to this equation, with the summation being replaced by an integral over a (Yetik and others 2000a).

A natural extension of the FFDD would be the linear canonical domain decomposition (LCDD) based on linear canonical transforms, as an extension of the work reported in Barshan, Kutay, and Ozaktas 1997.

10.8.2 Construction of the fractional Fourier domain decomposition

Denoting the lth columns of $\mathbf{F}_{N_{\text{out}}}^{-a_k}$ and $\mathbf{F}_{N_{\text{in}}}^{-a_k}$ as $\mathbf{w}_{l/N_{\text{out}}}^{-a_k}$ and $\mathbf{w}_{l/N_{\text{in}}}^{-a_k}$, the kth term of the summation in equation 10.55 can be written as an outer product $\sum_{l=1}^{N'} h_{kl} \, \mathbf{w}_{l/N_{\text{out}}}^{-a_k} \left(\mathbf{w}_{l/N_{\text{in}}}^{-a_k} \right)^{\mathrm{H}}$ so that equation 10.55 can be rewritten as

$$\mathbf{H} = \sum_{k=1}^{N} \sum_{l=1}^{N'} h_{kl} \, \mathbf{w}_{l/N_{\text{out}}}^{-a_k} \left(\mathbf{w}_{l/N_{\text{in}}}^{-a_k} \right)^{\mathrm{H}}. \tag{10.57}$$

To a certain extent, the inner summation resembles the outer product form of the SVD given in equation 10.54. The $N_{\text{out}} \times N_{\text{in}}$ matrices $\mathbf{w}_{l/N_{\text{out}}}^{-a_k} \left(\mathbf{w}_{l/N_{\text{in}}}^{-a_k} \right)^{\mathrm{H}}$ are of unit rank since they are the outer product of vectors. We will denote these matrices by $\mathbf{P}_{kl}$ so that

$$\mathbf{H} = \sum_{k=1}^{N} \sum_{l=1}^{N'} h_{kl} \mathbf{P}_{kl}. \tag{10.58}$$

This equation is simply an expansion of $\mathbf{H}$ in terms of the basis matrices $\mathbf{P}_{kl}$, $1 \le k \le N$, $1 \le l \le N'$, where the h_{kl} serve as the weighting coefficients of the expansion.

Equation 10.58 is a linear relation between the matrices $\mathbf{H}$ and h_{kl}, with the four-dimensional tensor $\mathbf{P}_{kl}$ representing the transformation between them. Let $\underline{\mathbf{H}}$ denote the $N_{\text{out}} N_{\text{in}} \times 1$ vector obtained by stacking the columns of $\mathbf{H}$ and let $\underline{\mathbf{h}}$ denote the $NN' \times 1$ vector obtained by stacking the column vectors $\mathbf{h}_1, \mathbf{h}_2, \ldots, \mathbf{h}_N$ on top of each other. Notice that we always have $NN' = \max(N_{\text{out}}, N_{\text{in}}) \min(N_{\text{out}}, N_{\text{in}}) = N_{\text{out}} N_{\text{in}}$. These column orderings determine a corresponding ordering which converts the four-dimensional tensor (or two-dimensional array of matrices) $\mathbf{P}_{kl}$ into a square matrix $\tilde{\mathbf{P}}$ of dimensions $N_{\text{out}} N_{\text{in}} \times N_{\text{in}} N_{\text{out}}$. (The vector obtained by stacking the columns of the matrix $\mathbf{P}_{kl}$, for a specific kl, goes into the $[(k-1)N' + l]$th column of the matrix $\tilde{\mathbf{P}}$.) With these conventions, we can write equation 10.58 as the square matrix equation $\underline{\mathbf{H}} = \tilde{\mathbf{P}}\underline{\mathbf{h}}$. This equation will have a unique solution for $\underline{\mathbf{h}}$ and thus h_{kl} if and only if the columns of $\tilde{\mathbf{P}}$ are linearly independent. Since the columns of $\tilde{\mathbf{P}}$ are merely column-stacked versions of the basis matrices $\mathbf{P}_{kl}$, this is the same as linear independence of these basis matrices. Recalling the definition of these matrices (just before equation 10.58), their linear independence follows from the fact that the inner product of any column of $\mathbf{F}^a$ with any column of $\mathbf{F}^{a'}$ ($a' \neq a$) is always nonzero. Thus *the FFDD always exists and is unique* (for given a_k).

Now, let $\mathbf{H}$ denote some linear matrix operator. Equation 10.55 represents a decomposition of this operator into N terms. Each term, taken by itself, corresponds to filtering in the a_kth fractional Fourier domain, where an a_kth order forward transform is followed by multiplication with a filter function $\mathbf{h}_k$ and concluded with an inverse a_kth order transform. All terms taken together, the FFDD can be represented by the block diagram shown in figure 10.37c and interpreted as the decomposition of an operator into fractional Fourier domain filters of different orders. An arbitrary linear operator $\mathbf{H}$ will in general not correspond to multiplicative filtering in the time or frequency domain or in any other single fractional Fourier domain. However, $\mathbf{H}$ can always be expressed as a combination of filtering operations in different fractional domains. *A sufficient number of different-ordered fractional Fourier domain filtering operations "span" the space of all linear operations.* The fundamental importance of the FFDD is that it shows how an arbitrary linear system can be decomposed into this complete set of domains in the time-frequency plane.

10.8.3 Pruning and sparsening

If $\mathbf{H}$ represents a shift-invariant system, all filter coefficients except those corresponding to $a_k = 1$ will be zero. More generally, different domains will make varying contributions to the decomposition. By eliminating domains for which the coefficients $h_{k1}, h_{k2}, \ldots, h_{kN'}$ are small, significant savings in storing and implementing $\mathbf{H}$ become possible. This procedure, which we refer to as *pruning* the FFDD, is the counterpart of truncating the SVD. An alternative to this selective elimination procedure will be referred to as *sparsening*, in which one simply employs a more coarsely spaced set of domains. In any event, the resulting smaller number of domains will be denoted by $M < N$. The upper limit of the summation in equation 10.55 is replaced by M and the equality is replaced by approximate equality. The equation $\underline{\mathbf{H}} = \tilde{\mathbf{P}}\underline{\mathbf{h}}$ is likewise replaced by $\underline{\mathbf{H}} \approx \tilde{\mathbf{P}}\underline{\mathbf{h}}$. If we solve this in the least-squares sense, minimizing $\|\underline{\mathbf{H}} - \tilde{\mathbf{P}}\underline{\mathbf{h}}\|$, we can find the filter coefficients resulting in the best *M-domain approximation* to $\mathbf{H}$. This procedure amounts to projecting $\mathbf{H}$ onto the subspace spanned by the MN' basis matrices, which now do not span the whole space.

The correspondence between the pruned FFDD and multichannel filtering should by now be evident. In other words, it is possible to interpret multichannel filtering configurations as pruned FFDDs. Since the fractional Fourier transform can be computed in $O(N \log N)$ time, implementation of the pruned version of figure 10.37c takes $O(MN \log N)$ time. If an acceptable approximation to **H** can be found with a relatively small value of M, this can be much smaller than the time $O(N_{\text{out}} N_{\text{in}})$ associated with direct implementation of the linear system. Likewise, optical implementation of the pruned FFDD requires a space-bandwidth product of $O(MN)$, as opposed to $O(N_{\text{out}} N_{\text{in}})$ for direct implementation.

The SVD of **H** can also be used to obtain an efficient approximate implementation if we keep only the M largest singular values and discard the others, by truncating equation 10.54 to M terms. Since each term is of outer product form, the overall implementation time of the linear system in this form is $O(MN)$.

The multichannel filtering examples discussed in section 10.4 can be viewed as illustrations of the use of the pruned or sparsened FFDD. For instance, let us reconsider the example on page 416 associated with figure 10.32. There we saw that the optimal general linear estimator **H** could be approximated with a normalized mean square Frobenius error of 5.2% by using $M = 6$ domains. On the other hand, if we approximate **H** by truncating equation 10.54 to M terms, $M = 6$ results in an error of 20%, demonstrating an instance where the FFDD yields better accuracy than the SVD. As another example, let us reconsider the example on page 416 associated with figure 10.31. There we saw that $M = 3$ results in a normalized mean square signal recovery error of $< 1\%$. Truncating the SVD to $M = 3$ terms, on the other hand, results in an error of 18%. Whereas the pruned FFDD gives good results for $M \geq P$, the SVD approach gives comparable results only when $M \geq Q$. These examples are not meant to imply that the FFDD is generally superior to the SVD; it is also possible to find many examples in which the SVD approach gives better results.

As we discussed on page 394, one way of using multichannel configurations or the FFDD is to take the M-domain decomposition as a constraint on the linear estimator and optimize over the filters $\mathbf{h}_k$ to minimize the mean square estimation error. One may construct similar constrained optimization problems by using the truncated SVD. However, this leads to a much more difficult nonlinear optimization problem because $\mathbf{u}_k$ and $\mathbf{v}_k$ in equation 10.54 are also unknowns, whereas the only unknowns in equation 10.55 are the $\mathbf{h}_k$, leading to a linear optimization problem.

We have not addressed the problem of optimally choosing the orders a_k, which corresponds to choosing the basis matrices. When $M = N$ the basis matrices form a complete set and any choice is acceptable. However, certain choices may offer better numerical stability. When $M < N$ the choice of a_k may reflect our knowledge about the ensemble of matrices **H** we wish to approximate. This knowledge may be statistical or in the form of restrictions on the set of matrices possible. It is also possible to choose the orders optimally for a given specific matrix.

Exploring the full range of properties and applications of the FFDD is beyond the scope of this book. Potential applications other than fast implementation of linear systems include data compression, optimal filtering and estimation, and regularization of ill-posed inverse problems, all of which may be based on the same basic idea of appropriately pruning or weighting the different domains.

10.9 Repeated filtering in the ordinary time and frequency domains

Here we show that any multistage fractional Fourier domain filtering configuration is equivalent to an appropriately chosen sequence of multiplicative filters inserted between appropriately scaled ordinary Fourier transform stages. Thus every operation that can be accomplished by repeated filtering in fractional Fourier domains can also be accomplished by repeated filtering alternately in the ordinary time and frequency domains. This section is closely related to section 9.8.

Ordinary Fourier domain filtering involves multiplication of the Fourier transform $F(\mu)$ of the input $f(u)$ with a filter function $H(\mu)$ to obtain the Fourier transform $G(\mu) = H(\mu)F(\mu)$ of the filtered output signal. This type of filtering allows the realization of shift-invariant (convolution-type) linear operations only: $g(u) = h(u) * f(u)$. The relation between the input and output of the system can also be expressed in operator notation as

$$g = \mathcal{F}^{-1}\Lambda_H \mathcal{F} f. \tag{10.59}$$

Now, let us consider the multistage filtering configuration shown in figure 10.38a. Each fractional Fourier transform stage transforms from one fractional domain to another, where a multiplicative filter is applied. In other words, the signal is repeatedly filtered in several consecutive fractional Fourier domains. Note that this configuration is slightly more general than the configuration in figure 10.4a since additional filters have been inserted just after the input and before the output. In operator notation, the relation between the input and output of a system with M stages can be expressed as

$$g = \Lambda_{h_{M+1}} \mathcal{F}^{a_M} \Lambda_{h_M} \cdots \Lambda_{h_3} \mathcal{F}^{a_2} \Lambda_{h_2} \mathcal{F}^{a_1} \Lambda_{h_1} \mathcal{F}^{a_0} \Lambda_{h_0} f. \tag{10.60}$$

Our claim is that any system of the form defined by this equation (figure 10.38a) is equivalent to a system composed of filters inserted between ordinary Fourier transform stages, appropriately scaled (figure 10.38b). Each time a Fourier transform is applied we alternate between the time and frequency domains, where multiplicative filters are applied. In operator notation,

$$g = \Lambda_{h'_{M+1}} \mathcal{F}_{\kappa_M} \Lambda_{h'_M} \cdots \Lambda_{h'_3} \mathcal{F}_{\kappa_2} \Lambda_{h'_2} \mathcal{F}_{\kappa_1} \Lambda_{h'_1} \mathcal{F}_{\kappa_0} \Lambda_{h'_0} f, \tag{10.61}$$

where $\mathcal{F}_{\kappa_k}$ is a scaled ordinary Fourier transform operator which maps a function $f(u)$ to $\int \exp(-i2\pi uu'/\kappa_k^2) f(u')\, du'$. The claim is that by appropriate choice of filters h'_k and scale factors κ_k, this relation between f and g can be made the same as that given in equation 10.60.

The proof is elementary. Upon examining the kernel of the fractional Fourier transform given in equation 4.4, we observe that calculating the fractional transform $f_a(u)$ amounts to multiplying $f(u)$ by a chirp function, taking its scaled ordinary Fourier transform, and multiplying the result by another chirp function. (Whereas this approach serves the purpose of the present argument, it is not necessarily the best way of decomposing the transform for the purpose of actual computation, as discussed in section 6.7.) The pre and post chirp multiplications can be absorbed into the multiplicative filters preceding and following the fractional Fourier transform stage, leaving us with a scaled ordinary Fourier transform.

(a)

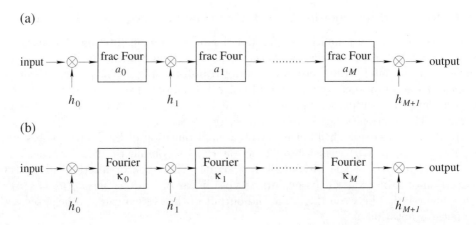

(b)

Figure 10.38. (a) A sequence of multiplicative filters inserted between fractional Fourier transform stages, each of which is characterized by its order a_k. (b) The system modeled as a sequence of multiplicative filters inserted between ordinary Fourier transform stages, each of which is characterized by its scale factor κ_k. (Ozaktas 1996)

Consecutive multiplicative filtering in the time and frequency domains is of course equivalent to alternately convolving and multiplying the signal with a sequence of functions. Such a chain of convolutions alternating with multiplications cannot be further reduced. This is because the order of convolution and multiplication cannot be interchanged. (If it could, such interchanges would allow the multiplications and convolutions to be segregated and reduced to a single convolution followed by a single multiplication, or the other way around.) To be more precise, let us assume that the output of a system $g(u)$ is related to its input $f(u)$ through the relation $g(u) = h_2(u)*[h_1(u)f(u)]$. Now, it is *not* possible to find any functions $h_3(u)$ and $h_4(u)$ such that the same system can be expressed in the form $g(u) = h_4(u)[h_3(u) * f(u)]$. Thus, the chain of convolutions and multiplications cannot be reduced, as claimed. (Proving that convolution and multiplication cannot be interchanged is not difficult. In the discrete case this is a consequence of the fact that diagonal and circulant matrices do not commute with each other and the DFT matrix.)

The result that repeated fractional Fourier domain filtering can be reduced to repeated ordinary Fourier domain filtering can be easily generalized. The ordinary Fourier transform is no more privileged than fractional transforms of other orders. The fractional Fourier transform stages in a repeated filtering configuration can be replaced with appropriately scaled fractional transforms of any desired order. Thus, by appropriate choice of scale factors and multiplicative filters, repeated filtering in any given sequence of fractional domains can be made equivalent to repeated filtering in any other desired sequence of fractional domains. In particular, we can choose to alternate between any two chosen domains. That is, applying multiplicative filters alternately in any two chosen domains (provided their orders do not differ by an integer multiple of 2) allows us to do everything that can be done by any configuration of the form given in figure 10.38a.

It is also possible to show that the scale factors κ_k appearing in equation 10.61 can be eliminated or made equal to each other. Since the Fourier transform of a scaled function is a scaled version of its Fourier transform, these scale factors can be migrated through the filters and transform stages and collected at either end of the system (by also replacing the filters with their appropriately scaled versions).

To summarize, we have shown that repeated filtering systems employing fractional transforms (as in figure 10.38a) can be reduced to repeated filtering systems employing only the ordinary Fourier transform (as in figure 10.38b). Applying multiplicative filters alternately in the time and frequency domains allows us to do everything that can be done by applying filters in fractional Fourier domains.

As discussed on page 376 in an optical context, this result does not compromise the conceptual and practical utility of the fractional Fourier transform. The fractional transform may be conceptually indispensable in devising an algorithm or designing an effective filter, even if the system is then reduced to one which does not employ fractional transforms. Furthermore, one would not necessarily engage in such a reduction, since the implementation of figure 10.38a is not more costly than figure 10.38b.

From a practical viewpoint, the implementation of the necessary filters may be much easier in certain domains, as compared to others. For instance, the filter required in one domain may be a simple window or mask, whereas the filter required in another domain may be a complex and highly oscillatory function. Furthermore, the precision with which filters must be realized may differ from domain to domain. In conclusion, the presented results should be used to increase the number of alternative realizations, not to reduce them to one. These alternative realizations provide additional degrees of freedom which may allow us to deal effectively with constraints arising from sampling and quantization.

10.10   Multiplexing in fractional Fourier domains

We now discuss the concept of multiplexing in fractional Fourier domains. First, let us review what it means to multiplex in the time or space domain (time-division or space-division multiplexing) and the frequency domain (frequency-division or wavelength-division multiplexing). Multiplexing in the u domain involves packing together signals whose representations are compact in this domain. We simply shift the several representations in this domain with respect to each other so that they do not overlap and can be easily separated later on. (If the original signals are of very large temporal or spatial extent, they can be cut into smaller pieces and interleaved with each other.) On the other hand, multiplexing in the frequency domain involves packing together signals which are compact in this domain. We simply shift the several representations in this domain with respect to each other so that they do not overlap.

It is instructive to view these processes in the u-μ plane. Let the total extent of the time or space "aperture" of our system be Δu_{sys} and the double-sided bandwidth of our system be $\Delta \mu_{\mathrm{sys}}$. This defines a region in the u-μ plane which we are free to utilize in packing signals. First, let us assume that the extents of our signals Δu are about equal to Δu_{sys} but that their bandwidths $\Delta \mu$ are smaller than $\Delta \mu_{\mathrm{sys}}$ (figure 10.39a). Then it is natural to pack several of these signals by using frequency-domain multiplexing, as shown in the figure. Second, let us assume that the bandwidths of our signals $\Delta \mu$ are about equal to $\Delta \mu_{\mathrm{sys}}$ but that their extents Δu are smaller than Δu_{sys} (figure 10.39b).

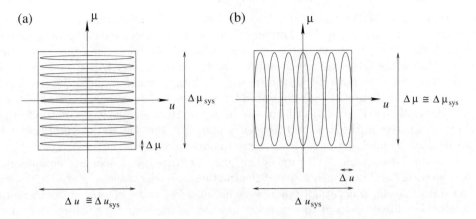

Figure 10.39. (a) Multiplexing in the frequency domain; (b) multiplexing in the time or space domain (Ozaktas and others 1994a).

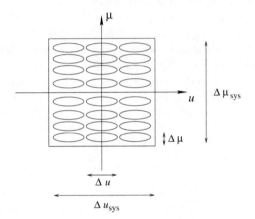

Figure 10.40. Multiplexing in both time (or space) and frequency (Ozaktas and others 1994a).

Then it is natural to employ time- or space-domain multiplexing.

If both Δu and $\Delta \mu$ are smaller than Δu_{sys} and $\Delta \mu_{\text{sys}}$ respectively, then both time-domain (or space-domain) and frequency-domain multiplexing can be employed together (figure 10.40). This is accomplished by shifting the Wigner distributions of the signals to be multiplexed by appropriate amounts in time (or space) and frequency.

Let us consider a set of signals which have Wigner distributions whose supports are of the oblique form illustrated in figure 10.41a. It is evident that the scheme in figure 10.41a, motivated by figure 10.40, is not the most efficient way of multiplexing such signals. The scheme illustrated in figure 10.41b is much more efficient. To pack signals in this manner, we transform them into the appropriate domain u_a and multiplex them as in figure 10.40 in that domain. This procedure can be generalized to signals whose Wigner distributions are not identical. An additional consideration is

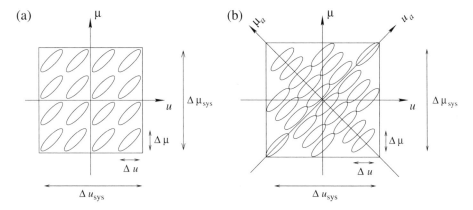

Figure 10.41. Inefficient (a) and efficient (b) multiplexing of signals with oblique Wigner
distributions (Ozaktas and others 1994a).

to take into account the time-frequency (or space-frequency) distribution of any noise
or interference and to avoid regions in which they are concentrated.

10.11 Historical and bibliographical notes

The optimal Wiener filtering problem in fractional Fourier domains (sections 10.2
and 10.6) was first treated in Kutay 1995, Kutay and others 1995, 1997 based on an
idea in Ozaktas and others 1994a, and independently in Zalevsky 1996, Zalevsky and
Mendlovic 1996a. Its applications to optics and image processing were discussed in
Kutay and Ozaktas 1998. The concept was generalized to filtering in linear canonical
transform domains in Barshan, Kutay, and Ozaktas 1997. The concept of repeated
(multistage) filtering in fractional and ordinary Fourier domains and its applications
to time- or space-variant filtering were first suggested in Ozaktas and Mendlovic 1993b
and Ozaktas and others 1994a, and were more fully treated in Erden 1997, Erden and
Ozaktas 1998, Ozaktas, Erden, and Kutay 1998, and Erden, Kutay, and Ozaktas 1999a.
The concept of parallel (multichannel) filtering and generalized filter circuits appeared
in Kutay and others 1998a, b, c and Kutay 1999. While section 10.5 is based mostly on
Ozaktas and others 1994a, the concept of fractional convolution was suggested earlier
in Mustard 1990, 1998, Mendlovic and Ozaktas 1993a, and Ozaktas and Mendlovic
1993a. The contents of section 10.7 appeared in Kutay and others 1998c, 2000. The
fractional Fourier domain decomposition was introduced in Kutay and others 1999.
The contents of section 10.9 first appeared in Ozaktas 1996. The potential for efficient
multiplexing and data compression was first noted in Ozaktas and others 1994a.

Brief reviews of the applications discussed in this chapter may be found in Erden,
Kutay, and Ozaktas 1999b and Ozaktas, Kutay, and Mendlovic 1998, 1999. Certain
parts of sections 10.2 and 10.3 have been derived from Ozaktas, Kutay, and Mendlovic
1998, 1999.

11

Applications of the Fractional Fourier Transform to Matched Filtering, Detection, and Pattern Recognition

11.1 Introduction

Estimation and detection are two basic applications of the theory of random variables and processes which are widely used in signal processing and communications. Chapter 10 dealt primarily with estimation problems—of which the prototype is the optimal Wiener filtering problem—and the synthesis of shift-variant systems. This chapter deals primarily with detection problems—of which the prototype is the matched filtering problem—and pattern recognition.

The applications of the fractional Fourier transform in signal analysis and processing are not limited to those we have discussed in chapter 10 and the present chapter. We discussed in chapter 9 the application of the transform to signal reconstruction and phase retrieval problems. The applications of the transform in the context of time- or space-frequency analysis were discussed in chapters 4 and 5, where we saw that the transform is not only related to well-established representations, but that it also allows the definition of new ones. An interesting and natural application of the fractional Fourier transform arises in the context of perspective projections in imaging (Yetik and others 2000b).

11.2 Fractional correlation

The correlation operation is often used to compare two signals, or to search for a given (usually smaller) pattern in a larger signal. It is essentially the same as the convolution operation, but interpreted differently. When a small signal is to be correlated with a larger one, it is usually more efficient to calculate it directly in the time or space domain. When large signals of similar extent are to be correlated, it is usually more efficient to employ the fast Fourier transform and multiply the Fourier transform of the first with the conjugate of the Fourier transform of the second. Optically, there are many ways of correlating two signals (Goodman 1996, VanderLugt 1964, Weaver and Goodman 1966, Lohmann and Werlich 1968); one common approach is to implement it just like a convolution by using a $4f$ spatial filtering system (page 247).

Essentially being the same as convolution, the correlation operation is intrinsically shift-invariant; shift of either of the input signals or patterns results in a corresponding shift in the output. This is often considered a desirable property, since a pattern can be detected regardless of where it is and, furthermore, the position of the peak at the output tells us where the pattern is. We will see that the correlation operation in fractional Fourier domains, or fractional correlation, in general exhibits a shift-variant dependence on the position of the pattern to be detected. Exploited properly, such a dependence finds many applications; for instance, when we wish to detect an object only if it is situated within a certain region but not otherwise, or when we wish the output to contain weighting information such that the weight is proportional to the closeness of the pattern to some designated point. Another situation is when we want to emphasize the central pixels of an image (which are considered more important) in the matching process in favor of the peripheral pixels. Indeed, the need for space-variant detection often arises in practice and several optical systems have been suggested for this purpose; see the references in Mendlovic, Ozaktas, and Lohmann 1995. With the use of fractional correlation, the degree of space variance becomes adjustable by varying the fractional transform order, including the special case of complete space invariance associated with ordinary correlation.

Let $g(u)$ denote the ordinary correlation of an input $f(u)$ with a reference $h(u)$:

$$g(u) = f(u) \star h(u) = \int f(u' + u)h^*(u')\, du' = \int f(u' + u/2)h^*(u' - u/2)\, du',$$
(11.1)

so that $G(\mu) = F(\mu)H^*(\mu)$. If the input $f(u)$ matches the reference $h(u)$, then at $u = 0$ we observe the correlation peak

$$g(0) = \int |h(u)|^2\, du.$$
(11.2)

There exist basically two different ways of fractionalizing the correlation operation. The first is to speak about "correlation in the ath order fractional Fourier domain," and thus write

$$g_a(u) = f_a(u) \star h_a(u).$$
(11.3)

This is consistent with the notion of "convolution in the ath order fractional Fourier domain" introduced in chapter 10. When $a = 0$ this simply corresponds to ordinary correlation in the time or space domain. When $a = 1$ it corresponds to correlation of the Fourier transforms, which implies multiplication of the original functions: $g(u) = f(u)h^*(u)$.

The second approach is to note that since $G(\mu) = F(\mu)H^*(\mu)$ represents a multiplication of ordinary Fourier transforms, the "fractional correlation" $g(u)$ of $f(u)$ and $h(u)$ can be defined by multiplying their ath order transforms as

$$g_a(u) = f_a(u)h_a^*(u).$$
(11.4)

When $a = 0$ this corresponds to multiplications of the two functions in the form $g(u) = f(u)h^*(u)$. When $a = 1$ it corresponds to $G(\mu) = F(\mu)H^*(\mu)$ associated with ordinary correlation. This approach is of course different than that represented by equation 11.3, but not essentially so, for the same reasons discussed on page 420. Indeed, the difference between the two approaches amounts to nothing more than replacement of the order $a \rightarrow a - 1$.

While these approaches represent the purest way of fractionalizing the correlation operation, many variations and extensions have been considered. Some of these will also be covered here since they have found a variety of interesting applications. Specifically, we will concentrate on the more general definition phrased as follows: the fractional correlation of $f(u)$ with $h(u)$ is obtained by forming the product $f_{a_1}(u)h_{a_1}^*(u)$ and then taking the inverse a_2th transform of this product. When $a_1 = a_2 = a$, this definition reduces to that given in equation 11.4. Although this definition is somewhat unnatural since the final result belongs to the $(a_1 - a_2)$th domain, rather than the original time (or space) domain, it provides an additional degree of flexibility which has been beneficially exploited in applications. Mathematically, this form of fractional correlation is defined through (Mendlovic, Ozaktas, and Lohmann 1995)

$$g_{a_2}(u) = f_{a_1}(u)h_{a_1}^*(u) \tag{11.5}$$

leading to

$$g(u) = \iint K_{\text{fc}}(u, u', u'')f(u')h^*(u')\, du'\, du'', \tag{11.6}$$

where

$$
\begin{aligned}
K_{\text{fc}}(u, u', u'') &= \int K_{-a_2}(u, u''')K_{a_1}(u''', u')K_{-a_1}(u''', u'')\, du''' \\
&= \frac{\text{sgn}(\cot \alpha_2)\sqrt{1 - i\tan \alpha_2}}{|\sin \alpha_1|} \exp\left[-i\pi(\cot \alpha_2\, u^2 + \cot \alpha_1(u''^2 - u'^2))\right] \\
&\quad \times \exp\left[i\pi \tan \alpha_2\, (\csc \alpha_1(u' - u'') - \csc \alpha_2\, u)^2\right].
\end{aligned}
\tag{11.7}
$$

As usual, $\alpha = a\pi/2$. When $a_2 = a_1 = a$, this reduces to the following result (Mendlovic, Ozaktas, and Lohmann 1995):

$$
\begin{aligned}
K_{\text{fc}}(u, u', u'') &= \int K_{-a}(u, u''')K_a(u''', u')K_{-a}(u''', u'')\, du''' \\
&= \frac{\text{sgn}(\cot \alpha)\sqrt{1 - i\tan \alpha}}{|\sin \alpha|} \exp\left[-i\pi \cot \alpha(u^2 + u''^2 - u'^2)\right] \\
&\quad \times \exp\left[i\pi \sec \alpha \csc \alpha\, (u' - u'' - u)^2\right].
\end{aligned}
\tag{11.8}
$$

When $a = 1$ this reduces to the kernel corresponding to the ordinary correlation operation. Alternatively, when we take $a_1 = a$ and $a_2 = \pm 1$, the kernel $K_{\text{fc}}(u, u', u'')$ given in equation 11.7 reduces to a form involving the delta function (Mendlovic, Ozaktas, and Lohmann 1995):

$$
\begin{aligned}
K_{\text{fc}}(u, u', u'') &= |\sin \alpha|^{-1}\, e^{i\pi \cot \alpha(u'^2 - u''^2)}\, \delta\left(\csc \alpha(u' - u'') \mp u\right) \\
&= e^{i\pi \cot \alpha(u'^2 - u''^2)}\, \delta(u' - u'' \mp \sin \alpha\, u),
\end{aligned}
\tag{11.9}
$$

and the correlation is given by

$$g(u) = e^{i\pi \cos \alpha \sin \alpha\, u^2} \int f(u' \pm \sin \alpha\, u) h^*(u')\, e^{\pm i2\pi \cos \alpha\, uu'}\, du'$$

$$= e^{-i\pi \cos \alpha \sin \alpha\, u^2} \int f(u') h^*(u' \mp \sin \alpha\, u)\, e^{\pm i2\pi \cos \alpha\, uu'}\, du'. \tag{11.10}$$

If we evaluate $g(u)$ at $u = 0$, for both $a_2 = +1$ and $a_2 = -1$ we obtain

$$g(0) = \int f(u') h^*(u')\, du'. \tag{11.11}$$

A definition of fractional correlation slightly more general than equation 11.5 has also been considered (Lohmann, Zalevsky, and Mendlovic 1996):

$$g_{a_3}(u) = f_{a_1}(u) h_{a_2}(u). \tag{11.12}$$

To obtain the fractional correlation $g(u)$ according to this definition, we multiply the a_1th transform of $f(u)$ with the a_2th transform of $h(u)$ and take the inverse a_3th transform of the product. If $f(u)$ and $h(u)$ are real, this reduces to equation 11.5 for $a_3 \to a_2$, $a_1 \to a_1$, $a_2 \to -a_1$, since $h_{-a_1}(u) = h^*_{a_1}(u)$ for real $h(u)$. The explicit expression for $g(u)$ is rather complicated and is not given here. However, it can be simplified considerably when

$$\cot(a_1 \pi/2) + \cot(a_2 \pi/2) = \cot(a_3 \pi/2). \tag{11.13}$$

Assuming this condition holds, the magnitude of the correlation can be expressed as

$$|g(u)| = \left| A_{\mathrm{fc}} \int f(u') h(u' - \sin \alpha\, u) \right.$$

$$\left. \times \exp\left[i\pi \left(u'^2 \left(\cot \alpha_1 + \frac{\cos \alpha_2 \sin \alpha_2}{\sin^2 \alpha_1} \right) - 2u'u \frac{\cos \alpha_2 \sin \alpha_2}{\sin \alpha_1 \sin \alpha_3} \right) \right] du' \right|, \tag{11.14}$$

$$A_{\mathrm{fc}} = \sqrt{1 - i \cot \alpha_1}\, \sqrt{1 - i \cot \alpha_2}\, \sqrt{1 + i \cot \alpha_3}\, |\sin \alpha_2|.$$

Several special cases of equation 11.13 have been discussed in Lohmann, Zalevsky, and Mendlovic 1996. The case $a_1 = -a_2 = a$ and $a_3 = \pm 1$, which we have already discussed, is merely one of these.

The fractional correlation operations defined above can be digitally computed in $O(N \log N)$ time since the fractional Fourier transform can be computed in $O(N \log N)$ time. Optical fractional correlation was experimentally demonstrated first in Mendlovic and others 1995a. Flexible practical fractional correlators allowing real-time control of the parameters were presented in Zalevsky, Mendlovic, and Caulfield 1997b. In the systems described in this work, one can change the order and thus the amount of space variance by adjusting only the axial location of the filter. The separations of the optical elements and the focal lengths of the lenses need not be changed. Neither is it necessary to recalculate or reencode the filter; the same filter can be used by simply adjusting its axial location. Clearly, these attributes are of great value in practice.

11.3 Controllable shift invariance

Correlation in fractional Fourier domains provides a flexible framework in which the degree of space variance can be adjusted through the fractional order parameter a. It is well known that the ordinary correlation operation is shift-invariant, meaning that a shift of either of the input functions results in a shifted version of the original correlation output. This is a direct consequence of the fact that the correlation operation is essentially a convolution, which is by definition a shift-invariant operation. In applications where an input object is correlated with a reference (matched filtering), this means that input objects matching the reference will be detected regardless of their location, with the position of the correlation peak obtained flagging the location of the detected object.

It should be evident that the fractional correlation operation will in general not be shift-invariant, unless the orders are chosen such that it reduces to ordinary correlation. However, since the fractional Fourier transform is continuous in the order, it is clear that as the orders deviate away from the values corresponding to ordinary correlation, the "amount" of shift invariance will also continuously decrease from complete shift invariance to no shift invariance. This possibility of controlling the amount of shift invariance (or shift variance) by adjusting the order, as will be seen, opens the doors to a variety of flexible processing schemes. (Since we are not defining a measure of the amount of shift invariance, statements referring to this notion should be regarded as being suggestive, rather than literal.)

Let us consider the magnitude of equation 11.10 (Lohmann, Zalevsky, and Mendlovic 1996):

$$|g(u)| = \left| \int f(u')h^*(u' - \sin \alpha \, u) \, e^{i2\pi \cos \alpha \, uu'} \, du' \right|$$

$$= \left| \int f(u' + \sin \alpha \, u/2)h^*(u' - \sin \alpha \, u/2) \, e^{i2\pi \cos \alpha \, uu'} \, du' \right|. \qquad (11.15)$$

Now, let the input $f(u)$ of this system be a shifted version $h_0(u - u_{\text{inp}})$ of the object $h_0(u)$ which we are trying to detect. Likewise, in the general case let $h(u)$ be a shifted version $h_0(u - u_{\text{ref}})$ of $h_0(u)$. The correlation peak will then be obtained at (Lohmann, Zalevsky, and Mendlovic 1996)

$$u_{\text{peak}} = (u_{\text{inp}} - u_{\text{ref}}) \csc(a\pi/2). \qquad (11.16)$$

This may be demonstrated by differentiating the correlation output and equating it to zero. Intuitively, u_{peak} is the value of u at which the arguments of the input and the reference appearing in equation 11.15 are matched so that their profiles are perfectly aligned, resulting in a maximally constructive contribution to the integral. The value of $|g(u)|$ at $u = u_{\text{peak}}$ is

$$|g(u_{\text{peak}})| = \left| \int |h_0(u)|^2 e^{i2\pi u(u_{\text{inp}} - u_{\text{ref}}) \cot(a\pi/2)} \, du \right|. \qquad (11.17)$$

When $a = 1$ (ordinary correlation) the peak is observed at $u_{\text{peak}} = u_{\text{inp}} - u_{\text{ref}}$ and $|g(u_{\text{peak}})| = \int |h_0(u)|^2 \, du$, representing the maximum value of the peak. In general, for an arbitrary value of a, the peak value will be equal to the maximum value of

$\int |h_0(u)|^2\, du$ if $u_{\text{inp}} = u_{\text{ref}}$ (the input is located at the same position as the reference). When $a = 1$ (ordinary correlation), even when $u_{\text{inp}} \neq u_{\text{ref}}$, we always obtain a peak of the same maximum height. However, for other values of a, the maximum peak will only be obtained when $u_{\text{inp}} = u_{\text{ref}}$; that is, when the position of the input object coincides with the position of the object in the reference template. As $|u_{\text{inp}} - u_{\text{ref}}|$ increases, the complex exponential in the integral oscillates more rapidly, and the peak becomes washed out. The frequency of oscillation depends on two factors: $(u_{\text{inp}} - u_{\text{ref}})$ and $\cot(a\pi/2)$. An increase in the magnitude of either means that the exponential will oscillate faster and wash out the peak. The value of the peak will decay with increasing $|u_{\text{inp}} - u_{\text{ref}}|$; that is, as the distance between the input and the reference increases. How quickly it decays depends on $\cot(a\pi/2)$, and thus the fractional transform order a. We see that by adjusting a we can control the sensitivity of the correlation peak to deviations of the position of the input from the position of the reference. To quantify this, let Δu denote the approximate support length of $|h_0(u)|^2$ (the extent over which it is significantly different than zero). Then, requiring that the argument of the exponential be less than π in order to prevent oscillations and washout, we obtain the following condition (Lohmann, Zalevsky, and Mendlovic 1996):

$$|u_{\text{inp}} - u_{\text{ref}}|\Delta u < \tan(a\pi/2). \tag{11.18}$$

This condition defines the maximum deviation of the input position from the reference position which can be tolerated to still get a peak comparable in size to the maximum. When the order $a = 1$ we see that the left-hand side can take on any value and we still get the full peak. However, for smaller values of a the peak decays with increasing deviation. The smaller the value of a, the faster the decay. The maximum correlation peak is obtained when $u_{\text{inp}} = u_{\text{ref}}$ and drops as u_{inp} moves away from u_{ref}, with the rate of decay being controlled by our choice of a. Thus, we are able to design systems which detect objects close to a certain point or within a certain region, but reject them far from this point or outside of the region, by properly choosing the order in conjunction with the detection threshold. Alternatively, we are able to design systems which not only detect the presence of an object, but also provide us with information as to its position encoded in the amplitude of the detection peak.

As a simple example, we consider the object shown in figure 11.1a. Figure 11.1b shows the autocorrelation peak when $a = 1$ and both input and reference are centered at the origin. As long as $a = 1$, the same correlation peak is obtained even when the input is shifted. Figure 11.1c shows the correlation peak when the input and reference are centered at the origin but $a = 0.5$. Finally, figure 11.1d shows the peak obtained when $a = 0.5$ and the input is shifted by a distance 0.8 away from the origin.

As a second example, consider the case where we wish to detect an object A only if it appears close to or within an interval around u_A and object B only if it appears close to or within an interval around u_B. The objects A and B centered at $u_A = -4$ and $u_B = 4$ are shown in figure 11.2a. Using the function shown in figure 11.2a as a reference, we obtain the correlation peak shown in figure 11.2b. On the other hand, if we interchange the positions of the objects as shown in figure 11.2c, we obtain the correlation pattern shown in figure 11.2d. The fractional order has been chosen as $a = 0.3$ in this example to ensure that the inequality given as equation 11.18 is comfortably violated when the positions of the objects are interchanged.

A two-dimensional version of this example will be discussed in section 11.7.

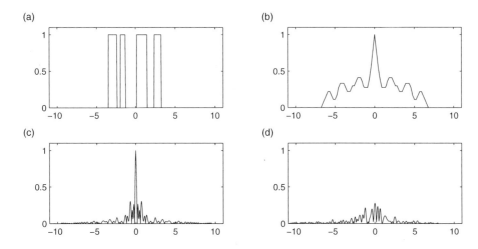

Figure 11.1. (a) Object to be detected; (b) ordinary autocorrelation of the object; (c) fractional correlation with $a = 0.5$ (magnitude); (d) fractional correlation with $a = 0.5$ when the input is shifted (magnitude). After Lohmann, Zalevsky, and Mendlovic 1996.

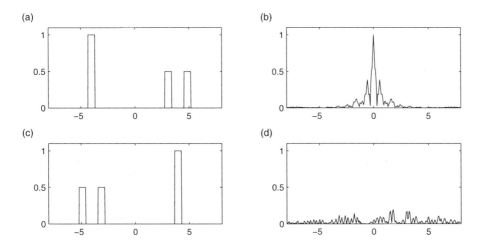

Figure 11.2. (a) The single taller pulse on the left is object A and the two shorter pulses on the right together constitute object B; (b) fractional correlation with a reference identical to the input shown in part a (magnitude); (c) new input with the positions of the two objects interchanged; (d) fractional correlation of the new input shown in part c with a reference identical to the function shown in part a (magnitude). After Lohmann, Zalevsky, and Mendlovic 1996.

11.4   Performance measures for fractional correlation

Here we will look at various performance measures commonly employed in optical pattern recognition for comparing ordinary correlation systems, and see that in the fractional case they are optimized by straightforward fractional generalizations of the filters which optimize them in the ordinary case (Bitran and others 1996). This section can be omitted without loss of continuity.

11.4.1   Fractional power filters

We consider the fractional correlation operation defined by equation 11.5 with $a_1 = a$ and $a_2 = 1$. Then the fractional correlation of $f(u)$ and $h(u)$, which we will denote by $g_{f,h}(u)$, can be expressed as follows:

$$g_{f,h}(u) = \int f_a(u_a) h_a^*(u_a) e^{i2\pi u_a u} \, du_a. \tag{11.19}$$

The function $h_a^*(u_a)$ appearing in this expression is the fractional generalization of the well-known ordinary matched filter (to which it reduces when $a = 1$). We will see that the matched filter is the filter which maximizes the signal-to-noise ratio. If it is desired to maximize performance measures other than the signal-to-noise ratio, it is necessary to use filters other than the matched filter. A one-parameter family of filter functions flexible enough to maximize a number of performance measures of interest is the family of *fractional power filters*, which are conventionally defined as follows (Kumar and Hassebrook 1990):

$$|H(\mu)|^{p-1} H^*(\mu) = |h_1(\mu)|^{p-1} h_1^*(\mu), \tag{11.20}$$

where p is a real parameter and $H(\mu) = h_1(\mu)$ as usual. The "fractional" of the fractional power filter has nothing to do with the "fractional" of the fractional Fourier transform. When $p = 1$ we obtain the matched filter. When $p = 0$ we obtain the so-called phase-only filter. When $p = -1$ we obtain the so-called inverse filter. We can generalize the family of filters defined by equation 11.20 to the fractional case in the same way as we have generalized the matched filter:

$$|h_a(u_a)|^{p-1} h_a^*(u_a). \tag{11.21}$$

Now, using this more general expression in equation 11.19 in place of $h_a^*(u_a)$, we obtain the more general system relation which will be employed in this section:

$$g_{f,h}(u) = \int f_a(u_a) \left[|h_a(u_a)|^{p-1} h_a^*(u_a) \right] e^{i2\pi u_a u} \, du_a. \tag{11.22}$$

We will speak of $f(u)$ as the input of the system, $h(u)$ as the reference, and $g_{f,h}(u)$ as the correlation output. The filtering system represented by this relation can be digitally implemented in $O(N \log N)$ time, and also has an efficient optical implementation. It provides a flexible scheme with two parameters a and p which can be adjusted to optimize a variety of objectives.

11.4.2   Performance measures and optimal filters

We now define three performance measures which are of interest in optical pattern recognition systems. The first of these is the *signal-to-noise ratio* (SNR) at the output. This measure allows us to compare how different systems reflect the same level of input noise to the output. The SNR will be defined as the absolute square of the expected value of the correlation peak at the origin divided by its variance, when the input is equal to the reference signal h plus additive noise n (that is, $f = h + n$):

$$\text{SNR} \equiv \frac{|\langle g_{h+n,h}(0)\rangle|^2}{\langle|g_{h+n,h}(0) - \langle g_{h+n,h}(0)\rangle|^2\rangle} = \frac{|\langle g_{h+n,h}(0)\rangle|^2}{\langle|g_{h+n,h}(0)|^2\rangle - |\langle g_{h+n,h}(0)\rangle|^2}. \quad (11.23)$$

The angle brackets denote ensemble averages. We will assume the noise n has zero mean and is independent of h. Larger SNR means a correlation peak which is less noisy.

In optical correlators it is important to make sure that as much as possible of the input light energy makes its way to the correlation plane since this improves the accuracy and speed of optical detection. Our second measure, the *Horner efficiency* (HE) was originally defined as the ratio of the total energy in the output plane to the total energy in the input plane (Horner 1982). However, it was later suggested that a more useful definition is (Horner 1992)

$$\text{HE} \equiv \frac{|g_{h,h}(0)|^2}{\int |h(u)|^2 \, du}, \quad (11.24)$$

since the correlation peak is often what one is primarily interested in, rather than the whole output plane.

It is also of interest to examine the sharpness of the correlation peak, which can be measured with the *peak-to-correlation energy* (PCE) (Kumar and Hassebrook 1990), defined as the ratio of the absolute square of the correlation peak to the total energy of the correlation output:

$$\text{PCE} \equiv \frac{|g_{h,h}(0)|^2}{\int |g_{h,h}(u)|^2 \, du}. \quad (11.25)$$

Larger PCE means a sharper correlation peak.

These three measures are often analytically tractable and thus advantageous over other common criteria such as *peak-to-maximum-sidelobe ratio* (PMSR) (Javidi 1989).

Several different filters have been proposed in the context of ordinary correlators ($a = 1$), which are optimal according to one of the performance measures above.

Best known among these is the *matched filter*, a shift-invariant operation which maximizes SNR. Since it is shift-invariant, it corresponds to multiplicative filtering in the frequency domain, with the filter being given by the familiar conjugate form

$$\propto \frac{H^*(\mu)}{S_{nn}(\mu)}, \quad (11.26)$$

where $H(\mu)$ is the ordinary Fourier transform of $h(u)$ and $S_{nn}(\mu)$ is the power spectral density of the noise (Papoulis 1977). For the special case of white noise, $S_{nn}(\mu)$ is a constant so proportionality 11.26 becomes

$$\propto H^*(\mu). \tag{11.27}$$

That this choice of $H(\mu)$ indeed maximizes SNR will follow as a special case of the fractional version of the same result to be discussed further below.

The *phase-only filter* is another shift-invariant operation which maximizes HE, and is characterized by the frequency-domain filter (Kumar and Hassebrook 1990)

$$\frac{H^*(\mu)}{|H(\mu)|}. \tag{11.28}$$

It is not surprising that this filter maximizes energy at the output since it has unit magnitude.

Finally, the *inverse filter* is a shift-invariant operation which maximizes PCE, and is characterized by the frequency-domain filter

$$\frac{H^*(\mu)}{|H(\mu)|^2}. \tag{11.29}$$

It can be seen that the inverse filter is designed with the intent of producing a delta function at the output. This filter is more fully discussed in Kumar and Hassebrook 1990.

For the white noise case, all three of these filters are special cases of the family of fractional power filters defined in equation 11.20, corresponding to $p = \pm 1, 0$.

11.4.3   Optimal filters for fractional correlation

We now turn our attention to the filtering system described by equation 11.22 and reexamine the three performance measures and the filters optimizing them for the fractional case.

First, we consider the signal-to-noise ratio as defined in equation 11.23. We assume that the noise n is independent of h, and that it is white with zero mean. Thus, the autocorrelation function of the noise is a delta function: $R_{nn}(u, u') = S_{nn}(0)\delta(u - u')$. Here $S_{nn}(0)$ represents the constant value of the power spectral density. It will be shown further below that the fractional Fourier transform $n_a(u_a)$ of zero-mean white noise is also zero-mean white noise. Thus we have $R_{n_a n_a}(u_a, u'_a) = S_{nn}(0)\delta(u_a - u'_a)$ for all values of a. Now, using equation 11.22 to explicitly write $g_{h+n,h}(0)$, it is possible to show that

$$\text{SNR} = \frac{|\langle g_{h+n,h}(0)\rangle|^2}{\langle |g_{h+n,h}(0)|^2\rangle - |\langle g_{h+n,h}(0)\rangle|^2} = \frac{|\langle g_{h,h}(0)\rangle|^2}{\langle |g_{n,h}(0)|^2\rangle} = \frac{\left|\int |h_a(u_a)|^{p+1}\, du_a\right|^2}{S_{nn}(0)\int |h_a(u_a)|^{2p}\, du_a}. \tag{11.30}$$

Setting the derivative of the above expression with respect to p equal to zero shows that SNR is maximized when $p = 1$, corresponding to the fractional generalization of the matched filter: $h_a^*(u_a)$. When $a = 1$ this reduces to the ordinary matched filter given by equation 11.27. We also note that the value of SNR is the same regardless of the value of a and equal to the value of SNR obtained when $a = 1$. Thus, the flexibility of choosing orders other than $a = 1$ does not entail any sacrifice in terms of SNR.

We now turn our attention to the Horner efficiency defined in equation 11.24:

$$\text{HE} \propto \frac{\left|\int h_a(u_a)|h_a(u_a)|^{p-1}h_a^*(u_a)\,du_a\right|^2}{\int |h(u)|^2\,du} = \frac{\left|\int |h_a(u_a)|^{p+1}\,du_a\right|^2}{\int |h(u)|^2\,du}. \tag{11.31}$$

Now, if $|h_a(u_a)|^{p-1}h_a^*(u_a)$ represents a passive optical filter (its magnitude never exceeds unity), then it is possible to show that the above expression is maximized when $p = 0$, corresponding to the fractional generalization of the phase-only filter: $h_a^*(u_a)/|h_a(u_a)|$. When $a = 1$ this reduces to the ordinary phase-only filter given by equation 11.28.

Finally, we examine the peak-to-correlation energy defined in equation 11.25:

$$\text{PCE} = \frac{\left|\int |h_a(u_a)|^{p+1}\,du_a\right|^2}{\int |h_a(u_a)|^{2(p+1)}\,du_a}. \tag{11.32}$$

Setting the derivative of this expression with respect to p equal to zero yields $p = -1$, corresponding to the fractional generalization of the inverse filter: $h_a^*(u_a)/|h_a(u_a)|^2$. When $a = 1$ this reduces to the ordinary inverse filter given by equation 11.29. It should be pointed out, however, that in the fractional case PCE does not have a meaningful interpretation. As noted before, PCE is a measure of peak sharpness. Since the fractional correlation operation is shift-variant, the shape of the peak is irrelevant. Fractional correlation cannot be used for localization of the input object; rather it tells us if the input object exists in some specific region.

In conclusion, we have shown that the filters optimizing three commonly employed performance measures for the fractional correlation output are straightforward fractional generalizations of the filters optimizing the same measures in the case of ordinary correlation. Furthermore, the resulting signal-to-noise ratio is independent of a, so that use of orders other than $a = 1$ does not entail a penalty in terms of signal-to-noise ratio. (Bitran and others 1996)

We finally prove our claim that the fractional Fourier transform of white noise is also white noise. The ensemble-averaged autocorrelation of white noise is given by $R_{nn}(u, u') = \langle n(u)n^*(u')\rangle = S_{nn}(0)\delta(u - u')$. The autocorrelation $R_{n_a n_a}(u_a, u_a')$ of the fractional Fourier transform $n_a(u_a)$ can be readily evaluated as

$$R_{n_a n_a}(u_a, u_a') = \langle n_a(u_a)n_a^*(u_a')\rangle = \left\langle \int K_a(u_a, u)n(u)\,du \int K_a^*(u_a', u')n^*(u')\,du'\right\rangle$$

$$= \iint K_a(u_a, u)K_a^*(u_a', u')R_{nn}(u, u')\,du\,du'$$

$$= \iint K_a(u_a, u)K_a^*(u_a', u')S_{nn}(0)\delta(u - u')\,du\,du'$$

$$= S_{nn}(0)\int K_a(u_a, u)K_a^*(u_a', u)\,du$$

$$= S_{nn}(0)\int K_a(u_a, u)K_{-a}(u, u_a')\,du$$

$$= S_{nn}(0)\,\delta(u_a - u_a'), \tag{11.33}$$

so that the transform is also white noise. It is also not difficult to show that if the mean of $n(u)$ is zero, then the mean of $n_a(u_a)$ is also zero. Thus the fractional Fourier transform of zero-mean white noise is also zero-mean white noise. The fact that the fractional Fourier transform of white noise is also white noise is not surprising when one considers the fact that white noise is uniformly distributed in the time- or space-frequency plane, and is thus not affected by rotation.

If we consider stationary zero-mean but colored noise, it is not difficult to show that the fractional Fourier transform is no longer stationary. Colored noise is nonuniformly distributed along the vertical μ axis in the time- or space-frequency plane, so that upon rotation it will exhibit a variation along the horizontal u axis and thus cannot be stationary. This is easily demonstrated mathematically by following a similar line of algebra as appears above but with R_{nn} given by an arbitrary function rather than a delta function. It is easy to see that $R_{n_a n_a}(u_a, u_a')$ cannot be expressed as a function of $(u_a - u_a')$ only. The colored noise case has been further discussed in Bitran and others 1996.

11.5 Fractional joint-transform correlators

The term "joint-transform correlator" does not refer to a mathematically distinct operation, but to a different way of optically implementing the correlation operation.

Conventional optical correlators are based on an input light distribution $f(u)$ being presented to an optical system housing a filter $h(u)$ to produce the correlation $f(u) \star h(u)$. When both of the functions to be convolved are input light distributions and one does not want to generate a complex filter corresponding to one of them, one can employ a joint-transform correlator. The two functions are both shifted by an amount $\Delta u/2$ in opposite directions with respect to the origin of the input plane so that they have no overlap or negligible overlap with each other:

$$f(u - \Delta u/2) + h(u + \Delta u/2). \tag{11.34}$$

Ordinarily, a Fourier transform stage is employed to obtain the Fourier transform of this distribution, and a square-law device (several approaches have been used to implement the square-law device, including photographic film and various kinds of spatial light modulators) is used to obtain an amplitude distribution proportional to the intensity distribution (Goodman 1996):

$$|F(\mu)e^{-i\pi\mu\Delta u} + H(\mu)e^{i\pi\mu\Delta u}|^2 = |F(\mu)|^2 + |H(\mu)|^2 + F(\mu)H^*(\mu)e^{-i2\pi\mu\Delta u}$$
$$+ F^*(\mu)H(\mu)e^{i2\pi\mu\Delta u}. \tag{11.35}$$

Now, by isolating the third term and ridding it of the harmonic carrier, we obtain $F(\mu)H^*(\mu)$. Then an inverse Fourier transform yields the desired correlation. Isolating this term is easy since the carrier $\exp(-i2\pi\mu\Delta u)$ corresponds to a shift in the space domain. Inverse Fourier transforming the above expression yields

$$\text{term centered around origin} + R_{fh}(u - \Delta u) + R_{hf}(u + \Delta u), \tag{11.36}$$

so that isolating $R_{fh}(u)$ is merely a matter of choosing Δu sufficiently large.

In the fractional joint-transform correlator, one employs a similar approach but replaces the ordinary Fourier transform with the fractional Fourier transform. It was first discussed in Lohmann and Mendlovic 1997 and is also discussed in Kuo and Luo 1998 and in Mendlovic, Zalevsky, and Ozaktas 1998. The problem of isolating and removing the undesired terms is less straightforward and requires somewhat greater care; see the cited references for a discussion in terms of Wigner distributions.

11.6 Adaptive windowed fractional Fourier transforms

11.6.1 Time- or space-dependent windowed transforms

The short-time or windowed Fourier transform was discussed in chapter 3 (equation 3.1). We can analogously define a windowed (or localized, or short-time) fractional Fourier transform as follows:

$$f^{(w)}_{a(\cdot)}(u, \xi) \equiv \int K_{a(\xi)}(u, u')[f(u')w^*(u' - \xi)]\, du', \tag{11.37}$$

where $w(u')$ is a suitably chosen lowpass unit-energy window function centered around the origin, which suppresses $f(u')$ outside an interval centered around ξ. The fractional order $a(\xi)$ is a function of the position ξ at which the window is centered. For this reason, the windowed fractional Fourier transform will also be referred to as the time- or space-dependent fractional Fourier transform. The windowed fractional Fourier transform at a given value of ξ is essentially the fractional Fourier transform of the function $f(u)$ over an interval around ξ. The order $a(\xi)$ may be different for each value of ξ. The transform is generalized to two dimensions in the obvious manner, with the two order parameters in the two dimensions each being a function of two variables.

The windowed fractional Fourier transform has many potential applications. For instance, considering the concept of noise filtering or multiplexing in fractional Fourier domains discussed in chapter 10, use of the windowed transform allows one to adapt the fractional order parameter according to the different signal and noise characteristics in different parts of the signal or image. This may allow the same filtering performance to be achieved with fewer stages, or it may allow more efficient multiplexing. In correlation, detection, and pattern recognition applications, the windowed transform allows control of the degree of shift variance throughout different regions of the signal or image. For these reasons, this transform has also been referred to as the adaptive fractional Fourier transform.

Since the properties of many signals and images are relatively slowly varying, the function $a(\xi)$ would normally also be chosen to be relatively slowly varying; in most applications there would not be a benefit in choosing the value of $a(\xi)$ for each value of ξ independently. In cases where there is knowledge of the underlying processes, an analytical model for $a(\xi)$ might be adopted. In other cases the function $a(\xi)$ would probably be chosen to be of parametric form, and the adaptation accomplished by optimizing over these parameters. One approach would be to restrict $a(\xi)$ to be a lowpass function of ξ, so that it can be fully characterized by a relatively small number of samples, compared to the number of samples characterizing the signals. A simpler approach is to restrict $a(\xi)$ to be piecewise constant and determine these constant values. Although this simpler approach is not the most elegant, and may result in block effects in certain circumstances, it is sufficient to illustrate the power of the transform in a straightforward manner.

In what follows we assume that the signal or image can be partitioned into nonoverlapping intervals or regions such that assigning a constant fractional Fourier order to each is sufficient. The intervals and regions need not be regular and can take arbitrary forms. An image $f(u, v)$ can be written as the sum of component images restricted to these regions as follows:

$$f(u, v) = \sum_j f(u, v) w_j(u, v), \tag{11.38}$$

where $w_j(u, v)$ is unity in region j and zero elsewhere. Now, the following expression may be used as a surrogate for the two-dimensional version of the windowed fractional Fourier transform defined in equation 11.37 (Mendlovic and others 1996b):

$$f_{a_1, a_2, \dots}(u, v) = \sum_j \omega_j(u, v) \left[\int K_{a_j}(u, v; u', v') f(u', v') w_j(u', v') \, du' \, dv' \right], \tag{11.39}$$

where the fractional transform orders in the two dimensions have been taken equal; more generally they can be different. The entity in square brackets is simply the a_jth fractional Fourier transform of that part of the image restricted to region j. The factor $\omega_j(u, v)$ is used to ensure that this transform is restricted to its designated region in the output plane, if necessary by cutting its corners. The index j plays the role of the variable ξ and its counterpart in the v dimension. It is actually each term $\omega_j(u, v)[\cdots]$ of the summation appearing on the right-hand side which corresponds to the windowed transform. However, since these terms are ensured to be nonoverlapping, we choose to deal with their sum (a patch-up of the nonoverlapping transforms), and thus eliminate j from the left-hand side.

While this might sound complicated, it amounts to nothing but partitioning the image into nonoverlapping parts, individually taking the fractional Fourier transforms of the parts with independently specified orders, and then assembling them back together side by side. This approach is much simpler to work with than the more elegant definition given in equation 11.37, although under certain circumstances it may lead to block effects.

An optical implementation of the described piecewise-constant windowed fractional Fourier transform has been proposed in Mendlovic and others 1996b. The underlying system is similar to the one we have already discussed on page 189. It consists of three lenses separated by two sections of free space such that any desired order can be realized by changing the focal lengths of the lenses while their separations remain constant. This property is important for being able to realize many different fractional Fourier transforms in parallel by using planar components. Based on this underlying system, Mendlovic and others have constructed the windowed fractional Fourier transforming configuration by using three diffractive optical elements separated by two sections of free space. Each of these diffractive elements is essentially a patchwork of diffractive lenses whose focal lengths are chosen so as to obtain the desired fractional Fourier order for that region. Spatial separation of the fractional transforms obtained for each region was accomplished by superimposing a periodic grating on each region, whose orientation and periodicity is chosen so as to direct the resulting fractional transforms to nonoverlapping spatial regions. (This amounts to encoding different regions with different spatial frequencies, an approach which can be readily adapted to digital signal processing.) Of course, the superimposed patchwork of lenses and gratings can be manufactured as a single diffractive element. Further details can be found in Mendlovic and others 1996b.

11.6.2 Applications

It is not difficult to think of potential applications of the adaptive windowed fractional Fourier transform. Let us first consider noise removal. We saw in chapter 10 that there was an optimal fractional Fourier domain in which the most effective separation of signal and noise could be accomplished. Now, if different regions of a signal or image are corrupted with noise or distortions of different characteristics, it will clearly be advantageous to adaptively choose the fractional Fourier order for these different regions, rather than working with a single order for the whole signal or image. Likewise, when multiplexing signals as described in chapter 10, it would be advantageous to be able to adapt the order to the local characteristics of the class of signals to be multiplexed. In correlation and detection applications, it may be desirable to have different degrees of shift invariance for different parts of the signal or image. For instance, we may be looking for objects characterized by a number of defining characteristics such that some of these are expected to be always in the same position, while others may appear in different positions. Being able to spatially adjust the degree of shift invariance can increase our discrimination capability in such circumstances.

The relationship between adaptive windowed fractional Fourier transforms and the generalized filtering configurations discussed in chapter 10 is not yet well understood. It seems that joint use of the two approaches may be complementary in certain circumstances but redundant in others. It also may be the case that use of the adaptive transform allows the accomplishment of certain tasks more efficiently than possible with generalized filtering configurations, or the other way around. On the other hand, adaptive systems are structurally related to multichannel filtering, so that in certain cases it may be possible to formulate one approach in terms of the other.

As an example of the use of the spatially adaptive windowed fractional Fourier transform, we will consider the problem of fingerprint recognition (Zalevsky 1996, Zalevsky, Mendlovic, and Caulfield 1997a). Since people press their fingers with varying degrees of force each time, the finger will be deformed by different amounts and the fingerprint will be distorted differently each time. The central region will undergo the least distortion, remaining more or less the same from time to time, while the outer regions will be subject to greater variation. Thus, it is natural to expect a processing system with spatially adjustable shift invariance to lead to better discrimination. In the central region one would desire smaller shift invariance, and in the outer regions one would desire greater shift invariance.

A detailed discussion of how the fingerprinting process can be modeled is found in Zalevsky, Mendlovic, and Caulfield 1997a. The piecewise-constant approach discussed above was employed for simplicity. Two fingerprints of the same finger of the same person were discretized into 256×256 images, one being used as the reference object and the other as the input object to be recognized (figure 11.3). The two fingerprints are, of course, not totally identical.

Figure 11.3. Fingerprint used as reference object (left); fingerprint used as input object (right) (Zalevsky 1996, Zalevsky, Mendlovic, and Caulfield 1997a).

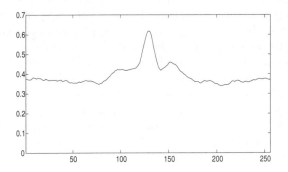

Figure 11.4. The central-line profile of the correlation peak obtained with ordinary matched filtering. After Zalevsky 1996, Zalevsky, Mendlovic, and Caulfield 1997a.

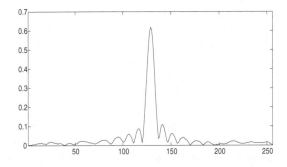

Figure 11.5. The central-line profile of the correlation peak obtained with matched filtering based on fractional correlation with $a = 0.9$. After Zalevsky 1996, Zalevsky, Mendlovic, and Caulfield 1997a.

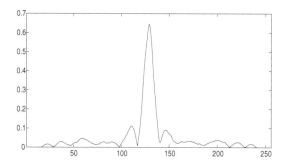

Figure 11.6.   The central-line profile of the correlation peak obtained for the inner area. After Zalevsky 1996, Zalevsky, Mendlovic, and Caulfield 1997a.

First, we examine what happens if we use an ordinary Fourier domain matched filter. We multiply the Fourier transform of the input by the conjugate of the Fourier transform of the reference, and finally take the inverse Fourier transform to obtain the correlation peak shown in figure 11.4. To obtain fair comparisons, the matched filter was normalized by the energy of the reference pattern in both this and the following cases. We observe that the correlation peak is not particularly sharp.

Second, we consider a (nonadaptive) matched filter based on fractional correlation with $a_1 = a$ and $a_2 = 1$ (page 445). The ath order fractional Fourier transform of the input is multiplied by the conjugate of the ath order fractional Fourier transform of the reference and followed by an ordinary inverse transform. The fractional Fourier order resulting in the sharpest correlation peak was found to be $a = 0.9$ (figure 11.5). We observe that the peak is much sharper than the peak in figure 11.4, which was obtained with ordinary matched filtering. Thus, we conclude that use of fractional correlation can increase discrimination capability, even when spatial adaptation is not employed.

Finally, we consider the use of the adaptive windowed fractional Fourier transform. Two regions were defined: the centrally located 128×128 inner square, and the remaining outer region. The degree of shift invariance can be adjusted by varying the fractional order: we can obtain less invariance in the inner region and greater invariance in the outer region by choosing the fractional order smaller for the inner region and larger for the outer region. It was found that best results are obtained by choosing the order equal to 0.86 for the inner region and 0.94 for the outer region. Two separate correlation peaks are obtained for the inner and outer regions and are shown in figures 11.6 and 11.7. Both of these peaks are very sharp and about 10% higher than the peaks shown in figures 11.5 and 11.4. However, we might observe that in this example most of the increase in performance comes from the use of fractional correlation, and relatively less from spatial adaptation. Nevertheless, the possibilities and potential room for improvement of the adaptive approach are far from exhausted. Presently, it is not known how to systematically determine or estimate the optimal partitioning into regions and the optimal transform orders for a given class of objects. Another issue is to formulate optimal decision or combination rules based on the correlation peaks corresponding to the several regions. Developments along these lines may be expected to offer further improvements.

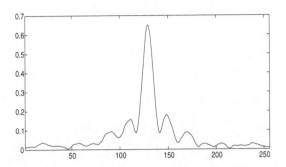

Figure 11.7. The central-line profile of the correlation peak obtained for the outer area.
After Zalevsky 1996, Zalevsky, Mendlovic, and Caulfield 1997a.

11.7   Applications with different orders in the two dimensions

The two-dimensional fractional Fourier transform was discussed in section 4.11, where
we saw that in general the order may be independently specified for the two dimensions
and represented as a vector order $\mathbf{a} = a_u \hat{\mathbf{u}} + a_v \hat{\mathbf{v}}$. The optical implementation of the
two-dimensional transform was discussed in section 9.11.3. General design formulas
for optical systems with arbitrarily specified parameters are given in Sahin, Ozaktas,
and Mendlovic 1998 and several practical issues are discussed in García and others
1996.

The nonseparable fractional Fourier transform is a further generalization in which
the orders can be specified along any oblique coordinate axes, and not necessarily the
conventional orthogonal u and v axes (Sahin, Kutay, and Ozaktas 1998). Although
the nonseparable transform cannot be reduced to the separable transform, and offers
further flexibility in applications, our discussion here will be restricted to the separable
transform.

Being able to choose the orders in the two dimensions differently provides us with
the additional flexibility of being able to treat the two dimensions differently and
thus beneficially exploit dimension-dependent characteristics. For instance, when the
noise and signal characteristics are not isotropic, and this applies in many image
processing problems, optimizing the order for each dimension independently allows
more effective separation of the noise from the signal. As a simple example, the reader
may wish to consider an image corrupted by a chirp whose chirp rate is not the same
in both dimensions. This and other types of anisotropic noise can be more effectively
eliminated if we are able to optimize the orders in the two dimensions independently.
Use of the two-dimensional separable transform for such problems was discussed in
section 10.4. (The best performance will be obtained, however, only if we use the
nonseparable transform mentioned above.)

Another application area in which being able to choose the orders in the two
dimensions independently may be of benefit is image multiplexing for transmission
or storage. The concept of multiplexing in fractional Fourier domains was discussed
in section 10.10. Essentially, multiplexing (or more generally data compression) is a
matter of efficiently packing signals in the time- or space-frequency hyperplane. If we
are dealing with images whose temporal/spatial or frequency content is of a different

nature in the two dimensions, then it may be advantageous to choose the orders in the two dimensions independently. Each image can be transformed with the appropriate orders, shifted, and superimposed with the others.

In the area of detection and pattern recognition, being able to choose the orders in the two dimensions differently will allow us to control the amount of shift invariance in the two dimensions independently. A particularly simple example arises if we wish to have complete shift invariance in one direction, but very little or none in the orthogonal direction. One situation where this might arise is when we have prior knowledge that the objects are constrained to a line (or with a suitable transformation, an arbitrary curve). For example, the objects we wish to recognize might be letters confined to a single line, or cars along an avenue. Not requiring shift invariance along one of the dimensions will relieve the design process so that other objectives might be better optimized. Another situation is when we wish to recognize objects regardless of their position along a certain line, but reject those which deviate from this line.

More concretely, let us consider the problem of designing a system that is able to recognize an object A or deformed versions of it only if they appear in region R_A, and an object B or deformed versions of it only if they appear in region R_B (García and others 1996). We will assume R_A is the upper half of the image plane and R_B is the lower half. Initially, let us assume that object A is placed at the origin, at the boundary of the two regions, and consider its a_vth fractional Fourier transform with respect to the vertical v axis. If object A is shifted upwards towards the inside of region R_A, this shift will result in a corresponding shift in the transform when $a_v = 0$. On the other hand, it will not result in any shift in the transform when $a_v = 1$. Since the fractional Fourier transform is continuous in the order a_v, this means that transforms of intermediate order will have shifts of intermediate amounts. For instance, if we consider $a_v = 0.5$, the transform will be shifted towards the inside of region R_A, but not as much as the original object A. Nevertheless, for values of a_v which are not very close to $a_v = 1$, it will be the case that a sizable shift of object A will bring not only object A, but also its fractional Fourier transform of order a_v totally into the region R_A. The same considerations apply to object B and its $a_v = 0.5$th order fractional Fourier transform. If object B is located considerably inside the region R_B, then its transform will also be found inside this region. Thus, the $a_v = 0.5$th fractional Fourier transform of the sum of object A located in R_A and object B located in R_B will consist of the two nonoverlapping transforms of the respective objects contained in their respective regions. By choosing the displacements of the original reference objects into the regions judiciously, it is possible to obtain a system which will recognize only object A in region R_A and only object B in region R_B. If object A appears in region R_B, it will not be detected, since use of the fractional order $a_v = 0.5$ along the direction perpendicular to the boundary of the regions will limit the amount of space invariance along this direction. On the other hand, in the direction parallel to the boundary of the two regions, we wish to obtain full space invariance; that is, detection regardless of position along this direction. Thus, the fractional transform order a_u is chosen equal to 1 along this direction.

The matched filter is obtained as follows. We situate object A in an appropriate position inside R_A and take its two-dimensional fractional Fourier transform with orders $a_u = 1$ and $a_v = 0.5$, and take its complex conjugate. The same process is repeated for object B and the final filter is obtained by adding the two results.

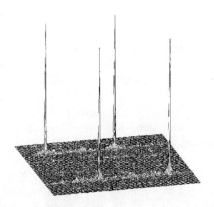

Figure 11.8. The input image: object A has been depicted by the letter Λ and object B has been depicted by the letter Y (left). The correlation output (right). Adapted from García and others 1996.

This filter will be multiplied by the corresponding two-dimensional fractional Fourier transform of the input, and then inverse ordinary Fourier transformed to obtain the correlation output. As things are, the correlation peak corresponding to A and that corresponding to B will both appear at the origin and overlap with each other. To separate them, the filters can be multiplied by appropriately chosen linear phase factors that will result in a shift of the correlation peaks into their respective regions. (García and others 1996)

Various forms of deformation invariance can be readily incorporated in this scheme. Different kinds of deformations for A and B can be specified. For instance, let us assume that one-dimensional scale invariance in the v direction is required for object A and one-dimensional scale invariance in the u direction is required for object B. An established approach for obtaining one-dimensional scale invariance is the use of the logarithmic harmonic decomposition (Mendlovic, Konforti, and Marom 1990). To achieve scale invariance, the filter is obtained by taking the fractional Fourier transforms of the harmonics determined with this technique, rather than the original objects. We note that this is merely an example and that other kinds of invariance can also be dealt with; for instance, rotation invariance for object A and two-dimensional scale invariance for object B.

We are considering an input image plane whose upper half is designated as region R_A and whose lower half is designated as region R_B, and we wish our system to recognize any instances of object A appearing in R_A and any instances of object B appearing in R_B. Thus the system is expected to be completely shift-invariant in the horizontal u direction, and moderately shift-invariant in the vertical v direction. Ideally, the shift invariance in the v direction should be such that deviations from the center of the upper region which are small enough that the object still remains in the upper region are tolerated, but deviations which are large enough that the object moves into the lower region (or out of the input plane) are not tolerated. This cannot be precisely achieved in the present system since the tolerance required would also depend on the position of the input object. Thus, the detection/rejection of objects

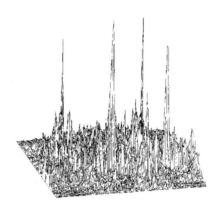

Figure 11.9.   The input image: object A has been depicted by the letter Λ and object B has been depicted by the letter Y (left). The correlation output (right). Adapted from García and others 1996.

close to the boundary of the two regions is a fuzzy one, with smaller and smaller peaks observed as the position of the input object approaches and crosses the boundary. As an example, assume that four instances of object A and four instances of object B appear in the input plane (left part of figure 11.8). The vertical positions of the two instances of object A appearing in the upper part, as well as the two instances of object B appearing in the lower part, were slightly shifted with respect to each other to demonstrate invariance to moderate v displacements. The filter was constructed by situating a single instance of object A in the center of the upper region, and a single instance of object B in the center of the lower region, and employing a fractional Fourier transform with $a_u = 1$, $a_v = 0.5$, as discussed above. The correlation output is shown in the right-hand part of figure 11.8. We observe that the upper two instances of object A and the lower two instances of object B are successfully detected with sharp correlation peaks.

Further simulations were undertaken to demonstrate the capability of the approach to deal with different kinds of deformation in the upper and lower parts of the image. In this example, we are no longer interested in recognizing object B anywhere; rather, we are interested in recognizing any vertically scaled version of object A in the upper half, and any horizontally scaled version of object A in the lower half. That is, the system must be v-scale-invariant in R_A and u-scale-invariant in R_B. The filter was obtained by situating a v-scale-invariant logarithmic harmonic of object A in the center of the upper region and a u-scale-invariant logarithmic harmonic of object A in the center of the lower region, and using a fractional Fourier transform with orders $a_u = 1$, $a_v = 0.5$. A sample result is shown in figure 11.9. A detection threshold which is 35% of the maximum intensity value was sufficient to correctly detect the true peaks among the background and false peaks (García and others 1996).

When the systems being discussed are implemented optically, it is crucial that the filter be produced at the correct scale, as determined by the scale of the input image and the scale factors associated with the optical fractional Fourier transformers (García and others 1996). If real-time operation is desired so that spatial light modulators

(SLMs) are employed for the input and/or the filter, the resolution and aperture of the SLMs may not allow the input or the filter to be generated at the necessary scale. Fortunately, this difficulty can be overcome by designing the fractional Fourier transformers to have specified input and output scale parameters, chosen so as to accommodate the limitations of the SLMs and/or the scale of the available input image. We saw in chapter 9 how to choose the focal lengths of the lenses and their separations so as to realize optical fractional Fourier transformers with specified order and scale parameters in one dimension. (For example, see equations 9.67 and 9.68 for the case of equal input and output scale parameters. In the more general case where the input and output scale parameters are not equal, equations 9.48, 9.49, and 9.50 tell us how the separation and focal lengths must be chosen.) These one-dimensional results immediately generalize to axially symmetric two-dimensional systems which realize fractional Fourier transforms with the same order in the two dimensions. We do not discuss implementation of two-dimensional transforms whose two orders are specified independently. However, general design formulas for such fractional Fourier transforming systems with specified orders and scale parameters can be found in Sahin, Ozaktas, and Mendlovic 1998. A further practical consideration is to make the scale of the fractional Fourier transforming configuration easily adjustable by moving some of the elements and/or the input transparency, without having to change the lenses. A scheme to achieve this has been described in Zalevsky 1996 and García and others 1997. In this scheme the input transparency (or SLM) is illuminated by a converging or diverging beam instead of a plane wave, so that moving the transparency results in the input being illuminated by a beam with varying wavefront radius. This is effectively the same as varying the focal length of a lens placed between an illuminating plane wave and the input transparency. The output plane must also be moved so that the overall condition for fractional Fourier transformation is still satisfied. This scheme allows the scale factor between the input and the filter to be adjusted as desired; such flexibility is especially important when working with SLMs. Furthermore, it allows different orders to be employed without changing the lenses.

For further details regarding optical implementation, as well as experimental results, the reader is referred to García and others 1996, 1997 and Zalevsky 1996, where excellent agreement is obtained with the numerical results presented above.

11.8   Historical and bibliographical notes

The purpose of this chapter, and especially the latter sections, has been not so much to describe specific systems, but rather to illustrate the flexibility afforded by use of the fractional Fourier transform. The application examples discussed represent a small sample of the many potential uses of the transform to design systems with performance and functionality exceeding those of conventional systems.

The concept of fractional correlation was first suggested in Ozaktas and Mendlovic 1993a, Mendlovic and Ozaktas 1993a, and first developed in Mendlovic, Ozaktas, and Lohmann 1995. This chapter is essentially based on Mendlovic, Ozaktas, and Lohmann 1995, Lohmann, Zalevsky, and Mendlovic 1996, Bitran and others 1996, Lohmann and Mendlovic 1997, Mendlovic and others 1996b, Zalevsky, Mendlovic, and Caulfield 1997a, and García and others 1996. A convenient starting point for further study is Mendlovic, Zalevsky, and Ozaktas 1998. Some of the results presented here also appear in Zalevsky 1996, where further details and results may be found. Fractional triple correlation and its applications are discussed in Mendlovic and others 1998. The application of fractional autocorrelation to the detection of linear FM signals is discussed in Akay and Boudreaux-Bartels 1998b.

Experimental results for optical fractional correlation were first reported in Mendlovic and others 1995a. Other works on optical fractional correlation include Granieri and others 1996, García and others 1997, Granieri, Arizaga, and Sicre 1997, Zalevsky, Mendlovic, and Caulfield 1997b, Zalevsky and others 1997, Almanasreh and Abushagur 1998, and Tripathi, Pati, and Singh 1999. Rotation-invariant versions of correlators with controllable space variance are discussed in Zhang and Gu 1998.

Correlation and matched filtering in Fresnel transform planes has recently received some attention (Hamam and Arsenault 1997). These approaches can also be formulated in the framework of fractional Fourier transforms.

A totally distinct approach to detection, feature extraction, and pattern recognition (based on the concept of filter circuits discussed in chapter 10) may be found in Güleryüz 1998.

Zalevsky 1999 is a brief review of the applications discussed in this chapter.

Bibliography on the Fractional Fourier Transform

(Abe and Sheridan 1994a) S. Abe and J. T. Sheridan. Generalization of the fractional Fourier transformation to an arbitrary linear lossless transformation: an operator approach. *J Phys A*, 27:4179–4187, 1994. Corrigenda in 7937–7938.

(Abe and Sheridan 1994b) S. Abe and J. T. Sheridan. Optical operations on wave functions as the Abelian subgroups of the special affine Fourier transformation. *Opt Lett*, 19:1801–1803, 1994.

(Abe and Sheridan 1995a) S. Abe and J. T. Sheridan. Almost-Fourier and almost-Fresnel transformations. *Opt Commun*, 113:385–388, 1995.

(Abe and Sheridan 1995b) S. Abe and J. T. Sheridan. Comment on 'The fractional Fourier transform in optical propagation problems.' *J Mod Opt*, 42:2373–2378, 1995.

(Abe and Sheridan 1995c) S. Abe and J. T. Sheridan. The Wigner distribution function, and the special affine Fourier transform: signal processing and optical imaging. *Zoological Studies*, 34:121–122, 1995.

(Abe and Sheridan 1997) S. Abe and J. T. Sheridan. An optical implementation for the estimation of the fractional-Fourier order. *Opt Commun*, 137:214–218, 1997.

(Agarwal and Simon 1994) G. S. Agarwal and R. Simon. A simple realization of fractional Fourier transforms and relation to harmonic oscillator Green's function. *Opt Commun*, 110:23–26, 1994.

(Akan and Chaparro 1996) A. Akan and L. F. Chaparro. Discrete rotational Gabor transform. In *Proc 1996 IEEE-SP Int Symp Time-Frequency Time-Scale Analysis*, IEEE, Piscataway, New Jersey, 1996. Pages 169–172.

(Akay and Boudreaux-Bartels 1997) O. Akay and G. F. Boudreaux-Bartels. Linear fractionally invariant systems: fractional filtering and correlation via fractional operators. In *Proc 31st Asilomar Conf Signals, Systems, Computers*, IEEE, Piscataway, New Jersey, 1997. Pages 1494–1498.

(Akay and Boudreaux-Bartels 1998a) O. Akay and G. F. Boudreaux-Bartels. Unitary and Hermitian fractional operators and their relation to the fractional Fourier transform. *IEEE Signal Processing Lett*, 5:312–314, 1998.

(Akay and Boudreaux-Bartels 1998b) O. Akay and G. F. Boudreaux-Bartels. Fractional autocorrelation and its applications to detection and estimation of linear FM signals. In *Proc 1998 IEEE-SP Int Symp Time-Frequency Time-Scale Analysis*, IEEE, Piscataway, New Jersey, 1998. Pages 213–216.

(Akay and Boudreaux-Bartels 1998c) O. Akay and G. F. Boudreaux-Bartels. Joint fractional representations. In *Proc 1998 IEEE-SP Int Symp Time-Frequency Time-Scale Analysis*, IEEE, Piscataway, New Jersey, 1998. Pages 417–420.

(Akay and Boudreaux-Bartels 1998d) O. Akay and G. F. Boudreaux-Bartels. Fractional Mellin transform: an extension of fractional frequency concept for scale. In *Proc 8th IEEE Int Digital Signal Processing Workshop*, IEEE, Piscataway, New Jersey, 1998. CD-ROM.

(Akay and Boudreaux-Bartels 1998e) O. Akay and G. F. Boudreaux-Bartels. Broadband interference excision in spread spectrum communication systems via fractional Fourier transform. In *Proc 32nd Asilomar Conf Signals, Systems, Computers*, IEEE, Piscataway, New Jersey, 1998. Pages 832–837.

(Alieva 1996a) T. Alieva. On the self-fractional Fourier functions. *J Phys A*, 29:L377–L379, 1996.

(Alieva 1996b) T. Alieva. Fractional Fourier transform as a tool for investigation of fractal objects. *J Opt Soc Am A*, 13:1189–1192, 1996.

(Alieva and Agulló-López 1995) T. Alieva and F. Agulló-López. Reconstruction of the optical correlation function in a quadratic refractive index medium. *Opt Commun*, 114:161–169, 1995. Erratum in 118:657, 1995.

(Alieva and Agullo-Lopez 1996a) T. Alieva and F. Agullo-Lopez. Optical wave propagation of fractal fields. *Opt Commun*, 125:267–274, 1996.

(Alieva and Agullo-Lopez 1996b) T. Alieva and F. Agullo-Lopez. Imaging in first-order optical systems. *J Opt Soc Am A*, 13:2375–2380, 1996.

(Alieva and Agullo-Lopez 1998) T. Alieva and F. Agullo-Lopez. Diffraction analysis of random fractal fields. *J Opt Soc Am A*, 15:669–674, 1998.

(Alieva and Barbé 1997) T. Alieva and A. M. Barbé. Self-fractional Fourier functions and selection of modes. *J Phys A*, 30:L211–L215, 1997.

(Alieva and Barbé 1998a) T. Alieva and A. M. Barbé. Fractional Fourier and Radon-Wigner transforms of periodic signals. *Signal Processing*, 69:183–189, 1998.

(Alieva and Barbé 1998b) T. Alieva and A. M. Barbé. About quantum state characterization. *J Phys A*, 31:L685–L688, 1998.

(Alieva and Barbé 1999) T. Alieva and A. M. Barbé. Self-fractional Fourier images. *J Mod Opt*, 46:83–99, 1999.

(Alieva and others 1994) T. Alieva, V. Lopez, F. Agullo-Lopez, and L. B. Almeida. The fractional Fourier transform in optical propagation problems. *J Mod Opt*, 41:1037–1044, 1994.

(Alieva and others 1995) T. Alieva, V. Lopez, F. Agullo-Lopez, and L. B. Almeida. Reply to the comment on the fractional Fourier transform in optical propagation problems. *J Mod Opt*, 42:2379–2383, 1995.

(Almanasreh and Abushagur 1998) A. M. Almanasreh and M. A. G. Abushagur. Fractional correlations based on the modified fractional order Fourier transform. *Opt Eng*, 37:175–184, 1998.

(Almeida 1993) L. B. Almeida. An introduction to the angular Fourier transform. In *Proc 1993 IEEE Int Conf Acoustics, Speech, Signal Processing*, IEEE, Piscataway, New Jersey, 1993. Pages III-257–III-260.

(Almeida 1994) L. B. Almeida. The fractional Fourier transform and time-frequency representations. *IEEE Trans Signal Processing*, 42:3084–3091, 1994.

(Almeida 1997) L. B. Almeida. Product and convolution theorems for the fractional Fourier transform. *IEEE Signal Processing Lett*, 4:15–17, 1997.

(Alonso and Forbes 1995) M. A. Alonso and G. W. Forbes. Fractional Legendre transformation. *J Phys A*, 28:5509–5527, 1995.

(Alonso and Forbes 1997) M. A. Alonso and G. W. Forbes. Uniform asymptotic expansions for wave propagators via fractional transformations. *J Opt Soc Am A*, 14:1279–1292, 1997.

(Andrés and others 1997) P. Andrés, W. D. Furlan, G. Saavedra, and A. W. Lohmann. Variable fractional Fourier processor: a simple implementation. *J Opt Soc Am A*, 14:853–858, 1997.

(Antosik, Mikusiński, and Sikorski 1973) P. Antosik, J. Mikusiński, and R. Sikorski. *Theory of Distributions: The Sequential Approach*. Elsevier, Amsterdam, 1973.

(Arıkan and others 1996) O. Arıkan, M. A. Kutay, H. M. Özaktaş, and Ö. K. Akdemir. The discrete fractional Fourier transformation. In *Proc 1996 IEEE-SP Int Symp Time-Frequency Time-Scale Analysis*, IEEE, Piscataway, New Jersey, 1996. Pages 205–207.

(Arrizón, Ibarra, and Ojeda-Castañeda 1996) V. Arrizón, J. G. Ibarra, and J. Ojeda-Castañeda. Matrix formulation of the Fresnel transform of complex transmittance gratings. *J Opt Soc Am A*, 13:2414–2422, 1996.

(Atakishiyev, Vicent, and Wolf 1999) N. M. Atakishiyev, L. E. Vicent, and K. B. Wolf. Continuous vs. discrete fractional Fourier transforms. *J Computational Applied Mathematics*, 107:73–95, 1999.

(Atakishiyev and Wolf 1997) N. M. Atakishiyev and K. B. Wolf. Fractional Fourier-Kravchuk transform. *J Opt Soc Am A*, 14:1467–1477, 1997.

(Atakishiyev and others 1998) N. M. Atakishiyev, E. I. Jafarov, S. M. Nagiyev, and K. B. Wolf. Meixner oscillators. *Revista Mexicana Fisica*, 44:235–244, 1998.

(Aytür and Ozaktas 1995) O. Aytür and H. M. Ozaktas. Non-orthogonal domains in phase space of quantum optics and their relation to fractional Fourier transforms. *Opt Commun*, 120:166-170, 1995.

(Bailey and Swarztrauber 1991) D. H. Bailey and P. N. Swarztrauber. The fractional Fourier transform and applications. *SIAM Review*, 33:389–404, 1991.

(Bailey and Swarztrauber 1994) D. H. Bailey and P. N. Swarztrauber. A fast method for the numerical evaluation of continuous Fourier and Laplace transforms. *SIAM J Sci Comput*, 15:1105–1110, 1994.

(Barbarossa 1995) S. Barbarossa, Analysis of multicomponent LFM signals by a combined Wigner-Hough transform. *IEEE Trans Signal Processing*, 43:1511–1515, 1995.

(Bargmann 1961) V. Bargmann. On a Hilbert space of analytic functions and an associated integral transform, Part I. *Comm Pure and Applied Mathematics*, 14:187–214, 1961.

(Barker 2000) L. Barker. The discrete fractional Fourier transform and Harper's equation. Technical Report BU-CE-0007, Bilkent University, Department of Computer Engineering, Ankara, January 2000. To appear in *Mathematika*.

(Barker and others 2000) L. Barker, Ç. Candan, T. Hakioğlu, M. A. Kutay, and H. M. Ozaktas. The discrete harmonic oscillator, Harper's equation, and the discrete fractional Fourier transform. *J Phys A*, 33:2209–2222, 2000.

(Barshan, Kutay, and Ozaktas 1997) B. Barshan, M. A. Kutay, and H. M. Ozaktas. Optimal filtering with linear canonical transformations. *Opt Commun*, 135:32–36, 1997.

(Bastiaans and van Leest 1998) M. J. Bastiaans and A. J. van Leest. From the rectangular to the quincunx Gabor lattice via fractional Fourier transformation. *IEEE Signal Processing Lett*, 5:203–205, 1998.

(Beck and others 1993) M. Beck, M. G. Raymer, I. A. Walmsley, and V. Wong. Chronocyclic tomography for measuring the amplitude and phase structure of optical pulses. *Opt Lett*, 18:2041–2043, 1993.

(Bernardo 1996) L. M. Bernardo. *ABCD* matrix formalism of fractional Fourier optics. *Opt Eng*, 35:732–740, 1996.

(Bernardo 1997) L. M. Bernardo. Talbot self-imaging in fractional Fourier planes of real and complex orders. *Opt Commun*, 140:195–198, 1997.

(Bernardo and Soares 1994a) L. M. Bernardo and O. D. D. Soares. Fractional Fourier transforms and optical systems. *Opt Commun*, 110:517–522, 1994.

(Bernardo and Soares 1994b) L. M. Bernardo and O. D. D. Soares. Fractional Fourier transforms and imaging. *J Opt Soc Am A*, 11:2622–2626, 1994.

(Bernardo and Soares 1996) L. M. Bernardo and O. D. D. Soares. Optical fractional Fourier transforms with complex orders. *Appl Opt*, 35:3163–3166, 1996.

(Bertrand and Bertrand 1987) J. Bertrand and P. Bertrand. A tomographic approach to Wigner's function. *Foundations Phys*, 17:397–405, 1987.

(Bitran and others 1995) Y. Bitran, D. Mendlovic, R. G. Dorsch, A. W. Lohmann, and H. M. Ozaktas. Fractional Fourier transform: simulations and experimental results. *Appl Opt*, 34:1329–1332, 1995.

(Bitran and others 1996) Y. Bitran, Z. Zalevsky, D. Mendlovic, and R. G. Dorsch. Fractional correlation operation: Performance analysis. *Appl Opt*, 35:297–303, 1996.

(de Bruijn 1973) N. G. de Bruijn. A theory of generalized functions, with applications to Wigner distribution and Weyl correspondence. *Nieuw Archief voor Wiskunde*, 21:205–280, 1973.

(Bultan 1999) A. Bultan. A four-parameter atomic decomposition of chirplets. *IEEE Trans Signal Processing*, 47:731–745, 1999.

(Candan 1998) Ç. Candan. *The Discrete Fractional Fourier Transform*. MS thesis, Bilkent University, Ankara, 1998.

(Candan, Kutay, and Ozaktas 1999) Ç. Candan, M. A. Kutay, and H. M. Ozaktas. The discrete fractional Fourier transform. In *Proc 1999 IEEE Int Conf Acoustics, Speech, Signal Processing*, IEEE, Piscataway, New Jersey, 1999. Pages 1713–1716.

(Candan, Kutay, and Ozaktas 2000) Ç. Candan, M. A. Kutay, and H. M. Ozaktas. The discrete fractional Fourier transform. *IEEE Trans Signal Processing*, 48:1329–1337, 2000.

(Cariolaro and others 1998) G. Cariolaro, T. Erseghe, P. Kraniauskas, and N. Laurenti. A unified framework for the fractional Fourier transform. *IEEE Trans Signal Processing*, 46:3206–3219, 1998.

(Condon 1937) E. U. Condon. Immersion of the Fourier transform in a continuous group of functional transformations. *Proc National Academy Sciences*, 23:158–164, 1937.

(Cong and Chen 1999) W. X. Cong and N. X. Chen. Optimization of Gaussian beam shaping in fractional Fourier transform domain. *Prog Natural Science*, 9:96–102, 1999.

(Cong, Chen, and Gu 1998a) W. X. Cong, N. X. Chen, and B. Y. Gu. Beam shaping and its solution with the use of an optimization method. *Appl Opt*, 37:4500–4503, 1998.

(Cong, Chen, and Gu 1998b) W. X. Cong, N. X. Chen, and B. Y. Gu. Recursive algorithm for phase retrieval in the fractional Fourier transform domain. *Appl Opt*, 37:6906–6910, 1998.

(Cong, Chen, and Gu 1998c) W. X. Cong, N. X. Chen, and B. Y. Gu. A new method for phase retrieval in the optical system. *Chinese Phys Lett*, 15:24–26, 1998.

(Cong, Chen, and Gu 1998d) W. X. Cong, N. X. Chen, and B. Y. Gu. A recursive method for phase retrieval in Fourier transform domain. *Chinese Science Bulletin*, 43:40–44, 1998.

(Dattoli, Torre, and Mazzacurati 1998) G. Dattoli, A. Torre, and G. Mazzacurati. An alternative point of view to the theory of fractional Fourier transform. *IMA J Appl Math*, 60:215–224, 1998.

(Deng, Caulfield, and Schamschula 1996) Z.-T. Deng, H. J. Caulfield, and M. Schamschula. Fractional discrete Fourier transforms. *Opt Lett*, 21:1430–1432, 1996.

(Deng and others 1996) X. Deng, Y. Li, Y. Qiu, and D. Fan. Diffraction interpreted through fractional Fourier transforms. *Opt Commun*, 131:241–245, 1996.

(Deng and others 1997) X. Deng, Y. Li, D. Fan, and Y. Qiu. A fast algorithm for fractional Fourier transforms. *Opt Commun*, 138:270–274, 1997.

(Dong and others 1997) B.-Z. Dong, Y. Zhang, B.-Y. Gu, and G.-Z. Yang. Numerical investigation of phase retrieval in a fractional Fourier transform. *J Opt Soc Am A*, 14:2709–2714, 1997.

(Dorsch 1995) R. G. Dorsch. Fractional Fourier transformer of variable order based on a modular lens system. *Appl Opt*, 34:6016–6020, 1995.

(Dorsch and Lohmann 1995) R. G. Dorsch and A. W. Lohmann. Fractional Fourier transform used for a lens design problem. *Appl Opt*, 34:4111–4112, 1995.

(Dorsch and others 1994) R. G. Dorsch, A. W. Lohmann, Y. Bitran, D. Mendlovic, and H. M. Ozaktas. Chirp filtering in the fractional Fourier domain. *Appl Opt*, 33:7599–7602, 1994.

(Dragoman 1996) D. Dragoman. Fractional Wigner distribution function. *J Opt Soc Am A*, 13:474–478, 1996.

(Dragoman 1998) D. Dragoman. The relation between light diffraction and the fractional Fourier transform. *J Mod Opt*, 45:2117–2124, 1998.

(Dragoman and Dragoman 1997a) D. Dragoman and M. Dragoman. Time-frequency modeling of atomic force microscopy. *Opt Commun*, 140:220–225, 1997.

(Dragoman and Dragoman 1997b) D. Dragoman and M. Dragoman. Near and far field optical beam characterization using the fractional Fourier tansform. *Opt Commun*, 141:5–9, 1997.

(Dragoman and Dragoman 1998) D. Dragoman and M. Dragoman. Temporal implementation of Fourier-related transforms. *Opt Commun*, 145:33–37, 1998.

(Dragoman and others 1998) D. Dragoman, K.-H. Brenner, M. Dragoman, J. Bahr, and U. Krackhardt. Hemispherical-rod microlens as a variant fractional Fourier transformer. *Opt Lett*, 23:1499–1501, 1998.

(Erden 1997) M. F. Erden. *Repeated Filtering in Consecutive Fractional Fourier Domains*. PhD thesis, Bilkent University, Ankara, 1997.

(Erden, Kutay, and Ozaktas 1999a) M. F. Erden, M. A. Kutay, and H. M. Ozaktas. Repeated filtering in consecutive fractional Fourier domains and its application to signal restoration. *IEEE Trans Signal Processing*, 47:1458–1462, 1999.

(Erden, Kutay, and Ozaktas 1999b) M. F. Erden, M. A. Kutay, and H. M. Ozaktas. Applications of the fractional Fourier transform to filtering, estimation and restoration. In *Proc 1999 IEEE-EURASIP Workshop Nonlinear Signal Image Processing*, Boğaziçi University, Bebek, Istanbul, 1999. Pages 481–485.

(Erden and Ozaktas 1998) M. F. Erden and H. M. Ozaktas. Synthesis of general linear systems with repeated filtering in consecutive fractional Fourier domains. *J Opt Soc Am A*, 15:1647–1657, 1998.

(Erden, Ozaktas, and Mendlovic 1996a) M. F. Erden, H. M. Ozaktas, and D. Mendlovic. Propagation of mutual intensity expressed in terms of the fractional Fourier transform. *J Opt Soc Am A*, 13:1068–1071, 1996.

(Erden, Ozaktas, and Mendlovic 1996b) M. F. Erden, H. M. Ozaktas, and D. Mendlovic. Synthesis of mutual intensity distributions using the fractional Fourier transform. *Opt Commun*, 125:288–301, 1996.

(Erden and others 1997) M. Fatih Erden, Haldun M. Ozaktas, Aysegul Sahin, and David Mendlovic. Design of dynamically adjustable anamorphic fractional Fourier transformer. *Opt Commun*, 136:52–60, 1997.

(Fonollosa 1996) J. R. Fonollosa. Positive time-frequency distributions based on joint marginal constraints. *IEEE Trans Signal Processing*, 44:2086–2091, 1996.

(Fonollosa and Nikias 1994) J. R. Fonollosa and C. L. Nikias. A new positive time-frequency distribution. In *Proc 1994 IEEE Int Conf Acoustics, Speech, Signal Processing*, IEEE, Piscataway, New Jersey, 1994. Pages IV-301–IV-304.

(García, Mas, and Dorsch 1996) J. García, D. Mas, and R. G. Dorsch. Fractional-Fourier-transform calculation through the fast-Fourier-transform algorithm. *Appl Opt*, 35:7013–7018, 1996.

(García and others 1996) J. García, D. Mendlovic, Z. Zalevsky, and A. Lohmann. Space-variant simultaneous detection of several objects by the use of multiple anamorphic fractional-Fourier-transform filters. *Appl Opt*, 35:3945–3952, 1996.

(García and others 1997) J. García, R. G. Dorsch, A. W. Lohmann, C. Ferreira, and Z. Zalevsky. Flexible optical implementation of fractional Fourier transform processors. Applications to correlation and filtering. *Opt Commun*, 133:393–400, 1997.

(Gómez-Reino, Bao, and Pérez 1996) C. Gómez-Reino, C. Bao, and M. V. Pérez. GRIN optics, Fourier optics and optical connections. In *17th Congress ICO: Optics for Science and New Technology, SPIE Proc 2778*, SPIE, Bellingham, Washington, 1996. Pages 128–131.

(Gori, Santarsiero, and Bagini 1994) F. Gori, M. Santarsiero, and V. Bagini. Fractional Fourier transform and Fresnel transform. *Atti della Fondazione Giorgio Ronchi*, IL:387–390, 1994.

(Granieri, Arizaga, and Sicre 1997) S. Granieri, R. Arizaga, and E. E. Sicre. Optical correlation based on the fractional Fourier transform. *Appl Opt*, 36:6636–6645, 1997.

(Granieri, Sicre, and Furlan 1998) S. Granieri, E. E. Sicre, and W. D. Furlan. Performance analysis of optical imaging systems based on the fractional Fourier transform. *J Mod Opt*, 45:1797–1807, 1998.

(Granieri, Trabocchi, and Sicre 1995) S. Granieri, O. Trabocchi, and E. E. Sicre. Fractional Fourier transform applied to spatial filtering in the Fresnel domain. *Opt Commun*, 119:275–278, 1995.

(Granieri and others 1996) S. Granieri, M. del Carmen Lasprilla, N. Bolognini, and E. E. Sicre. Space-variant optical correlator based on the fractional Fourier transform: implementation by the use of a photorefractive $Bi_{12}GeO_2$ (BGO) holographic filter. *Appl Opt*, 35:6951–6954, 1996.

(Granieri and others 1997) S. Granieri, W. D. Furlan, G. Saavedra, and P. Andrés. Radon-Wigner display: a compact optical implementation with a single varifocal lens. *Appl Opt*, 36:8363–8369, 1997.

(Guinand 1956) A. P. Guinand. Matrices associated with fractional Hankel and Fourier transformations. *Proc Glasg Math Assoc*, 2:185–192, 1956.

(Güleryüz 1998) Ö. Güleryüz. *Feature Extraction with the Fractional Fourier Transform*. MS thesis, Bilkent University, Ankara, 1998.

(Hamam and Arsenault 1997) H. Hamam and H. H. Arsenault. Fresnel transform-based correlator. *Appl Opt*, 36:7408–7414, 1997.

(Howe 1988) R. Howe. The oscillator semigroup. In *The Mathematical Heritage of Hermann Weyl: Proc Symp Pure Mathematics 48*, American Mathematical Society, Providence, Rhode Island, 1988. Pages 61–132.

(Hua and Liu 1997) J. Hua and L. Liu. Optical dual fractional Fourier and Fresnel self-imaging transform. *Optik*, 105:47–50, 1997.

(Hua, Liu, and Li 1997a) J. Hua, L. Liu, and G. Li. Observing the fractional Fourier transform by free-space Fresnel diffraction. *Appl Opt*, 36:512–513, 1997.

(Hua, Liu, and Li 1997b) J. Hua, L. Liu, and G. Li. Scaled fractional Fourier transform and its optical implementation. *Appl Opt*, 36:8490–8492, 1997.

(Hua, Liu, and Li 1997c) J. Hua, L. Liu, and G. Li. Extended fractional Fourier transforms. *J Opt Soc Am A*, 14:3316–3322, 1997.

(Hua, Liu, and Li 1997d) J. Hua, L. Liu, and G. Li. Performing the fractional Fourier transform by one Fresnel diffraction and one lens. *Opt Commun*, 137:11–12, 1997.

(Hua, Liu, and Li 1997e) J. W. Hua, L. R. Liu, and G. Q. Li. Imaginary angle fractional Fourier transform and its optical implementation. *Science in China Series E: Technological Sciences*, 40:374–378, 1997.

(Huang and Suter 1998) Y. Huang and B. Suter. The fractional wave packet transform. *Multidimensional Systems Signal Processing*, 9:399–402, 1998.

(Jagoszewski 1998a) E. Jagoszewski. Fractional Fourier transform in optical setups. *Optica Applicata*, 28:227–237, 1998.

(Jagoszewski 1998b) E. Jagoszewski. Conventional and fractional optical Fourier transform. *Optica Applicata*, 28:257–259, 1998.

(James and Agarwal 1995) D. F. V. James and G. S. Agarwal. Generalized Radon transform for tomographic measurement of short pulses. *J Opt Soc Am A*, 12:704–708, 1995.

(James and Agarwal 1996) D. F. V. James and G. S. Agarwal. The generalized Fresnel transform and its applications to optics. *Opt Commun*, 207–212, 1996.

(Jiang 1995) Z. Jiang. Scaling laws and simultaneous optical implementation of various order fractional Fourier transforms. *Opt Lett*, 20:2408–2410, 1995.

(Jiang, Lu, and Zhao 1997) Z. Jiang, Q. Lu, and Y. Zhao. Sensitivity of the fractional Fourier transform to parameters and its application in optical measurement. *Appl Opt*, 8455–8458, 1997.

(Karasik 1994) Y. B. Karasik. Expression of the kernel of a fractional Fourier transform in elementary functions. *Opt Lett*, 19:769–770, 1994.

(Karp 1995) D. B. Karp. The fractional Hankel transform and its applications in mathematical physics. *Russian Acad Sci Dokl Math*, 50:179–185, 1995.

(Kerr 1988a) F. H. Kerr. A distributional approach to Namias' fractional Fourier transforms. *Proc Royal Soc Edinburgh*, 108A:133–143, 1988.

(Kerr 1988b) F. H. Kerr. Namias' fractional Fourier transforms on L^2 and applications to differential equations. *J Math Anal Applic*, 136:404–418, 1988.

(Kerr 1991) F. H. Kerr. A fractional power theory for Hankel transforms in $L^2(\mathbf{R}^+)$. *J Math*

Anal Applic, 158:114–123, 1991.

(Kerr 1992) F. H. Kerr. Fractional powers of Hankel transforms in the Zemanian spaces. *J Math Anal Applic*, 166:65–83, 1992.

(Khare 1974) R. S. Khare. Fractional Fourier analysis of defocused images. *Opt Commun*, 12:386–388, 1974.

(Kober 1939) H. Kober. Wurzeln aus der Hankel-, Fourier- und aus anderen stetigen Transformationen. *Quart J Math (Oxford)*, 10:45–59, 1939.

(Kong and Lü 1997) F. Kong and B. Lü. Implementation of the fractional Fourier transform with a paraboloidal reflective mirror. *Optik*, 106:42–43, 1997.

(Kraniauskas, Cariolaro, and Erseghe 1998) P. Kraniauskas, G. Cariolaro, and T. Erseghe. Method for defining a class of fractional operations. *IEEE Trans Signal Processing*, 46:2804–2807, 1998.

(Kuo and Luo 1998) C. J. Kuo and Y. Luo. Generalized joint fractional Fourier transform correlators: a compact approach. *Appl Opt*, 37:8270–8276, 1998.

(Kutay 1995) M. A. Kutay. *Optimal Filtering in Fractional Fourier Domains*. MS thesis, Bilkent University, Ankara, 1995.

(Kutay 1999) M. A. Kutay. *Generalized Filtering Configurations with Applications in Digital and Optical Signal and Image Processing*. PhD thesis, Bilkent University, Ankara, 1999.

(Kutay and Ozaktas 1998) M. A. Kutay and H. M. Ozaktas. Optimal image restoration with the fractional Fourier transform. *J Opt Soc Am A*, 15:825–833, 1998.

(Kutay and others 1995) M. A. Kutay, H. M. Ozaktas, L. Onural, and O. Arıkan. Optimal filtering in fractional Fourier domains. In *Proc 1995 IEEE Int Conf Acoustics, Speech, Signal Processing*, IEEE, Piscataway, New Jersey, 1995. Pages 937–940.

(Kutay and others 1997) M. A. Kutay, H. M. Ozaktas, O. Arıkan, and L. Onural. Optimal filtering in fractional Fourier domains. *IEEE Trans Signal Processing*, 45:1129–1143, 1997.

(Kutay and others 1998a) M. A. Kutay, M. F. Erden, H. M. Ozaktas, O. Arıkan, Ö. Güleryüz, and Ç. Candan. Space-bandwidth-efficient realizations of linear systems. *Opt Lett*, 23:1069–1071, 1998.

(Kutay and others 1998b) M. A. Kutay, M. F. Erden, H. M. Ozaktas, O. Arıkan, Ç. Candan, and Ö. Güleryüz. Cost-efficient approximation of linear systems with repeated and multi-channel filtering configurations. In *Proc 1998 IEEE Int Conf Acoustics, Speech, Signal Processing*, IEEE, Piscataway, New Jersey, 1998. Pages 3433–3436.

(Kutay and others 1998c) M. A. Kutay, H. Özaktaş, M. F. Erden, H. M. Ozaktas, and O. Arıkan. Solution and cost analysis of general multi-channel and multi-stage filtering circuits. In *Proc 1998 IEEE-SP Int Symp Time-Frequency Time-Scale Analysis*, IEEE, Piscataway, New Jersey, 1998. Pages 481–484.

(Kutay and others 1999) M. A. Kutay, H. Özaktaş, H. M. Ozaktas, and O. Arıkan. The fractional Fourier domain decomposition. *Signal Processing*, 77:105–109, 1999.

(Kutay and others 2000) M. A. Kutay, H. Özaktaş, M. F. Erden, and H. M. Ozaktas. Optimization and cost analysis of multi-stage and multi-channel filtering configurations. Technical Report BU-CE-0004, Bilkent University, Department of Computer Engineering, Ankara, January 2000.

(Labunets and Labunets 1998) E. V. Labunets and V. G. Labunets. Fast fractional Fourier transform. In *9th European Signal Processing Conference: Signal Processing IX: Theories Applications*, Typorama Publications, Patras, Greece, 1998. Pages 1757–1760.

(Lakhtakia 1993) A. Lakhtakia. Fractal self-Fourier functions. *Optik*, 94:51–52, 1993.

(Lakhtakia and Caulfield 1992) A. Lakhtakia and H. J. Caulfield. On some mathematical and optical integral transforms of self-similar (fractal) functions. *Optik*, 91:131–133, 1992.

(Lancis and others 1997) J. Lancis, T. Szoplik, E. Tajahuerce, V. Climent, and M. Fernández-Alonso. Fractional derivative Fourier plane filter for phase-change visualization. *Appl Opt*, 36:7461–7464, 1997.

(Larkin 1995a) K. G. Larkin. A beginner's guide to the fractional Fourier transform, part 1. *Australian Optical Society News*, June 1995, pages 18–21.

(Larkin 1995b) K. G. Larkin. A beginner's guide to the fractional Fourier transform, part 2: a brief history of time frequency distributions. *Australian Optical Society News*, December 1995, pages 13–17.

(Lee and Szu 1994) S.-Y. Lee and H. H. Szu. Fractional Fourier transforms, wavelet transforms, and adaptive neural networks. *Opt Eng*, 33:2326–2330, 1994.

(Liu, Wu, and Fan 1998) Z. Y. Liu, X. Y. Wu, and D. Y. Fan. Collins formula in frequency-domain and fractional Fourier transforms. *Opt Commun*, 155:7–11. 1998.

(Liu, Wu, and Li 1995) S. Liu, J. Wu, and C. Li. Cascading the multiple stages of optical fractional Fourier transforms under different variable scales. *Opt Lett*, 20:1415–1417, 1995.

(Liu, Zhang, and Zhang 1997) S. Liu, J. Zhang, and Y. Zhang. Properties of the fractionalization of a Fourier transform. *Opt Commun*, 133:50–54, 1997.

(Liu and others 1995) S. Liu, J. Xu, Y. Zhang, L. Chen, and C. Li. General optical implementation of fractional Fourier transforms. *Opt Lett*, 20:1053–1055, 1995.

(Liu and others 1997a) S. Liu, J. Jiang, Y. Zhang, and J. Zhang. Generalized fractional Fourier transforms. *J Phys A*, 30:973–981, 1997.

(Liu and others 1997b) S. Liu, H. Ren, J. Zhang, and Z. Zhang. Image-scaling problem in the optical fractional Fourier transform. *Appl Opt*, 36:5671–5674, 1997.

(Lohmann 1993a) A. W. Lohmann. Concept and implementation of the fractional Fourier transform. In *Optical Computing 1993, Tech Digest Ser 7*, OSA, Washington DC, 1993. Pages 18–21.

(Lohmann 1993b) A. W. Lohmann. Image rotation, Wigner rotation, and the fractional order Fourier transform. *J Opt Soc Am A*, 10:2181–2186, 1993.

(Lohmann 1994) A. W. Lohmann. Fourier curios. In *Current Trends in Optics*, Academic Press, London, 1994. Chapter 11, pages 149–161.

(Lohmann 1995) A. W. Lohmann. A fake zoom lens for fractional Fourier experiments. *Opt Commun*, 115:437–443, 1995.

(Lohmann and Mendlovic 1994b) A. W. Lohmann and D. Mendlovic. Fractional Fourier transform: photonic implementation. *Appl Opt*, 33:7661–7664, 1994.

(Lohmann and Mendlovic 1997) A. W. Lohmann and D. Mendlovic. Fractional joint transform correlator. *Appl Opt*, 36:7402–7407, 1997.

(Lohmann, Mendlovic, and Shabtay 1999) A. W. Lohmann, D. Mendlovic, and G. Shabtay. Coherence waves. *J Opt Soc Am A*, 16:359–363, 1999.

(Lohmann, Mendlovic, and Zalevsky 1996) A. W. Lohmann, D. Mendlovic, and Z. Zalevsky. Fractional Hilbert transform. *Opt Lett*, 21:281–283, 1996.

(Lohmann, Mendlovic, and Zalevsky 1998) A. W. Lohmann, D. Mendlovic, and Z. Zalevsky. Fractional transformations in optics. In *Progress in Optics XXXVIII*, Elsevier, Amsterdam, 1998. Chapter IV, pages 263–342.

(Lohmann and Soffer 1993) A. W. Lohmann and B. H. Soffer. Relationships between two transforms: Radon-Wigner and fractional Fourier. In *1993 OSA Annual Meeting Tech Digest*, OSA, Washington DC, 1993. Page 109.

(Lohmann and Soffer 1994) A. W. Lohmann and B. H. Soffer. Relationships between the Radon-Wigner and fractional Fourier transforms. *J Opt Soc Am A*, 11:1798–1801, 1994.

(Lohmann, Tepichín, and Ramírez 1997) A. W. Lohmann, E. Tepichín, and J. G. Ramírez. Optical implementation of the fractional Hilbert transform for two-dimensional objects. *Appl Opt*, 36:6620–6626, 1997.

(Lohmann, Zalevsky, and Mendlovic 1996) A. W. Lohmann, Z. Zalevsky, and D. Mendlovic. Synthesis of pattern recognition filters for fractional Fourier processing. *Opt Commun*, 128:199–204, 1996.

(Lohmann and others 1996a) A. W. Lohmann, R. G. Dorsch, D. Mendlovic, Z. Zalevsky, and C. Ferreira. Space-bandwidth product of optical signals and systems. *J Opt Soc Am A*, 13:470–473, 1996.

(Lohmann and others 1996b) A. W. Lohmann, D. Mendlovic, Z. Zalevsky, and R. G. Dorsch. Some important fractional transformations for signal processing. *Opt Commun*, 125:18–20,

1996.

(Lohmann and others 1998) A. W. Lohmann, Z. Zalevsky, R. G. Dorsch, and D. Mendlovic. Experimental considerations and scaling property of the fractional Fourier transform. *Opt Commun*, 146:55–61, 1998.

(Lü and Kong 1997) B. Lü and F. Kong. Synthesis and decomposition of fractional Fourier optical systems written in terms of $ABCD$ matrix. *Optik*, 106:155–158, 1997.

(Lü, Kong, and Zhang 1997) B. Lü, F. Kong, and B. Zhang. Optical systems expressed in terms of fractional Fourier transforms. *Opt Commun*, 137:13–16, 1997.

(Ludwig 1988) L. F. Ludwig. General thin-lens action on spatial intensity distribution behaves as non-integer powers of Fourier transform. In *Spatial Light Modulators and Applications 1988*, OSA, Washington DC, 1988. Pages 173–176.

(Mann and Haykin 1992) S. Mann and S. Haykin. 'Chirplets' and 'warblets': novel time-frequency methods. *Electronics Lett*, 28:114–116, 1992.

(Mann and Haykin 1995) S. Mann and S. Haykin. The chirplet transform: physical considerations. *IEEE Trans Signal Processing*, 43:2745–2761, 1995.

(Marhic 1995) M. E. Marhic. Roots of the identity operator and optics. *J Opt Soc Am A*, 12:1448–1459, 1995.

(McAlister and others 1995) D. F. McAlister, M. Beck, L. Clarke, A. Meyer, and M. G. Raymer. Optical phase-retrieval by phase-space tomography and fractional-order Fourier transforms. *Opt Lett*, 20:1181–1183, 1995.

(McBride and Kerr 1987) A. C. McBride and F. H. Kerr. On Namias's fractional Fourier transforms. *IMA J Appl Math*, 39:159–175, 1987.

(Mendlovic 1999) D. Mendlovic. The fractional Fourier transform: a tutorial. In *Proc 1999 IEEE-EURASIP Workshop Nonlinear Signal Image Processing*, Boğaziçi University, Bebek, Istanbul, 1999. Pages 476–480.

(Mendlovic and Ozaktas 1993a) D. Mendlovic and H. M. Ozaktas. Fractional Fourier transforms and their optical implementation: I. *J Opt Soc Am A*, 10:1875–1881, 1993. Reprinted in *Selected Papers on Fourier Optics, SPIE Milestone Ser 105*, SPIE, Bellingham, Washington, 1995. Pages 427–433.

(Mendlovic and Ozaktas 1993b) D. Mendlovic and H. M. Ozaktas. Fourier transforms of fractional order and their optical interpretation. In *16th Congress ICO: Optics as a Key to High Technology, SPIE Proc 1983*, SPIE, Bellingham, Washington, 1993. Pages 387–388.

(Mendlovic and Ozaktas 1999) D. Mendlovic and H. M. Ozaktas. Fractional Fourier transform in optics. In *18th Congress ICO: Optics for the Next Millennium, SPIE Proc 3749*, SPIE, Bellingham, Washington, 1999. Pages 40–41.

(Mendlovic, Ozaktas, and Lohmann 1993) D. Mendlovic, H. M. Ozaktas, and A. W. Lohmann. Fourier transforms of fractional order and their optical interpretation. In *Optical Computing 1993, Tech Digest Ser 7*, OSA, Washington DC, 1993. Pages 127–130.

(Mendlovic, Ozaktas, and Lohmann 1994a) D. Mendlovic, H. M. Ozaktas, and A. W. Lohmann. Graded-index fibers, Wigner-distribution functions, and the fractional Fourier transform. *Appl Opt*, 33:6188–6193, 1994.

(Mendlovic, Ozaktas, and Lohmann 1994b) D. Mendlovic, H. M. Ozaktas, and A. W. Lohmann. Self Fourier functions and fractional Fourier transforms. *Opt Commun*, 105:36–38, 1994.

(Mendlovic, Ozaktas, and Lohmann 1995) D. Mendlovic, H. M. Ozaktas, and A. W. Lohmann. Fractional correlation. *Appl Opt*, 34:303–309, 1995.

(Mendlovic, Zalevsky, and Konforti 1997) D. Mendlovic, Z. Zalevsky, and N. Konforti. Computation considerations and fast algorithms for calculating the diffraction integral. *J Mod Opt*, 44:407–414, 1997.

(Mendlovic, Zalevsky, and Ozaktas 1998) D. Mendlovic, Z. Zalevsky, and H. M. Ozaktas. Applications of the fractional Fourier transform to optical pattern recognition. In *Optical Pattern Recognition*, Cambridge University Press, Cambridge, 1998. Chapter 4, pages 89–125.

(Mendlovic and others 1995a) D. Mendlovic, Y. Bitran, R. G. Dorsch, and A. W. Lohmann. Optical fractional correlation: experimental results. *Appl Opt*, 34:1665–1670, 1995.

(Mendlovic and others 1995b) D. Mendlovic, Y. Bitran, R. G. Dorsch, C. Ferreira, J. Garcia, and H. M. Ozaktaz. Anamorphic fractional Fourier transform: optical implementation and applications. *Appl Opt*, 34:7451–7456, 1995.

(Mendlovic and others 1995c) D. Mendlovic, Z. Zalevsky, N. Konforti, R. G. Dorsch, and A. W. Lohmann. Incoherent fractional Fourier transform and its optical implementation. *Appl Opt*, 34:7615–7620, 1995.

(Mendlovic and others 1995d) D. Mendlovic, Z. Zalevsky, R. G. Dorsch, Y. Bitran, A. W. Lohmann, and H. Ozaktas. New signal representation based on the fractional Fourier transform: definitions. *J Opt Soc Am A*, 12:2424–2431, 1995.

(Mendlovic and others 1996a) D. Mendlovic, R. G. Dorsch, A. W. Lohmann, Z. Zalevsky, and C. Ferreira. Optical illustration of a varied fractional Fourier-transform order and the Radon-Wigner display. *Appl Opt*, 35:3925–3929, 1996.

(Mendlovic and others 1996b) D. Mendlovic, Z. Zalevsky, A. W. Lohmann, and R. G. Dorsch. Signal spatial-filtering using the localized fractional Fourier transform. *Opt Commun*, 126:14–18, 1996.

(Mendlovic and others 1997) D. Mendlovic, Z. Zalevsky, D. Mas, J. García, and C. Ferreira. Fractional wavelet transform. *Appl Opt*, 36:4801–4806, 1997.

(Mendlovic and others 1998) D. Mendlovic, D. Mas, A. W. Lohmann, Z. Zalevsky, and G. Shabtay. Fractional triple correlation and its applications. *J Opt Soc Am A*, 15:1658–1661, 1998.

(Mihovilovic and Bracewell 1991) D. Mihovilovic and R. N. Bracewell. Adaptive chirplet representation of signals on time-frequency plane. *Electronics Lett*, 27:1159–1161, 1991.

(Mihovilovic and Bracewell 1992) D. Mihovilović and R. N. Bracewell. Whistler analysis in the time-frequency plane using chirplets. *J Geophysical Research*, 97:17199–17204, 1992.

(Miller 1968) W. Miller, Jr. *Lie Theory and Special Functions*. Academic Press, New York, 1968.

(Mustard 1987a) D. Mustard. Lie group imbeddings of the Fourier transform. School of Mathematics Preprint AM87/13, The University of New South Wales, Kensington, Australia, 1987.

(Mustard 1987b) D. A. Mustard. The fractional Fourier transform and a new uncertainty principle. School of Mathematics Preprint AM87/14, The University of New South Wales, Kensington, Australia, 1987.

(Mustard 1987c) D. Mustard. Lie group imbeddings of the Fourier transform and a new family of uncertainy principles. In *Proc Miniconf Harmonic Analysis Operator Algebras*, Australian National University, Canberra, 1987. Pages 211–222.

(Mustard 1989) D. Mustard. The fractional Fourier transform and the Wigner distribution. School of Mathematics Preprint AM89/6, The University of New South Wales, Kensington, Australia, 1989.

(Mustard 1990) D. Mustard. Fractional convolution. School of Mathematics Preprint AM90/26, The University of New South Wales, Kensington, Australia, 1990.

(Mustard 1991) D. Mustard. Uncertainty principles invariant under the fractional Fourier transform. *J Australian Mathematical Society B*, 33:180–191, 1991.

(Mustard 1996) D. Mustard. The fractional Fourier transform and the Wigner distribution. *J Australian Mathematical Society B*, 38:209–219, 1996.

(Mustard 1998) D. Mustard. Fractional convolution. *J Australian Mathematical Society B*, 40:257–265, 1998.

(Namias 1980a) V. Namias. The fractional order Fourier transform and its application to quantum mechanics. *J Inst Maths Applics*, 25:241–265, 1980.

(Namias 1980b) V. Namias. Fractionalization of Hankel transforms. *J Inst Maths Applics*, 26:187–197, 1980.

(Onural, Erden, and Ozaktas 1997) L. Onural, M. F. Erden, and H. M. Ozaktas. Extensions to common Laplace and Fourier transforms. *IEEE Signal Processing Lett*, 4:310–312, 1997.

(Onural 1993) L. Onural. Diffraction from a wavelet point of view. *Opt Lett*, 18:846–848, 1993.

(Onural and Kocatepe 1995) L. Onural, M. Kocatepe. Family of scaling chirp functions, diffraction, and holography. *IEEE Trans Signal Processing*, 43: 1568–1578, 1995.

(Onural, Kocatepe, and Ozaktas 1994) L. Onural, M. Kocatepe, and H. M. Ozaktas. A class of wavelet kernels associated with wave propagation. In *Proc 1994 IEEE Int Conf Acoustics, Speech, Signal Processing*, IEEE, Piscataway, New Jersey, 1994. Pages III-9–III-12.

(Owechko 1998) Y. Owechko. Time-frequency distributions using rotated-window spectrograms. In *Proc 1998 IEEE-SP Int Symp Time-Frequency Time-Scale Analysis*, IEEE, Piscataway, New Jersey, 1998. Pages 301–304.

(Ozaktas 1994) H. M. Ozaktas. The fractional Fourier-transform in optics. In *1994 OSA Annual Meeting Prog*, OSA, Washington DC, 1994. Page 173.

(Ozaktas 1996) H. M. Ozaktas. Repeated fractional Fourier domain filtering is equivalent to repeated time and frequency domain filtering. *Signal Processing*, 54:81–84, 1996.

(Ozaktas and Aytür 1995) H. M. Ozaktas and O. Aytür. Fractional Fourier domains. *Signal Processing*, 46:119–124, 1995.

(Ozaktas, Barshan, and Mendlovic 1994) H. M. Ozaktas, B. Barshan, and D. Mendlovic. Convolution and filtering in fractional Fourier domains. *Optical Review*, 1:15–16, 1994.

(Ozaktas and Erden 1997) H. M. Ozaktas and M. F. Erden. Relationships among ray optical, Gaussian beam, and fractional Fourier transform descriptions of first-order optical systems. *Opt Commun*, 143:75–86, 1997.

(Ozaktas, Erden, and Kutay 1998) H. M. Ozaktas, M. F. Erden, and M. A. Kutay. Signal processing with repeated filtering in fractional Fourier domains. *J Optoelectronics. Laser, supplement: Optics for Information Infrastructure*, 9:160–162, 1998.

(Ozaktas, Erkaya, and Kutay 1996) H. M. Ozaktas, N. Erkaya, and M. A. Kutay. Effect of fractional Fourier transformation on time-frequency distributions belonging to the Cohen class. *IEEE Signal Processing Lett*, 3:40–41, 1996.

(Ozaktas and Kutay 2000a) H. M. Ozaktas and M. A. Kutay. Time-order signal representations. Technical Report BU-CE-0005, Bilkent University, Department of Computer Engineering, Ankara, January 2000. Also in *Proc First IEEE Balkan Conf Signal Processing, Communications, Circuits, Systems*, Bilkent University, Ankara, 2000. CD-ROM.

(Ozaktas, Kutay, and Mendlovic 1998) H. M. Ozaktas, M. A. Kutay, and D. Mendlovic. Introduction to the fractional Fourier transform and its applications. Technical Report BU-CEIS-9802, Bilkent University, Department of Computer Engineering and Information Science, Ankara, January 1998.

(Ozaktas, Kutay, and Mendlovic 1999) H. M. Ozaktas, M. A. Kutay, and D. Mendlovic. Introduction to the fractional Fourier transform and its applications. In *Advances in Imaging and Electron Physics 106*, Academic Press, San Diego, California, 1999. Pages 239–291.

(Ozaktas and Mendlovic 1993a) H. M. Ozaktas and D. Mendlovic. Fourier transforms of fractional order and their optical interpretation. *Opt Commun*, 101:163–169, 1993.

(Ozaktas and Mendlovic 1993b) H. M. Ozaktas and D. Mendlovic. Fractional Fourier transforms and their optical implementation: II. *J Opt Soc Am A*, 10:2522–2531, 1993. Reprinted in *Selected Papers on Fourier Optics, SPIE Milestone Ser 105*, SPIE, Bellingham, Washington, 1995. Pages 434–443.

(Özaktaş and Mendlovic 1993) H. M. Özaktaş and D. Mendlovic. Kesirli Fourier dönüşümleri, Wigner dağılımları ile ilgisi, ve optik sinyal işlemeye uygulamaları. In *Sinyal İşleme ve Uygulamaları*, Boğaziçi University, Bebek, Istanbul, 1993. Pages 189–191. (In Turkish: Fractional Fourier transforms, their relationship to Wigner distributions, and applications to optical signal processing. In *Signal Processing and its Applications*.)

(Ozaktas and Mendlovic 1994) H. M. Ozaktas and D. Mendlovic. Fractional Fourier transform as a tool for analyzing beam propagation and spherical mirror resonators. *Opt Lett*,

19:1678–1680, 1994.

(Ozaktas and Mendlovic 1995a) H. M. Ozaktas and D. Mendlovic. Fractional Fourier optics. *J Opt Soc Am A*, 12:743–751, 1995.

(Ozaktas and Mendlovic 1995b) H. M. Ozaktas and D. Mendlovic. The fractional Fourier transform and its applications in optics and signal processing. *Optical Processing and Computing: SPIE Int Tech Working Group Newsletter*, October 1995, page 5.

(Ozaktas and Mendlovic 1996a) H. M. Ozaktas and D. Mendlovic. Every Fourier optical system is equivalent to consecutive fractional-Fourier-domain filtering. *Appl Opt*, 35:3167–3170, 1996.

(Ozaktas and Mendlovic 1996b) H. M. Ozaktas and D. Mendlovic. Applications of the fractional Fourier transform in optics and signal processing—a review. In *17th Congress ICO: Optics for Science and New Technology, SPIE Proc 2778*, SPIE, Bellingham, Washington, 1996. Pages 414–417.

(Ozaktas, Mendlovic, and Lohmann 1993) H. M. Ozaktas, D. Mendlovic, and A. W. Lohmann. About the Wigner distribution of a graded index medium and the fractional Fourier transform operation. In *16th Congress ICO: Optics as a Key to High Technology, SPIE Proc 1983*, SPIE, Bellingham, Washington, 1993. Pages 407–408.

(Ozaktas and Miller 1996) H. M. Ozaktas and D. A. B. Miller. Digital Fourier optics. *Appl Opt*, 35:1212–1219, 1996.

(Ozaktas and others 1994a) H. M. Ozaktas, B. Barshan, D. Mendlovic, and L. Onural. Convolution, filtering, and multiplexing in fractional Fourier domains and their relation to chirp and wavelet transforms. *J Opt Soc Am A*, 11:547–559, 1994.

(Ozaktas and others 1994b) H. M. Ozaktas, B. Barshan, L. Onural, and D. Mendlovic. Filtering in fractional Fourier domains and their relation to chirp transforms. In *Proc 7th Mediterranean Electrotechnical Conference*, IEEE, Piscataway, New Jersey, 1994. Pages 77–79.

(Ozaktas and others 1995) H. M. Ozaktas, B. Barshan, D. Mendlovic, and H. Urey. Space-variant filtering in fractional Fourier domains. In *Optical Computing, Inst Phys Conf Ser 139*, Institute of Physics Publishing, Bristol, UK, 1995. Pages 285–288.

(Ozaktas and others 1996) H. M. Ozaktas, O. Arikan, M. A. Kutay, and G. Bozdağı. Digital computation of the fractional Fourier transform. *IEEE Trans Signal Processing*, 44:2141–2150, 1996.

(Özdemir and Arıkan 2000) A. K. Özdemir and O. Arıkan. Fast computation of the ambiguity function and the Wigner distribution on arbitrary line segments. *IEEE Trans Signal Processing*, 48, 2000.

(Palma and Bagini 1997) C. Palma and V. Bagini. Extension of the Fresnel transform to *ABCD* systems. *J Opt Soc Am A*, 14:1774–1779, 1997.

(Patterson 1959) A. L. Patterson. *Z Krist*, 112:22, 1959.

(Pei and Yeh 1997) S.-C. Pei and M.-H. Yeh. Improved discrete fractional Fourier transform. *Opt Lett*, 22:1047–1049, 1997.

(Pei and Yeh 1998a) S.-C. Pei and M.-H. Yeh. Two-dimensional discrete fractional Fourier transform. *Signal Processing*, 67:99–108, 1998.

(Pei and Yeh 1998b) S.-C. Pei and M.-H. Yeh. Discrete fractional Hilbert transform. In *Proc 1998 IEEE Int Symp Circuits Systems*, IEEE, Piscataway, New Jersey, 1998. Pages IV-506–IV-509.

(Pei and Yeh 1999) S.-C. Pei and M.-H. Yeh. Discrete fractional Hadamard transform. In *Proc 1999 IEEE Int Symp Circuits Systems*, IEEE, Piscataway, New Jersey, 1999. Pages III-179–III-182.

(Pei, Yeh, and Tseng 1999) S.-C. Pei, M.-H. Yeh, and C.-C. Tseng. Discrete fractional Fourier transform based on orthogonal projections. *IEEE Trans Signal Processing*, 47:1335–1348, 1999.

(Pei and others 1998) S.-C. Pei, C.-C. Tseng, M.-H. Yeh, and J.-J. Shyu. Discrete fractional Hartley and Fourier transforms. *IEEE Trans Circuits Systems II*, 45:665–675, 1998.

(Pellat-Finet 1994) P. Pellat-Finet. Fresnel diffraction and the fractional-order Fourier transform. *Opt Lett*, 19:1388–1390, 1994.

(Pellat-Finet 1995) P. Pellat-Finet. Transfert du champ électromagnétique par diffraction et transformation de Fourier fractionnaire. *C R Acad Sci Paris*, 320:91–97, 1995.

(Pellat-Finet and Bonnet 1994) P. Pellat-Finet and G. Bonnet. Fractional order Fourier transform and Fourier optics. *Opt Commun*, 111:141–154, 1994.

(Pellat-Finet and Torres 1997) P. Pellat-Finet and Y. Torres. Image formation with coherent light: the fractional Fourier transform approach. *J Mod Opt*, 44:1581–1594, 1997.

(Pons and others 1999) A. M. Pons, A. Lorente, C. Illueca, D. Mas, and J. M. Artigas. Fresnel diffraction in a theoretical eye: a fractional Fourier transform approach. *J Mod Opt*, 46:1043–1050, 1999.

(Rashid 1989) M. A. Rashid. A simple proof of the result that the Wigner transformation is of finite order. *J Math Phys*, 30:1999–2000, 1989.

(Raveh and Mendlovic 1999) I. Raveh and D. Mendlovic. New properties of the Radon transform of the cross Wigner ambiguity distribution function. *IEEE Trans Signal Processing*, 47:2077–2080, 1999.

(Raymer, Beck, and McAlister 1994a) M. G. Raymer, M. Beck, and D. F. McAlister. Complex wave-field reconstruction using phase-space tomography. *Phys Rev Lett*, 72:1137–1140, 1994.

(Raymer, Beck, and McAlister 1994b) M. G. Raymer, M. Beck, and D. McAlister. Spatial and temporal optical field reconstruction using phase-space tomography. In *Quantum Optics VI*, Springer, Berlin, 1994.

(Richman and Parks 1997) M. S. Richman and T. W. Parks. Understanding discrete rotations. In *Proc 1997 IEEE Int Conf Acoustics, Speech, Signal Processing*, IEEE, Piscataway, New Jersey, 1997. Pages 2057–2060.

(Ristic and Boashash 1993) B. Ristic and B. Boashash. Kernel design for time-frequency signal analysis using the Radon transform. *IEEE Trans Signal Processing*, 41,1996–2008, 1993.

(Ruiz and Rabal 1997) B. Ruiz and H. Rabal. Fractional Fourier transform description with use of differential operators. *J Opt Soc Am A*, 14:2905–2913, 1997.

(Sahin 1996) A. Sahin. *Two-dimensional Fractional Fourier Transform and Its Optical Implementation*. MS thesis, Bilkent University, Ankara, 1996.

(Sahin, Kutay, and Ozaktas 1998) A. Sahin, M. A. Kutay, and H. M. Ozaktas. Nonseparable two-dimensional fractional Fourier transform. *Appl Opt*, 37:5444–5453, 1998.

(Sahin, Ozaktas, and Mendlovic 1995) A. Sahin, H. M. Ozaktas, and D. Mendlovic. Optical implementation of the two-dimensional fractional Fourier transform with different orders in the two dimensions. *Opt Commun*, 120:134–138, 1995.

(Sahin, Ozaktas, and Mendlovic 1998) A. Sahin, H. M. Ozaktas, and D. Mendlovic. Optical implementations of two-dimensional fractional Fourier transforms and linear canonical transforms with arbitrary parameters. *Appl Opt*, 37:2130–2141, 1998.

(Sang, Williams, and O'Neill 1996) T.-H. Sang, W. J. Williams, and J. C. O'Neill. An algorithm for positive time-frequency distributions. In *Proc 1996 IEEE-SP Int Symp Time-Frequency Time-Scale Analysis*, IEEE, Piscataway, New Jersey, 1996. Pages 165–168.

(Santhanam and McClellan 1995) B. Santhanam and J. H. McClellan. The DRFT—a rotation in time-frequency space. In *Proc 1995 IEEE Int Conf Acoustics, Speech, Signal Processing*, IEEE, Piscataway, New Jersey, 1995. Pages 921–924.

(Santhanam and McClellan 1996) B. Santhanam and J. H. McClellan. The discrete rotational Fourier transform. *IEEE Trans Signal Processing*, 44:994–998, 1996.

(Seger 1993) O. Seger. *Model Building and Restoration with Applications in Confocal Microscopy*. PhD thesis, Linköping University, Sweden, 1993.

(Seger and Lenz 1992) O. Seger and R. Lenz. Modelling of the point-spread function of laser scanning microscopes using canonical transforms. *J Visual Communication Image Representation*, 3:364–380, 1992.

(Shamir and Cohen 1995) J. Shamir and N. Cohen. Root and power transformations in optics.
    J Opt Soc Am A, 12:2415–2423, 1995.
(Sheppard 1998) C. J. R. Sheppard. Free-space diffraction and the fractional Fourier
    transform. *J Mod Opt*, 45:2097–2103, 1998.
(Sheppard and Larkin 1998) C. J. R. Sheppard and K. G. Larkin. Similarity theorems for
    fractional Fourier transforms and fractional Hankel transforms. *Opt Commun*, 154:173–178,
    1998.
(Sheridan 1996) J. T. Sheridan. General fractal preserving linear and non-linear
    transformations. *Optik*, 103:27–30, 1996.
(Shih 1995a) C.-C. Shih. Optical interpretation of a complex-order Fourier transform. *Opt
    Lett*, 20:1178–1180, 1995.
(Shih 1995b) C.-C. Shih. Fractionalization of Fourier transform. *Opt Commun*, 118:495–498,
    1995.
(Shin and others 1998) S. G. Shin, S. I. Jin, S. Y. Shin, and S. Y. Lee. Optical neural network
    using fractional Fourier transform, log-likelihood, and parallelism. *Opt Commun*, 153:218–
    222, 1998.
(Simon and Mukunda 1998) R. Simon and N. Mukunda. Iwasawa decomposition in first-
    order optics: universal treatment of shape-invariant propagation for coherent and partially
    coherent beams. *J Opt Soc Am A*, 15:2146–2155, 1998.
(Smithey and others 1993) D. T. Smithey, M. Beck, M. G. Raymer, and A. Faridani.
    Measurement of the Wigner distribution and the density matrix of a light mode using
    optical homodyne tomography: application to squeezed states and the vacuum. *Phys Rev
    Lett*, 70:1244–1247, 1993.
(Song and others 1997a) S. H. Song, S. Park, E.-H. Lee, P. S. Kim, and C. H. Oh.
    Planar optical implementation of multichannel fractional Fourier transforms. *Opt Commun*,
    137:219–222, 1997.
(Song and others 1997b) S. H. Song, J.-S. Jeong, S. Park, and E.-H. Lee. Planar optical
    implementation of fractional correlation. *Opt Commun*, 143:287–293, 1997.
(Stankovic and Djurovic 1998) L. Stankovic and I. Djurovic. Relationship between the
    ambiguity function coordinate transformations and the fractional Fourier transform. *Annals
    Telecommunications*, 53: 316–319, 1998.
(*Status Report on the Fractional Fourier Transform* 1995) *Status Report on the Fractional
    Fourier Transform*. A. W. Lohmann, D. Mendlovic, and Z. Zalevsky, editors. Tel-Aviv
    University, Faculty of Engineering, Tel-Aviv, Israel, 1995.
(Taylor 1984) M. E. Taylor. Noncommutative microlocal analysis, part I. *Memoirs AMS*,
    volume 52, number 313, 1984.
(Tripathi, Pati, and Singh 1999) R. Tripathi, G. S. Pati, and K. Singh. A simple technique
    for space-variant target detection using fractional correlation. In *International Conference
    on Optics and Optoelectronics '98, SPIE Proc 3729*, SPIE, Bellingham, Washington, 1999.
    Pages 45–50.
(Tu and Tamura 1997) J. Tu and S. Tamura. Wave field determination using tomography of
    the ambiguity function. *Phys Rev E*, 55:1946–1949, 1997.
(Tu and Tamura 1998) J. H. Tu and S. Tamura. Analytic relation for recovering the mutual
    intensity by means of intensity information. *J Opt Soc Am A*, 15:202–206, 1998.
(Tucker, Ojeda-Castañeda, and Cathey 1999) S. B. Tucker, J. Ojeda-Castañeda, and W. T.
    Cathey. Matrix description of near-field diffraction and the fractional Fourier transform. *J
    Opt Soc Am A*, 16: 316–322, 1999.
(Turski 1998) J. Turski. Harmonic analysis on $SL(2,c)$ and projectively adapted pattern
    representation. *J Fourier Anal Appl*, 4:67–91, 1998.
(Uozumi and Asakura 1994) J. Uozumi and T. Asakura. Fractal optics. In *Current Trends in
    Optics*. Academic Press, London, 1994. Chapter 6, pages 83–93.
(Várilly and Gracia-Bondía 1987) J. C. Várilly and J. M. Gracia-Bondía. The Wigner
    transformation is of finite order. *J Math Phys*, 28:2390–2392, 1987.

(Vilenkin 1965) N. Ja. Vilenkin. *Special Functions and the Theory of Group Representations.* Izd. Nauka., Moscow, 1965 (in Russian). English translation: AMS Translations, vol. 22, American Mathematical Society, Providence, Rhode Island, 1968.

(Vogel and Risken 1989) K. Vogel and H. Risken. Determination of quasiprobability distributions in terms of probability distributions for the rotated quadrature phase. *Phys Rev A*, 40:2847–2849, 1989.

(Vogel and Schleich 1991) W. Vogel and W. Schleich. Phase distribution of a quantum state without using phase states. *Phys Rev A*, 44:7642–7646, 1991.

(Wang, Chan, and Chui 1998) M. Wang, A. K. Chan, and C. K. Chui. Linear frequency-modulated signal detection using Radon-ambiguity transform. *IEEE Trans Signal Processing*, 46:571–586, 1998.

(Wang and Zhou 1998) X. E. Wang and J. Zhou. Scaled fractional Fourier transform and optical systems. *Opt Commun*, 147:341–348, 1998.

(Wawrzyńczyk 1990) A. Wawrzyńczyk. On the infinitesimal generator of the Fourier transform. *An Inst Mat Univ Nac Autónoma México*, 30:83–88, 1990.

(Weissler 1979) F. B. Weissler. Two-point inequalities, the Hermite semigroup, and the Gauss-Weierstrass semigroup. *J Functional Analysis*, 32:102–121, 1979.

(Weyl 1927) H. Weyl. Quantenmechanik und Gruppentheorie. *Ztsch Physik*, 46:1–47, 1927.

(Weyl 1930) H. Weyl. *The Theory of Groups and Quantum Mechanics*, second edition. Dover, New York, 1930.

(Wiener 1929) N. Wiener. Hermitian polynomials and Fourier analysis. *J Math Phys (MIT)*, 8:70–73, 1929.

(Wolf 1979) K. B. Wolf. *Integral Transforms in Science and Engineering*. Plenum Press, New York, 1979.

(Wolf and Krötzsch 1999) K. B. Wolf and G. Krötzsch. Metaxial correction of fractional Fourier transformers. *J Opt Soc Am A*, 16:821–830, 1999.

(Wolf and Rivera 1997) K. B. Wolf and A. L. Rivera. Holographic information in the Wigner function. *Opt Commun*, 144:36–42, 1997.

(Wood and Barry 1994a) J. C. Wood and D. T. Barry. Tomographic time-frequency analysis and its application toward time-varying filtering and adaptive kernel design for multicomponent linear-FM signals. *IEEE Trans Signal Processing*, 42:2094–2104, 1994.

(Wood and Barry 1994b) J. C. Wood and D. T. Barry. Linear signal synthesis using the Radon-Wigner transform. *IEEE Trans Signal Processing*, 42:2105–2111, 1994.

(Wood and Barry 1994c) J. C. Wood and D. T. Barry. Radon transformation of time-frequency distributions for analysis of multicomponent signals. *IEEE Trans Signal Processing*, 42:3166–3177, 1994.

(Xia 1996) X.-G. Xia. On bandlimited signals with fractional Fourier transform. *IEEE Signal Processing Lett*, 3:72–74, 1996.

(Xia and others 1996) X.-G. Xia, Y. Owechko, B. H. Soffer, and R. M. Matic. On generalized-marginal time-frequency distributions. *IEEE Trans Signal Processing*, 44:2882–2886, 1996.

(Yetik and others 2000a) İ. Ş. Yetik, M. A. Kutay, H. Özaktaş, and H. M. Ozaktas. Continuous and discrete fractional Fourier domain decomposition. In *Proc 2000 IEEE Int Conf Acoustics, Speech, Signal Processing*, IEEE, Piscataway, New Jersey, 2000. Pages I-93–I-96.

(Yetik and others 2000b) İ. Ş. Yetik, H. M. Ozaktas, B. Barshan, and L. Onural. Perspective projections in the space-frequency plane and fractional Fourier transforms. *J Opt Soc Am A*, 17, 2000.

(Yoshimura and Iwai 1997) H. Yoshimura and T. Iwai. Properties of the Gaussian Schell-model source field in a fractional Fourier plane. *J Opt Soc Am A*, 14:3388–3393, 1997.

(Yoshimura and Iwai 1998) H. Yoshimura and T. Iwai. Effect of lens aperture on the average intensity in a fractional Fourier plane. *Pure Appl Opt*. 7:1133–1141, 1998.

(Yu and others 1998a) L. Yu, Y. Y. Lu, X. M. Zeng, M. C. Huang, M. Z. Chen, W. D. Huang, and Z. Z. Zhu. Deriving the integral representation of a fractional Hankel transform from a fractional Fourier transform. *Opt Lett*, 23:1158–1160, 1998.

(Yu and others 1998b) L. Yu, W. D. Huang, M. C. Huang, Z. Z. Zhu, Z. M. Zeng, and W. Ji. The Laguerre-Gaussian series representation of two-dimensional fractional Fourier transform. *J Phys A*, 31:9353–9357, 1998.

(Yurke, Schleich, and Walls 1990) B. Yurke, W. Schleich, and D. F. Walls. Quantum superpositions generated by quantum nondemolition measurements. *Phys Rev A*, 42:1703–1711, 1990.

(Zalevsky 1996) Z. Zalevsky. *Unconventional Optical Processors for Pattern Recognition and Signal Processing*. PhD thesis, Tel-Aviv University, Tel-Aviv, 1996.

(Zalevsky 1999) Z. Zalevsky. Applications of the fractional Fourier transform to correlation, feature extraction and pattern recognition. In *Proc 1999 IEEE-EURASIP Workshop Nonlinear Signal Image Processing*, Boğaziçi University, Bebek, Istanbul, 1999. Pages 579–584.

(Zalevsky and Mendlovic 1996a) Z. Zalevsky and D. Mendlovic. Fractional Wiener filter. *Appl Opt*, 35:3930–3936, 1996.

(Zalevsky and Mendlovic 1996b) Z. Zalevsky and D. Mendlovic. Fractional Radon transform: definition. *Appl Opt*, 35:4628–4631, 1996.

(Zalevsky and Mendlovic 1997) Z. Zalevsky and D. Mendlovic. Light propagation analysis in graded index fiber—review and applications. *Fiber and Integrated Optics*, 16:55–61, 1997.

(Zalevsky, Mendlovic, and Caulfield 1997a) Z. Zalevsky, D. Mendlovic, and J. H. Caulfield. Localized, partially space-invariant filtering. *Appl Opt*, 36:1086–1092, 1997.

(Zalevsky, Mendlovic, and Caulfield 1997b) Z. Zalevsky, D. Mendlovic, and J. H. Caulfield. Fractional correlator with real-time control of the space-invariance property. *Appl Opt*, 36:2370–2375, 1997.

(Zalevsky, Mendlovic, and Dorsch 1996) Z. Zalevsky, D. Mendlovic, and R. G. Dorsch. Gerchberg-Saxton algorithm applied in the fractional Fourier or the Fresnel domain. *Opt Lett*, 21:842–844, 1996.

(Zalevsky, Mendlovic, and Lohmann 1998) Z. Zalevsky, D. Mendlovic, and A. W. Lohmann. The *ABCD*-Bessel transformation. *Opt Commun*, 147:39–41, 1998.

(Zalevsky, Mendlovic, and Ozaktas 2000) Z. Zalevsky, D. Mendlovic, and H. M. Ozaktas. Energetic efficient synthesis of general mutual intensity distribution. *J Opt A*, 2:83–87, 2000.

(Zalevsky and others 1997) Z. Zalevsky, I. Raveh, G. Shabtay, D. Mendlovic, and J. Garcia. Single-output color pattern recognition using a fractional correlator. *Opt Eng*, 36:2127–2136, 1997.

(Zayed 1996) A. I. Zayed. On the relationship between the Fourier and fractional Fourier transforms. *IEEE Signal Processing Lett*, 3:310–311, 1996.

(Zayed 1998a) A. I. Zayed. A convolution and product theorem for the fractional Fourier transform. *IEEE Signal Processing Lett*, 5:101–103, 1998.

(Zayed 1998b) A. I. Zayed. Hilbert transform associated with the fractional Fourier transform. *IEEE Signal Processing Lett*, 5:206–208, 1998.

(Zayed 1998c) A. I. Zayed. Fractional Fourier transform of generalized functions. *Integral Transforms Special Functions*, 7:299–312, 1998.

(Zhang and Gu 1998) Y. Zhang and B.-Y. Gu. Rotation-invariant and controllable space-variant optical correlation. *Appl Opt*, 37:6256–6261, 1998.

(Zhang, Gu, and Yang 1998) Y. Zhang, B.-Y. Gu, and G.-Z. Yang. Generation of self-fractional Hankel functions. *J Phys A*, 31:9769–9772, 1998.

(Zhang and others 1997) Y. Zhang, B.-Y. Gu, B.-Z. Dong, G.-Z. Yang, H. Ren, X. Zhang, and S. Liu. Fractional Gabor transform. *Opt Lett*, 22:1583–1585, 1997.

(Zhang and others 1998a) Y. Zhang, B.-Y. Gu, B.-Z. Dong, and G.-Z. Yang. Optical implementations of the Radon-Wigner display for one-dimensional signals. *Opt Lett*, 23:1126–1128, 1998.

(Zhang and others 1998b) Y. Zhang, B.-Z. Dong, B.-Y. Gu, and G.-Z. Yang. Beam shaping in the fractional Fourier transform domain. *J Opt Soc Am A*, 15:1114–1120, 1998.

(Zhang and others 1998c) Y. Zhang, B.-Y. Gu, B.-Z. Dong, and G.-Z. Yang. A new kind of windowed fractional transforms. *Opt Commun*, 152:127–134, 1998.

(Zhang and others 1999a) Y. Zhang, B.-Y. Gu, B.-Z. Dong, and G.-Z. Yang. New optical configurations for implementing Radon-Wigner display: matrix analysis approach. *Opt Commun*, 160:292–300, 1999.

(Zhang and others 1999b) Y. Zhang, B.-Y. Gu, B.-Z. Dong, and G.-Z. Yang. Rotationally symmetric beam shaping in the fractional Fourier transform domain. *Optik*, 110: 61–65, 1999.

Other Cited Works

(*Advanced Topics in Shannon Sampling and Interpolation Theory* 1993) *Advanced Topics in Shannon Sampling and Interpolation Theory*. R. J. Marks II, editor. Springer, New York, 1993.

(Agarwal 1992) R. P. Agarwal. *Difference Equations and Inequality Theory: Methods and Applications*. Marcel Dekker, New York, 1992.

(Akansu and Haddad 1992) A. N. Akansu and R. A. Haddad. *Multiresolution Signal Decomposition: Transforms, Subbands, and Wavelets*. Academic Press, London, 1992.

(Aldrovandi and Galetti 1990) R. Aldrovandi and D. Galetti. On the structure of quantum phase space. *J Math Phys*, 31:2987–2995, 1990.

(Alonso and Forbes 1995) M. A. Alonso and G. W. Forbes. Generalization of Hamilton's formalism for geometrical optics. *J Opt Soc Am A*, 12:2744–2752, 1995.

(Arnaud 1973) J. A. Arnaud. Hamiltonian theory of beam mode propagation. In *Progress in Optics XI*, Elsevier, Amsterdam, 1973. Chapter VI, pages 247–304.

(Arnaud 1976) J. A. Arnaud. *Beam and Fiber Optics*. Academic Press, New York, 1976. Chapter 4, pages 220–324.

(Atakishiev and Suslov 1991) N. M. Atakishiev and S. K. Suslov. Difference analogs of the harmonic oscillator. *Theoretical Mathematical Physics*, 85:1055–1062, 1991.

(Atakishiyev, Chumakov, and Wolf 1998) N. M. Atakishiyev, S. M. Chumakov, and K. B. Wolf. Wigner distribution function for finite systems. *J Math Phys*, 39:6247–6261, 1998.

(Athanasiu and Floratos 1994) G. G. Athanasiu and E. G. Floratos. Coherent states in finite quantum mechanics. *Nuclear Phys B*, 425:343–364, 1994.

(Bacry and Cadilhac 1981) H. Bacry and M. Cadilhac. Metaplectic group and Fourier optics. *Phys Rev A*, 23:2533–2536, 1981.

(Balian and Itzykson 1986) R. Balian and C. Itzykson. Observations sur la mechanique quantique finie. *Compte Rendus Acad Sci*, 303:773–778, 1986.

(Bamler and Glünder 1983) R. Bamler and H. Glünder. The Wigner distribution function of two-dimensional signals: Coherent optical generation and display. *Optica Acta*, 30:1789–1803, 1983.

(Barrett 1984) H. H. Barrett. The Radon transform and its applications. In *Progress in Optics XXI*, Elsevier, Amsterdam, 1984. Chapter 3, pages 217–286.

(Bartelt, Brenner, and Lohmann 1980) H. O. Bartelt, K.-H. Brenner, and A. W. Lohmann. The Wigner distribution function and its optical production. *Opt Commun*, 32:32–38, 1980.

(Bastiaans 1978) M. J. Bastiaans. The Wigner distribution function applied to optical signals and systems. *Opt Commun*, 25:26–30, 1978.

(Bastiaans 1979a) M. J. Bastiaans. Wigner distribution function and its application to first-order optics. *J Opt Soc Am*, 69:1710–1716, 1979.

(Bastiaans 1979b) M. J. Bastiaans. The Wigner distribution function and Hamilton's characteristics of a geometric-optical system. *Opt Commun*, 30:321–326, 1979.

(Bastiaans 1979c) M. J. Bastiaans. Transport equations for the Wigner distribution function. *Optica Acta*, 26:1265–1272, 1979.

(Bastiaans 1979d) M. J. Bastiaans. Transport equations for the Wigner distribution function in an inhomogeneous and dispersive medium. *Optica Acta*, 26:1333–1344, 1979.

(Bastiaans 1980) M. J. Bastiaans. Gabor's expansion of a signal into Gaussian elementary signals. *Proc IEEE*, 68:538–539,1980.

(Bastiaans 1981a) M. J. Bastiaans. A sampling theorem for the complex spectrogram, and Gabor's expansion of a signal in Gaussian elementary signals. *Opt Eng*, 20:594–598, 1981.

(Bastiaans 1981b) M. J. Bastiaans. The Wigner distribution function of partially coherent light. *Optica Acta*, 28:1215–1224, 1981.

(Bastiaans 1982a) M. J. Bastiaans. Gabor's signal expansion and degrees of freedom of a signal. *Optica Acta*, 28:1223–1229, 1982.

(Bastiaans 1982b) M. J. Bastiaans. Gabor's signal expansion and its relation to sampling of the sliding-window spectrum. In *Advanced Topics in Shannon Sampling and Interpolation Theory*, Springer, New York, 1993. Chapter 1, pages 1–35.

(Bastiaans 1985) M. J. Bastiaans. On the sliding-window representation in digital signal processing. *IEEE Trans on Acoustics, Speech, and Signal Processing*, 33:868–873, 1985.

(Bastiaans 1986a) M. J. Bastiaans. Application of the Wigner distribution function to partially coherent light. *J Opt Soc Am A*, 3:1227–1238, 1986.

(Bastiaans 1986b) M. J. Bastiaans. Uncertainty principle and informational entropy for partially coherent light. *J Opt Soc Am A*, 3:1243–1246, 1986.

(Bastiaans 1989) M. J. Bastiaans. Propagation laws for the second-order moments of the Wigner distribution function in first-order optical systems. *Optik*, 82:173–181, 1989.

(Bastiaans 1991a) M. J. Bastiaans. Gabor's signal expansion applied to partially coherent light. *Opt Commun*, 86:14–18, 1991.

(Bastiaans 1991b) M. J. Bastiaans. Second-order moments of the Wigner distribution function in first-order optical systems. *Optik*, 88:163–168, 1991.

(Bastiaans 1994) M. J. Bastiaans. Gabor's signal expansion and the Zak transform. *Appl Opt*, 33:5241–5255, 1994.

(Bastiaans 1997) M. J. Bastiaans. Applications of the Wigner distribution function in optics. In *The Wigner Distribution: Theory and Applications in Signal Processing*, Elsevier, Amsterdam, 1997. Pages 375–426.

(Basu and Wolf 1982) D. Basu and K. B. Wolf. The unitary irreducible representations of $SL(2, R)$ in all subgroup reductions. *J Math Phys*, 23:189–205, 1982.

(Beckner 1975) W. Beckner. Inequalities in Fourier analysis. *Annals Mathematics*, 102:159–182, 1975.

(Bendinelli and others 1974) M. Bendinelli, A. Consortini, L. Ronchi, and B. R. Frieden. Degrees of freedom. and eigenfunctions, for the noisy image. *J Opt Soc Am*, 64:1498–1502, 1974.

(Bertrand and Bertrand 1992) J. Bertrand and P. Bertrand. A class of affine Wigner functions with extended covariance properties. *J Math Phys*, 33:2515–2527, 1992. Erratum, 34:885, 1993.

(Birkhoff and Rota 1989) G. Birkhoff and G. C. Rota. *Ordinary Differential Equations*. Wiley, New York, 1989.

(Boashash 1988) B. Boashash. Note on the use of the Wigner distribution for time-frequency signal analysis. *IEEE Trans Acoustics, Speech, and Signal Processing*, 36:1518–1521, 1988.

(Boone 1997) B. G. Boone. *Signal Processing Using Optics*. Oxford University Press, New York, 1997.

(Born and Wolf 1980) M. Born and E. Wolf. *Principles of Optics*, sixth edition. Pergamon Press, Oxford, 1980.

(Boyd 1983) R. W. Boyd. *Radiometry and the Detection of Optical Radiation*. Wiley, New York, 1983.

(Bracewell 1986) R. N. Bracewell. *The Fourier Transform and its Applications*, second edition. McGraw-Hill, New York, 1986.

(Bracewell 1995) R. N. Bracewell. *Two-dimensional Imaging*. Prentice Hall, Englewood Cliffs, New Jersey, 1995.

(Bracewell 1999) R. N. Bracewell. *The Fourier Transform and its Applications*, third edition. McGraw-Hill, New York, 1999.

(Butterweck 1977) H. J. Butterweck. General theory of linear, coherent, optical data-processing systems. *J Opt Soc Am*, 67:60–70, 1977.

(Butterweck 1981) H. J. Butterweck. Principles of optical data-processing. In *Progress in Optics XIX*, Elsevier, Amsterdam, 1981. Chapter 4, pages 211–280.

(Byun 1993) D.-W. Byun. Inversions of Hermite semigroup. *Proc Amer Math Soc*, 118:437–445, 1993.

(Campbell 1994) C. E. Campbell. Ray vector fields. *J Opt Soc Am A*, 11:618–622, 1994.

(Caola 1991) M. J. Caola. Self-Fourier functions. *J Phys A*, 24:L1143–L1144. 1991.

(Caratheodory 1965) C. Caratheodory. *Calculus of Variations and Partial Differential Equations of the First Order*. Chelsea Publishing, New York, 1965.

(Carmichael 1999) H. J. Carmichael. *Statistical Methods in Quantum Optics I: Master Equations and Fokker-Planck Equations*. Springer, Berlin, 1999.

(Cathey 1974) W. T. Cathey. *Optical Information Processing and Holography*. Wiley, New York, 1974.

(Choi and Williams 1989) H. I. Choi and W. J. Williams. Improved time-frequency representation of multicomponent signals using exponential kernels. *IEEE Trans Acoustics, Speech, and Signal Processing*, 37:862–871, 1989.

(Choudhury, Puntambekar, and Chakraborty 1995) D. Choudhury, P. N. Puntambekar, A. K. Chakraborty. Optical synthesis of self-Fourier functions. *Opt Commun*, 119:279–282, 1995.

(Chui 1992) C. K. Chui. *An Introduction to Wavelets*. Academic Press, London, 1992.

(Cincotti, Gori, and Santarsiero 1992) G. Cincotti, F. Gori, and M. Santarsiero. Generalized self-Fourier functions. *J Phys A*, 25:L1191–L1194, 1992.

(Claasen and Mecklenbräuker 1980a) T. A. C. M. Claasen and W. F. G. Mecklenbräuker. The Wigner distribution—a tool for time-frequency signal analysis. Part I: continuous-time signals. *Philips J Res*, 35:217–250, 1980.

(Claasen and Mecklenbräuker 1980b) T. A. C. M. Claasen and W. F. G. Mecklenbräuker. The Wigner distribution—a tool for time-frequency signal analysis. Part II: discrete-time signals. *Philips J Res*, 35:276–300, 1980.

(Claasen and Mecklenbräuker 1980c) T. A. C. M. Claasen and W. F. G. Mecklenbräuker. The Wigner distribution—a tool for time-frequency signal analysis. Part III: relations with other time-frequency signal transformations. *Philips J Res*, 35:372–389, 1980.

(Claasen and Mecklenbräuker 1983) T. A. C. M. Claasen and W. F. G. Mecklenbräuker. The aliasing problem in discrete-time Wigner distributions. *IEEE Trans Acoustics, Speech, and Signal Processing*, 31:1067-1072, 1983.

(Coffey 1994) M. W. Coffey. Self-reciprocal Fourier functions. *J Opt Soc Am A*, 11:2453–2455, 1994.

(Cohen 1966) L. Cohen. Generalized phase-space distribution functions. *J Math Phys*, 7:781–786, 1966.

(Cohen 1976) L. Cohen. Quantization problem and variational principle in the phase space formulation of quantum mechanics. *J Math Phys*, 17:1863–1866, 1976.

(Cohen 1989) L. Cohen. Time-frequency distributions—a review. *Proc IEEE*, 77:941–981, 1989.

(Cohen 1995) L. Cohen. *Time-Frequency Analysis*. Prentice Hall, Englewood Cliffs, New Jersey, 1995.

(Cohen-Tannoudji, Diu, and Laloë 1977) C. Cohen-Tannoudji, B. Diu, and F. Laloë. *Quantum Mechanics*. Wiley, New York, 1977 (2 volumes).

(Daubechies 1990) I. Daubechies. The wavelet transform, time-frequency localization and signal analysis. *IEEE Trans Information Theory*, 36:961–1006, 1990.

(Daubechies 1992) I. Daubechies. *Ten Lectures on Wavelets*. Society for Industrial and Applied Mathematics, Philadelphia, Pennsylvania, 1992.

(Debnath and Mikusiński 1990) L. Debnath and P. Mikusiński. *Introduction to Hilbert Spaces with Applications*. Academic Press, San Diego, California, 1990.

(van Dekker and van den Bos 1997) A. J. van Dekker and A. van den Bos. Resolution: a survey. *J Opt Soc Am A*, 14:547–557, 1997.

(Deschamps 1972) G. A. Deschamps. Ray techniques in electromagnetics. *Proc IEEE*, 60:1022–1035, 1972.

(Dickinson and Steiglitz 1982) B. W. Dickinson and K. Steiglitz. Eigenvectors and functions of the discrete Fourier transform. *IEEE Trans Acoustics, Speech, and Signal Processing*, 30:25–31, 1982.

(Dragoman 1994) D. Dragoman. Higher-order moments of the Wigner distribution function in first-order optical systems. *J Opt Soc Am A*, 11:2643–2646, 1994.

(Dragoman 1997) D. Dragoman. The Wigner distribution function in optics and optoelectronics. In *Progress in Optics XXXVII*, Elsevier, Amsterdam, 1997. Chapter I, pages 1–56.

(Dragt and Finn 1976) A. J. Dragt and J. M. Finn. Lie series and invariant functions for analytic symplectic maps. *J Math Phys*, 17:2215–2227, 1976.

(Dragt 1982) A. J. Dragt. Lie algebraic theory of geometrical optics and optical aberrations. *J Opt Soc Am*, 72:372–379, 1982.

(Dudgeon and Mersereau 1984) D. E. Dudgeon and R. M. Mersereau. *Multidimensional Digital Signal Processing*. Prentice Hall, Englewood Cliffs, New Jersey, 1984.

(Dym and McKean 1972) H. Dym and H. P. McKean. *Fourier Series and Integrals*. Academic Press, San Diego, California, 1972.

(Easton, Ticknor, and Barrett 1984) R. L. Easton, Jr., A. J. Ticknor, and H. H. Barrett. Application of the Radon transform to optical production of the Wigner distribution function. *Opt Eng*, 23:738–744, 1984.

(Engheta 1995) N. Engheta. A note on fractional calculus and the image method for dielectric spheres. *J Electromagnetic Waves Applic*, 9:1179–1188, 1995.

(Engheta 1996a) N. Engheta. On fractional calculus and fractional multipoles in electromagnetism. *IEEE Trans Antennas Propagation*, 44:554–566, 1996.

(Engheta 1996b) N. Engheta. Electrostatic "fractional" image methods for perfectly conducting wedges and cones. *IEEE Trans Antennas Propagation*, 44:1565–1574, 1996.

(Engheta 1996c) N. Engheta. Use of fractional integration to propose some "fractional" solutions for the scalar Helmholtz equation. *Prog Electromagnetics Research*, 12:107–132, 1996.

(Engheta 1997) N. Engheta. On the role of fractional calculus in electromagnetic theory. *IEEE Antennas and Propagation Magazine*, August 1997, pages 35–46.

(Engheta 1998) N. Engheta. Fractional curl operator in electromagnetics. *Microwave and Optical Tech Lett*, 17:86–91, 1998.

(Erden and Ozaktas 1997) M. F. Erden and H. M. Ozaktas. Accumulated Gouy phase shift in Gaussian beam propagation through first-order optical systems. *J Opt Soc Am A*, 14:2190–2194, 1997.

(Ferraro and Caelli 1988) M. Ferraro and T. M. Caelli. Relationship between integral transform invariances and Lie group theory. *J Opt Soc Am A*, 5:738–742, 1988.

(Flandrin 1993) P. Flandrin. *Temps-fréquence*. Hermès, Paris, 1993.

(Folland 1989) G. B. Folland. *Harmonic Analysis in Phase Space*. Princeton University Press, Princeton, New Jersey, 1989.

(Friberg 1981) A. T. Friberg. Phase-space methods for partially coherent wavefields. In *Optics in Four Dimensions—1980*, American Institute of Physics, New York, 1981. Pages 313–331.

(Friberg 1986) A. T. Friberg. Energy transport in optical systems with partially coherent light. *Appl Opt*, 25:4547–4556, 1986.

(Frieden 1971) B. R. Frieden. Evaluation, design, and extrapolation methods for optical signals, based on use of the prolate functions. In *Progress in Optics IX*, Elsevier, Amsterdam, 1971. Chapter 8, pages 311–407.

(Gabor 1961) D. Gabor. Light and information. In *Progress in Optics I*, Elsevier, Amsterdam,

1961. Chapter 4, pages 109–153.

(Galetti and de Toledo Piza 1988) D. Galetti and A. F. R. de Toledo Piza. An extended Weyl-Wigner transformation for special finite spaces. *Physica A*, 149:267–282, 1988.

(García-Calderón and Moshinsky 1980) G. García-Calderón and M. Moshinsky. Wigner distribution functions and the representation of canonical transformations in quantum mechanics. *J Phys A*, 13:L185–L188, 1980.

(Gardiner 1991) C. W. Gardiner. *Quantum Noise*. Springer, Berlin, 1991.

(Gaskill 1978) J. D. Gaskill. *Linear Systems, Fourier Transforms, and Optics*. Wiley, New York, 1978.

(Ghatak and Thyagarajan 1980) A. Ghatak and K. Thyagarajan. Graded index optical waveguides: a review. In *Progress in Optics XVIII*, Elsevier, Amsterdam, 1980. Chapter 1, pages 1–126.

(Gilmore 1974) R. Gilmore. Baker-Campbell-Hausdorff formulas. *J Math Phys*, 15:2090–2092, 1974.

(Goldstein 1980) H. Goldstein. *Classical Mechanics*, second edition. Addison-Wesley, Reading, Massachusetts, 1980.

(Goodman 1985) J. W. Goodman. *Statistical Optics*. Wiley, New York, 1985.

(Goodman 1996) J. W. Goodman. *Introduction to Fourier Optics*, second edition. McGraw-Hill, New York, 1996.

(Gori 1993) F. Gori. Sampling in optics. In *Advanced Topics in Shannon Sampling and Interpolation Theory*, Springer, New York, 1993. Chapter 2, pages 37–83.

(Gori 1994) F. Gori. Why is the Fresnel transform so little known? In *Current Trends in Optics*, Academic Press, London, 1994. Chapter 10, pages 139–148.

(Gori and Guattari 1971) F. Gori and G. Guattari. Effects of coherence on the degrees of freedom of an image. *J Opt Soc Am*, 61:36–39, 1971.

(Gori and Guattari 1973) F. Gori and G. Guattari. Shannon number and degrees of freedom of an image. *Opt Commun*, 7:163–165, 1973.

(Gori, Santarsiero, and Guattari 1993) F. Gori, M. Santarsiero, and G. Guattari. Coherence and the spatial distribution of intensity. *J Opt Soc Am A*, 10:673–679, 1993.

(Grünbaum 1982) F. A. Grünbaum. The eigenvectors of the discrete Fourier transform: a version of the Hermite functions. *J Math Anal Applic*, 88:355–363, 1982.

(Guillemin and Sternberg 1981) V. Guillemin and S. Sternberg. The metaplectic representation, Weyl operators, and spectral theory. *J Funct Anal*, 42:128–225, 1981.

(Guillemin and Sternberg 1984) V. Guillemin and S. Sternberg. *Symplectic Techniques in Physics*. Cambridge University Press, Cambridge, 1984.

(Hakioğlu 1998) T. Hakioğlu. Finite-dimensional Schwinger basis, deformed symmetries, Wigner function, and an algebraic approach to quantum phase. *J Phys A*, 31:6975–6994, 1998.

(Hakioğlu 1999) T. Hakioğlu. Linear canonical transformations and quantum phase: a unified canonical and algebraic approach. *J Phys A*, 32:4111–4130, 1999.

(Hazak 1992) G. Hazak. Comment on "Wave field determination using three-dimensional intensity information." *Phys Rev Lett*, 69:2874, 1992.

(Hecht, Zajac, and Guardino 1997) E. Hecht, A. Zajac, and K. Guardino. *Optics*, third edition. Addison-Wesley, Reading, Massachusetts, 1997.

(Hildebrand 1965) F. B. Hildebrand. *Methods of Applied Mathematics*, second edition. Prentice Hall, Englewood Cliffs, New Jersey, 1965.

(Hildebrand 1968) F. Hildebrand. *Finite-Difference Equations and Simulations*. Prentice Hall, Englewood Cliffs, New Jersey, 1968.

(Hille and Phillips 1957) E. Hille and R. S. Phillips. *Functional Analysis and Semi-Groups*, revised edition. American Mathematical Society, Providence, Rhode Island, 1957.

(Hillery and others 1984) M. Hillery, R. F. O'Connell, M. O. Scully, and E. P. Wigner. Distribution functions in physics: fundamentals. *Phys Rep*, 106:121–167, 1984.

(Hlawatsch and Boudreaux-Bartels 1992) F. Hlawatsch and G. F. Boudreaux-Bartels. Linear and quadratic time-frequency signal representations. *IEEE Signal Processing Magazine*, April 1992, pages 21–67.

(Hlawatsch and Kozek 1993) F. Hlawatsch and W. Kozek. The Wigner distribution of a linear signal space. *IEEE Trans Signal Processing*, 41:1248–1258, 1993.

(Horner 1982) J. L. Horner. Light utilization in optical correlators. *Appl Opt*, 21:4511–4514, 1982.

(Horner 1992) J. L. Horner. Clarification of Horner efficiency. *Appl Opt*, 31:4629, 1992.

(Iizuka 1987) K. Iizuka. *Engineering Optics*, second edition. Springer, Berlin, 1987.

(*Image Recovery: Theory and Applications* 1987) *Image Recovery: Theory and Applications*. H. Stark, editor. Academic Press, Orlando, Florida, 1987.

(Ivanov, Sivokon, and Vorontsov 1992) V. Yu. Ivanov, V. P. Sivokon, and M. A. Vorontsov. Phase retrieval from a set of intensity measurements: theory and experiment. *J Opt Soc Am A*, 9:1515—1524, 1992.

(Jain 1989) A. K. Jain. *Fundamentals of Digital Image Processing*. Prentice Hall, Englewood Cliffs, New Jersey, 1989.

(Janssen 1988) A. J. E. M. Janssen. The Zak transform: A signal transform for sampled time-continuous signals. *Philips J Res*, 43:23–69, 1988.

(Javidi 1989) B. Javidi. Nonlinear joint power spectrum based optical correlation. *Appl Opt*, 28:2358–2367, 1989.

(Javidi and Horner 1994) B. Javidi and J. L. Horner. *Real-Time Optical Information Processing*, Academic Press, San Diego, California, 1994.

(Jenkins and White 1976) F. A. Jenkins and H. E. White. *Fundamentals of Optics*, fourth edition. McGraw-Hill, New York, 1976.

(Kauderer 1990) M. Kauderer. Fourier-optics approach to the symplectic group. *J Opt Soc Am A*, 7:231–239, 1990.

(Kelley and Peterson 1991) W. G. Kelley and A. C. Peterson. *Difference Equations: An Introduction with Applications*. Academic Press, San Diego, California, 1991.

(Klein and Furtak 1986) M. V. Klein and T. E. Furtak. *Optics*, second edition. Wiley, New York, 1986.

(Kline and Kay 1965) M. Kline and I. W. Kay. *Electromagnetic Theory and Geometrical Optics*. Wiley, New York, 1965.

(Kolner and Nazarathy 1989) B. H. Kolner and M. Nazarathy. Temporal imaging with a time lens. *Opt Lett*, 14:630–632, 1989. Erratum, 15:655, 1990.

(Kumar and Hassebrook 1990) B. V. K. V. Kumar and L. Hassebrook. Performance measures for correlation filters. *Appl Opt*, 29:2997–3006, 1990.

(Lakhtakia 1993) A. Lakhtakia. Fractal self-Fourier functions. *Optik*, 94:51–52, 1993.

(Lanczos 1970) C. Lanczos. *The Variational Principles of Mechanics*, fourth edition. Dover, New York, 1970.

(Landau 1993) H. J. Landau. On the density of phase-space expansions. *IEEE Trans Information Theory*, 39:1152–1156, 1993.

(Landau and Pollak 1961) H. J. Landau and H. O. Pollak. Prolate spheroidal wave functions, Fourier analysis and uncertainty—II. *Bell System Technical J*, 40:65–84, 1961.

(Landau and Pollak 1962) H. J. Landau and H. O. Pollak. Prolate spheroidal wave functions, Fourier analysis and uncertainty—III: the dimension of the space of essentially time- and band-limited signals. *Bell System Technical J*, 41:1295–1336, 1962.

(Leonhardt 1996) U. Leonhardt. Discrete Wigner function and quantum-state tomography. *Phys Rev A*, 53:2998–3015, 1996.

(Lewis 1986) F. L. Lewis. *Optimal Estimation*. Wiley, New York, 1986.

(Li and Sheng 1998) Y. Li and Y. Sheng. Wavelets, optics, and pattern recognition. In *Optical Pattern Recognition*, Cambridge University Press, Cambridge, 1998. Chapter 3, pages 64–88.

(*Lie Methods in Optics* 1986) *Lie Methods in Optics*. J. S. Mondragón and K. B. Wolf, editors. Springer, Berlin, 1986.

(*Lie Methods in Optics* 1989) *Lie Methods in Optics II*. K. B. Wolf, editor. Springer, Berlin, 1989.

(Lipson 1993) S. G. Lipson. Self-Fourier objects and other self-transform objects: comment. *J Opt Soc Am A*, 10:2088–2089, 1993.

(Lipson and Lipson 1981) S. G. Lipson and H. Lipson. *Optical Physics*, second edition. Cambridge University Press, Cambridge, 1981.

(Lohmann 1954) A. Lohmann. Ein neues Dualitätsprinzip in der Optik. *Optik*, 11:478–488, 1954. An English version appeared as: Duality in Optics. *Optik*, 89:93–97, 1992.

(Lohmann 1986) A. W. Lohmann. *Optical Information Processing*, lecture notes. Optik+Info, Post Office Box 51, Uttenreuth, Germany, 1986.

(Lohmann and Mendlovic 1992a) A. W. Lohmann and D. Mendlovic. Self-Fourier objects and other self-transform objects. *J Opt Soc Am A*, 9:2009–2012, 1992.

(Lohmann and Mendlovic 1992b) A. W. Lohmann and D. Mendlovic. An optical self-transform with odd cycles. *Opt Commun*, 93:25–26, 1992.

(Lohmann and Mendlovic 1992c) A. W. Lohmann and D. Mendlovic. Temporal filtering with time lenses. *Appl Opt*, 31:6212–6219, 1992.

(Lohmann and Mendlovic 1994a) A. W. Lohmann and D. Mendlovic. Image formation of a self-Fourier object. *Appl Opt*, 33:153–157, 1994.

(Lohmann and Werlich 1968) A. W. Lohmann and H. W. Werlich. Incoherent matched filter with Fourier holograms. *Appl Opt*, 7:561–563, 1968.

(Luis and Peřina 1998) A. Luis and J. Peřina. Discrete Wigner function for finite-dimensional systems. *J Phys A*, 31:1423–1441, 1998.

(Luneburg 1964) R. K. Luneburg. *Mathematical Theory of Optics*. University of California Press, Berkeley, California, 1964.

(Mallat 1989) S. G. Mallat. A theory for multiresolution signal decomposition: The wavelet representation. *IEEE Trans on Pattern Analysis and Machine Intelligence*, 11:674–693, 1989.

(Mallat 1998) S. Mallat. *A Wavelet Tour of Signal Processing*. Academic Press, London, 1998.

(Mandel and Wolf 1995) L. Mandel and E. Wolf. *Optical Coherence and Quantum Optics*. Cambridge University Press, Cambridge, 1995.

(Marcuse 1982) D. Marcuse. *Light Transmission Optics*, second edition. Krieger Publishing, Malabar, Florida, 1982.

(Marks 1991) R. J. Marks II. *Introduction to Shannon Sampling and Interpolation Theory*. Springer, New York, 1991.

(McClellan 1973) J. H. McClellan. Comments on "Eigenvalue and eigenvector decomposition of the discrete Fourier transform." *IEEE Trans Audio Electroacoustics*, 21:65, 1973.

(McClellan and Parks 1972) J. H. McClellan and T. W. Parks. Eigenvalue and eigenvector decomposition of the discrete Fourier transform. *IEEE Trans Audio Electroacoustics*, 20:66–74, 1972.

(McLachlan 1964) N. W. McLachlan. *Theory and Applications of Mathieu Functions*. Dover, New York, 1964.

(Mendlovic, Konforti, and Marom 1990) D. Mendlovic, N. Konforti, and E. Marom. Shift and projection invariant pattern recognition using logarithmic harmonics. *Appl Opt*, 29:4784–4789, 1990.

(Mendlovic and Lohmann 1997) D. Mendlovic and A. W. Lohmann. Space-bandwidth product adaptation and its application to superresolution: fundamentals. *J Opt Soc Am A*, 14:558–562, 1997.

(Mendlovic, Lohmann, and Zalevsky 1997) D. Mendlovic, A. W. Lohmann, and Z. Zalevsky. Space-bandwidth product adaptation and its application to superresolution: examples. *J Opt Soc Am A*, 14:563–567, 1997.

(Mendlovic and Ozaktas 1993c) D. Mendlovic and H. M. Ozaktas. Optical-coordinate transformation methods and optical-interconnection architectures. *Appl Opt*, 32:5119–5124, 1993.

(Mendlovic and Zalevsky 1997) D. Mendlovic and Z. Zalevsky. Definition, properties and applications of the generalized temporal-spatial Wigner distribution function. *Optik*, 107:49–56, 1997.

(Miller 1998) D. A. B. Miller. Spatial channels for communicating with waves between volumes. *Opt Lett*, 23:1645–1647, 1998.

(Miller and Ross 1989) K. S. Miller and B. Ross. Fractional difference calculus. In *Univalent Functions, Fractional Calculus, and their Applications*, Ellis Horwood, Chichester, UK, 1989. Chapter 12, pages 139–152.

(Möller 1988) K. D. Möller. *Optics*. University Science Books, Mill Valley, California, 1988.

(Moshinsky and Quesne 1971) M. Moshinsky and C. Quesne. Linear canonical transformations and their unitary representations. *J Math Phys*, 12:1772–1780, 1971.

(Moshinsky, Seligman, and Wolf 1972) M. Moshinsky, T. H. Seligman, and K. B. Wolf. Canonical transformations and the radial oscillator and Coulomb problem. *J Math Phys*, 13:901–907, 1972.

(Navarro-Saad and Wolf 1986a) M. Navarro-Saad and K. B. Wolf. The group theoretical treatment of aberrating systems. I. Aligned lens systems in third aberration order. *J Math Phys*, 27:1449–1457, 1986.

(Navarro-Saad and Wolf 1986b) M. Navarro-Saad and K. B. Wolf. Factorization of the phase-space transformation produced by an arbitrary refracting surface. *J Opt Soc Am A*, 3:340–346, 1986.

(Naylor and Sell 1982) A. W. Naylor and G. R. Sell. *Linear Operator Theory in Engineering and Science*. Springer, New York, 1982.

(Nazarathy and Shamir 1980) M. Nazarathy and J. Shamir. Fourier optics described by operator algebra. *J Opt Soc Am*, 70:150–159, 1980.

(Nazarathy and Shamir 1981) M. Nazarathy and J. Shamir. Holography described by operator algebra. *J Opt Soc Am*, 71:529–541, 1981.

(Nazarathy and Shamir 1982a) M. Nazarathy and J. Shamir. First-order optics—a canonical operator representation: lossless systems. *J Opt Soc Am*, 72:356–364, 1982.

(Nazarathy and Shamir 1982b) M. Nazarathy and J. Shamir. First-order optics—operator representation for systems with loss or gain. *J Opt Soc Am*, 72:1398–1408, 1982.

(Nazarathy, Hardy, and Shamir 1982) M. Nazarathy, A. Hardy, and J. Shamir. Generalized mode propagation in first-order optical systems with loss or gain. *J Opt Soc Am*, 72:1409–1420, 1982.

(Nazarathy, Hardy, and Shamir 1986) M. Nazarathy, A. Hardy, and J. Shamir. Misaligned first-order optics: canonical operator theory. *J Opt Soc Am A*, 3:1360–1369, 1986.

(Nugent 1992) K. A. Nugent. Wave field determination using three-dimensional intensity information. *Phys Rev Lett*, 68:2261–2264, 1992.

(Ojeda-Castañeda and Noyola-Isgleas 1988) J. Ojeda-Castañeda and A. Noyola-Isgleas. Differential operators for scalar wave propagation. *J Opt Soc Am A*, 5:1605–1609, 1988.

(O'Neill, Flandrin, and Williams 1999) J. C. O'Neill, P. Flandrin, and W. J. Williams. On the existence of discrete Wigner distributions. *IEEE Signal Processing Lett*, 6:304–306, 1999.

(Onural and Özgen 1992) L. Onural and M. T. Özgen. Extraction of three-dimensional object location information directly from in-line holograms using Wigner analysis. *J Opt Soc Am A*, 9:252–260, 1992.

(Oppenheim and Shafer 1989) Alan V. Oppenheim and Ronald W. Shafer. *Discrete-Time Signal Processing*. Prentice Hall, Englewood Cliffs, New Jersey, 1989.

(*Optical Pattern Recognition* 1998) *Optical Pattern Recognition*. F. T. S. Yu and S. Jutamulia, editors. Cambridge University Press, Cambridge, 1998.

(Osipov 1992) V. F. Osipov. *Pochti periodicheskie funktsii Bora-Frenelya*. University of St. Petersburg, St. Petersburg, 1992. (In Russian: Bohr-Fresnel Almost-Periodic Functions.)

(Ozaktas, Brenner, and Lohmann 1993) H. M. Ozaktas, K.-H. Brenner, and A. W. Lohmann. Interpretation of the space-bandwidth product as the entropy of distinct connection patterns in multifacet optical interconnection architectures. *J Opt Soc Am A*, 10:418–422,

1993.

(Ozaktas and Kutay 2000b) H. M. Ozaktas and M. A. Kutay. Wigner distributions, linear canonical transforms, and phase-space optics. Technical Report BU-CE-0006, Bilkent University, Department of Computer Engineering, Ankara, January 2000.

(Ozaktas and Mendlovic 1993c) H. M. Ozaktas and D. Mendlovic. Multi-stage optical interconnection architectures with least possible growth of system size. *Opt Lett*, 18:296–298, 1993.

(Ozaktas and Nuss 1996) H. M. Ozaktas and M. C. Nuss. Time-variant linear pulse processing. *Opt Commun*, 131:114–118, 1996.

(Ozaktas and Urey 1993) H. M. Ozaktas and H. Urey. Space-bandwidth product of conventional Fourier transforming systems. *Opt Commun*, 104:29–31, 1993.

(Ozaktas, Urey, and Lohmann 1994) H. M. Ozaktas, H. Urey, and A. W. Lohmann. Scaling of diffractive and refractive lenses for optical computing and interconnections. *Appl Opt*, 33:3782–3789, 1994.

(Papoulis 1968) A. Papoulis. *Systems and Transformations with Applications in Optics.* McGraw-Hill, New York, 1968.

(Papoulis 1974) A. Papoulis. Ambiguity function in Fourier optics. *J Opt Soc Am*, 64:779–788, 1974.

(Papoulis 1977) A. Papoulis. *Signal Analysis.* McGraw-Hill, New York, 1977.

(Papoulis 1991) A. Papoulis. *Probability, Random Variables, and Stochastic Processes*, third edition. McGraw-Hill, New York, 1991.

(Papoulis 1994) A. Papoulis. Pulse-compression, fiber communications, and diffraction—a unified approach. *J Opt Soc Am A*, 11:3–13, 1994.

(Patorski 1989) K. Patorski. The self-imaging phenomenon and its applications. In *Progress in Optics XXVII*, Elsevier, Amsterdam, 1989. Chapter 1, pages 1–108.

(Pegis 1961) R. J. Pegis. The modern development of Hamiltonian optics. In *Progress in Optics I*, Elsevier, Amsterdam, 1961. Chapter VI, pages 1–29.

(Perelomov 1986) A. Perelomov. *Generalized Coherent States and Their Applications.* Springer, Berlin, 1986.

(Peyrin and Prost 1986) F. Peyrin and R. Prost. A unified definition for the discrete-time, discrete-frequency, and discrete-time/frequency Wigner distributions. *IEEE Trans Acoustics, Speech, and Signal Processing*, 34:858–866, 1986.

(Qian and Chen 1996) S. Qian and D. Chen. *Joint Time-Frequency Analysis.* Prentice Hall, Englewood Cliffs, New Jersey, 1996.

(Quesne and Moshinsky 1971) C. Quesne and M. Moshinsky. Canonical transformations and matrix elements. *J Math Phys*, 12:1780–1783, 1971.

(Rammal and Bellissard 1990) R. Rammal and J. Bellissard. An algebraic semi-classical approach to Bloch electrons in a magnetic field. *J Phys France*, 51:1803–1830, 1990.

(Ramo, Whinnery, and Van Duzer 1994) S. Ramo, J. R. Whinnery, and T. Van Duzer. *Fields and Waves in Communication Electronics*, third edition. Wiley, New York, 1994.

(Reardon and Chipman 1990) P. J. Reardon and R. A. Chipman. Maximum power of refractive lenses: a fundamental limit. *Opt Lett*, 15:1409-1411, 1990.

(Reynolds and others 1989) G. O. Reynolds, J. B. DeVelis, G. B. Parrent, Jr., and B. J. Thompson. *The New Physical Optics Notebook: Tutorials in Fourier Optics.* SPIE, Bellingham, Washington, 1989.

(Rhodes 1994) W. T. Rhodes. Simple procedure for the analysis of coherent imaging systems. *Opt Lett*, 19:1559–1561, 1994.

(Rhodes 1998) W. T. Rhodes. The fallacy of single-lens coherent imaging theory. In *1998 OSA Annual Meeting Prog*, OSA, Washington DC, 1998. Page 117.

(Richman, Parks, and Shenoy 1998) M. S. Richman, T. W. Parks, and R. G. Shenoy. Discrete-time, discrete-frequency, time-frequency analysis. *IEEE Trans Signal Processing*, 46:1517–1527, 1998.

(Rioul and Vetterli 1991) O. Rioul and M. Vetterli. Wavelets and signal processing. *IEEE*

Signal Processing Magazine, October 1991, pages 14–38.

(Roman 1992) S. Roman. *Advanced Linear Algebra*. Springer, New York, 1992.

(Saleh and Subotic 1985) B. E. A. Saleh and N. S. Subotic. Time-variant filtering of signals in the mixed time-frequency domain. *IEEE Trans Acoustics, Speech, and Signal Processing*, 33:1479–1485, 1985.

(Saleh and Teich 1991) B. E. A. Saleh and M. C. Teich. *Fundamentals of Photonics*. Wiley, New York, 1991.

(Santhanam and Tekumalla 1976) T. S. Santhanam and A. R. Tekumalla. Quantum mechanics in finite dimensions. *Foundations Physics*, 6:583-587, 1976.

(Schleich, Mayr, and Krähmer 1999) W. P. Schleich, E. Mayr, and D. Krähmer. *Quantum Optics in Phase Space*. Wiley, New York, 1999.

(Segal 1963) I. E. Segal. Transforms for operators and symplectic automorphisms over a locally compact Abelian group. *Math Scand*, 13:31–43, 1963.

(Sekiguchi and Wolf 1987) T. Sekiguchi and K. B. Wolf. The Hamiltonian formulation of optics. *Am J Phys*, 55:830–835, 1987.

(Shamir 1979) J. Shamir. Cylindrical lens systems described by operator algebra. *Appl Opt*, 18:4195–4202, 1979.

(Siegman 1986) A. E. Siegman. *Lasers*. University Science Books, Mill Valley, California, 1986.

(Simon and Wolf 2000) R. Simon and K. B. Wolf. Structure of the set of paraxial optical systems. *J Opt Soc Am A*, 17:342–355, 2000.

(Slepian 1964) D. Slepian. Prolate spheroidal wave functions, Fourier analysis and uncertainty—IV: extensions to many dimensions; generalized prolate spheroidal functions. *Bell System Technical J*, 43:3009–3057, 1964.

(Slepian 1978) D. Slepian. Prolate spheroidal wave functions, Fourier analysis and uncertainty—V. *Bell System Technical J*, 57:1371–1430, 1978.

(Slepian and Pollak 1961) D. Slepian and H. O. Pollak. Prolate spheroidal wave functions, Fourier analysis and uncertainty—I. *Bell System Technical J*, 40:43–63, 1961.

(Solimeno, Crosignani, and Di Porto 1986) S. Solimeno, B. Crosignani, and P. Di Porto. *Guiding, Diffraction, and Confinement of Optical Radiation*. Academic Press, Orlando, Florida, 1986.

(Stavroudis 1972) O. N. Stavroudis. *The Optics of Rays, Wavefronts, and Caustics*. Academic Press, New York, 1972.

(Stoler 1981) D. Stoler. Operator methods in physical optics. *J Opt Soc Am*, 71:334–341, 1981.

(Strang 1988) G. Strang. *Linear Algebra and its Applications*, third edition. Harcourt Brace Jovanovich, New York, 1988.

(Strang and Nguyen 1996) G. Strang and T. Nguyen. *Wavelets and Filter Banks*. Wellesley-Cambridge Press, Wellesley, Massachusetts, 1996.

(Suter 1997) B. W. Suter. *Multirate and Wavelet Signal Processing*. Academic Press, London, 1997.

(Synge 1937) J. L. Synge. *Geometrical Optics: An Introduction to Hamilton's Method*. Cambridge University Press, Cambridge, 1937.

(Synge 1954) J. L. Synge. *Geometrical Mechanics and de Broglie Waves*. Cambridge University Press, Cambridge, 1954.

(Tanaka 1986) K. Tanaka. Paraxial theory in optical design in terms of Gaussian brackets. In *Progress in Optics XXIII*, Elsevier, Amsterdam, 1986. Chapter 2, pages 63–111.

(Testorf and Ojeda-Castañeda 1996) M. Testorf and J, Ojeda-Castañeda. Fractional Talbot effect: analysis in phase space. *J Opt Soc Am A*, 13:119–125, 1996.

(Toraldo di Francia 1955) G. Toraldo di Francia. Resolving power and information. *J Opt Soc Am*, 45:497–501, 1955.

(Toraldo di Francia 1969) G. Toraldo di Francia. Degrees of freedom of an image. *J Opt Soc Am*, 59:799–804, 1969.

(*The Transforms and Applications Handbook* 2000) *The Transforms and Applications Handbook*, second edition. A. D. Poularikas, editor. CRC Press, Boca Raton, Florida, 2000.

(Vakman 1968) D. E. Vakman. *Sophisticated Signals and the Uncertainty Principle in Radar*. Springer, Berlin, 1968.

(VanderLugt 1964) A. VanderLugt. Signal detection by complex spatial filtering. *IEEE Trans Information Theory*, 10:139–146, 1964.

(VanderLugt 1992) A. VanderLugt. *Optical Signal Processing*. Wiley, New York, 1992.

(Vetterli and Kovacevic 1995) M. Vetterli and J. Kovacevic. *Wavelets and Subband Coding*. Prentice Hall, Englewood Cliffs, New Jersey, 1995.

(Walls and Milburn 1994) D. F. Walls and G. J. Milburn. *Quantum Optics*. Springer, Berlin, 1994.

(Walter 1994) G. G. Walter. *Wavelets and Other Orthogonal Systems with Applications*. CRC Press, Boca Raton, Florida, 1994.

(Walther 1967) A. Walther. Gabor's theorem and energy transfer through lenses. *J Opt Soc Am*, 57:639–644, 1967.

(Walther 1968) A. Walther. Radiometry and coherence. *J Opt Soc Am*, 58:1256–1259, 1968.

(*Wavelets: Mathematics and Applications* 1993) *Wavelets: Mathematics and Applications*. J. J. Benedetto and M. W. Frazier, editors. CRC Press, Boca Raton, Florida, 1993.

(Weaver and Goodman 1966) C. S. Weaver and J. W. Goodman. A technique for optically convolving two functions. *Appl Opt*, 5:1248–1249, 1966.

(Weil 1964) A. Weil. Sur certains groupes d'opérateurs unitaires. *Acta Mathematica*, 111:143–211, 1964.

(Wiegmann and Zabrodin 1995) P. B. Wiegmann and A. V. Zabrodin. Algebraization of difference eigenvalue equations related to $U_q(sl_2)$. *Nuclear Phys B*, 451:699–724, 1995.

(Wiener 1933) N. Wiener. *The Fourier Integral and Certain of its Applications*. Cambridge University Press, Cambridge, 1933.

(Wigner 1932) E. Wigner. On the quantum correction for thermodynamic equilibrium. *Phys Rev*, 40:749–759, 1932.

(*The Wigner Distribution: Theory and Applications in Signal Processing* 1997) *The Wigner Distribution: Theory and Applications in Signal Processing*. W. Mecklenbräuker and F. Hlawatsch, editors. Elsevier, Amsterdam, 1997.

(*Wigner Distributions and Phase Space in Optics* 2000) *Wigner Distributions and Phase Space in Optics*. G. W. Forbes, V. I. Man'ko, H. M. Ozaktas, R. Simon, and K. B. Wolf, editors. Feature issue of *J Opt Soc Am A*, 17 (12), December 2000.

(Wilcox 1967) R. M. Wilcox. Exponential operators and parameter differentiation in quantum physics. *J Math Phys*, 8:962–982, 1967.

(Wilkinson 1988) J. H. Wilkinson. *The Algebraic Eigenvalue Problem*. Oxford University Press, Oxford, 1988.

(Winthrop 1971) J. T. Winthrop. Propagation of structural information in optical wave fields. *J Opt Soc Am*, 61:15–30, 1971.

(Wolf 1974a) K. B. Wolf. Canonical transforms. I. Complex linear transforms. *J Math Phys*, 15:1295–1301, 1974.

(Wolf 1974b) K. B. Wolf. Canonical transforms. II. Complex radial transforms. *J Math Phys*, 15:2102–2111, 1974.

(Wolf 1976) K. B. Wolf. Canonical transforms. separation of variables, and similarity solutions for a class of parabolic differential equations. *J Math Phys*, 17:601–613, 1976.

(Wolf 1977) K. B. Wolf. On self-reciprocal functions under a class of integral transforms. *J Math Phys*, 18:1046–1051, 1977.

(Wolf 1979) K. B. Wolf. *Integral Transforms in Science and Engineering*. Plenum Press, New York, 1979.

(Wolf 1986a) K. B. Wolf. Symmetry in Lie optics. *Annals of Physics*, 172:1–25, 1986.

(Wolf 1986b) K. B. Wolf. The group theoretical treatment of aberrating systems. II. Axis-symmetric inhomogeneous systems and fiber optics in third aberration order. *J Math Phys*,

27:1458–1465, 1986.

(Wolf 1991) K. B. Wolf. Fourier transform in metaxial geometric optics. *J Opt Soc Am A*, 8:1399—1403, 1991.

(Wolf 1993) K. B. Wolf. Relativistic aberration of optical phase space. *J Opt Soc Am A*, 10:1925–1934, 1993.

(Wolf 1996) K. B. Wolf. Wigner distribution function for paraxial polychromatic optics. *Opt Commun*, 132:343–352, 1996.

(Wolf and Krötzsch 1995) K. B. Wolf and G. Krötzsch. Geometry and dynamics in refracting systems. *Eur J Phys*, 16:14–20, 1995.

(Wolf and Kurmyshev 1993) K. B. Wolf and E. V. Kurmyshev. "Squeezed states" in Helmholtz optics. *Phys Rev A*, 47:3365–3370, 1993.

(Wright and Garrison 1987) E. M. Wright and J. C. Garrison. Path-integral derivation of the complex *ABCD* Huygens integral. *J Opt Soc Am A*, 4:1751–1755, 1987.

(Yamamoto and İmamoğlu 1999) Y. Yamamoto and A. İmamoğlu. *Mesoscopic Quantum Optics*. Wiley, New York, 1999.

(Yariv 1989) A. Yariv. *Quantum Electronics*, third edition. Wiley, New York, 1989.

(Yariv 1997) A. Yariv. *Optical Electronics in Modern Communications*, fifth edition. Oxford University Press, Oxford, 1997.

(Yarlagadda 1977) R. Yarlagadda. A note on the eigenvectors of DFT matrices. *IEEE Trans Acoustics, Speech, and Signal Processing*, 25:586–589, 1977.

(Yosida 1984) K. Yosida. *Operational Calculus: A Theory of Hyperfunctions*. Springer, New York, 1984.

(Yu 1983) F. T. S. Yu. *Optical Information Processing*. Wiley, New York, 1983.

(Yu and Jutamulia 1992) F. T. S. Yu and S. Jutamulia. *Optical Signal Processing, Computing, and Neural Networks*. Wiley, New York, 1992.

(Zhu, Peyrin, and Goutte 1989) Y. M. Zhu, F. Peyrin, and R. Goutte. Equivalence between two-dimensional analytic and real signal Wigner distribution. *IEEE Trans Acoustics, Speech, and Signal Processing*, 37:1631–1634, 1989.

Credits

We are grateful to the following for granting permission to use/adapt material for our figures, tables, and text. Individual sources are indicated in the captions or text.

© Academic Press: text.

Bilkent University: figures, tables, and text.

© Çağatay Candan: figures, tables, and text.

© Elsevier Science: figures 8.6, 8.8, 9.3, 9.4, 9.12, 10.37, 10.38, 11.1, 11.2 and text, adapted/reprinted from *Optics Communications* and *Signal Processing*, with permission from Elsevier Science.

© M. Fatih Erden: figures and text.

© The Institute of Electrical and Electronics Engineers (IEEE): figures 3.1, 4.4, 6.1, 6.2, 10.1, 10.3, 10.4, 10.6, 10.7, 10.8, 10.14, 10.15, 10.16, 10.17, 10.18, 10.19, 10.20, 10.21, 10.22, 10.23, 10.28, 10.36, 10.37, tables 3.2, 3.4, 3.5, 3.6, 6.1, 6.3, 6.4, and text.

© International Society for Optical Engineering (SPIE): text.

© Optical Society of America (OSA): figures 5.5, 9.5, 9.6, 9.7, 9.10, 9.13, 10.1, 10.2, 10.4, 10.10, 10.11, 10.12, 10.13, 10.24, 10.25, 10.26, 10.27, 10.33, 10.34, 10.35, 10.37, 10.39, 10.40, 10.41, 11.3, 11.4, 11.5, 11.6, 11.7, 11.8, 11.9 and text.

© Ayşegül Şahin: text.

© Taylor & Francis (www.tandf.co.uk/journals): figures 3.1, 3.2.

Index